6100553805

KU-542-758

SPECIAL TOPICS IN HETEROCYCLIC CHEMISTRY

This is the thirtieth volume in the series

THE CHEMISTRY OF HETEROCYCLIC COMPOUNDS

THE CHEMISTRY OF HETEROCYCLIC COMPOUNDS

A SERIES OF MONOGRAPHS

ARNOLD WEISSBERGER and EDWARD C. TAYLOR

Editors

SPECIAL TOPICS IN HETEROCYCLIC CHEMISTRY

Edited by

Arnold Weissberger

Research Laboratories
Eastman Kodak Company
Rochester, New York

Edward C. Taylor

Princeton University
Princeton, New Jersey

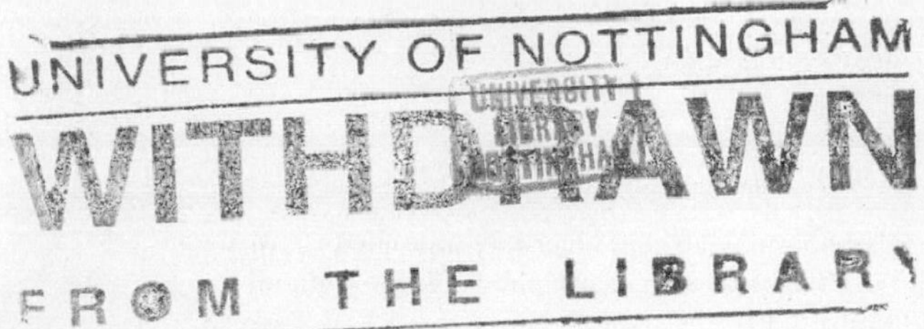
UNIVERSITY OF NOTTINGHAM
WITHDRAWN
FROM THE LIBRARY

AN INTERSCIENCE® PUBLICATION

JOHN WILEY & SONS

NEW YORK • LONDON • SYDNEY • TORONTO

An Interscience ® Publication

Copyright © 1977 by John Wiley & Sons, Inc.

All rights reserved. Published simultaneously in Canada.

No part of this book may be reproduced by any means, nor transmitted, nor translated into a machine language without the written permission of the publisher.

Library of Congress Cataloging in Publication Data:

Main entry under title:

Special topics in heterocyclic chemistry.

(The Chemistry of heterocyclic compounds; v. 30)
"An Interscience publication."
Bibliography: p.
1. Heterocyclic compounds—Addresses, essays, lectures.
I. Weissberger, Arnold, 1898–. II. Taylor, Edward Curtis, 1923–
QD400.S512 547'.59 76-10672
ISBN 0-471-67253-X

Printed in the United States of America

10 9 8 7 6 5 4 3 2 1

Contributors

H. L. Blewitt, *University of Alabama, Tuscaloosa, Alabama*

E. Campaigne, *Ft. Lewis College, Durango, Colorado*

Albert J. Fritsch, *Center for Science in the Public Interest, Washington, D.C.*

R. D. Hamilton, *Ft. Lewis College, Durango, Colorado*

Georges Maury, *Department de Chimie, Faculté des Sciences, Rabat, Morocco*

John P. Paolini, *Merrell-National Laboratories, Cincinnati, Ohio*

K. T. Potts, *Department of Chemistry, Rensselaer Polytechnic Institute, Troy, New York*

David M. Sturmer, *Research Laboratories, Eastman Kodak Company, Rochester, New York*

Alfred Taurins, *Department of Chemistry, McGill University, Montreal, Canada*

The Chemistry of Heterocyclic Compounds

The chemistry of heterocyclic compounds is one of the most complex branches of organic chemistry. It is equally interesting for its theoretical implications, for the diversity of its synthetic procedures, and for the physiological and industrial significance of heterocyclic compounds.

A field of such importance and intrinsic difficulty should be made as readily accessible as possible, and the lack of a modern detailed and comprehensive presentation of heterocyclic chemistry is therefore keenly felt. It is the intention of the present series to fill this gap by expert presentations of the various branches of heterocyclic chemistry. The subdivisions have been designed to cover the field in its entirety by monographs which reflect the importance and the interrelations of the various compounds, and accommodate the specific interests of the authors.

In order to continue to make heterocyclic chemistry as readily accessible as possible, new editions are planned for those areas where the respective volumes in the first edition have become obsolete by overwhelming progress. If, however, the changes are not too great so that the first editions can be brought up-to-date by supplementary volumes, supplements to the respective volumes will be published in the first edition.

ARNOLD WEISSBERGER

Research Laboratories
Eastman Kodak Company
Rochester, New York

EDWARD C. TAYLOR

Princeton University
Princeton, New Jersey

Concerning **Special Topics in Heterocyclic Chemistry**

This volume is the first in the series "The Chemistry of Heterocyclic Compounds" to contain collected treatments of topics not necessarily related to each other and not comprising sufficient pages to be issued individually. Since the series' inception about 25 years ago, volumes have contained exhaustive discussions of syntheses, reactions, properties, structure, physical chemistry, and so on, of compounds belonging to a specific ring system (such as pyridines, thiophenes, pyrimidines, and indoles). This series has become the basic reference collection for information on heterocyclic compounds.

The series "General Heterocyclic Chemistry," initiated in 1971, is devoted to those disciplines of heterocyclic chemistry that are of *general* significance and application, and that are of interest to all organic chemists as well as to those whose particular concern is heterocyclic chemistry. Each volume in this series surveys the entire field of heterocyclic chemistry rather than a particular ring system.

We have long felt a need for an additional forum for discussions of topics of more limited scope whose treatment in a separate monograph might not be appropriate. We hope that readers and research workers in the field will comment on the usefulness of this new forum; we welcome suggestions for improvements and contributions to future volumes.

ARNOLD WEISSBERGER
EDWARD C. TAYLOR

Contents

SPECIAL TOPICS IN HETEROCYCLIC CHEMISTRY

This is the thirtieth volume in the series

THE CHEMISTRY OF HETEROCYCLIC COMPOUNDS

CHAPTER I

5,5-Systems with a Bridgehead Nitrogen Atom

JOHN P. PAOLINI, PH.D.

Merrell-National Laboratories
Cincinnati, Ohio

I. Introduction

The systems we discuss in this chapter are those 5,5 fused ring systems with a bridgehead nitrogen atom which have π-electrons in mutual cyclic conjugation or for which the tautomeric potential for this condition exists.[1] These systems contain 10 π-electrons and as the bridgehead nitrogen atoms of such molecules contribute two π-electrons, the systems will of necessity be anions (e.g., **1**) or contain one or two other heteroatoms that can contribute two π-electrons. The introduction of one other heteroatom contributing two π-electrons (e.g., **2**) gives a neutral molecule and the introduction of a second heteroatom contributing two π-electrons gives a cation (e.g., **3**). Polycyclic systems containing these 5,5 groupings will also be considered. Cyclazines, in which the nitrogen

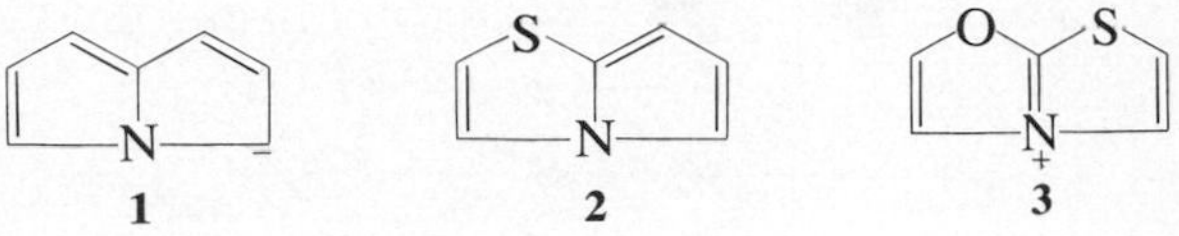

1 **2** **3**

atom is common to three rings, and mesoionic systems are discussed in other chapters.

No preference is given to any one definition of aromaticity,[2a,b] and the term is used in its broadest sense.

The term pyridine type nitrogen refers to a ring nitrogen atom contributing one π-electron to the system, and a pyrrole type nitrogen is one that contributes two π-electrons to the system.

Chemical Abstracts has been covered through December, 1972; later selected references are also included.

II. Synthesis

The formation of fused ring systems with bridgehead nitrogen atoms usually involves the building of one ring onto an existing ring, although there are examples in which both rings are formed during the course of the reaction. We deal with the former type first, discussing those ring closures that produced 5,5 systems.

For convenience, the following notation is used in designating systems. The bridgehead nitrogen is "1" and the other bridgehead atom "2" and nitrogen is given preference in nomenclature. The example shown would be a [1,4,3]thiadiazole.

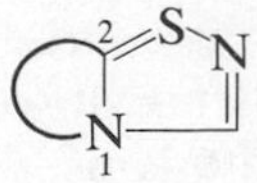

A. Pyrroles

1. The reaction of a 2-alkyl system (**4**) with an α-halo carbonyl compound gives a quaternary salt (**5**) which cyclizes, in base, to a pyrrole ring (**6**). Sodium bicarbonate,[3–6] sodium carbonate,[7,8] sodium ethoxide,[9,10] and triethylamine[11] and sodium acetate-acetic anhydride have all been used as bases.

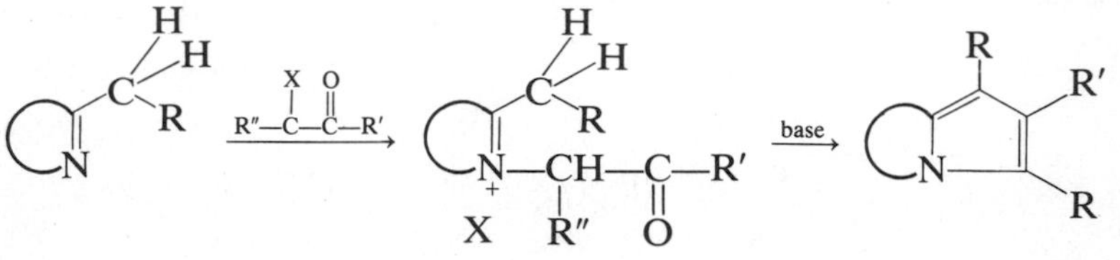

4 **5** **6**

R = H, alkyl, aryl or NH—CO—C_6H_5
R′ = alkyl or aryl
R″ = H or alkyl
X = Cl or Br

This procedure has been used extensively for the preparation pyrrolo-[1,2-*a*]imidazoles[3,5] (**7**), pyrrolo[1,2-*a*]benzimidazoles (**8**),[3,4,7,8,10] pyrrolo[2,1-*b*]thiazoles (**9**),[11,12,14] pyrrolo[2,1-*b*]benzothiazoles (**10**),[13,14] 7H-naptho[2,1-*d*]pyrrolo[1,2-*a*]imidazoles (**11**)[9] and pyrrolo[1,2-*b*]-1,2,4-triazole (**12**)[6].

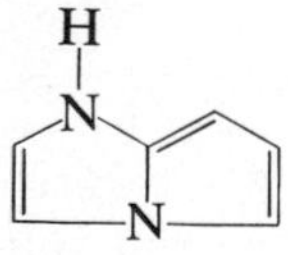

7 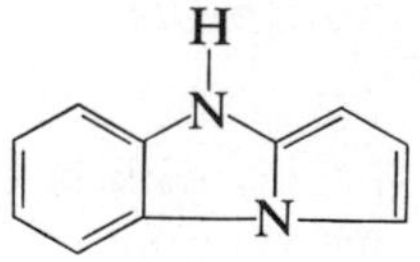 **8** 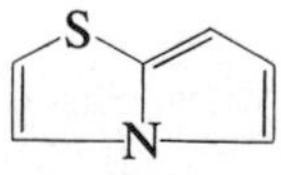 **9**

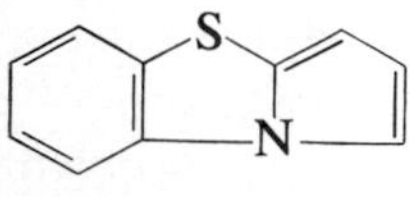

10 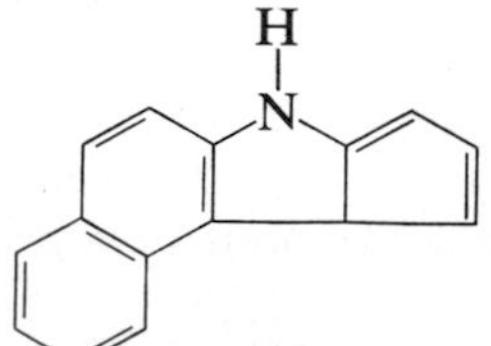 **11** 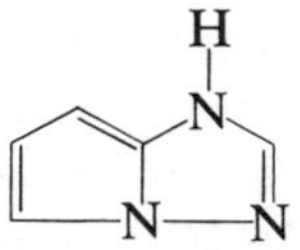 **12**

2. Boekelheide and Fedoruk[15] prepared pyrrolo[1,2-*a*]imidazoles (**7**) from 1-alkyl imidazoles (**13**) phenacylbromide and ethylpropiolate as shown here. Excess ethylpropiolate is used and serves not only to react

with the ylid (**15**) but also to dehydrogenate the apparent intermediate (**16**).

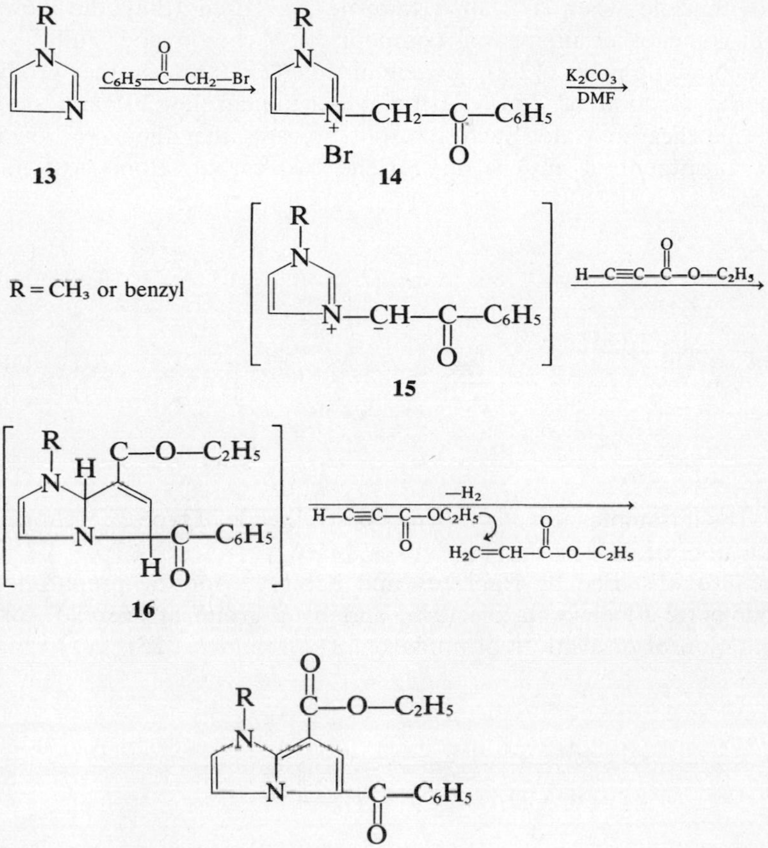

13

C_6H_5—C(=O)—CH_2—Br

14

K_2CO_3 DMF

R = CH_3 or benzyl

15

H—C≡C—C(=O)—O—C_2H_5

16

$-H_2$

H—C≡C—C(=O)—OC_2H_5

H_2C=CH—C(=O)—O—C_2H_5

17

3. Dimethylacetylene dicarboxylate reacts with compounds with pyridine type nitrogens. Thiazole and benzothiazole (**18**) reacted with two moles of diethylacetylene dicarboxylate to give the corresponding triester

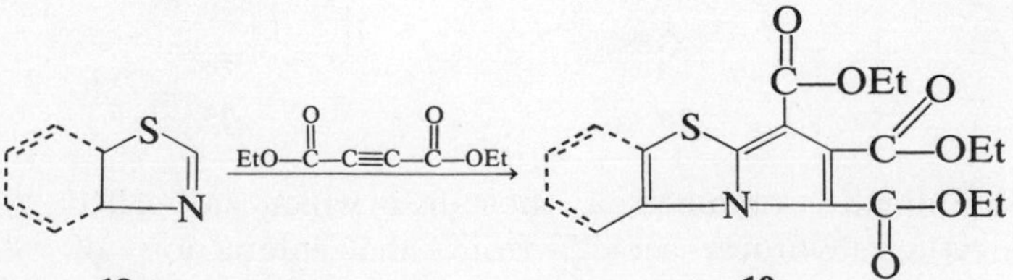

EtO—C(=O)—C≡C—C(=O)—OEt

18

19

(**19**).[16] The benzothiazole was obtained in better yield (40%) than pyrrolo[2,1-*b*]thiazole (8.5%). While these workers did not obtain any pyrrolothiazole when a 2-alkyl thiazole was used (they did however obtain some other interesting compounds), Acheson and Tully[17] were able to obtain a pyrrolo[2,1-*b*]benzimidazole (**21**) (among other products) from the reaction of 2-ethyl-1-methylbenzimidazole (**20**) and diethyl-acetylene dicarboxylate. Such a result indicates that the 2-carbon (and alkyl substituent, if any) is one of the two carbon atoms lost in this process.

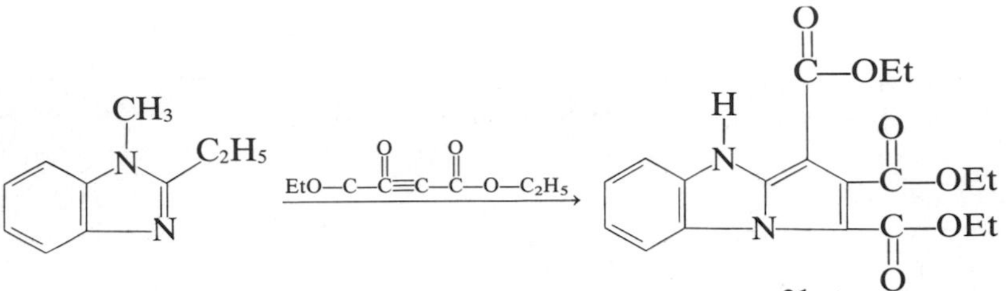

20 **21**

4. The intramolecular alkylation of benzylhalide of type **22** (followed by abstraction of HX with base) gives fused pyrroles of type **23**. This procedure was used by Babichev and Kibirev[18] for the preparation of isoindolo[1,2-*b*]benzothiazole (**24**), and by a group at Merck[19] for the preparation of derivatives of imidazo[2,1-*a*]isoindole (**25**).

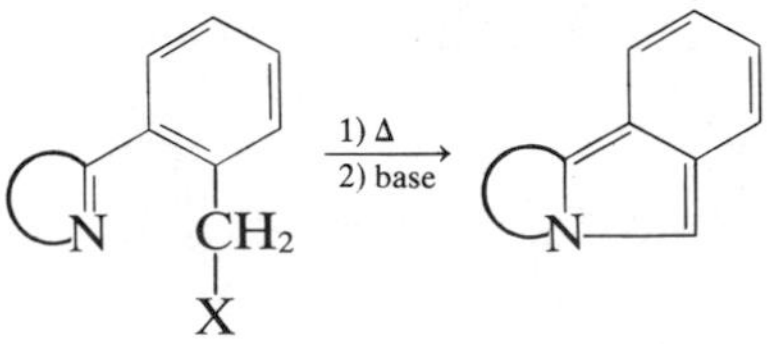

22 **23**

X = Cl or Br

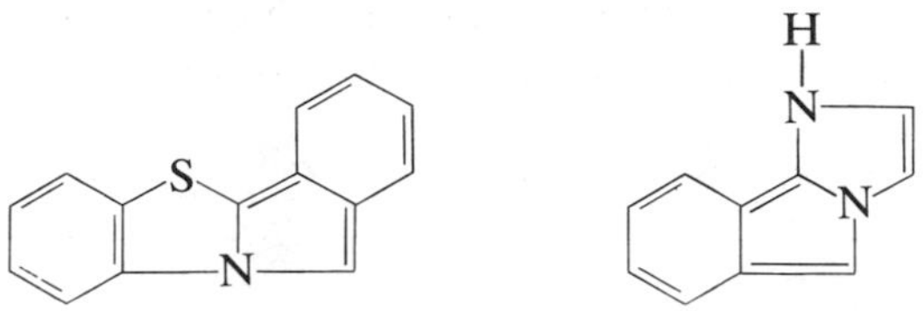

24 **25**

5. Manjunath[20] obtained a substance which he called "8,9-(1,2-cyclohexyl)tetrahydrocarbazole" from the interaction of 9-nitroso-1,2,3,4-tetrahydrocarbazole (**26**; R = H), cyclohexanone, zinc, and acetic

acid. Lions and Ritchie[21] pointed out that Manjunath had reported an incorrect structural formula. They designated structure **27**, a normal product of the Fisher indole synthesis, as being the correct structure.

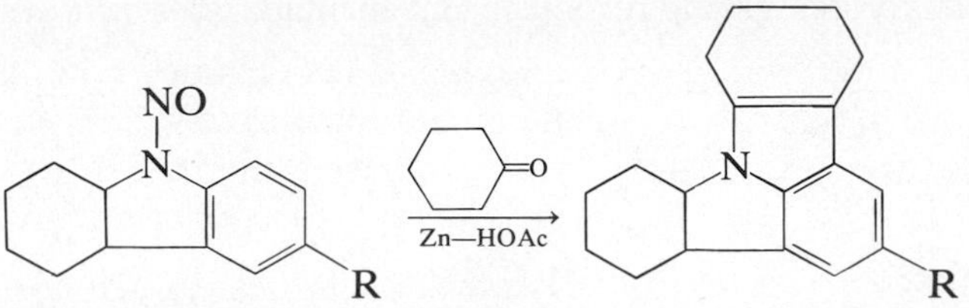

NO N R Zn—HOAc O N R

26 **27**

[They prepared a 6-methyl derivative (**27**; R = CH_3) using the same approach.] 8-Methyl-9-nitrosocarbazole (**28**); in which the ortho position is blocked, failed to give a similar substance, which is to be expected if structure **27** is correct.

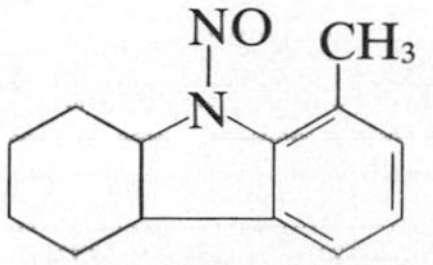

NO CH_3 N

28

Manjunath[20] also prepared the hydrazone (**29**) but was unable to effect a Fisher ring closure with it. Preston and Tucker[22] accomplished this by treating this hydrazone with dry HCl in tetralin at 160° and obtained the tetrahydro indolo[3,2,1-*j,l*]carbozole (**30**).

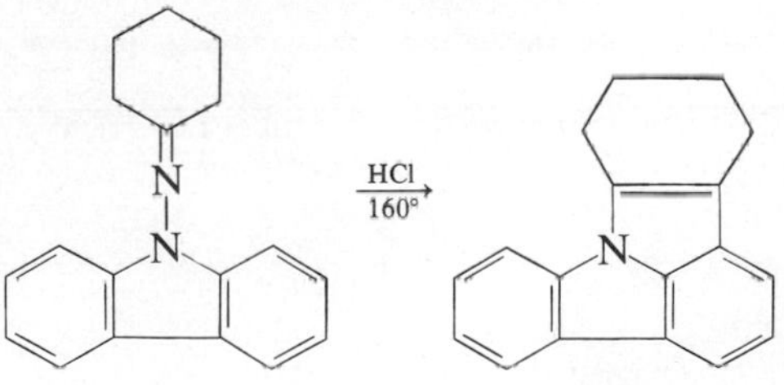

N N HCl 160° N

29 **30**

6. The Pschorr reaction was used by Preston and Tucker[22] to prepare indolo[3,2,1-*j,k*]carbazole (**22**) from 1-amino-9-phenyl carbazole (**31**).

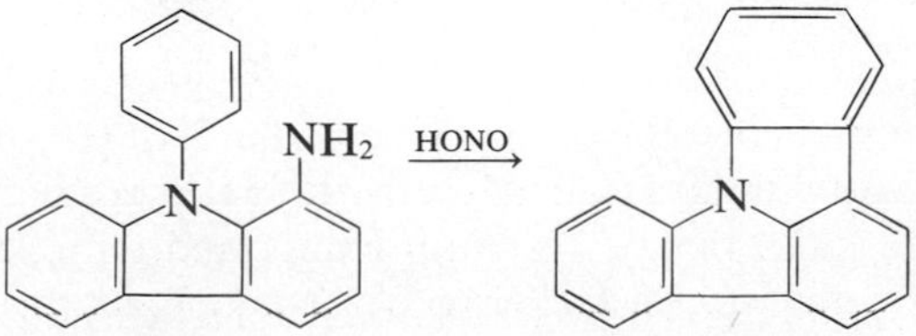

NH_2 N HONO N

31 **32**

7. Kovtinenko and Babichev[23] alkylated 1-alkyl-2-methyl benzimidazoles (**33**) with an alkyl halide having the halogen atom on a methylene group, to give quaternary salts of type **34**. Treatment of these salts with anhydrides gave pyrrolo[1,2-*a*]benzimidazoles of type **35**.

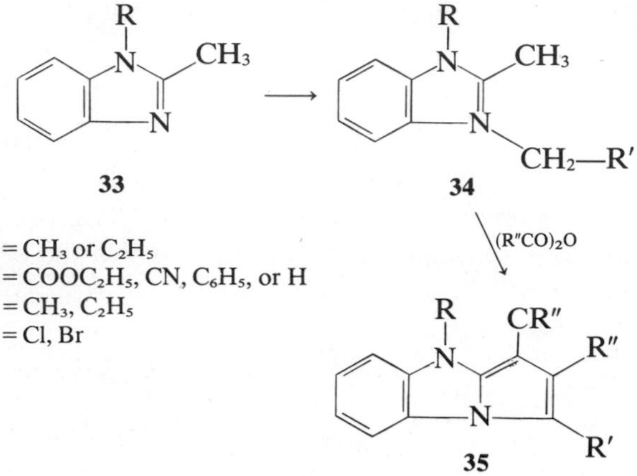

33 **34** **35**

$(R''CO)_2O$

R = CH_3 or C_2H_5
R′ = $COOC_2H_5$, CN, C_6H_5, or H
R″ = CH_3, C_2H_5
X = Cl, Br

8. In their study on diynes, Muller and Zountsas[24] prepared fused pyrroloindoles. Tris(triphenylphosphine)rhodium chloride acted on the

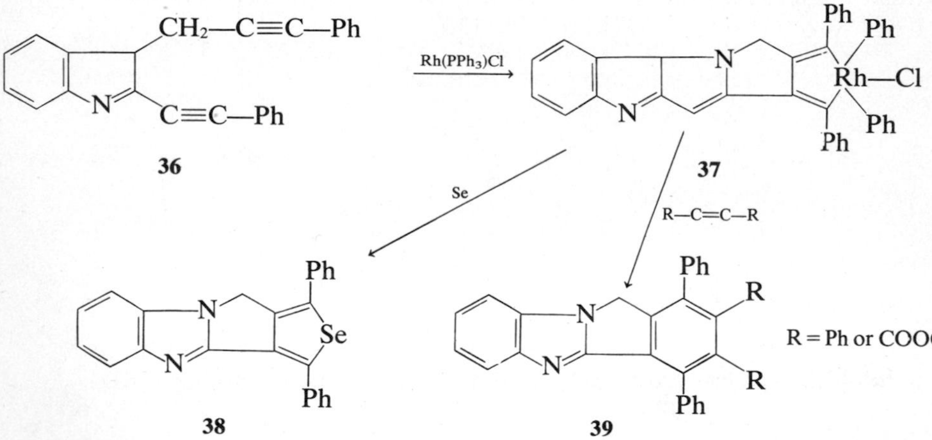

36 $Rh(PPh_3)Cl$ **37**

Se R—C≡C—R

38 **39** R = Ph or $COOC_2H_5$

diacetylenic benzimidazole (**36**) to give chloro 1*H*-pyrrolo[1,2-*a*]-benzimidazole-2,3-diylidenbis (phenylmethylidyne)bistriphenylphosphine rhodium (**37**). Treatment of this rhodium compound with selenium gave 1,3-diphenyl-10*H*-seleno[3′,4′:3,4]pyrrolo[1,2-*a*]benzimidazole (**38**). The rhodium compound (**37**) reacted with disubstituted aceytenes to form the 11*H*-isoindolo[2,1-*a*]benzimidazole (**39**). (These compounds are written in the "methylene" form but the potential for tautomerism to the N–H form exists—see tautomerism.)

9. Likhitskaya and Babichev[25] used the reaction of 2-(2-hydroxymethylphenyl)benzoselenazole (**40**) and phosphorous tribromide to prepare isoindolo[1,2-*b*]benzoselenazole as the hydrobromide salt (**41**). Treatment of this salt with sodium hydroxide gives the free base **42**. Pyrido-[2,3-*c*]benzoselenazolo-[3,2-*a*]pyrrole (**43**).

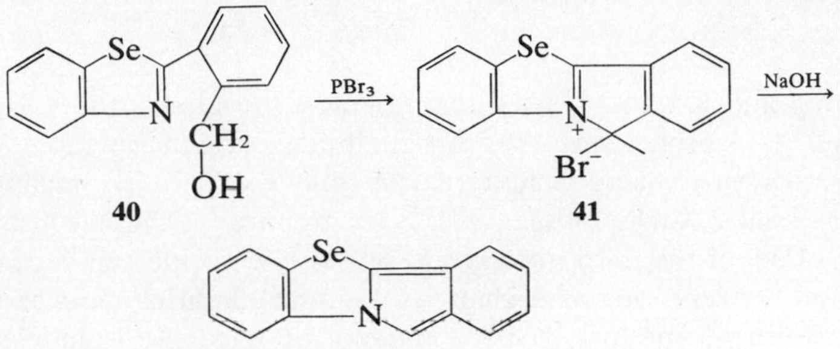

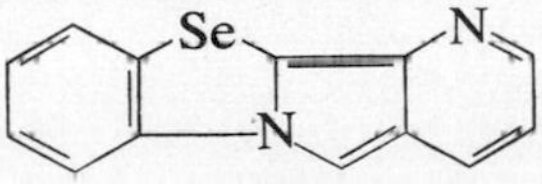

$$\mathbf{40} \xrightarrow{PBr_3} \mathbf{41} \xrightarrow{NaOH} \mathbf{42}$$

43

10. Quarternary salts of heterocyclic systems having an active *N*-methylene group and an open carbon α to the nitrogen (e.g., **44**) react with picryl chloride (**45**) to form a pyrrole ring.

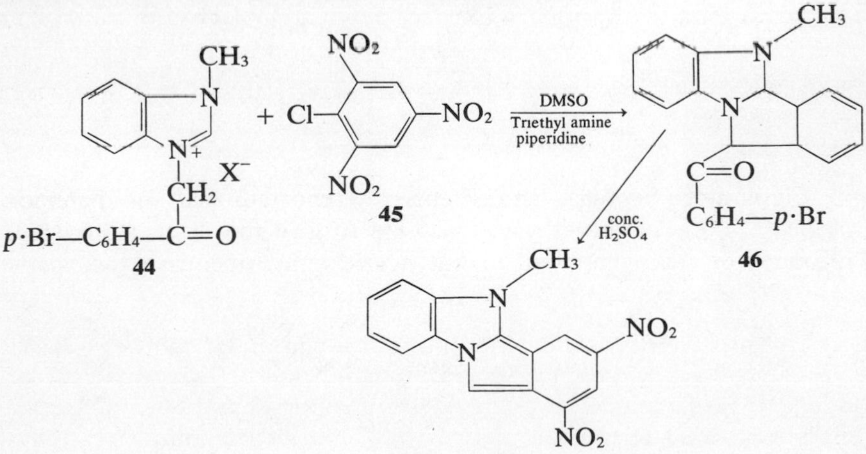

$$\mathbf{44} + \mathbf{45} \xrightarrow[\text{piperidine}]{\text{DMSO, Triethyl amine}} \mathbf{46} \xrightarrow{\text{conc. } H_2SO_4} \mathbf{47}$$

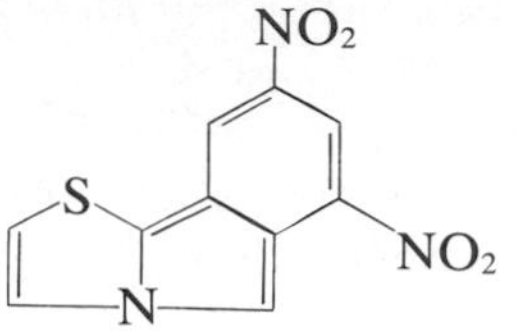

48

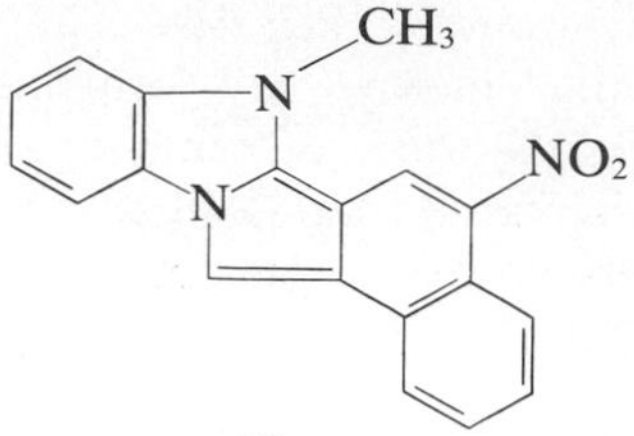

49

Reuschling and Krohnke[86] used this reaction to prepare the 1,3-dinitro-5-methyl-11-(4-bromobenzoyl)-5*H*-benzimidazo[2,1-*a*]isoindole (**46**) which was debenzoylated in sulfuric acid (giving **47**). In like manner, 6,8-dinitrothiazolo[2,3-*a*]isoindole (**48**) was prepared. The electron withdrawing effect of the third nitro group of picryl chloride can be simulated by a fused benzene ring as 1-chloro-2,4-dinitronaphthalone was used to prepare 5-nitro-7-methyl-7*H*-benzimidazo[2,1-*a*]benz[*e*]isoindole (**49**).

B. Pyrazoles

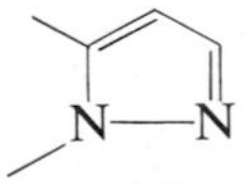

1,5

1. Becker and Bottcher[26] used the Wolff reaction to effect the ring contraction of **50** to a pyrazole (**51**).

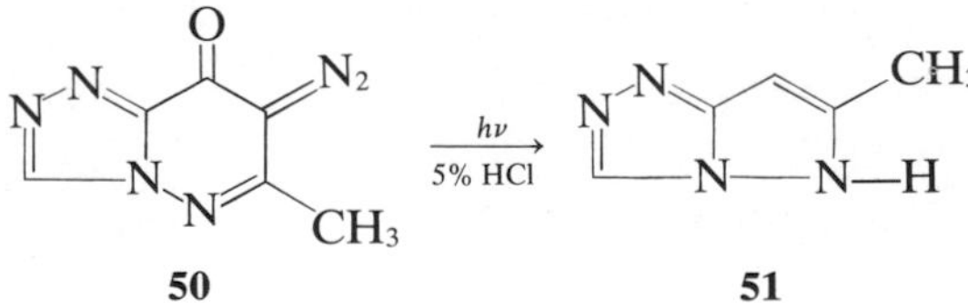

$\xrightarrow[\text{5\% HCl}]{h\nu}$

50 51

2. Claramunt, Fabrega, and Elguero[27] reported that 3,4-diamino-triazole (**52**) reacts with ethylacetoacetate to give the diazepinone (**53**). Treatment of this compound in hot acetic anhydride gave the acetyl

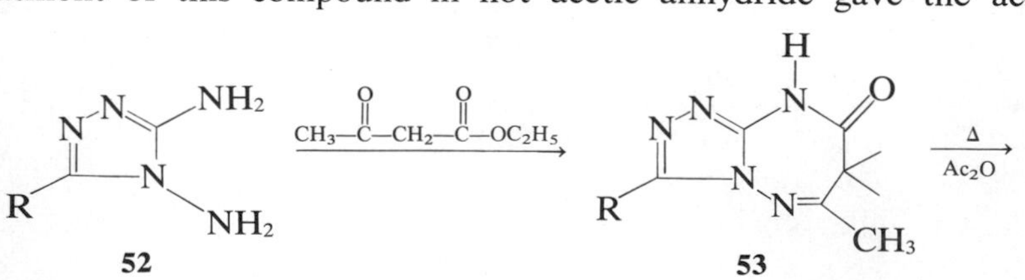

$\xrightarrow{CH_3-CO-CH_2-CO-OC_2H_5}$ $\xrightarrow[Ac_2O]{\Delta}$

52 53

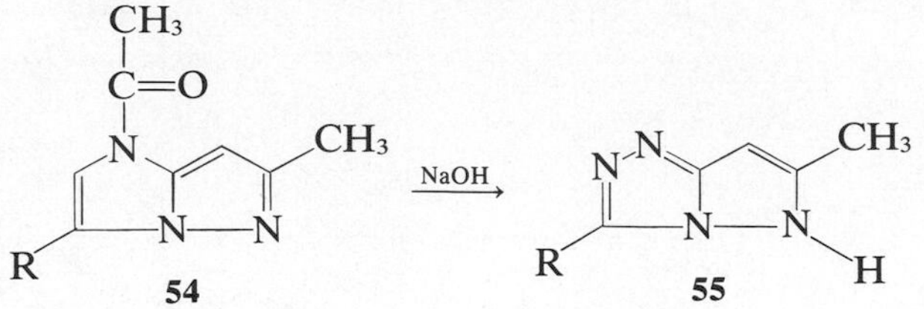

54 $\xrightarrow{NaOH}$ 55

R = H, CH_3, C_2H_3, and benzyl

compound **54** which, in base, yielded the 5*H*-pyrazolo[*s*,1-*c*]-*s*-triazole (**55**). This series of reactions had been reported earlier by Gehlen and Drohla[28] to yield initially the triazepinone (**56**) and ultimately either the

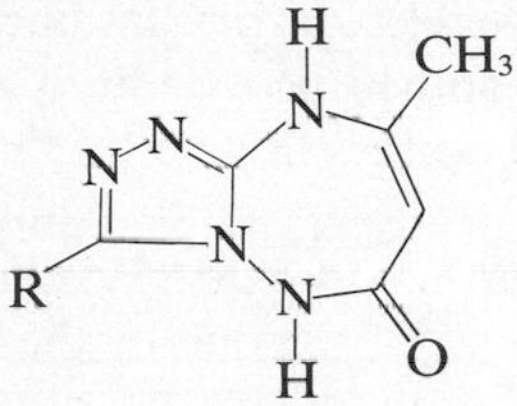

56 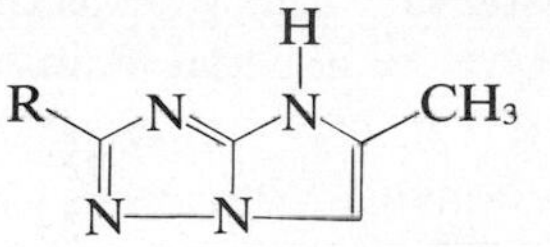 57 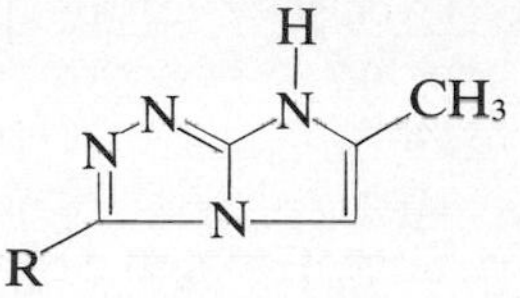 58

imidazo[1,2-*b*-*c*]-*s*-triazole (**57**) or the imidazo[2,1-*c*-*b*]-*s*-triazole (**58**). Claramunt et al.[27] ruled out structures **57** and **58** as possible base hydrolysis products on the basis of their nmr studies. The nmr spectra showed hydrogens on carbon atoms which was not attached to a nitrogen (δ's of 5.55, 5.50, and 5.46 for R = H, CH_3, and C_2H_5, respectively). Such a condition suggested a pyrazole ring and structure **55** was accordingly assigned. [Compound **55** (R = CH_3) prepared by a different route had been previously reported by Bailey, Knott, and Marr.[29]]

C. Imidazoles

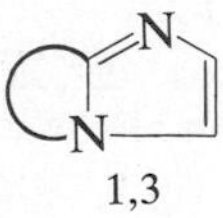

1,3

1. The most commonly used method for the synthesis of fused 1,3 imidazoles is the alkylation of amines of type **59** with α-halo carbonyl or related compounds.

a. The reaction of an α-haloketone and an amine of type **59** gives quaternary salts (**60**) which can then by cyclized to form fused imidazoles (**61**).

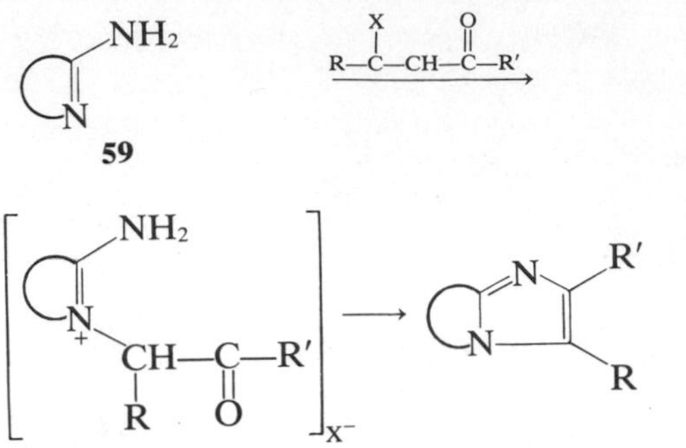

X = Cl or Br

Cyclization of the quaternary salts (**60**) to the fused imidazole has been effected in aqueous base,[30–32] aqueous acid,[33–34] and in hot water.[35] A few examples of the direct conversion (i.e., no excess amine in the reaction mixture or no aqueous base used in the work up which would remove HX and promote cyclization) of the amino compound (**59**) and the bromoketone to imidazoles have been reported.[30–31] In one of these the ketone used had two phenyl groups[30] and the amine in the other example had strong electron withdrawing substituents on the ring.[31] Thus electronic and/or steric factor appears to be involved.

Kickhöfen and Kröhnke,[35] using the following sequence of reactions, showed that 2-aminothiazole (**62**) gave 6-phenylimidazo[2,1-*b*]thiazole (**63**) rather than the 5-phenyl compound (**64**). Kröhnke, Kickhöfen, and

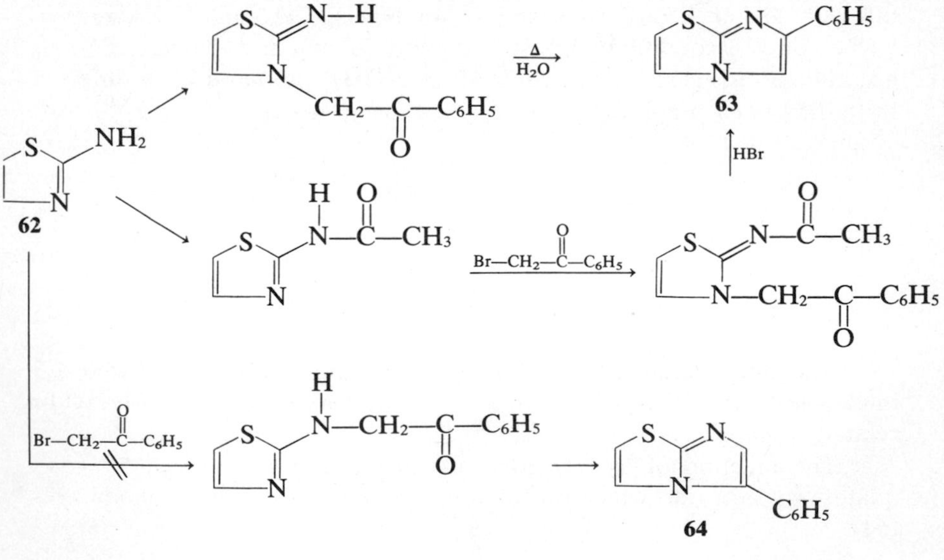

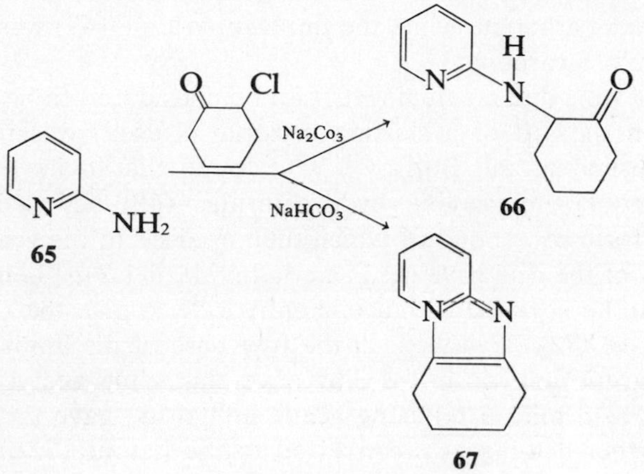

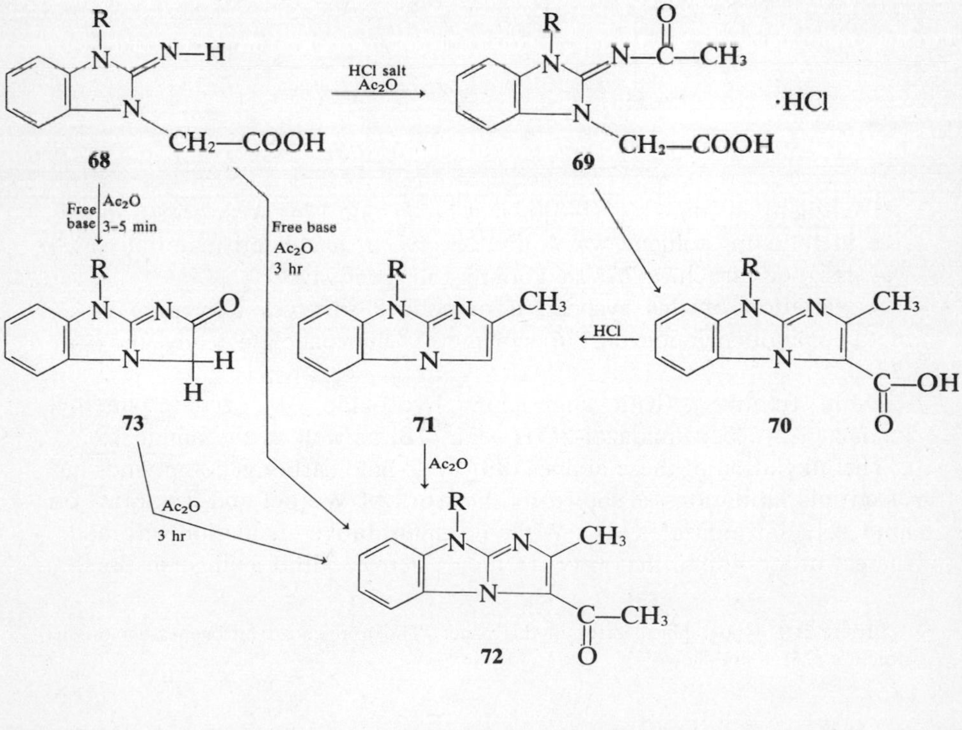

Thoma[36] had shown earlier that phenacyl bromide alkylated 2-aminopyridine on the ring nitrogen. By analogy with these two examples, it seems reasonable to conclude that alkylation of 2-amino compounds of type **59** can be considered to proceed as shown in most instances; however, the possibility that this pattern could be altered by medium or by large steric or electronic factors should not be ruled out. A steric effect was reported by Rietman[37] from a study on a 6,5 system. 2-Aminopyridine (**65**) and α-chlorocyclohexanone gave the exocyclic substituted product **66** rather than the ring substituted material **67**. This work was confirmed by Campbell and McCall[38] who further found that pH was a factor in this reaction. The cyclohexanone (**66**) was obtained in aqueous sodium carbonate while the imidazopyridine (**67**) was formed in aqueous sodium bicarbonate.

An effect of time and medium (HCl can be considered to be part of the medium when the salt of a starting material is used) was reported by Simonov, Anisimova, and Borisov[39] who found that 1-alkyl-3-carboxymethyl-3-iminobenzimidazoline hydrochlorides (**68**) acylated in acetic anhydride to form compound **69** which then cyclized to the acid (**70**). Decarboxylation of the acid gave the 2-methylimidazo[1,2-*a*]-benzimidazole (**71**) which can be acylated with acetic anhydride to give the 2-methyl-3-acetyl derivative (**72**). Treatment of the free base of the imidazoline (**68**) in refluxing acetic anhydride for 3 hr gave this same acetyl compound (**72**), while 3 to 5 min in refluxing acetic anhydride gave the lactam **73** which, on further heating, was converted to the ketone (**72**).

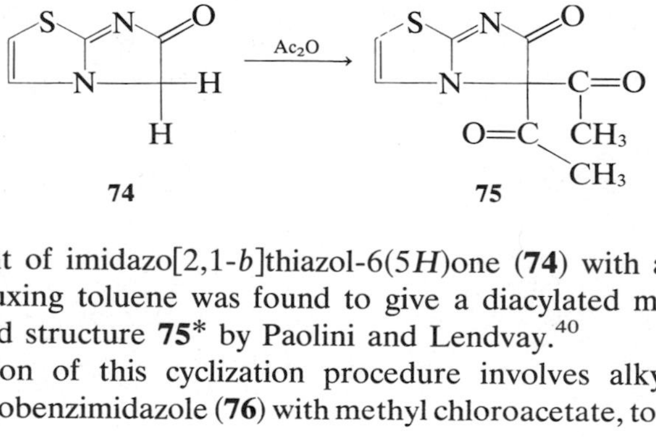

74 $\xrightarrow{Ac_2O}$ **75**

Treatment of imidazo[2,1-*b*]thiazol-6(5*H*)one (**74**) with acetic anhydride in refluxing toluene was found to give a diacylated material which was assigned structure **75*** by Paolini and Lendvay.[40]

A variation of this cyclization procedure involves alkylation of 2-methylaminobenzimidazole (**76**) with methyl chloroacetate, to give the ester (**77**).

Upon treatment with ammonium hydroxide, it formed 1-methyl imidazo[1,2-*a*]benzimidazol-2(3*H*)-one (**78**) as well as the amide **79**.[41]

The alkylation of these amines (**80**) by α-halo carbonyl compounds has electronic limitations as shown by the work of Werbel and Zamora[31] on imidazo[2,1-*b*]thiazoles (**81**). While no quantitative evaluation was made, these workers found that very strongly electron withdrawing substituents

* There is a typographical error in this paper. The bridgehead nitrogen atom of this structure (**74**) is not shown.

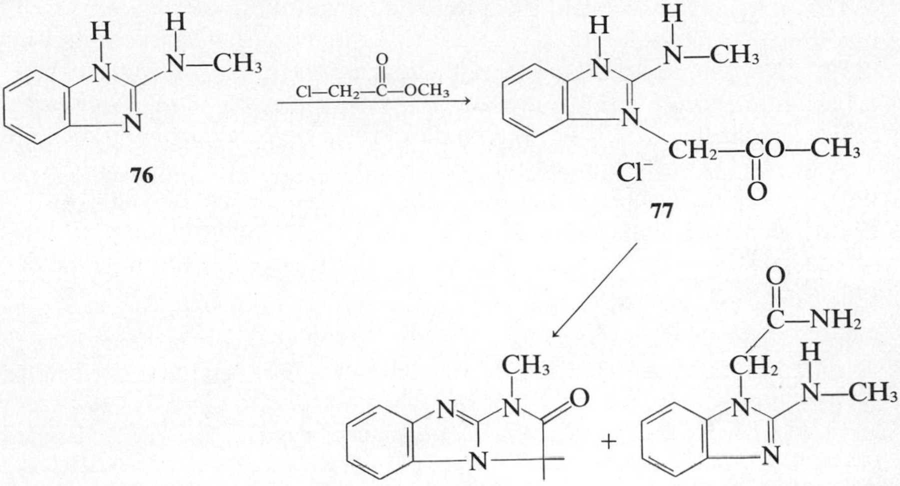

$Cl-CH_2-C(=O)-OCH_3$

76 → 77 → 78 + 79

in the 5-position of a 2-aminothiazole (**80**) will inhibit its capacity to react with phenacylbromide.

The acetals of α-haloaldehydes are more stable than the free aldehydes and are often used in reactions of this type.[42]

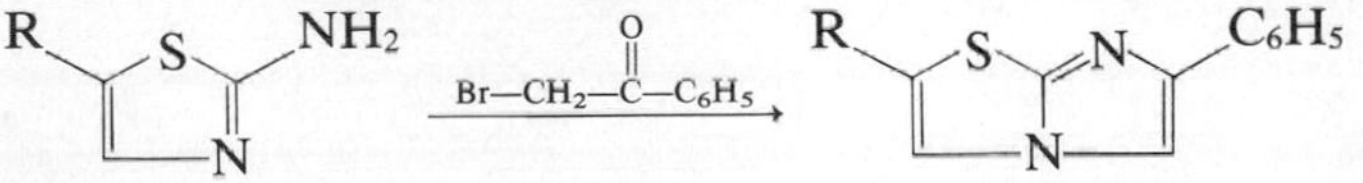

$Br-CH_2-C(=O)-C_6H_5$

80 → 81

R	Result
H	reaction
Cl	reaction
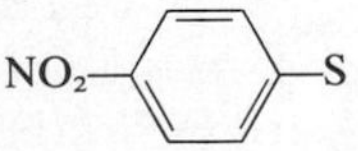 $NO_2-C_6H_4-S$	reaction
NO_2	no reaction
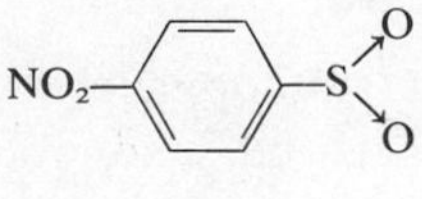 $NO_2-C_6H_4-SO_2$	no reaction

This approach has been used for the preparation of derivatives of imidazo[2,1-*b*]thiazole (**82**),[30,31,33,35,43–52] imidazo[2,1-*b*]benzothiazole (**83**),[31,48,53,55] imidazo[2,1-*b*][1,3,4]thiadiazole (**84**),[31,55–60] imidazo[1,2-*d*][1,2,4]thiadiazole (**85**),[61] imidazo[2,1-*b*]benzoxazole (**86**),[31] imidazo-[1,2-*a*]imidazole (**87**),[62–65] 9*H*-imidazo[1,2-*a*]benzimidazole (**88**),[61–68] 1*H*-imidazo[1,2-*a*]benzimidazole (**89**),[64] imidazo[2,1-*b*][1,3,4]oxidiazole (**90**),[32] 1*H*-imidazo[2,1-*f*]purine (**91**),[70] 1*H*-imidazo[1,2-*d*]tetrazolo (**92**),[51] and 1*H*-imidazo[1,2-*b*]-*s*-triazole (**93**),[51] imidazo[2,1-*b*]-1,3,4-selenadiazoles (**94**),[71,72] and imidazo[2,1 : 2,7]thiazolo[4,5-*b*]quinoxaline (**95**).[73]

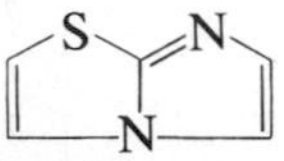

82 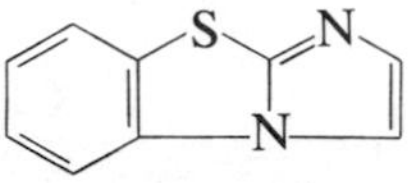 **83** 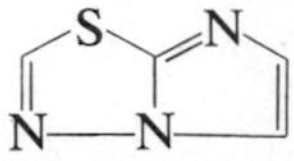 **84**

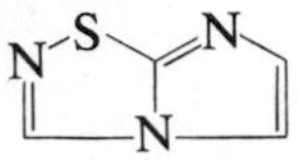

85 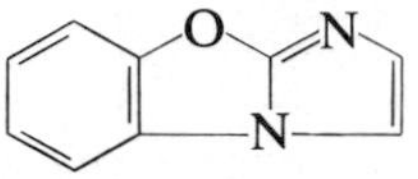 **86** 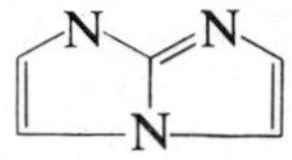 **87**

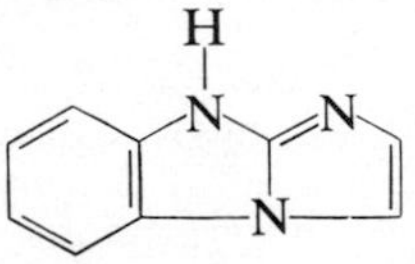

88 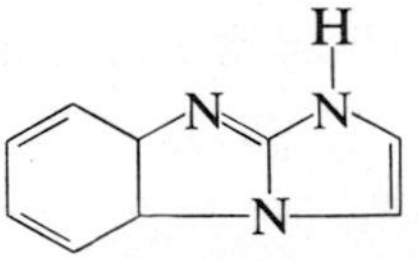 **89** 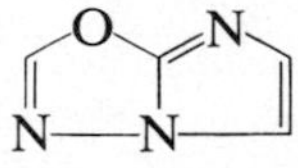 **90**

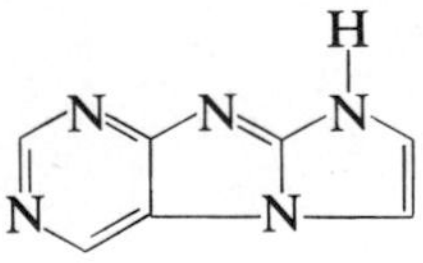

91 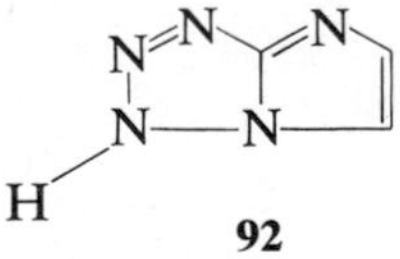 **92** 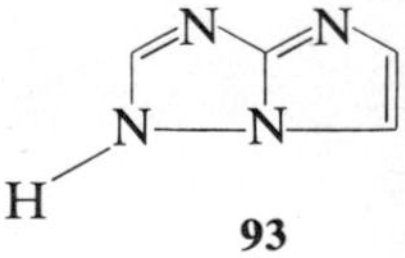 **93**

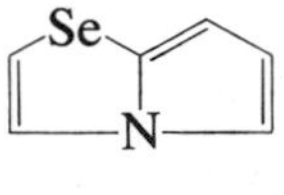

94 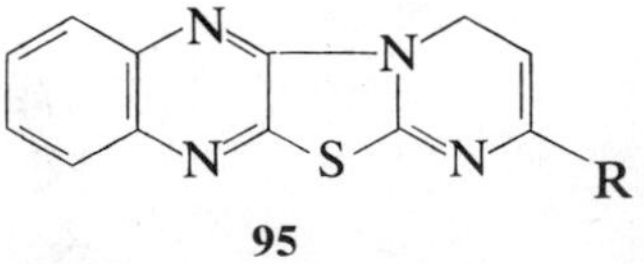 **95**

b. Iwai and Hiroaka[53] treated 2-aminothiazole (**62**) with propargyl bromide, the anhydride of bromoacetone, and obtained the propynyl compound (**96**) which on treatment in base gave-6-methylimidazo[2,1-*b*]-thiazole (**97**). Propargyl bromide was found to be superior to bromoacetone for the preparation of compound **97**.

62 $\xrightarrow{\text{Br—CH}_2\text{—C≡C—H}}$ [thiazolium: 2-NH_2, N⁺—CH_2—C≡C—H, Br^-] **96** $\xrightarrow{\text{NaOEt}}$ [imidazothiazole with CH_3] **97**

c. The reaction of 2-imino-3-carboxymethylthiazoline (**98**) with phosphoryl chloride was used by Paolini and Lendvay to prepare 6-chloroimidazo[2,1-*b*]thiazole (**99**). Parrick and Pearson[74] used this approach to prepare 2-chloroimidazo[2,1-*b*]benzothiazole (**99**).

98 [S, =NH, N—CH_2—C(=O)—OH] $\xrightarrow{\text{POCl}_3}$ **99** [S, N, Cl]

d. A group at Janssen Pharmaceutica[34] reported the isolation of 6-(2-thienyl)imidazo[2,1-*b*]thiazole (**100**) from the feces of chickens who had been fed 2-acetylamino-3-[2-hydroxy-(2-thienyl)ethyl]thiazoline (**101**). No further reports concerning the possible synthetic utility of this procedure has been put forward.

100 [S, N—C(=O)—CH_3, N—CH_2—C(H)(OH)—thienyl (S)] ⟶ **101** [S, N, N, thienyl (S)]

e. Alkylation of theophylline salts having a good leaving group at the 8-position (**102**) (bromine[75] and methylthio[76] have been used) with phenacyl bromide gave ketone **103** which on treatment with ammonia gave the 6,8-dimethyl-5,7-dioxo-2-phenylimidazo[2,1-*f*]purine (**105**). The amine (**104**) is an apparent intermediate, and this method may be looked upon as a variation of the first method discussed, that is alkylation of a ring nitrogen diposed 1,3 to an exocyclic amine.

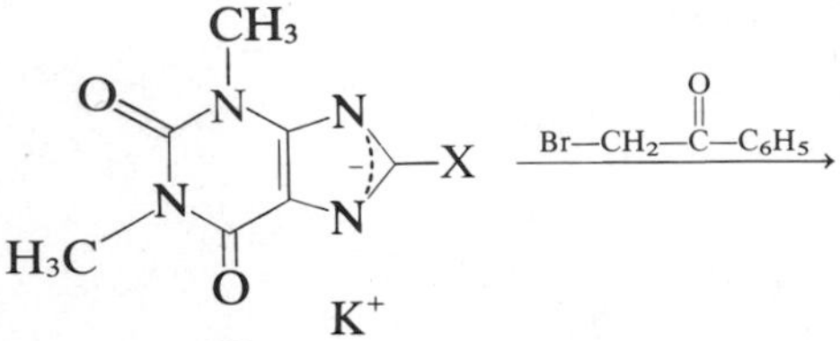

$Br—CH_2—C(=O)—C_6H_5$ →

102

$X = Br$ or SCH_3

NH_3 140–160°

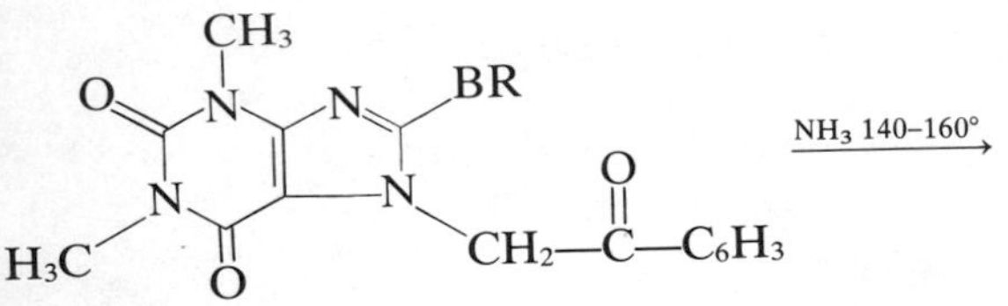

103

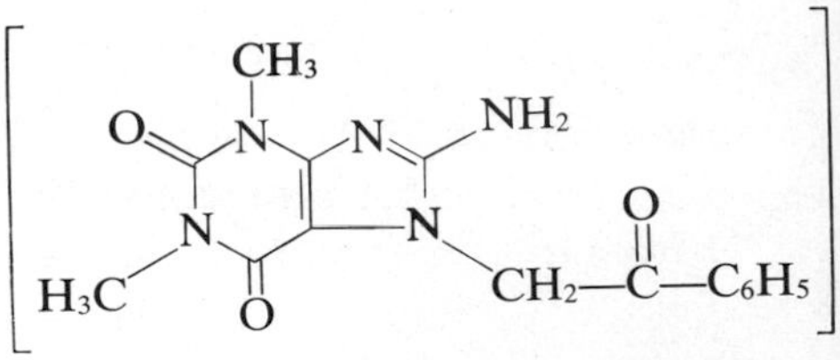

104

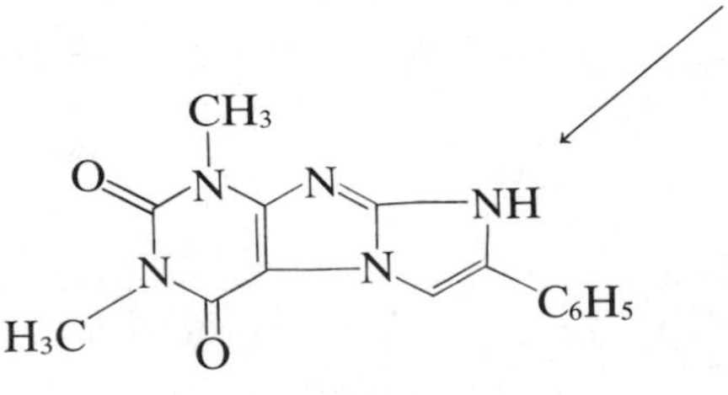

105

This reaction is general, many such imidazo purines having been prepared by Kochergin and co-workers.[75,76] Additionally, imidazo[1,2-*a*]-imidazole (**77**) and 1*H*-naphtho[1,2-*d*]imidazo[3,2-*d*]imidazoles (**106**) have been prepared in this manner.[78]

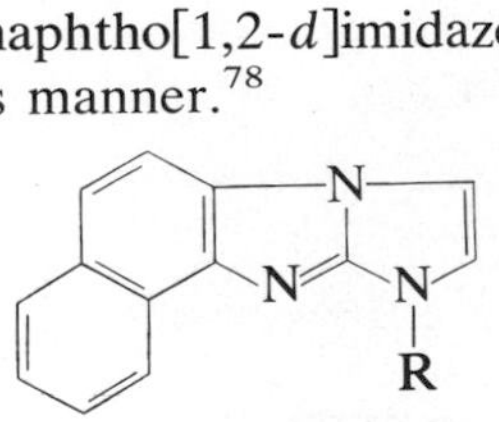

106

2. The reaction of 2-amino-6-ethoxybenzothiazole (**106**) and 1,4-naphthaquinone was reported by Rudner[79] to give a polycyclic system possessing one of two structures, **108** or **109**.

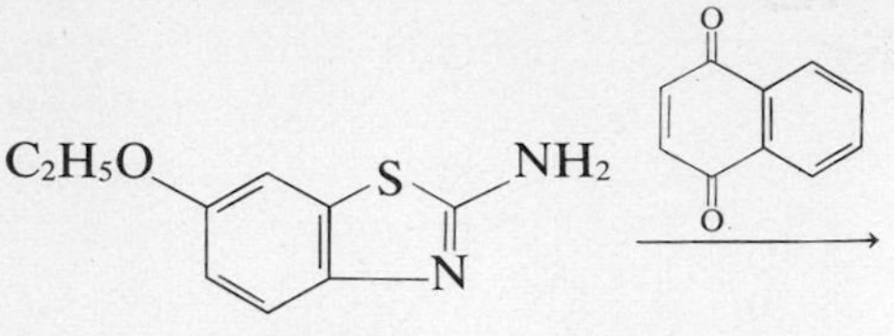

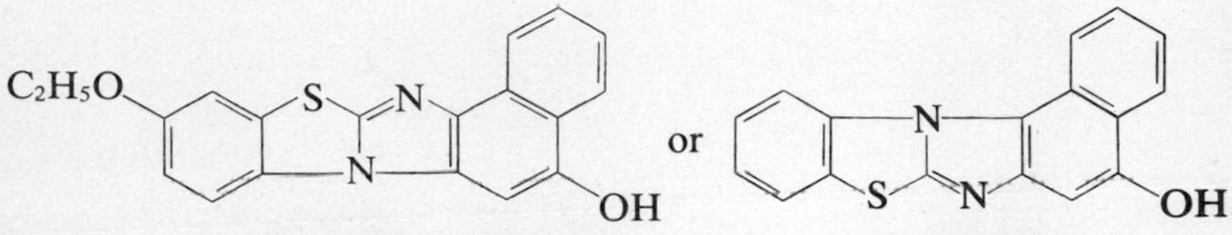

3. Menzel and co-workers[80–82] reported the synthesis of a series of pyrazolo[2,3-*a*]benzimidazoles (**109**) as shown in this reaction scheme.

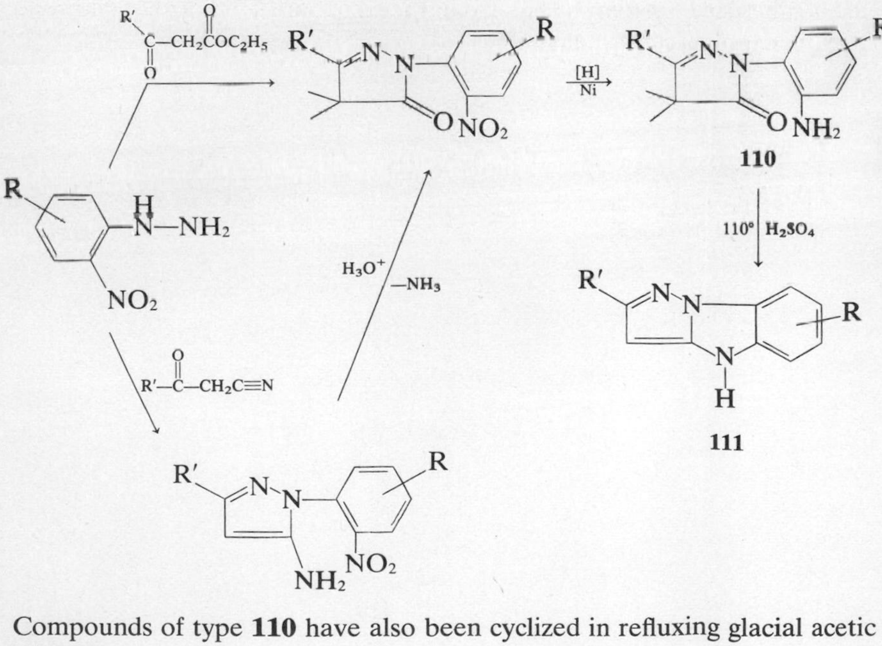

Compounds of type **110** have also been cyclized in refluxing glacial acetic acid.

4. Ochiai and Nisigawa[83] prepared 7,9-dinitrobenzimidazo[2,1-*b*]-benzothiazole (**114**) from the reaction of 2-aminobenzothiazole (**112**) and picryl chloride. The intermediate amine (**113**) was isolated and characterized.

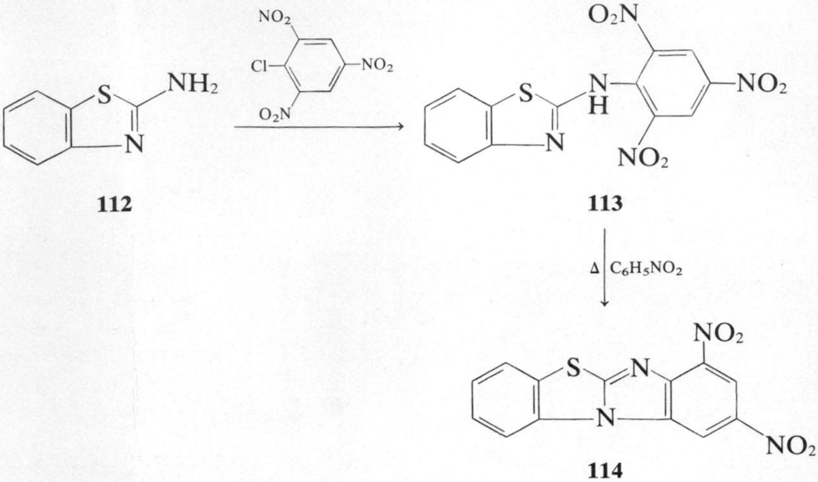

112 **113** **114**

5. Scheyer and Schwamberger[84] reported the formation of polycyclic fused imidazole systems (**116**) from the reaction of thiazolo anthrone (**115**) and polycyclic *o*-haloamines.

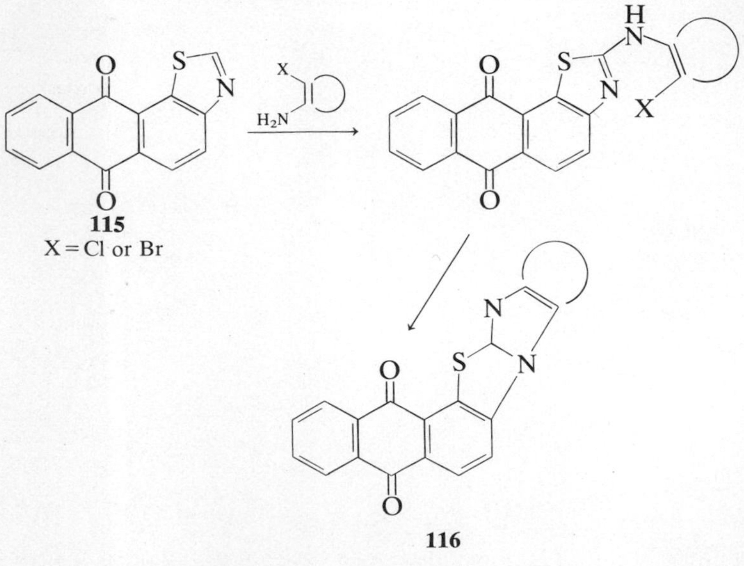

115
X = Cl or Br

116

6. Hubert and Reimlinger[85] studied the photolysis and pyrolysis of benzotriazoles of type **117**. A molecule of nitrogen was lost from the

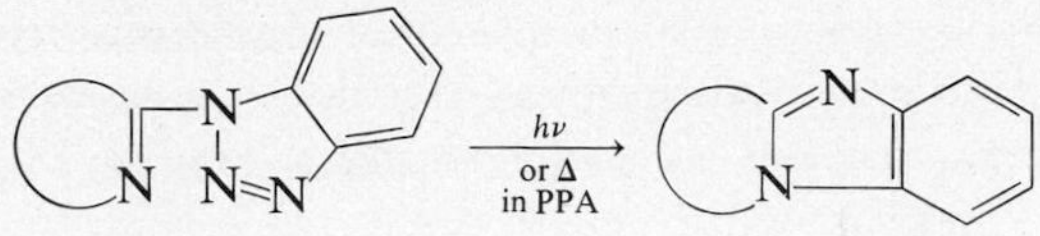

117 $\xrightarrow[\text{or } \Delta \text{ in PPA}]{h\nu}$ 118

TABLE I

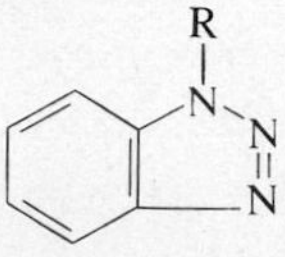

119

Starting R of benzotriazole	Product	Process	Ref.
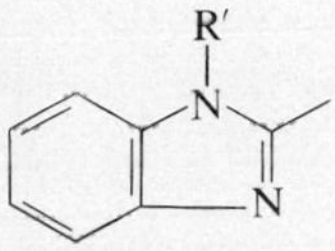 $R' = H, CH_3$, or $CH_2—C_6H_5$	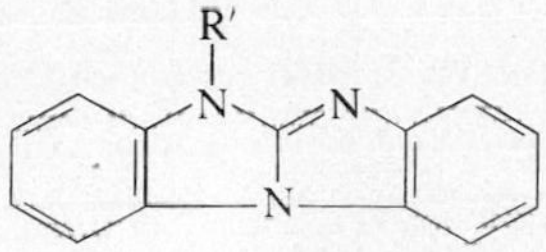	$h\nu$	85
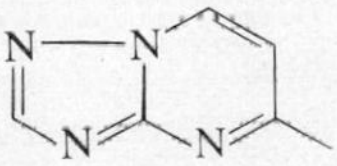	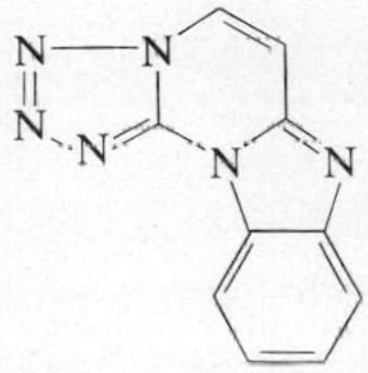	Δ In PPA	85
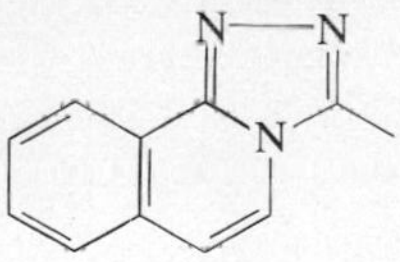	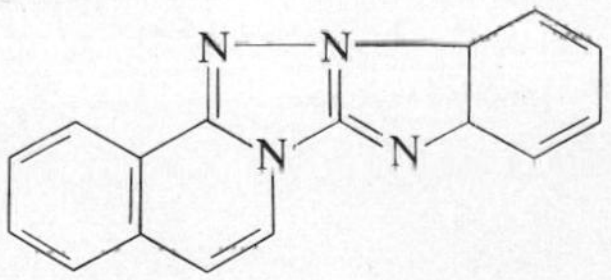	Δ In PPA	85
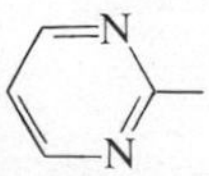	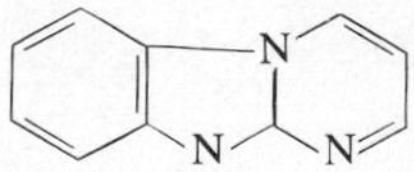	Δ In PPA	85
HO	OH	Δ In PPA	85
		$h\nu$	89
			90
		Δ Gas phase	90

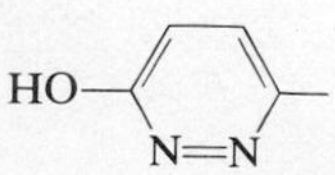

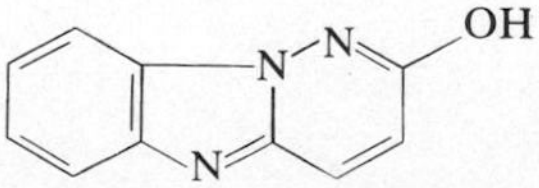

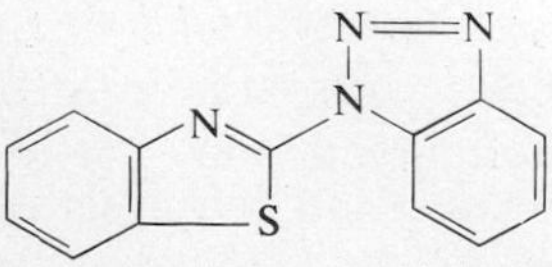

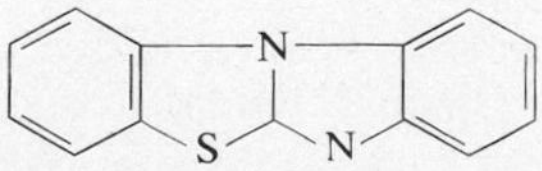

triazole ring and the reactive species formed closed onto the pyridine type nitrogen of the other nucleus to give fused imidazoles of type **118**. These reactions were carried out in polyphosphoric acid, a medium that was essential to the success of the reaction.

Table I summarizes the results of the successful reaction of triazoles of type **119**.

The photolysis of 1-(2-benzothiazolyl)benzotriazole was found by Hubert[89] with confirmation by Lin and DeJongh[90] to give benzimidazo-[2,1-*b*]benzothiazole (see Table I). The results of pyrolysis were dependent on the conditions and/or the medium. Hubert[89] found that pyrolysis in PPA gave *o,o'*-diamino phenyl sulfide while Lin and DeJongh[90] found that pyrolysis in the gas phase, at 750°, gave benzimidazo[2,1-*b*]-benzothiazole—the same as was obtained in photolysis.

7. Glushkov and Magidson[87] treated 1-ethoxyisoindole (**120**) with aminocyanoacetamide and obtained the imidazoisoindole (**121**). These workers state that formation of product **121** "... demonstrated that the reaction of lactim esters with α-amino-α-cyanoacetamide is applicable to partially aromatized lactim esters although the reactivity of such aromatized systems is considerably reduced."

OC_2H_5 N **120** $\xrightarrow{HCl}$ [O, C, H_2N, C, H, N≡C, N, H—N] → O, H_2N—C, N, H_2N, N **121**

8. The reaction of α-aminoaldehydes and cyanamide was shown by Lawson[63] to give the aminoketones (**122**) which on treatment with concentrated hydrochloric acid yielded 2,5-dialkylimidazo[1,2-*a*]-imidazoles (**123**).

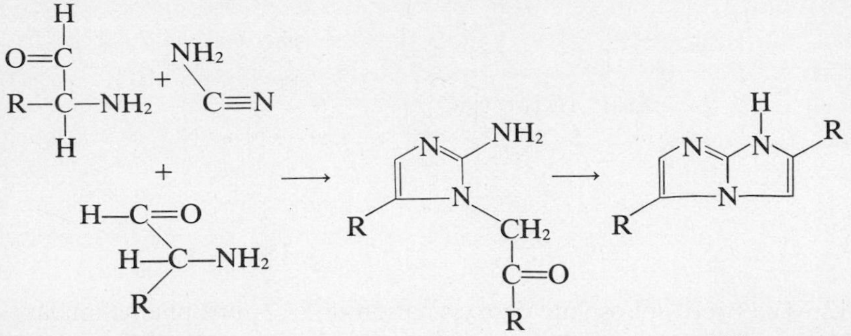

R = alkyl

122

123

9. Keutzberger and Meyer[88] treated 3,5-diaminotriazole (**124**) with acyloins in acetic acid, using *p*-toluenesulfonic acid as the catalyst and obtained imidazo[1,2-*b*]-*s*-triazoles.

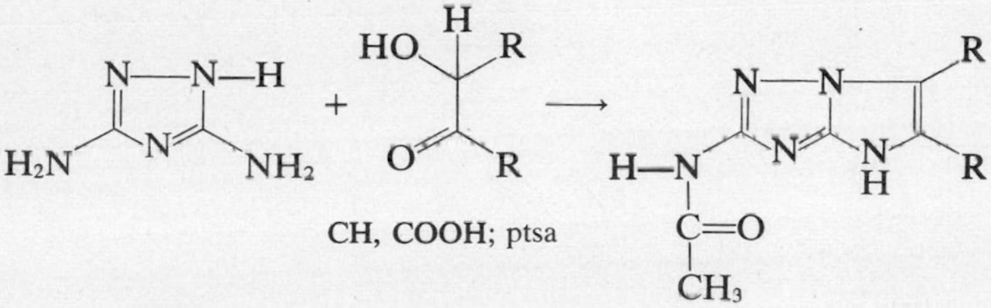

CH, COOH; ptsa

124

125

10. The reaction of 2,3-dichloroquinoxaline (**126**) with thiourea gave **2-aminothiazolo[4,5-*b*]quinoxaline** (**127**) which reacted with a second molecule of 2,3-dichloroquinoxaline to give Quinoxalino[2,3 : 4′5′]-imidazo[2′,1′ : 2,3]thiazolo[4,5-*b*]quinoxaline (**128**).[91]

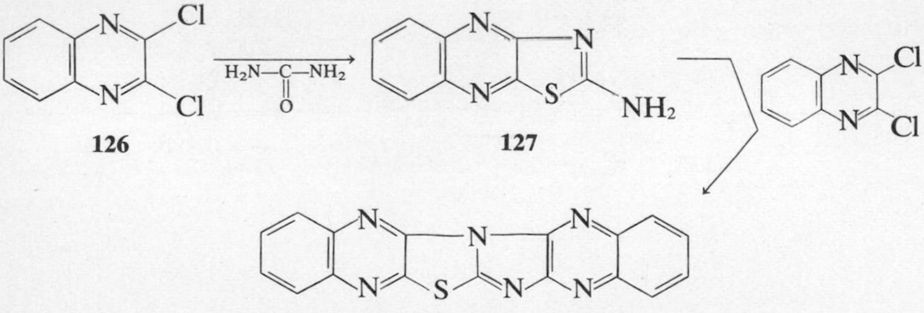

126

127

128

11. Mignonac-Mondon, Elguero, and Lazaro[92] treated 1-(2-chlorophenyl)-3-methyl-5-aminopyrazole **129** with potassium in ammonia to obtain 2-methylpyrazolo[2,3-*a*]benzimidazole **130**.

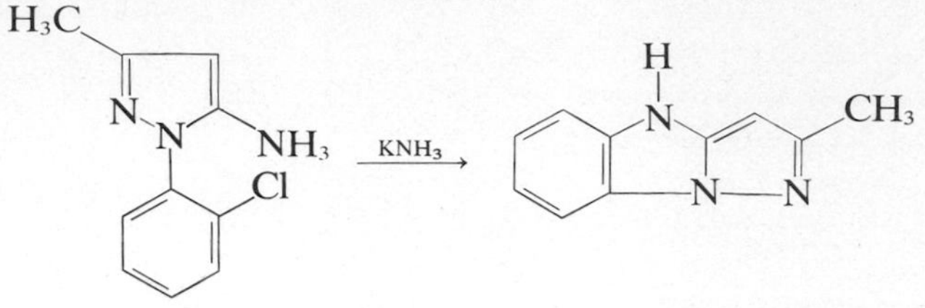

H_3C NH$_3$ Cl N N KNH$_3$ → H N CH_3 N N

129 **130**

12. The triethylphosphite deoxygenation of 2-(2-nitrophenyl)imidazole (**131**) was reported by Tsuge and Samura[120] to give the inidazolo-benzimidazole (**132**).

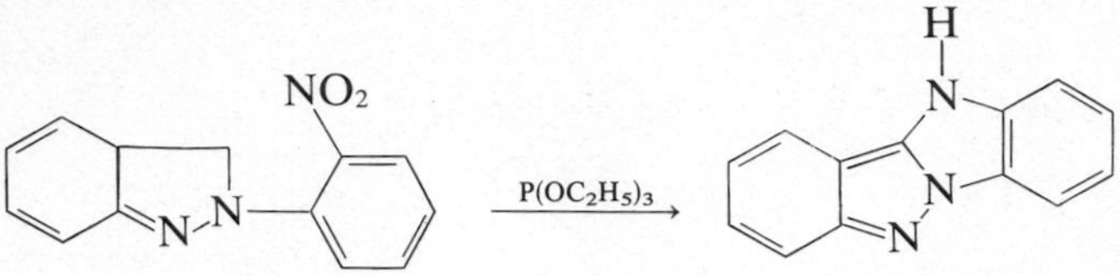

NO_2 N N $P(OC_2H_5)_3$ → H N N N

131 **132**

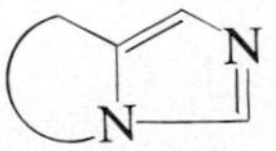

N N

1,4

Imidazoles of the 1,4 type are prepared *via* the cyclization of amides, ureas, or thioureas of type **133**. The cyclization of the amides is effected in phosphoryl chloride,[93–101] to give compounds of type **134** while the ureas and thioureas are cyclized thermally[96,102,103] to give oxo or thio imidazoles (**135**).

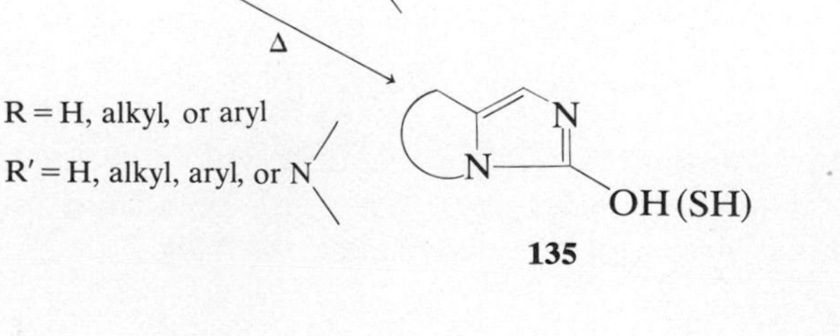

R H N N—H C=O(S) R′ **133** where R = H, alkyl or aryl $POCl_3$ → R N N R′ **134**

where R = N Δ

R = H, alkyl, or aryl

R′ = H, alkyl, aryl, or N

N N OH (SH)

135

Ring closures of these types have been used to prepare derivatives of imidazo[5,1-*b*]thiazoles (**136**),[94,103] imidazo[5,1-*b*]benzothiazoles (**137**),[94,101,102] 4*H*-imidazo[1,5-*a*]benzimidazoles (**138**),[96–99] and imidazo[5,1-*b*]benzoxazoles (**139**).[95,100]

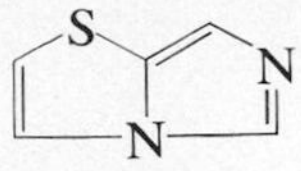

136

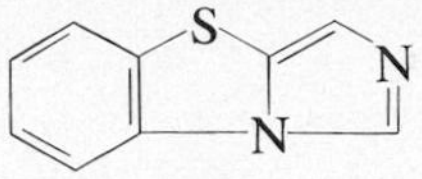

137

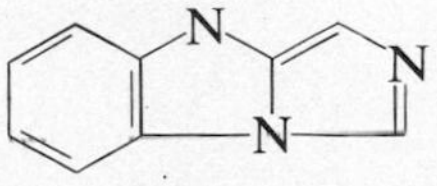

138

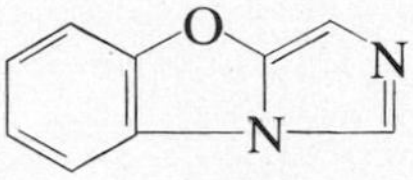

139

D. Triazoles

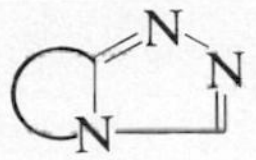

1,3,4

The following four procedures are related in that they all involve hydrazide intermediates, although these intermediates are not always isolated.

1a. The reaction of acylating agents with hydrazines of type **140** has been used to form fused 1,3,4 triazoles (**142**). Direct triazole formation from hydrazines (**150**) was achieved when formic acid,[105] formamide,[106]

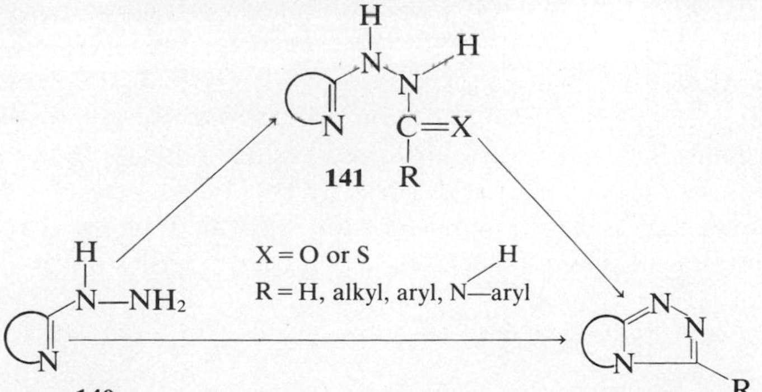

141

X = O or S
R = H, alkyl, aryl, NH—aryl

140

142

R = H, alkyl, aryl, NH_2, OH, SH

and orthoesters[107–110] were used. When anhydrides,[112] acid chlorides,[105,110] and acids[105] other than formic acid were used, hydrazides [**141**; R = alkyl or aryl, X = 0] were obtained which were ring closed in phenol,[105] acetic anhydride–phosphoryl chloride,[111] and phosphoryl chloride[110] and polyphosphoric acid.[113]

Potts and Hussain[110] found that electronic and/or steric factors were involved in these ring closures. 2-Hyrazino-4-methylthiazole (**143**, R = H) when heated under reflux with formic, acetic, or propionic acid, for 6 to

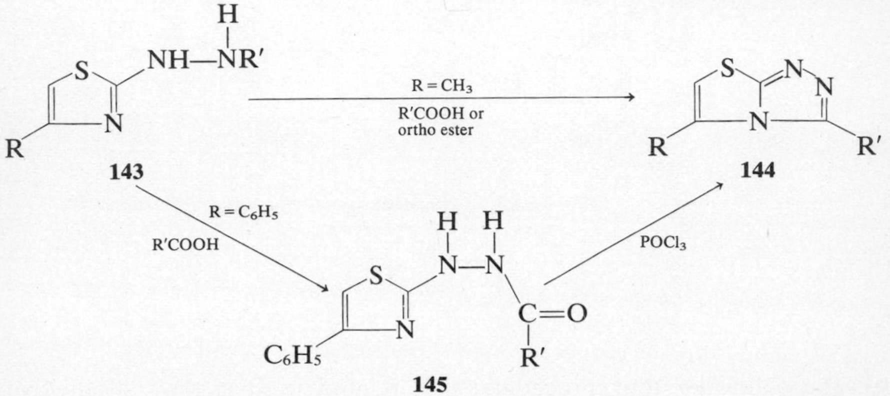

R = CH_3
R'COOH or ortho ester

143

144

R = C_6H_5
R'COOH

$POCl_3$

145

8 hr, gave the thiazolo-[2,3-*c*]triazole (**144**) directly. When 2-hydrazino-4-phenylthiazole was the starting material, these same conditions gave the hydrazides (**145**) which were then closed in phosphoryl chloride.

Treatment of hydrazines (**140**) with carbon disulfide in basic medium gave fused triazoles wherein R is thiol (**142**; R = SH).[112,114–119] Fused 3-aminotriazoles (**143**; R = NH_2) were obtained by treating the hydrazines (**140**) with cyanogen bromide.[112,115–118]

The reaction of hydrazines with isocyanates and isothiocyanates in refluxing 1,2,4-trichlorobenzene gave hydroxy and thiol derivatives (**142**; R = OH and SH) directly.[105] In lower boiling solvents, benzene, semicarbazide and thiosemicarbazide respectively (**141**; X = O or S, R = C_6H_5) were obtained.[119] The thiosemicarbazide could then be ring closed to the thio-triazole (**142**; R = SH) either in refluxing pyridine or acetic anhydride.[119]

The use of hydrazines and acylating agents to prepare fused triazoles

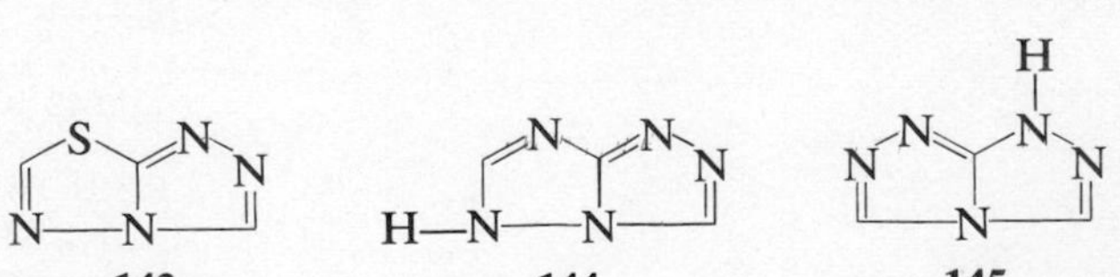

143 **144** **145**

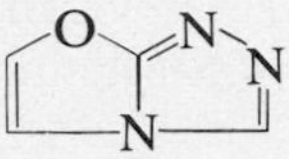

146

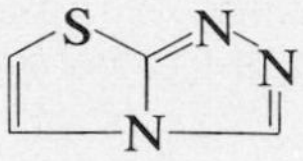

147

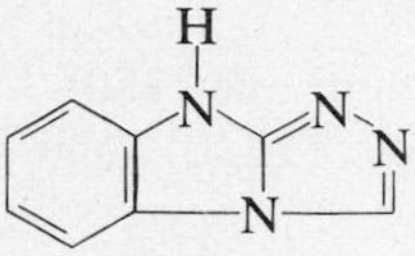

148

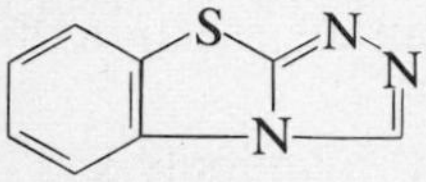

149

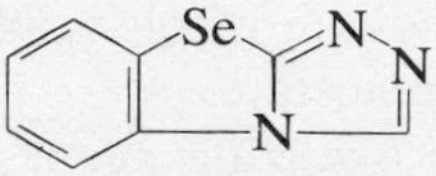

150

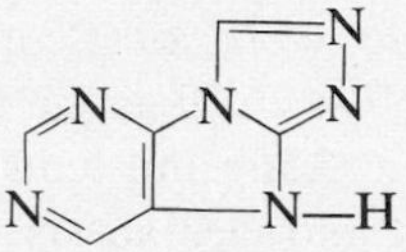

151

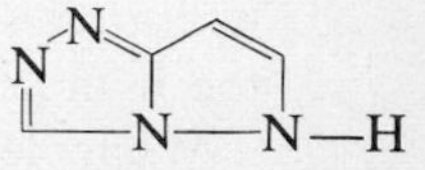

152

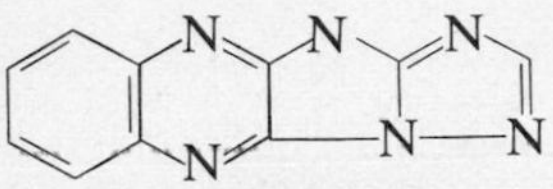

153

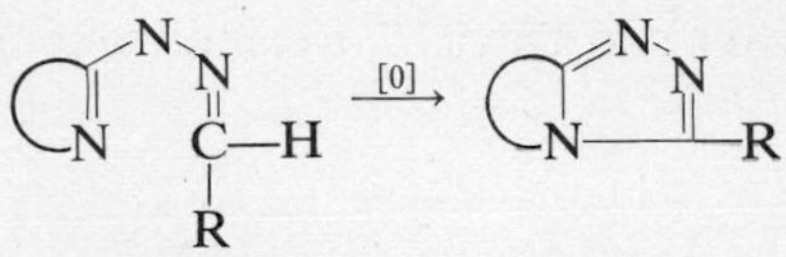

154 **155**

has been used for the preparation of derivatives of *s*-triazolo[3,4-*b*]-[1,3,4]thiadiazole (**142**),[107,108] *s*-triazolo[4,3-*b*]-*s*-triazole (**144**)[117,118] *s*-triazolo[3,4-*c*]-*s*-triazole (**145**),[119] oxazolo[2,3-*c*]-*s*-triazole (**146**),[119] thiazolo[2,3-*c*]-*s*-triazole (**147**),[119] *s*-triazolo[4,3-*a*]benzimidazole

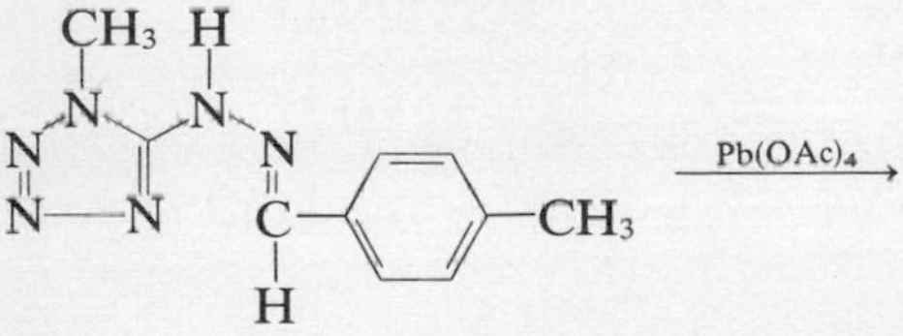

156

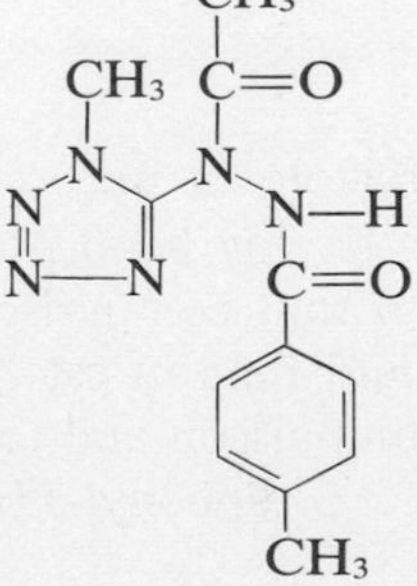

157

+

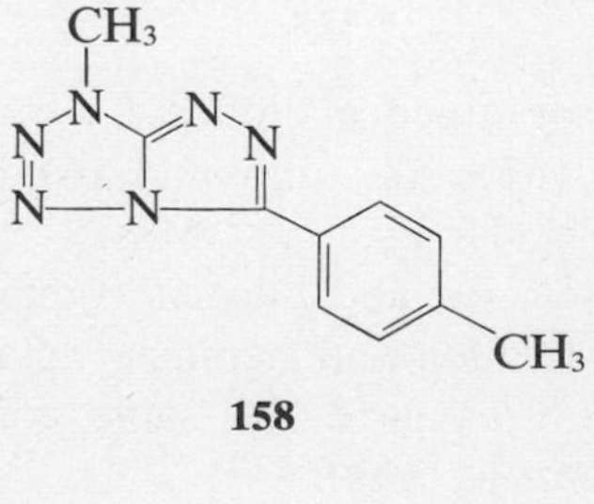

158

(**148**),[105,107,108] *s*-triazolo[3,4-*b*]benzothiazole (**149**),[75,79,81] *s*-triazolo[3,4-*b*]benzoselenazole (**150**),[105] triazolo[4,3-*e*]purine (**151**),[106] pyrazolo[5,1-*c*]-*s*-triazole (**152**),[29] and *s*-triazolo[3′,2′ : 2,3]thiazolo[4,5-*b*]quinoxaline (**153**).[73]

b. The oxidation of hydrazones of type **154** is used to give fused 1,3,4-triazoles of type **155**. Hydrazides are intermediates in this process as shown by the work of Scott and Butler[121] who isolated the acetyl hydrazide (**157**) as well as the triazolo tetrazole (**158**) from the reaction of the hydrazone (**156**) and lead tetraacetate.

While lead tetraacetate is the oxidizing agent most commonly employed,[114,121–123] ferric chloride[124] and hydrogen peroxide[119] have also been used.

Bower and Doyle[114] treated compound **159** with lead tetraacetate and

obtained a diphenyl-*s*-triazolo-*s*-triazole of uncertain structure (**160** or **162**). As compound **161** is known, having been prepared unequivocally by Hoggarth,[125] Mosby[126] suggested that comparison of Hoggarth's and Bower and Doyle's compounds would help in establishing the structure. Gehlen and Lemme[127] made the comparison and found that both procedures gave the same compound, 3,6-diphenyl-1*H*-*s*-triazolo[4,3-*b*]triazole (**161**).

Butler and O'Sullivan and Scott[128] reported that there is a limitation to the use of lead tetraacetate which is that acetoxylation is apparently favored by the presence of a sulfur atom at a potential cyclization site. These workers report that the triazolobenzothiazoles (**165**) (with a molecule of solvated acetic acid) as reported by Bower and Doyle[114] as having been obtained from the action of lead tetraacetate on the benzothiazolylhydrazones (**163**), are actually acylated hydrazides (**164**). Their

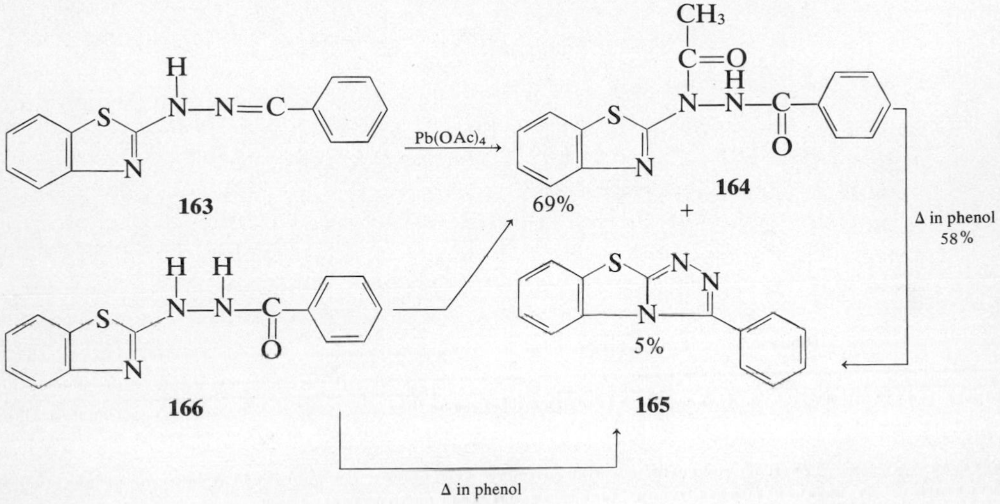

best effort at obtaining a triazolobenzothiazole (**165**) was a 5% yield along with a 69% yield of the diacylhydrazide (**164**) which was prepared independently by acylation of the hydrazide (**166**). 3-Phenyltriazolo[3,4-*b*]benzothiazole (**165**) had been reported earlier by Reynolds and Van Allan.[105] This was obtained by heating the hydrizide (**166**) in refluxing phenol. The Scott group[128] confirmed this work and found, additionally, that the *N*-acetylhydrazide (**164**) in refluxing phenol, was converted to the triazino benzothiazole (**165**).

The oxidation of hydrazones of type **154** has been used to prepare derivatives of *s*-triazolo[4,3-*b*]-*s*-triazole (**144**),[119] oxazolo[2,3-*c*]-*s*-triazole (**146**)[119] thiazolo[2,3-*c*]-*s*-triazole (**147**),[119] *s*-triazolo[3,4-*b*]-benzothiazole (**149**),[114–124] 1*H*-*s*-triazolo[4,3-*d*]tetrazole (**166**),[123] and 3*H*-*s*-triazolo[4,3-*d*]tetrazole (**167**).[121,122]

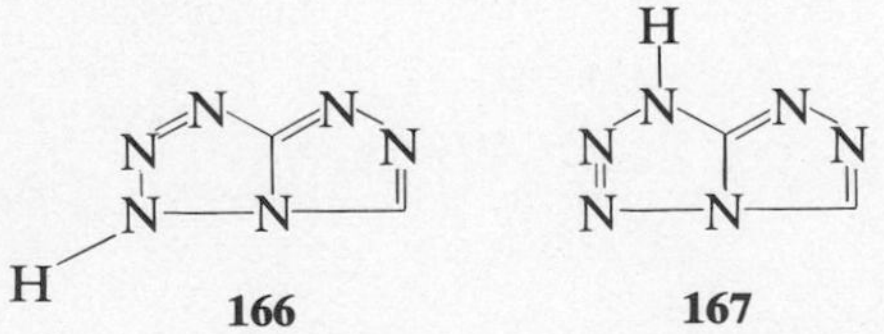

c. The reaction of hydrazones of type (**154**) with bromine in acetic acid gives hydrazidic bromides (**168**) which on treatment with aqueous sodium acetate give fused triazoles of type **155**.

154 $\xrightarrow{Br_2}$ **168** $\xrightarrow[NaOAc]{H_2O}$ **155** R = alkyl or aryl

R = alkyl or aryl

168

This procedure has been used to prepare derivatives of *s*-triazole[3,4-*d*]-tetrazole (**167**),[123] *s*-triazolo[4,3-*b*]-*s*-triazole (**144**),[129] and 3*H*-*s*-triazolo[4,3-*d*]tetrazole (**166**).[130]

d. Halo compounds of type (**166** react with aryl hydrazides in phenol, using sodium phenoxide as the base, to give fused triazoles of type **170** (R = aryl).

169 $\xrightarrow[C_6H_5OH;\ C_6H_5ONa]{H_2N—NH—CO—\phi}$ **170**

This procedure has been used to prepare derivatives of *s*-triazolo[3,4-*b*]benzothiazole (**170**).[105]

e. Reimlinger and Merenyi[131] reported obtaining the parent pyrazolo-[5,1-*c*]-*s*-triazole (**172**) in 1% yield from the reaction of diazomethane and pyrazol-3-diazonium chloride (**171**).

171 $\xrightarrow{CH_2N_2}$ **172**

f. The reaction of 2-ethoxy-1-methyl-1-phenyl-diazenium tetrafluoroborate (**174**) and benzothiazole (**173**) was reported by Eicker, Hunig, Hansen, and Nickalaus[132] to give 2-phenyl-*s*-triazolo[3,4-*b*]benzothiazolium tetrafluoroborate (**175**). The diazenium salt did not react with

173 + $C_2H_5—O—N{=}\overset{+}{N}(CH_3)—C_6H_5\ BF_4^-$ (**174**) ⟶ **175** (BF_4^-)

benzoxazole, 1-methylbenzimidazole, 1-phenylpyrazole, indole, or imidazole.

E. Triazolium

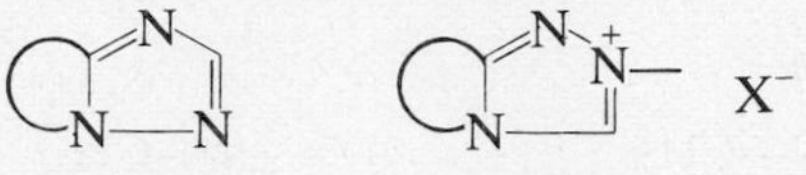

1,3,5 1,3,4 X^-

1. N-Amino-imino compounds of type **176** react with acylating agents

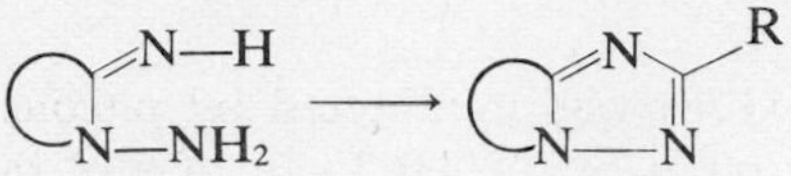

176 **177**

R = H or phenyl

to give fused 1,3,4-tetrazoles (**177**). Benzoyl chloride[125] and formic acid have been used in this reaction.

This procedure has been used for the preparation of derivatives of 5*H*-*s*-triazolo[4,3-*b*]-*s*-triazole (**144**)[125] and thiazolo[3,2-*b*]triazole (**178**),[133] and *s*-triazolo[1,5-*a*]benzimidazoles[134] (**179**).

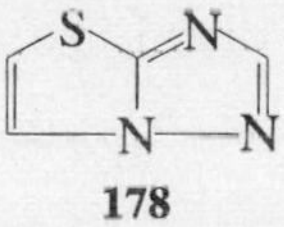

178 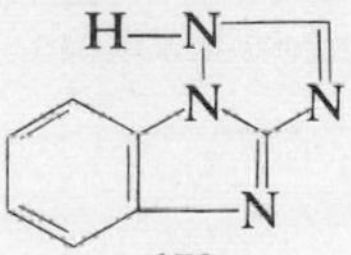 **179**

A variation of this procedure, used to obtain the parent thiazolo[3,2-*b*]triazole (**178**) involved the preparation of the *N*-amino compound

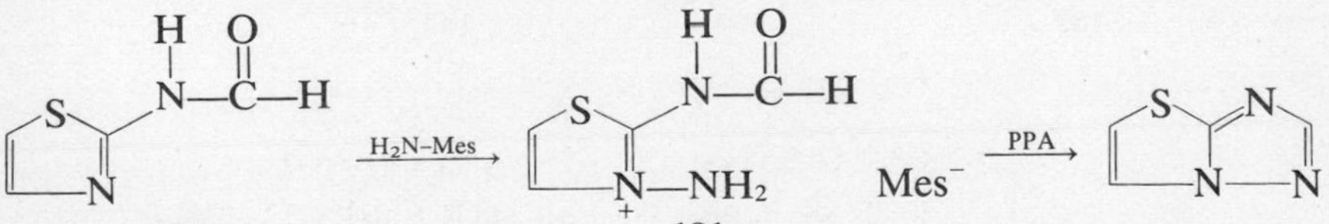

H_2N-Mes PPA

180 **181** Mes^-

$Mes = {}^-O_3S$-(2,4,6-trimethylphenyl) (CH_3, CH_3, CH_3)

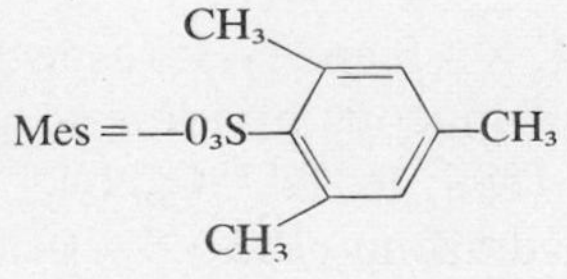

(**181**) from the amide (**180**) followed by cyclization in polyphosphoric acid.[135]

2. Näf[136] reported obtaining thiazolo[3,2-*b*]-*s*-triazole (**178**) from 2-methylaminothiazole (**182**) as shown. This reaction sequence has been questioned by Mosby.[137] Tamura and co-workers[135] reported that

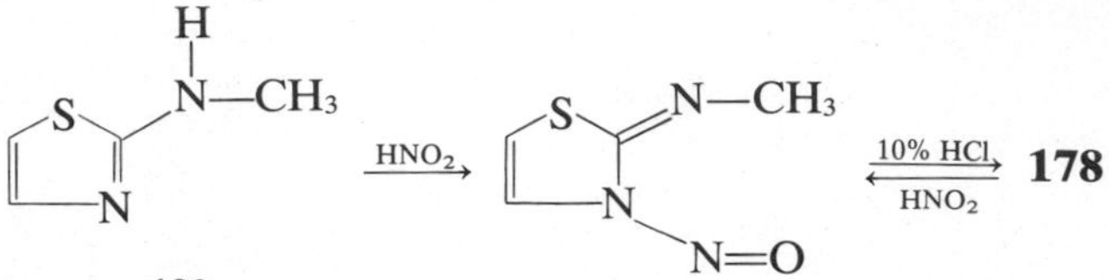

HNO₂ → ; 10% HCl ⇄ HNO₂ **178**

182

thiazolo[3,2-*b*]thiazole (**178**) was unaffected by nitrous acid. They also report that the melting point of the HCl salt of **178** they obtained (see above) is slightly different than the HCl salt of Näf's compound.

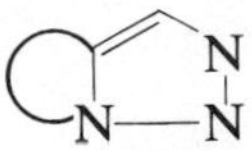

1,4,5

Messmer and Gelleri[138] treated hydrazones of type **183** with *N*-bromosuccinimide in ethylacetate at room temperature and obtained 1,3-diaryl-*v*-triazolo[5,1-*b*]benzothiazolium and 1,3-diaryl-*v*-triazolo[1,5-*a*]benzimidazolium salts (**184**; X = S and NH, respectively). The benzimidazolium compound lost HBr in aqueous pyridine to give 1,3-diphenyl-*v*-triazolo-[1,5-*a*]benzimidazole (**185**; R, $R^1 = C_6H_5$).

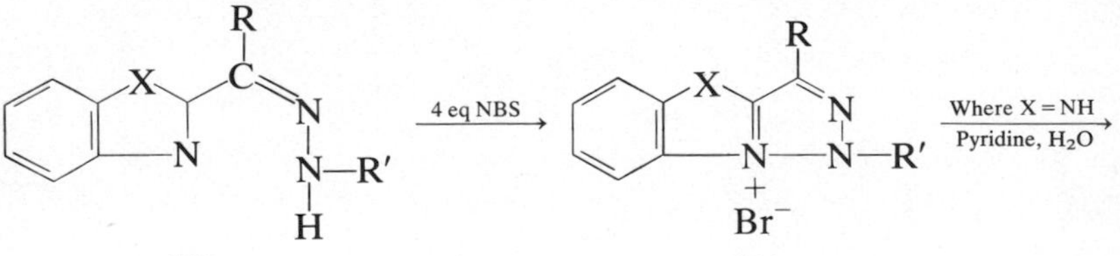

4 eq NBS → ; Where X = NH, Pyridine, H_2O →

183 **184**

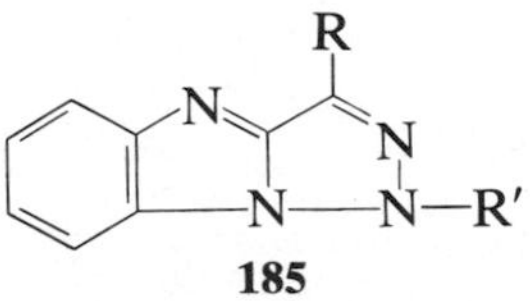

185

X = S or NH
R and R′ = aryl

The benzoxazole hydrazone (**183**; X = 0) failed to cyclize.

F. Tetrazoles

Fused tetrazoles of type **182** and open chain azides of type **186** are tautomers. In this synthesis section we consider the tetrazole form to be the favored form and the azide to be an intermediate.

1. The reaction of hydrazines of type **140** gives tetrazoles of type **187**. This procedure has been found to be rather general. While 2-hydrazino-4-

methylthiazole (**188**; R = CH_3) reacts in this manner with nitrous acid, to give the corresponding thiazolo tetrazole (**189**).[139–143] 2-Hydrazinothiazole (**188**; R = H) reacts in an anomalous manner with nitrous acid giving instead of the thiazolo tetrazole, a dark, light-sensitive compound to which Reynolds, Van Allan, and Tinker[142] assigned the triazene

structure **191**. Additionally Beyer and co-workers[140–141] found that the thiadiazine (**190**), an isomer of the hydrazinothiazole (**188;** R = CH_3) reacts with nitrous acid to give the thiazolotetrazole (**189**) (see ref. 141 for a discussion of the structure of compound **190**).

This approach has been used for the preparation of derivatives of thiazolo[3,2-*d*]tetrazole (**192**),[139–142] 1*H*-pyrazolo[2,3-*d*]tetrazole (**193**)[143] tetrazolo[5,1-*b*]benzothiazole (**194**),[105,114,144] tetrazolo[1,5-*a*]benzimidazole (**195**),[105,114] tetrazolo[5,1-*b*]benzoxazole (**196**),[105] tetrazolo[5,1-*b*]benzoselenazole (**197**),[105] tetrazolo[1,5-*b*]purine (**198**),[106] tetrazolo-[5,1-*b*]naphtho[1′,2′-*d*]thiazole (**199**),[145,146] tetrazolo[5,1-*b*]naphtho-[2′,1′-*d*]thiazole (**200**),[146] and 1,3,4-triazolo[4,3-*d*]tetrazole (**201**).[241]

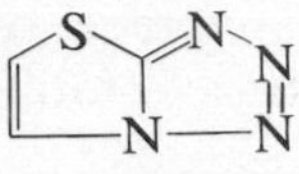

192 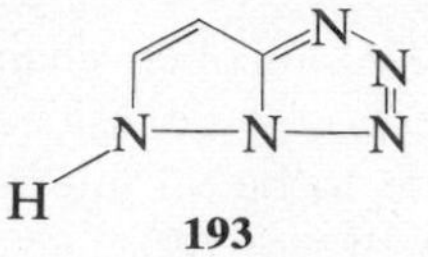 **193** 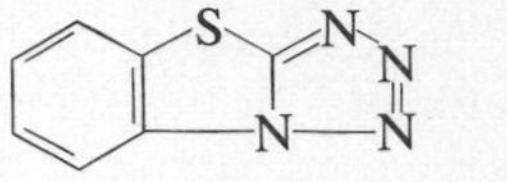 **194**

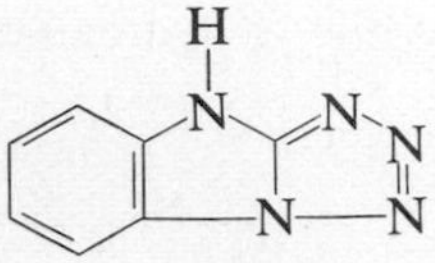

195 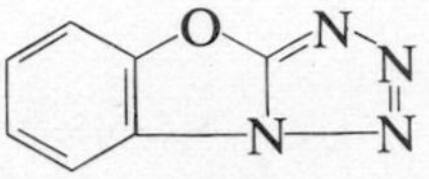 **196** 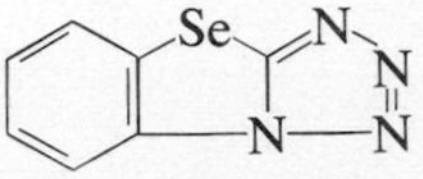 **197**

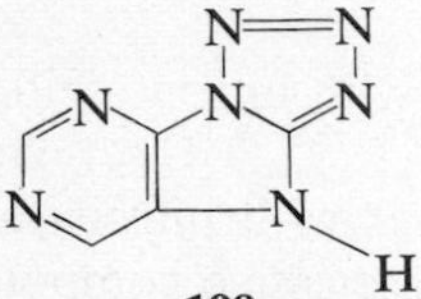

198 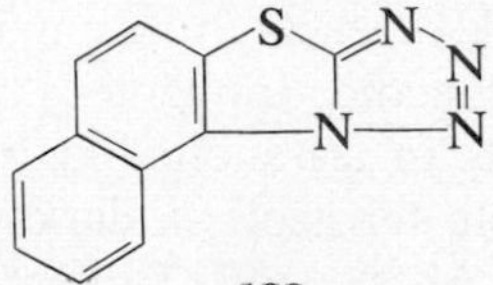 **199** 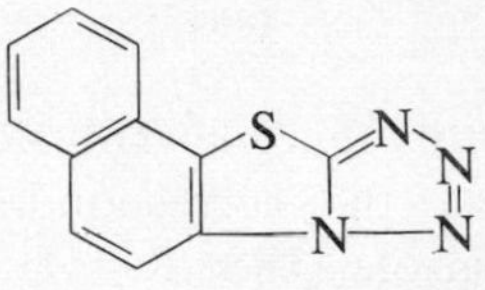 **200**

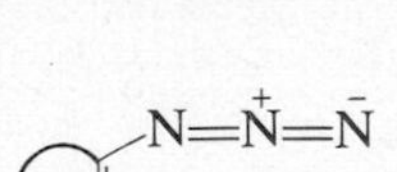

201

2. Chloro compounds of type **169** react with sodium azide in acetic acid,[145] ethanol,[147] or aqueous methanol[148] to give tetrazoles (**187**).

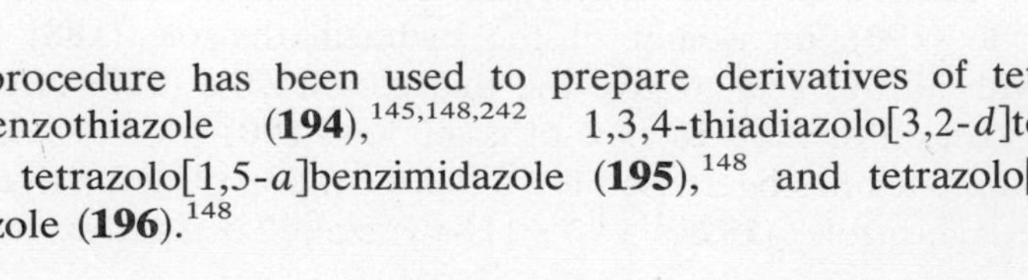

NaN_3

169 **1&6** **187**

This procedure has been used to prepare derivatives of tetrazolo-[5,1-*b*]benzothiazole (**194**),[145,148,242] 1,3,4-thiadiazolo[3,2-*d*]tetrazole (**202**),[147] tetrazolo[1,5-*a*]benzimidazole (**195**),[148] and tetrazolo[5,1-*b*]-benzoxazole (**196**).[148]

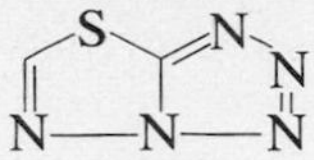

202

3. Fused tetrazoles can also be prepared from amines of type **59**. In this procedure the diazonium salt is prepared and treated *in situ* with sodium azide to give the tetrazole (**203**).

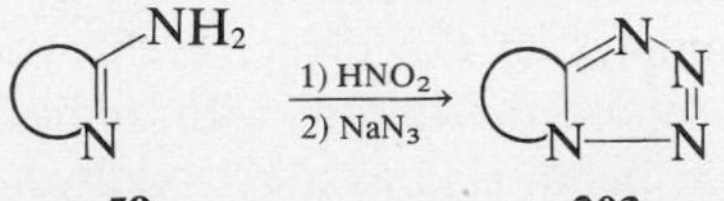

1) HNO_2
2) NaN_3

59 203

This procedure has been used to prepare derivatives of tetrazolo[5,1-*b*]benzothiazole (**194**),[145,149,150] and tetrazolo[5,1-*b*]naphtho[2′,1′-*d*]-thiazole (**199**).[145]

G. Thiazoles

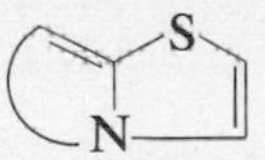

1,3

1. The most commonly used method for the preparation of fused thiazoles involves the reaction of compounds of type **204** with an α-halo

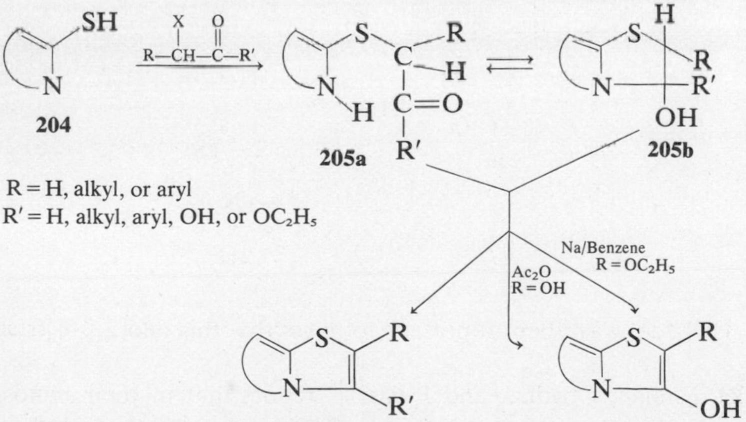

204 205a 205b

R = H, alkyl, or aryl
R′ = H, alkyl, aryl, OH, or OC_2H_5

Na/Benzene $R = OC_2H_5$

Ac_2O R = OH

206 207

carbonyl or α-halo carboxyl compound to give *S*-alkylated products, which can be shown to exist in the acyclic (**205a**) and cyclic alcohol

(**205b**) forms[112,151,152,243,246] which can then be dehydrated to give fused thiazoles (**206** or **207**).

Kochergin and Shchukina[153] have reported that acid or strong dehydrating agents are necessary to cyclize aldehydes and ketones of type (**205**; R^1 = H, alkyl or aryl) to fused triazoles (**206**; R and R^1 = H, alkyl or aryl). Phosphoryl chloride is the cyclizing agent usually employed,[112,152–159] although sulfuric acid,[156,158,160] hydrochloric acid,[153] acetic acid,[161] polyphosphoric acid,[244,245] and ethanolic HCl[244] have also been used. These are, however, some examples of the direct cyclization of thiols of type **204** to triazoles (**206**) upon treatment with α-halo ketones.[112,162,163]. Acetic anhydride has also been used as a cyclodehydrating agent for ring closures of this type,[247] but it usually enters into the reaction (see Section 2).

Potts and Hussain[110] reported that treatment of 2-mercapto-5-phenyl-*s*-triazole (**208**; R = C_6H_5) with chloroacetone or phenacyl bromide in refluxing ethanol for 4 hr gave the thiazolo[3,2-*b*]-*s*-triazole (**209**) directly. The analogous thione (**208**; R = methyl) was reported to give the uncyclized ketone (**210**) after 4 hr of heating but cyclized product (**209**) after 24 hr of heating. Ring closure of the uncyclized ketone in phosphoryl chloride was then reported to give the thiazolo[2,3-*c*]triazole (**211**).

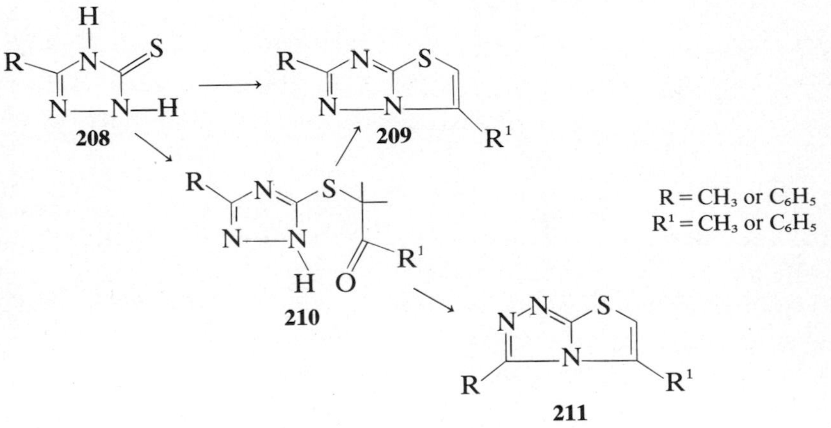

R = CH_3 or C_6H_5
R^1 = CH_3 or C_6H_5

Dhaka, Muhan, Chadha, and Pujari[248] report that in their hands, 3-mercapto-5-phenyl-*s*-triazole gave only the uncyclized ketone (**210**) after 4 hr in refluxing ethanol. Also, they report the cyclization of the ketone in phosphoryl chloride or polyphosphoric acid gave only the thiazolo[3,2-*b*]-*s*-triazole (**209**) and no thiazolo[2,3-*c*]-*s*-triazole (**211**).

Kochergin[125] studied the influence of 4(5) substituents, in imidazole rings, on the direction of ring closure and found that the *p*-nitrophenyl compound (**212**) which could form either the 5- or 6-aryl isomer (**213** or **214**, respectively), cyclized to give only the 6-*p*-nitroimidazothiazole (**214**). Steric influence would appear to be important in this reaction. The structure of compound **214** was proven by independent synthesis from 2-amino-5-acetyl-4-methylthiazole (**215**) and *p*-nitrophenacylbromide (see the imidazole section for a discussion of the direction of ring closure of 2-aminothiazoles and α-halocarbonyl compounds). Other work on the direction of cyclization of 2-alkylthio-4(5) substituted mercaptoimidazoles is somewhat confusing as Kochergin[155] reported that treatment of 4(5)-carbomethoxy-2-mercaptoimidazole (**216**, with bromoacetaldehyde gave a compound "... which, by analogy with 3-hydroxy-6-phenylimidazo-[2,1-*b*]thiazoline, probably has the structure of the methyl ester of 3-hydroxyimidazo[2,1-*b*]thiazoline-5-carboxylic acid [(**217b**)]. . . ."

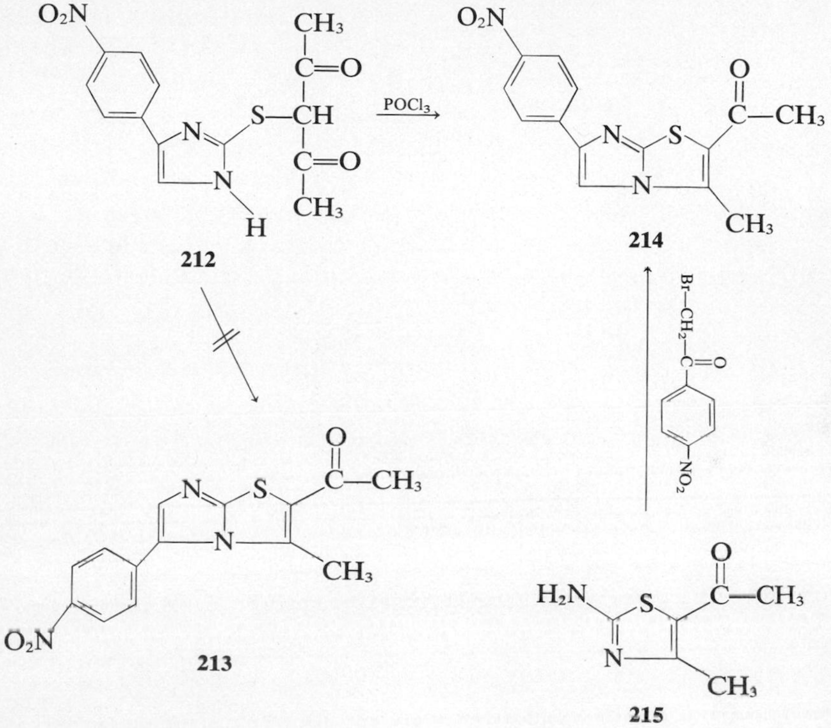

The 3-hydroxy-6-phenyl compound referred to is structure **218b** in the following sequence from a paper by Kochergin and Shchukina.[152]

This comparison is not altogether clear.

216 → ($Br-CH_2-C(=O)-H$) → [217a] ⇌ 217b → ($POCl_3$)

→ ($Br-CH_2-C(=O)-H$) → [218a] ⇌ 218b → ($POCl_3$)

Mohan and Pujari[247] found that 5-methoxy-2-benzimidazothiolacetic acid (**219**) in acetic anhydride gave only the 6-methoxy isomer (**221**) and none of the 7-methoxy isomer (**220**).

219 $\xrightarrow{Ac_2O}$ **221** (not **220**)

222 $\xrightarrow[\text{Base}]{Br—CH_2—C≡C—H}$ **223**

223 $\xrightarrow{R—O^-}$ **224**

223 $\xrightarrow[\text{HMPT}]{\Delta}$ **225** → **226** + **227**

227 $\xrightarrow{NaOC_2H_5}$ **226**

3-Hydroxy(-one) derivatives have been prepared by treatment of the ester (**205a**; R = OC_2H_5) with sodium in benzene[165,166] or by treating the acid (**205a**; R = OH) with acetic anhydride.[167,169]

b. The 5-propargyl compound (**223**), obtained from 2-mercaptobenzimidazole (**222**) and propargyl bromide, cyclized in the presence of alkoxide ion to give 3-methylthiazolo[3,2-*a*]benzimidazole (**224**).[170] It is interesting to note that that propargyl compound (**223**), a "ketone anhydride," cyclizes in basic medium while ketones cyclize in acid.[161]

Balasubramanian and Venugopalan[249] have confirmed this basic ring closure and found additionally that the propargyl compound (**223**) undergoes a Claisen type rearrangement, to the allenic intermediate (**225**) on heating in hexamethylphosphorous triamide, as 2-methylthiazolo[3,2-*a*] benzimidazole (**226**) is formed. The methylene-thiazoline (**228**) is also obtained and this compound, on treatment with base, is converted to the 2-methyl derivative.

c. Kochergin and co-workers[152,171] used acetals in this type of synthesis. The reaction of 6-phenyl-2-mercaptoimidazole (**228**) with α-bromodimethyl (or diethyl) acetal gave the *S*-alkylated compound (**229**) which cyclized to the 2,3-dihydrothiazoline ether (**230**) in phosphoryl chloride. Treatment of these ethers (**230**) with sulfuric acid gave 6-phenyl-imidazo[2,1-*b*]thiazole (**231**).

C_6H_5, SH, N, N, H — **228** —($Br—CH_2—CH(OR)_2$)→ C_6H_5, N, S, CH_2, $C(OR)_2$, H — **229** —($POCl_3$)→ C_6H_5, N, S, H, H, OR, H — **230** —(H_2SO_4)→ C_6H_5, N, S, N — **231**

Gordon[172] reported that 2,6-diamino-8-mercaptopurine (**232**) reacted with chloroacetaldehyde to form the thiazolopurine (**233**) directly.

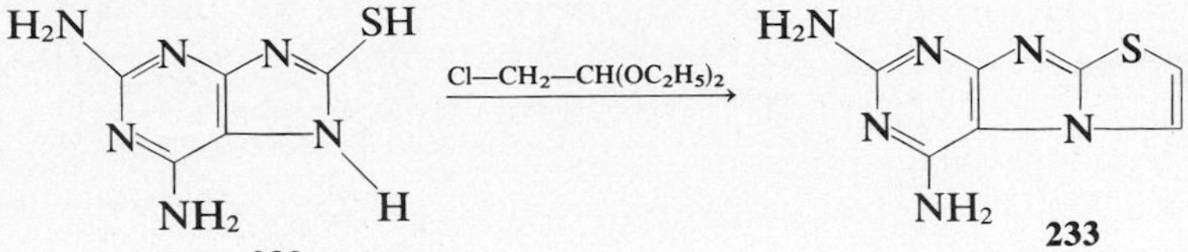

This approach has been used for the preparation of derivatives of imidazo[2,1-*b*]thiazole (**234**),[133–155,157,158,160,161,173] thiazolo-[3,2-*b*]-*s*-triazole (**235**),[174] thiazolo[3,2-*a*]benzimidazole (**236**),[151–153,156,160,162,165–168,170,174,243] naphthol[2′,3′ : 4,5]imidazo[2,1-*b*]-thiazole (**237**),[159,169] naptho[1,′2′ : 4,5]imidazo [2,1-*b*]thiazole (**238**), thiazolo[2,3-*b*]purine (**239**),[172] and thiazolo[3,2-*a*]indole (**240**).[175]

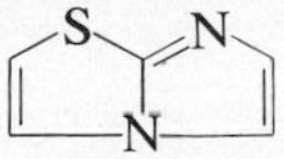

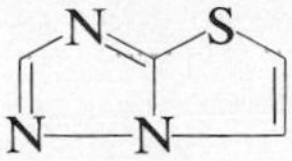

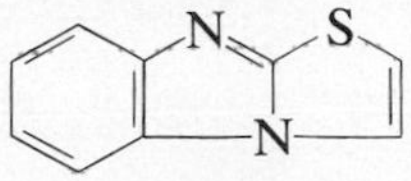

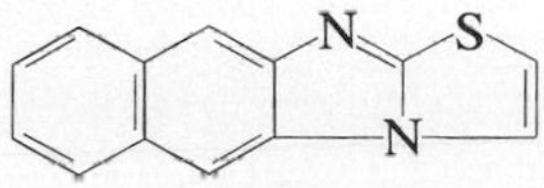

2. In addition to effecting cyclodehydration acetic anhydride can enter into the reaction. Ochiai[154] treated the *S*-acetonyl imidazole (**241**) with acetic anhydride and obtained 2-acetyl-3,6-dimethyl-5-carbethoxy-imidazo[2,1-*b*]-thiazole (**242**). Kochergin[164] found that heating phenyl-*S*-acetonylimidazole (**243**; R = CH_3) with acetic anhydride for 5 to 7 min gave the *N*-acetyl compound (**244**; R = CH_3) which on longer heating

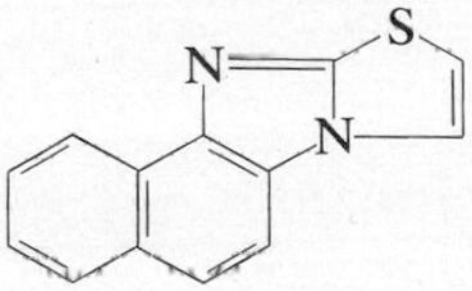

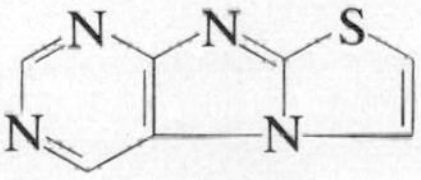

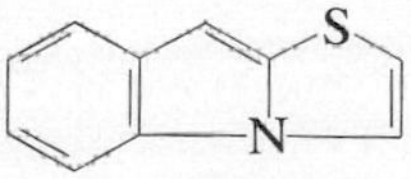

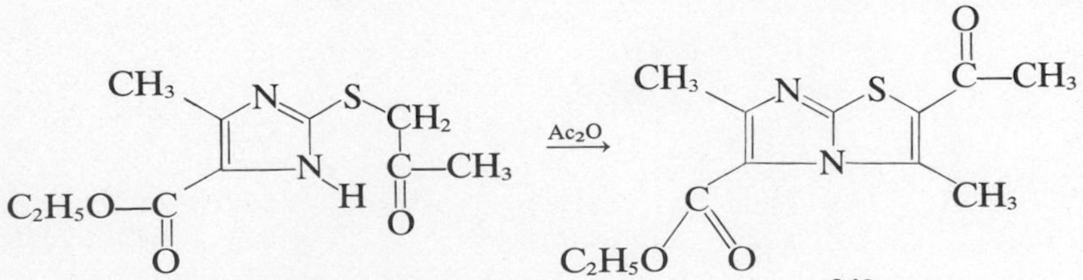

C_6H_5 N S CH_2 N H C=O R **243** → (Ac_2O, Reflux 5–7 min) → C_6H_5 N S–CH_2 C=O R N C O CH_3 **244**

→ (Ac_2O, NaOAc, Reflux 3 hr) → C_6H_5 N S C–R (O) N CH_3 **245**

R = CH_3, C_6H_5 or C_6H_4-$p$$NO_2$

N S–CH_2 N H C=O H **246** → (Ac_2O) → N S–CH_3 C O H N C O CH_3 **247** ⇸ N S C–H (O) N CH_3 **248**

247 → (Ac_2O, Pyridine) → [N S H C–H (O) N OH CH_3 **249**] ⇌ [N S C–H (O) C OH N H CH_3 **250**] → N S C–CH_3 (O) N **251**

with acetic anhydride–sodium acetate, gave 2-acetyl-3-methylimidazo-[2,1-*b*]thiazole (**245**; R = CH_3). In general, higher temperatures and basic conditions favor cyclization.[250–252] The ring closure was viewed as involving the methylene carbon of the acetonyl group and the carbonyl carbon of the amide portion of **244**. Alper and Taurins[151] in their study on thiazolo[3,2-*a*]benzimidazoles, found that the *N*-acetyl compound (**247**), obtained from the aldehyde (**241**) on heating with acetic anhydride in pyridine, did not give the formyl-methyl compound (**248**) which would result if the reaction followed Kochergin's mechanism. Instead, 2-acetylthiazolo[3,2-*a*]benzimidazole (**251**) was obtained. Alper and Taurins[151] suggest that a dihydrothiazoline ring is formed (**249**) which undergoes cleavage of the C–N bond giving intermediate **250** which then cyclizes to give the acetyl compound (**251**). The acyclic intermediate **247** also appears to have the potential to form the aldehyde **248** but none of this compound was reported as having been isolated.

Kochergin and co-workers have since reported the preparation of imidazo[2,1-*b*]thiazoles[252] and thiazolo[3,2-*a*]benzimidazoles[253] as follows.

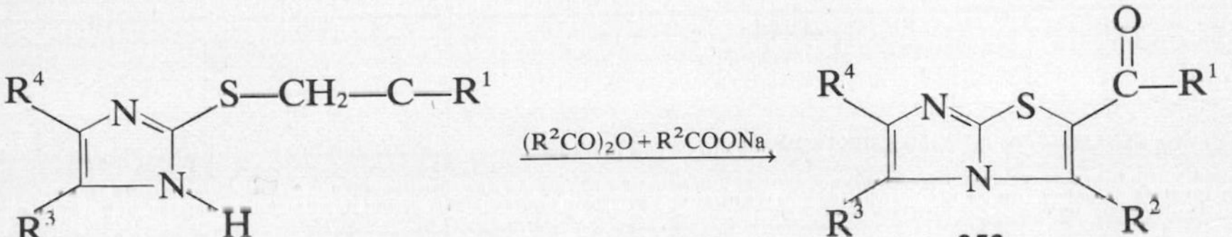

$(R^2CO)_2O + R^2COONa$

252

R^3 = H or aryl

R^4 = H or aryl or $R^3 - R^4$ = benzene ring residue

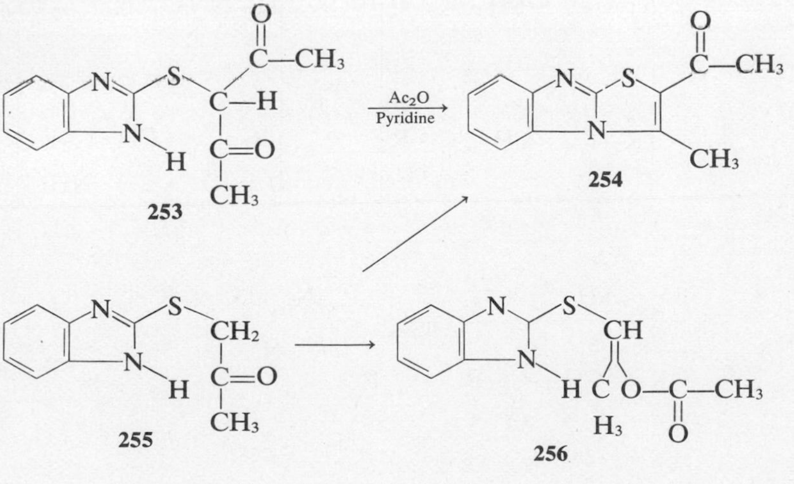

Ac_2O / Pyridine

253 **254**

255 **256**

The products obtained (**252**) were assigned structures according to the Kochergin rather than the Alper–Taurins mechanism.

D'Amico[176] reported that the pentanedione (**253**) on treatment with acetic anhydride gave the expected acetyl methyl compound (**254**) but that the *S*-acetonyl compound (**255**) on treatment with acetic anhydride–pyridine, gave the propenol acetate (**256**). Alper and Taurins[151] reported that the *S*-acetonyl compound (**255**) on treatment with acetic anhydride in pyridine gave 2-acetyl-3-methylthiazolo[3,2-*a*]benzimidazole (**254**)—as does the pentanedione (**253**)—and not the porpenol acetate (**256**) as reported by D'Amico.[176]

4. Lawson and Morley[177,178] prepared 2,5-disubstituted imidazo[2,1-*b*]thiazoles (**258**) by effecting the cyclization of 1-substituted-2-mercapto-imidazoles in hydrochloric acid. The interesting chemistry involved in the preparation of the precursors is also shown.

R—CH(NH₂)—C(=O)—OH + R′—CH(NH₂)—C(=O)—OH $\xrightarrow[\text{pH 2}]{\text{Na(Hg)}}$ [R—C(=O)—N=CH—CH(NH₂)—R′] ⇌ [R—C(=O)—NH—CH=C(NH₂)—R′]

R = alkyl or aryl
R′ = alkyl

↓ KSCN, pH 4 (pH 2.8) where R + R′ = phenyl)

[H—N=C(SH)—N(C(=O)—R)(CH=C(NH₂)—R′)] $\xrightarrow[-\text{NH}_3]{\text{H}_2\text{O}}$ [H—N=C(SH)—N(C(=O)—R)(CH₂—C(=O)—R′)]

↓

257 $\xrightarrow{\text{HCl}}$ **258**

5. 2,3-Dichloroquinoxaline (**126**) with two active halogen atoms has been used to prepare fused thiazoles (**257**).[91,113]

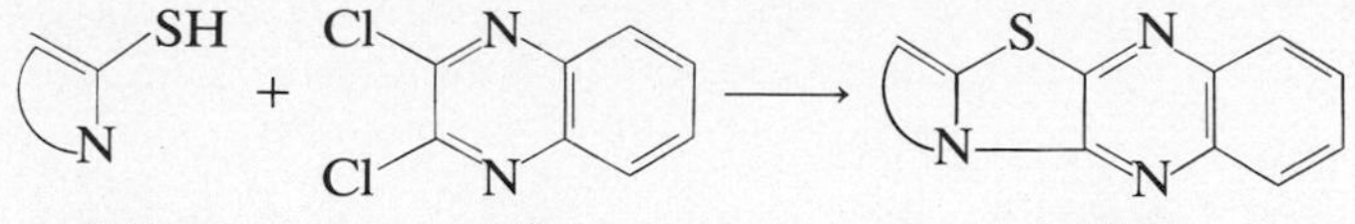

204 **126** **257**

This procedure was used to prepare derivatives of benzimidazo-[2′,1′ : 2,3]thiazolo[4,5-*b*]quinoxalines (**258**)[91] *s*-triazolo[3′,2′ : 2,3]-thiazolo[4,5-*b*]quinoxaline (**259**).[113]

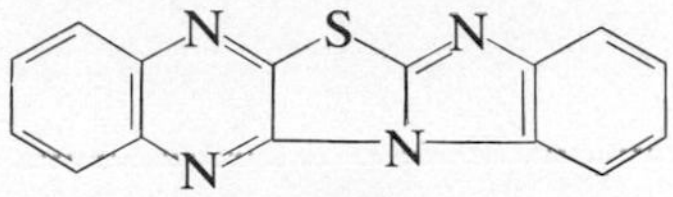

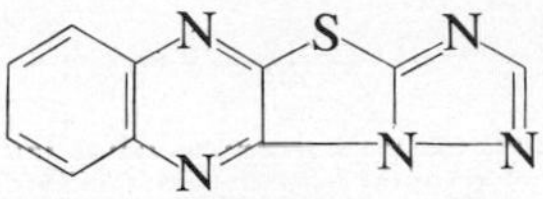

258 **259**

6. Kohner[254] in studying the reaction of 2,5-dithio-1,3,4-thiadiazole (**260**) and aromatic amines (aniline is used in the example shown) obtained compounds which he identified as 3-mercapto-1,2,4-triazolo-[3,4-*b*]benzothiazoles (**261**). In working to elucidate the mechanism, he

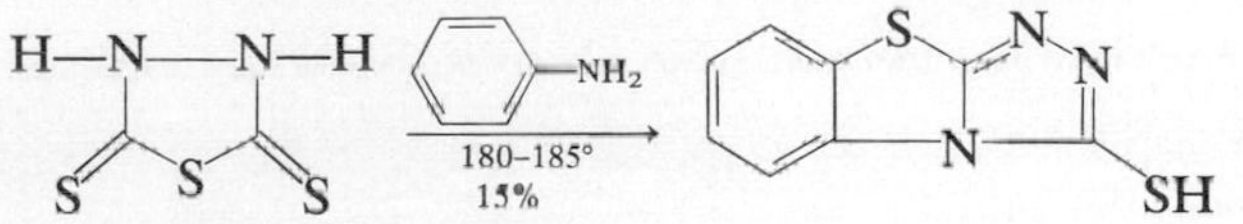

180–185°
15%

260 **261**

6–8% −2H

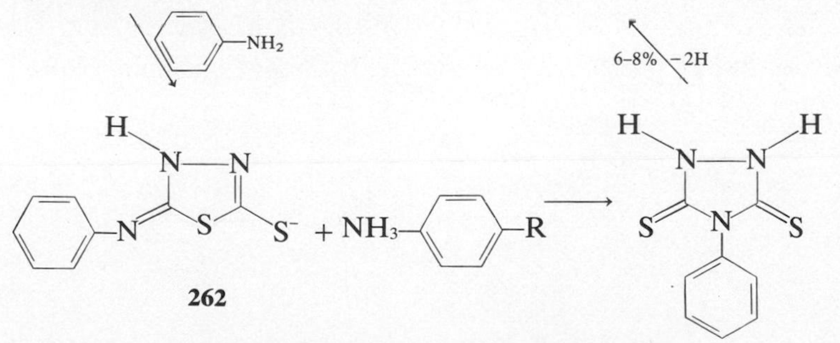

262

prepared the amine salt (**262**) which then proceeded to give the fused system (**261**), but in lower yield than originally obtained.

H. Thiazolium

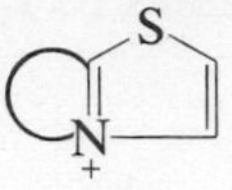

1,3

A number of cationic-fused thiazoles have been prepared. The synthesis of such systems was developed by Bradsher, Lohr, and Jones[179–182] and involves treating a thiol of type **263** with an α-halo-ketone to give an

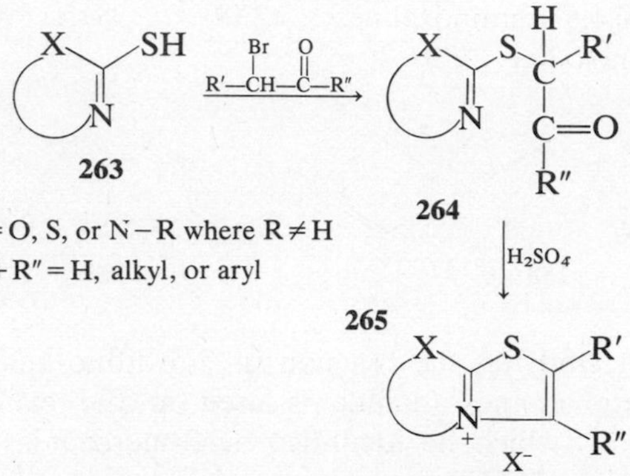

263

X = O, S, or N − R where R ≠ H
R′ + R″ = H, alkyl, or aryl

264

H_2SO_4

265

265
X = $\frac{1}{2}SO_4^{2-}$ or ClO_4^-

intermediate α-thioketone (**264**) which cyclizes to the thiazolium salt (**270**; X = $\frac{1}{2}SO_4$) which is a water soluble compound. The aqueous solution is then treated with perchloric acid to give the more insoluble perchlorate salts (**265**; X = ClO_4^-).

This procedure was used to prepare derivatives of the thiazolo[2,3-*b*]thiazolium (**266**),[179,180] oxazolo[2,3-*b*]thiazolium (**267**),[181] thiazolo[2,3-*b*]benzothiazolium (**268**),[182] thiazolo[2,3-*b*]benzoxazolium (**269**),[183] and thiazolo[3,2-*d*]tetrazolium (**270**)[183,255] cations.

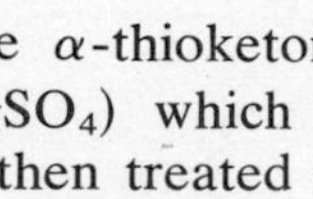

266 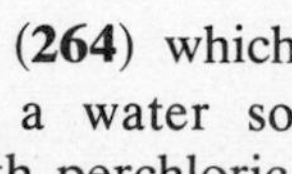 **267** 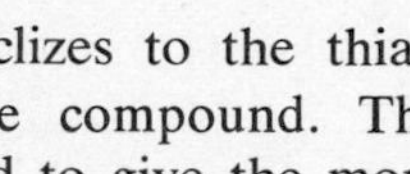 **268**

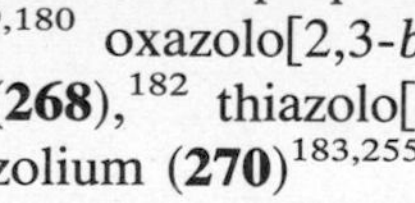

269 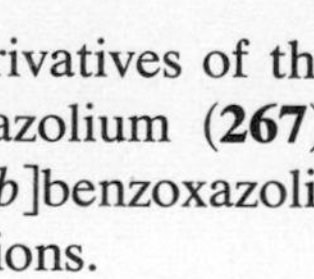 **270**

I. Oxadiazoles

1,5,4

The only example of an oxadiazole ring being fused onto another 5-membered ring was reported by Beyer and Hetzheim.[184] The synthesis of 2-methyl-6-phenylimidazo[2,1-*b*][1.3.4]oxadiazole (**274**) starts with an oxadiazole (**271**) which gives compound **271** on treatment with phenacyl bromide. In aqueous sodium carbonate, the quaternary salt (**272**) rearranges to the imidazolone (**273**) which is isolated.[185] This imidazolone (**273**) on treatment with phosphoryl chloride gives the imidazooxadiazole (**274**).

J. Thiadiazoles

1,5,4

The only reported examples of the fusion of a 1,5,4 thiadiazole ring onto another 5-membered ring involve treatment of a mercapto-amino compound of type **275** with an acylating agent.

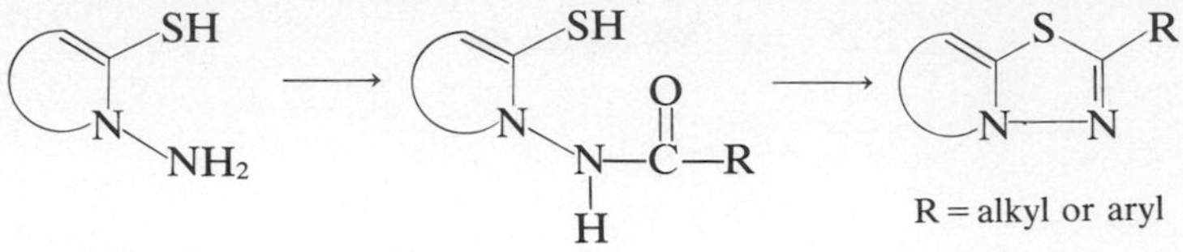

SH / NH₂ → SH / N—C(=O)—R / H → S R / N—N

R = alkyl or aryl

275 **276** **277**

When the acylating agent is an acid chloride[186,256] or anhydride,[187] an acylamino group is obtained which cyclizes in phosphoryl chloride to the fused thiadiazole (**277**).

When the acylating agent is carbon disulfide or cyanogen bromide, cyclization is spontaneous. Potts and Huseby[188] reported that the reaction of 5-alkyl-4-amino-3-mercapto-*s*-triazoles (**278**) with carbon disulfide in alcoholic potassium hydroxide gave 5-alkyl-2-mercapto-*s*-triazolo[3,4-*b*][1,3,4]thiadiazole (**279**). Reaction did not take place when the thio-amine(**278**) was aryl substituted. The nature of the reaction medium is

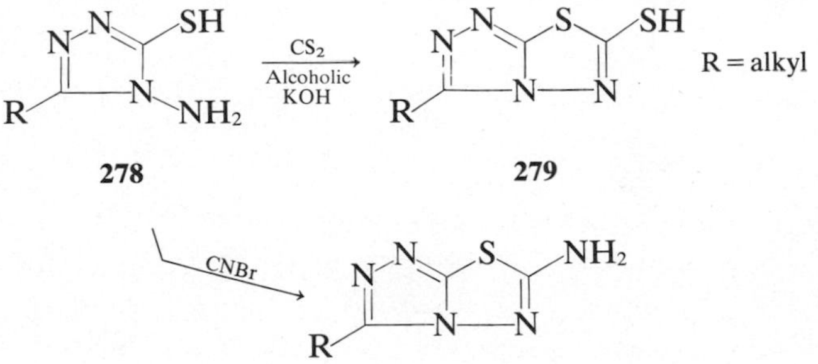

CS_2, Alcoholic KOH

R = alkyl

278 **279**

CNBr

280

apparently important as Kanaoka[189] reported that compounds of type **278** would not react with carbon disulfide in refluxing pyridine. Alkyl and aryl derivatives of the amino thiol (**278**) reacted with cyanogen bromide to give 2-aminotriazolothiadiazoles (**280**).

Formation of a thiadiazole ring has been used to prepare derivatives of imidazo[2,1-*b*][1.3.4]thiadiazole (**281**)[187] and *s*-triazolo[3,4-*b*][1.3.4]-thiadiazole (**282**).[186,188]

281 **282**

K. Thiadiazolium

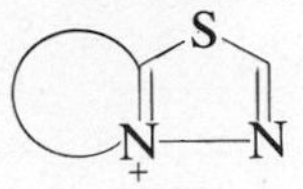

1,5,3

The first examples of a thiadazolium salt (**284**) were reported by Ege[191] who prepared these compounds by treating the thione (**283**) with concentrated sulfuric acid. The sulfate salts (**284**; $X=\frac{1}{2}SO_4^-$) are water

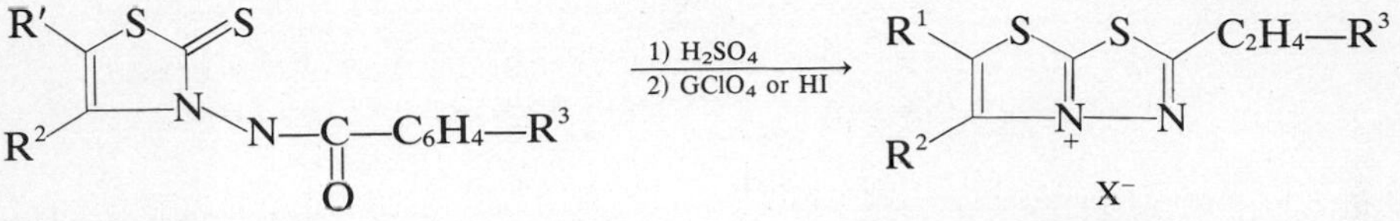

283

$R^1 = H, CH_3$

$R^2 = CH_3$ or C_6H_5

$R^3 = H$ or NO_2

1) H_2SO_4
2) $GClO_4$ or HI

284

$X=\frac{1}{2}SO_4^{2-}, ClO_4^-, I^-$

soluble, and the products are isolated as the rather insoluble perchlorates or iodides (**284**; $R = ClO_4^-$ or I^-, respectively). This procedure is similar to that used by Bradsher and co-workers,[179–182] for the preparation of thiazolium cations (see thiazolium).

L. Thiadiazole

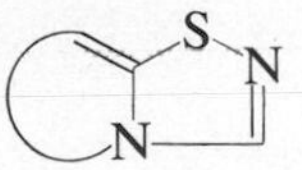

1,4,3

The only example of the fusion of a 1,4,3 thiadiazole ring onto another 5-membered ring was reported by Stanovnik and Tishler,[190] who heated the benzimidazole sulfonamide (**285**) with ethyl orthoformate and obtained the 1,1-dioxo-[1.2.4]thiadiazolo[4,5-*a*]benzimidazole (**287**) as shown. On shorter heating time, the intermediate, **286**, was obtained and on further heating converted to the fused structure.

$$\text{285} \xrightarrow[\Delta\text{-5 min}]{HC(OC_2H_5)_3} \text{286}$$

285

286

$HC(OC_2H_5)_3$, 30 min; Δ

287

M. Thiadiazolium

1,3,5

Barnikow and Bodeker[257] treated 2-aminothiazole (**72**) with benzoylisothiocyanate to obtain the benzoylthiourea (**288**) which upon hydrolysis with base, gave the thiazolo thiourea (**290**). Bromine and this

72 → (S=C=N—C(O)—C_6H_5) 288

OH^-

Br^-

Br^-

291

290

thiourea (**290**) gave 2-amino thiazolo[3,2-*b*]-1,2,4-thiadiazolium bromide (**291**).

N. Simultaneous Formation of Two Rings

There are several reports concerning the formation of both rings of the 5,5 system during the course of the reaction. Reactions of type 2 and 4 appear to be more limited in scope than those shown as types 1 and 3.

1. The reaction of a 1,2-dinucleophile and a 1,4-dielectrophile has been used to prepare several 5,5-systems.

a. Thiele and Falk treated[192] *o*-phthalodicarboxaldehyde (**292**) with phenylenediamine (**293**) and obtained 11*H*-isoindolo[2,1-*a*]benzimidazole (**294a**). [This compound can also be written in the 6*H* form (**294b**)

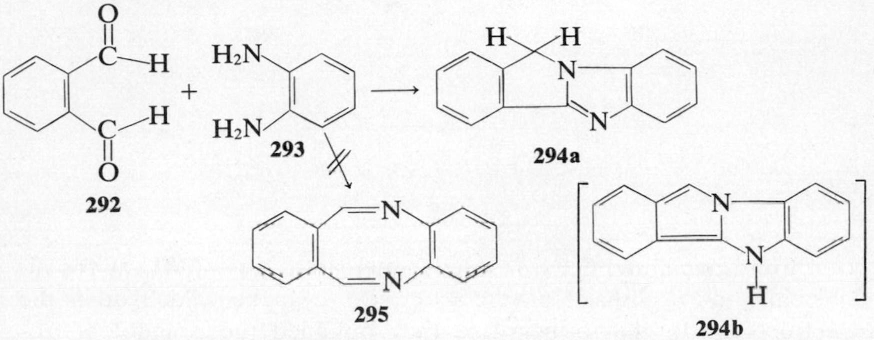

(see tautomerism section).] Betrabet and Chakravarti[193] reported that the compound obtained by Thiele and Falk[192] was really the dibenzodiazocine (**295**). Ried and Bodem[194] prepared a series of compounds from several aromatic 1,2-diamines and 1,2-dialdehydes and assigned diazacine structures to them. Unequivocal support for the isoindolobenzimidazole structure (**294a**) was provided by the nmr spectral work of Amos and Gillis[195] and chemical proofs were offered by Bistrzyki and Schmutz[196] and by Sparatori and Bignardi,[197] the latter being shown here.[198]

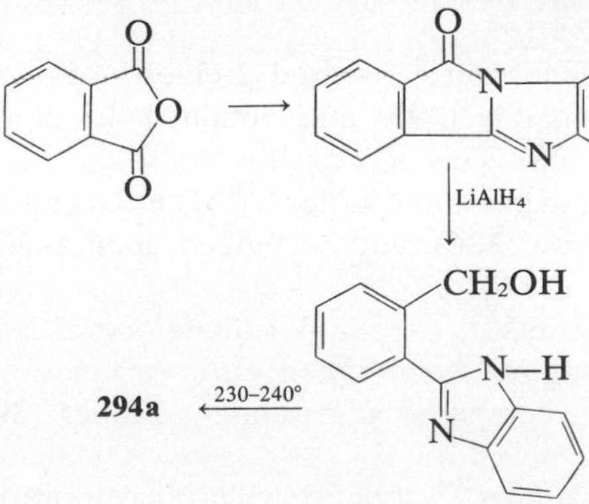

$\xrightarrow{\text{LiAlH}_4}$ ... **294a** $\xleftarrow{230\text{–}240°}$

Perlmutter and Knapp[199] showed that the compounds prepared by Ried and Bodem[194] did not have the diazacine structure but were the following systems: 12*H*-benzo[4,5]isoindolo[2,1-*a*]benzimidazole (**296**), 13*H*-naphth[2′,3′ : 4,5]imidazo[2,1-*a*]isoindole (**297**), 14*H*-benzo[*f*]-naphth[2′,3′ : 4,5]imidazo[2,1-*a*]isoindole (**298**) and either 8*H*-benzo[*f*]-naphth[1′,2′ : 4,5]imidazo[2,1-*a*]isoindole (**299**) or 14*H*-benzo[*f*]naphth-[2′,1′ : 4,5]imidazo[2,1-*a*]isoindole (**300**).

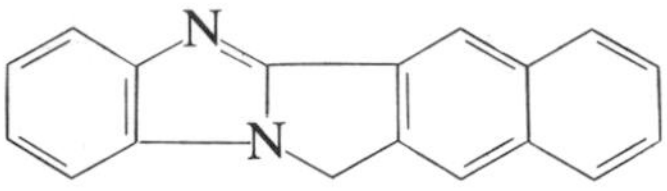

296 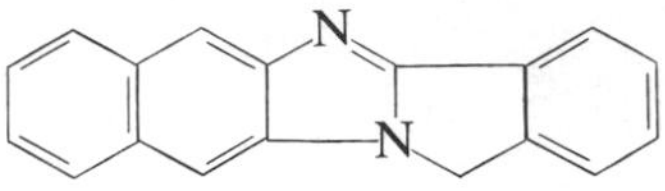 **297**

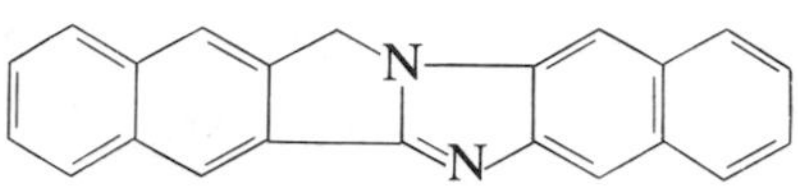

298

b. Oliver, Dann, and Gates[200] used 2-mercaptoaniline (**301**) as the dinucleophile and *o*-phthalaldehydic acid and *o*-cyanobenzaldehyde as the dielectrophiles. In the former case they obtained the isoindolobenzothiazolone (**302a**) [which can also be written in the ol form (**302b**)] and in

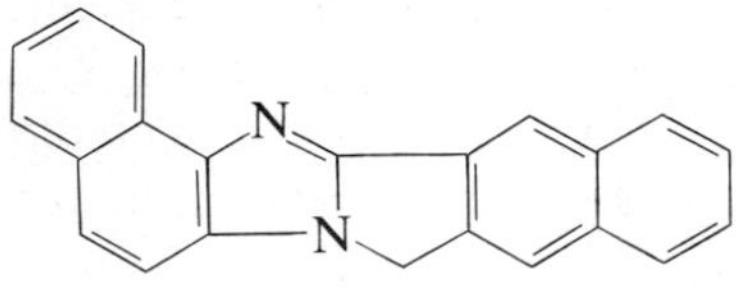

299 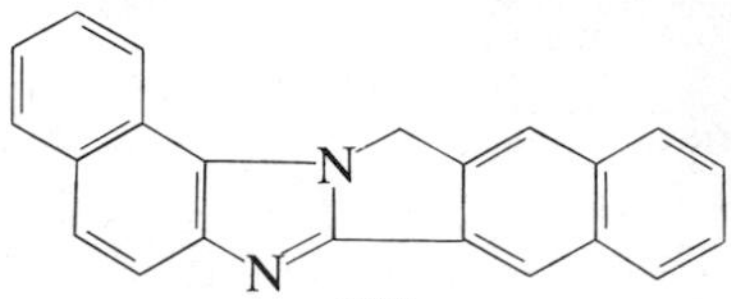 **300**

the latter case they obtained the imine (**303a**) [which can also be written in the amine form (**303b**)].

c. Gompper and Effenberger[201] treated 2-chloro-4,5-diphenyloxazole (**304**) with phenylenediamine (**293**) and obtained 2,3-diphenylimidazo-[1,2-*a*]benzimidazole (**305**). This approach was also used to prepare 1,2-diisopropyl-imidazo[1,2-*a*]benzimidazoles.[258] The oxazole (**304**), prepared from the oxazolone (**306**) may be looked upon as a derivative of formamidodesoxybenzoin (**307**).

2. Rowe and co-workers[202] prepared isoindolobenzimidazole (**294a**) *via* the rather interesting reaction sequence shown below.

3. Grob and Ankli[203] prepared pyrroloimidazolones (**308**) as shown here.

4. Middleton and Metzger[204] treated trifluoroacetonitrile (**309**) with

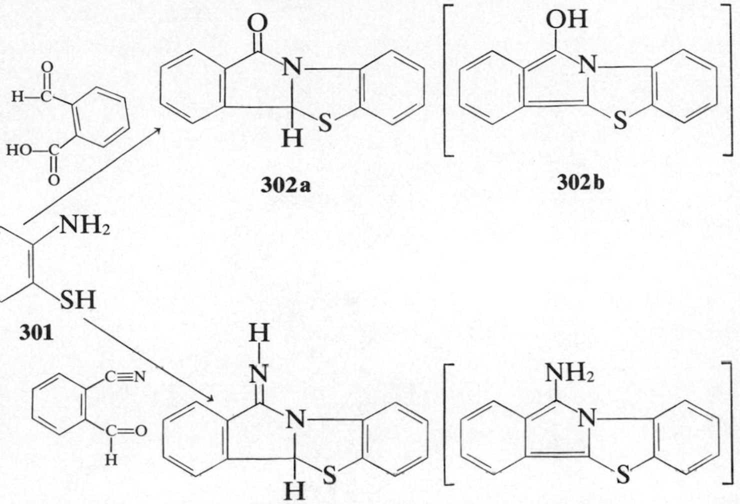

302a

302b

301

303a

303b

sodium cyanide and obtained 3H-2,5,7-trifluoromethyl imidazo[3,4-b]-s-triazole (**310**). They proposed the mechanism shown below.

5. Malaviya and Dutt[205] reported that photolysis of 1,8-diaminonaphthalene (**311**) in dilute hydrochloric acid gave peri-dinaphthaleneazotide (**312**).

6. Wikel and Paget[259] reported the synthesis of s-triazolo[3,4-b]-benzothiazoles from 2-halophenylisothiocyanates (**313**) and hydrazones

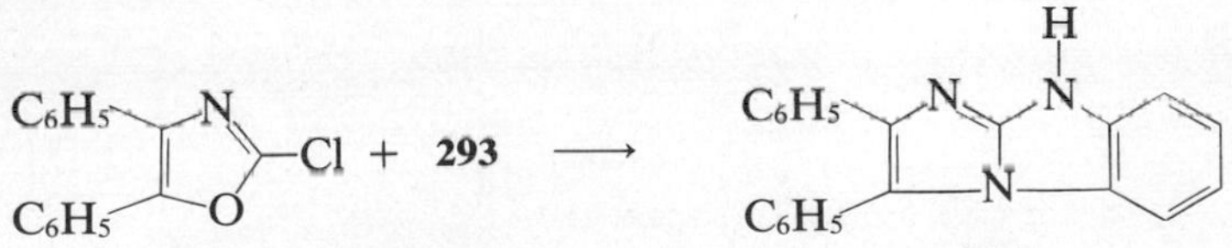

304 + 293 ⟶ 305

306 $\xrightarrow[Et_3N]{POCl_3}$ 304

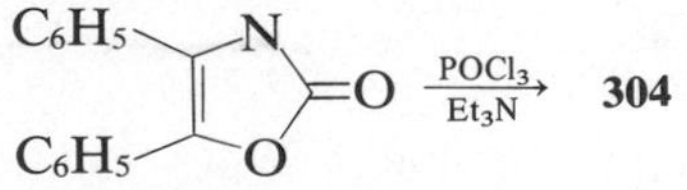

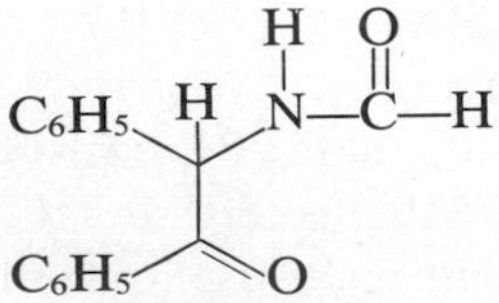

307

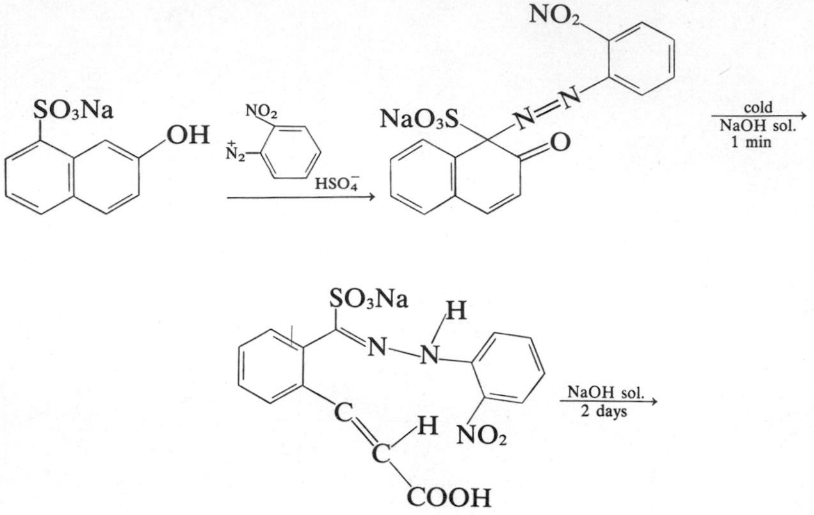

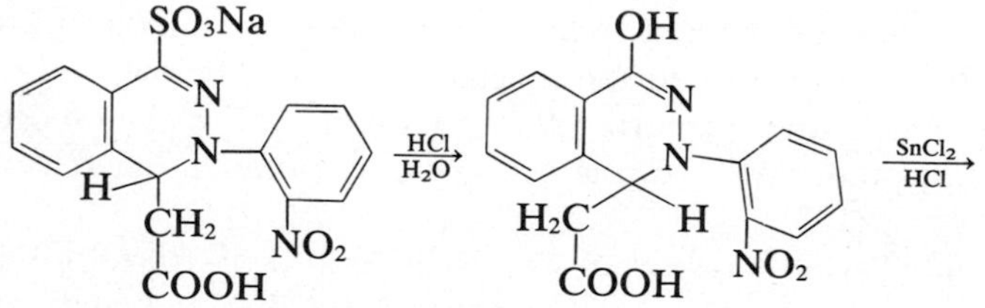

294a

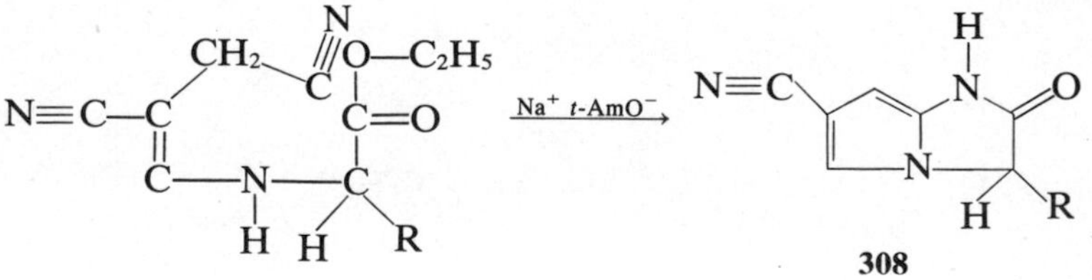

308

R = H or CH_3

or hydrazides in dimethylformamide with sodium hydride. While this reaction can be effected in one step, these workers prepared the intermediates (**314** and **315**) which were then cyclized to triazolobenzothiazole (**316**).

7. Golgolab, Lalezari, and Hosseini-Gohari[256] prepared triazolo[3,4-*b*]-1,3,4-thiadiazoles (**319**) from thiocarbohydrazide (**318**) and a carboxylic acid (**317**) in the presence of phosphoryl chloride. For the preparation of symmetrically substituted compounds, this apparently facile procedure

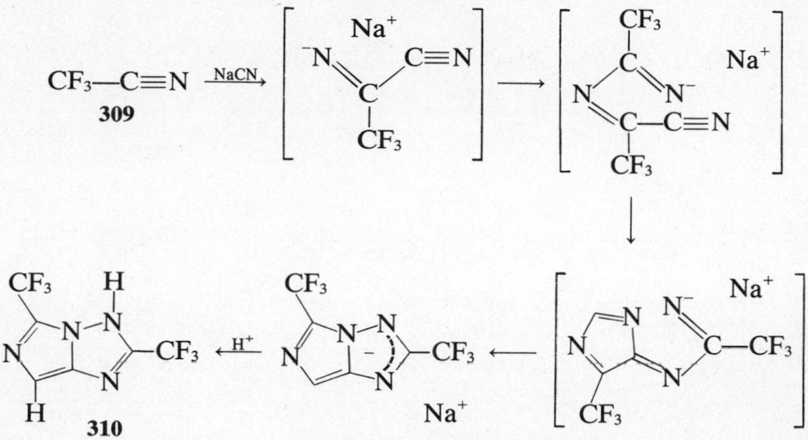

CF$_3$—C≡N $\xrightarrow{\text{NaCN}}$ [Na$^+$ $^-$N=C(CF$_3$)—C≡N] → [CF$_3$ Na$^+$ ... C—C≡N CF$_3$]

309

310

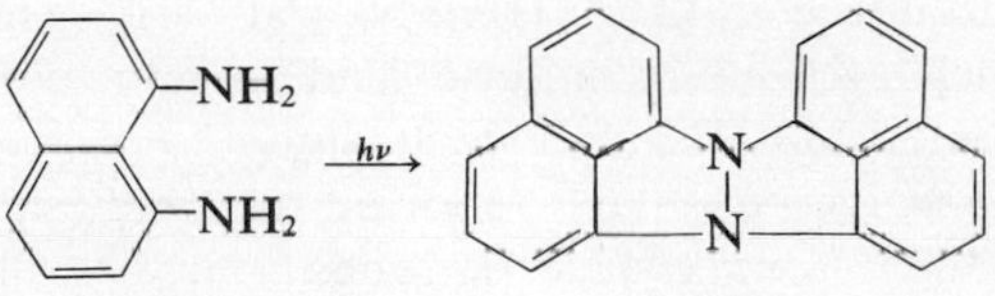

Na$^+$ CF$_3$

311 $\xrightarrow{h\nu}$ 312

obviated the need for preparing the intermediate aminomercaptotriazoles.

8. 5*H*-Tetrazolo[1,5-*a*]isoindole was prepared by Babichev and Romanov[260] in a two-step procedure from 2-chloromethylbenzonitrile (**320**).

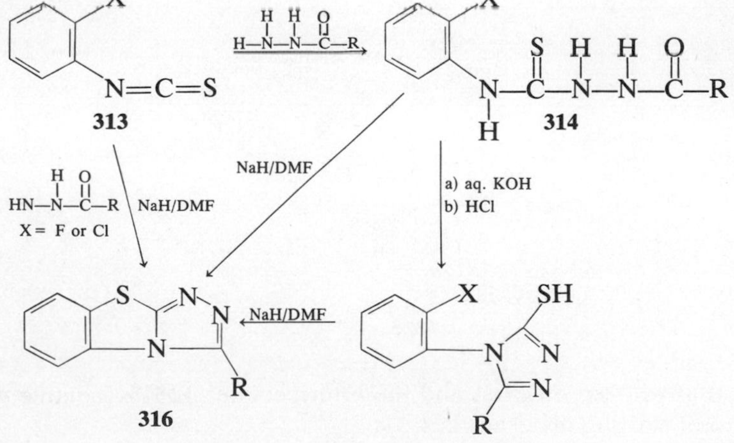

313 → 314

H—N(H)—N(H)—C(=O)—R

NaH/DMF

HN—N(H)—C(=O)—R NaH/DMF

X = F or Cl

a) aq. KOH
b) HCl

NaH/DMF

316

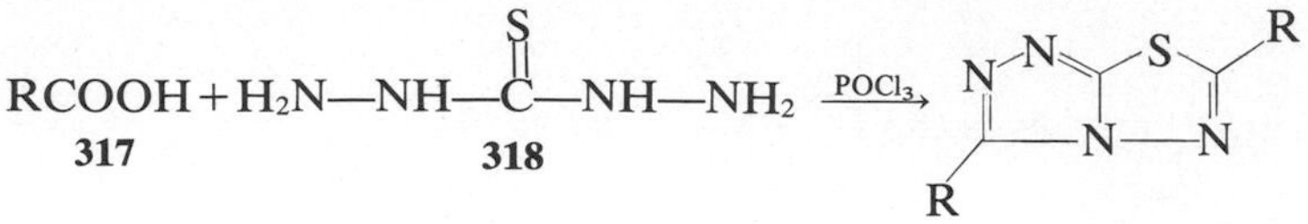

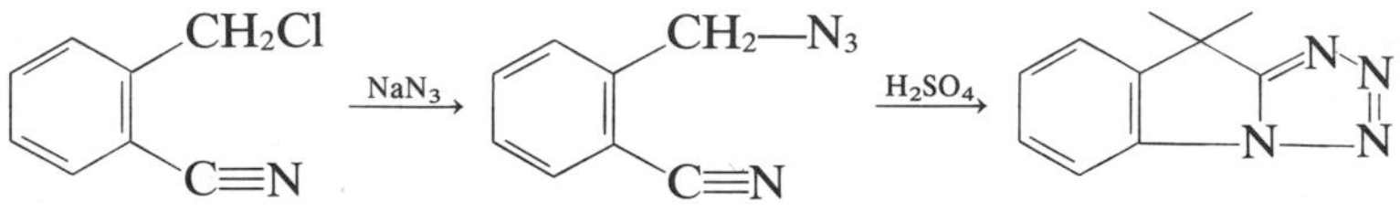

$RCOOH + H_2N—NH—C(=S)—NH—NH_2 \xrightarrow{POCl_3}$ 319

317 318 319

320 $\xrightarrow{NaN_3}$ 321 $\xrightarrow{H_2SO_4}$ 322

The reaction of this chloro compound (**320**) with sodium azide gave 2-azidomethylbenzonitrile (**321**) (IR $-N_3$ at $2100\ cm^{-1}$; C≡N at $2230\ cm^{-1}$) which cyclized in sulfuric acid to tetrazolo[1,5-*a*]isoindole (**312**) (see tautomerism—NH-methylene).

9. Dziomko and Ivashchenko[65] reported the synthesis of imidazo[1,2-*a*]imidazoles (**324**) from α-halo ketones and benzalaminoguanidine (**323**).

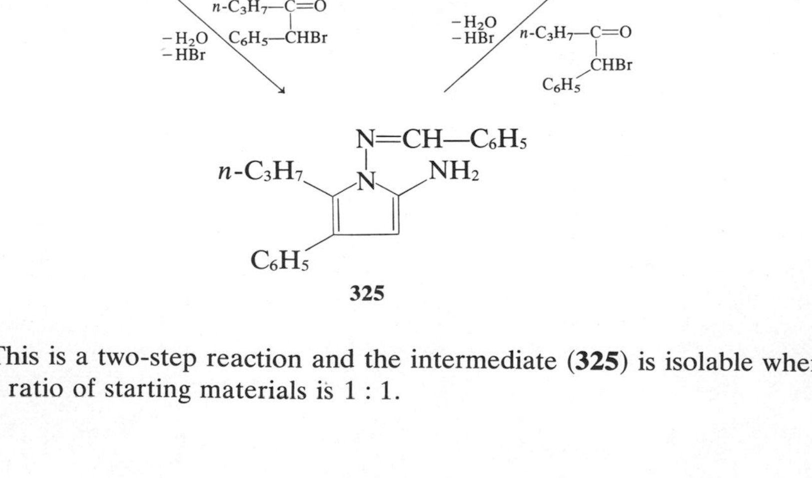

323 $\xrightarrow[-2H_2O,\ -2HBr]{n\text{-}C_3H_7—C(=O)—CHBrC_6H_5\ (2)}$ 324

323 $\xrightarrow[-H_2O,\ -HBr]{n\text{-}C_3H_7—C(=O)—CHBr—C_6H_5}$ 325 $\xrightarrow[-H_2O,\ -HBr]{n\text{-}C_3H_7—C(=O)—CHBr—C_6H_5}$ 324

This is a two-step reaction and the intermediate (**325**) is isolable when the ratio of starting materials is 1 : 1.

III. Reactions

In examining their reactions we should consider what role, if any, the aromaticity then these 5,5 systems are aromatic.

In examining their reactions we should consider what role, if any, the bridgehead nitrogen atom plays. The canonical structures (e.g., **326a**) are more helpful in assessing this than are the delocalized structures (e.g., **326b**). Any canonical form that can be written for any of these systems shows a pyrrole nitrogen common to each ring. In the imidazo[2,1-*b*]-thiazole (**326a**), for example, we see that the imidazo portion contains a

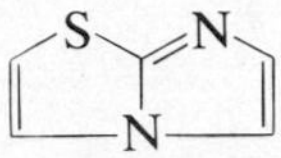

326a  **326b**

pyrrole and a pyridine nitrogen as does an imidazole ring. However, the thiazole portion has a "pyrrole" nitrogen unlike an isolated thiazole nucleus which has a pyridine-type nitrogen. While no quantitative studies have been carried out, it will be seen that the imidazole portion of this system is easily susceptible to electrophilic attack as is an imidazole ring, and unlike an isolated thiazole ring, the thiazole portion of the system is also readily susceptible to electrophilic attack. Thus the "pyrrole" nitrogen behaves as a pyrrole nitrogen donating electrons to both parts of the system.

A. Electrophilic Attack

1. *Pyrrolo[2,1-b]thiazole*

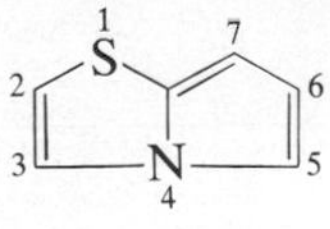

This system is π-excessive and although it has no basic nitrogen atom, it can still be protonated. The nmr spectra of the perchlorate salts have been studied by Malloy, Reid, and McKenzie.[206] They found that protonation takes place mainly at the 5-position (**327**) with some of the 7-substituted derivatives (**328**) also being formed.

Electrophilic substitution follows this same pattern, that is, the 5-position is the favored site for electrophilic attack and 5-formyl (*via* the Vilsmeier–Haack reaction), acetyl, trifluoroacetyl, and nitroso derivatives have been prepared.[207] (Attempted nitration even with a mild agent—cupric nitrate in acetic anhydride—resulted in decomposition of the system.) The site of attack was proven by means of nmr. A chemical proof was also offered. Lithium aluminum hydride–aluminum chloride

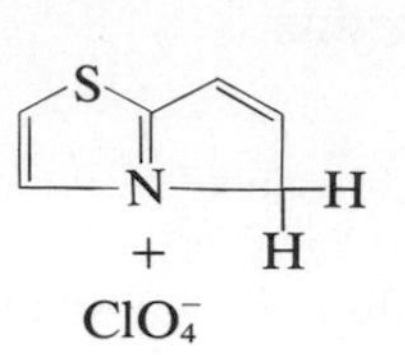

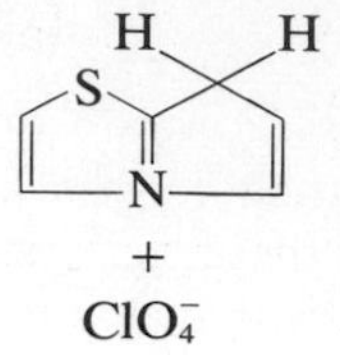

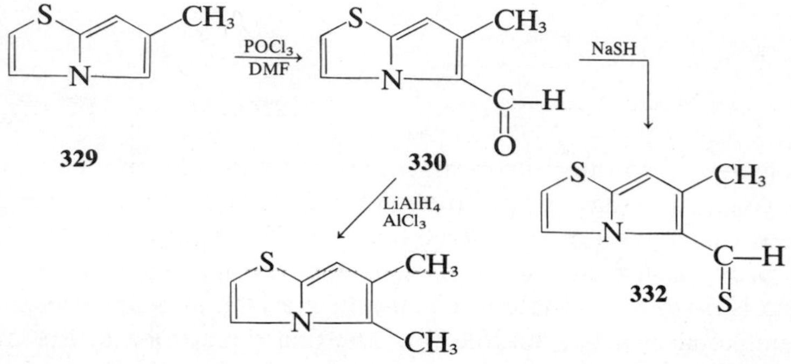

reduction of the formyl compound (**330**) obtained from 6-methylpyrrolo-[2,1-*b*]thiazole (**329**) gave the known 5,6-dimethylpyrrolo[2,1-*b*]thiazole (**331**) showing that the formyl group occupied the 5-position.

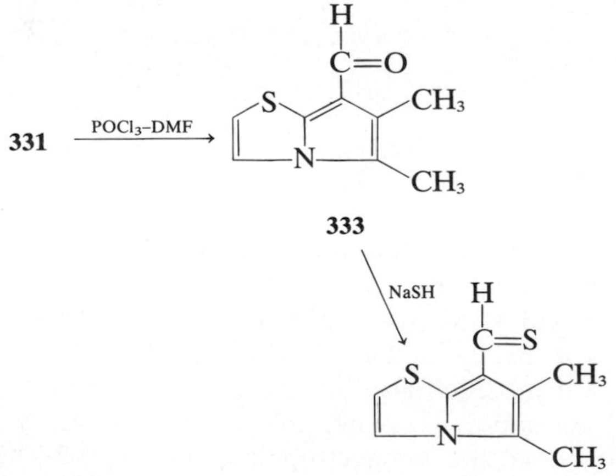

When the 5-position is blocked, the 7-position is subject to attack thus formylation of the 5,6-dimethyl compound (**331**) gave the 7-formyl compound (**330**). A small quantity (2.3%) of diacetyl derivative (**335**) was obtained from the acetylation of 6-methylpyrrolo[2,1-*b*]thiazole (**329**), although the 5-acetyl compound (**335**) was the major product (97.7%).

329 ⟶ **335** + **336**

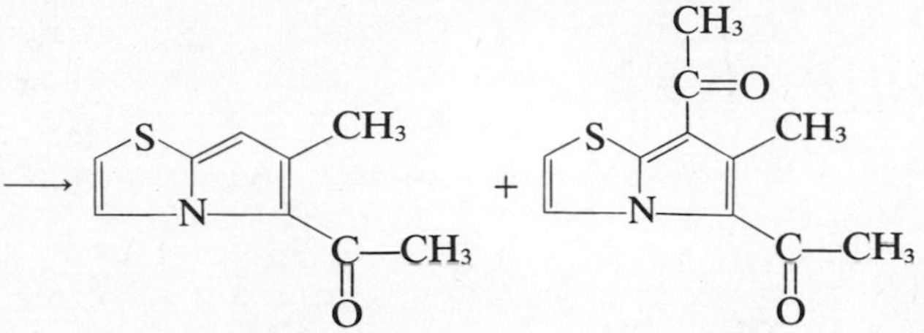

Mackie, McKenzie, Reid, and Webster[261] treated the aldehydes (**330** and **333**) with sodium hydrosulfide and obtained thioaldehydes **332** and **334**. The nmr spectrum of the 5-thioaldehyde (**332**) showed that it could exist in either the *syn* or *anti* form. The 7-thioaldehyde (**334**) existed only in the *syn* conformation.

Tritylation of the 6-methyl derivative (**339**) with triphenylmethyl perchlorate gave a mixture of the 5,7-ditrityl (**337**), 5-trityl (**338**), and

329 ⟶ **337** (27%) + **338** (20%) + **339** (5%)

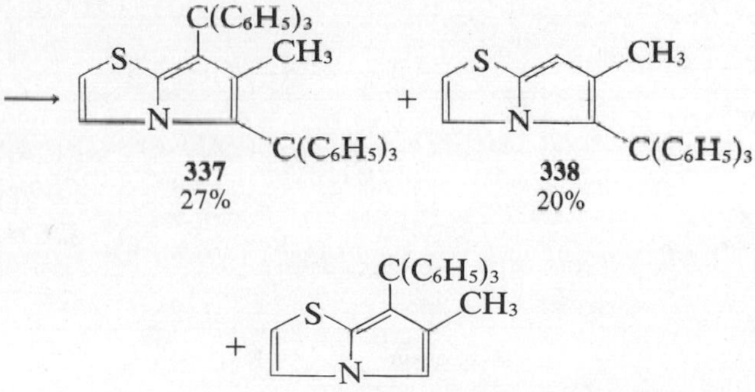

7-trityl (**339**) derivatives. This was the only reaction tried which gave a 7-substituted product while the 5-position was unblocked. A study of models indicated that the trityl group has a more hindered approach to the electronically more favorable 5-position.

Ceder and Beijer[262] metallated 3-methyl-6-phenyl (**340**) and 2,6-dimethyl pyrrolo[2,1-*b*]thiazole (**342**) with butyl lithium. Treatment of

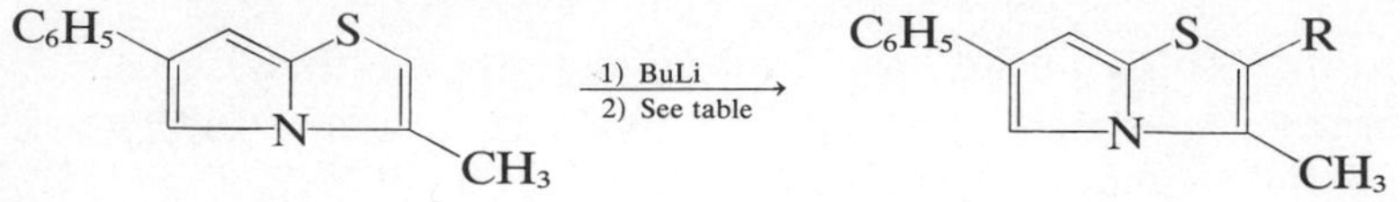

340

1) BuLi
2) See table

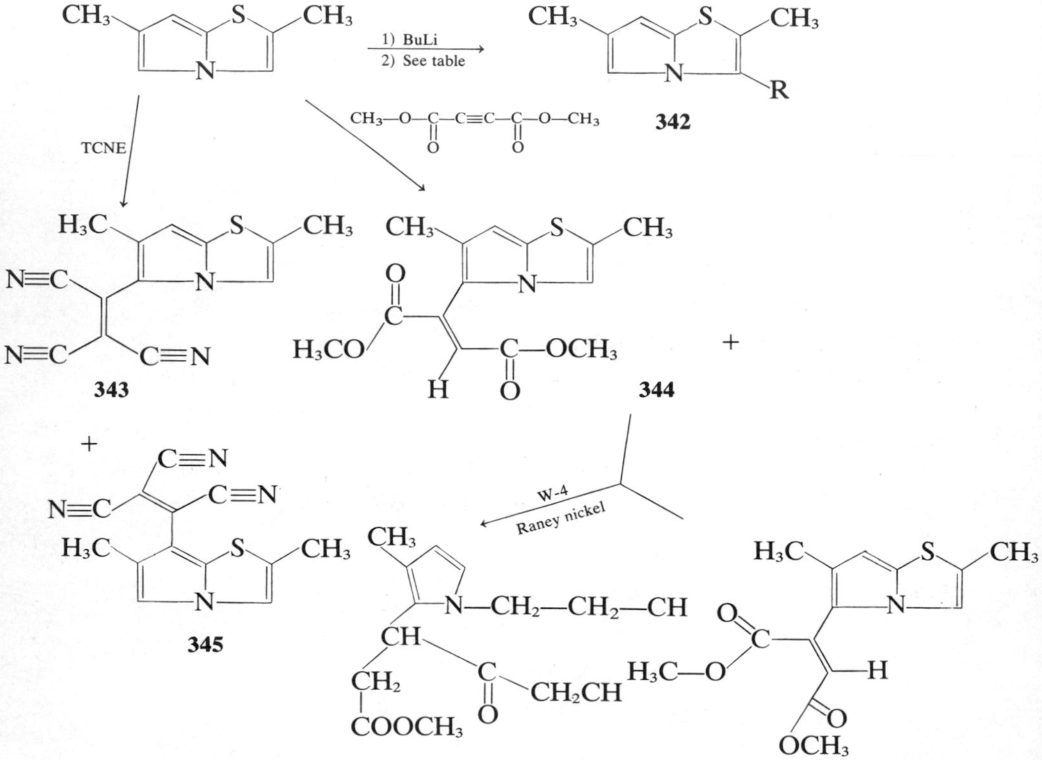

341

1) BuLi
2) See table

342

TCNE

$CH_3—O—C(=O)—C≡C—C(=O)—O—CH_3$

343

344

+

+

345

W-4
Raney nickel

346

347

Reagent	R
	H
$H—C(=O)—N(CH_3)_2$	—C=O
D_2O	D
CO_2	COO^-
(CDV only)	

the metallated materials with dimethyl formamide, to obtain ring substituted compounds, **341** and **342** (see table for substituents), as shown by nmr.

2,6-Dimethylpyrrolo[2,1-*b*]thiazole reacted with dimethylacetylene dicarboxylate to give both the E and Z isomer (**344** and **342**, respectively). That these were not positional isomers was demonstrated by their desulfurization with W-4 Raney nickel to give the pyrrole.

Reaction of **342** with TCNE gave two isomers. As the tricyanoethylene group has no potential for isomer formation, these compounds had to be positional isomers (**343**+**345**).

An unusual product was obtained from the reaction of the 6-methyl (**329**) compound in tetranitromethane. This material possesses either structure **348** or **349**.

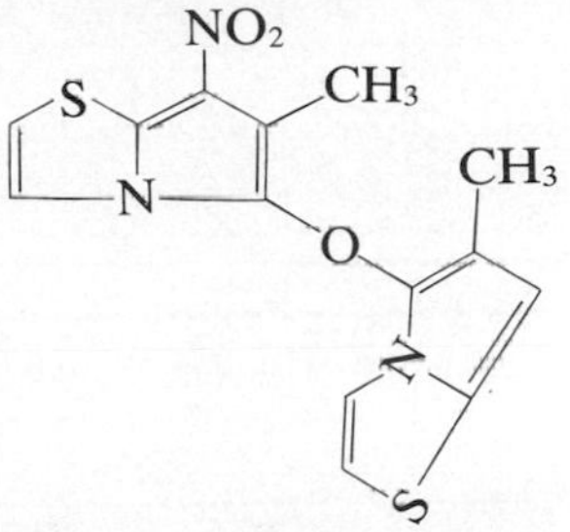

348 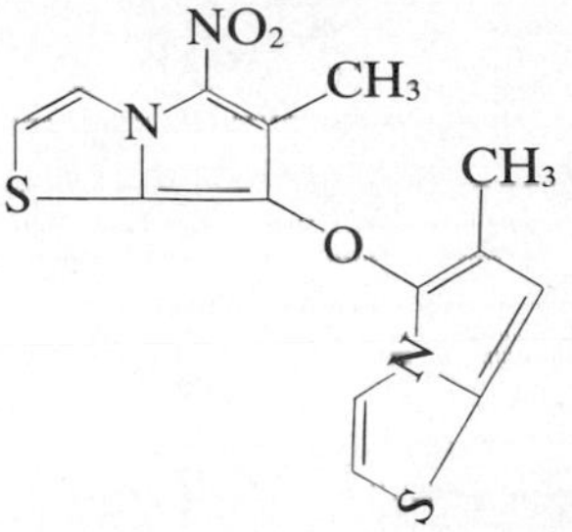 **349**

Pyl, Gille, and Nusch[12] acetylated derivatives of 6-phenylpyrrolo[2,1-*b*]thiazole (**350**) and as reaction took place when the 2,7 or para position

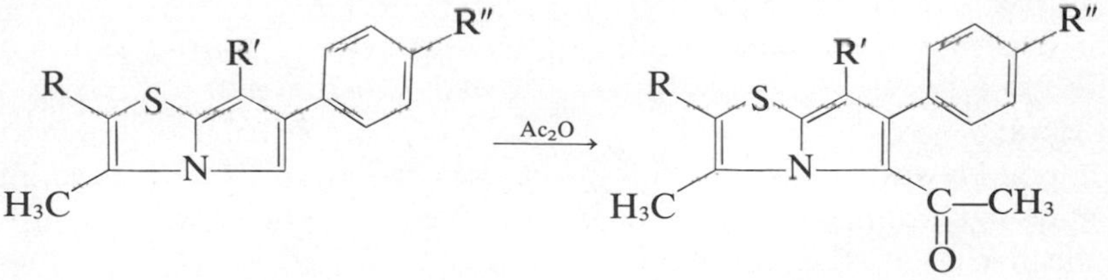

$\xrightarrow{Ac_2O}$

350 **351**

R = H or CH_3 R′ = H or CH_3 R″ = H or Br

were blocked these workers concluded that substitution went at the 5-position. While this is not a vigorous structure proof as it ignores the possibility that any one of several positions might be subject to attack and that blocking of one position might merely result in attack taking place at another position. The work of McKenzie, Malloy, and Reid[207] indicate that the structural assignment of compound **351** is correct.

2. *Imidazo[1,2-a]imidazole*

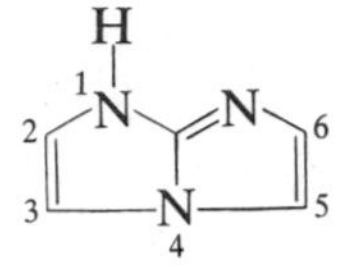

Miller and Bambury[272] treated 1,6-dimethyl imidazo[1,2-*a*]imidazole (**352**) with several electrophiles and found that substitution took place at

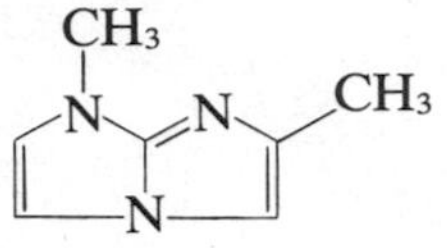

352 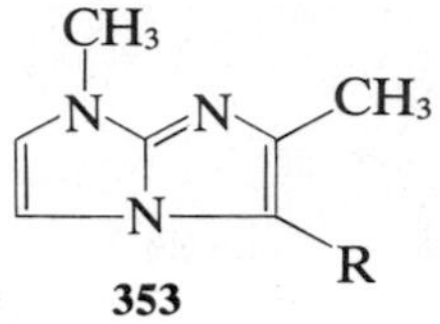 **353**

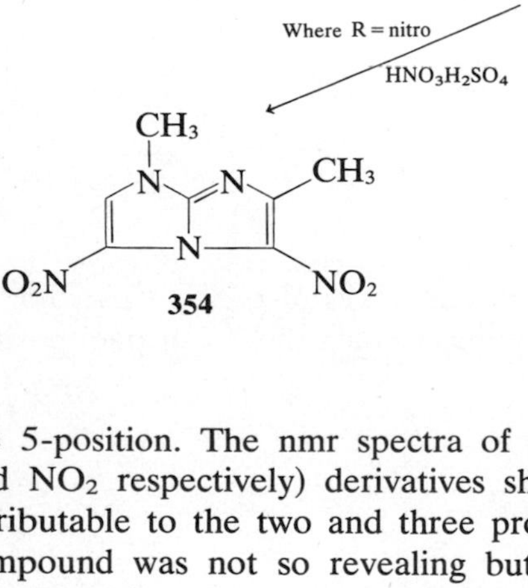

Where R = nitro

$HNO_3H_2SO_4$

354

Reagent	R
HNO_3–H_2SOR	NO_2
Br_2–$CHCl_3$	Br
$POCl_3$–DMF	H–C=O

the 5-position. The nmr spectra of the formyl and nitro **353**–R=C(H)=O and NO_2 respectively) derivatives showed two doublets (A,B pattern) attributable to the two and three protons. The spectrum of the bromo compound was not so revealing but conversion of it to a piperidino compound (see displacement reactions) gave a compound whose nmr was indicative of a 5,6 disubstituted compound.

Nitration of (**353**; R = H) proved to be tricky and the best results (66% yield) were obtained by treating a sulfuric acid solution of **353** with one equivalent of ethyl nitrate. When two equivalents of ethyl nitrate were used, a dinitro compound was obtained. Using nmr and nuclear Overhauser effect, these workers were unable to determine the site of the second nitro group. Their structural assignment—1,6-dimethyl-3,5-dinitro imidazo[1,2-*a*]imidazole (**354**)—was based on their Hückel Molecular Orbital (HMO) calculations (see theoretical section).

3. *Pyrrolo[1,2-a]benzimidazole*

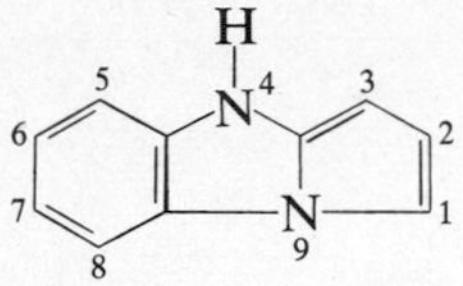

Kornilov, Dyadyusha, and Babichev[208] found, by means of nmr, that 1-methylpyrrolo[1,2-*a*]benzimidazole protonates on the 1-position (**355**). Babichev and co-workers[10,208] using nmr showed that the 1-position

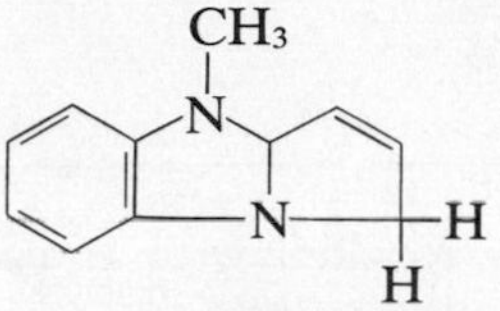

355

(**352**) is also the favored position for electrophilic substitution as shown here. (see Table II).

TABLE II

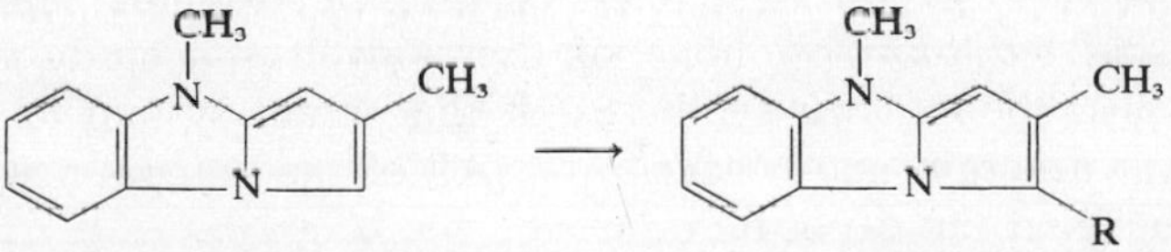

356 ⟶ **357**

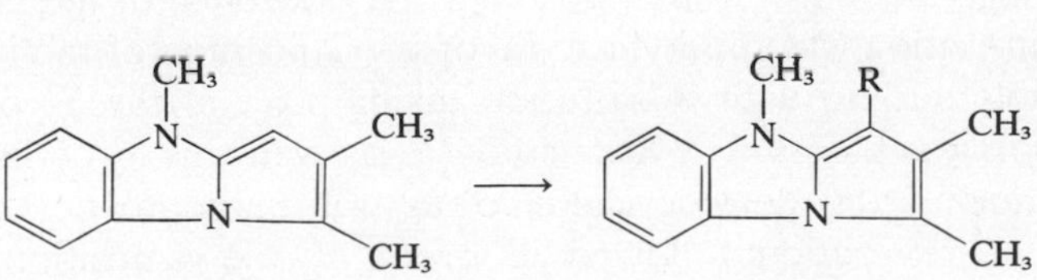

358 ⟶ **359**

Reagent	R	Ref.
$POCl_3$–DMF	$\mathrm{-CH{=}O}$	10
Ac_2O	$\mathrm{-C(=O)-CH_3}$	8,10

TABLE II (Continued)

Reagent	R	Ref.
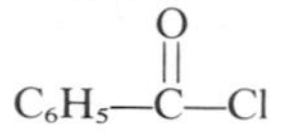 C_6H_5—C(=O)—Cl	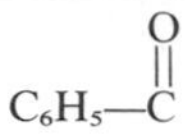 C_6H_5—C(=O)—	10
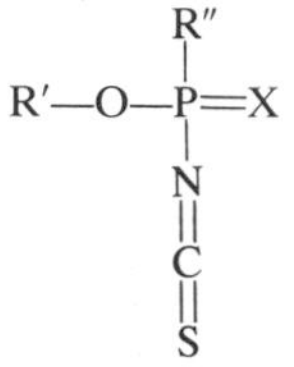 R′—O—P(R″)(=X)—N=C=S R′ = alkyl R′ = H or CH_3 X = S or O	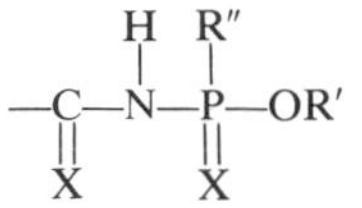 —C(=X)—N(H)—P(R″)(=X)—OR′	209

When the 1-position was blocked with a methyl group (**353**), the 3 position (**359**) accepted the electrophile.[10,209]

4. *Imidazo*[*2,1-b*]*thiazole*

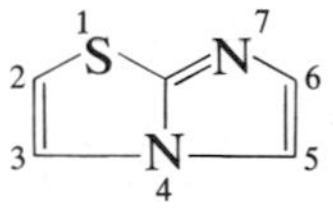

While there are several reports concerning electrophilic substitution onto imidazo[2,1-*b*]thiazoles, none of these involves a study with the parent system although two syntheses of the parent system have been reported by Kochergin and Mazur.[210,211] All of the compounds studied had substituents at the 6-position.

Pyl and co-workers studied the nitration, nitrosation, bromination, and diazo coupling reaction of 6-arylimidazo[2,1-*b*]thiazoles. In the first publication[30] nitrosation was reported to favor the 2-position of the imidazo-[2,1-*b*]thiazole nucleus with substitution taking place at the 5-position if the 2-position was blocked. A later study[212] showed this to be incorrect. The 5-position of the heterocyclic nucleus was found to be the most susceptible to nitrosation followed in order by the 4-position of the benzene ring and the 2-position of the heterocyclic system. This is based on the following sequence of reactions. 6-Phenyl, 2-methyl-6-phenyl, 3-methyl-6-phenyl, and 2,3-dimethyl-6-phenylimidazo[2,1-*b*]thiazole (**360a–d**) with potassium permanganate gave benzoic acid (**362**). Nitrosation of 6(4-bromophenyl)imidazo[2,1-*b*]thiazole (**363**) gave a mononitroso compound which on oxidation with permanganate gave 4-bromobenzoic acid (**365**). A mononitroso compound was obtained from the 5-methyl-6-phenyl compound (**366**) and oxidation of this nitroso compound

R, S, N, R′, N — HONO → R, S, N, R′, N, NO — $KMnO_4$ → H—O—C(=O)—C_6H_5

360 **361** **362**

$KMnO_4$

a) $R = R' = H$
b) $R - CH_3$, $R' = H$
c) $R = H$, $R' = CH_3$
d) $R = R' = CH_3$

Br, S, N, N — HONO → Br, S, N, N, NO — $KMnO_4$ → H—O—C(=O)—C_6H_4—Br

363 **364** **365**

S, N, N, CH_3 — HONO → NO, S, N, N, CH_3 — $KMnO_4$ → H—O—C(=O)—C_6H_4—NO_2

366 **367** **368**

Br, S, N, N, CH_3 — HONO → NO_2, Br, S, N, N, CH_3 — $KMnO_4$ → **365**

369 **370**

Br, S, N, N, H_3C, CH_3 — HONO → Br, ON, S, N, N, H_3C, CH_3

371 **372**

gave 4-nitrobenzoic acid (**368**). Treatment of 6-(4-bromophenyl)-5-methylimidazo[2,1-*b*]thiazole (**369**) with nitrous acid resulted in the formation of a *nitro* compound (**370**) which on oxidation with potassium permanganate gave 4-bromobenzoic acid (**375**). The 3,5-dimethyl-6-*p*-bromophenyl derivative (**371**) gives a mononitroso compound which is apparently the 2-nitroso derivative (**372**). These reactions along with an earlier report that 2,5-dimethyl-6-phenylimidazo[2,1-*b*]thiazole does not react with nitrous acid[30] were used to assign the structures of compounds **361a–d**, **364**, **372**, and **370**.[212] [Note: These reactions do not appear to completely rule out reaction at the 3-position. That nitrous acid is capable of putting a nitroso group on the benzene ring of 5-methyl-6-phenylimidazo[2,1-*b*]thiazole (**366**) (giving compound **367**),[212] but does not react with 2,5-dimethyl-6-phenylimidazo[2,1-*b*]thiazole[30] is indeed a curious phenomenon. Furthermore, although the 2-nitro compound (**370**) was reported as being known[30] it was prepared by nitration of compound **369** which does not make for a valid comparison.]

Using this same approach, that is, treatment of 6-arylimidazo[2,1-*b*]thiazoles and various methyl derivatives, Pyl, Giebelman, and Beyer[30] reported that bromination in chloroform takes place only at the 5-position of the heterocyclic nucleus and that the bromo compounds were isolated as perbromide hydrobromides which lose bromine and hydrogen bromide on treatment with sodium acetate in aqueous methanol.

Nitration is sulfuric acid[46] was reported to favor the 4-position (**364**) of the benzene ring of 6-phenylimidazo[2,1-*b*]thiazole (**360a**) followed by the 5- and the 2-positions of the heterocyclic nucleus. Diazo coupling takes place only at the 5-position of the heterocyclic nucleus.[212]

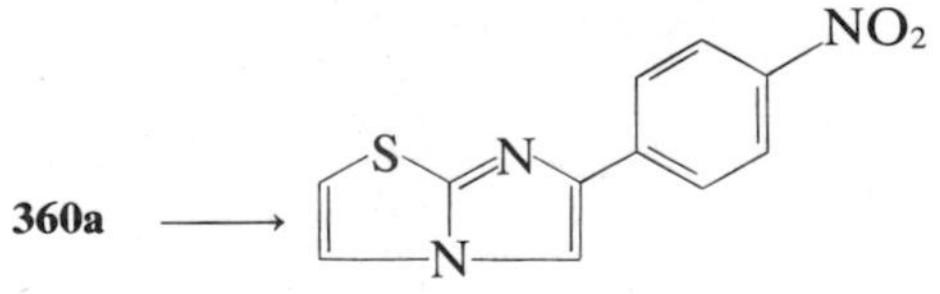

360a ⟶ [structure with S, N, N, NO_2]

374

The work of the Pyl group on nitration, nitrosation, bromination, and diazo coupling is summarized in Table III.

Pyl, Wünch, and Beyer[212] reported that 6-phenylimidazo[2,1-*b*]thiazole (**360a**) undergoes diazo coupling only when diazo compounds containing strong electron withdrawing substituents were used. However Pentimalli and Passalaqua[213] found this reaction to be more general, taking place even when benzenediazonium chloride was used as the electrophile. Pentimalli, Cogo, and Guerra,[214] using nmr and chemical studies, essentially confirmed the work of the Pyl group with respect to

TABLE III

	Position substituted in order of favor		
Nitrosation[211]	5IT[a]	4B[b]	2IT
Nitration (in H_2SO_4)[46]	4B	5IT	2IT
Bromination[16]	5IT	—	—
Diazo coupling[211]	5IT	—	—

[a] IT—imidazothiazole nucleus.
[b] B—benzene ring.

nitration, nitrosation, and diazo coupling. In addition, they showed that 6-phenyl-imidazo[2,1-*b*]thiazole (**360a**) undergoes reactions with some dienophiles, namely, maleic anhydride and diethyldiazodicarboxylate, to give the acid (**375**) and hydrazine (**376**) respectively.

The structures of these compounds were proven chemically. Neither dienophile reacted with 2,5-dimethyl-6-phenylimidazo[2,-1-*b*]thiazole

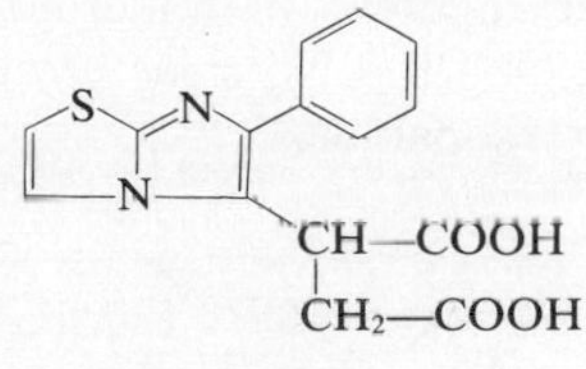

375

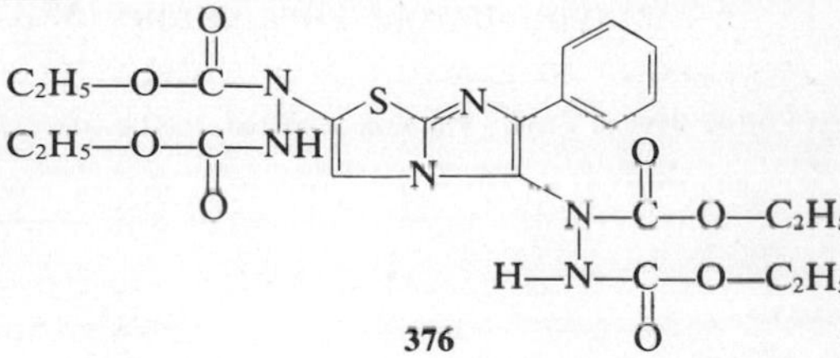

376

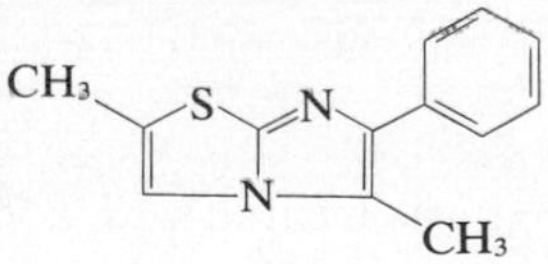

377

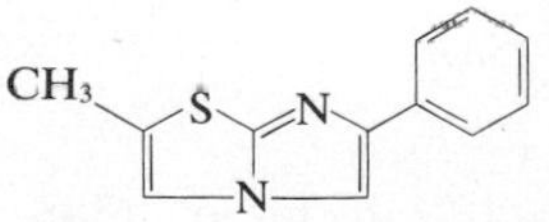

378

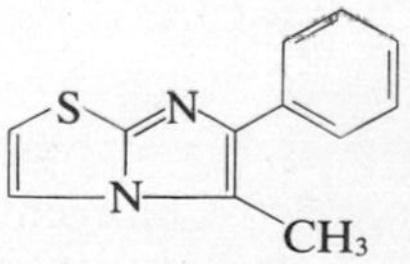

379

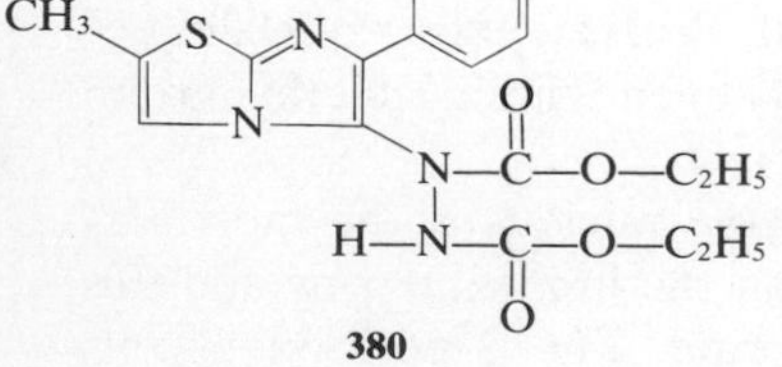

380

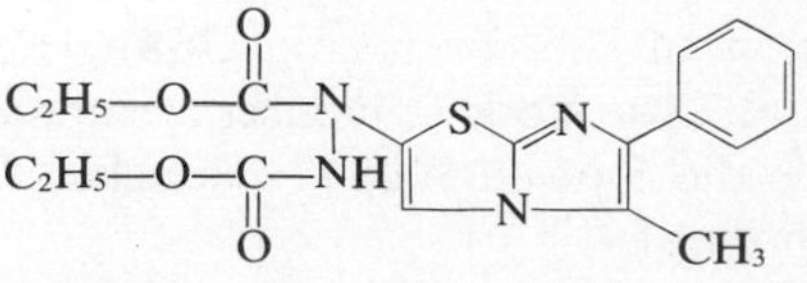

381

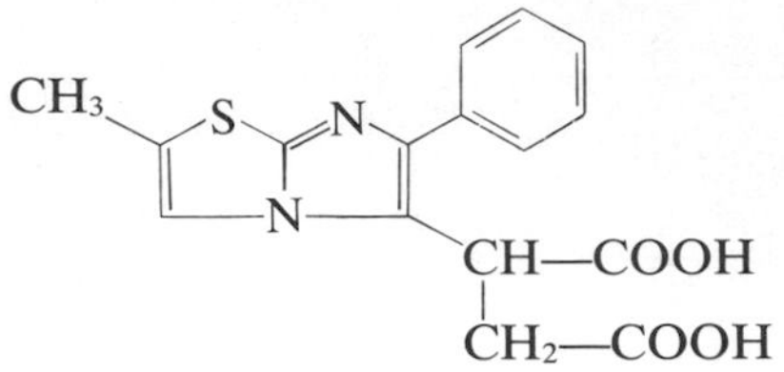

382

(**377**) showing the 2- and 5-positions to be the reactive sites. Diethylazodicarboxylate reacted with both 2-methyl-6-phenyl (**378**) and 5-methyl-6-phenylimidazo[2,1-*b*]thiazole (**379**), to give monoaddition products **380** and **381**, respectively, leaving little doubt that the *bis* product is the 2,5-dihydrazine (**376**). Maleic anhydride which gives only a monosubstitution product (**382**), reacted with 2-methyl-6-phenylimidazo[2,1-*b*]thiazole (**377**) but not with 5-methyl-6-phenylimidazo[2,1-*b*]thiazole (**379**) showing that the succinic acid derivative is indeed compound (**375**). *N*-Phenylmaleimide was reported to give no product with 6-phenylimidazao[2,1-*b*]thiazole (**360a**).

6-Methylimidazo[2,1-*b*]thiazole (**383**) was reported to undergo thiocyanation to give the 5-substituted product (**384**).[215] No structure proof was given but the positional assignment is reasonable.

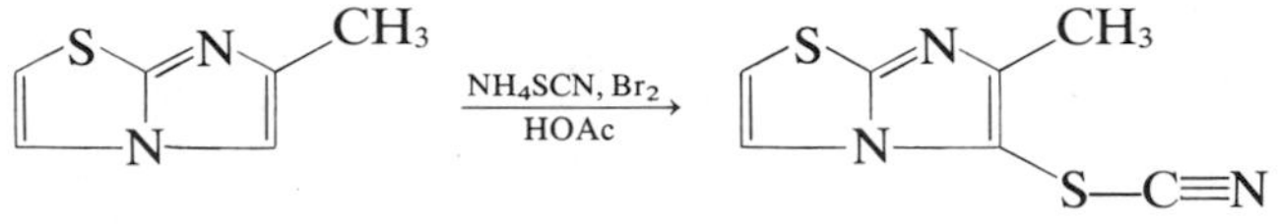

$\xrightarrow[\text{HOAc}]{\text{NH}_4\text{SCN, Br}_2}$

383 **384**

6-Chloroimidazo[2,1-*b*]thiazole (**385**; R = H) was treated with a variety of nucleophiles to give compounds, shown by nmr, to be substituted at the 5-position (Table III).[140]

6-(2-Furyl)imidazo[2,1-*b*]thiazole (**386**) is reported to bromiate at the 5-position (**387**) when there is no substituent at the 3-position, and at the 5-position of the furan ring (**388**) when there is a substituent at the 3-position. The steric influences are apparent. With an extra equivalent of bromine the 5-position can be attacked (**389**) even with a 3 methyl group being present.[263]

The 5-position of the imidazo[2,1-*b*]thiazole nucleus is the most susceptible to electrophilic attack. This site is on the imidazole ring and this is not unusual behavior for an imidazole ring. The 2-position of this

TABLE IV

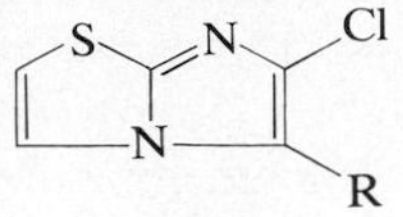

385

Reagent	R
HNO_3–H_2SO_4	NO_2
CH_2O; acetic acid, H—N(R′)(R′)	CH_2—N(R′)(R′) (R′ = alkyl or cycloalkyl ring residue)
KSCN, Br_2, acetic acid	S—C≡N
$POCl_3$—HC(=O)—$N(CH_3)_2$	C(=O)—H
N-Bromosuccinimide	Br
N-Chlorosuccinimide	Cl

system, in many cases, also undergoes electrophilic substitution with relative ease, and this is not typical behavior for a thiazole ring—compare the relative ease of nitration of the thiazole portion of 6-(4-bromophenyl)-5-methylimidazo[2,1-*b*]thiazole (**369**) (in cold nitric acid)[46] with the more drastic condition required for the nitration of mono and

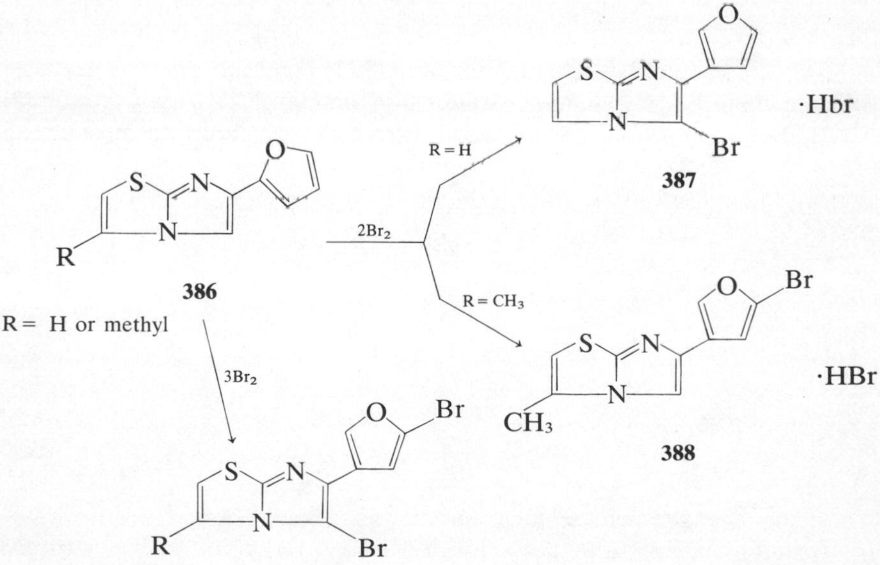

R = H or methyl

386 —$2Br_2$→ (R = H) 387 ·Hbr; (R = CH_3) 388 ·HBr

386 —$3Br_2$→ 389

dimethyl thiazoles.[216] This difference in reactivity of the thiazole ring is apparently due to the bridgehead nitrogen atom, a pyrrole-type nitrogen that can donate electrons to the system rather than the pyridine-type nitrogen that is found in an isolated thiazole ring.

Pentimalli and Milani[113] reported the methylation of the 5-diazophenyl coupled compounds (**390**) with dimethyl sulfate to give **391**.

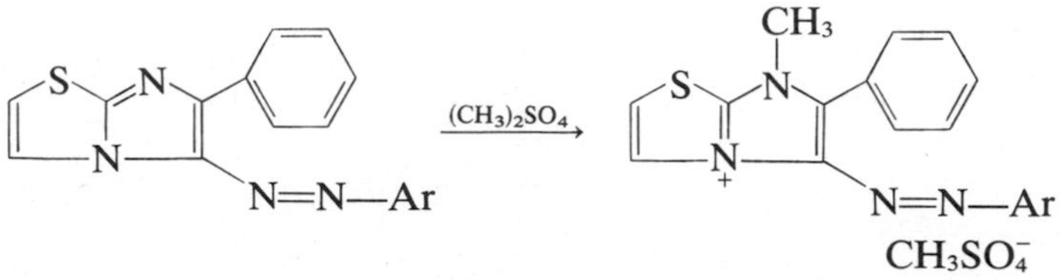

$(CH_3)_2SO_4$

N=N—Ar

CH_3

N=N—Ar
$CH_3SO_4^-$

390 **391**

5. *Pyrrolo[2,1-b]benzothiazole*

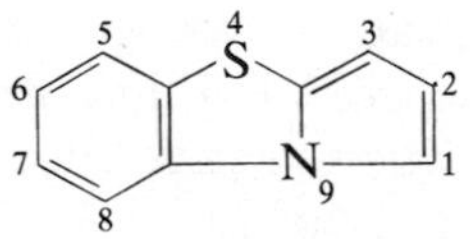

2-Phenylpyrrolo[2,1-*b*]benzothiazole (**392**) is reported to undergo diazo coupling.[213] By analogy with 2-phenylimidazo[2,1-*b*]thiazole (**360a**) the carbon atom α to the bridgehead nitrogen (1-position) was designated

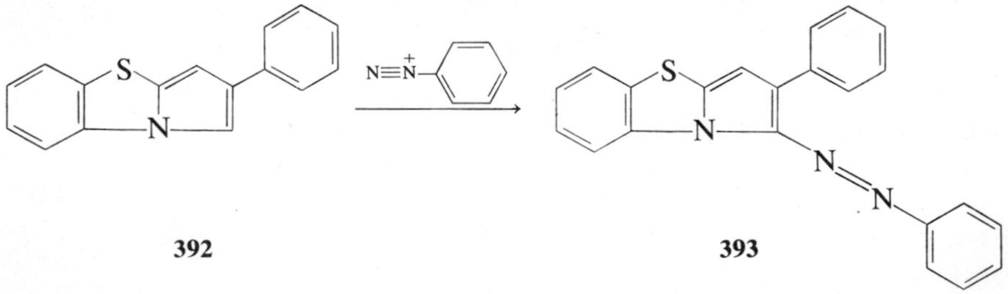

392 **393**

as being the site where substitution took place (**393**). Even weakly electrophillic diazo salts, those formed from *p*-toluidine and *p*-anisidine were used in this reaction.

6. *Imidazo[2,1-b]benzothiazole*

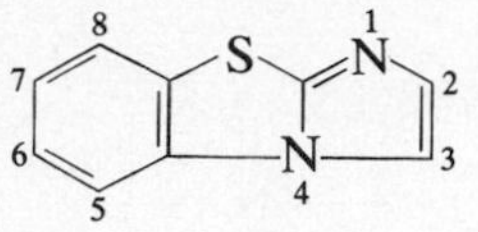

2-Phenylimidazo[2,1-*b*]benzothiazole was found by Pentimalli and co-workers[213,217] to be susceptible to attack by electrophilic reagents (nitrous acid, maleic anhydride, diethylazodicarboxylate, and some benzenediazonium salts).

Nitration gave a mixture containing a mono and a dinitro compound. The mononitro compound was 2-(4-nitrophenylimidazo[2,1-*b*]-benzothiazole (**395**), as shown from independent synthesis from 2-aminobenzothiazole (**92**) and *p*-nitrophenacyl bromide. The dinitro compound

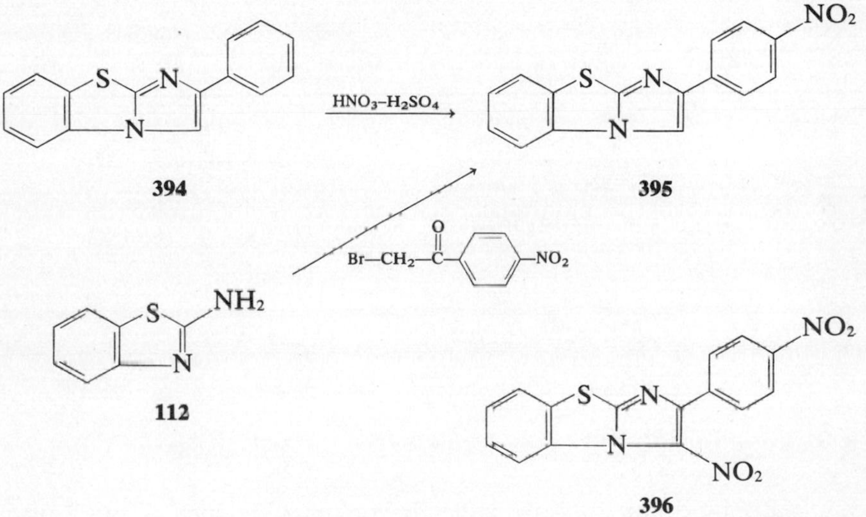

HNO_3–H_2SO_4

394 **395**

Br—CH_2—C(=O)—C_6H_4—NO_2

112 **396**

was assumed to be 2-(4-nitrophenyl)-3-nitroimidazo[2,1-*b*]benzothiazole (**396**).

The other electrophilic reagents used were, by analogy with the chemistry observed with 2-phenylimidazo[2,1-*b*]thiazole (**360a**) assumed to give 3-substituted derivatives (**397**—Table V).

By way of comparison with the similar 6-phenylimidazo[2,1-*b*]thiazole (**360a**) and 2-phenylpyrrolo[2,1-*b*]benzothiazole (**394**) which reacted with weakly electrophilic diazonium salts (those formed from *p*-toluidine and *p*-anisidine), 2-phenylimidazo[2,1-*b*]thiazole (**360a**) only

TABLE V

397

Reagent	R
HNO_2	NO
Maleic anhydride (O=C–CH=CH–C=O, ring-closed through O)	$—CH(—COOH)—CH_2—COOH$
$C_2H_5—OC(=O)—N{=}N—C(=O)—O—C_2H_5$	$—N(—C(=O)—O—C_2H_5)—N(H)—C(=O)—O—C_2H_5$
$N{\equiv}\overset{+}{N}—C_6H_4R$ (R = NO_2 or Cl)	$—N{\equiv}N—C_6H_4R$

reacted with strongly electrophilic diazonium salts such as *p*-Cl or *p*-nitrobenzenediazonium chloride. While the imidazo thiazole (**360a**) and the pyrrolobenzothiazole (**392**) each have two electron withdrawing components, pyridine nitrogen and a phenyl substituent in the former and a fused benzene ring and a phenyl substituent in the latter, the electron density about the position α to the pyrrole nitrogen is still large enough to permit electrophilic attack from weak electrophiles. 2-Phenylimidazo-[2,1-*b*]benzothiazole (**394**) has three electron-withdrawing components, a fused benzene ring, a pyridine nitrogen, and a phenyl substituent, which lower the reactive potential of the 3-position so as to permit attack by only strongly electrophilic diazonium salts.

7. *Imidazo[5,1-b]benzothiazole*

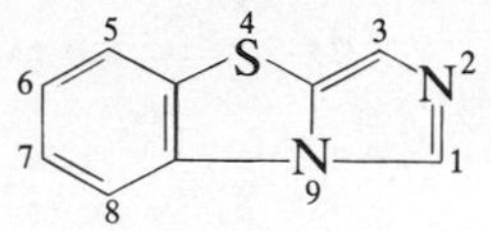

Avidon and Shchukina[218] treated 3-phenyl-imidazo[5,1-*b*]benzothiazole with a variety of electrophilic reagents and obtained products shown by nmr to be 1-substituted-3-phenylimidazo[5,1-*b*]benzothiazoles (**398**—Table VI).

TABLE VI

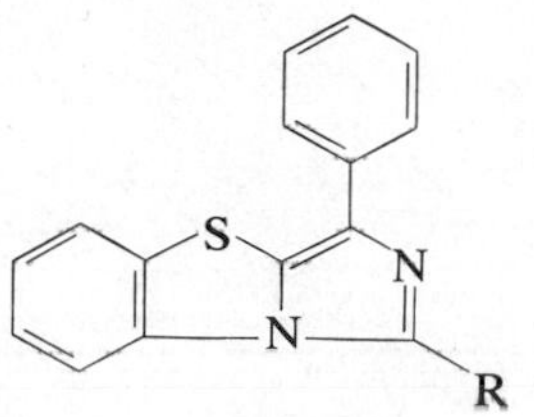

398

Reagent	R
H—C(=O)—N(CH_3)$_2$—$POCl_3$	C(=O)—H
Hg(CH_3COO)$_2$	Hg—0—C(=O)—CH_3
CH_2O, H—N(CH_3)$_2$·HCl	CH_2—N(CH_3)$_2$
HNO_3—HOAc	NO_2
Br_2	Br

8. *Imidazo[5,1-b]benzimidazole*

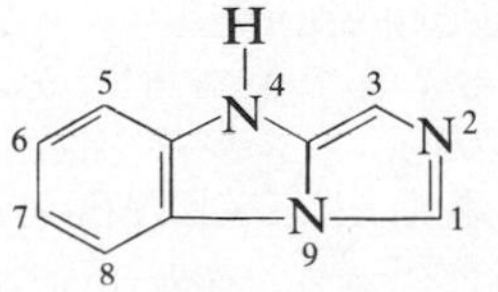

The imidazo[5,1-*b*]benzimidazole was studied by Aryuzina and Shchukina[265,266] who reported that substitution takes place at the 1-position when the 3-position is blocked[265] (**399** to **400**) and at the 3-position if the 1-position is blocked[266] (**401** to **402**). Compound **399** also undergoes diazo coupling and acetylates with acetic anhydride.[266]

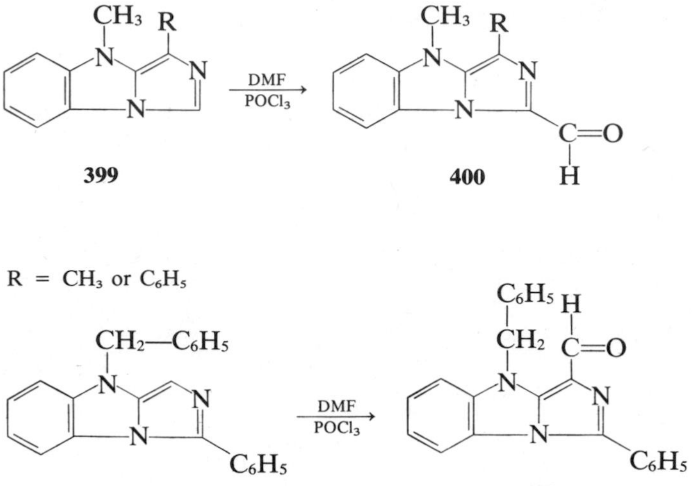

399 $\xrightarrow[\text{POCl}_3]{\text{DMF}}$ **400**

R = CH_3 or C_6H_5

401 $\xrightarrow[\text{POCl}_3]{\text{DMF}}$ **402**

9. *Thiazolo[3,2-b]-1,3,4-triazole*

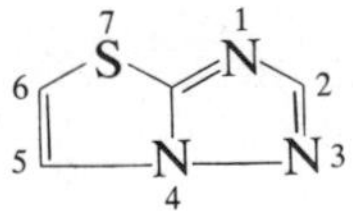

5-Methyl (**403**; $-R = H$, $R^1 = CH_3$), 2,5-dimethyl (**403**; $-R$; $R = R^1 = CH_3$)[267] and 5-methyl-2-phenyl Thiazolotriazoles (**403**; $-R = C_6H_5$, $R^1 = CH_3$)[135] react with *N*-bromosuccinimide to give 6-bromo derivatives (**404**). The unsubstituted, as well as the 2-methyl and 2-phenyl thiazolo[3,2-*b*]-1,3,4-triazoles (**403**; $-R^1 = H$ and $R = H$, CH_3 and C_6H_5, respectively), failed to give a bromo compound (**405**) on treatment with NBS. Tamura and co-workers[135] suggested that "... thiazolo[3,2-*b*]-*s*-triazole is resistant to electrophilic substitution but the introduction of a 5-methyl substituent greatly activates the C-6 position presumably due to stabilization of the transition state through hyperconjugation."

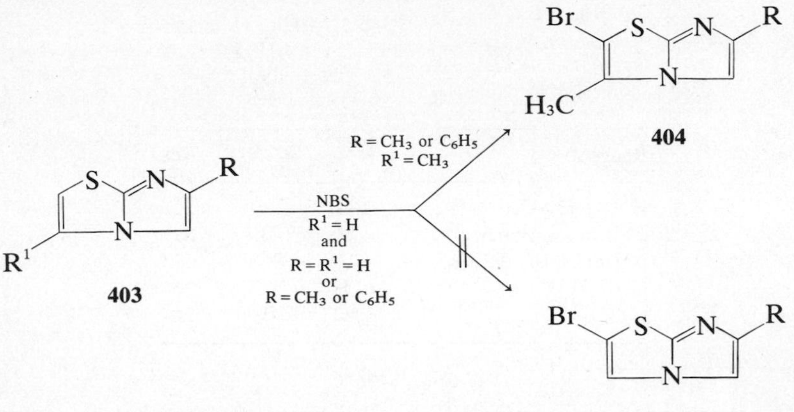

403

NBS
$R^1 = H$
and
$R = R^1 = H$
or
$R = CH_3$ or C_6H_5

$R = CH_3$ or C_6H_5
$R^1 = CH_3$

404

405

10. *Imidazo[2,1-b]-1,3,4-thiadiazole*

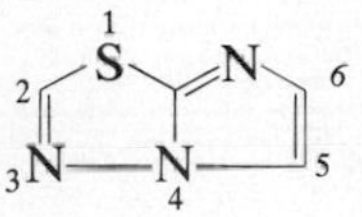

2-Methyl-6-(4-nitrophenyl)imidazo[2,1-*b*]-1,3,4-thiadiazole (**406**) is reported to undergo the Vilsmeier–Haack reaction (with $POCl_3$ and DMF) to give the 5-formyl-derivative (**407**).[60]

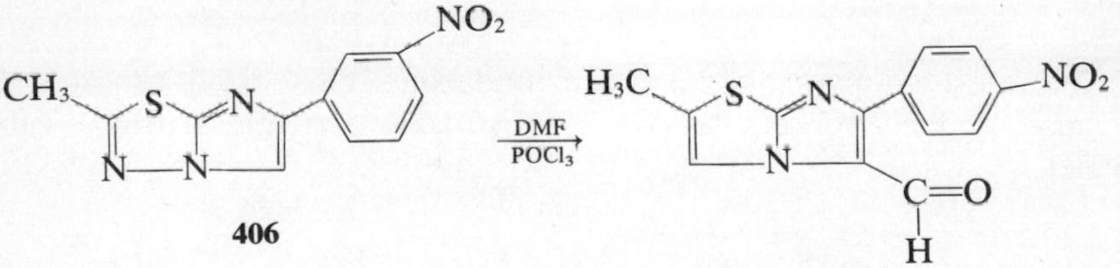

DMF
$POCl_3$

406

407

11. *Thiazolo[3,2-d]tetrazole*

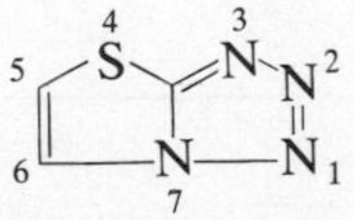

Facile electrophilic substitution at the 5-position of 6-substituted thiazole[3,2-*d*]tetrazoles has been reported by Avramenko and co-workers[219] as well as by Skolenko and co-workers[264] (**408**–Table VII).

TABLE VII

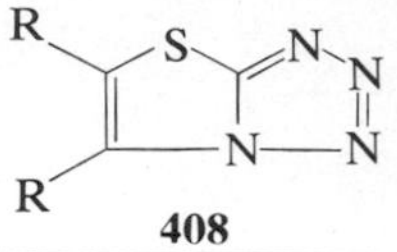

408

Reagent	R	R′	Ref.
HNO_3–H_2SO_4(−10°)	CH_3	NO_2	219
N-Bromosuccinimide	CH_3	Br	219
NH_4SCN, Br_2, HOAc	CH_3	SCN	219
$ClSO_3H$	C_6H_5	Cl–SO_2	264
H_2SO_4	C_6H_5	SO_3H	264

Here again, as with imidazo[2,1-*b*]thiazole, we see that the thiazole portion of the system nitrates under considerably milder conditions than mono or dimethyl thiazoles[216]—this despite its fusion to a system containing three pyridine nitrogens.

12. *Pyrazolo[2,3-a]benzimidazole*

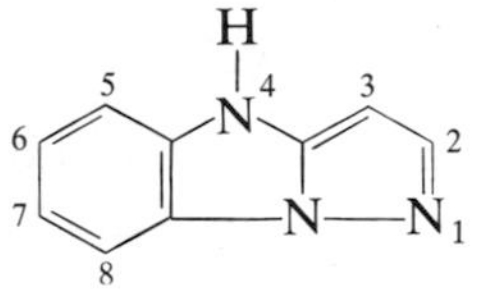

2-Substituted pyrazolo[2,3-*a*]benzimidazoles have been reported by Menzel and co-workers to give 3-substituted derivatives (**409**) (Table VIII).

TABLE VIII

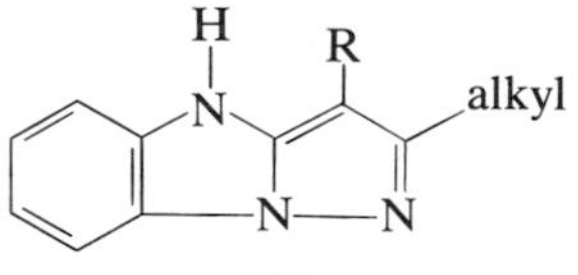

409

Reagent	R	Ref.
SO_2Cl_2, HOAc, NaOAc	Cl	81
H_2SO_4	SO_3H	81
$N{\equiv}N^+$-aryl	—N=N-aryl	220–222

13. *Isoindolo[1,2-b]benzothiazole*

Babichev and Kibirev[213] reported that isoindolo[1,2-*b*]benzothiazoles react with a variety of nucleophiles to give 11-substituted derivatives (**410**—Table IX)

TABLE IX

410

Reagent	R
$HC(=O)—N(CH_3)_2—POCl_3$	$—C(=O)H$
Ac_2O	$—C(=O)—CH_3$
$Cl—C(=O)—C_6H_5$, pyridine	$—C(=O)—C_6H_5$
1-Cl-2,4,6-$(NO_2)_3C_6H_2$, Et_3N	2,4,6-$(NO_2)_3C_6H_2$—
$Cl—C{=}C—C(=O)—C_6H_5$	$—C{=}C—C(=O)—C_6H_5$
Cl^- $N{\equiv}N^+—C_6H_5$	$—N{=}N—C_6H_5$

A nitroso compound was also prepared by reaction of the parent system with nitrous acid. This compound was isolated as the perchlorate salt and the infrared spectrum shows no evidence of an NO band but has strong absorption at 3200 cm^{-1} showing that the product exists as the oxime of the thiazolium salt (**409**).

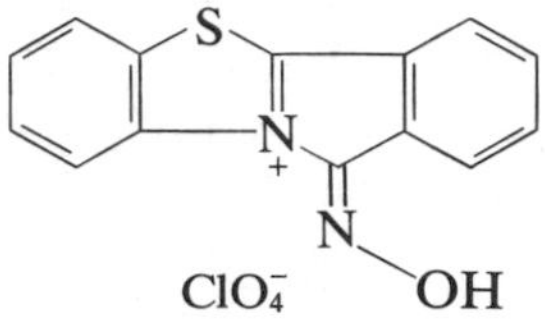
S
N
+
N
ClO_4^-
OH

411

14. *1,3,4-triazolo[3,4-b]benzothiazole*

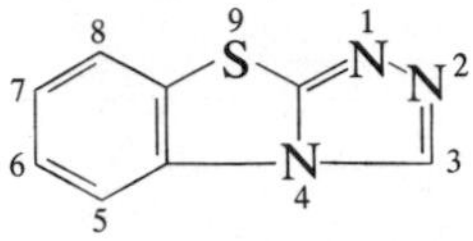
8
9
1
S
N
N
2
7
6
N
3
4
5

It has been reported by Sicheva and co-workers[268] as well as by Paget[269] that aminoalkylation, halogenation, and formylation of this system (**412**) takes place at the 3-position (**414**—Table X).

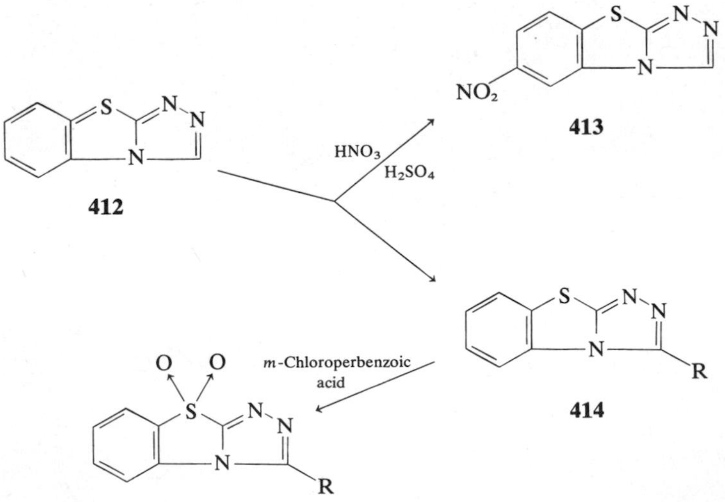
S
N
N
N
412
HNO_3
H_2SO_4
NO_2
413
414
R
m-Chloroperbenzoic acid
O
O
415

TABLE X

Reagent	R	Ref.
$CH_2O_3HN(CH_3)_2 \cdot HCl$	$CH_2N(CH_3)_2$	268,269
DMF·$POCl_3$	CHO	268
Br_2 or NBS	Br	268,269
NCS[a] (?)	Cl	269

[a] Reagent not specified.

Nitration with nitric acid–sulfuric acid is reported by Sicheva, *et al.*[270] to take place at the 6-position (**413**). NMR and some synthesis was used to make this structural assignment.

The sulfur atom of the 3-chloro derivative (**414**) was oxidized to the 7,9 dioxo compound (**415**) by *m*-chloroperbenzoic acid.

B. Nucleophilic Attack

1. *Pyrrolo[1,2-a]imidazole*

1-Alkyl-5-benzoylpyrroloimidazoles (**416**) were debenzoylated in aqueous hydrochloride. Only the 1-benzyl compound (**417**; R=CH_2–C_6H_5) was stable.[15]

HCl / H_2O

416 **417**

2. *Pyrrolo[2,1-b]thiazole*

Pyl, Dinse, and Sietz[11] reported that the acid hydrolysis of 7-benzamido-6-phenylpyrrolo[2,1-*b*]thiazole (**417**) gave the 7-hydroxy compound (**409**).

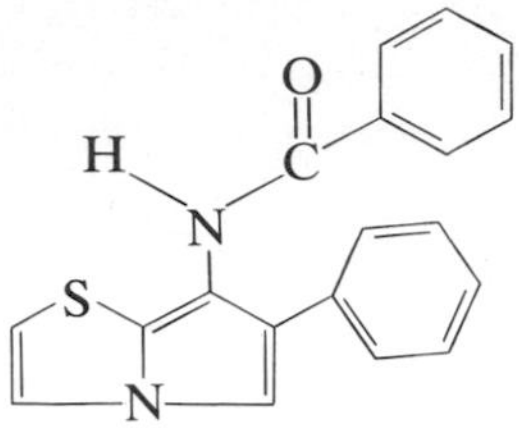

418

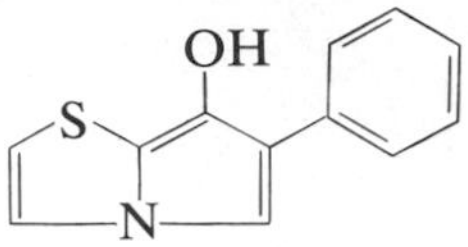

419

3. *Thiazolo[2,3-c]-s-triazole*

The exocyclic sulfur atom of 5,6-diphenyl-3-mercaptothiazolo[3,4-*b*]-*s*-triazole (**410**) can be removed with either Raney nickel or with peroxide to give 5,6-diphenylthiazolo[3,4-*b*]-*s*-triazole (**411**).[119]

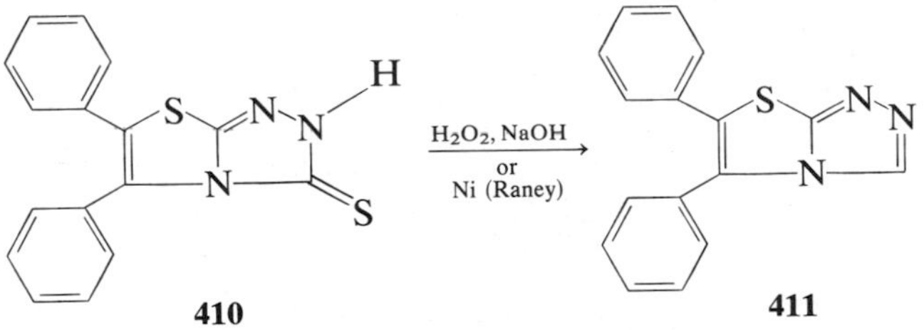

H_2O_2, NaOH
or
Ni (Raney)

410

411

4. *Imidazo[2,1-b]benzothiazole*

Parrick and Pearson[46] displaced the chlorine atom of 2-chloroimidazo-[2,1-*b*]thiazole (**412**) with phosphorous and hydroiodic acid and obtained the parent imidazo[2,1-*b*]benzothiazole (**413**). Attempted displacements using palladium on charcoal with either hydrazine or hydrogen failed.

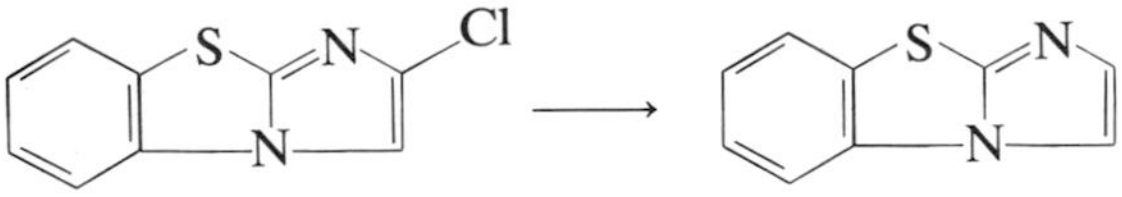

412

413

5. *Imidazo[5,1-b]benzothiazole*

1-Cyano-3-phenylimidazo[5,1-*b*]benzothiazole (**415**) was prepared by heating 1-bromo-3-phenylimidazo[5,1-*b*]thiazole (**414**) with potassium cyanide in dimethylformamide.[101]

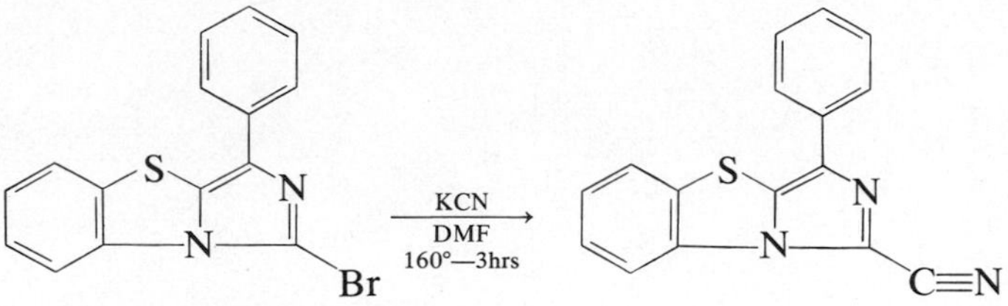

414 $\xrightarrow[\text{160°—3hrs}]{\text{KCN, DMF}}$ **415**

6. *Imidazo[1,2-a]benzimidazole*

Siminov and Anisimova[80] displaced the bromine of 3-bromo-9-methyl-2-phenylimidazo[1,2-*a*]benzimidazole **416** with several nucleophiles in refluxing dimethylformamide (Table XI).[101]

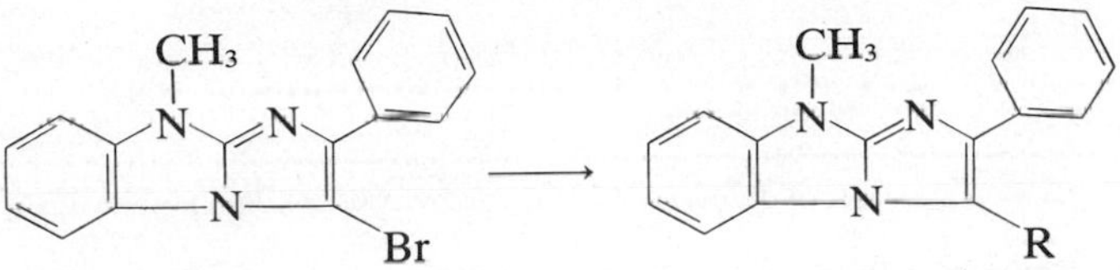

416 ⟶ **417**

TABLE XI

Reagent	R
$NaNO_2$	NO_2
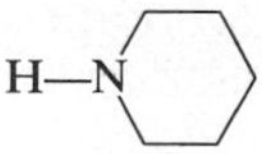 H—N (piperidine)	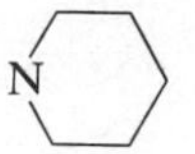 N (piperidino)
 HN (morpholine)	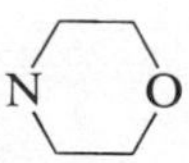 N O (morpholino)

Debenzylation of 9-benzyl-2-phenylimidazo[1,2-*a*]benzimidazole **418** with sodium and ammonia gave the expected 2-phenylimidazo[1,2-*a*]-benzimidazole **419** as well as the 2,3-dihydro derivative **420**.[271]

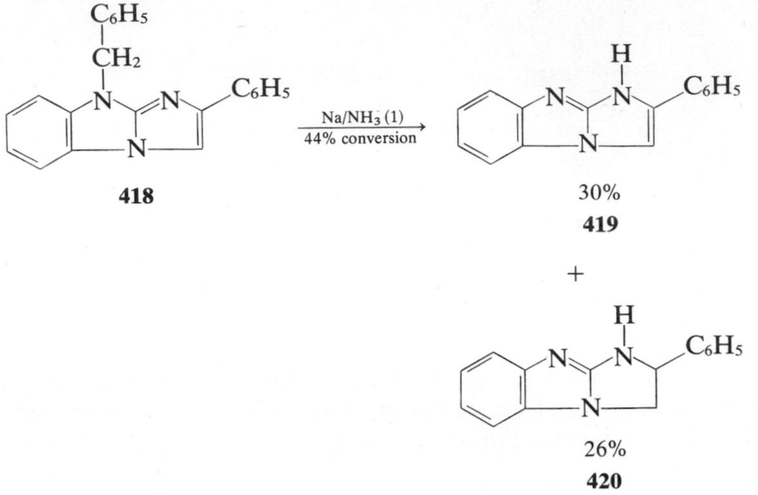

C_6H_5 CH_2 C_6H_5 Na/NH_3 (1) 44% conversion H C_6H_5

418

30%
419

+

H C_6H_5

26%
420

7. *Imidazo[5,1-b]benzimidazole*

Aryuzina and Schukukina[71] reported the displacement of the bromine atom of compound **421** by nitrite to give the nitro compound **422**. They

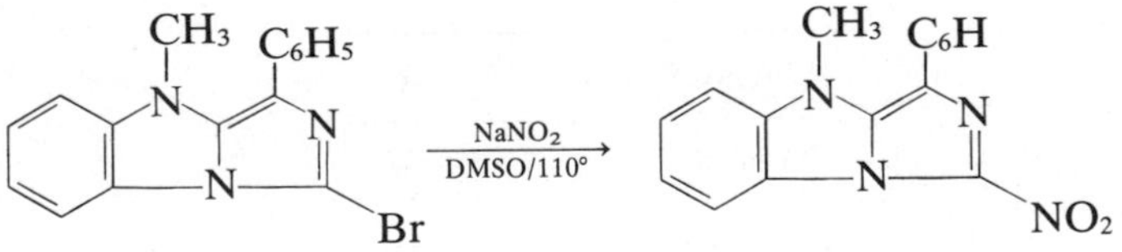

CH_3 C_6H_5 $NaNO_2$ DMSO/110° CH_3 C_6H Br NO_2

421 **421**

were, however, unable to prepare an ethyl ether by treating **421** with sodium ethoxide.

8. *Isoindolobenzothiazole*

The 11-acyl derivatives of this system **422** are deacylated in aqueous acid to give the parent system **423**.[223]

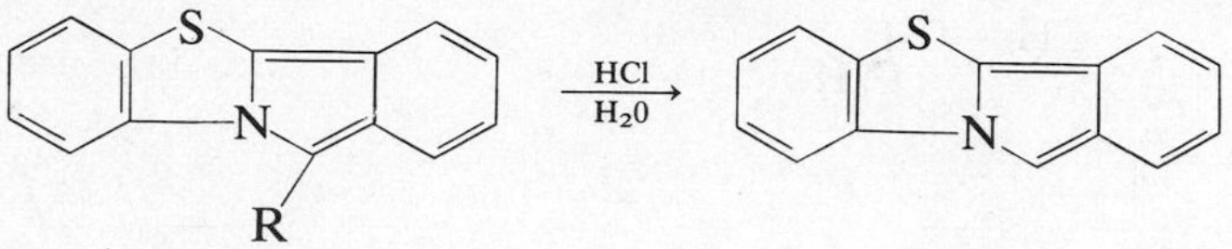

422 → (HCl, H_2O) → 423

R = CH_3—C(=O), H—C(=O), or C_6H_5—C(=O)

9. *Imidazo[1,2-a]imidazole*

Miller and Bambury[272] reported the displacement of the 5-bromo atom of **424** by piperidine to give compound **425**.

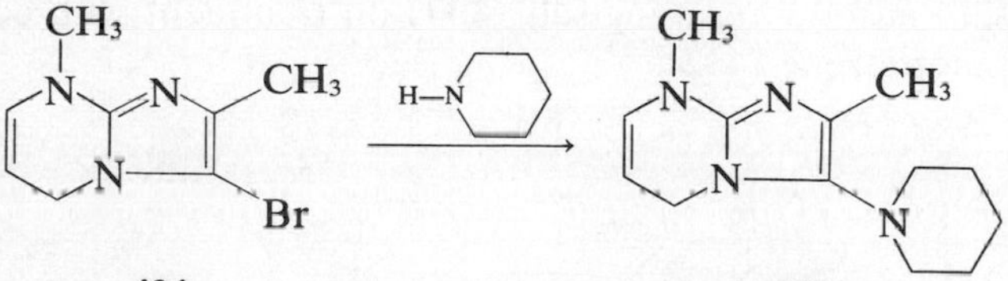

424 → (H—N piperidine) → 425

10. *Pyrrolo[1,2-a]benzimidazole*

Kochergin and co-workers[273] reported the debenzylation of phenylpyrrolo[1,2,-*a*]imidazole and pyrrolo[1,2-*a*]benzimidazoles gave the expected products **427**. The structural assignment, with respect to location of the methylene group, was done on the basis of their nmr studies.

426 → (Na/NH_3) → 427

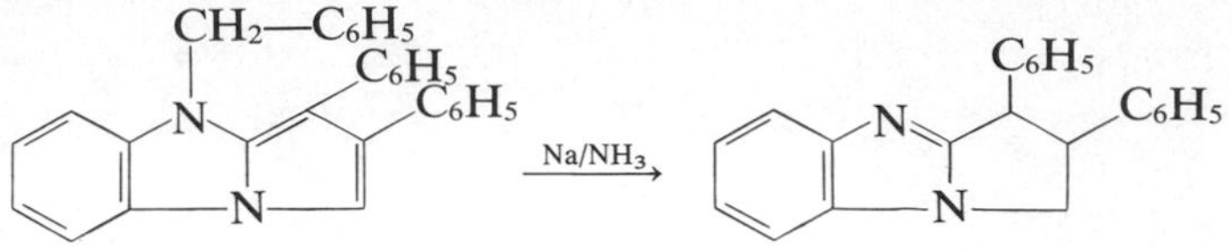

$\xrightarrow{Na/NH_3}$

428 **429**

When the diphenyl-benzyl compound **428** was treated with sodium and ammonia, not only did debenzylation occur but saturation of the pyrrole ring also took place giving compound **429**.

C. Ring Openings

1. *Thiazolo[3,2-d]tetrazole*

Sodium cyanide opens the tetrazole ring of 6-methylthiazolo[3,2-*d*]-tetrazole (**430**) and the resulting salt (**431**) on treatment with acetic acid gave the cyanotriazene (**432**).[224]

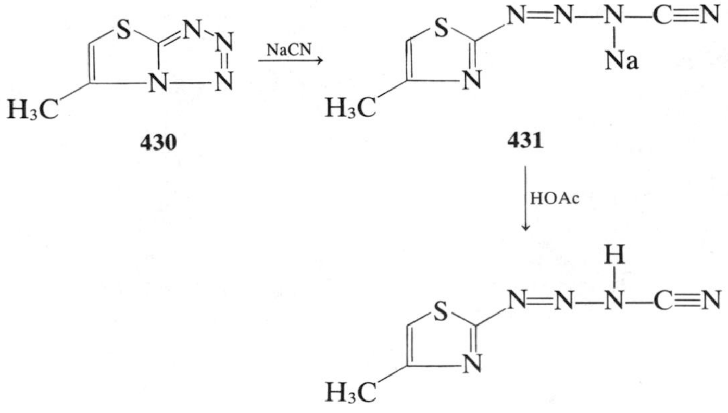

$\xrightarrow{NaCN}$

430 **431**

$\downarrow$ HOAc

432

2. *Thiazolo[2,3-b][1.3.4]thiadiazolium*

The thiazolothiadiazolium perchlorate (**433**) prepared by Ege[191] is converted to the dithiadiazacine (**434**) in aqueous base. The mechanism shown was proposed by Ege.

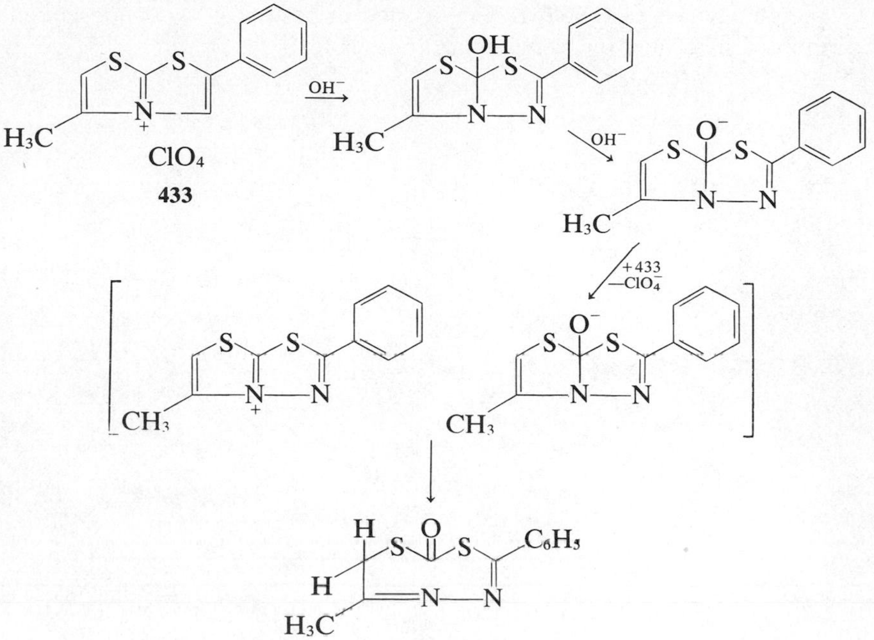

H_3C ClO$_4$ **433** $\xrightarrow{OH^-}$ ... $\xrightarrow{OH^-}$ ... $\xrightarrow{+433, -ClO_4^-}$ [...] $\rightarrow$ **434**

3. *Imidazo[1,2-a]benzimidazole*

Reduction of the nitro compound **435** gave the amine as a stannic chloride dihydrochloride (**436**) which on heating in ethanol underwent ring cleavage to the phenylglycine derivative (**437**).[225]

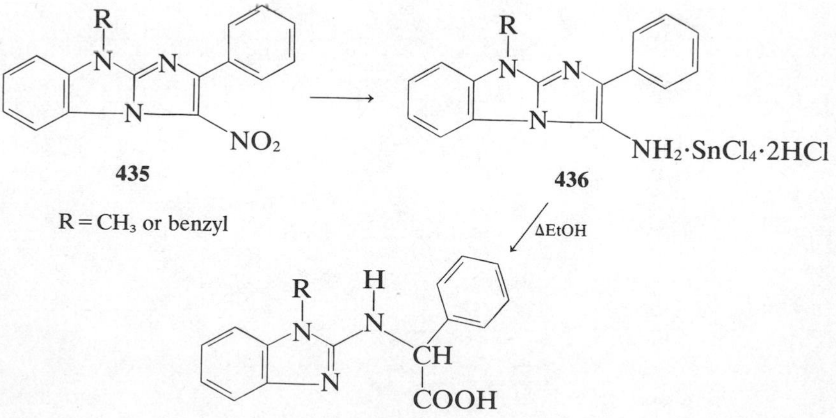

435 $\longrightarrow$ **436** ($NH_2{\cdot}SnCl_4{\cdot}2HCl$) $\xrightarrow{\Delta EtOH}$ **437**

R = CH_3 or benzyl

The quaternary salt (**438**) is ring opened in base (**439**) with subsequent deamination to give the benzimidazolone (**440**)[68]

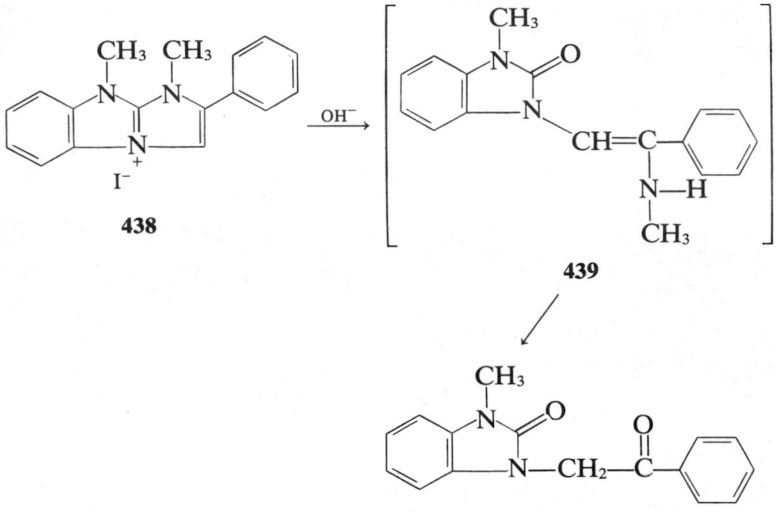

The permanganate oxidation of 9-methylimidazo[1,2-*a*]benzimidazol 2(3*H*)-one (**441**) gave the bis azo compound (**442**) while base hydrolysis of this lactam (**441**) gave 1-carboxymethyl-3-methylbenzimidazol-2-one (**445**).[41]

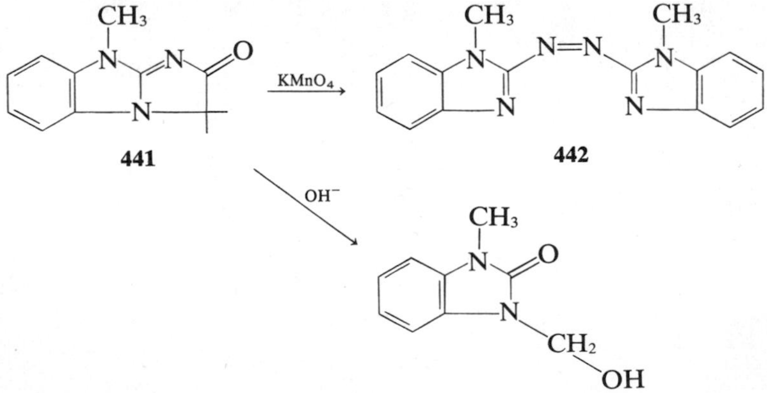

4. *Naphtho[1′,2′ : 4,5]imidazo[2,1-b]thiazole*

The desulfurization of this system was studied by Knish, Krasovskii, and Kochergin.[245,246] These workers found that the thiazole ring could be reductively opened. The 3-methyl derivative of this system (**444**) on

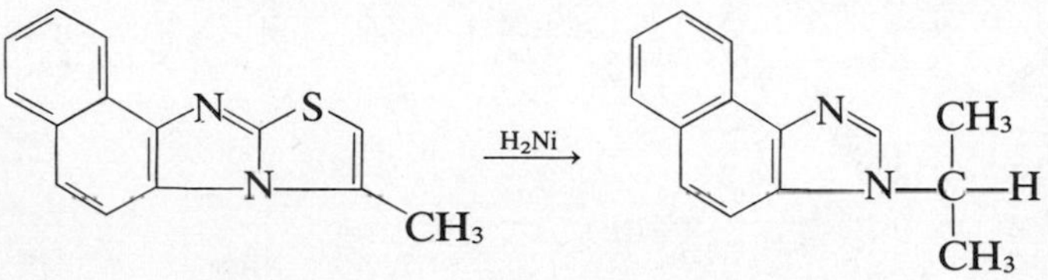

$$\xrightarrow{H_2Ni}$$

444 **445**

treatment with Raney-Nickel and hydrogen gave 3-isopropylnaphtho[1,2-*d*]imidazole (**445**).

5. *Triazolo[3,4-b]benzothiazole*

Sicheva, Kisleva, and Shuchukina[268] found that the triazole ring of the triazolo[3,4-*b*]benzothiazole system can be opened in base, with

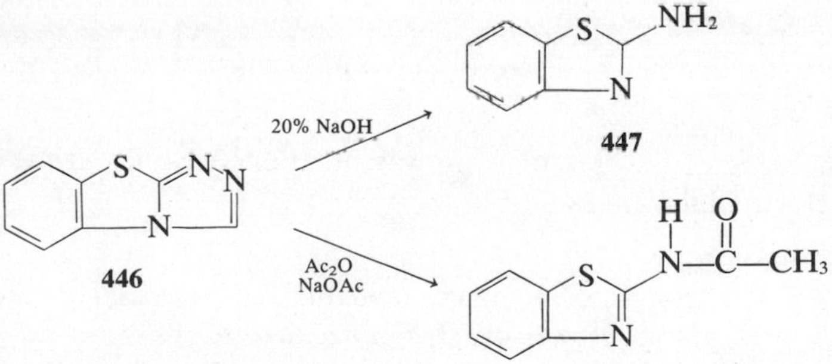

20% NaOH; Ac_2O, NaOAc

446 **447** **448**

nitrogen–nitrogen cleavage as well as carbon–nitrogen cleavage. In aqueous sodium hydroxide **446** is degraded to 2-aminobenzothiazole (**447**). With sodium acetate in acetic anhydride, this same cleavage apparently occurs followed by acylation of the amino group to give **448**.[274]

6. *5H-Tetrazolo[1,5-a]isoindole*

Alkylation of 5*H*-tetrazolo[1,5-*a*]isoindole (**449**) with dimethylsulfate gave a quarternary salt (**450**). The site of alkylation was demonstrated as shown here.[260]

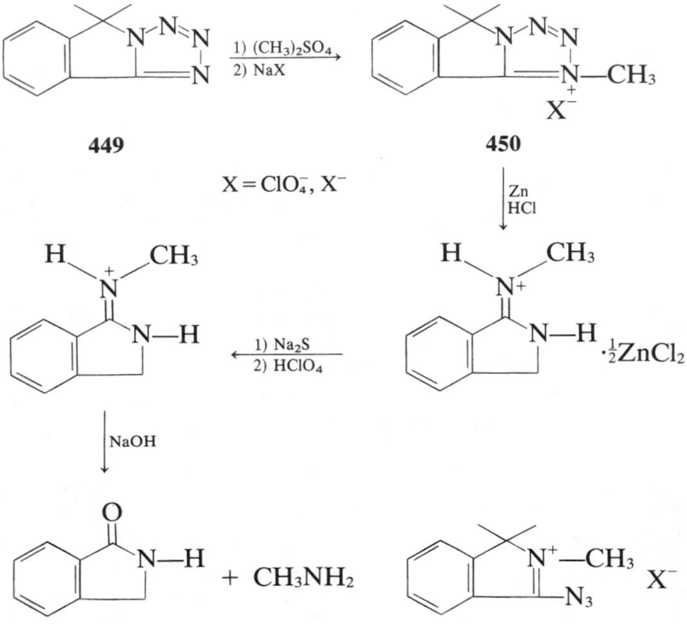

1) $(CH_3)_2SO_4$
2) NaX

449 **450**

X = ClO_4^-, X^-

Zn
HCl

$\cdot \frac{1}{2}ZnCl_2$

1) Na_2S
2) $HClO_4$

NaOH

+ CH_3NH_2

7. *Pyrrolo[2,1-b]thiazole*

The opening of the thiazole ring was mentioned earlier in the discussion of electrophilic attack.[262]

8. *Tetrazolo[1,5-a]benzimidazole, tetrazolo[5,1-b]benzothiazole, and tetrazolo[5,1-b]benzoxazole*

In the following examples, the starting materials exist as the azide—a situation that the experimenters set out to exploit. As a result of reaction, the potential for tetrazole formation was lost.

Shiokawa and Ohki[277] attempted cycloaddition reactions with 1-methyl-2-azidobenzimidazole (9-methyltetrazolobenzimidazole) (**451**).

Diphenyl ketone reacted exothermically with the loss of nitrogen to give a compound assigned structure **452**. Dimethylacetylene dicarboxylate, in both acetonitrile and benzene, added across the 2–3 double bond to give the azido diester (**453**).

451 → $(C_6H_5)C{=}C{=}O$ → 452 (94%)

451 → $CH_3OOC{-}C{\equiv}C{-}COOCH_3$ in CH_3CN → 453 (87%)

451 → $CH_3OOC{-}C{\equiv}C{-}COOCH_3$ in benzene → 81% + 454 (4%)

451 → $H{-}C{\equiv}C{-}COOCH_3$ → 455 (37%) + 456 (5.5%)

In benzene, dimethylacetylene dicarboxylate also adds to the azide group (**454**). Methylpropargylate, in contrast to the diester gave more azide addition product (**455**) than azetidine (**456**). (Note. Structures **453** and **454** are depicted as having 5-membered rather than 4-membered rings in the paper. This is inconsistent with the molecular formulae in the paper as well as the chemistry.)

Van Allen and Reynolds[278] added triphenylphosphine to azidobenzimidazoles as well as azidobenzothiazole and azidobenzoxazole (**457**) to give complexes (**458**) which were converted thermally to the phosphinimines—**459**.

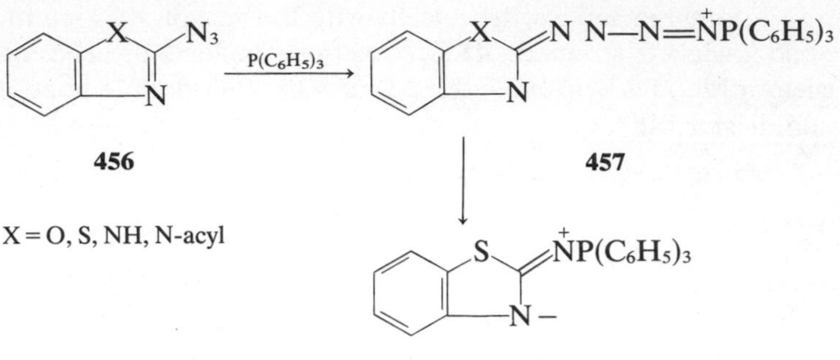

456

457

X = O, S, NH, N-acyl

459

D. Condensations

Several systems have been reported to undergo condensation at an endocyclic site. Such systems possess the potential for *N-H*-methylene tautomerism (see tautomerism section).

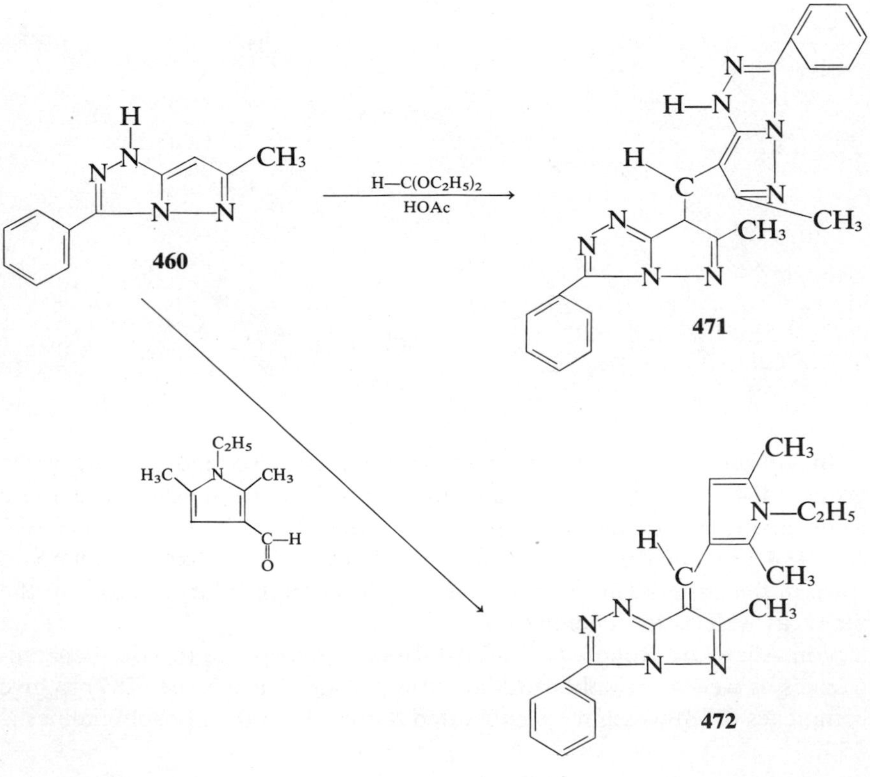

460

471

472

1. *Pyrazolo[3,2-c]-s-triazole*

Bailey, Knott, and Marr[274] reported the 3-phenyl-6-methyl[3,2-*c*]-*s*-triazole (**460**) undergoes condensation at the 7-position. Ethyl orthoformate in acetic acid reacts with two molecules of the pyrazolotriazole condensing with one and electrophilically attacking the other to give compound **471**. Condensation with 1-ethyl-2,5-dimethylpyrrole-3-aldehydes gave compound **472**.

Stein and Machiele[275] effected oxidative anil formation (**474**) by treating the pyrazolotriazole (**473**) with 3,5-dichloro-4-hydroxyaniline and potassium ferricyanide, in the presence of sodium carbonate.

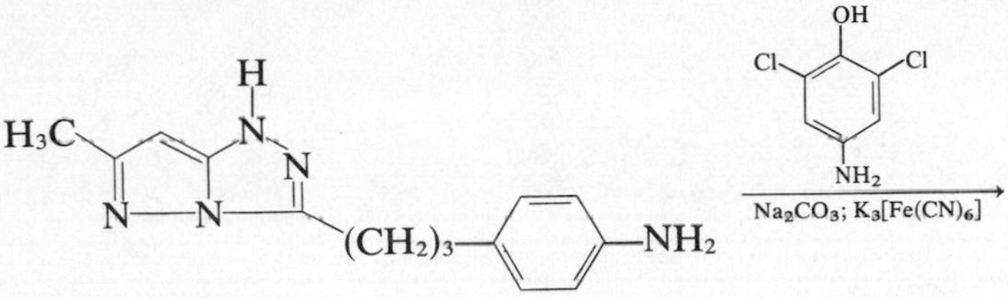

Na_2CO_3; $K_3[Fe(CN)_6]$

473

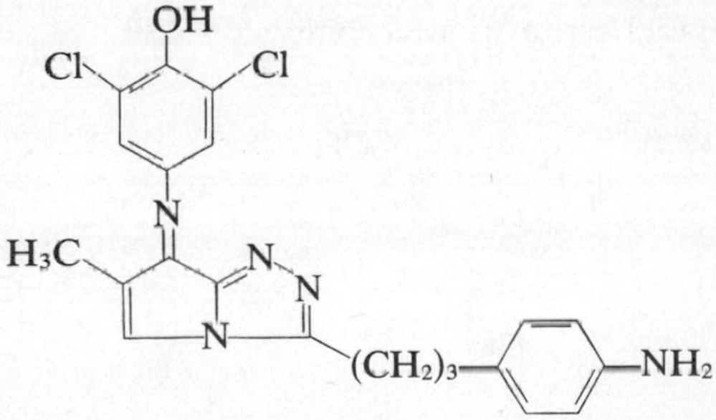

474

2. *Pyrrolo[2,1-b]thiazole*

Babichev, Kibirov, and Chulko[220] reported that the perchlorate salt of 7-methyl-6-phenylpyrrolo[2,1-*b*]thiazole (**475**) condenses with the formyl pyrrolothiazole (**476**) to give compound **477**. McKenzie, Malloy, and Reid[207] reported that the perchlorate salt of 6-methylpyrrolothiazole (**478**) condenses with ethyl orthoformate to give the ethoxymethylene

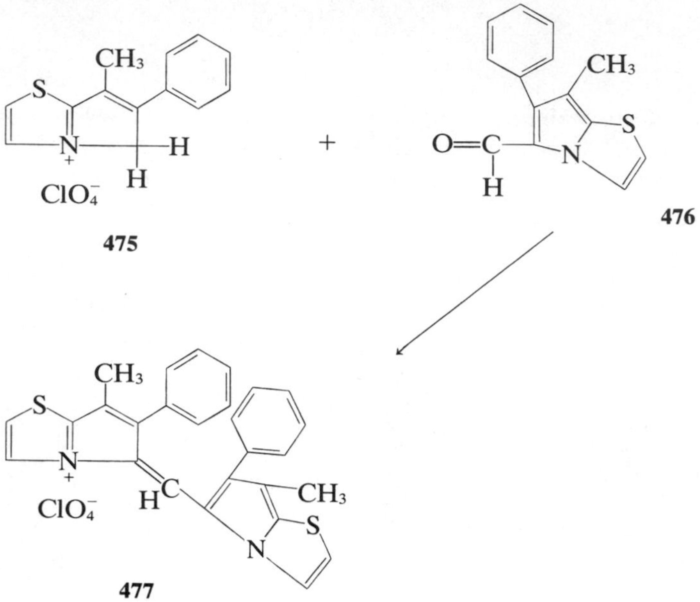

CH3 S N+ H H ClO4− 475 + O=C H CH3 S N 476

CH3 S N+ ClO4− HC CH3 S N 477

material (**479**) which then condenses with 2,3-dimethylbenzothiazolium perchlorate to give compound **480**.

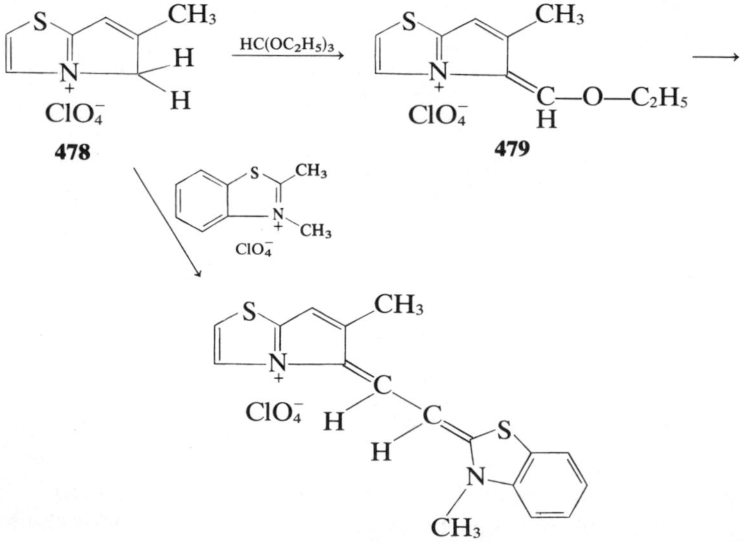

S N+ CH3 H H ClO4− 478 $\xrightarrow{HC(OC_2H_5)_3}$ S N+ CH3 C—O—C_2H_5 H ClO4− 479 ⟶

S CH3 N+ CH3 ClO4−

S N+ CH3 ClO4− H C C H S N CH3 480

3. *Thiazolo[3,2-a]indole*

The bromide salts of 3-methylthiazolo[3,2-*a*]indole **481** condenses with *p*-dimethylaminobenzaldehyde, reaction being reported to take place at the 9-position of the heterocyclic nucleus. The resulting product is isolated as the perchlorate salt **482**.[175]

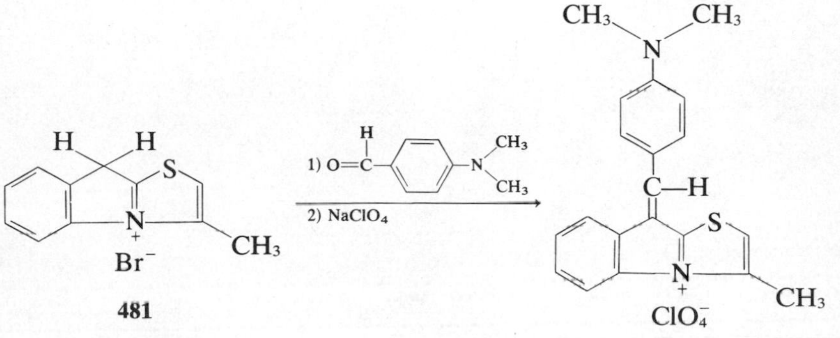

H H S N + Br^- CH_3

481

1) O=C(H)–C₆H₄–N(CH_3)₂

2) $NaClO_4$

CH_3 N CH_3 C—H S N + ClO_4^- CH_3

482

4. *Pyrazolo[2,3-a]benzimidazole*

2-Heptadecylpyrazolo[2,3-*a*]benzimidazole (**483**) condenses with *p*-dimethylaminobenzaldehyde, in the presence of base to give compound **484**.[82]

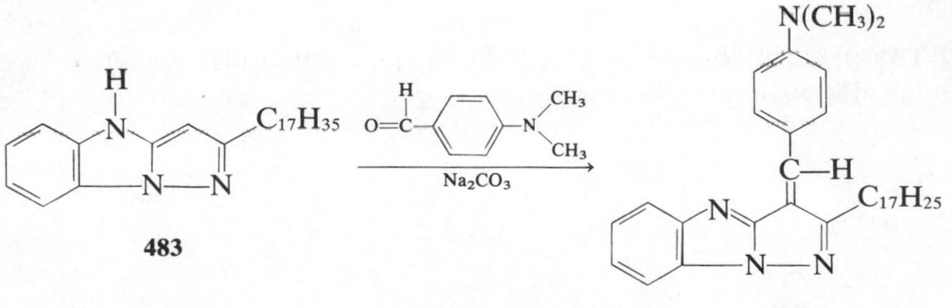

H N $C_{17}H_{35}$ N—N

483

O=C(H)–C₆H₄–N(CH_3)₂

Na_2CO_3

$N(CH_3)_2$ C—H $C_{17}H_{25}$ N N—N

484

The acetyl amino compound (**485**) undergoes oxidative coupling with *p*-dimethylaminoaniline to give the diamino derivative (**486**). Treatment of compound (**486**) with sodium methoxide gives the anil (**487**).[227]

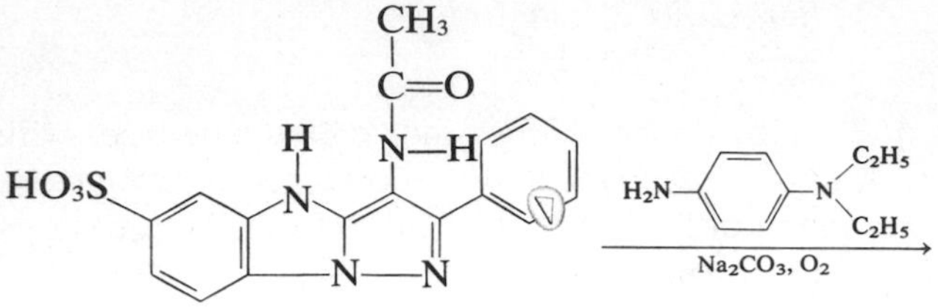

485

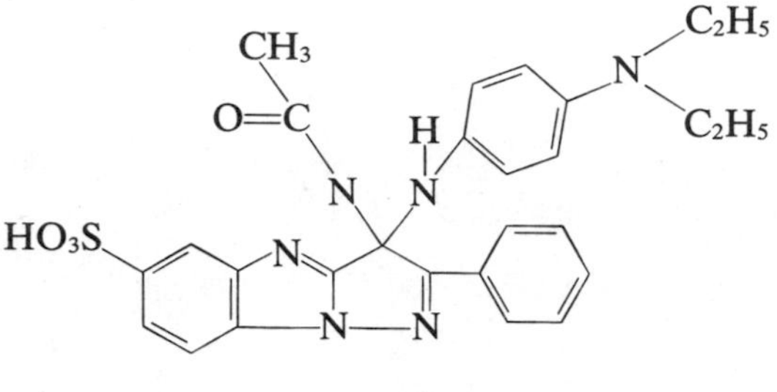

H_2N–C_6H_4–$N(C_2H_5)_2$

Na_2CO_3, O_2

486

$NaOCH_3$

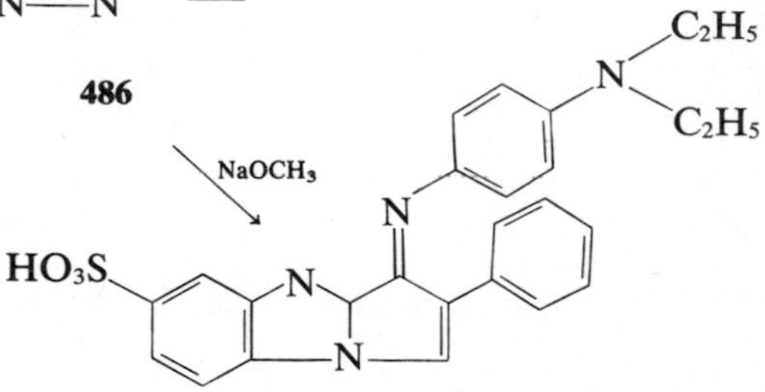

487

5. *Pyrrolo[1,2-a]benzimidazole*

The perchlorate salts of pyrrolo[1,2-*a*]benzimidazoles condense with aldehydes and aldehyde acetals, the 1- and 3-positions being susceptible

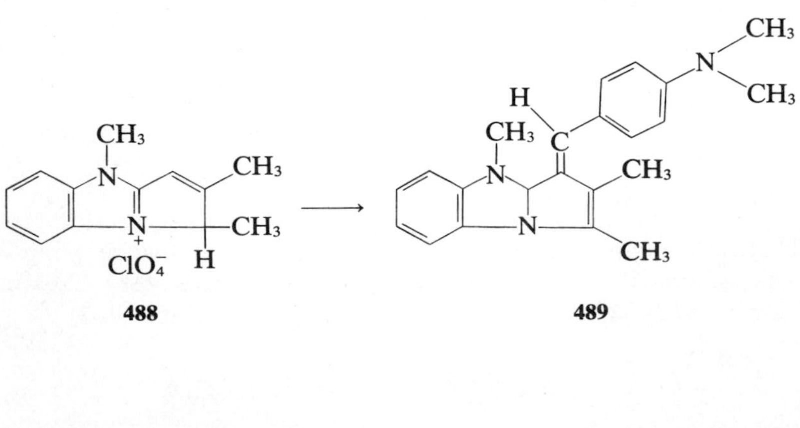

488

489

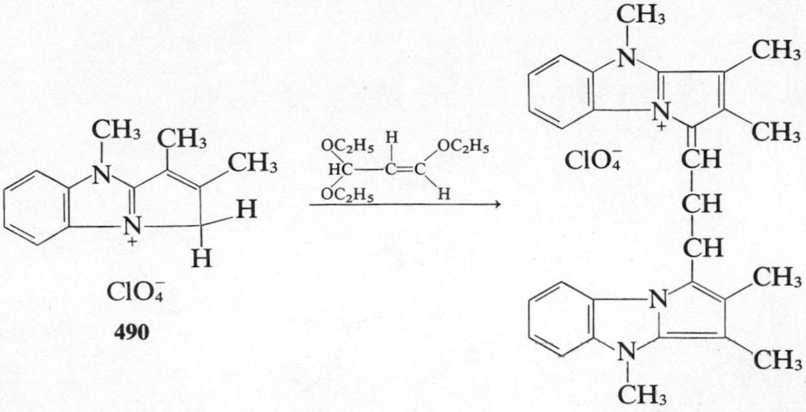

to attack. When the 1-position was substituted with a methyl group (**488**) condensation took place at the 3-position (**489**), and when the 3-position was blocked with a methyl group (**490**) substitution took place at the 1-position of two molecules to give (**491**).[8]

6. *Isoindolo[1,2-b]benzothiazole*

The hydroiodide salt of isoindolo[1,2-*b*]benzothiazole (**492**) condenses with *p*-dimethylaminobenzaldehyde in acetic anhydride to give compound **493**.[223]

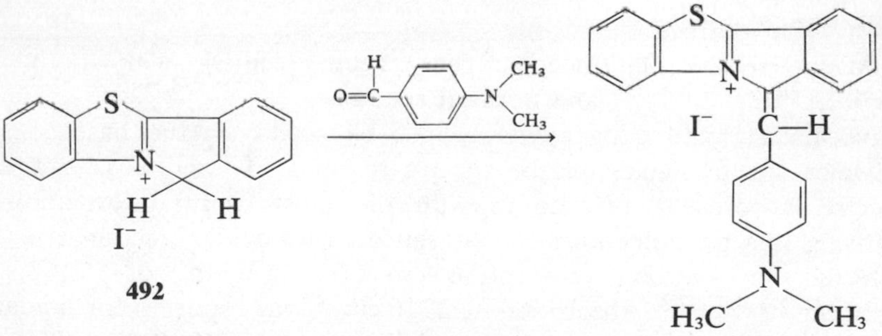

7. *Pyrrolizinium anion*

The pyrrolizinium anion (**495**) prepared by Okamura and Katz[228] from 3*H*-pyrrolo[1,2-*a*]pyrrole (**494**) condenses with benzophenone to give the benzhydrilidene (**496**).

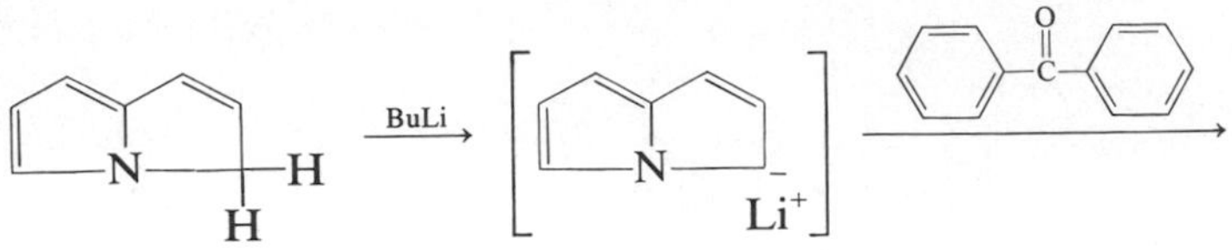

IV. Tautomerism

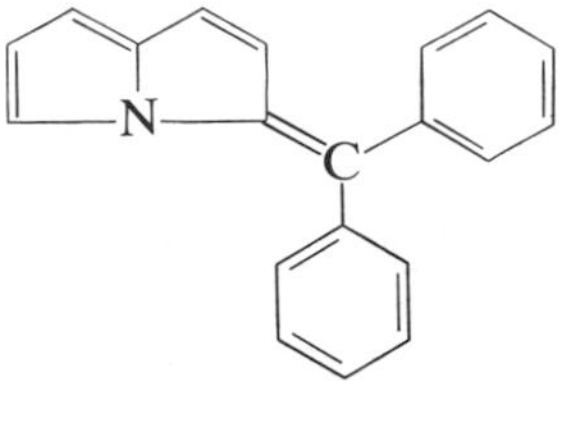

1. *Azido-tetrazole Tautomerism*

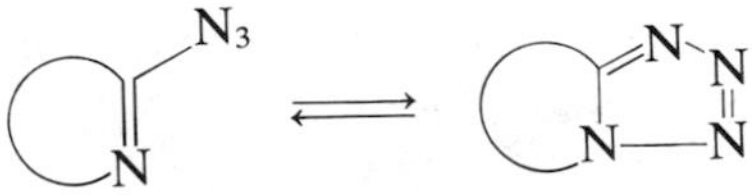

Azido-tetrazole tautomerism has been studied with both ultraviolet[105,142,149,229,190] and infrared spectra.[131,142,146,150,219,229–232] Unequivocal structural assignments cannot be made on the bases of ultraviolet spectra (compare the spectra in refs. 149 and 229). Infrared spectra, at present, provide us with the most useful information in studying this phenomenon. The presence of an azide function is easily detected by strong absorption in the region of 2160 to 2120 cm^{-1}.[131,233,276] Absorption at 2210 cm^{-1} was reported for imidazo-[3,4-*d*]tetrazole,[241] as well as a weak band in the 1340 to 1180 cm^{-1} region,[233] the former being more easily recognizable. Tetrazole rings show bands at 1080, 975, and 735 cm^{-1}.

Electronic factors, phase medium, and temperature are among the parameters found to influence the direction of tautomerization. Reynolds, Van Allan, and Tinker[142] were the first to report that an azide group attached to a 5-membered ring could cyclize to a tetrazole. Using infrared spectra (phase not given) these investigators observed azidobenzimidazole (**497a**), 2-azidobenzoxazole (**498a**), and 2-azido-1,1-dioxobenzothiazole (**499a**) were favored over their respective ring-closed forms while tetra-

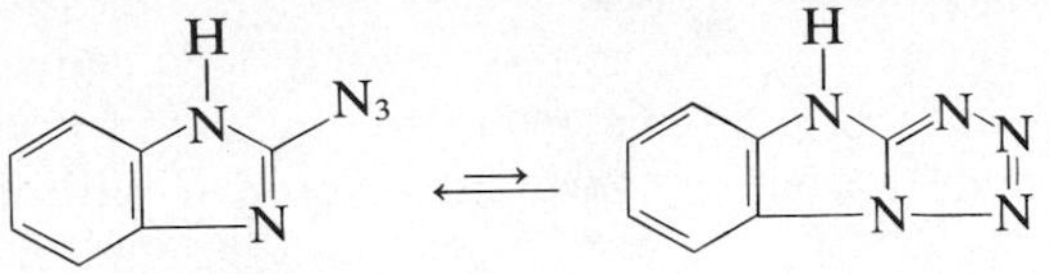

497a ⇌ **497b**

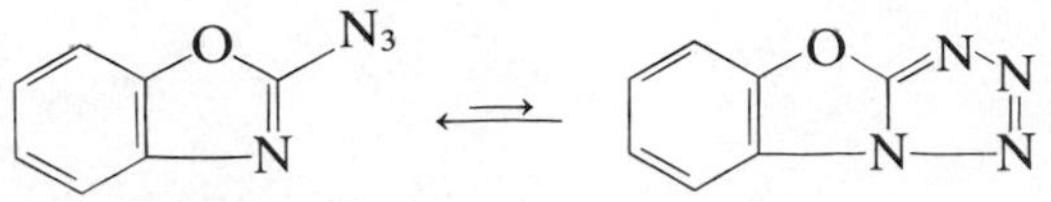

498a ⇌ **498b**

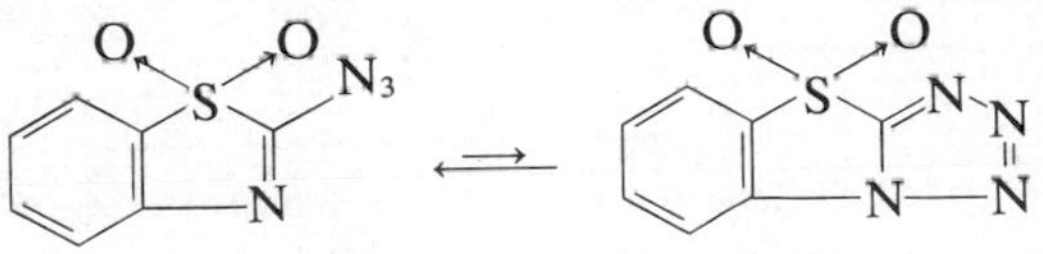

499a ⇌ **499b**

zolobenzothiazole (**500b**) and tetrazolobenzoselenazole (**501b**) were favored over the ring-open azides. The workers suggested that electronic effects are important in determining whether the system is to exist in the azide or tetrazole form. The electronic influence on tetrazolo benzothiazoles has been reported by Avramenko and co-workers[150] and Skolenko and co-workers[242] who reported that with electron withdrawing substituents (NO_2, COOH) stablize the azide form while electron-donating substituents (CH_3, OCH_3) favored the tetrazole form. Chlorine in the 4-position favored the azide form while chlorine in the 6-position favored tetrazole form. This is probably electronic rather than steric as methyl in the 4- or 6-positions favor the tetrazole form.[150] The work of Boyer and Miller[230] showed that this phenomenon could be dependent on phase as well as electronic environment. Spectral studies showed that 2-azido-benzothiazole (**500a**) predominated in chloroform

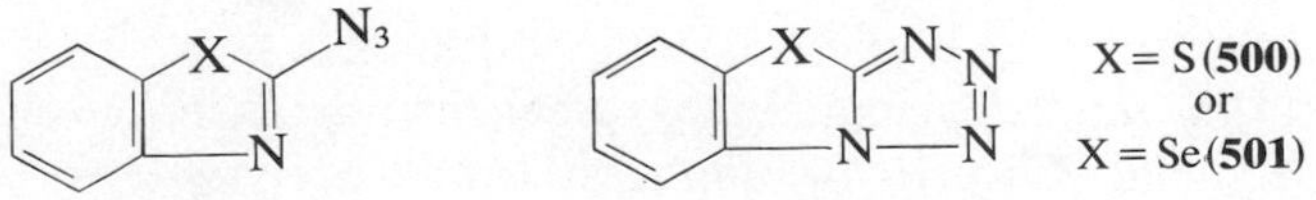

500a **500b** X = S(**500**) or X = Se(**501**)

solution while the ring closed tetrazolobenzothiazole (**500b**) was favored in the solid state. As 3-azido-1,1-dioxybenzoisothiazole did not cyclize in either chloroform solution or in the solid state. Other workers have essentially confirmed these observations.[219,229]

Sheinker and co-workers[232] demonstrated thermal dependancy. These workers showed that in the solid-state tetrazolobenzothiazole (**500b**) tautomerizes to the azide (**500a**) at 120° and reverts to tetrazole form (**500b**) on cooling.

Avramenko, Pochinok, and Rozum[231] reported that the methiodide of tetrazolo[5,1-*b*]thiazole existed, in the solid state, as the tetrazole. This is in contrast to the ethylfluoroborate salt which is reported to exist in the azide form.[234]

2. *Lactam–Lactim Tautomerism*

N—H =O   N —OH

As a result of many studies[235] the assumption is usually made that compounds with lactam–lactim potential exist in the lactam form. Where 5,5 systems are concerned very few spectral data have been reported to confirm this.

Grob and Ankli[203] reported that the infrared spectra (phase not reported) of the pyrroloimidazole (**502**) had bands at 2.93 μ(N–H) and 5.80

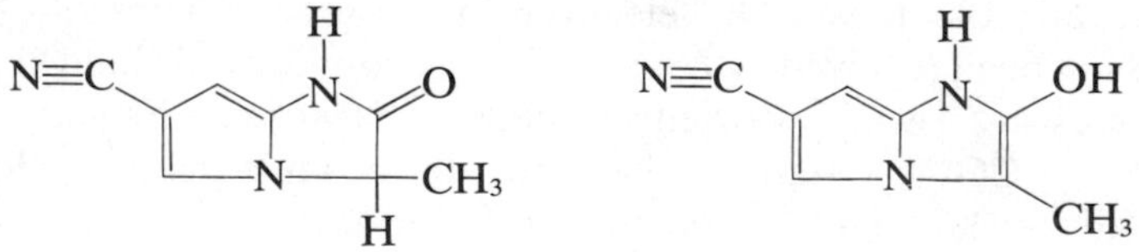

502a **502b**

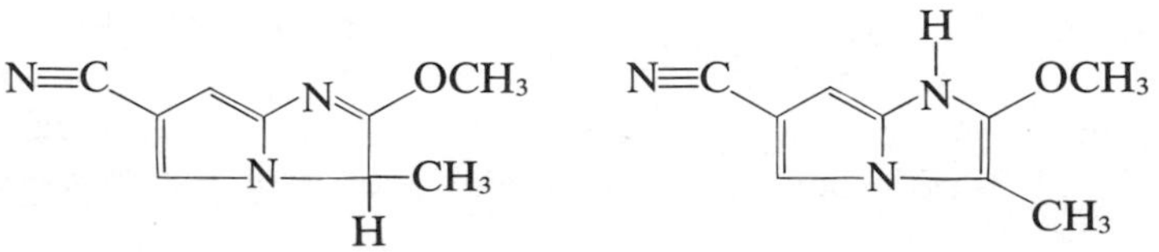

503a **503b**

μ(C=O) showing the lactam (**502a**) rather than the lactim form (**502b**) to be favored. The *O*-methyl compound (**503**), prepared from **502** and diazomethane, has absorption at 6.29 μ(C=N) and no N–H absorption showing that this compound exists as the 3-*H* isomer (**503a**) and not as the closed conjugated system (**503b**). (This tautomeric form has been questioned by Katritzky and Lagowski.[236])

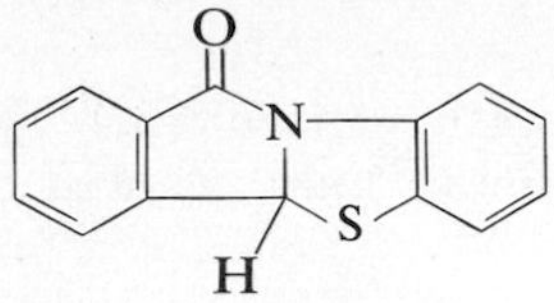

504

The lactam (**504**) was reported to have been obtained from the reaction 2-mercaptoaniline and phthalaldehydic acid[200] (see Section II.N). This study, however, established that cyclization had occurred and lactam *versus* lactim structures were not considered.

3. *Thiol–Thione Tautomerism*

Spinner[237] suggested that mercapto compounds, whether they exist in the thiol or thione form, should be referred to as thiols. Most investigators follow this convention. Spectral data of the 5,5 systems studied show the thione form to prevail.

Kanaoka, Okuda, and Shiho[116] reported that the ultraviolet spectra of 2-alkyl-5-mercapto-*s*-triazolo[3,4-*b*][1.3.4]thiadiazole (**505**) show absorption at 243 nm (log $\epsilon = 3.34–3.79$) and 3.04 to 3.05 nm (3.74–3.93) while the spectra of the corresponding *S*-methyl compounds (**506**) show

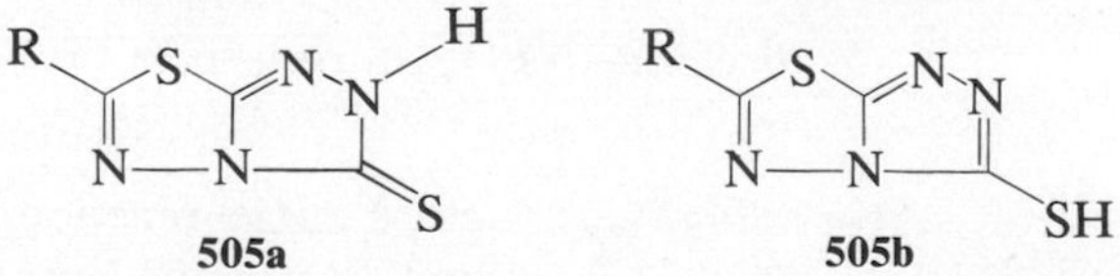

505a **505b**

R = alkyl

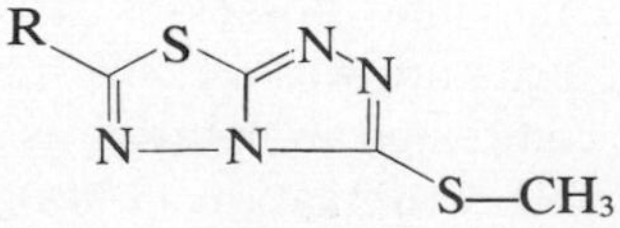

506

single bands in the region of 267.5 to 268.5 nm (3.52–3.66) showing that the mercapto compounds (**505**) exist predominantly in the thione form (**a**). The infrared absorption of these compounds, 2680 cm^{-1} (DSH), 1113 to 1180 (νC=S), and 1527 to 1546 (δ NH), showed a mixture of tautomers **505a** and **b**).

Ultraviolet spectra of other *s*-triazolo[3,4-*b*][1.3.4]thiadiazoles[112,115] and *s*-triazolo[5,1-*c*]-*s*-triazoles[118] also showed the thione form to predominate.

4. *Amine–Imine Tautomerism*

Where amine–imine tautomerism exists the amine form is usually assumed to predominate as a result of work on other systems.[235] One example of an imine has been reported. Oliver Dann and Gates[200] prepared the lactamidine hydrochloride (**507a**). This compound was isolable only as the salt and evidence that the free base, in solution, would behave as the imine was not presented. (There also exists the possibility that the salt might exist with proton and amine on the same carbon atom (**507b**—see section below.)

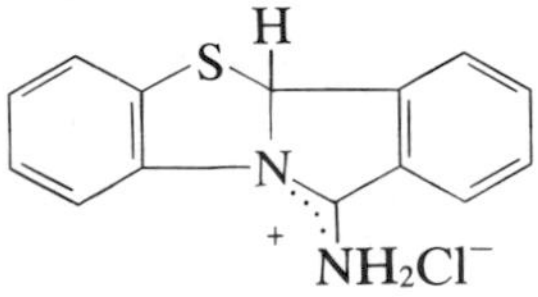

507a 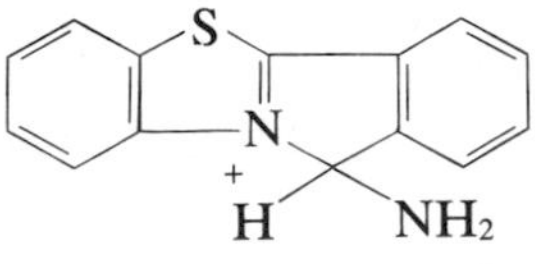 **507b**

5. *Methylene–NH Tautomerism*

The controversy surrounding the structure of the reaction product of *o*-phthalodicarboxaldehyde and 1,2-phenylenediamine was touched on briefly in the synthesis section (Section II.14) and the reaction scheme confirming that this compound possesses structure **294** was shown. An nmr spectral proof of structure was put forth by Amos and Gillis.[198] The favored form of this compound in solution is not the closed shell (**294b**) but the form with broken conjugation (**294a**). The nmr spectrum of this compound (**294**) showed a band at τ5.15 (two protons). As the azomethine proton of benzylidene aniline absorbs at $\tau 1.53$, there can be little doubt that the band at $\tau 5.15$ is a methylene signal. Perlmutter and Knapp,[199] using nmr, studied some related derivatives with additional

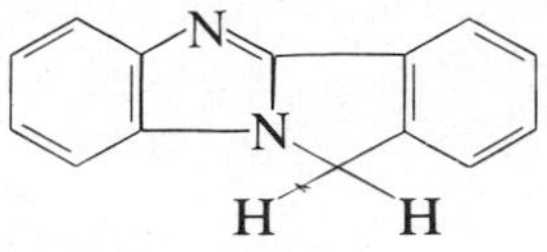

294a

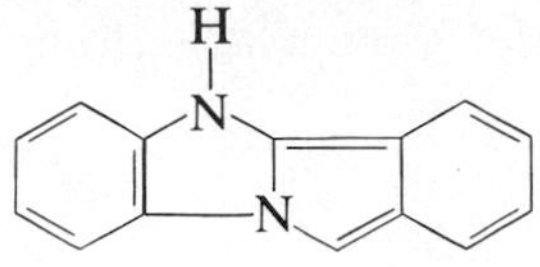

294b

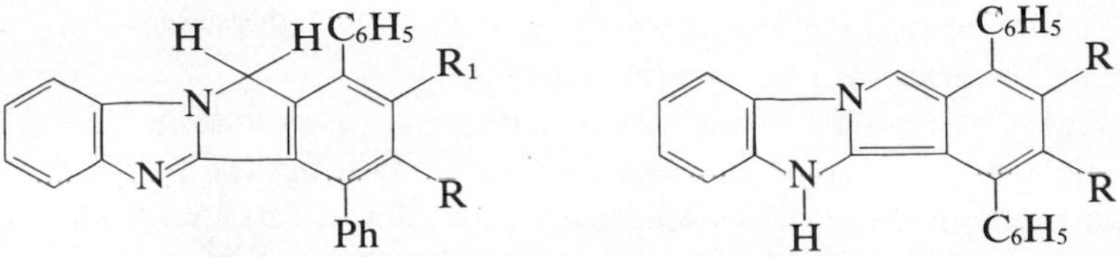

508

fused benzene rings and also found the methylene group. Müller and Zountsas[24] found that compound **508** was favored as the methylene having a peak at $\tau 5.08$ and 5.17 (for R = C_6H_5 or $COOC_2H_5$, respectively).

The location of the maverick proton on the pyrrolo[1,2-*a*]imidazole has still not been settled absolutely. Grob and Ankli[203] on the basis of ultraviolet spectra and basicitier assigned structures (**509** and **510**) the 2-chloro-3-methyl-6-cyano and 3-methoxy-3-methyl-6-cyano derivative respectively.

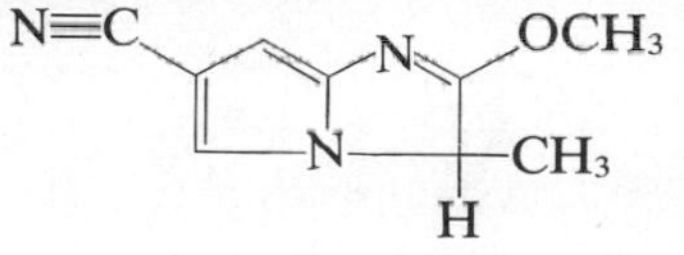

509

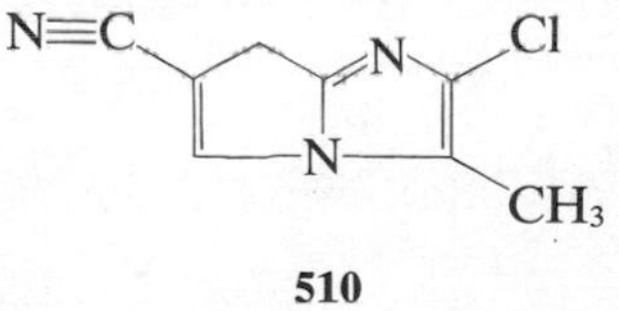

510

Kochergin and co-workers[273] then prepared to study the nmr spectra of pyrrolo[1,2-*a*]imidazoles and pyrrolo[1,2-*a*]benzimidazoles. The nmr

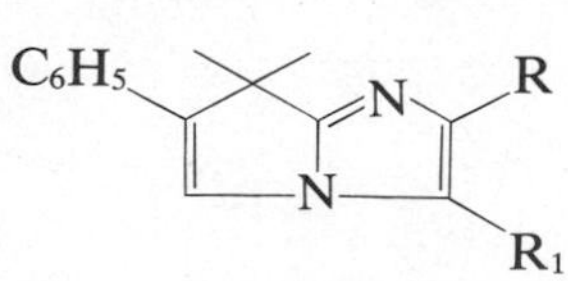

511
R^1 = H or C_6H_5
R = H or C_6H_5

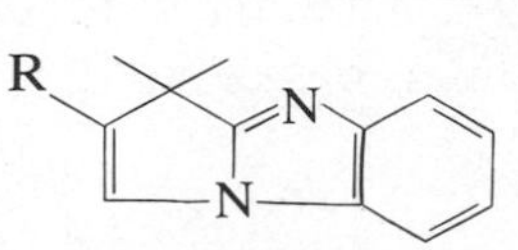

512
R = C_6H_5 or $C_6H_4OCH_3p$

spectra of these compounds showed methylene groups and the 7- and 1-position respectively (**451** and **512**) designated as being the bearer of the itinerant proton in these systems.

The nmr spectrum of 5*H*-tetrazolo[1,5-*a*]isoindole (**513**) shows the presence of a methylene group (a band at $\delta 5.72$).

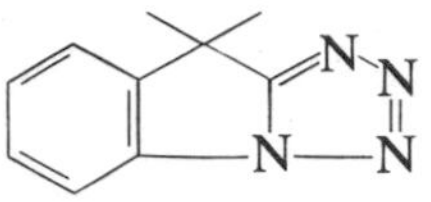

513

The perchant for the methylene form is also reflected in the nature of the reactivity of the molecule, for example, its behavior as an active methylene compound (see condensations).

Salmond[238] prepared two derivatives of 5-methylisoindolo[1,2-*a*]benzimidazole (**514**) which do possess the totally conjugated imidazopyrrole nucleus. Both of these compounds are locked into this situation by having a methyl group on the 5-nitrogen and have electron-withdrawing substituents, acetyl or carboxymethyl, at the 11-position. The 11-unsubstituted derivative was isolable only as the perchlorate salt (**515**).

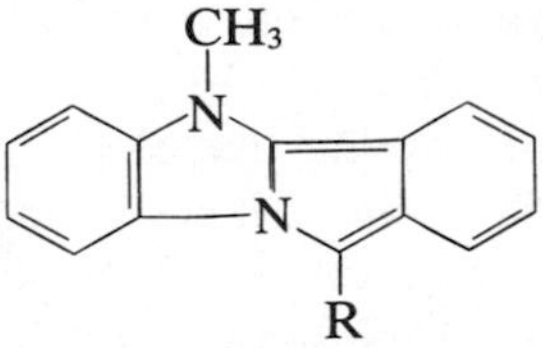

514 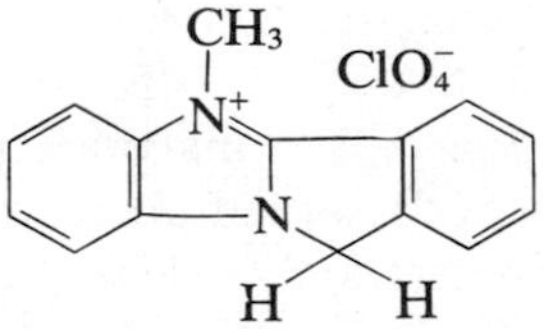 **515**

$R = \overset{O}{\overset{\|}{C}}—CH_3, \overset{O}{\overset{\|}{C}}—O—CH_3$

Similarly the 5-benzoyl derivative of 1-methylpyrrolo[1,2-*a*]imidazole (**516**; $R = H$, $R^1 = \overset{O}{\overset{\|}{C}}–C_6H_5$) is stable but removal of the electron withdrawing benzoyl group results in the formation of a rather unstable compound (**516**; $R = H$, $R^1 = H$). Interestingly enough, the 1-benzylpyrrolo[1,2-*a*]imidazole (**516**; $R = C_6H_5$, $R' = H$) is a stable compound. No explanation was offered for this phenomenon.[15]

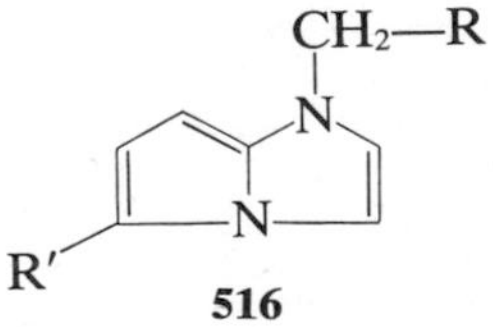

516

V. NMR-Aromaticity

Elvidge and Jackman[239] defined aromaticity as the ability of a system to sustain a ring current in a magnetic field. While there is considerable nmr data reported on 5,5-systems, there are no studies concerning deshielding due to ring current.

While one usually looks at the $\delta 6.5$ to 9 region of the spectrum to see aromatic protons, and adjustment in thinking is required when considering these 5,5-systems, especially those containing a pyrrole ring because of the high π-electron density possible at each position. These systems have 10 π-electrons distributed over 8 atoms, and ignoring all other effects, this is 1.25 π-electrons per atom, which is greater than the 1.20 π-electrons possible at each position of an isolated pyrrole, oxazole, or cyclopentenylanion. The signals of ring protons have been observed as high at 4.94 for the 3-proton of 1,2,4-trimethylpyrrolo[1,2-*a*]-benzimidazole,[8] a nonionic species. Higher field signals can be found for anions.

Okamura and Katz[228] assumed a ring current and used corrected line positions to calculate the change densities of pyrrolizinium anion (**495**). The values so obtained (Table XII) for the β- and γ-positions showed rather good correlation with HMO calculations. The nmr spectra of the lithium, potassium, and sodium salts of pyrrolizinium anion all showed three bands integrating for two-protons each. (The use of double resonance permitted assignment of these bands.) And while these bands are at relatively high field, the equivalence of both α-, both β-, and both γ-positions clearly demonstrates delocalization in this system.

TABLE XII

Position	Chemical shifts (τ) Pyrrolizinium anion (Li salt in THF)	Charge densities obtained from nmr data	Charge densities obtained from HMO calculations
α	3.57	−0.17	−0.26
β	3.97	−0.18	−0.14
γ	5.25	−0.32	−0.30

VI. Theoretical

Very little has been reported on the theoretical aspects of the 5,5-systems under discussion. This leaves an appreciable amount of work to those so inclined.

Despite the paucity of theoretical studies, the pyrrolizinium anion (**495**) enjoyed the attention of two groups of workers. Okamura and Katz,[228] using the HMO approximation, and Galasso and De Alti[240] using ASMO SCF CI calculations obtained the data shown in Table XIII.

TABLE XIII

	Electron densities	
Position	HMO	ASMO SCF CI
α	1.26	1.391
β	1.14	1.088
γ	1.30	1.270

Both methods show the β-position to have the lowest electron density, the HMO calculations show the γ-position to have the highest electron density while the ASMO SCF CI calculation favors the α-position.

While one should be judicious in using data for systems in the ground state to predict chemical reactivity, Galasso and De Alti[240] chose to make such a correlation and stated that the data show the α-position to be the most disposed to cationoid attack. This is in agreement with the finding of Okamura and Katz[228] (see Section III. D).

Galasso and De Alti[240] also calculated the singlet transition energies for pyrrolizinium anion (**495**) and a comparison of their data with the findings of Okamura and Katz[228] is shown in Table XIV.

TABLE XIV. SINGLET TRANSITION ENERGIES FOR PYRROLIZINIUM ANION

	Calculated	Found
$'A_1 \rightarrow 'A_1$	4.694 eV	4.20 eV (295 nm)
$'A_1 \rightarrow 'B_1$	5.865 eV	5.90 eV (210 nm)

While the data for the second band, due to the jump of an electron from the highest occupied molecular orbital to the highest unoccupied molecular orbital, is in good agreement with experimental findings, the calculated value for the first band, excitation of an electron from the highest bonding to the lowest antibonding orbital, does not approach the found value so closely.

k_i			
	−2.00		
	−1.81	−1.67	−1.69
			−1.50
	−1.41	−1.41	
		−1.19	−1.22
	0.00	0.00	
	+0.47		+0.29
	+1.00	+0.92	+0.66
	+1.41	+1.41	+1.38
		+1.69	
	+2.34		+2.08
		+3.26	+3.02
		N, N	N, N–R
$\sum k_i\beta$	10.44β	14.55β	14.86β

Fig. 1. Energy level diagrams, $E_i = \alpha + k_{i},\beta$. [Reproduced from *J. Am. Chem. Soc.*, **90**, 3830 (1964).]

Boekelheide and Fedoruk[15] compared the HMO data of pyrrolo[1,2-*a*]imidazole with that of pentadienyl dianion and the mesoionic pyrazolo-[1,2-*a*]pyrazole systems (Fig. 1). The energy diagrams of the dianion and the pyrazole systems with symmetrical charge distribution show four bonding, one nonbonding, and three antibonding orbitals while the pyrrolo[1,2-*a*]imidazole, with unsymmetrical charge distribution has five bonding and three antibonding orbitals. An appreciable difference between the delocalization energies (4.42β) of the dianion and the pyrroloimidazole can be seen.

Galasso[279] carried out CNDO/2-SCFMO calculations on pyrrolizinium cation as well as several other bridgehead nitrogen systems and reported that: "... the hybridization on the nitrogen atom is $S^a p6^b$ with $a \simeq 1.20$ and $b \simeq 2.36 - 2.51$ a.u., drastically different from the classical image of the S^1p6^2 trigonal hybrid ...."

Concerning the skeletal carbon atoms, he reported that the population on the 2s atomic orbital was 1.00 ± 0.3 a.u. The populations on the 2p6 orbitals fell into two categories: a population of 1.75–1.80 a.u. for those

Fig. 2. Reproduced from *Gazz. Chim. Ital.* **99,** 1078 (1969).

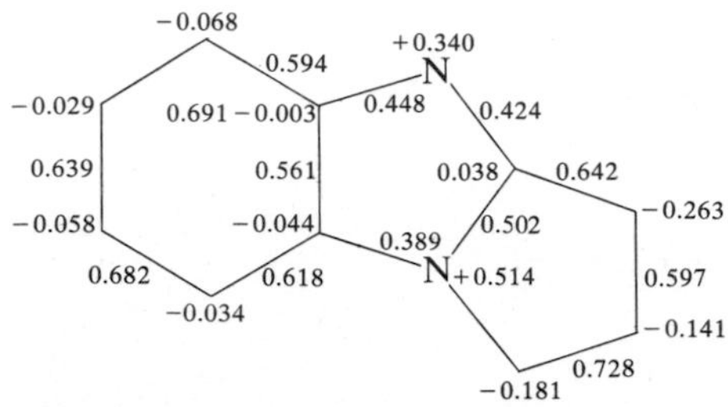

Fig. 3. Reproduced from *Khim. Geterotsikl. Soedin.*, 905 (1968).

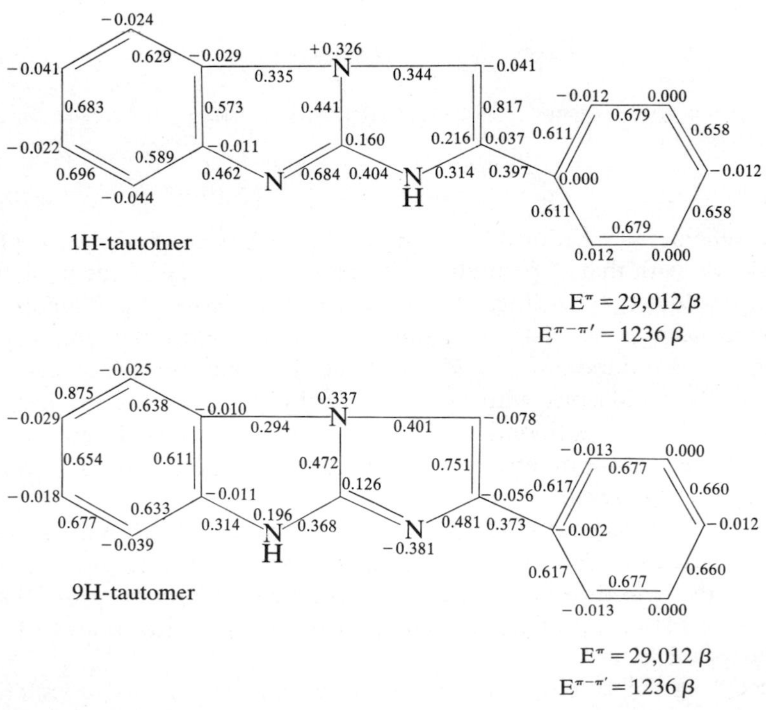

Fig. 4

carbon atoms linked directly to the bridgehead nitrogen and 1.93–2.01 a.u. for the other carbon atoms.

The sigma and Π net atomic charges for the pyrrolizinium cation according to Galasso[279] is shown in Fig. 2.

Kornilov, Dyadusha, and Babichev[208] carried out HMO calculations for pyrrolo[1,2-*a*]benzimidazole. The charge densities and bond orders are shown in Fig. 3.

Both of these sets of HMO calculations used the same parameters, that is, $\alpha_N = \alpha_C + \beta_{C1}$ and $\beta_{CN} = \beta_{C1}$. Okamura and Katz state that these parameters are between the values $h = 0.5$, $k = 1$, $\delta = 0.17$, and $h = 1.5$, $k = 0.8$, $\delta = 0.01$, which have been used by other workers and that the results of calculation obtained from the three sets of parameters were not greatly affected.

HMO data for 2-phenylimidazo-1,2-*a* benzimidazole (Fig. 4) has also been reported by Anisomova, Simonov, and Pozharskii[104] who reported that the data for the 9*H*-tautomer agrees with the experimental finding that electrophilic substitution takes place at the 3-position.

TABLE XV. HMO DERIVED QUANTITIES

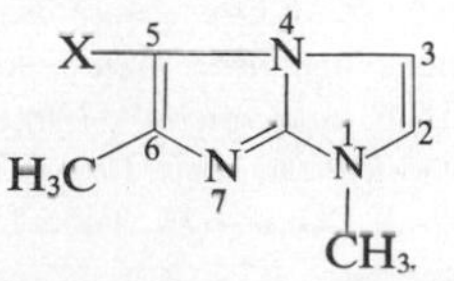

1, X = H; **5**, X = NO_2

Compound	S^E			Effective charge		
	2-position	3-position	5-position	2-position	3-position	5-position
1	1.26795	1.50860	1.64927	−0.03733	−0.08638	−0.12586
1 Protonated 1-position	1.34897	1.45185	1.72875	−0.03621	−0.06999	−0.12390
1 Protonated 4-position	1.20402	1.52460	1.65697	−0.02787	−0.08271	−0.11638
1 Protonated 7-position	1.20146	1.31850	1.38532	−0.03407	−0.07023	−0.09709
1 Triprotonated	1.17410	1.24761	1.37118	−0.02063	−0.04977	−0.08757
5	1.25970	1.43224		−0.03566	−0.07080	
5 Protonated 1-position	1.28494	1.30901		−0.03194	−0.05500	
5 Protonated 4-position	1.16463	1.34144		−0.02669	−0.05920	
5 Protonated 7-position	1.20271	1.29557		−0.03444	−0.06193	
5 Triprotonated	1.17523	1.23434		−0.02094	−0.04356	

Miller and Bambury[272] carried out HMO calculations for 6-methyl and 6-methyl-5-nitroimidazo[1,2-*a*]imidazole (Table XV) and used these data to assign the site of nitration of the 6-methyl-5-nitro derivative (see electrophilic substitution).

VII. References

1. For a more general review on the 5,5 systems with a bridgehead nitrogen see W. L. Mosby, "Heterocyclic Systems with Bridgehead Nitrogen Atoms—Part One," Interscience Publishers, New York, 1961, pp. 63–237.
2. For a critical discussion of various definitions of aromaticity see (a) D. Lloyd, "Carbocyclic Non-Benzenoid Aromatic Compounds," Elsevier Publishing Co., Amsterdam, 1966, pp. 1–14; (b) G. M. Badger, "Aromatic Character and Aromaticity," Cambridge University Press, New York, 1969.
3. P. M. Kochergin, A. A. Druzhnina, and R. M. Palei, *Khim. Geterotsikl. Soedin.*, 149 (1966).
4. A. A. Druzhnina and P. M. Kochergin, *Khim. Geterotsikl. Soedin.*, 532 (1967).
5. R. M. Palei and P. M. Kochergin, *Khim. Geterotsikl. Soedin.*, 536 (1967).
6. H. G. O. Becker, H. D. Steinleitner, and H. J. Timpe, *Synthesis*, 415 (1973).
7. F. S. Babichev and A. F. Babicheva, *Khim. Geterotsikl. Soedin.*, 187 (1967).
8. F. S. Babichev and A. F. Babicheva, *Khim. Geterotsikl. Soedin.*, 917 (1967).
9. G. Kutrov, P. Koslovskaya, and F. S. Babichev, *Ukr. Khim. Zh.*, **33,** 738 (1969).
10. F. S. Babichev, G. P. Kutrov, and M. Yu. Kornilov, *Ukr. Khim. Zh.*, **34,** 1020 (1968).
11. T. Pyl, H.-D. Dinse, and O. Sietz, *Ann.*, **676,** 141 (1964).
12. T. Pyl, H. Gille, and D. Nusch, *Ann.*, **679,** 139 (1964).
13. F. Kröhnke and W. Friedrich, *Ber.*, **96,** 1195 (1963).
14. B. B. Molloy, D. H. Reid, and F. S. Skelton, *J. Chem. Soc.*, 653 (1965).
15. V. Boekelheide and N. A. Fedoruk, *J. Am. Chem. Soc.*, **90,** 3830 (1968).
16. D. H. Reid, F. S. Skelton, and W. Bonthrone, *Tetrahedron Lett.*, 1797 (1964).
17. R. M. Acheson and W. R. Tully, *J. Chem. Soc.*, 1623 (1968).
18. F. S. Babichev and V. K. Kibirev, *Zh. Obshch. Khim.*, **33,** 2000 (1963); English translation, **33,** 1946 (1963).
19. Belgian Patent 660836.
20. B. L. Manjunath, *Quart. J. Indian Chem. Soc.*, **4,** 271 (1927)*, *Chem. Abstr.*, **21,** 3198 (1927).
21. F. Lions and E. Ritchie, *J. Am. Chem. Soc.*, **61,** 1927 (1939).
22. W. G. Preston and S. H. Tucker, *J. Chem. Soc.*, 659 (1943).
23. V. A. Kovtinenko and F. S. Babichev, *Ukr. Khim. Zh.*, **38,** 1244 (1972).
24. E. Muller and G. Zountsas, *Ber.*, **105,** 2529 (1972).
25. V. S. Likhitskaya and F. S. Babichev, *Ukr. Khim. Zh.*, **35,** 746 (1969).
26. H. G. O. Becker and H. Botcher, *J. Prakt. Chem.*, **314,** 55 (1972).
27. R.-M. Claramunt, J.-M. Fabrega, and J. Elguero, *J. Het. Chem.*, **11,** 751 (1974).
28. H. Gehlen and R. Drohla, *Arch. Pharm.*, **309,** 709 (1970); *Chem. Abstr.*, **73,** 130946f (1970).

* C. A. gives this year as 1917.

29. J. B. Bailey, E. B. Knott, and P. A. Marr., German Patent 1810462, *Chem. Abstr.*, **76,** 47395s; British Patent 1252418, *Chem. Abstr.*, **76,** 87186f (1972).
30. T. Pyl, R. Giebelman, and H. Beyer, *Ann.*, **643,** 145 (1961).
31. L. M. Werbel and M. L. Zamora, *J. Het. Chem.*, **2,** 287 (1965).
32. G. Westphal and P. Henklein, *Z. Chem.*, **9,** 25 (1969); *Chem. Abstr.*, **70,** 87686v (1969).
33. T. Pyl, K. Wunch, L. Bulling, and H. Beyer, *Ann.*, **657,** 113 (1962).
34. A. H. M. Raemaeker *et al.*, *J. Med. Chem.*, **9,** 545 (1966).
35. B. Kickhofen and F. Krohnke, *Ber.*, **88,** 1109 (1955).
36. F. Kröhnke, B. Kickhofen, and C. Thoma, *Ber.*, **88,** 117 (1955).
37. J. Reitman, U.S. Patent 2,057,978; British Patent 360,027; *Chem. Abstr.*, **27,** 514 (1933); German Patent 547,985; *Chem. Abstr.*, **26,** 3514 (1932).
38. N. Campbell and E. B. McCall, *J. Chem. Soc.*, 2411 (1951).
39. A. M. Simonov, V. A. Anisimova, and T. A. Borisov., *Khim. Geterotsikl. Soedin.*, 111 (1973).
40. J. P. Paolini and L. J. Lendvay, *J. Med. Chem.*, **12,** 1031 (1969).
41. T. A. Borisov, A. M. Simonov, and V. A. Anisimova, *Khim. Geterotsikl. Soedin.*, 803 (1973).
42. H. Ogura, *et al.*, *J. Med. Chem.*, **19,** 923 (1972).
43. H. K. Kondo and G. Nagasawa, *Yakugaku Zasshi*, **57,** 1050 (1937); *Chem. Abstr.*, **32,** 4161 (1938).
44. T. Matsukawa and S. Ban, *Yakugaku Zasshi*, **71,** 756 (1951); *Chem. Abstr.*, **46,** 8094 (1952).
45. T. Matsukawa and S. Ban, *Yakugaku Zasshi*, **72,** 884 (1952); *Chem. Abstr.*, **47,** 6410 (1953).
46. T. Pyl, L. Bulling, K. Wunch, and H. Beyer, *Ann.*, **643,** 153 (1961).
47. F. Dallacker, H. Pauling, and M. Lipp, *Ann.*, **663,** 58 (1963).
48. N. P. Buu-Hoi, N. D. Xuong, and T. Thu-Cuc, *Bull. Soc. Chim. Fr.*, 205 (1966).
49. G. Kempter *et al.*, *Z. Chem.*, **8,** 339 (1968).
50. N. O. Saldabol, G. Ya. Zarina, and S. A. Hillers, *Khim. Geterotsikl. Soedin.*, 178 (1968).
51. C. Altimirante *et al.*, *J. Med. Chem.*, **9,** 29 (1966).
52. G. Kempter *et al.*, *J. Prakt. Chem.*, **313,** 977 (1971).
53. I. Iwai and T. Hiraoka, *Chem. Pharm. Bull.*, **12,** 813 (1964).
54. B. Rudner, U.S. Patent 2,790,172.
55. S. Ban, *Yakugaku Zasshi*, **74,** 658 (1954).
56. N. O. Saldabol, S. Hillers, L. N. Alekseeva, and B. Brizga, *Khim.-Farm. Zh.*, **1,** 27 (1967); *Chem. Abstr.*, **68,** 2856m (1968).
57. A. Sitte, H. Paul, and G. Hilgetag, *Z. Chem.*, **7,** 341 (1967).
58. T. Pyl, F. Waschk, and H. Beyer, *Ann.*, **663,** 113 (1963).
59. Netherlands Patent 7,108,313
60. J. W. Carpenter, J. D. Mee, and D. W. Heseltine, U.S. Patent 3,615,639.
61. J. Goerdler and W. Roth, *Ber.*, **96,** 534 (1963).
62. P. M. Kochergin and B. A. Priimenko, *Khim. Geterotsikl. Soedin*, 176 (1969).
63. A. Lawson, *J. Chem. Soc.*, 307 (1956).
64. L. F. Miller and R. E. Bambury, Abstracts of the Third Central Regional Meeting-American Chemical Society, Cincinnati, Ohio, 1971.
65. V. M. Dziomko and A. V. Ivashchenko, *Khim. Geterotsikl. Soedin.*, **119,** (1973).
66. A. M. Siminov and P. M. Kochergin, *Khim. Geterotsikl. Soedin.*, 316 (1965).
67. P. M. Kochergin and A. M. Siminov, *Khim. Geterotsikl. Soedin.*, 133 (1967).

68. A. M. Siminov and V. A. Anisimova, *Khim. Geterotsikl. Soedin.*, 1102 (1968).
69. V. S. Ponomar and P. M. Kochergin, *Khim. Geterotsikl. Soedin.*, 253 (1972).
70. T. P. Sycheva, Z. A. Pankina, and M. N. Shchukina, *Khim. Geterotsikl. Soedin.*, 440 (1970); *Chem. Abstr.*, **73,** 87850f (1970).
71. I. Lalezari and A. Shafiee, *J. Het. Chem.*, **8,** 835 (1971).
72. E. Bulka and D. Ehlers, *J. Prakt. Chem.*, **315,** 510 (1973).
73. K. S. Dhaka *et al.*, *Indian J. Chem.*, **12,** 966 (1974).
74. J. Parrick and K. Pearson, *Chem. Ind.*, 1261 (1970).
75. P. M. Kochergin, A. A. Tkachenko, and M. V. Povstyanoi, U.S.S.R. Patent 213,881.
76. A. A. Tkachemko, P. M. Kochergin, and F. A. Zubkov, *Khim. Geterotsikl. Soedin.*, 682 (1971).
77. A. A. Tkachemko, P. M. Kochergin, and G. F. Panchemko, *Khim. Geterotsikl. Soedin.*, 686 (1971).
78. V. A. Priimenko and P. M. Kochergin, *Khim. Geterotsikl. Soedin.*, 1243 (1971).
79. B. Rudner, U.S. Patent 2,790, 172.
80. K. Löffler and K.-H. Menzel, Belgian Patent 621241.
81. K.-H. Menzel and R. Putter, U.S. Patent 3,369,897.
82. K.-H. Menzel, O. Wahl, and W. Pelz, German Patent 1,070,030.
83. E. Ochiai and T. Nisizawa, *Yakugaku Zasshi*, **60,** 127 (1940); *Chem. Abstr.*, **34,** 5082 (1940).
84. H. Scheyer and E. Schwamberger, U.S. Patent 2,108,413; French Patent 803,279; *Chem. Abstr.*, **31,** 2832 (1937); German Patent 642,339; *Chem. Abstr.*, **31,** 3709 (1937).
85. A. J. Hubert and H. Reimlinger, *Ber.*, **103,** 2828 (1970).
86. D. B. Reuschling and F. Krohnke, *Ber.*, **104,** 2103 (1971).
87. R. G. Glushkov and O. Yu. Magidson, *Khim. Geterotsikl. Soedin.* 85 (1965).
88. A. Kreutzberger and B. Meyer, *Chem. Ber.* **105,** 1810 (1972).
89. A. J. Hubert, *J. Chem. Soc.*, **C,** 1334 (1969).
90. D. C. K. Lin and D. C. DeJongh, *J. Org. Chem.*, **39,** 1780 (1974).
91. K. S. Ohaka, J. Mohan, V. K. Chada, and H. K. Pujan, *Indian J. Chem.*, **12,** 966 (1974).
92. S. Mignonac-Mondon, J. Elguero, and R. Lazaro, *CR. Soc. C.*, **276** 1533 (1973); *Chem. Abstr.*, **79,** 31985f (1973).
93. T. Pyl, O. Sietz, and K. Staege, *Ann.*, **679,** 144 (1964).
94. V. V. Avidon and M. N. Shchukina, *Khim. Geterotsikl. Soedin.*, 349 (1965).
95. T. P. Sycheva, I. D. Kiseleva, and M. N. Shchukina, *Khim. Geterotsikl. Soedin.*, 690 (1966).
96. V. M. Aryuzina and M. N. Shchukina, *Khim. Geterotsikl. Soedin.*, 605 (1966).
97. V. M. Aryuzina and M. N. Shchukina, *Khim. Geterotsikl. Soedin.*, 506 (1968).
98. V. M. Aryuzina and M. N. Shchukina, *Khim. Geterotsikl. Soedin.*, 509 (1968).
99. V. M. Aryuzina and M. N. Shchukina, *Khim. Geterotsikl. Soedin.*, 525 (1970); *Chem. Abstr.*, **73,** 87845h (1970).
100. T. P. Sycheva, Z. A. Pankina, and M. N. Shchukina, *Khim. Geterotsikl. Soedin.*, 440 (1970); *Chem. Abstr.*, **73,** 87850f (1970).
101. V. V. Avidon and M. N. Shchukina, *Khim. Geterotsikl. Soedin.*, 719 (1968). *Note:* As the result of a typographical error in the experimental section of this paper, *Chem. Abstr.*, **70,** 37716j (1969) reports that cyclization is effected using the procedure of van Es. Avidon and Shchukina[94] is the intended reference.
102. V. M. Avidon and M. N. Shchukina, *Khim. Geterotsikl. Soedin.*, 64 (1965).
103. B. G. Ermolaeva and M. N. Shchukina, *Khim. Geterotsikl. Soedin.*, 84 (1967).

104. V. A. Anisimova, A. M. Simonov, and A. F. Pozharski., *Khim. Geterotsikl. Soedin.*, 797 (1973).
105. G. A. Reynolds and J. A. Van Allan, *J. Org. Chem.*, **24,** 1478 (1959).
106. R. G. Glushkov, E. S. Golovchinskaya, and O. Yu. Magidson, *Zh. Obshch. Khim.*, **29,** 3742 (1959); English Translation, **29,** 3700 (1959).
107. M. Kanaoka, *Yakugaku Zasshi*, **76,** 113 (1956); *Chem. Abstr.*, **51,** 3597 (1957).
108. M. Kanaoka, *Pharm. Bull.*, **5,** 385 (1957); *Chem. Abstr.*, **52,** 5390 (1958).
109. J. deMendoza and J. Elguero, *Bull. soc. Chim. Fr.*, 1675 (1974).
110. K. T. Potts and S. Hussain, *J. Org. Chem.*, **36,** 10 (1971).
111. L. G. S. Brooker and J. Van Lare, U.S. Patent 2,786,054.
112. K. T. Potts and R. M. Huseby, *J. Org. Chem.*, **31,** 3528 (1966).
113. Pentimalli, L. and Milani, G., *Ann. Chim. (Rome)* **61,** 672 (1971).
114. J. D. Bower and F. P. Doyle, *J. Chem. Soc.*, 727 (1957).
115. J. Sandstrom, *Acta Chem. Scand.*, **15,** 1295 (1961).
116. M. Kanaoka, T. Okuda, and D. Shiho, *Yakagaku Zasshi*, **87,** 119 (1967).
117. K. T. Potts and C. Hirsch, *Chem. Ind.*, 2168 (1966).
118. K. T. Potts and C. Hirsch, *J. Org. Chem.*, **33,** 143 (1968).
119. H. Beyer, E. Bulker, and K. Dittrich, *J. Prakt. Chem.*, **30,** 280 (1963).
120. *O. Tsuge and H. Samura, Org. Prep. Proc. Int.*, **6,** 161 (1974).
121. F. L. Scott and R. N. Butler, *J. Chem. Soc.*, 1202 (1966).
122. R. N. Butler and F. L. Scott, *J. Chem. Soc.*, 1319 (1966).
123. F. L. Scott and R. N. Butler, *J. Chem. Soc.*, 1711 (1968).
124. V. R. Rao and V. R. Srinivasan, *Experientia*, **20,** 200 (1964).
125. E. Hoggarth, *J. Chem. Soc.*, 614 (1950).
126. See ref. 1, p. 214.
127. H. Gehlen and F. Lemme, *Ann.*, **703,** 116 (1967).
128. R. N. Butler, P. O'Sullivan, and F. L. Scott, *J. Chem. Soc.* 2265 (1971).
129. F. L. Scott and J. B. Aylward, *Tetrahedron Lett.* 847 (1965).
130. R. N. Butler and F. L. Scott, *J. Chem. Soc.*, 239 (1967).
131. H. Reimlinger and R. Merenyi, *Ber.*, **103,** 3284 (1970).
132. T. Eicher, S. Hunig, H. Hansen, and P. Nikolaus, *Ber.*, **102,** 3159 (1969).
133. J. Hadacek and J. Slotova-Trnkova, *Spisy. Prirodevecke Fak. Univ. Brne*, **22,** 39 (1960–61).
134. R. I.-F. Ho and A. R. Day, *J. Org. Chem.*, **38,** 3084 (1073).
135. Y. Tamura *et al.*, *J. Het. Chem.*, **11,** 459 (1974).
136. E. Näf, *Ann.*, **265,** 108 (1891).
137. See ref. 1, p. 203.
138. A. Messmer and A. Gelleri, *Angew. Chem. Int. Ed.*, **6,** 261 (1967).
139. G. Fodor and G. Wilheim, *Acta Chim. Acad. Sci. Hung.*, **2,** 189 (1952); *Chem. Abstr.*, **48,** 3346 (1954).
140. H. Beyer, W. Lassig, and G. Ruhlig, *Ber.*, **86,** 764 (1953).
141. H. Beyer, W. Lassig, and E. Bulka, *Ber.*, **87,** 1385 (1954).
142. G. A. Reynolds, J. A. Van Allan, and J. F. Tinker, *J. Org. Chem.*, **24,** 1205 (1959).
143. H. Beyer, G. Wolter, and H. Lemke, *Ber.*, **89,** 2350 (1956).
144. M. Colonna and R. Andrisano, *Publ. 1st. Chim. Ind. Univ. Bologna* No. 5, 3; No. 6, 3 (1943); *Chem. Abstr.*, **41,** 754 (1947).
145. V. Ya. Pochinok, S. D. Zaitseva, and R. G. El'gort, *Ukr. Khim. Zh.*, **17,** 509 (1951); *Chem. Abstr.*, **48,** 11392 (1954).
146. I. Ya. Postovskii, G. N. Tyorenkova, and L. F. Pilatova, *Dokl. Akad. Nauk.* **179,** 111 (1968).

147. T. Bacchetti, A. Alemagna, and B. Danieli, *Ann. Chim. (Roma)*, **55,** 615 (1965).
148. British Patent 1,069,771; French Patent 1,392,310.
149. V. Ya. Pochinok and L. F. Avramenko, *Ukr. Khim. Zh.*, **28,** 511 (1962).
150. L. F. Avramenko, V. Ya. Pochinok, and Yu. S. Rozum, *Zh. Obsch. Khim.*, **33,** 980 (1963); English Translation **33,** 968 (1963).
151. A. E. Alper and A. Taurins, *Can. J. Chem.* **45,** 2903 (1963).
152. P. M. Kochergin and M. N. Shchukina, *Zh. Obshch. Khim.*, **26,** 2905 (1956).
153. P. M. Kochergin and M. N. Shchukina, *Zh. Obshch. Khim.*, **26,** 458 (1956).
154. E. Ochiai, *Ber.*, **69,** 1650 (1936).
155. P. M. Kochergin, *Zh. Obshch. Khim.*, **30,** 1529 (1960); English Translation, **30,** 1542 (1960).
156. P. M. Kochergin and A. N. Krasovskii, *Khim. Geterotsikl. Soedin.*, 945 (1969).
157. P. M. Kochergin, A. M. Tsyganova, and L. M. Viktorova, *Khim. Geterotsikl. Soedin.*, 93 (1967).
158. P. M. Kochergin and A. M. Viktorova, *Khim. Geterotsikl. Soedin.*, 313 (1965).
159. P. M. Kochergin, USSR Patent 232,976; *Chem. Abstr.*, **70,** 87809n (1969).
160. P. M. Kochergin, *Zh. Obshch. Khim.*, **26,** 2493 (1956).
161. H. Andersag and K. Westphal, *Ber.*, **70,** 2035 (1937).
162. A. R. Todd, E. Bergel, and Karimullah, *Ber.*, **69,** 217 (1936).
163. R. C. Elderfield and R. N. Prasad, *J. Org. Chem.*, **24,** 1410 (1959).
164. P. M. Kochergin, *Zh. Obshch. Khim.*, **26,** 2916 (1956).
165. H. W. Stephen and F. J. Wilson, *J. Chem. Soc.*, 2531 (1926).
166. F. J. Wilson, W. Baird, R. Burns, A. M. Munro, and H. W. Stephan, *J. Roy. Tech. Coll.* [*Glasgow*, **2,** No. 1 (1929); *Chem. Abstr.*, **23,** 5164 (1929)].
167. J. D. Kendall and G. F. Duffin, U.S. Patent 2,527,265.
168. J. D. Duffin and F. G. Kendall, *J. Chem. Soc.*, 361 (1956).
169. D. J. Brown, *J. Chem. Soc.*, 1974 (1958).
170. I. Iwai and T. Hiraoka, Japanese Patent 41-5099.
171. A. N. Krasovskii and P. M. Kochergin, *Khim. Geterotsikl. Soedin.*, 849 (1967).
172. M. Gordon, *J. Am. Chem. Soc.*, **73,** 984 (1951).
173. G. DeStevens and A. Halamandaris, *J. Am. Chem. Soc.*, **79,** 5710 (1957).
174. J. D. Kendall and F. G. Duffin, British Patents 634,951/2; *Chem. Abstr.*, **44,** 9287 (1950); *U.S. Patents* 2,527,265/6.
175. A. I. Kiprianov and V. O. Khilya, *Zh. Obshch. Khim.*, **2,** 1474 (1966); English Translation, **2,** 1457 (1966).
176. J. J. D'Amico, U.S. Patent 3,225,059; J. J. D'Amico, R. N. Campbell, and E. C. Guinn, *J. Org. Chem.*, **29,** 865 (1964).
177. A. Lawson and H. V. Morley, *J. Chem. Soc.*, 1695 (1955).
178. A. Lawson and H. V. Morley, *J. Chem. Soc.*, 566 (1957).
179. C. K. Bradsher, D. F. Lohr, and W. J. Jones, *Tetrahedron Lett.*, 1723 (1965).
180 C. K. Bradsher and W. J. Jones, *J. Org. Chem.*, **32,** 2074 (1967).
181. C. K. Bradsher and W. J. Jones, *J. Org. Chem.*, **32,** 2079 (1967).
182. C. K. Bradsher and D. F. Lohr, *J. Het. Chem.*, **4,** 75 (1967).
183. R. Neidlein and J. Tauber, *Tetrahedron Lett.*, 6287 (1968).
184. H. Beyer and A. Hetzheim, *Z. Chem.*, **2,** 153 (1962).
185. H. Beyer and A. Hetzheim, *Z. Chem.*, **2,** 152 (1962).
186. M. Kanaoka, *Yakugaku Zasshi*, **76,** 113 (1956); *Chem. Abstr.*, **51,** 3579 (1957).
187. T. Pyl, F. Waschk, and H. Beyer, *Ann.*, **663,** 113 (1963).
188. K. T. Potts and R. M. Huseby, *Chem. Ind.*, 1919 (1964).
189. M. Kanaoka, *Yakugaku Zasshi*, **87,** 119 (1967).

190. B. Stanovnik and M. Tishler, *Archiv. Pharmazie*, **300,** 322 (1967).
191. G. Ege, *Angew. Chem.* (*Int. Ed.*), **6,** 629 (1967).
192. J. Thiele and F. G. Falk, *Ann.*, **347,** 114 (1906).
193. M. V. Betrabet and G. C. Chakravarti, *J. Indian Chem. Soc.*, **7,** 495 (1930); *Chem. Abstr.*, **25,** 701 (1931).
194. W. Reid and H. Bodem, *Ber.*, **89,** 708 (1956).
195. D. Amos and R. G. Gillis, *Aust. J. Chem.*, **17,** 1440 (1964).
196. A. Bistrzyki and W. Schmutz, *Ann.*, **415,** 1 (1918).
197. F. Sparatore and G. Bignardi, *Gayn. Chim. Ital.*, **92,** 606 (1962).
198. For a discussion of related chemistry see K. Hoffman, "Imidazole and Derivatives," part I, Interscience, New York, 1953, pp. 267–271.
199. H. D. Perlmutter and P. S. Knapp, *J. Org. Chem.*, **32,** 2350 (1967).
200. G. L. Oliver, J. R. Dann, and J. W. Gates, *J. Am. Chem. Soc.*, **80,** 702 (1958); U.S. Patent 2,860,985; Belgian Patent 564,592.
201. R. Gompper and F. Effenberger, *Ber.*, **92,** 1928 (1959).
202. F. M. Rowe *et al.*, *J. Chem. Soc.*, 1796 (1935).
203. C. A. Grob and P. Ankli, *Helv. Chim. Acta*, **33,** 273 (1950).
204. W. J. Middleton and D. Metzger, *J. Org. Chem.*, **35,** 3985 (1970).
205. B. K. Malaviya and S. Dutt, *Proc. Acad. Sci. United Provinces Agra Oudh, India*, **4,** 319 (1935); *Chem. Abstr.*, **30,** 1057 (1936).
206. B. B. Malloy, D. H. Reid, and S. McKenzie, *J. Chem. Soc.*, 4368 (1965).
207. S. McKenzie, B. B. Malloy and D. H. Reid, *J. Chem. Soc.*, 1908 (1966).
208. M. Yu. Kornilov, G. G. Dyadyusha, and F. S. Babichev, *Khim. Geterotsikl. Soedin.*, 905 (1968).
209. N. Phonh, Z. M. Ivanova, G. I. Derkoxh, and F. S. Babichev, *Zh. Obshch. Khim.*, **41,** 319 (1971).
210. P. M. Kochergin and I. A. Mazur, USSR Patent 196,868; Derwent Basic 39,591.
211. P. M. Kochergin and I. A. Mazur, USSR Patent 196,872; Derwent Basic 32,594.
212. T. Pyl, L. Wünch, and H. Beyer, *Ann.*, **657,** 108 (1962).
213. L. Pentimalli and V. Passalaqua, *Bull. Sci. Fac. Chim. Ind. Bologna*, **24,** 205 (1966). [Note ref. 5 of this paper is in error. The article cited should be that of Pyl, Wunch, and Beyer.[173]]
214. L. Pentimalli, G. Cogo, and A. M. Guerra, *Gazz. Chim. Ital.*, **97,** 488 (1967).
215. S. Kano, D. Takiguchi, and T. Noguchi, Japanese Patent 24187/68; Derwent basic 34,614.
216. J. M. Sprague and A. H. Land, in "Heterocyclic Compounds" Vol. 5, R. C. Elderfield, Ed, Wiley, New York, 1957, p. 634.
217. L. Pentimalli and A. M. Guerra, *Gazz. Chim. Ital.*, **97,** 1286 (1967).
218. V. V. Avidon and M. N. Shchukina, *Khim. Geterotsikl. Soedin.*, 292 (1966).
219. L. F. Avramenko, T. A. Zacharova, V. Ya. Pochinok, and Yo. S. Rozum, *Khim. Geterotsikl. Soedin.*, 423 (1968).
220. K.-H. Menzel, R. Putter, and G. Wolfrom, British Patent 951,113.
221. G. Wolfrom, R. Putter, and K.-H. Menzel, German Patent 1,234,891.
222. P. Schmitz, K. G. Kleb, H.-A. Dortman, K.-H. Menzel, and R. Putter, British Patent 927,614.
223. F. S. Babichev and V. K. Kibirev, *Zh. Obshch. Khim. SSSR*, **33,** 1946 (1963).
224. V. Ya. Pochinok and S. D. Zaitseva, *Ukr. Khim. Zh.*, **26,** 351 (1960).
225. A. M. Siminov, B. A. Anisimova, and Yu. V. Koschienko, *Khim. Geterotsikl. Soedin.*, 184 (1969).
226. F. S. Babichev, V. K. Kibirev, and L. G. Chulko, *Ukr. Khim. Zh*, **32,** 64 (1966).

227. K.-H. Menzel, French Patent 658,107.
228. W. H. Okamura and T. J. Katz, *Tetrahedron,* **23,** 2941 (1967).
229. V. Ya. Pochinok and L. F. Avramenko, *Acta. Phys. Chim. Debrecina,* **12,** 131 (1966).
230. J. H. Boyer and E. J. Miller, *J. Am. Chem. Soc.,* **81,** 4671 (1959).
231. L. F. Avramenko, V. Ya. Pochinok and Yu. S. Rozum, *Zh. Obshch. Khim.,* **34,** 278 (1964); English Translation, **34,** 275 (1964).
232. Yu. N. Sheinker, I. Ya. Postovskii, N. P. Bednyagina, L. B. Senyavina, and L. F. Pilatova, *Dokl. Akad. Nauk.* **141,** 1388 (1961).
233. L. J. Bellamy, "The Infra-red Spectra of Complex Molecules," 2nd ed., Wiley, New York, 1958, p. 263.
234. H. Balli and F. Kersting, *Ann.,* **647,** 1 (1961).
235. A. R. Katritsky and J. M. Lagowski, in "Advances in Heterocyclic Chemistry," Vol. 2, A. R. Katritsky, Ed., Academic Press, New York, 1963, pp. 1–81.
236. *Ibid.,* p. 52.
237. E. Spinner, *J. Chem. Soc.,* 1237 (1960).
238. W. G. Salmond, *Tetrahedron Lett.,* 4689 (1967).
239. J. A. Elvidge and L. M. Jackman, *J. Chem. Soc.,* 859 (1961).
240. V. Galasso and G. DeAlti, *Theor. Chim. Acta,* **11,** 411 (1968).
241. V. I. Nifonter *et al., Khim. Geterotsikl. Soedin.,* 94 (1973).
242. V. N. Skopenko *et al., Ukr. Khim. Zh.,* **39,** 215 (1973).
243. H. Singh and S. Singh, *Indian. J. Chem.,* **11,** 311 (1973).
244. S. Kamo and T. Noguchi, Japanese Patent 71-37,836.
245. E. G. Knish, A. N. Krasonsku, and P. M. Kochergin, *Khim. Geterotsikl. Soedin.,* 25 (1972).
246. E. G. Knish, A. N. Krasovsku, and P. M. Kochergin, *Khim. Geterotsikl. Soedin.,* 1128 (1971).
247. J. Mohan and H. K. Pujari, *Indian J. Chem.,* **10,** 274 (1972).
248. K. S. Khaka, J. Morham, V. K. Chadka, and H. K. Pujari, *Indian J. Chem.,* **12,** 485 (1974).
249. K. K. Balasubramanian and B. Venugepalen, *Tetrahedron Lett.,* 2643 (1974).
250. E. G. Knish, A. N. Krasovskii, and P. M. Kochergin, *Khim. Geterotsikl. Soedin.,* 30 (1972).
251. A. N. Krasovskii, P. M. Kochergin, and T. E. Kozlovskaya, *Khim. Geterotsikl. Soedin.,* 393 (1971).
252. I. A. Mazur, P. M. Kochergin, and V. G. Tromsa, *Khim. Geterotsikl. Soedin.,* 389 (1971).
253. British Patent 1,241,069.
254. M. V. Kohner, *Indian J. Chem.,* **11,** 321 (1973).
255. H. Alper and R. W. Stout, *J. Het. Chem.,* **10,** 5 (1973).
256. H. Golgolab, I. Lalezari, and L. Hosseini-Gobari, *J. Het. Chem.,* **10,** 387 (1973).
257. G. Barnikow and J. Bodeker, *J. Prakt. Chem.,* **313** (1971).
258. V. M. Dzimoko and A. V. Ivashchenko, *Zh. Obshch. Khim.,* **43,** 1330 (1973); English Translation, **43,** 1322 (1973).
259. J. H. Wikel and C. J. Paget, *J. Org. Chem.,* **39,** 3506 (1974).
260. F. S. Babichev and N. N. Romanov, *Ukr. Khim. Zh.,* **39,** 49 (1973).
261. R. K. Mackie, S. McKenna, D. H. Reid, and R. G. Webster, *J. Chem. Soc. Perkins Trans.,* **I,** 657 (1973).
262. O. Cedar and B. Beijer, *Tetrahedron,* **28,** 4341 (1972).
263. N. O. Saldabol *et al., Khim. Geterotsikl. Soedin.,* 1353 (1972).
264. V. N. Skolenko *et al., Ukr. Khim. Zh.,* **39,** 60 (1973).

265. V. M. Aryozina and M. N. Shchukina, *Khim. Farm. Zh.*, **6,** 22 (1972).
266. V. M. Aryozina and M. N. Shchukina, *Khim. Geterotsikl. Soedin.*, 395 (1973).
267. S. Kano and T. Naguchi, Japanese Patent 71,26,498; *Chem. Abstr.*, **75,** 140864h (1970).
268. T. P. Sicheva, I. D. Kisleva, and M. N. Shchukina, *Khim. Geterotsikl. Soedin.*, 916 (1970).
269. C. J. Paget, German Patent 2,250,077.
270. T. P. Sicheva, I. D. Kisleva, G. P. Sirova, and M. N. Shchukina, *Khim. Geterotsikl. Soedin.*, 913 (1970).
271. V. A. Anisimera, A. M. Siminov, and T. A. Borisova, *Khim. Geterotsikl. Soedin.*, 791 (1973).
272. L. F. Miller and R. E. Bambury, *J. Org. Chem.*, **38,** 1955 (1973).
273. P. M. Kochergin *et al.*, *Khim. Geterotsikl. Soedin.*, 826 (1971).
274. J. Bailey, E. B. Kratt, and P. A. Morr, German Patent 1,810,962; British Patent 1,253,933.
275. M. H. Stern and E. D. Machiele, German Patent 2,260,202.
276. E. Alcalde, J. deMendoza, and J. Elguero, *J. Het. Chem.*, **11,** 921 (1974).
277. Y. Shickawa and S. Ohki, *Chem. Pharm. Bull.*, **21,** 981 (1973).
278. J. A. Van Allan and G. A. Reynolds, *J. Het. Chem.*, **5,** 471 (1968).
279. V. Galasso, *Gazz. Chim. Ital.*, **99,** 1078 (1969).

CHAPTER II

Indolizine and Aza Derivatives with Additional Nitrogens in the 5-Membered Ring

H. L. BLEWITT

University of Alabama, Tuscaloosa, Alabama

I. Introduction

The diazaindenes have long been of interest to both the organic chemist and biochemist because of their close structural relationship to purine (**1**). Hundreds of these compounds have been prepared for

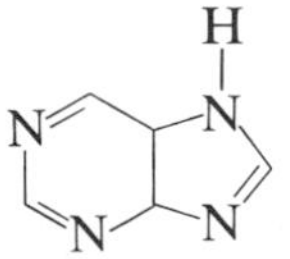

1

physiological testing. Although organic chemists have vigorously pursued the synthesis and testing of these compounds, it was not until 1965 that rather extensive works were begun to elucidate the chemical and physical properties of these systems.

The diazaindenes discussed in this chapter include indolizine and azaderivatives with additional nitrogens in the 5-membered ring. Several alternate names (see Table I) and numbering systems have been used for these compounds. The nomenclature (shown below) used in this Chapter is the same as that currently used by *Chemical Abstracts.* The numbering of these systems has often caused confusion, but in recent years the numbering, illustrated below, has become widely used[1–16] and will therefore be followed in this chapter.

The chemistry of these diazaindenes has been extensively reviewed by Mosby[17] through 1956 (Volume 50 of *Chemical Abstracts*). Other limited reviews, each reviewing a single diazaindene, covering the literature from 1957 are also available.[4,8,9,11,121] Colonna and Monti[160] have also published a review article dealing with the synthesis of indolizine derivatives with azodicarbonic esters.

TABLE I. ALTERNATE NAMES ENCOUNTERED IN THE LITERATURE

Imidazo[1,2-*a*] pyridine[a]	Imidazo[1,5-*a*]-pyridine[a]	Pyrazolo[1,5-*a*]-pyridine[a]	Indolizine[a]
1-Azaindolizine	2-Azaindolizine	3-Azaindolizine	Pyrrocoline
1,4-Imidazopyridine	2,3*a*-Diazaindene	3-Azapyrrocoline	Pyrindole
Pyrimidazole	2,4-Imidazopyridine	1,7*a*-Diazaindene	Pyrrodine
Pyridino(1′ : 2′-1 : 2)-glyoxaline		Polyazaindene	8-Pyrrolopyridine
3*a*-Azaindole		Pyrazolo[2,3-*a*]-pyridine	Pyrrolo[1,2-*a*]-pyridine
9-Azainolenine			Indolicine[b]
1,3*a*-Diazaindene			
3,7*a*-Diazaindene			

[a] Name currently used by *Chemical Abstracts.*
[b] Name often encountered in the Russian literature.

2
Imidazo[1,2-*a*]pyridine

3
Imidazo[1,5-*a*]pyridine

4
Pyrazolo[1,5-*a*]pyridine

5
Indolizine

Derivatives of these systems have been known for a long time, but the parent substances are comparatively new. In 1925 Tschitschibabin[18] synthesized imidazo[1,2-*a*]pyridine (**2**) and only 2 years later he successfully prepared the unsubstituted indolizine (**5**) utilizing the following sequences.

$$\text{2-aminopyridine} + BrCH_2CHO \xrightarrow[150\text{–}200°C]{\text{Sealed tube}} \mathbf{2}$$

$$\text{2-methylpyridine} + BrCH_2CHO \longrightarrow \mathbf{5}$$

In the mid-1950s Bowers and Ramage[20,21] successfully synthesized the parent substances imidazo[1,5-*a*]pyridine (**3**) and pyrazolo[1,5-*a*]pyridine (**4**). The reaction sequences are shown below. These syntheses of parent

materials (**2**, **3**, **4**, and **5**) do not necessarily afford the highest yields. Several alternate methods are known and are discussed in the synthesis section of this chapter.

The major purpose of this review is to report on advances in the field of diazaindene chemistry that have been accomplished since 1956.

II. Physical Properties

A. Melting and Boiling Point Trends of Some Simple Diazaindenes

The parent substances are either liquids or low-melting solids. The addition of a high molecular weight substituent (e.g., bromine) or a polar substituent, such as nitroso or nitro, generally results in the expected high melting solid (see Table II).

Some limited studies have even been carried out by Mattu and Margoniu[53] on the crystal forms of certain substituted imidazo[1,2-*a*]-pyridines.

B. Quantum Chemical Calculations

During the past few years several workers have carried out quantum mechanical calculations on these diazaindenes and have attempted to correlate the results with various chemical and physical properties.[1,2,4,7–9,11,54–64,159]

Frontier and total π-electron densities* for several diazaindenes have

* The frontier electrondensity is the electronic distribution to a given ring position of those electrons in the highest occupied molecular orbital. The total π-electron density is the sum of the electron densities contributed by all of the occupied π-molecular orbitals.

been calculated using several sets of parameters.[1,2,4,8,9,11,63,221,222,225] Correlations between both frontier and total π-electron densities with electrophilic and nucleophilic substitution have been made with some success.[1,4,8,9,221,222] These correlations work especially well in predicting electrophilic substitution. Typical electron densities for some of the

TABLE II. PROPERTIES OF SOME REPRESENTATIVE DIAZAINDENES

Compound	M.P.&/or B.P. (°C)	Ref.
Imidazo[1,2-*a*]pyridine	$b_{27}=153–155$, $b_3=114–113$	4, 18, 22
2-Methyl-	$b_{15}=134–136$, $m=45–46$	4
3-Methyl-	$b_{1.3}=84–89$, $m=64–65.5$	4
5-Methyl-	$b_{0.5}=104–105$	4
3-Bromo-	$m=92.9–93.3$	4
3-Nitroso-	$m=160–160.5$	4
3-Nitro-	$m=203–204.5$	4
Imidazo[1,5-*a*]pyridine	$b_3=120–125$, $m=54–55$	8, 20
1-Methyl-	$b_{1.4}=98–99$, $m=64–65$	8, 20
3-Methyl-	$b_{0.5}=90–91$, $m=55$	8, 20
5-Methyl-	$b_{0.25}=149–152$	8
1,3-Dibromo-	$m=125–126$	8
1-Nitro-	$m=244–245$	8
Indolizine	$b_{760}=205$, $m=75$	19, 25–38
1-Methyl-	$m=44$	36, 39
3-Methyl-	$b_{760}=230$	27, 29, 36, 51
5-Methyl-	$b_{34}=124$	49
Pyrazolo[1,5-*a*]pyridine	$b_{25}=108$	9, 21
1-Bromo-	$m=67–67.5$	9
1-Nitroso-	$m=139–140$	9
1-Nitro-	$m=181–182$	9

parent diazaindenes (**2**, **3**, **4**, and **5**) are shown in Tables III and V. More recently CNDO/2 calculations on these same compounds have been used to obtain electron densities, dipole moments, bond angles, and bond distances. These CNDO/2 electron densities appear to be a reasonably good index for predicting electrophilic substitution (see Table III).[220]

A number of successful correlations have been made between the transition energies, relative peak positions, and intensities in the ultraviolet spectra of some diazaindenes and the energy of the lowest lying unoccupied molecular orbital.[4,8,56,61–63,159] Paudler and Blewitt[4] used correlations of this type to predict ultraviolet spectral peak positions of substituted imidazo[1,2-*a*]pyridines relative to the parent substance (see

TABLE III[a]. CNDO/2, FRONTIER, AND TOTAL π-ELECTRON DENSITIES FOR THE PARENT DIAZAINDENES

2 3 4 5

Ring position	Compound 2[b]	Compound 3[c]	Compound 4[d]	Compound 5[e]
	Total π-electron densities			
1	1.46 (1.36)	1.13 (0.99)	1.14 (1.19)	1.18 (1.13)
2	1.04 (0.89)	1.31 (1.26)	1.01 (0.86)	1.08 (1.02)
3	1.11 (1.28)	1.05 (1.10)	1.42 (1.48)	1.16 (1.24)
4	1.48 (1.37)	1.49 (1.40)	1.49 (1.34)	— (1.39)
5	0.95 (1.11)	0.98 (1.13)	0.95 (1.08)	0.98 (1.12)
6	1.02 (0.99)	1.02 (0.96)	1.02 (1.01)	1.03 (0.98)
7	0.98 (1.02)	1.00 (1.07)	0.98 (1.01)	1.01 (1.05)
8	1.02 (0.97)	0.99 (0.93)	1.01 (0.98)	1.01 (0.95)
9	0.95 (1.02)	1.03 (1.16)	0.98 (1.06)	— (1.12)
	Frontier π-electron densities [CNDO/2 electron densities][f]			
1	0.59 [5.171]	0.51 [4.012]	0.46 [4.067]	— [4.080]
2	0.52 [3.894]	0.00 [5.162]	0.02 [3.939]	— [4.018]
3	0.51 [4.067]	0.50 [3.887]	0.61 [5.166]	— [3.967]
4	0.01 [5.051]	0.02 [5.035]	0.01 [4.894]	— [5.015]
5	0.31 [3.927]	0.34 [3.903]	0.33 [3.896]	— [3.899]
6	0.07 [4.006]	0.10 [4.025]	0.10 [4.031]	— [4.032]
7	0.19 [3.974]	0.22 [3.984]	0.18 [3.966]	— [3.998]
8	0.23 [4.019]	0.24 [4.019]	0.26 [4.026]	— [4.023]
9	0.04 [3.857]	0.08 [3.924]	0.04 [3.896]	— [3.906]

[a] Values in parentheses are taken from Ref. 63.
[b] Values from Ref. 1 and 4.
[c] Values from Ref. 8.
[d] Values taken from Ref. 7 and 9.
[e] Values from Ref. 8 and 182.
[f] Values in brackets are CNDO/2 electron densities; taken from Ref. 220.

Table IV). Paudler and Brumbaugh's[55] improved methods yielded greater success in working with a related system, *s*-triazolo[1,5-*a*]pyrimidine (**6**).

6

S-Triazdo[1,5-*a*]pyrimidine

TABLE IV[a]. ULTRAVIOLET SPECTRAL SHIFTS OF SUBSTITUTED IMIDAZO-[1,2-*a*-] PYRIDINES RELATIVE TO THE PARENT

Compound	Longest UV wavelength (mμ)	Δm[b]	Predicted[c] shift	Observed[c] shift
Parent	336	1.185	—	—
2-Methyl	339	1.199	H	B
3-Methyl	348	1.119	B	B
5-Methyl	326	1.212	H	H
6-Methyl	340	1.177	B	B
7-Methyl	335	1.209	H	H
8-Methyl	328	1.200	H	H
5,7-Dimethyl	334	1.251	H	H
3-Bromo	343	1.163	B	B
2-Methyl-3-bromo	346	1.194	H	B
3-Bromo-5-methyl	338	1.209	H	B
3-Bromo-7-methyl	333	1.205	H	H
3-Bromo-5,7-dimethyl	338	1.248	H	B
6-Chloro	344	1.184	B	B

[a] Table taken from ref. 4.
[b] The value $\Delta m[\Delta m = \beta_0(\text{eigenvalue})]$ represents the difference between the bonding energy coefficients (eigenvalues) of the highest occupied and lowest unoccupied molecular orbitals.
[c] The H and B, as they appear in the table, represent hypsochromic and bathochromic shifts, respectively.

TABLE V. FRONTIER π-ELECTRON DENSITIES FOR NUCLEOPHILIC SUBSTITUTION[A]

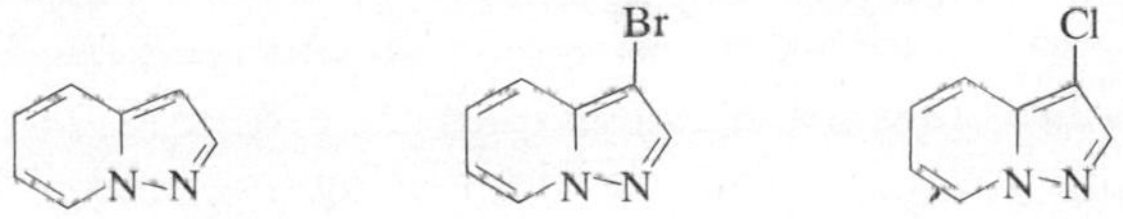

Ring position[b]	Electron densities (parent)	Electron densities (Br)	Electron densities (Cl)
1	0.116	0.118	0.118
2	0.096	0.101	0.102
3	0.011	0.010	0.009
4	0.144	0.144	0.144
5	0.650	0.647	0.647
6	0.102	0.100	0.100
7	0.308	0.310	0.310
8	0.572	0.567	0.567
9	0.002	0.001	0.001
10	—	0.002	0.002

[a] These electron densities are found for the sixth HMO—for addition of two electrons to the molecule.
[b] The halogen derivatives are numbered the same as the parent molecules with the halogen atoms receiving number 10.

TABLE VI[a]. π-ELECTRON DENSITY CORRELATION WITH CHEMICAL SHIFT POSITIONS OF THE 5-MEMBERED RING PROTONS IN IMIDAZO[1,2-*a*]PYRIDINE

Substituent	H_2(c/s)[b]	qr(hetero)[c]	H_3(c/s)[b]	qr(hetero)[c]
2-Methyl	—	—	431	1.174
5-Methyl	458	1.041	441	1.115
6-Methyl	463	1.038	455	1.115
7-Methyl	452	1.041	447	1.119
8-Methyl	457	1.038	452	1.112
5,7-Dimethyl	445	1.046	426	1.121

[a] Data taken from Ref. 1 and 4.
[b] Chemical shift positions are given in cycles per second relative to tetramethylsilane.
[c] qr = total π-electron densities. These calculations are based on the hetero-model LCAO method and the parameters used were suggested by A. Streitwieser, Jr. (see Ref. 183, p. 135).

Attempts have been made to correlate nuclear magnetic resonance (nmr) chemical shift positions of the different protons of several imidazo-[1,2-*a*]pyridines with the total π-electron densities of the respective ring positions.[1] These attempts were limited in their scope and application and only worked well for H_2 and H_3 in substituted imidazo[1,2-*a*]pyridines (see Table VI). It should be noted that these are ground state calculations, and as a result one would not necessarily expect an excellent correlation.

Recently more sophisticated techniques have been devised that accurately predict nmr chemical shift positions using quantum chemical calculations.[1,8,9,59,60,225]

Many other calculations and attempted correlations have been made including dipole moments, ionization potentials, molecular geometry, bond order superdelocalizability (for both electrophilic and nucleophilic attack), and free valancy index.[8,56,61–64,220–224] Many of the recent difficulties encountered seem to stem from a lack of understanding of the extent of sigma electron polarization.[59,60]

C. Spectra

1. *Infrared Spectra*

There appears to have been little attention given to the infrared spectra of these diazaindenes (**2**, **3**, **4**, and **5**). Several authors have routinely

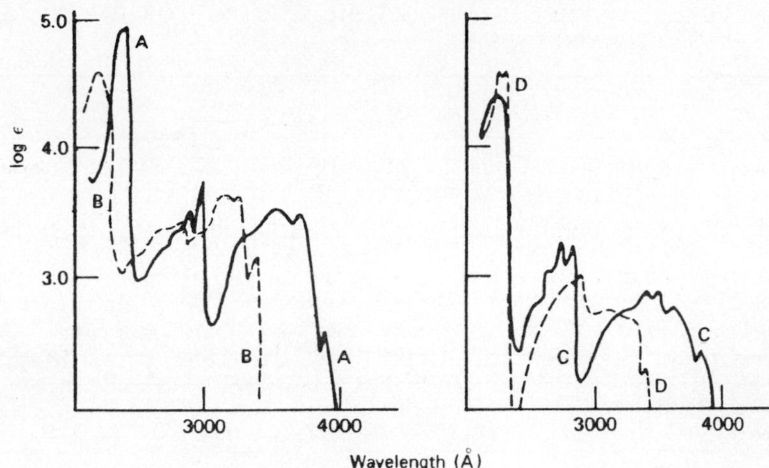

Fig. 1. Ultraviolet spectra of the parent diazaindenes: indolizine (A), imidazo[1,2-*a*]pyridine (B), imidazo[1,5-*a*]pyridine (C), and pyrazolo[1,5-*a*]pyridine (D).

recorded and reported[65,161–166,178] the infrared spectra of these compounds, but little if any effort has gone into correlating the position of the substituents with the infrared absorption spectra of these substances.

2. *Ultraviolet Spectra*

The ultraviolet spectra of imidazo[1,2-*a*]pyridine (**2**), imidazo[1,5-*a*]-pyridine (**3**), pyrazolo[1,5-*a*]pyridine (**4**), and indolizine (**5**) have been the subject of a great deal of research and are shown in Fig. 1. The spectra of indolizine (**5**) and imidazo[1,5-*a*]pyridine strongly resemble one another, exhibiting three major absorption maxima at approximately the same wavelengths.[167] The ultraviolet spectra of imidazo[1,2-*a*]pyridine (**2**) and pyrazolo[1,5-*a*]pyridine (**4**) also strongly resemble each other but are quite different from the observed spectra of imidazo[1,5-*a*]pyridine (**4**) and indolizine (**5**).[167]

Imidazo[1,2-*a*]pyridine (**2**) and pyrazolo[1,5-*a*]pyridine exhibit two major absorption maxima both centered at about the same wavelength and having similar log ϵ values[167] (see Table VII).

Detailed analysis and comparisons of the ultraviolet spectra of the parent compounds (**2**, **3**, **4**, and **5**) have appeared[66,167] as well as numerous other spectra.[2,21,122,154,156,168–170,172–175,178] These ultraviolet spectra have been utilized in structural proof[154,156,161,165,166,168,171,172] and extensively in quantum chemical calculations. Ionization and ultraviolet

TABLE VII[a]. ULTRAVIOLET ABSORPTION SPECTRA FOR THE PARENT DIAZAINDENES

Compound	Conc. (10^{-5} M)	Absorption max. (A°) and log ε (in parentheses)								
Indolizine	2.08	3815	3610	3465	3360[b]	2945	2825	2755	—	2375
		(2.92)	(3.24)	(3.29)	(3.23)	(3.56)	(3.45)	(3.34)	—	(4.51)
Imidazo[1,2-*a*]-pyridine	3.40	3360	3200	3100	—	2800	2725	2530	2285[b]	2245
		(3.10)	(3.38)	(3.42)	—	(3.41)	(3.38)	(3.15)	(4.29)	(4.41)
Imidazo[1,5-*a*]-pyridine	3.40	3785	3580	3440	3330[b]	2845	2735	2650[b]	—	2180
		(2.89)	(3.20)	(3.27)	(3.23)	(3.66)	(3.72)	(3.55)	—	(4.46)
Pyrazolo[1,5-*a*]-pyridine	3.22	3400	3230[b]	3100	3005	2890	2810[b]	2270	2225	
		(2.72)	(3.14)	(3.25)	(3.26)	(3.52)	(3.46)		(4.54)	(4.56)

[a] Data from Ref. 4 and 167.
[b] Inflection.

spectral measurements have also been used to indicate which ring position(s) undergoes protonation in an acidic medium.[154,156] This is discussed in the section dealing with ionization properties. A detailed discussion of the ultraviolet spectra of these diazaindenes and their numerous substituted derivatives is beyond the scope of this review.

3. *Nuclear Magnetic Resonance Spectra*

In recent years several groups of workers have studied the nuclear magnetic resonance characteristics of the parent substances (**2**, **3**, **4**, and **5**) and several of the substituted diazaindenes. Prior to 1965 work on the chemistry of these ring systems posed many varied and difficult problems. However the use of nmr spectroscopy in the structure determination of a large number of these diazaindenes has greatly facilitated investigations involving cyclization, substitution reactions (electrophilic and nucleophilic), deuterium exchange, and protonation of the ring systems.[4,7,107,155,176,221,222,302]

In general the agreement in interpretation of the chemical shifts and coupling constants among workers has been good with only minor differences encountered in determining accurate spin–spin coupling constants.[1,4,5,7,226]

The nuclear magnetic resonance spectra of the unsubstituted imidazo[1,2-*a*]pyridine (**2**), imidazo[1,5-*a*]pyridine (**3**), pyrazolo[1,5-*a*]pyridine (**4**), and indolizine (**5**) have been thoroughly studied. The vast majority of these compounds and their derivatives can be interpreted by first-order splitting rules including cross-ring and para spin-spin coupling. The

TABLE VIII. NMR SPECTRA OF THE PARENT DIAZAINDENES

Compound	Chemical shift (τ)							Solvent	Ref.
	H_1	H_2	H_3	H_5	H_6	H_7	H_8		
Indolizine	3.72	3.36	2.86	2.24	3.69	3.50	2.75	CCl_4	5
Imidazo[1,2-*a*]pyridine	—	2.23	2.40	1.95	3.45	3.03	2.03	$CHCl_3$	1, 4, 5, 6
Imidazo[1,5-*a*]pyridine	2.73	—	2.03	2.12	2.59	3.42	2.66	CCl_4	5
Pyrazolo[1,5-*a*]pyridine	3.62	2.20	—	1.61	3.38	3.03	2.56	$CHCl_3$	5, 7, 9

Coupling constants $J(H_2)$

Compound	1,2	1,3	1,5	2,3	2,7	2,8	3,5	3,8	5,6	5,7	5,8	6,7	6,8	7,8
Indolizine	3.9	1.2	1.0	2.7	0.0	0.0	0.0	0.5	6.8	1.0	1.2	6.4	1.0	8.9
Imidazo[1,2-*a*]pyridine	—	—	—	1.2	0.0	0.0	0.0	0.8	6.8	1.3	1.3	6.6	1.3	9.0
Imidazo[1,5-*a*]pyridine	—	0.0	0.5	—	—	—	—	1.0	7.1	0.9	1.1	6.4	1.1	9.2
Pyrazolo[1,5-*a*]pyridine	2.18	—	0.9	—	0.5	0.0	—	—	6.9	1.0	1.0	6.8	1.2	8.9

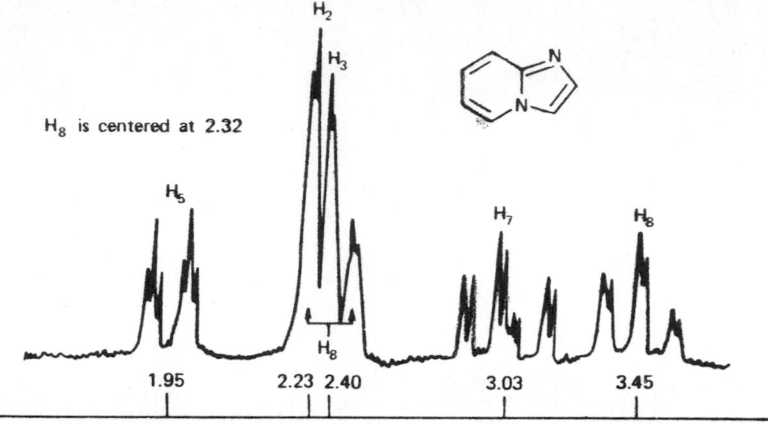

Fig. 2. Nmr spectrum of imidazo[1,2-*a*]pyridine.

chemical shifts and coupling constants for the parent compounds are listed in Table VIII. The nmr spectra of these compounds appear in Figs. 2 to 5. These systems also lend themselves admirably to establishing perianisotropy effects of the hetero atoms.[3,4]

The nmr spectra of the many diazaindenes that have been reported are too numerous to list individually. The references in this section and Tables VIII and IX can be used as guides leading to more complete

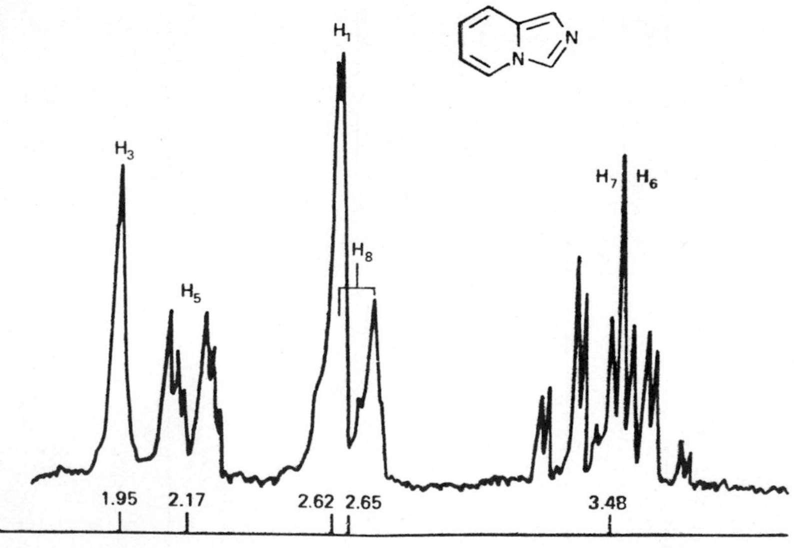

Fig. 3. Nmr spectrum of imidazo[1,5-*a*]pyridine.

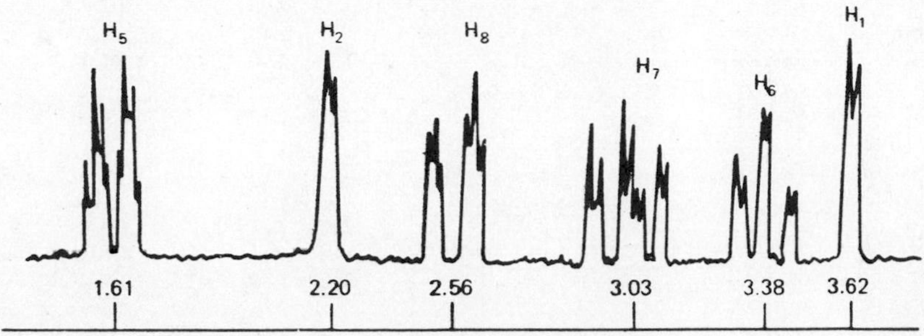

Fig. 4. Nmr spectrum of pyrazolo[1,5-*a*]pyridine.

listings. The chemical shift positions of some selected diazaindenes (substituted) appear in Table IX.

4. *Mass Spectra*

The vast majority of aromatic nitrogen heterocyclic systems that have been described may be outlined in the scheme shown below.[8] The loss of HCN seems to be the major pathway in many heterocyclic systems such

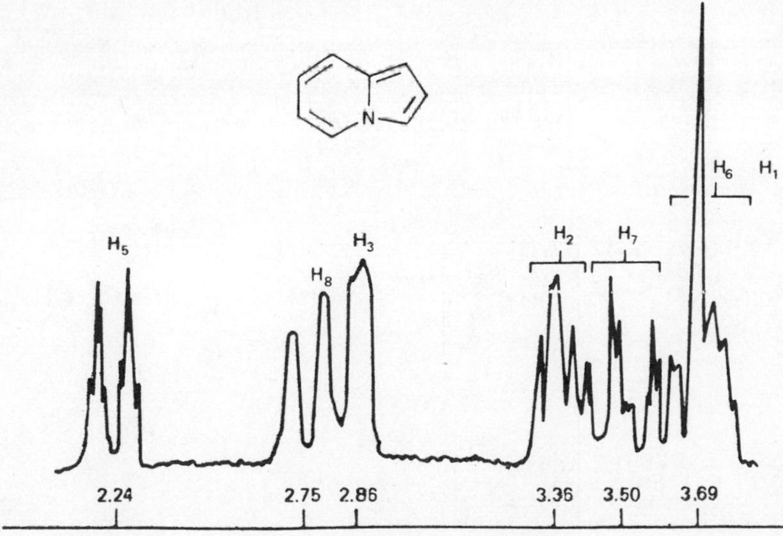

Fig. 5. Mnr spectrum of indolizine.

TABLE IX. NMR SPECTRAL DATA FOR SOME SUBSTITUTED DIAZAINDENES

Compound	Chemical shift (τ)							Solvent	Ref.
	H_1	H_2	H_3	H_5	H_6	H_7	H_8		
2-Methylimidazo[1,2-*a*]-pyridine	—	—	2.82	2.12	3.53	3.08	2.58	$CDCl_3$	1,4,6
3-Bromoimidazo[1,2-*a*]-pyridine	—	2.38	—	2.10	3.25	2.90	2.42	$CDCl_3$	1,2
3-Nitroimidazo[1,2-*a*]-pyridine	—	0.87	—	0.33	2.00	1.63	1.63	D_2SO_4	4
5-Methylimidazo[1,2-*a*]-pyridine	—	2.37	2.68	—	3.65	3.05	2.52	$CDCl_3$	1,4
1-Bromopyrazolo[1,5-*a*]-pyridine	—	2.08	—	1.62	2.88	2.72	2.52	$CDCl_3$	9
1-Nitropyrazolo[1,5-*a*]-pyridine	—	1.83	—	1.74	3.32	2.80	2.41	$CDCl_3$	9
3-Methylpyrazolo[1,5-*a*]-pyridine	2.78	1.50	—	1.05	2.30	2.04	1.79	$CDCl_3$	9
1-Methylimidazo[1,5-*a*]-pyridine	—	—	2.07	2.30	3.58	3.58	2.77	$CDCl_3$	8
1-Nitro-3-methylimidazo-[1,5-*a*]pyridine	—	—	—	1.65	2.70	2.27	1.77	$CDCl_3$	8
5-Methylimidazo[1,5-*a*]-pyridine	2.53	—	2.03	—	3.75	3.40	2.70	$CDCl_3$	8
1-Methylindolizine	—	[a]	2.95	2.12	[a]	[a]	[a]	CCl_4	5
2,6-Dimethylindolizine	4.02	—	3.16	2.52	—	3.71	2.99	CCl_4	5

[a] See Ref. 5; the multiplet was insufficiently resolved at low concentration to permit accurate determination of the chemical shift position.

$$\left[(CH)_n\text{-}CH_3\right]^{+\cdot}\ \mathbf{7}\ \xleftarrow{-HCN}\ (CH)_n(CH_3)CH{=}N^{+\cdot}\ \xrightarrow{-H}\ \left[(CH)_n(CH_2)CH{=}N\right]^{+\cdot} \longrightarrow (CH)_{n+1}CH{=}N^{+}\ \mathbf{8}\ \xrightarrow{-HCN}\ \left[(CH)_{n+1}\right]^{+}$$

$$(CH)_n(CH_3)CH{=}N^{+\cdot}\ \xrightarrow{-CH_3}\ \left[(CH)_nCH{=}N\right]^{+}\ \mathbf{9}$$

as pyridines,[179] quinolines,[180] indoles,[179] and pyrazines.[179,181] The methyl derivatives of many of these systems may either lose a hydrogen or methyl radical to yield species **8** or **9** respectively. Species **8**, a ring-expanded fragment, loses HCN when structurally possible. The presence of a nitrogen atom at the bridgehead of imidazo[1,2-*a*]pyridine (**2**), imidazo[1,5-*a*]pyridine (**3**), pyrazolo[1,5-*a*]pyridine (**4**), and indolizine (**5**) provides a structural variation of considerable interest in terms of its influence upon the fragmentation patterns of these compounds.

The mass spectra of these diazaindenes and their methyl derivatives have been thoroughly studied and in many ways are quite similar. This similarity can be easily discerned by studying Fig. 6 and 7. Each of the diazaindenes is discussed individually in the following paragraphs.

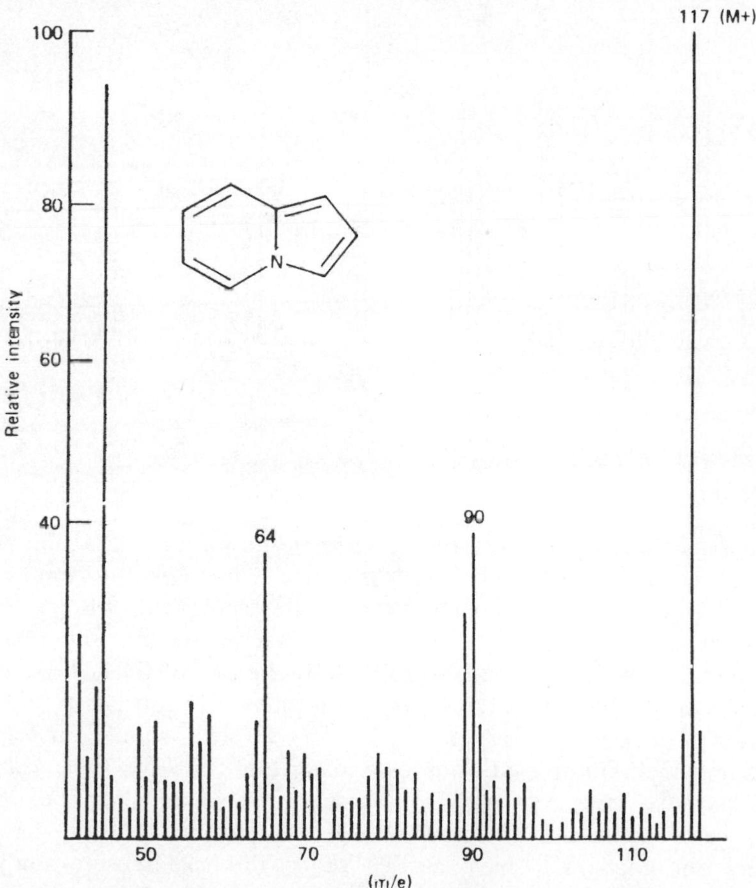

Fig. 6. Mass spectrum of indolizine.

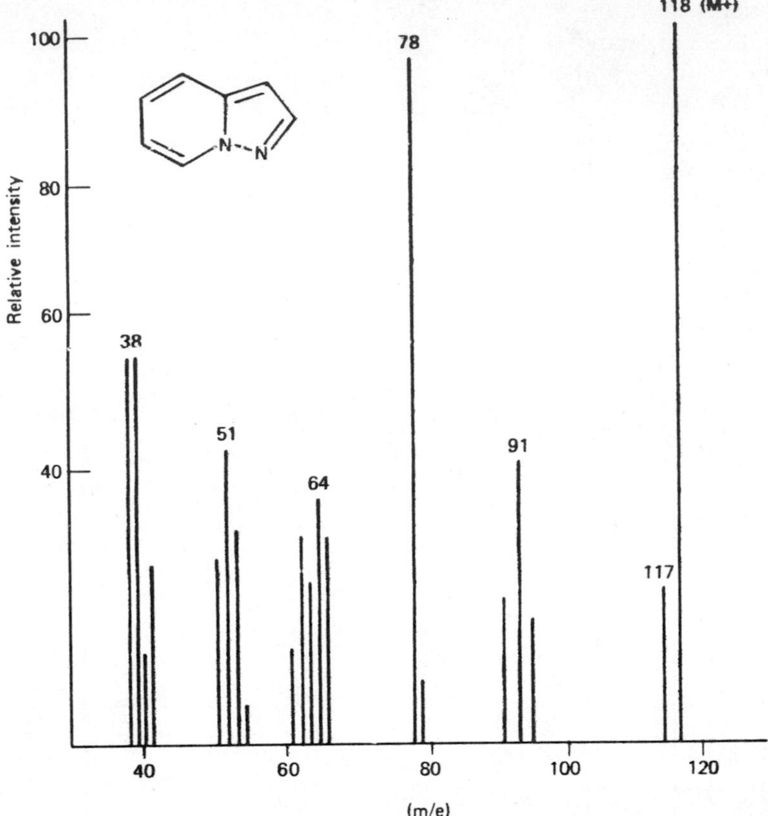

Fig. 7. Mass spectrum of pyrazolo[1,5-*a*]pyridine.

The mass spectrum of pyrazolo[1,5-*a*]pyridine indicates that the primary decomposition pathway is the loss of HCN from the molecular ion, m/e 118 (100%).[138] Another major fragment results from the loss of an additional molecule of HCN, followed by acetylene. An ion, m/e 78, corresponding to C_5H_4N was also formed by the loss of $C_2H_2N\cdot$ from the molecular ion. This type of fragmentation pattern is consistent with the behavior of related ring systems.[13,139,140,142–149] It is significant to note that Dunham's[9] interpretation is essentially the same for the pyrazolo-[1,5-*a*]pyridine ring system. The fragmentation scheme for the parent substance is shown on the following page (see Scheme 1).

Jones and Stanyer[150] have carried out a rather extensive study on indolizine and several of its methyl and dimethyl derivatives. The mass spectra is shown in Fig. 6. The molecular ion is also the base peak (m/e

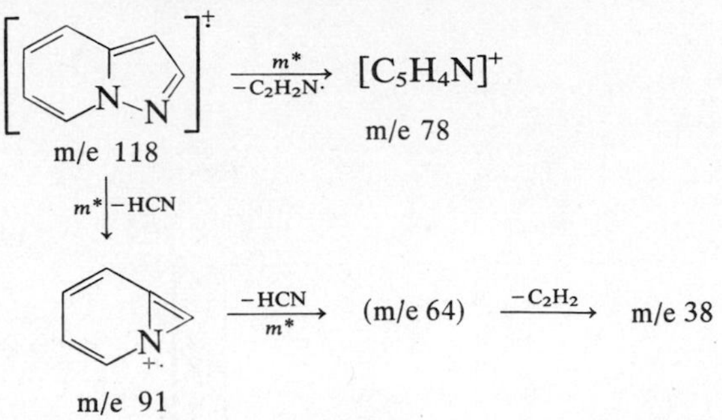
m/e 118
m^*
$-C_2H_2N\cdot$
$[C_5H_4N]^+$
m/e 78
m^* −HCN
−HCN
m^*
(m/e 64)
$-C_2H_2$
m/e 38
m/e 91

Scheme 1

117) with the first obvious loss being HCN yielding a peak m/e 90 and a second peak at m/e 89 indicating a loss of H_2CN. The next ion of high relative intensity is at m/e 64 corresponding to a loss of C_2H_2. Of much more interest are the mass spectra of the methylindolizines, as here the opportunity for ring expansion exists. The principal fragmentation pattern is shown on the following page in Scheme 2.[150]

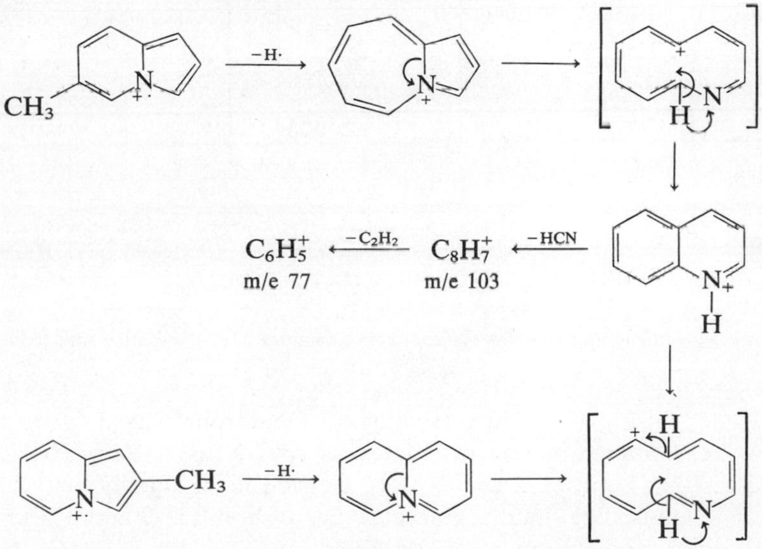
CH_3
−H·
−HCN
$-C_2H_2$
$C_6H_5^+$
m/e 77
$C_8H_7^+$
m/e 103
CH_3
−H·

Scheme 2

Scheme 3[8] shows the paths which are common to the imidazo[1,5-*a*]-pyridine and imidazo[1,2-*a*]pyridine ring systems. The loss of HCN seems to occur quite readily in these compounds. Imidazo[1,5-*a*]pyridine seems to lose HCN in two ways. The preferred direction has been determined

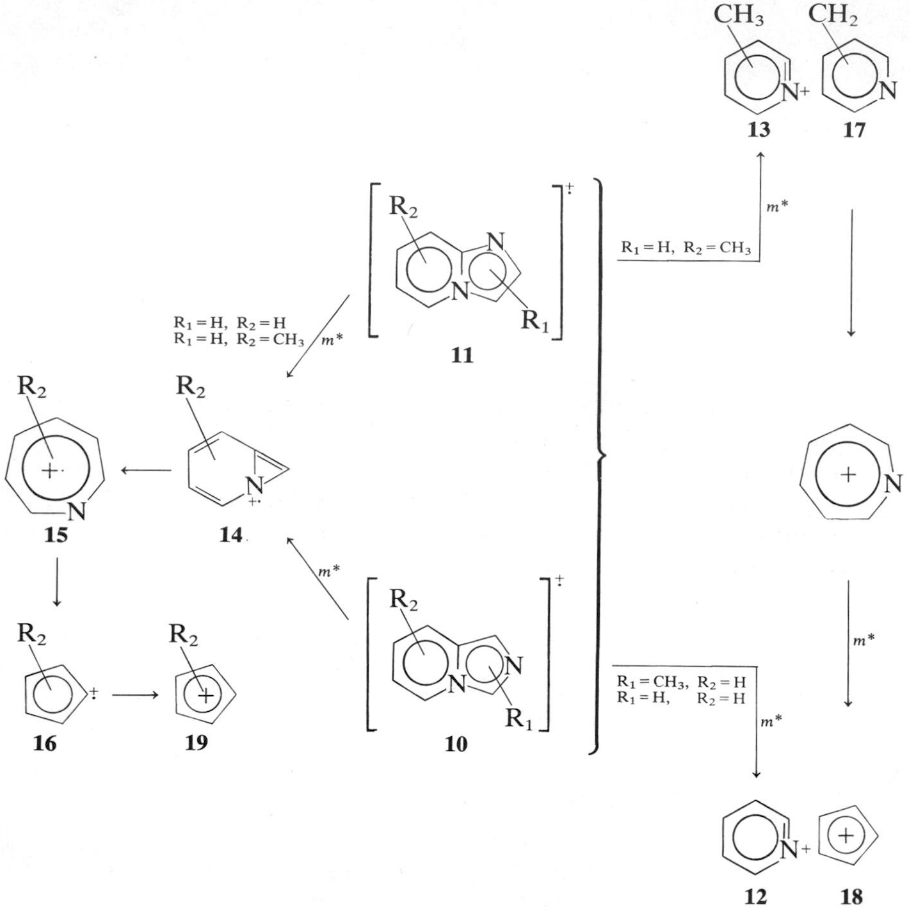

Scheme 3 Fragmentation pathways common to imidazo[1,5-*a*]pyridine and to imidazo-[1,2-*a*]pyridine

by Kuder[8] using deuterium labeling. 3-Deuteroimidazo[1,5-*a*]pyridine loses DCN approximately two times as readily as it loses HCN. This indicates that bonds 1–2 and 3–4 are cleaved more readily than the 1–9 and 2–3 bonds. The simultaneous cleavage of bonds 1–9 and 3–4 in these compounds yields fragments **12** or **13**.

D. Ionization Properties

1. *Basic Strength*

The pK_a values of the weakly basic diazaindenes vary over a rather wide range (see Table X). The basic strength of these compounds seems to depend on both the number and position of the ring nitrogens. It has been suggested[156] that resonance forms **2a**, **2b**, **3a**, **3b**, **4a**, and **4b** make significant contributions to the overall resonance hybrid of these

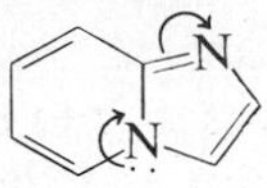

2 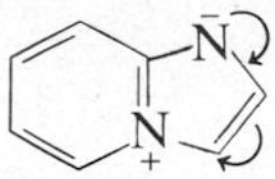 **2a** 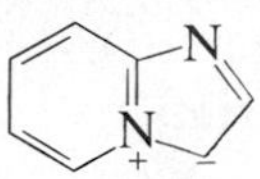 **2b**

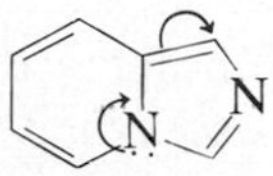

3 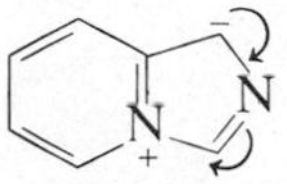 **3a** 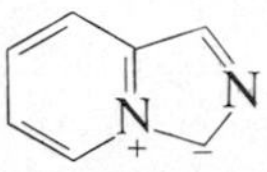 **3b**

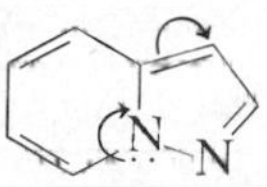

4 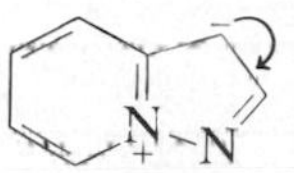 **4a** 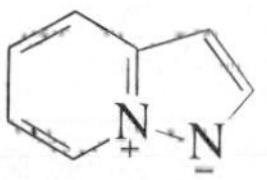 **4b**

heteroaromatic systems. Data by Grant and co-workers tend to strongly support this thesis.[226] Imidazo[1,2-*a*]pyridine (**2**) is a stronger base than imidazo[1,5-*a*]pyridine because of the increased availability of electrons on N_1 as shown by **2a**. The fact that pyrazolo[1,5-*a*]pyridine is the weakest base must be rationalized in terms of structure **4b** which seems to imply a rather high electron density at N_3. This can however be easily justified by considering the close proximity of a positively charged nitrogen which undoubtedly exerts a strong inductive effective on the electrons on N_3 in pyrazolo[1,5-*a*]pyridine. Indolizine must be considered as a separate case because of the protonation which occurs not on nitrogen but on carbon in the 5-numbered ring. The protonation of indolizine is discussed in the section dealing with protonation and *N*-methylation. In general substituted ring systems vary in basic character precisely as one would predict depending on the electron donating or withdrawing power of the substituent (see Table X).

2. *Site of Protonation and N-Methylation*

Protonation and *N*-methylation has been the subject of a significant amount of research in the diazaindene series. In fact several studies of this type have been devoted to demonstrating the aromatic character and planarity of the ring system.

Protonation and *N*-methylation in imidazo[1,2-*a*]pyridine occur at the 1-position, a fact that is predicted by frontier electron density calculations (see Table III).[3,4] Paudler and Blewitt[3] were able to establish the position of protonation and methylation by utilizing nmr spectroscopy to study the peri-type interaction that exists between the 8-position and the methyl group substituted on N-1. Work published by Bradsher, Litzinger, and Zinn[72] also confirmed the position of methylation in the imidazo[1,2-*a*]pyridine series. These workers showed that the cation, **22**, produced from 2-methylaminopyridine (**20**) by the reaction with bromoacetone was identical with that obtained by methylation of 2-methylimidazo[1,2-*a*]pyridine (**21**).[151]

NH—CH_3 + $BrCH_2COCH_3$

20

CH_3Br

21

22

The research described above shows the correctness of earlier conjecture[151–153,156] concerning the position which quaternization occurs in the imidazo[1,2-*a*]pyridine system.

TABLE X. IONIZATION CONSTANTS OF SOME DIAZAINDENES

Compound	pK_a	Ref.
Imidazo[1,2-*a*]pyridine	$5.06^a (6.79)^c$	3,156
5-Methylimidazo[1,2-*a*]pyridine	5.55^a	3
5,7-Dimethylimidazo[1,2-*a*]pyridine	5.96^a	3
Imidazo[1,5-*a*]pyridine	5.54^c	150
Pyrazolo[1,5-*a*]pyridine	1.43^c	156
Indolizine	$2.63^b (3.94)^c$	36,156
1-Methylindolizine	3.60^b	36
2,7-Dimethylindolizine	5.26^b	36

[a] In 90% ethanol–10% H_2O, obtained by extropolating several determinations to zero concentration.
[b] Determined in 60% ethyl alcohol.
[c] See Ref. 156.

Treatment of pyrazolo[1,5-*a*]pyridine with methyliodide in a variety of solvents failed to yield a crystalline methiodide.[9] Heating an equimolar mixture of pyrazolo[1,5-*a*]pyridine and methyl iodide in the absence of solvent (sealed tube) at 100°C yielded a nearly quantitative amount of the pure methiodide (**23**).

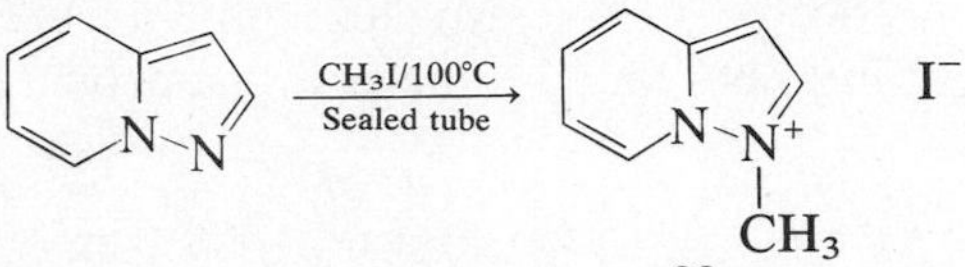

CH_3I/100°C
Sealed tube

I^-

CH_3

23

The difficulty encountered in the preparation of the methiodide is undoubtedly due to the relatively weakly basic nature of pyrazolo[1,5-*a*]-pyridine.[154] Protonation is also reported to occur at the 3-position.[154]

Armarego[154,156] concluded on the basis of ultraviolet spectral studies that protonation of imidazo[1,5-*a*]pyridine occurs at the 2-position (non-bridgehead nitrogen). Kuder[8] using nuclear magnetic resonance spectroscopy confirmed this hypothesis. Kuder demonstrated that imidazo[1,5-*a*]-pyridine formed an *N*-methyl derivative (**24**) at the 2-position, by showing that the nuclear magnetic resonance spectrum of the parent methiodide (**24**) in D_2O was essentially superimposable upon that of the parent compound (**3**) in acid solution.

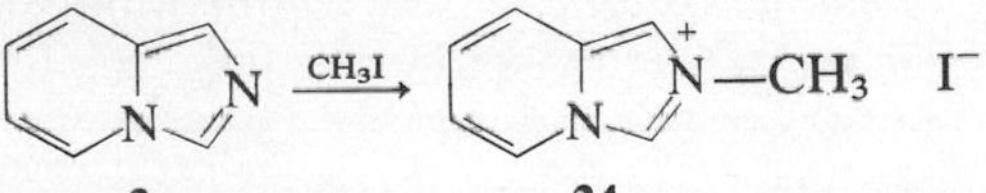

CH_3I

N—CH_3 I^-

3 **24**

Treatment of indolizine with methyliodide can yield a wide variety of products substituted in the 5-membered ring depending both on the conditions used and the substituents present in the 1,2, or 3 ring positions of indolizine. This type of reaction for indolizine has been thoroughly reviewed by Mosby.[17] The structure of indolizinium perchlorate and several of its methyl derivatives has been determined by Fraser, Melera, Malloy, and Reid[155,158] utilizing nmr spectroscopy in trifluoroacetic acid. Protonation of all indolizines studied occurred preferentially at position 3, as illustrated by compound **25**.

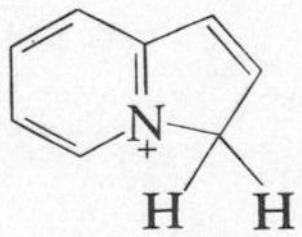

H H

25

Armarego[156,157] also found that indolizine was protonated predominately at the 3-position. However, apparently for steric reasons, the 3-methyl-, 2,3-dimethyl-, and 3,7-dimethyl indolizines were protonated on C-1. In 3,5-dimethyl indolizine steric repulsion of a proton seems to be overcome by the steric crowding of the 3 and 5 methyl groups. The protonation thus occurs on C-3, which results in a significant decrease in the steric crowding between the methyl group on C-3 and the methyl group on C-5.

III. Syntheses

A. Preparation of Imidazo[1,2-*a*]Pyridines

The unsubstituted compound was first prepared by Tschitsbabin[18] by reacting bromoacetaldehyde with 2-aminopyridine in a sealed tube at 150–200°C. Imidazo[1,2-*a*]pyridine may also be synthesized in high yield under less drastic conditions by refluxing 2-aminopyridine and bromo- or chloroacetaldehyde in aqueous-alcoholic or dioxane solution in the presence of sodium bicarbonate.

The best general method of preparation of substituted imidazo[1,2-*a*]-pyridines is the condensation of substituted 2-aminopyridines with α-haloaldehydes or α-haloketones. Nearly quantitative yields are often obtained. The general scheme is shown below.

$$\text{2-aminopyridine } (NH_2) + R'\text{—}\underset{\displaystyle X}{\underset{|}{C}}HCOR'' \longrightarrow \text{imidazo[1,2-}a\text{]pyridine (2-}R''\text{, 3-}R'\text{)}$$

(X = Br or Cl)

Schmidt and Grundig[53] had reported earlier that 2,6-diaminopyridine failed to yield the 5-aminoimidazo[1,2-*a*]pyridine. However more recently Paolini and Robins[6] have successfully condensed 2,6-diaminopyridine with α-bromoacetaldehyde to yield the 5-amino compound. There has been only one substituted 2-aminopyridine reported which does not undergo this very facile condensation. Attempts to prepare 5-hydroxyimidazo[1,2-*a*]pyridine by the condensation of 2-amino-6-hydroxypyridine with bromoacetaldehyde were unsuccessful.[6] This is probably due to the decreased basic character of the ring nitrogen caused by the keto-enol (**26** $\rightleftharpoons$ **27**) equilibrium.

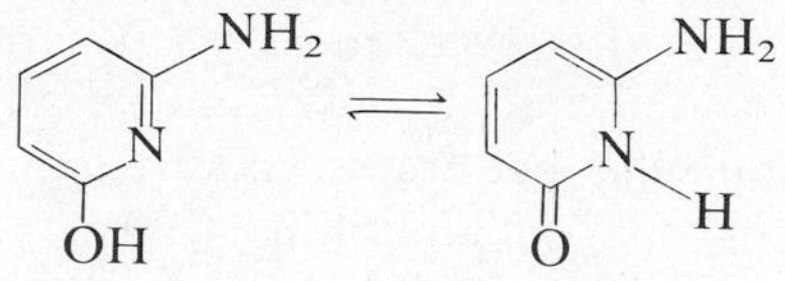

26 27

Steric effects also appear to play a prominent role in this condensation, although this problem has never been thoroughly investigated. When 2-aminopyridine is condensed with phenacyl bromide, 3-bromobutanone, or desyl chloride (PhCOCHClPh) the compounds produced were 3-phenylimidazo[1,2-*a*]pyridine, 2,3-dimethylimidazo[1,2-*a*]pyridine, and 2,3-diphenylimidazo[1,2-*a*]pyridine in yields of 88.0% 41.4%, and 38.0%, respectively.[6,69]

In an earlier volume of the series by Mosby,[17] several other synthetic methods for substituted imidazo[1,2-*a*]pyridines are discussed in detail. These methods are summarized below (Schemes 4, 5, and 6).

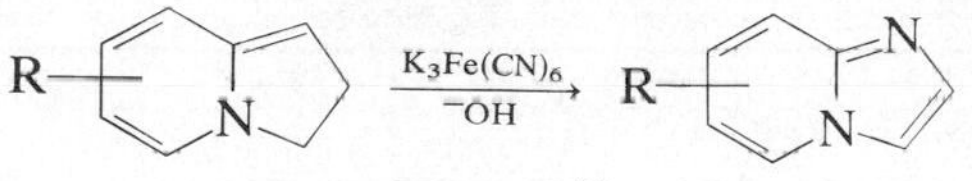

$K_3Fe(CN)_6$, ^-OH

Scheme 4[22]

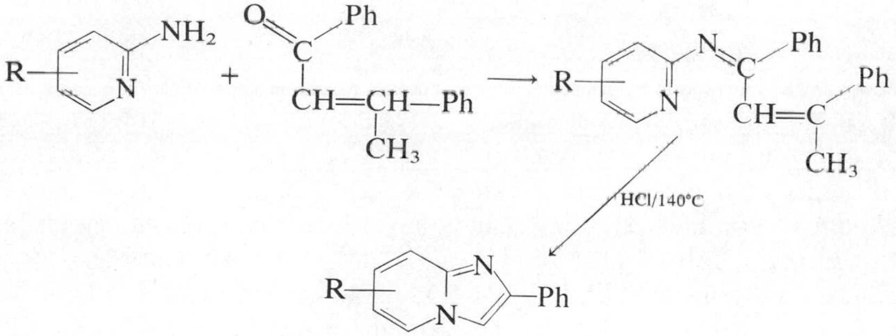

HCl/140°C

Scheme 5[70]

$2CH_3O_2CC{\equiv}CCO_2CH_3$ +

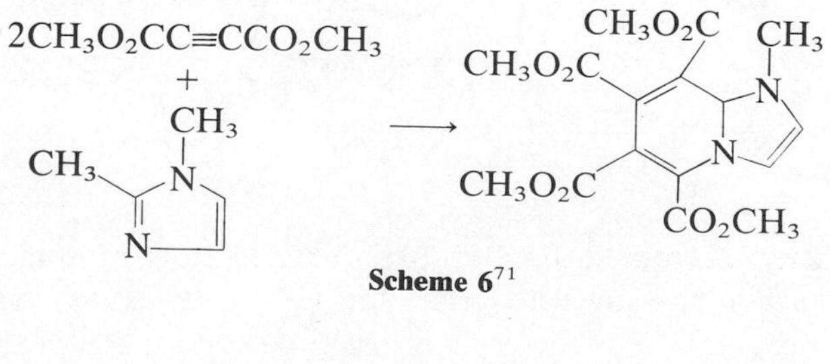

Scheme 6[71]

In 1965 Bradsher[72] and co-workers devised a rather straightforward one-step synthesis of imidazo[1,2-*a*]pyridinium salts. This method provides access to several salts that are essentially impossible to produce by any other method. When 2-anilopyridine (**28**) is allowed to react with bromoacetone in refluxing acetone, 1-phenyl-2-methylimidazo[1,2-*a*]-pyridinium (**29**) is produced. To demonstrate the general nature of the

NH—Ph + $BrCH_2COCH_3$ ⟶ Ph, CH_3

28 **29**

reaction Bradsher condensed 2-methylaminopyridine (**30**) and 2-benzylaminopyridine (**31**) with α-haloketones and was able to produce several imidazo[1,2-*a*]pyridinium cations (**32**) in yields ranging from 60 to 97%.

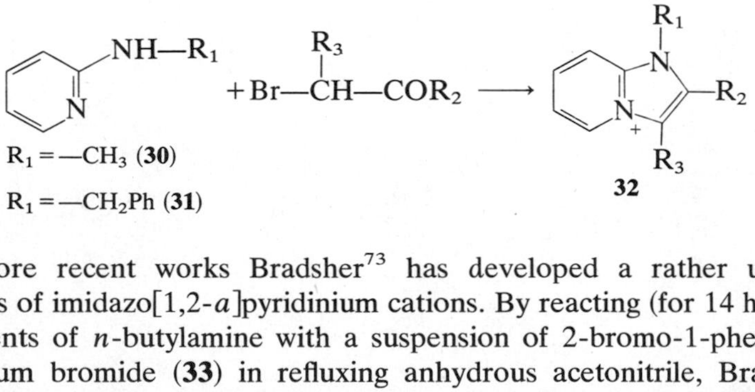

$NH—R_1$ + Br—CH(R_3)—COR_2 ⟶ R_1, R_2, R_3

R_1 = —CH_3 (**30**)

R_1 = —CH_2Ph (**31**)

32

In more recent works Bradsher[73] has developed a rather unique synthesis of imidazo[1,2-*a*]pyridinium cations. By reacting (for 14 hr) two equivalents of *n*-butylamine with a suspension of 2-bromo-1-phenacyl-pyridinium bromide (**33**) in refluxing anhydrous acetonitrile, Bradsher obtained the 1-butyl-2-phenylimidazo[1,2-*a*]pyridinium cation (**34**). It is significant to note that if the reaction is interrupted after 10 min of refluxing, 2-phenyloxazolo[3,2-*a*]pyridinium cation (**35**) and 1-butyl-2-

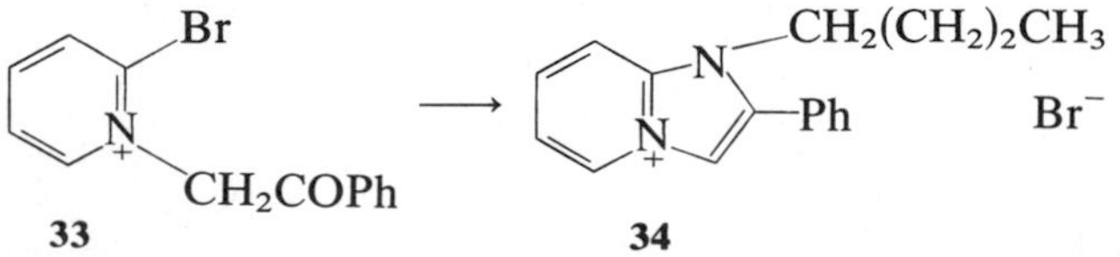

Br, CH_2COPh ⟶ $CH_2(CH_2)_2CH_3$, Ph, Br^-

33 **34**

hydroxy-2-phenyl-2,3-dihydroimidazo[1,2-*a*]pyridinium bromide (**36**) are obtained. Bradsher has suggested that the failure to isolate any of the

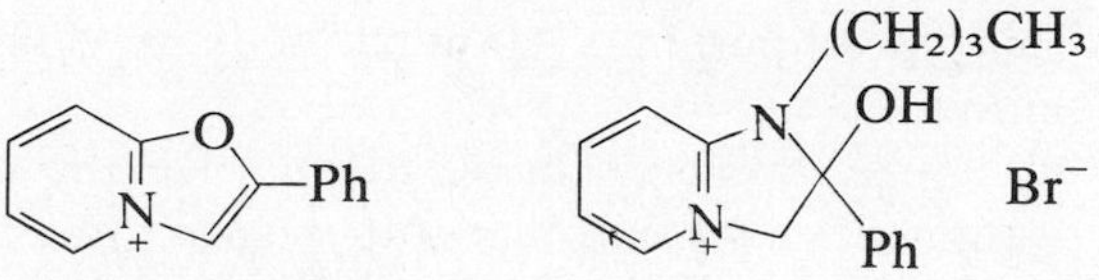

35

36

oxazolo[3,2-*a*]pyridinium cation (**35**), when the reaction was allowed to proceed for 14 hr, is apparently due to ring opening that occurs when **35** is heated with *n*-butylamine. Improved yields of oxazolopyridinium salts are obtained when tertiary amines are substituted for *n*-butylamine.

Recently a rather unusual synthesis of substituted imidazo[1,2-*a*]-pyridines was carried out by Paudler, Van Dahm, and Park. 1,4-Diaza-cycl[3,2,2]azines are decomposed in acid solution to produce substituted imidazo[1,2-*a*]pyridines.[227] Other syntheses of imidazo[1,2-*a*]pyridines have recently been reported. However their usefulness as general methods of preparation seem to be quite limited and thus the details are not discussed in this Chapter.[303,308]

B. Preparation of Imidazo[1,5-*a*]Pyridines

The synthesis of the parent substance and several derivatives was first carried out by Bower and Ramage[20] by the cyclization of substituted 2-aminomethylpyridines (**37**) in the presence of phosphorylchloride.

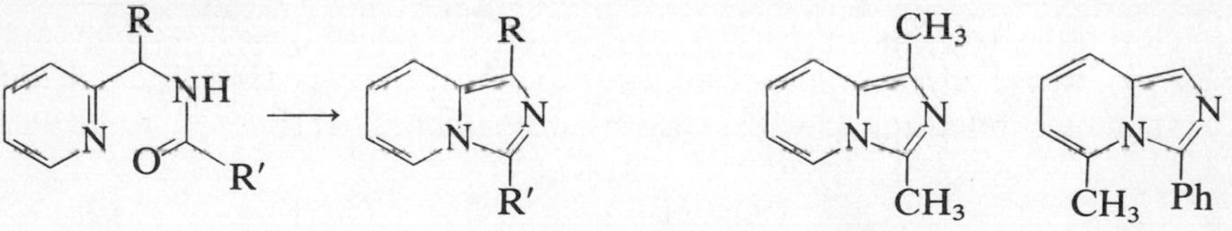

37

38

39

Mosby[17] has reviewed this synthesis as well as the synthesis of several related, partially saturated, systems. More recently Paudler and Kuder[8] have synthesized several alkylderivatives via this method and generally obtained yields ranging from 58 to 83%. Only compounds **38** and **39** of these derivatives attempted were obtained in yields less than 50%. Their respective yields were 8% and 37%. No reasons were given for the rather low yields in these syntheses. One might speculate that compound **39** poses synthetic difficulties because of the "peri-type" steric interaction that exists between the methyl group on the 5-position and the 3-phenyl substituent. This type of peri interaction has been previously documented in related compounds.[3]

Very few attempts have been made to improve the synthesis of the imidazo[1,5-*a*]pyridines and only one other successful method has recently been employed. Boyer and Wolford[117] succeeded in synthesizing 1-phenylimidazo[1,5-*a*]pyridine utilizing the Leuckart reduction.

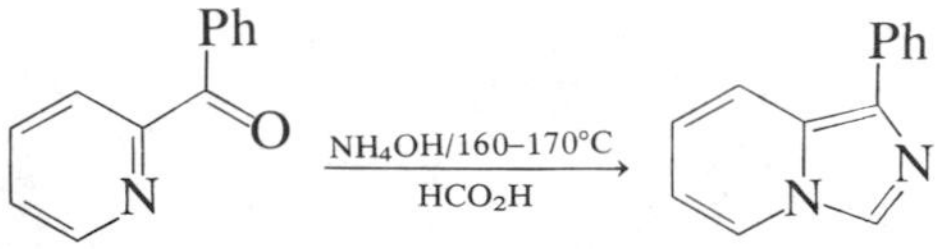

$$\xrightarrow[\text{HCO}_2\text{H}]{\text{NH}_4\text{OH}/160\text{–}170°\text{C}}$$

C. Preparation of Pyrazolo[1,5-*a*]Pyridines

The first synthesis of pyrazolo[1,5-*a*]pyridine was accomplished by Bower and Ramage.[21] Oxidation of 2-(2-aminoethyl)pyridine (**40**) with alkaline ferricyanide solution produced the parent substance (**4**) in approximately 24% yield. This reaction has been briefly reviewed by Mosby.[17]

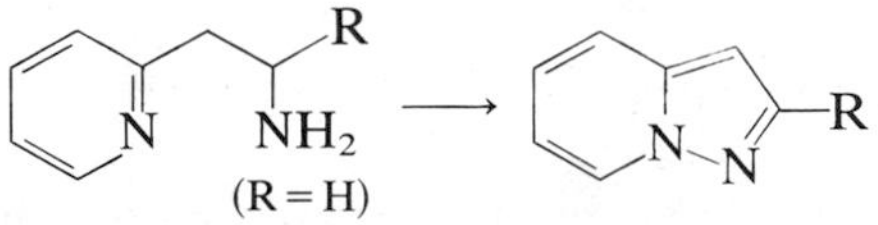

(R = H)

40 → **4**

Several other less practical syntheses of pyrazolo[1,5-*a*]pyridines are also known. Huisgen and co-workers[74] synthesized two derivatives of the system by condensing **41** with substituted acetylenes (**42**).

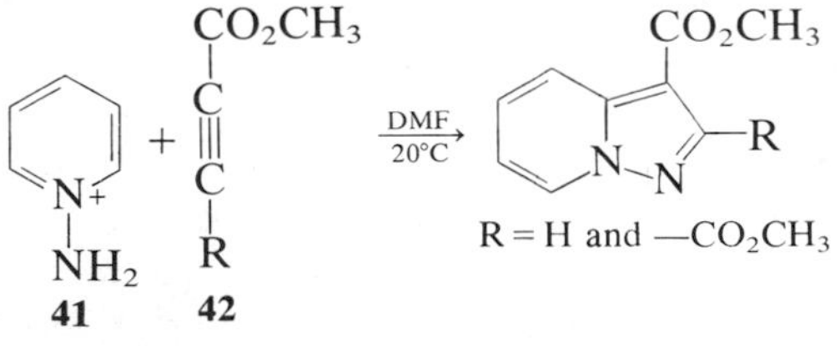

41 + **42** $\xrightarrow[20°\text{C}]{\text{DMF}}$

R = H and $-CO_2CH_3$

A similar synthesis has been carried out by Sasaki, Kanematsu, and Yukimoto.[76] The reaction of 1-aminopyridinium iodide (**41**) with cyanoacetylene in dimethyl formamide at room temperature produced 1-cyanopyrazolo[1,5-*a*]pyridine.

Rees and Yelland[75] found that oxidation of 4-substituted-3-phenyl-pyrazolin-5-ones (**43**) with lead tetra-acetate in methylene chloride containing tetraphenylcyclopentadiene gave good yields of the Diels–Alder adducts (**45**) and of the unstable pyrazolones (**44**). When the Diels–Alder adducts (**44**) were heated at 210°C/1 mm for 16 hr, carbon dioxide was evolved, and the pyrazolo[1,5-*a*]pyridines (**46**) were formed in 30 to 45% yield.[228] This represents a rather unusual rearrangement, since it involves

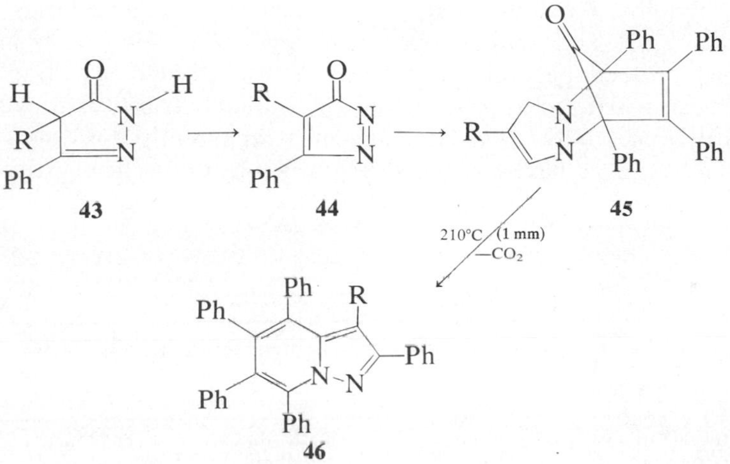

43 → **44** → **45**

210°C (1 mm), $-CO_2$

46

the elimination of carbon dioxide from two nonadjacent carbonyl groups.

Taylor and Hartke[77] have also been successful in synthesizing a pyrazolo[1,5-*a*]pyridine derivative. The reaction of 3-cyanomethyl-4-cyano-5-aminopyrazole (**47**) with acetylacetone in the presence of potassium hydroxide yielded 2-cyanomethyl-3-cyano-5,7-dimethyl-pyrazolo-[2,3-*a*]pyrimidine (**48**) in 40% yield and a minor product (20%) identified as 1,8-dicyano-2-amino-5,7-dimethylpyrazolo[1,5-*a*]pyridine (**49**). This reaction also seems to work equally well with ethyl acetoacetate.

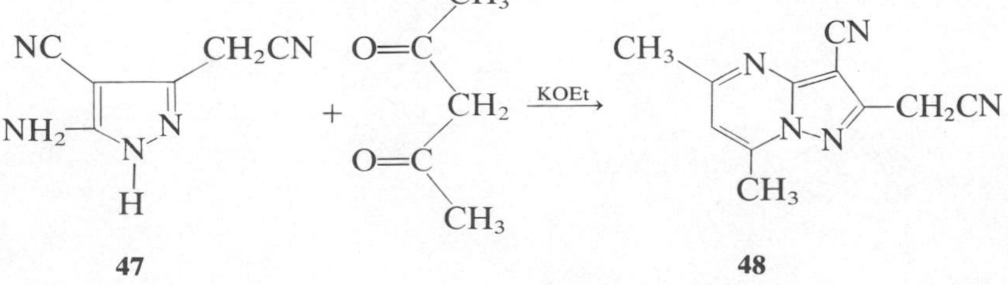

47 + $CH_3COCH_2COCH_3$ $\xrightarrow{KOEt}$ **48**

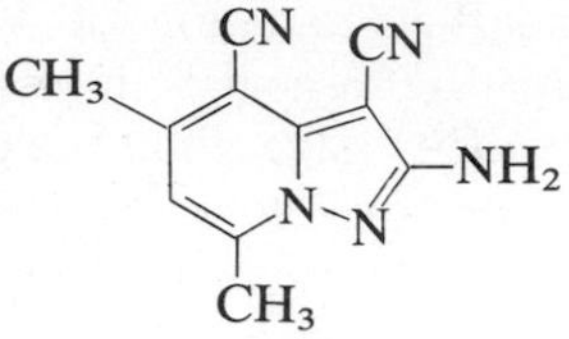

49

Duerr and Sergio have been successful in obtaining certain pyrazolo-[1,5-*a*]pyridine derivatives through a sequence of rather unusual reactions.[230,305–307] The spirocyclopentadienepyrazoles (**49a**), prepared by reacting diazocyclopentadienes with certain disubstituted acetylenes, once formed can rearrange to produce the appropriately substituted pyrazolo-[1,5-*a*]pyridines (**49b**). This rearrangement apparently does not always occur and seems to have a rather strong dependence on the substituents.

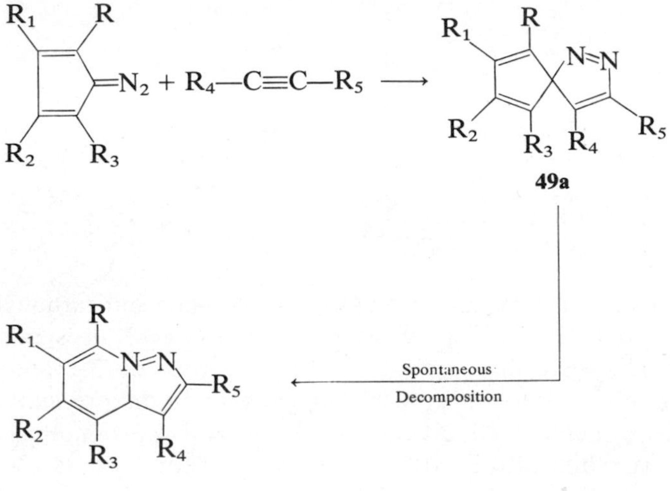

$=N_2 + R_4—C{\equiv}C—R_5 \longrightarrow$

49a

Spontaneous Decomposition

In another rather unusual synthesis Michalska found that the acetate, tosylate or diphenylphosphoryl ester of isonitrosoflavanone (**49c**) gave a substituted pyrazolo[1,5-*a*]pyridine (**49d**) when heated with pyridine. The yield varied from 30 to 85% depending on the ester used.[231]

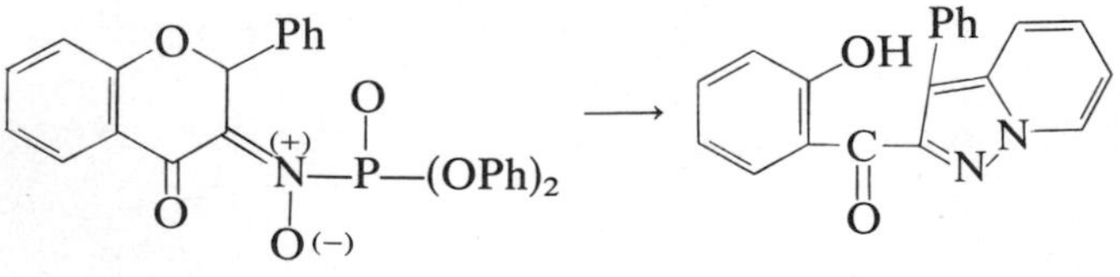

49c **49d**

Tamura and co-workers found that N-(β-acylvinyl)iminopyridium betaines cyclize in refluxing toluene to produce pyrazolo[1,5-*a*]pyridines. This intramolecular cyclization of ylides was also extended to the synthesis of indolizine derivatives.[232,299] Other workers have also been successful in utilizing a ylide in achieving the synthesis of pyrazolo[1,5-*a*]-pyridines.[233]

Recently Chaplea and Dreiding have also succeeded in preparing pyrazolo[1,5-*a*]pyridine via a new method.[305] When bicyclo[4.1.0]hept-3-ene-2,5-dione monoethylene acetal was treated with HCl and the resulting homo-*p*-quinone (**49e**) was treated with *p*-$MeC_6H_6SO_2NHNH_2$, **49f** was produced. When **49** was heated with sodium methoxide pyrazolo-[1,5-*a*] was produced.

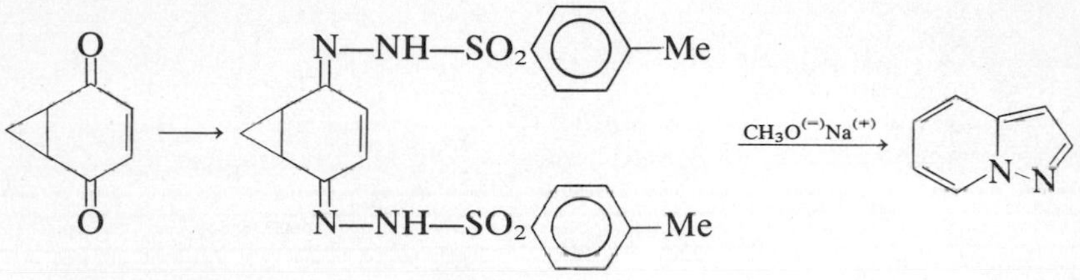

$CH_3O^{(-)}Na^{(+)}$

49e **49f**

D. Preparation of Indolizines

The most widely employed synthesis of indolizines[17] is that first devised by Tschitschibabin.[19] This scheme which employs the condensation of 2-picoline (**50**) or its derivatives with haloketones (**51**) is similar to Tschitschibabin's first synthesis of imidazo[1,2-*a*]pyridine.

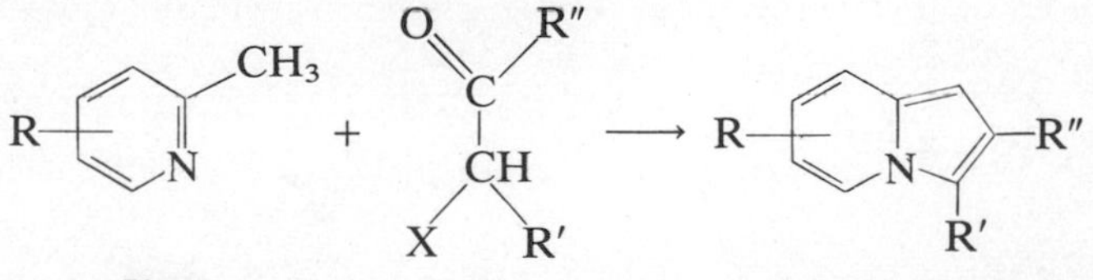

50 **51**

In an earlier volume of the series[17] several other synthetic methods for indolizines are discussed in detail. The most important of these synthetic methods are summarized below (Schemes 7–11).

Several other syntheses of indolizines have also been developed. Bragg and Wibberley[92] have synthesized ethyl alkyl and arylindolizine-1-carboxlates by the reaction of α-haloketones with ethyl 2-pyridylacetate

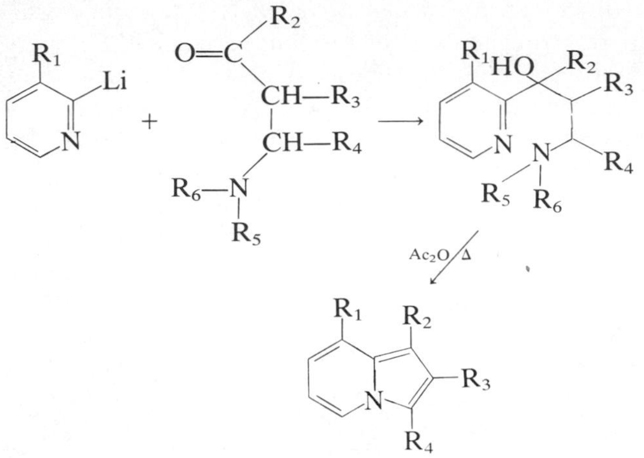

Scheme 7[39,78–80]

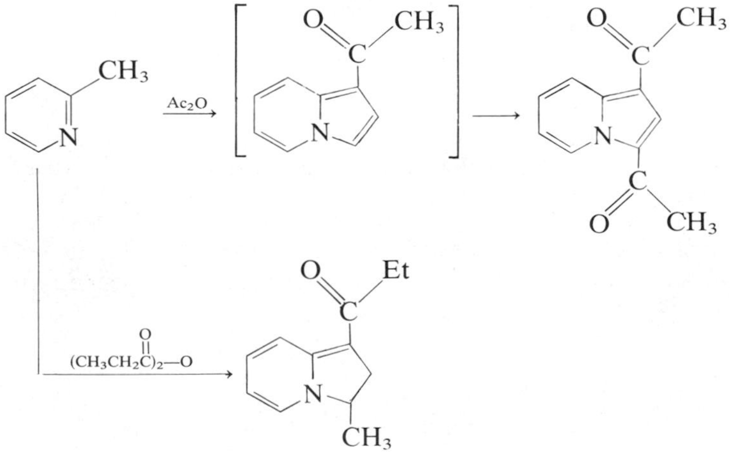

Scheme 8[27–29,50,80–84]

(**52**). When this reaction is carried out under refluxing conditions in ethylether, ethyl 2-pyridylacetate hydrobromide (**53**) separates instead of the quaternary salt and after filtration and concentration of the filtrate ethyl 2-phenylindolizine-1-carboxylate (**54**) crystallizes. The authors suggested that part of the ester may behave as a base, removing hydrogen bromide and causing ring closure. The structure of the products were confirmed by hydrolysis and decarboxylation to alkyl- and arylindolizines. The yield was greatly improved by using 2 moles of ester and still further

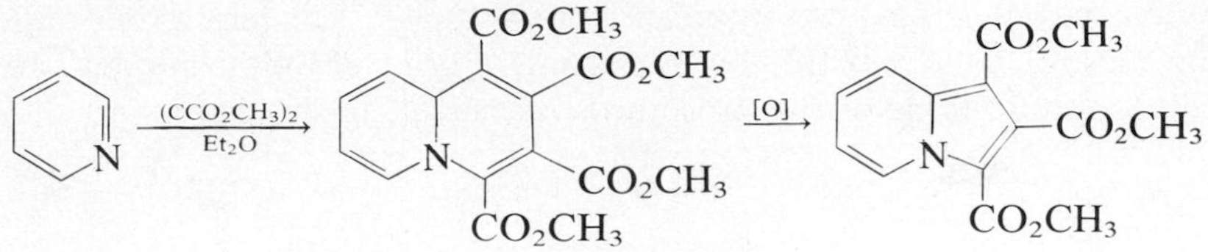

CO_2CH_3 CO_2CH_3 $(CCO_2CH_3)_2$ Et_2O [O] CO_2CH_3 CO_2CH_3 CO_2CH_3 CO_2CH_3 CO_2CH_3

Scheme 9[28,30,51,85–88]

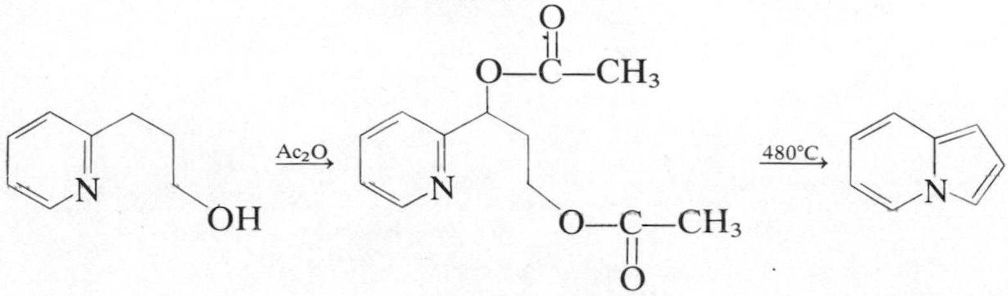

Ac_2O OH 480°C

Scheme 10[31]

$(CH_3COCH_2)_2$ $CH_3CH_2OH/HCl/0°C$ $HOAc/Zn(OAc)_2$ Δ

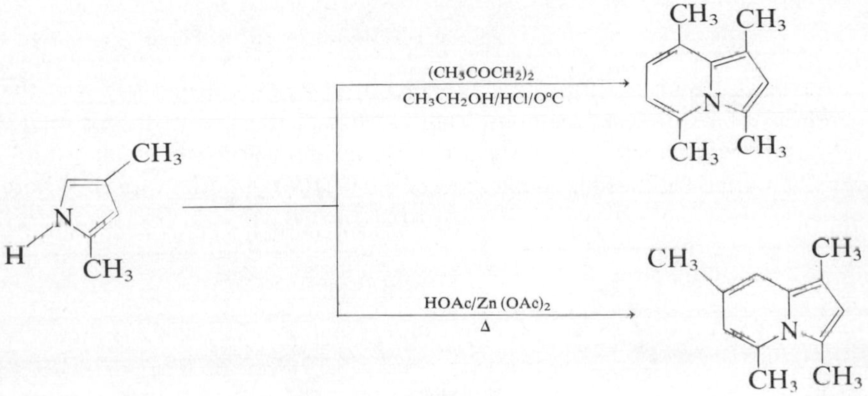

Scheme 11[89–91]

by using acetone. These same workers[92,93] have also extended the synthesis of indolizidnes to the use of other 2-picolyl derivatives and α-haloketones. They synthesized directly indolizines with ester, acyl, and cyano groups in positions 1 and/or 3. The first detailed mechanism for the formation of indolizines was also proposed. This proposed mechanism is shown on the following page. Danis has prepared some very unusual

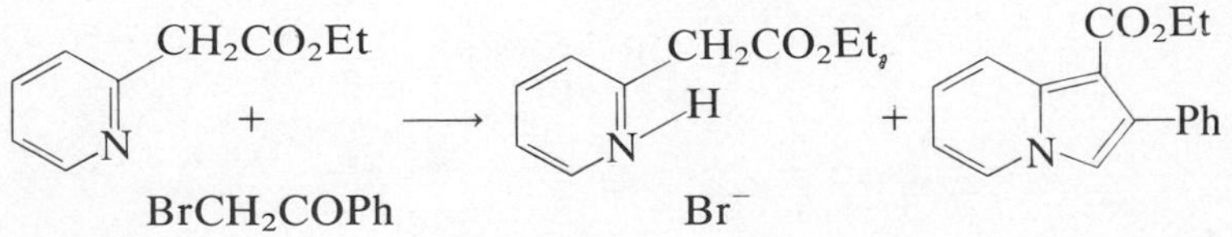

CH_2CO_2Et + $BrCH_2COPh$ ⟶ CH_2CO_2Et H Br^- + CO_2Et Ph

52 **53** **54**

indolizines utilizing a closely related method.[234–236] Several other authors have also been successful in synthesizing a series of indolizine derivatives using modifications of the procedures discussed in this paragraph.[237–239]

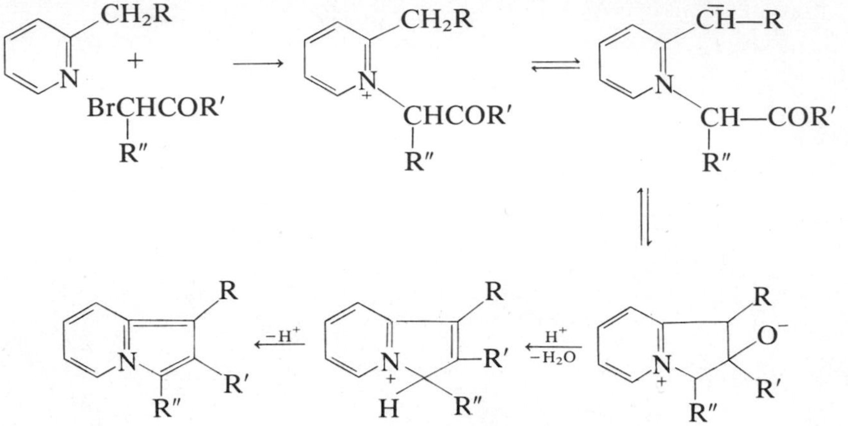

Wibberley and Melton[95,96] have recently studied a method for the synthesis of indolizines involving ring closure at the 2–3 positions of the indolizine ring system. The cyclization involved an intramolecular "aldol-type" condensation of a suitable acyl ethoxycarbonylmethine (**59**). Where

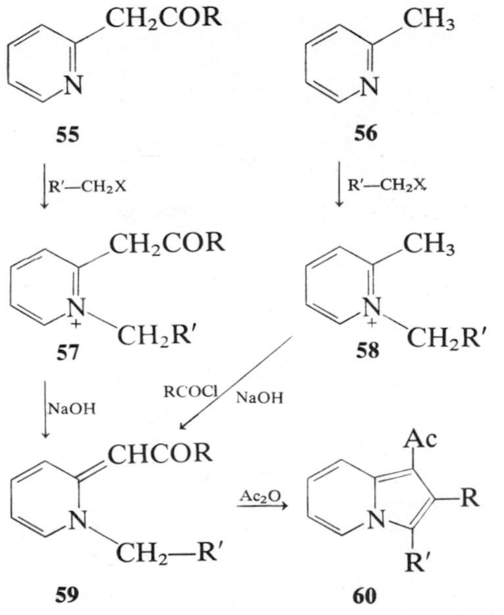

55 **56**

57 **58**

59 **60**

the required methine could be prepared by simultaneous acylation and dehydrahalogenation of a suitable 3-methylpyridinium bromide (**58**) the overall process gave a rapid synthesis from α-picoline (**56**). This route to the methines was not applicable in the majority of cases thus necessitating an approach via the quaternary salts of the less readily accessible substituted picolines. The overall reaction scheme is shown below.

Several authors[97–99,240,241,301] have been successful in using pyridinium ylides in the synthesis of indolizines. Treatment of pyridinium phenacylid (**61**) in refluxing toluene in the presence of Pd/C for 20 hr with dimethylacetylenedicarboxylate yields dimethyl-3-benzoylindolizine-1,2-dicarboxylate (**62**) in 18% yield.[98] Better yields (32%) were obtained by generating the ylid from the salt in dimethylformamide by the addition of sodium hydride,[100,101] and then adding the ester when the exothermic reaction begins.

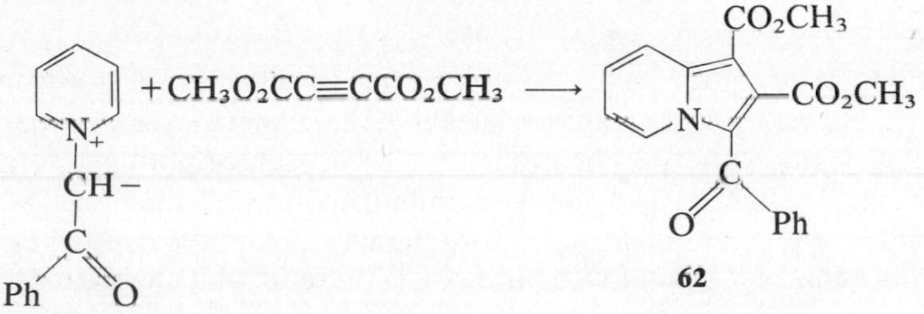

$$\mathbf{61} + CH_3O_2CC{\equiv}CCO_2CH_3 \longrightarrow \mathbf{62}$$

Several other workers[102–104] have also studied the synthesis of indolizines using reactions of substituted acetylenes, such as diethylacetylenedicarboxylate, cyanoacetylenes, and cyanochloro acetylenes, with pyridinium phenacylides.

Acheson and Robinson[105] have also successfully synthesized several indolizines. They discovered that pyridine and its 3- and 4-methyl and 3,5-dimethyl derivatives when condensed with methyl propiolate gave indolizines (**63**) and cyclic (3,2,2) azine esters (**64**) of the type shown below.

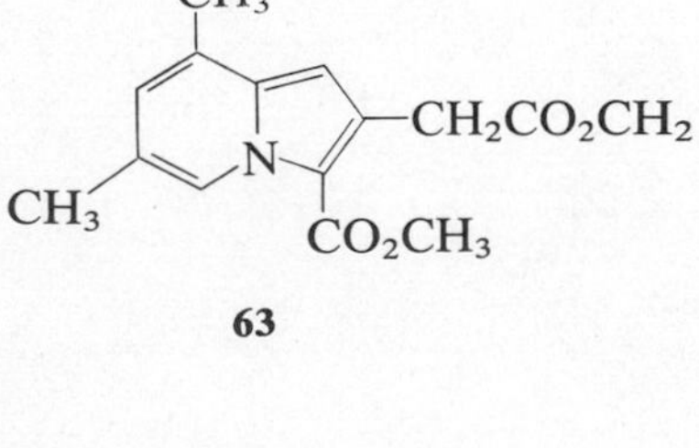

63 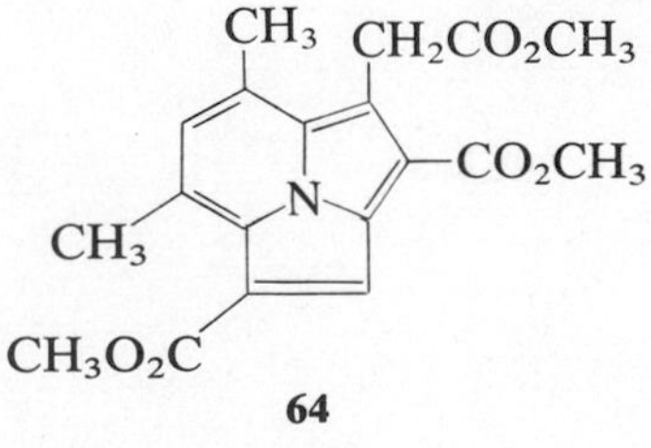 **64**

Acheson and Bridson[106] also synthesized a rather unusual bisindolizine derivative utilizing a unique method. When 2-phenylethynylpyridine (**65**) was condensed with dimethyl acelylenedicarboxylate the product was tetramethyloxybis-1-phenylmethyleneindolizine-2,3-dicarboxylate (**66**).

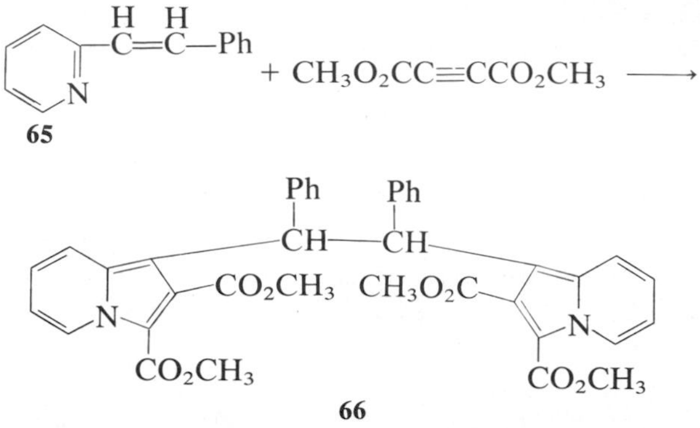

$$\text{(65: 2-pyridyl-CH=CH-Ph)} + CH_3O_2CC{\equiv}CCO_2CH_3 \longrightarrow \textbf{66}$$

Lown and Matsumoto have recently prepared a series of substituted indolizines utilizing a rather unique method.[243] Diphenylcyclopropenone (**66a**) reacts with a variety of heteroaromatic nitrogen compounds to give in good yield either 1,2-diphenyl-3-indolizinyl-cis-1,2-diphenyl acrylates (**66b**) and aza-analogs, 5,6-diphenyl-7-hydroxypyrrolo[1,2-*b*]-pyridazines (**66c**) and similar structures or adducts in which cycloaddition takes place to an N=N bond. These same authors have noted that diphenylcyclopropenone in certain of its reactions behaves like a more reactive form of diphenylacetylene. This analogy is useful here since pyridine, pyridazine, and pyrazinium ylide react with activated acetylenes to afford indolizine structures and aza-analogs of the indolizine skeleton.

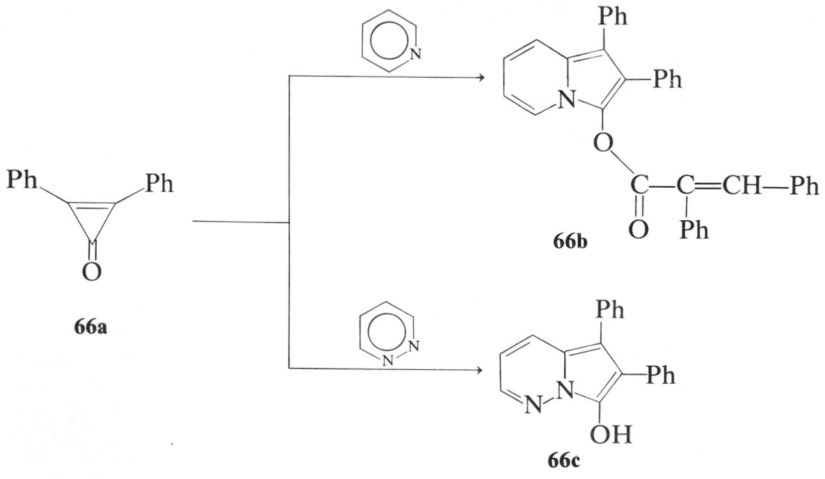

66a → (pyridine) → **66b**; **66a** → (pyridazine) → **66c**

Many other workers have recently synthesized rather wide variety of substituted indolizines mainly for the purpose of testing for biological activity.[244–250,277,279,281,286,300,304]

IV. General Reactions

Of the four parent substances discussed in this chapter only indolizine (**5**) had been extensively studied prior to 1957. An excellent review of the chemistry of indolizine has appeared in an earlier volume of the series by Mosby.[17] The nature of the chemistry of indolizine therefore is not discussed in detail except where their nature is important for comparison purposes.

Very little attention, prior to 1965, had been devoted to the chemistry of imidazo[1,2-*a*]pyridine and in fact only 11 papers had appeared that touched upon this subject. In some of this work, several structures were only postulated and other structure proofs left much to be desired. The chemistry of the other two diazaindenes, imidazo[1,5-*a*]pyridine and pyrazolo[1,5-*a*]pyridine, was virtually unexplored prior to 1957.

A. Electrophilic Substitution

Since 1956 a great deal of effort has been put forth in order to understand the process and nature of electrophilic substitution in the diazaindenes. The vast majority of the studies dealing with indolizine were carried out prior to 1957 and are discussed in *Mosby's Review.*[17]

1. *Halogenation*

Two groups of researchers, Paolini and Robins[107] and Paudler and Blewitt[2,4] have carried out rather detailed halogenation studies on imidazo[1,2-*a*]pyridine. In all cases when the 3-position is unsubstituted, using either bromine water, *N*-bromosuccinimide, or *N*-chlorosuccinimide, the 3-substituted halogenated compound is obtained. When 3-methylimidazo[1,2-*a*]pyridine (**67**) was treated with *N*-bromosuccinimide the 5-bromo-3-methyl (**68**) compound is produced.[107] When the same compound was treated with bromine water only starting material was recovered.[2,4] The yields in these halogenation reactions are generally good and often run as high as 90%. In all cases discussed above the structures of the products were established by comparing the nuclear magnetic resonance spectra of the starting material and the brominated product. Saldabols and Popelis have also studied the bromination of imidazo[1,2-*a*]pyridine. Their studies were performed on 2-(2-furyl)-imidazo[1,2-*a*]pyridine and showed that bromination could occur on either the 3-position of the imidazo[1,2-*a*]pyridine or on the 5-position of the furan ring, depending on reaction conditions, reagents used or position of substituents.[251]

Imidazo[1,5-*a*]pyridine when treated with Br_2/H_2O yielded a 1,3-dibromo derivative. No monobromo compound was isolated. *N*-Bromosuccinimide has also been successfully used to produce 1-methyl-3-bromoimidazo[1,5-*a*]pyridine by brominating the 1-methyl derivative.[8]

Pyrazolo[1,5-*a*]pyridine also reacts smoothly with bromine–water to produce 1-bromopyrazolo[1,5-*a*]pyridine.[7–9]

In respect to halogenation, indolizine is by far the most unusual of these diazaindenes. Indolizine reacts with bromine or iodine to yield unstable indefinite products.[27] These results have apparently discouraged further work in this area, although 3-acetyl-1-iodo-2-phenylindolizine (**70**) and 1,3-diiodo-2-phenylindolizine (**72**) were obtained by direct iodination of 2-phenyl-3-acetylindolizine (**69**) and 2-phenylindolizine (**71**), respectively.[252]

2. *Nitrosation*

Several workers[4,9,109–114,253] have successfully nitrosated the diazaindenes. In the majority of these cases the reactions proceed as would be predicted producing 1- or 3-substituted products depending mainly on the position of the ring nitrogen. Imidazo[1,2-*a*]pyridines and several of its derivatives were found to react smoothly with nitrous acid to produce in high yields the 3-nitroso products.[4,109,115]

A series of methylimidazo[1,2-*a*]pyridines have been nitrosated with some surprising results.[115] The 6-, 7-, and 8-monomethyl compounds do react smoothly to produce the respective 3-nitroso derivatives. However 5-methylimidazo[1,2-*a*]pyridine appears to react only with great difficulty and all attempts to isolate 3-nitroso-5-methylimidazo[1,2-*a*]pyridine have been unsuccessful. The reaction of 5,7-dimethylimidazo[1,2-*a*]pyridine has been somewhat more successful and a small yield of the 3-nitroso derivative has been isolated.[115] It is interesting to note that these results are in distinct conflict with those observed in the nitration of similar methyl derivatives (see following section on nitration). Thus far, no hypotheses has been put forth to explain this interesting experimental observation.

Paudler and Dunham[9] have studied the nitrosation of pyrazolo[1,5-*a*]-pyridine and were able to obtain in almost quantitative yield 1-nitrosopyrazolo[1,5-*a*]pyridine. The structure of the product was established by reduction followed by acetylation to yield the acetylamino compound and then comparing the nmr spectrum to that of the known 1-bromopyrazolo-[1,5-*a*]pyridine.

Indolizine readily undergoes this electrophilic substitution reaction and derivatives of the system, unsubstituted in positions 1 or 3, readily undergo nitrosation.[110–114,254]

Attempted nitrosation of imidazo[1,5-*a*]pyridine did not produce the expected nitroso derivative but instead yielded a very unusual rearranged product.[8,10] When imidazo[1,5-*a*]pyridine was treated with a cold solution of sodium nitrate in hydrochloric acid, a white crystalline solid was obtained. This is quite unusual, since nitroso derivatives of these diazaindenes are usually deep blue-green solids. Nitrosation of the 3-methyl and

N N R → N N O N R

73

R = H, —CH_3, and Ph

3-phenyl compounds also yielded white crystalline solids. Pyrolysis, alkaline hydrolysis, as well as mass, ultraviolet, and nmr spectral studies were used to establish the structures of the rearranged products as 3-(2-pyridyl)-1,2,4-oxadizoles (**73**).

3. *Nitration*

Prior to 1957 little was known concerning the nitration characteristics of the diazaindenes. Since that time the nitration of these heterocyclic molecules has been explored in some detail. Blewitt[4] was able to prepare 3-nitroimidazo[1,2-*a*]pyridine in approximately 80% yield by treating the parent substance with a solution of cold HNO_3/H_2SO_4. It also appears that the mono methylimidazo[1,2-*a*]pyridines readily undergo nitration in the 3-position in good yield regardless of the position of the methyl substituent.[115]

Nitration of imidazo[1,5-*a*]pyridine is also easily accomplished but less rigorous conditions must be utilized. Attempts to nitrate using a solution of cold concentrated nitric acid in sulfuric acid seem to catalyze decomposition and only a low yield of 2-picolinamide (**72**) was isolated.[8]

$\xrightarrow[\text{Cold}]{HNO_3/H_2SO_4}$ (N, N, NH₂, O labels in structure)

74

However under milder conditions it is possible to form a nitro derivative Thus the treatment of imidazo[1,5-*a*]pyridine with cupric nitrate in cold acetic anhydride resulted in the formation of 1-nitroimidazo[1,5-*a*]-pyridine. The position of ring nitration was established using nuclear magnetic resonance spectroscopy.[8]

Pyrazolo[1,5-*a*]pyridine also readily undergoes nitration easily. Paudler and Dunham, using a cold concentration solution of HNO_3 in H_2SO_4, were able to isolate 1-nitropyrazolo[1,5-*a*]pyridine in good yield.[9] Lynch and Lem have also done a rather detailed study on the nitration of pyrazolo[1,5-*a*]pyridine.[309] Using a solution of cold (O°C) concentrated H_2SO_4 and fuming HNO_3 they isolated the 1-nitro derivative. If, however, they maintained similar reaction conditions using excess fuming HNO_3, the 1,8-dinitro compound was the only product reported. The most unusual product was obtained when pyrazolo[1,5-*a*]pyridine was treated with fuming HNO_3 (O°C) in acetic anhydride. In this case, in addition to the 1-nitro compound, di(3-pyrazolo 1,5-*a*-pyridyl)oxidaommonium nitrate (**74a**) also isolated.

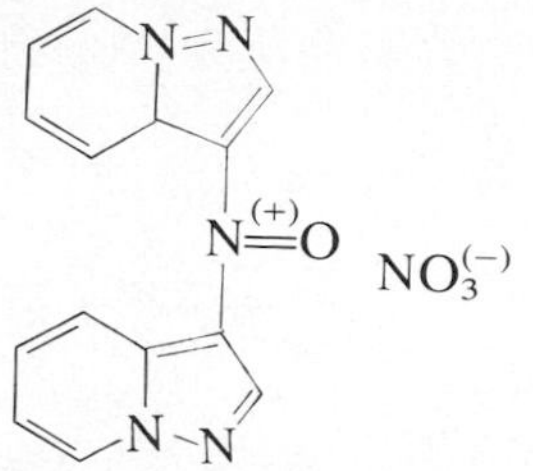

$NO_3^{(-)}$

74a

Earlier workers[43,116] claimed that indolizine could not be nitrated because of oxidation. However, it has since been determined that by varying the reaction conditions the oxidation could be minimized and nitroindolizines can be obtained. Nitration in the indolizines is rather unusual because of the preference for the 1-position. Nitration of 2-methylindolizine produced a 62% yield of the 1-nitro isomer and only 1.5% yield of the 3-nitro derivative.[43] Recently Hickman and Wibberley have been able to successfully nitrate indolizine in the 3-position. Treating either indolizine or 2-methylindolizine with HNO_3 in an excess of acetic anhydride at −70°C produced the corresponding 3-nitroindolizines in moderate yields.[310,311]

This reaction has been reviewed in more detail in an earlier volume of the series by Mosby.[17]

4. *Friedel–Crafts Reaction*

A relatively small amount of research has been carried out on the susceptibility of the diazaindenes to undergo the Friedel–Crafts reactions. Only one reference has been made to this type of reaction on imidazo-[1,5-*a*]pyridine (**3**). Bower and Ramage[20] found that imidazo[1,5-*a*]-pyridine acetylates under Friedel–Crafts conditions in the 1-position and if this is blocked in the 3-position.

Friedel–Crafts acylation of indolizine has been widely studied and is known to yield the 3-substituted product.[42,117,118] If more rigorous conditions are used the 1,3-diacetylated compound is obtained.[42,118]

Neither imidazo[1,5-*a*]pyridine nor pyrazolo[1,5-*a*]pyridine have successfully undergone Friedel–Crafts acylation. Dunham[9] has reported an unsuccessful attempt to acetylate pyrazolo[1,5-*a*]pyridine and no reaction

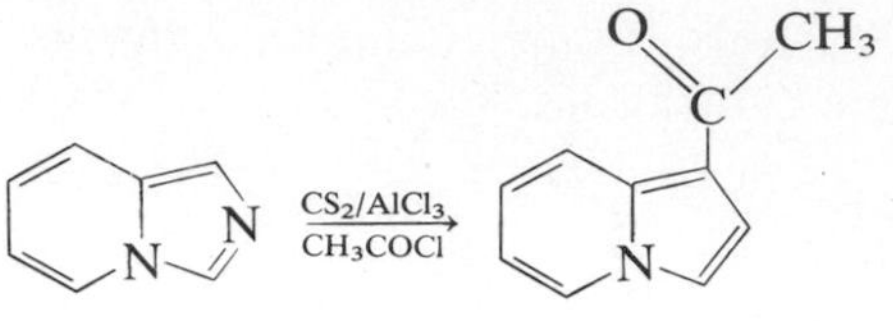

3

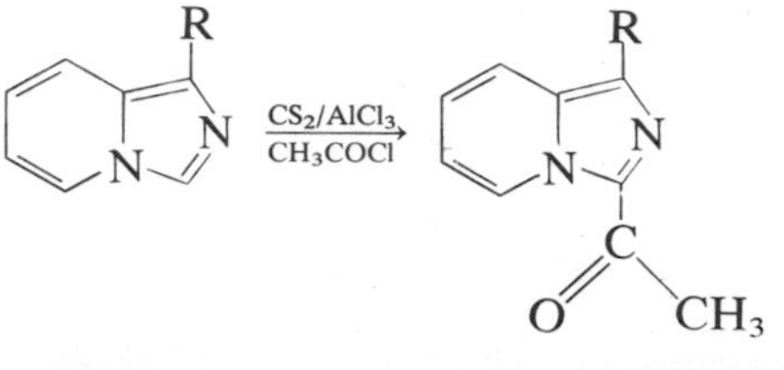

of this type has apparently been attempted on imidazo[1,2-*a*]pyridine. Since these systems all seem to undergo electrophilic substitution readily, one would suspect that the proper choice of reaction conditions should lead to Friedel–Crafts products in both imidazo[1,2-*a*]pyridine and pyrazolo[1,5-*a*]pyridine.

5. *Miscellaneous Electrophilic Substitution*

Treatment of pyrazolo[1,5-*a*]pyridine with benzene diazonium chloride afforded a mixture of seven products.[9] Column chromatography (alumina) permitted the isolation of one of these seven compounds. This product was tentatively identified as 1-phenyldiazopyrazolo[1,5-*a*]pyridine (**75**).

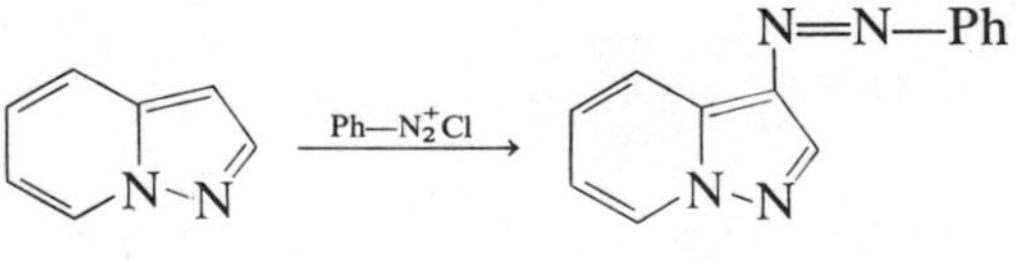

75

In a rather unusual electrophilic substitution reaction, Hill and Battiste[120] found that diphenylcyclopropenone in the presence of HCl condensed readily with 2-phenyl-7-methylindolizine (**76**) to yield **77**.

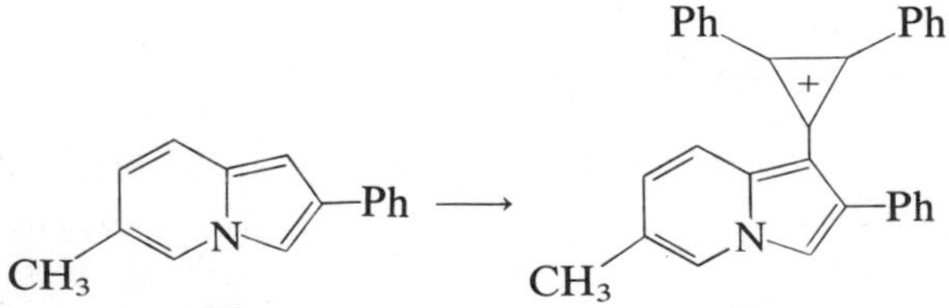

76 **77**

B. Nucleophilic Substitution and Displacement Reactions

Treatment of pyrazolo[1,5-*a*]pyridine with the following reagents afforded no new products.[9]

1. *n*-Butyl lithium in hexane at room temperature
2. *n*-Butyl lithium in refluxing *p*-xylene
3. Potassium amide in *N,N*-dimethyl aniline at 140°C
4. Potassium amide in refluxing *p*-cymene

Although unchanged starting materials were recovered in many instances, not one pure product was isolated from any of the reactions above. The frontier π-electron densities (Table V)[9] for nucleophilic substitution indicate that substitution should occur rather easily at the 5- and/or 8-positions in pyrazolo[1,5-*a*]pyridine. Therefore these results are not fully understood.

Treatment of 1-bromopyrazolo[1,5-*a*]pyridine with potassium *t*-butoxide in refluxing *t*-butanol for 48 hr also resulted in the recovery of unchanged starting material.[9]

The reaction of 1,3-dibromoimidazo[1,5-*a*]pyridine with sodium methoxide in refluxing methanol for 48 hr likewise afforded only recovered starting material. Similarly treatment with potassium *t*-butoxide did not effect the displacement of bromide ions from the molecules.[8] This inertness to nucleophilic displacement parallels the reported behavior of imidazo[1,2-*a*]pyridine and pyrazolo[1,5-*a*]pyridine with halogen substituents in the 5-membered ring.

Treatment of 2-chloro-3-nitroimidazo[1,2-*a*]pyridine (**78**) with dimethylamine yielded compound **79**. This result is certainly not surprising because of the strong electron withdrawing effect of the adjacent nitro group.[107]

$$\mathbf{78}\ (\text{2-Cl, 3-}NO_2) \xrightarrow{(CH_3)_2NH} \mathbf{79}\ (\text{2-}N(CH_3)_2\text{, 3-}NO_2)$$

Paolini and Robins[6] and Paudler and Good[121] have both studied the nucleophilic displacement of chlorine from the 6-membered ring in imidazo[1,2-*a*]pyridine. Paolini and Robins[6] found that the chlorine was easily displaced from 5-chloroimidazo[1,2-*a*]pyridine (**80**) using several organic oxide ions. Paudler and Good[121] also observed a similar phenomena. They found that when perchloroimidazo[1,2-*a*]pyridine (**81**)

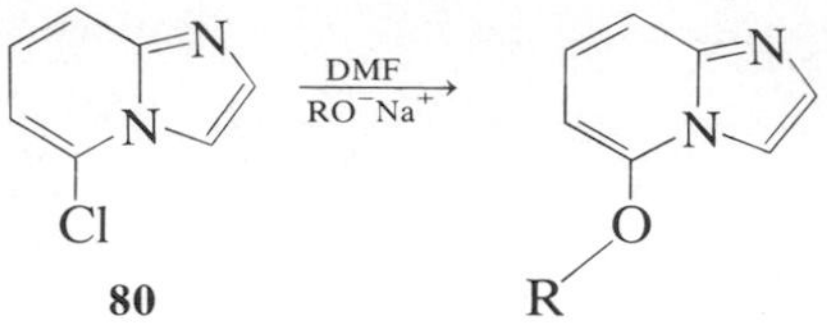

DMF, RO^-Na^+

80

$(R = —CH_2Ph, —CH_3, \text{and} —C_2H_5)$

was treated with sodium methoxide only the 5-chloro substituent was displaced to yield 2,3,6,7,8-pentachloro-5-methoxyimidazo[1,2-*a*]-pyridine (**82**).

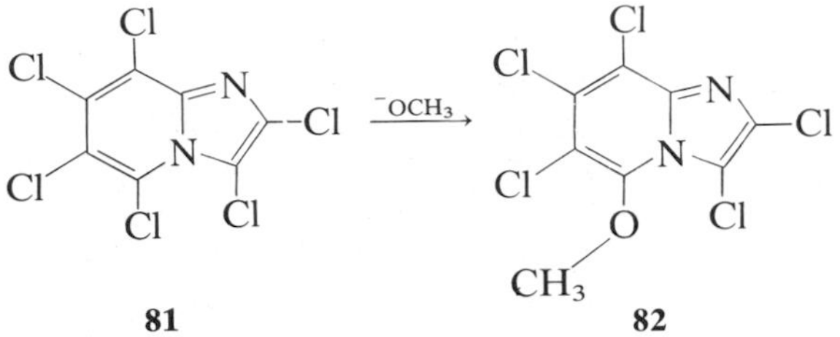

$^-OCH_3$

81 **82**

C. Mannich Reaction

There have been no reported Mannich reactions run on either imidazo-[1,5-*a*]pyridine or pyrazolo[1,5-*a*]pyridine. However, applications of the Mannich Reaction to both imidazo[1,2-*a*]pyridine and indolizine have been widely studied. Imidazo[1,2-*a*]pyridine has been found to undergo the Mannich reaction at the 3-position. Lombardino[122] demonstrated this for both the parent and the 2-methyl derivative. He also examined the

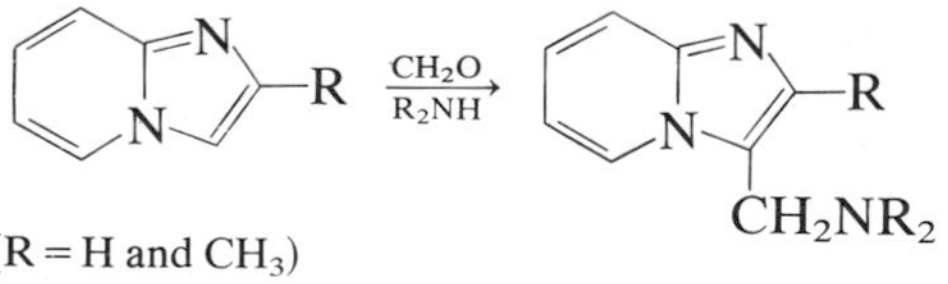

CH_2O, R_2NH

$(R = H \text{ and } CH_3)$

reactivity of the methyl group in the 2-position of imidazo[1,2-*a*]pyridine in an attempt to determine if it would behave similarly to the methyl groups in other heterocyclic systems, such as, the 2- and 4-picolines. When 2-methylimidazo[1,2-*a*]pyridine is treated with chloral, compound **83** was produced which on hydrolysis yielded the acrylic acid (**84**). Other workers attempting Mannich reactions have also obtained similar substitution products when studying related 2-substituted imidazo[1,2-*a*]-pyridines.[255]

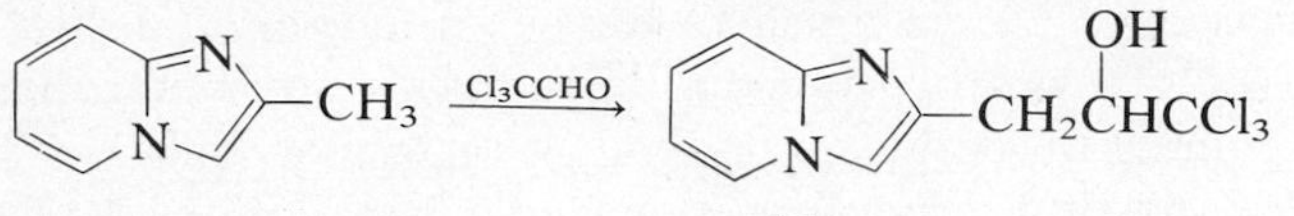

Paudler and Shin[15] reported a ylide-like intermediate in the reaction of imidazo[1,2-*a*]pyridine with phenyl lithium. If the reaction was carried out in the presence of cyclohexanone compound **85** resulted. The reaction

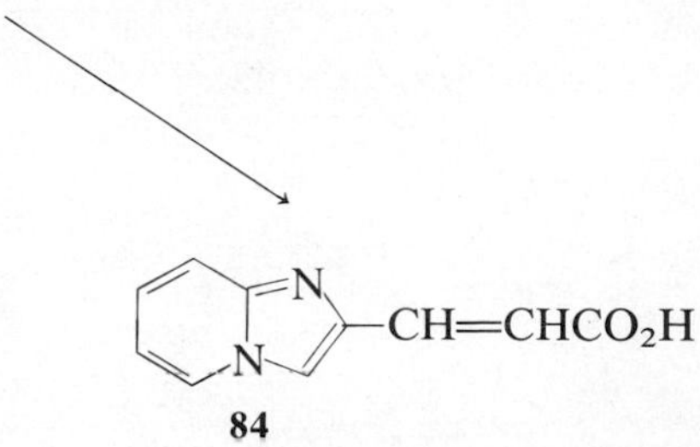

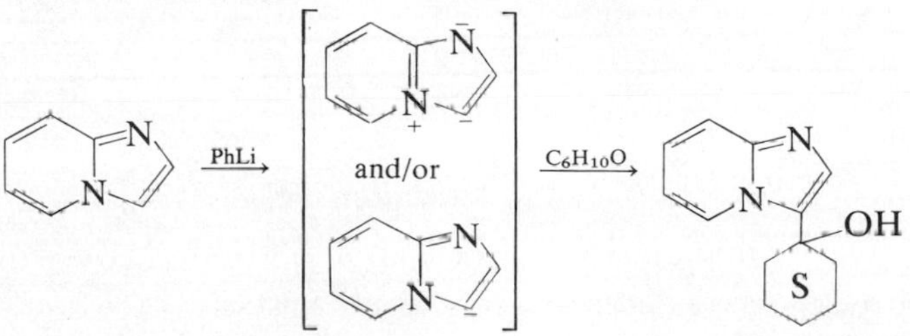

also will proceed with methyl substituents present at the 2-position in the 5-membered ring or in the 6-, 7-, or 8-positions of the 6-membered ring. In the case of 5-methylimidazo[1,2-*a*]pyridine, the methyl protons are acidic enough to react with phenyl lithium and they condense with cyclohexanone to yield **86**. Imidazo[1,5-*a*]pyridine also reacts with phenyl lithium and cyclohexanone to yield 3-substituted products analogous to compound **85**.[256]

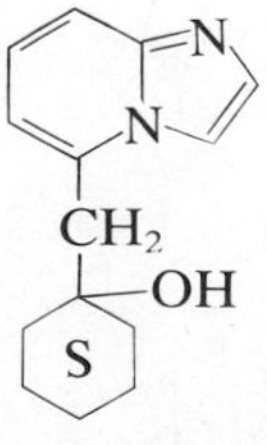

Harrell and co-workers[123–126] in a search for new therapeutic derivatives have synthesized a series of Mannich derivatives from indolizine. The reaction was found to occur quite readily on either the 1- or 3-position. The 2,3-disubstituted indolizines yielded a 1-substituted Mannich produce (**87**)[123] and the 1,2-disubstituted indolizines yielded a 3-substituted product (**88**).[124]

CH_2NR_2 R_1 $\xrightarrow[R_2NH]{CH_2O}$ R_1 R_2 R_2

87

R_1 R_1 R_2 $\xrightarrow[R_2NH]{CH_2O}$ R_2 CH_2NR_2

88

D. Oxidation

All of these diazaindenes are oxidatively unstable. However very little effort has been devoted to determining the nature of their oxidation products. Bowers and Brown[127] studied the potassium ferricyanide oxidation of methyl-2-hydroxy-3-methyl and methyl-2-hydroxy-3-phenylindolizine-1-carboxylates. When these compounds were treated under mild oxidative conditions 2,3-dihydro-3-hydroxy-2-oxo-3-substituted indolizines (**89**) were produced. In other studies on indolizine acrylonitrile was achieved under a variety of oxidizing conditions.[259] Under slightly different conditions Huenig and Linhart have successfully achieved oxidative dimerization at the 3-position of 1,2-dimethyl indolizine using potassium ferricyanide.[265]

CO_2R_1 OH $\xrightarrow{K_3Fe(CN)_6}$ CO_2R_1 =O HO R_2 R_2

R_1 = —CH_2 and —CH_2CH_3

R_2 = —CH_2 and —Ph

89

When imidazo[1,5-*a*]pyridine was treated with a fourfold molar excess of potassium permanganate, the major product was picolinic acid (**90**).[8] This product accounted for 62% of the starting material. An additional 29% was accounted for as picolinamide (**91**). The oxidative stability of

$$\text{imidazo[1,5-}a\text{]pyridine} \xrightarrow{KMnO_4} \text{2-pyridyl-}CO_2H \;(\mathbf{90}) + \text{2-pyridyl-}C(=O)NH_2 \;(\mathbf{91})$$

indolizine has been extensively studied. An excellent review of its oxidative properties has appeared in a volume of the series by Mosby.[17]

E. Diazonium Salt Reactions

There are no reported studies on diazonium salts of imidazo[1,5-*a*]-pyridine and imidazo[1,2-*a*]pyridine. There has been, however, some work done on both pyrazolo[1,5-*a*]pyridine and indolizine.

Diazotization of 1-aminopyrazolo[1,5-*a*]pyridine and subsequent treatment with copper (I) cyanide yielded a rather unusual product, 1-pyrazolo[1,5-*a*]pyridinediazocarboxamicle (**92**). This result is not surprising,

$$\text{1-}NH_2\text{-pyrazolo[1,5-}a\text{]pyridine} \xrightarrow[\text{2) } CuCN/H_2O]{\text{1) HONO}} \text{1-}(N{=}N{-}C(=O)NH_2)\text{-pyrazolo[1,5-}a\text{]pyridine} \;(\mathbf{92})$$

ing, especially when one considers the unusual stability of this type of system reported by Tedder and co-workers.[128,129] Aminoindolizines are reported to be unstable in air[81] and the diazonium compounds were therefore prepared from the corresponding nitroso compounds. 1-Nitroso-2-phenyl-3-methylindolizine and 1-nitroso-2-phenyl-3-acetyl-indolizine both yielded the corresponding diazonium nitrates when treated with nitric oxide (NO). Indolizine-1-diazonium salts (nitrates and chlorides) proved to be exceptionally stable both in solution and in the crystalline state. They were found to couple with phenoxide ions yielding the predicted azo dyes. Interestingly enough, the 3-nitrosoindolizines seem to be inert to treatment with nitric oxide and were recovered unchanged.

F. Hydrogen–Deuterium Exchange

These diazaindenes readily undergo hydrogen–deuterium exchange under either acidic or basic conditions. This exchange under acid conditions generally occurs at the same positions where electrophilic substitution normally takes place. The base catalyzed exchanges of these compounds usually occur at positions adjacent to nitrogen atoms which possess a partial positive charge as indicated by major contributing resonance structures (see Section II.D). Deuterium exchange reactions may be conveniently studied using an nmr sample tube as the reaction vessel. The spectrum of the sample is determined periodically and the site of deuterium exchange is indicated by the disappearance of a given peak in the spectrum.[8]

Imidazo[1,2-*a*]pyridine is quite typical of this type of system. Paudler and Helmick[11,12,14] showed that acid catalyzed exchange in this parent substance occurs at the 3-position in 3*M* deutero sulfuric acid. It is significant to note that under the conditions used by these researchers the exchange rate of the methiodides is considerably slower than that of the free bases. This suggests that hydrogen–deuterium exchange occurs on the free base rather than the protonated species.

Base catalyzed hydrogen–deuterium exchange also occurs easily on the imidazo[1,2-*a*]pyridine system. In view of the postulated contribution of structure **93** to the overall resonance hybrid of the molecule one would predict base catalyzed deuterium exchange to occur on positions 3 and 5.

2 ⟷ **93**

Paudler and Helmick[11,12,14] have observed that imidazo[1,2-*a*]pyridine rapidly undergoes a hydrogen–deuterium exchange ($NaOCH_3$ in CH_3OD) at the 3- and 5-positions. These workers also observed that the methiodides of imidazo[1,2-*a*]pyridines undergo deuterium exchange at positions 2, 3, and 5, which is predictable due to the contributing resonance forms, **94** and **95**.

Pyrazolo[1,5-*a*]pyridine undergoes hydrogen–deuterium exchange rapidly in 5.0*N* deuterosulfuric at room temperature.[9] The nuclear magnetic resonance spectrum shows that all five remaining protons are

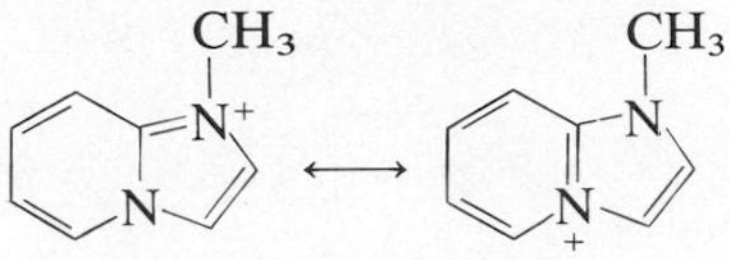

94 95

deshielded with H_2 appearing as a sharp singlet, in contrast to the doublet (for H-2) noted in all pyrazolo[1,5-*a*]pyridines studied. Dunham[9] thus concluded from the nmr spectrum (described previously) that H-1 must be absent and the exchange therefore occurred at the 1-position. No studies of hydrogen–deuterium exchange in basic media have apparently been carried out on pyrazolo[1,5-*a*]pyridine or its methiodides. However one would expect exchange to occur in the same fashion as imidazo[1,2-*a*]pyridine.

Kuder[8] showed that imidazo[1,5-*a*]pyridine and several of its methyl derivatives undergo deuterium exchange in 7*N* D_2SO_4 at the 1-position. When imidazo[1,5-*a*]pyridine is dissolved in a 0.2*N* solution of sodium methoxide in CH_3OD hydrogen–deuterium exchange occurs at the 3-position. Deuterium exchange also occurred at the 3-position under neutral conditions when a dilute solution of the parent substance in D_2O is allowed to stand for 24 hr. No deuterium exchange was noted under similar conditions when the methiodide of imidazo[1,5-*a*]pyridine was treated with D_2O.[8]

Very little effort has been devoted to studying the hydrogen–deuterium exchange in the indolizines. This is probably due to the nature of the stable salts (**94**) formed in dilute acid solution.

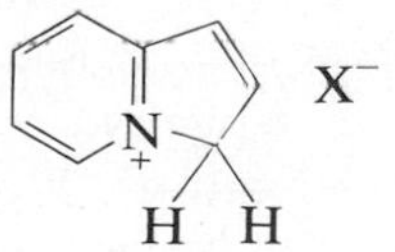

X^-

H H

94a

Fraser, Melera, Malloy, and Reid[155] did however demonstrate that salts (**94a**) of the type shown above do not undergo hydrogen–deuterium exchange in deuterotrifluoroacetic acid solution. Recently, Engewald, Muehlstaedt, and Weiss have shown the hydrogen deuterium exchange will occur on indolizine. If indolizine is warmed in a solution of D_2O/Dioxane at 50°C, H–D exchange does occur with the 3-position being the most reactive.[221,222]

G. Miscellaneous Reactions

1. *Reactions with Dienophiles*

These diazaindenes seem to be inert to Diels–Alder reaction. This is undoubtedly due to their high degree of aromatic character. Dunham[9] subjected pyrazolo[1,5-*a*]pyridine to a wide variety of dienophiles under essentially neutral conditions and was not able to isolate anything other than unchanged starting materials. Successful Diels–Alder reactions have not been reported for these or similar systems. However, Galbraith, Small, Barnes, and Boekelheide[49,130,131] have observed a related reaction of indolizine and dimethyl acetylenedicarboxylate affords a 68% yield of the substituted cyclo[3.3.2]azine (**95**).

CO_2CH_3–C≡C–CO_2CH_3 + indolizine → (Pd, Refluxing toluene) → **95** (CH_3O_2C, CO_2CH_3)

Indolizine does however react under acidic conditions with dienophiles to yield typical electrophilic substituted products.[132,133]

2. *Catalytic Reduction*

Catalytic hydrogenation has been carried out on all four of these diazaindenes.[8,9,134–136,257] The course of the reductions are all very similar and are characterized by the reduction of the pyridine ring yielding the respective 5,6,7,8-tetrahydro derivatives. It is of interest to note that the reduction of the 6-membered ring leaves an aromatic sextet of electrons in the 5-membered ring whereas if the reduction had occurred in the 5-membered ring the resulting product would not be aromatic. Complete saturation of the indolizines occur with great facility under a variety of conditions.[28–31,51,91,134,135,137]

3. *Acid Hydrolysis*

When imidazo[1,5-*a*]pyridine is heated in dilute hydrochloric acid the compound is hydrolyzed to 2-aminomethylpyridine (**96**) and formic acid.[9]

Some limited work has also been carried out on indolizine. Hydrolysis of 3-amino-2-phenyl- and 3-amino-1,2-diphenylindolizine with HCl produced 3-(2-pyridyl)propionic acids.[259] This type of sensitivity to acid hydrolysis has not been reported for any of the other diazaindenes.

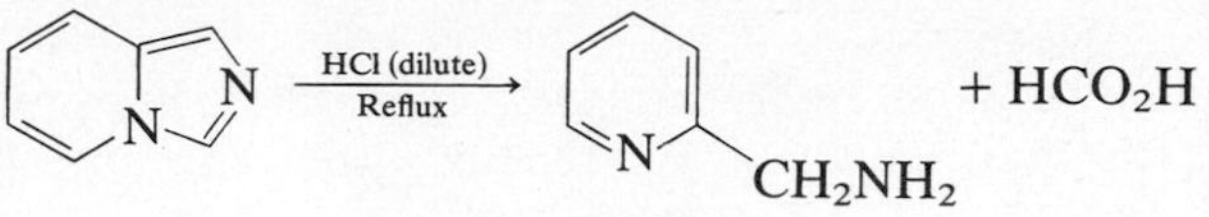

HCl (dilute), Reflux → CH_2NH_2 + HCO_2H

96

4. *Perhalogenation*

Perchloromidazo[1,2-*a*]pyridine (**97**) has been obtained in high yield by the treatment of imidazo[1,2-*a*]pyridine with phosphorous pentachloride at 275°C in a sealed tube.[16] The fact that no skeletal rearrangement occurs is adequately confirmed by the fact that hydrogenation with $Pd/CaCo_3/H_2$ regenerates imidazo[1,2-*a*]pyridine.

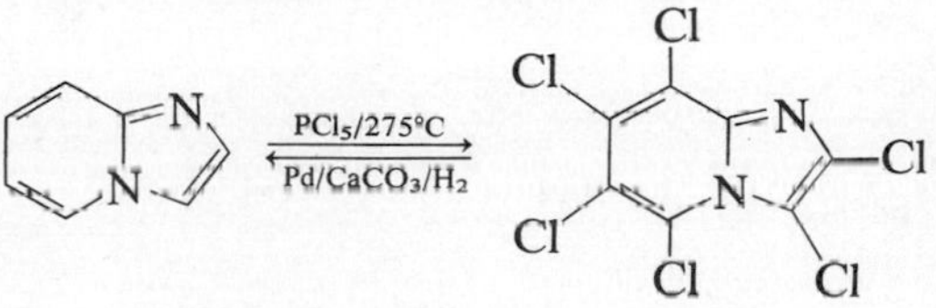

PCl_5/275°C; $Pd/CaCO_3/H_2$

97

5. *Coupling Reactions*

Recently a great deal of interest seems to have been generated in the coupling reactions of indolizines and other diazaindenes. Several authors have found that when certain 1,2-disubstituted indolizines are treated with tosylazide 3,3′-azaindolizines (**97a**) are formed. If the starting indolizines are substituted in the 2- and 3-positions then 1,1′-azoindolizines are formed.[260–263]

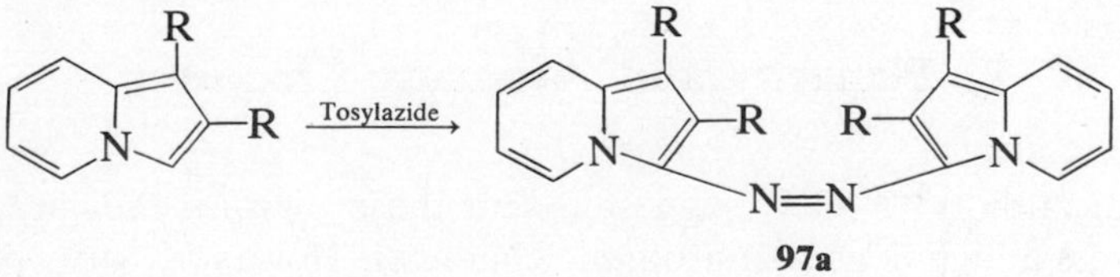

Tosylazide →

97a

Glover and York have also demonstrated a related synthesis of a coupled diazo-compound with imidazo[1,2-*a*]pyridine.[264] When 1-amino-1-*H*-imidazo[1,2-*a*]pyridin-4-ium salts (**97b**) were treated with bromine, 1,1′-azoimidazo[1,2-*a*]pyridinium salts (**97c**) were produced.

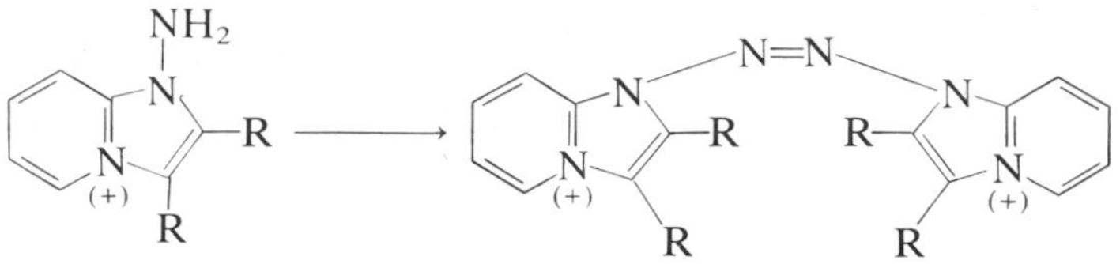

97b **97c**

In a rather unusual reaction Bogdanowicz-Szwed and Kawalek[266] have recently shown that a coupling reaction occurs when 2-methyl and 2,7-dimethylindolizine are treated with *m*- and *p*-phenylene diisocyanates to yield compound **97d**.

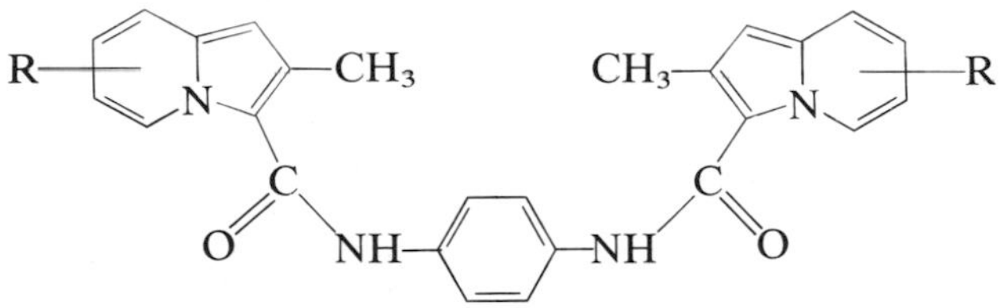

97d

6. *Rearrangements*

Jacquier, Lopez, and Maury in their studies on the diazaindenes have shown that the Dimroth rearrangement does occur readily on the appropriately substituted imidazo[1,2-*a*]pyridine.[312] When 3-methyl-6-nitroimidazo[1,2-*a*]pyridine is treated with dilute base the rearranged product was identified as 2-methyl-6-nitroimidazo[1,2-*a*]pyridine. This rearrangement can occur with a nitro group in the 6- or 8-position and with a variety of different substituents in the 2-position.

V. Planarity and Aromatic Character

Several authors[3,4,6] have suggested that these diazaindenes are planar and possess a high degree of aromatic character. However, only one study

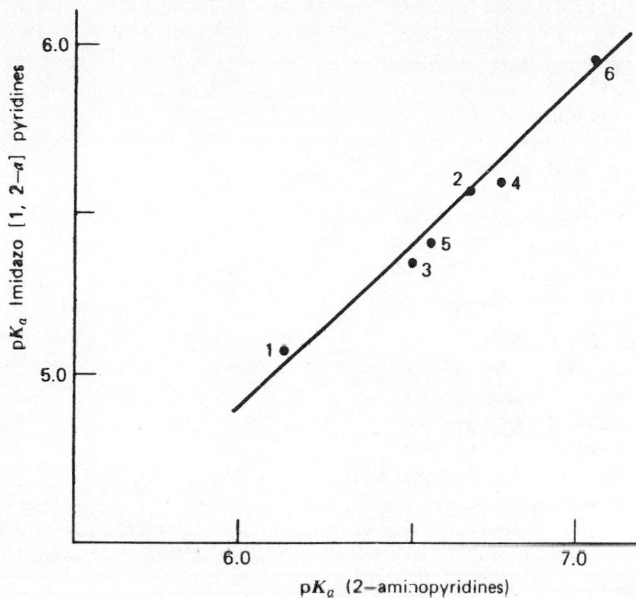

Fig. 8. Correlation of the basicities of 2-aminopyridines with the corresponding imidazo[1,2-*a*]pyridines. The numbers refer to compounds listed and numbered in Table XI.

has been carried out in an attempt to demonstrate the planarity of this type of ring system. Paudler and Blewitt[3] compared the basicities of imidazo[1,2-*a*]pyridines with the corresponding 2-aminopyridines. It would be quite a coincidence if the free-energy changes involved in the protonation of the two systems were the same without involving similar structural changes in the compounds. This linear free-energy relationship is clearly shown by an inspection of Fig. 8 and Table XI. These same authors also compared the chemical shift positions of ring protons and methyl substituents in some imidazo[1,2-*a*]pyridines with the "equivalent" substituents in the corresponding 2-aminopyridines. Table XII shows that nearly all of the protons in the various imidazo[1,2-*a*]pyridines are at least as deshielded as those in the corresponding 2-aminopyridines. These studies coupled with the typical aromatic reactions that imidazo[1,2-*a*]-pyridine undergoes (see Section IV) leads one to conclude that this system possesses a high degree of aromatic character. Recently x-ray data on the crystal structure of 1,1-azobis(arylimidazo[1,2-*a*]pyridinium)-bromides showed that this ring system is nearly planar.[267,268]

Although no such studies have been carried out on imidazo[1,5-*a*]-pyridine, pyrazolo[1,5-*a*]pyridine, or indolizine, their similarity to

TABLE XI. BASICITIES OF VARIOUS IMIDAZO[1,2-*a*]-PYRIDINES AND 2-AMINOPYRIDINES

Compound	pK_a[a]	Compound Number[b]
Imidazo[1,2-*a*]pyridine	5.06	1
5-Methylimidazo[1,2-*a*]pyridine	5.55	2
6-Methylimidazo[1,2-*a*]pyridine	5.33	3
7-Methylimidazo[1,2-*a*]pyridine	5.59	4
8-Methylimidazo[1,2-*a*]pyridine	5.40	5
5,7-Dimethylimidazo[1,2-*a*]pyridine	5.96	6
2-Aminopyridine	6.14	1
2-Amino-6-methylpyridine	6.69	2
2-Amino-5-methylpyridine	6.50	3
2-Amino-4-methylpyridine	6.76	4
2-Amino-3-methylpyridine	6.52	5
2-Amino-4,6-dimethylpyridine	7.12	6

[a] In 90% ethanol–10% water, obtained by extrapolating several determinations at different concentrations, to zero concentration.
[b] Compound numbers refer to those used in Fig. 8.

imidazo[1,2-*a*]pyridine both in general reactivity (see Section IV) and spectral characteristics strongly indicates the existence of a high degree of aromatic character in these closely related systems.

VI. Uses

Many derivatives of these diazaindenes have been screened, with limited success, for their biological[124,125,191,202,218,219,279,282,285,286,288,295] activity. The vast majority of the syntheses and subsequent screening have concentrated on substituted indolizines and imidazo-[1,2-*a*]pyridines. Imidazo[1,2-*a*]pyridines have been screened for their activity as hypertensive,[184] antipyretic,[185–187] antiinflammatory,[187–189] analgesic,[186,187,190] antifungal,[192,193,275,276] antiphlogistic,[186] anthelmintic,[275,276] and anticonvulsant[186,187] agents. Certain imidazo[1,2-*a*]-pyridines have also been shown to provide respiratory stimulant activity.[194,280] The indolizines have also been studied for their psychotropic,[195] antiinflammatory,[196,277,281,284] analgesic,[196,281] antimicrobial,[197,278,283] antiexudative,[198] and hypoglycemic[277] activity. Substituted indolizines have been found to have central nervous system depressant activity.[123,126] Certain imidazo[1,2-*a*]pyridines have also been found to act as powerful selective neuromuscular blocking agents.[267,268,287] Recently a great deal

TABLE XII. NMR SPECTRA OF SOME 2-AMINOPYRIDINES AND RELATED IMIDAZO[1,2-*a*]PYRIDINES

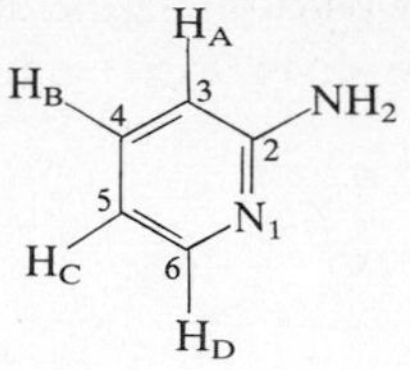

1 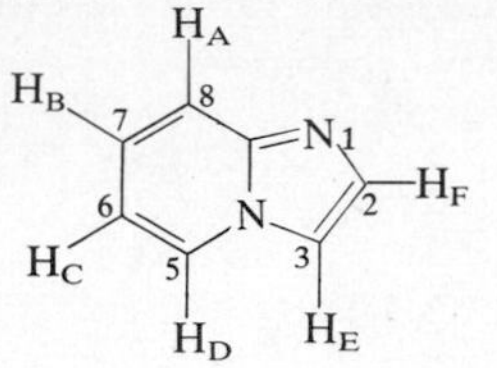 **2**

Compound		Chemical shift						
Number	Substituent	H_A	H_B	H_C	H_D	H_E	H_F	Substituent
I[a]	None	2.72	3.48	3.62	1.96	—	—	—
II[a]	None	2.44	2.79	3.20	1.80	—	—	—
I[b]	5-Methyl	2.63	3.49	—	1.99	—	—	7.81
II[b]	6-Methyl	2.40	3.01	—	2.10	2.48	2.35	7.78
I[a]	4,6-Dimethyl	3.65	—	3.65	—	—	—	7.68 7.87
I[b]	4,6-Dimethyl	3.58	—	3.76	—			7.63(6-CH_3) 7.81(4-CH_3)
IIb[b]	5,7-Dimethyl	2.59	—	3.50	—	2.55	2.31	7.51(5-CH_3) 7.68(7-CH_3)
I[b]	3-Methyl	—	2.67	3.35	1.97	—	—	7.88
II[b]	8-Methyl	—	3.02	3.32	1.96	2.40	2.32	7.41
I[b]	4-Methyl	3.60	—	3.50	2.04	—	—	7.76
I[a]	4-Methyl	3.52	—	3.48	2.20	—	—	7.82
II[a]	7-Methyl	2.85	—	3.54	2.08	2.48	2.43	7.80
II[b]	7-Methyl	2.64	—	3.45	2.04	2.52	2.48	7.65

[a] 1.0 *M* solution in D_2O.
[b] 1.0 *M* solution in $CDCl_3$.

of success has been obtained in utilizing derivatives of imidazo[1,2-*a*]-pyridines as nondepolarizing muscle relaxants and in the treatment of ulcers.[288–293]

Limited medical testing has been carried out on both imidazo[1,5-*a*]-pyridine[294,295] and pyrazolo[1,5-*a*]pyridine. Some derivatives of imidazo-[1,5-*a*]pyridines have been found to inhibit the secretion of gastric juice and pepsin.[295] Pyrazolo[1,5-*a*]pyridines have been tested as blood vessel dialating agents and also found to be of some use as an antimicrobials.[296–298]

Both substituted indolizines[128,172,203–213,271] and imidazo[1,2-*a*]-pyridines[199–201,272] have been extensively studied as azo dyes. Indolizines have been incorporated into polymers[163,214,269] and also used as photographic sensitizing[215,216,273,274] and fabric brightening agents.[217] There has also been some effort to utilize imidazo[1,2-*a*]pyridines as electron acceptors and spectral sensitizers for direct positive silver halide emulsions.[270]

VII. References

1. W. W. Paudler and H. L. Blewitt, *Tetrahedron,* **21** (2), 356 (1965).
2. W. W. Paudler and H. L. Blewitt, *J. Org. Chem.*, **30,** 4081 (1965).
3. W. W. Paudler and H. L. Blewitt, *J. Org. Chem.*, **31,** 1295 (1966).
4. H. L. Blewitt, Ph.D. Dissertation, Ohio University, Athens, Ohio, 1966.
5. P. J. Black, M. L. Heffernan, L. M. Jackman, Q. N. Porter and G. R. Underwood, *Aust. J. Chem.*, **17,** 1128 (1964).
6. J. P. Paolini and R. K. Robins, *J. Heterocycl. Chem.*, **2,** 53 (1965).
7. W. W. Paudler and D. E. Dunham, *J. Heterocycl. Chem.*, **2,** 410 (1965).
8. J. E. Kuder, Ph.D. Dissertation, Ohio University, Athens, Ohio, 1968.
9. D. E. Dunham, Ph.D. Dissertation, Ohio University, Athens, Ohio, 1967.
10. W. W. Paudler and J. E. Kuder, *J. Org. Chem.*, **32,** 2430 (1967).
11. L. S. Helmick, Ph.D. Dissertation, Ohio University, Athens, Ohio, 1968.
12. W. W. Paudler and L. S. Helmick, *Chem. Commun.*, 377 (1967).
13. W. W. Paudler, J. E. Kuder, and L. S. Helmick, *J. Org. Chem.*, **33,** 1379 (1968).
14. W. W. Paudler and L. S. Helmick, *J. Org. Chem.*, **33,** 1087 (1968).
15. W. W. Paudler and H. G. Shin, *J. Org. Chem.*, **33,** 1638 (1968).
16. W. W. Paudler, D. J. Pokorny, and J. J. Good, *J. Heterocycl. Chem.*, **8,** 37 (1971).
17. W. L. Mosby, "Heterocyclic Systems with Bridgehead Nitrogen Atoms," Vol. XV (Part 1), Interscience, New York, 1961.
18. A. E. Tschitschibabin, *Ber.*, **58,** 1704 (1925).
19. A. E. Tschitschibabin, *Ber.*, **60,** 1607 (1927).
20. Bower and Ramage, *J. Chem. Soc.*, 2834 (1955).
21. Bower and Ramage, *J. Chem. Soc.*, 4506 (1957).
22. Bower, *J. Chem. Soc.*, 4511 (1957).
23. Rath, *Ber.*, **58,** 346 (1925).
24. A. E. Tschitschibabin, *Ber.*, **58,** 1707 (1925).
25. Scholtz, *Ber.*, **45,** 734 (1912).
26. Scholtz, *Ber.*, **45,** 1718 (1912).
27. Scholtz and Fraude, *Ber.*, **46,** 1069 (1913).
28. Diels, Alder, Kashimoto, Freidrichsen, Eckhardt, and Klare, *Ann.*, **498,** 16 (1932).
29. Ochiai and Tsuda, *Ber.*, **67,** 1011 (1934).
30. Diels, Alder, Friedrichsen, Klare, Winkler, and Schrum, *Ann.*, **505,** 103 (1933).
31. Boekelheide and Freely, *J. Org. Chem.*, **22,** 589 (1957).
32. Brand and Reuter, *Ber.*, **72,** 1668 (1939).
33. Wilson, *J. Chem. Soc.*, 63 (1945).
34. Krumholtz, *Selecta Chim.*, No. 8, 3 (1949): *Chem. Abstr.*, **44,** 3992 (1950).
35. Bower, *J. Chem. Soc.*, 4510 (1957).
36. Miller and Brown, American Chemical Society, 130th National Meeting, Atlantic City, N.J., September 16–21, 1956; *Abstr.*, p. 49.
37. Longuet-Higgens and Coulson, *Trans. Faraday Soc.*, **43,** 87 (1947).

38. Fukui, Yonezawa, Nogata, and Shingu, *J. Chem. Phys.*, **22,** 1433 (1954).
39. Barrett and Chambers, *J. Chem. Soc.*, 338 (1958).
40. A. E. Tschitschibabin, German Patent 464,481; *Frdl.*, **16,** 2652 (1931).
41. Holland and Nayler, *J. Chem. Soc.*, 1083 (1946).
42. Barrows, Holland, and Kenyon, *J. Chem. Soc.*, 1069 (1946).
43. Barrows, Holland, and Kenyon, *J. Chem. Soc.*, 1077 (1946).
44. A. E. Tschitschibabin and Stepanow, *Ber.*, **62,** 1068 (1929).
45. Rossiter and Saxton, *J. Chem. Soc.*, 3654 (1953).
46. Konds and Osawa, *J. Pharm. Soc. Japan*, **56,** 73 (1936); *Chem. Zent.*, **I,** 4158 (1936); *Chem. Abstr.*, **30,** 3431 (1936).
47. Leonard and Hay, *J. Am. Chem. Soc.*, **78,** 1984 (1956).
48. Konda and Kokeguchi, *J. Pharm. Soc. Japan*, **57,** 573 (1937); *Chem. Abstr.*, **31,** 6230 (1937); *Chem. Zent.*, **II,** 3747 (1937).
49. Boekelheide and Windgassen, *J. Am. Chem. Soc.*, **80,** 2020 (1958).
50. Scholtz, *Arch. Pharm.*, **251,** 666 (1913); *Chem. Zent.*, **I,** 1284 (1914); *Chem. Abstr.*, **8,** 2698 (1914).
51. Diels and Schrum, *Ann.*, **530,** 68 (1937).
52. A. E. Tschitschibabin and Stepanow, *Ber.*, **63,** 470 (1930).
53. F. Mattu and E. Marongiu, *Chimica (Milan)*, **41** (1), 9 (1965).
54. W. W. Paudler and J. E. Kuder, *J. Org. Chem.*, **31,** 809 (1966).
55. R. Brumbaugh, Ph.D. Dissertation, Ohio University, Athens, Ohio, 1970.
56. C. Aussems, S. Jaspers, L. Georges, and F. van Remoortere, *Bull. Soc. Chim. Belg.*, **78,** 479 (1969).
57. A. Streitwieser, *J. Am. Chem. Soc.*, **82,** 4123 (1960).
58. J. A. Berson, M. Evleth, and S. L. Manott, *J. Am. Chem. Soc.*, **87,** 2901 (1965).
59. P. J. Black, R. D. Brown, and M. L. Heffernan, *Aust. J. Chem.*, **20** (7), 1325 (1967).
60. P. J. Black, R. D. Brown, and M. L. Heffernan, *Aust. J. Chem.*, **20** (7), 1305 (1967).
61. J. Fettelson, *J. Chem. Phys.*, **43** (7), 2511 (1965).
62. A. Gamba and G. Favini, *Gazz. Chim. Ital.*, **98** (2), 167 (1968).
63. V. Galasso, G. DeAlti, and A. Bigotto, *Theor. Chim. Acta*, **9** (3), 222 (1968).
64. S. Hillers, F. B. Mazheika, and I. I. Grandberg, *Khim. Geterotsikl. Soedin.*, 884 (1967). (5); *Chem. Abstr.*, **68,** 86719e (1967).
65. W. W. Paudler and H. L. Blewitt, unpublished results.
66. A. R. Katritsky, "Physical Methods in Heterocyclic Chemistry," Vol. II, Academic Press, New York, 1963.
67. Schmidt and Grundig, *Monatsh.*, **84,** 491 (1953).
68. Matveev, *Bull. Acad. Sci. USSR, Classe Sci. Math. Nat., Ser. Chim.*, 533 (1936); *Chem. Zent.*, **I,** 1125 (1938); *Chem. Abstr.*, **31,** 6654 (1937); *Brit. Chem. Abstr.*, **AII,** 263 (1937).
69. Kaye, Parris, and Burlant, *J. Am. Chem. Soc.*, **75,** 746 (1953).
70. Schmidt and Bangler, *Ber.*, **59,** 1360 (1926).
71. Diels, Alder, Winckler, and Peterson, *Ann.*, **498,** 1 (1932).
72. C. K. Bradsher, E. F. Litzinger, and M. F. Zinn, *J. Heterocycl. Chem.*, **2,** 331 (1965).
73. C. K. Bradsher, R. D. Brandau, J. E. Bobek, and T. L. Hough, *J. Org. Chem.*, **34,** 2129 (1969).
74. R. Huisgen, R. Grashey, and R. Krischke, *Tetrahedron Lett.*, **9,** 387 (1962).
75. C. W. Rees and M. Yelland, *J. Chem. Soc. (D)*, 377 (1969).
76. T. Sasaki, K. Kanematsu, and Y. Yukimoto, *J. Chem. Soc. (C)*, 481 (1970).
77. E. C. Taylor and K. S. Hartke, *J. Am. Chem. Soc.*, **81,** 2452 (1959).
78. Barrett, British Patent 765,874; *Chem. Abstr.*, **51,** 14829 (1957).

79. Barrett. *J. Chem. Soc.*, 325 (1958).
80. Adamson, Barrett, Billinghurst, and Jones, *J. Chem. Soc.*, 2315 (1957).
81. Borrows and Holland, *Chem. Rev.*, **42,** 611 (1948).
82. Scholtz, *Ber.*, **45,** 734 (1912).
83. A. E. Tschitschibabin and Stepanow, *Ber.*, **62,** 1068 (1929).
84. A. E. Tschitschibabin and Stepanow, *Ber.*, **63,** 470 (1930).
85. Diels, Alder, Friedrichsen, Peterson, Brodersen, and Kech, *Ann.*, **510,** 87 (1934).
86. Diels and Meyer, *Ann.*, **513,** 129 (1934).
87. Diels and Moeller, *Ann.*, **516,** 45 (1935).
88. Diels and Pistor, *Ann.*, **530,** 87 (1937).
89. Plancher, *Ber.*, **35,** 2606 (1902); *Atti Reale. Accad. Lincei*, **11** (5), **II,** 210 (1902); *Chem. Zent.*, **II,** 1472 (1902).
90. Plancher and Crusa, *Atti Reale. Accad. Lincei*, **15** (5), **II,** 453 (1906); *Chem. Zent.*, **II,** 1847 (1906).
91. Saxton, *J. Chem. Soc.*, 3239 (1951).
92. D. R. Bragg and D. G. Wibberley, *J. Chem. Soc.*, 2627 (1962).
93. D. R. Bragg and D. G. Wibberley, *J. Chem. Soc.* (*C*), 3277 (1966).
94. D. R. Bragg and D. G. Wibberley, *J. Chem. Soc.* (*C*), 2120 (1966).
95. J. Hurst, T. Melton, and D. G. Wibberley, *J. Chem. Soc.*, 2948 (1965).
96. T. Melton and D. G. Wibberley, *J. Chem. Soc.* (*C*), 983 (1967).
97. C. A. Henrick, E. Ritchie, and W. C. Taylor, *Aust. J. Chem.*, **20,** 2467–2477 (1967).
98. V. Boekelheide and K. Fahrenholtz, *J. Am. Chem. Soc.*, **83,** 458 (1961).
99. F. Krohnke and H. H. Stevernagel, *Chem. Ber.*, **97,** 1118 (1964).
100. C. A. Henrick, E. Ritchie, and W. C. Taylor, *Aust. J. Chem.*, **20,** 2441 (1967).
101. C. A. Henrick, E. Ritchie, and W. C. Taylor, *Aust. J. Chem.*, **20,** 2455 (1967).
102. V. Boekelheide and N. A. Fedariek, *J. Org. Chem.*, **33,** 2062 (1968).
103. T. Okamoto, M. Hirobe, Y. Tamai, and E. Yabe, *Chem. Pharm. Bull. Japan*, **14,** 506 (1966).
104. T. Sasaki, K. Kanematsu, and Y. Yukimoto, *J. Chem. Soc.* (*C*), 481–485 (1970).
105. R. M. Acheson and D. A. Robinson, *J. Chem. Soc.* (*C*), 1633–1638 (1968).
106. R. M. Acheson and J. N. Bridson, *J. Chem. Soc.* (*C*), 1143–1146 (1969).
107. J. P. Paolini and R. K. Robins, *J. Org. Chem.*, **30,** 4085 (1965).
108. Borrows and Holland, *J. Chem. Soc.*, 670 (1947).
109. L. Pentimalli and S. Bozzini, *Boll. Sci. Fac. Chem. Ind. Bologna*, **23** (2–3), 181–187 (1965) (Ital.); *Chem. Abstr.*, **63,** 14848e (1965).
110. Borrows, Holland, and Kenyon, *J. Chem. Soc.*, 1075 (1946).
111. Kondo and Nishizawa, *J. Pharm. Soc. Japan*, **56,** 1 (1936); *Chem. Zent.*, **I,** 4158 (1936); *Chem. Abstr.*, **30,** 3431 (1936).
112. Buu-Hoi and Hoan, *Rec. Trav. Chim.*, **68,** 781 (1949).
113. Buu-Hoi, Hoan, and Royer, *Bull. Soc. Chim. France*, 489 (1950).
114. Buu-Hoi, Binh, Loc, Xuong, and Jacquignon, *J. Chem. Soc.*, 3126 (1957).
115. H. L. Blewitt, S, Taylor, F. Moorsreiner, J. Graue, and M. Serra, unpublished results from the Tarkio College Research Laboratories.
116. Scholtz, *Ber.*, **45,** 1718 (1912).
117. D. E. Ames, T. F. Grey, and W. A. Jones, *J. Chem. Soc.*, 620–622 (1959).
118. A. E. Tschitschibabin and Stepanow, *Ber.*, **63,** 470 (1929).
119. Holland and Naylor, *J. Chem. Soc.*, 1504 (1955).
120. J. H. M. Hill and M. A. Battiste, *Tetrahedron Lett.*, **53,** 5537–5540 (1968).
121. J. J. Good, M.Sc. Dissertation, Ohio University, Athens, Ohio, 1970.
122. J. G. Lombardino, *J. Org. Chem.*, **30,** 2403 (1965).

123. W. B. Harrell and R. F. Doerge, *J. Pharm. Sci.*, **56,** 225 (1967).
124. W. B. Harrell and R. F. Doerge, *J. Pharm. Sci.*, **56,** 1200 (1967).
125. W. B. Harrell and R. F. Doerge, *J. Pharm. Sci.*, **57,** 1989 (1968).
126. W. B. Harrell, *J. Pharm. Sci.*, **59,** 275 (1970).
127. R. J. Bowers and A. G. Brown, *J. Chem. Soc.* (*C*), 1434 (1970).
128. J. M. Tedder and K. H. Todd, *Chem. Commun.*, 424 (1967).
129. J. M. Tedder, K. H. Todd, and W. K. Gileson, *J. Chem. Soc.* (*C*), 1279 (1969).
130. R. J. Windgassen, W. H. Saunders, and V. Boekelheide, *J. Am. Chem. Soc.*, **81,** 1459 (1959).
131. A. Galbraith, T. Small, R. A. Barnes, and V. Boekelheide, *J. Am. Chem. Soc.*, **83,** 453–458 (1961).
132. L. Pentimalli and S. Bozzini, *Boll. Sci. Fac. Chim. Ind. Bologna*, **23,** 179 (1965); *Chem. Abstr.*, **65,** 13652d.
133. L. Pentimalli and S. Bozzini, *Ann. Chim.* (*Rome*), **56,** 752 (1966); *Chem. Abstr.* **63,** 14848d (1965).
134. Barrows, Holland, and Kenyon, *J. Chem. Soc.*, 1083 (1946).
135. Ochiai and Kobayashi, *J. Pharm. Soc. Japan*, **56,** 376 (1936); *Chem. Zent.*, **II,** 1938 (1936); *Chem. Abstr.*, **30,** 6364 (1936).
136. Kondo and Mitsugi, *J. Pharm. Soc. Japan*, **57,** 397 (1937); *Chem. Zent.*, **II,** 991 (1937); *Chem. Abstr.*, **33,** 2139 (1939).
137. Clemo, Fox, and Raper, *J. Chem. Soc.*, 4173 (1953).
138. K. T. Potts and V. P. Singlr, *Org. Mass. Spectrom*, **3,** 433 (1970).
139. J. H. Bowie, P. F. Donaghue, H. J. Rodda, R. G. Cooks, and D. H. Williams, *Org. Mass Spectrom.*, **1,** 13 (1938).
140. P. M. Draper and D. B. Maclean, *Can. J. Chem.*, **46,** 1487 (1938).
141. S. O. Lawesson, G. Schroll, J. H. Bowie, and R. G. Cooks, *Tetrahedron*, **24,** 1875 (1968).
142. R. Lawrence and E. S. Waight, *J. Chem. Soc.* (*B*), 1 (1968).
143. M. Marx and C. Djerassi, *J. Am. Chem. Soc.*, **90,** 678 (1968).
144. F. G. Holliman, R. A. W. Johnstone, and B. J. Millard, *J. Chem. Soc.* (*C*), 2351 (1967).
145. T. Nishiwaki, *J. Chem. Soc.* (*C*), 885 (1967).
146. W. A. Cole, D. H. Williams, and A. N. H. Yeo, *J. Chem. Soc.* (*B*), 1284 (1968).
147. D. M. Forkey, *Org. Mass Spectrom.*, **2,** 309 (1969).
148. K. T. Potts and C. Hirsch, *J. Org. Chem.*, **33,** 143 (1968).
149. K. T. Potts, V. P. Singh, and J. Bhattachyra, *J. Org. Chem.*, **33,** 3766 (1968).
150. G. Jones and J. Stanyer, *Org. Mass. Spectrom.*, **3,** 1489 (1970).
151. K. Shillings, F. Krohnke, and B. Kickhofen, *Chem. Ber.*, **88,** 1093 (1955).
152. A. E. Tschitschibabin, *Ber.*, **59,** 2048 (1926).
153. A. E. Tschitschibabin, *J. Russ. Phys. Chem. Soc.*, **58,** 1159 (1926).
154. W. L. F. Armarego, *J. Chem. Soc.*, 2778 (1965).
155. M. Fraser, A. Melera, B. B. Malloy, and D. H. Reid, *J. Chem. Soc.*, 3288 (1962).
156. W. L. F. Armarego, *J. Chem. Soc.*, 4226 (1964).
157. W. L. F. Armarego, *J. Chem. Soc.*, *Phys. Org.* (2), 191 (1966).
158. M. Fraser, S. McKenzie, and D. H Reid, *J. Chem. Soc.*, *Phys. Org.* (1), 44 (1966).
159. E. M. Evleth, *Theor. Chim. Acta*, **16,** 22 (1970); *Chem. Abstr.*, **72,** 60988z (1970).
160. M. Colonna and A. Monti, *Atti. Accad. Sci. Ist. Bologna, Classe Sci. Frs., Rend.*, **251,** 13 (1962); *Chem. Abstr.*, **62,** 16177f (1962).
161. I. I. Grandberg, S. B. Nikitina, V. A. Mosknieako, and V. I. Minken, *Khim. Geterotsikl. Soedin*, 1076 (1967); *Chem. Abstr.*, **69,** 52067w (1968).

162. U. Mikstais and A. Arens, *Khiom. Geterotsikl. Soedin.*, 116 (1968); *Chem. Abstr.*, **70,** 77908e (1968).
163. G. Yu. Turchinovich, *Tr. Krevsk. Politekhn. Inst.*, **43,** 91 (1963); *Chem. Abstr.*, **62,** 10559a, e & f (1963).
164. S. McKenzie and D. H. Reid, *Chem. Commun.*, 401 (1966).
165. M. Cardellini, S. Ottolina, and P. Tafara, *Ann. Chim.* (*Rome*), **58,** 1206 (1968); *Chem. Abstr.*, **70,** 77753a (1968).
166. B. S. Thyagarajan and P. V. Gopalakrishnan, *Tetrahedron*, **21,** 3305 (1965).
167. J. D. Bower, *J. Chem. Soc.*, 4510 (1957).
168. S. McKenzie and D. H. Reid, *J. Chem. Soc.* (*C*), 145 (1970).
169. O. Dann and P. Nickel, *Ann.*, **667,** 101 (1963).
170. J. Mirek and J. Mazurek, *Rocz. Chem.*, **42,** 79 (1968).
171. N. J. Leonard, K. Conrow, and R. W. Fulmer, *J. Org. Chem.*, **22,** 1445 (1957).
172. J. Bailey, *Ind. Chim. Belge*, **32,** 90 (1967); *Chem. Abstr.*, **70,** 79138c (1968).
173. J. Mirek, *Rocz. Chem.*, **41,** 307 (1967).
174. E. A. Kochetkova, *Zh. Obshch. Khim.*, **33,** 1201 (1963),
175. N. Saldabols and I. Mazeika, *Khim. Geterotsikl. Soedin.*, 118 (1969); *Chem. Abstr.*, **70,** 115069 (1968).
176. J. P. Paolini, *Diss. Abstr.*, **26,** 2482 (1965).
177. J. H. Boyer and L. T. Wolford, *J. Org. Chem.*, **23,** 1053 (1958).
178. V. Boekelheide and W. Feely, *J. Org. Chem.*, **22,** 589 (1957).
179. H. Budzikiewicz, C. Djerassi, and D. H. Williams, "Interpretation of Mass Spectra," Holden-Day, Inc., San Francisco, 1964.
180. S. D. Sample, D. A. Lightner, O. Buchardt, and C. Djerassi, *J. Org. Chem.*, **32,** 997 (1967).
181. A. L. Jennings and J. E. Boggs, *J. Org. Chem.*, **29,** 2065 (1964).
182. W. W. Paudler and J. E. Kuder, *J. Heterocycl. Chem.*, **3,** 33 (1966).
183. A. Streitwieser, Jr., "Molecular Orbital Theory for Organic Chemists," Wiley, New York, 1961.
184. British Patent 1,058,258 (Cl. C 07d), February 8, 1967; *Chem. Abstr.*, **66,** PC 95047v (1967).
185. U. M. Teotino, L. Polo-Friz, A. Grandini, and D. Della Bella, *Farmaco* (*Pavia*) *Ed. Sci.*, **17,** 988 (1962); *Chem. Abstr.*, **64,** 9677e (1966).
186. L. Almirante and W. Murmann, British Patent 1,076,089 (Cl. C 07d), July 19, 1967: Ital. Appl. Nov. 9, 1965; *Chem. Abstr.*, **68,** 8725a (1968).
187. L. Almirante, A. Mugnaini, P. Rugarli, A. Gamba, E. Zefelippa, N. DeToma, and W. Murmann, *J. Med. Chem.*, **12,** 122 (1969).
188. D. Lednicer, U.S. Patent 3,455,924 (Cl. 260-256.4; C 07d, A61h), 15 Jul. 1969, Appl. 08 Feb. 1967; *Chem. Abstr.*, **71,** 101878p (1969).
189. B. J. Northover, *Brit. J. Pharmacol. Chemother.*, **31,** 483 (1967); *Chem. Abstr.*, **68,** 28263r (1968).
190. R. Aries, French Patent 1,536,351 (Cl. C 07d, A61k), 27 Aug. 1968, Appl. 03 Jul. 1967; *Chem. Abstr.*, **71,** 124437h (1969).
191. W. B. Harrell, S. Kuang, and C. O'Dell, *J. Pharm. Sci.*, **59,** 721 (1970).
192. S. N. Godovikova and A. Ya. Malkina, *Vestn. Dermatol Venerol.*, **42,** 76 (1968); *Chem. Abstr.*, **69,** 50038p (1968).
193. A. Ya. Malkina, S. N. Godovikova, V. F. Gladkikh, E. V. Sazonona, and G. L. Astakhova, *Vestn. Dermatol. Venerol.*, **41,** 44 (1967); *Chem. Abstr.*, **67,** 72351w (1967).
194. C. Casagrande, A. Invernizzi, and G. Ferrari, *Farmaco. Ed. Sci.*, **23,** 1141 (1968); *Chem. Abstr.*, **70,** 57734s (1969).

195. V. S. Venturella, *J. Pharm. Sci.*, **52,** 868 (1963).
196. J. Nayler, British Patent 1,174,124 (Cl. C 07d), December 10, 1969, Appl. 30 Jun. 1967; *Chem. Abstr.* **72,** 55285 (1970).
197. N. Saldabols, L. N. Alekseeva, B. Brizga, L. Kruzmetra, and S. Hillers, *Khim. Farm. Zh.*, **4,** 20 (1970); *Chem. Abstr.*, **73,** 77136p (1970).
198. W. Engel, E. Seeger, H. Teufel, and A. Eckenfels, German Offen. 1,922,191 (Cl. C 07d, A61k), November 12, 1970, Appl. April 30, 1969; *Chem. Abstr.*, **74,** 12995u (1971).
199. J. M. Straley and J. G. Fisher, Def. Publ., U.S. Patent Office 663,484, (Cl. 260-152), December 3, 1968, Appl. August 28, 1967; *Chem. Abstr.*, **71,** 40199u (1969).
200. L. Pentimalli and V. Passalacqua, *Boll. Sci. Fac. Chim. Ind. Bologna*, **24,** 25 (1966); *Chem. Abstr.*, **66,** 11837s (1967).
201. L. Pentimalli and V. Passalacqua, *Boll. Sci. Fac. Chim. Ind. Bologna*, **24,** 205 (1966); *Chem. Abstr.*, **66,** 105869f (1967).
202. G. Ferrari and C. Casagrande, U.S. Patent 3,539,582 (Cl. 260-295); C 08d), November 10, 1970, Belg. Appl. January 18, 1967; *Chem. Abstr.*, **74,** 100042m (1971).
203. J. Bailey, British Patent 1,156,496 (Cl. C 09b), June 25, 1969, Appl. July 23, 1965; *Chem. Abstr.*, **71,** 92638f (1969).
204. E. A. Kochetkova, *Sb. Nauch. Tr. Vladimir. Politekh. Inst.*, (7), 146 (1969); *Chem. Abstr.*, **74,** 88647c (1971).
205. L. Pentimalli, *Boll. Sci. Fac. Chim. Ind. Bologna*, **23,** 15 (1965); *Chem. Abstr.*, **64,** 6789b (1965).
206. F. N. Stepanov and N. A. Aldanova, *Khim. Teckhnol. Promenenie Proizvodnykh Piridina Khinolina*, Materialy Soveshchanya, Inst. Khim., Acad. Nauk Latv. S.S.R., Riga, 1957, 203 (Publ. 1960); *Chem. Abstr.*, **55,** 23529e, f, g (1960).
207. F. N. Stepanov and L. I. Lukashina, *Zhur. Obshchei Khim.*, **30,** 2850 (1960); *Chem. Abstr.*, **55,** 15481d (1960).
208. J. M. Straley and J. G. Fisher, Def. Publ., U.S. Patent Office 663,484 (Cl. 260-152), December 3, 1968, Appl. August 28, 1967; *Chem. Abstr.*, **71,** 40199u (1969).
209. D. W. Heseltine and G. S. Leslie, British Patent 756,226, September 5, 1956; *Chem. Abstr.*, **51,** 4187g (1956).
210. R. S. Long and R. J. Boyle, U.S. Patent 2,877,230, March 10, 1959; *Chem. Abstr.*, **53,** 10781h (1959).
211. R. S. Long and R. J. Boyle, U.S. Patent 2,863,714, December 9, 1958; *Chem. Abstr.*, **53,** 6633b (1958).
212. F. N. Stepanov and N. A. Aldanova, *Angew. Chem.*, **71,** 125 (1959).
213. F. N. Stepanov and L. I. Lukashina, *Zh. Obshch. Khim.*, **33,** 2364 (1963); *Chem. Abstr.*, **60,** 695a (1963).
214. F. N. Stepanov and G. Yu. Turchinovich, *Ukr. Khim. Zh.*, **30,** 738 (1964); *Chem. Abstr.*, **61,** 13480e (1964).
215. R. F. Posse, W. Saleck, W. Laessig, and M. Heilmann, German Patent 1,183,371 (Cl. G 03c), December 10, 1964, Appl. May 10, 1963; *Chem. Abstr.*, **62,** 3568D, 1964.
216. Belg. Patent, 564,823, February 28, 1958; *Chem. Abstr.*, **53,** 14798h (1958).
217. L. E. Craig, U.S. Patent 2,785,133, March 12, 1957; *Chem. Abstr.*, **51,** 9175a (1957).
218. H. Euler, H. Hasselquist, and O. Heidenberger, *Arkiv Kemi* **14,** 419 (1959); *Chem. Abstr.*, **54,** 1215d (1960).
219. L. Petit, N. P. Buu-Hoi, and N. D. Xuong, *Grasas y aceites* (Seville, Spain), **10,** 243 (1959); *Chem. Abstr.*, **54,** 17410i (1960).
220. W. W. Paudler and J. N. Casman, *J. Heterocycl. Chem.*, **10** (4), 499 (1973).
221. W. Engewald, M. Muehlstaedt, and C. Weiss, *Tetrahedron*, **27** (4), 851 (1971).

222. W. Engewald, M. Muehlstaedt, and C. Weiss, *Tetrahedron*, **27,** (17), 4171 (1971).
223. C. Glier, F. Dietz, M. Scholz, and G. Fischer, *Tetrahedron*, **28,** 5779 (1972).
224. B. A. Hess, Jr., L. J. Schaad, and C. W. Holyoke, Jr., *Tetrahedron*, **28,** 3657 (1972).
225. E. Kleinpeter, R. Borsdorf, G. Fischer, and H. J. Hofmann, *J. Prakt. Chem.*, **314** (3–4), 515 (1972).
226. D. M. Grant, R. J. Pugmire, M. J. Robins, and R. K. Robins, *J. Am. Chem. Soc.*, **93** (8), 1887 (1971).
227. W. W. Paudler, R. Van Dahm, and Y. N. Park, *J. Heterocycl. Chem.*, **9** (1), 81 (1972).
228. C. W. Rees and M. Yelland, *J. Chem. Soc., Perkin Trans.*, **1** (3), 221–225 (1973).
229. F. Ya. Perveev and I. I. Afonina, *Zh. Org. Khim*, **7** (2), 420 (1971): *Chem. Abstr.*, **74,** 14164oy (1971).
230. H. Duerr and R. Sergio, *Tetrahedron Lett.*, **33,** 3479 (1972).
231. Maria Michalska, *Tetrahedron Lett.*, **28,** 2667 (1971).
232. Y. Tamura, N. Tsuyimato, Y. Sumida, and M. Ikeda, *Tetrahedron*, **28** (1), 21 (1972).
233. T. Sasaki, K. Kanematsu, and A. Kakehl, *Tetrahedron Lett.*, **51,** 5245 (1972).
234. I. Danis, *Aust. J. Chem.*, **25** (5), 1003 (1972).
235. I. Danis, *Aust. J. Chem.*, **25** (5), 1025 (1972).
236. I. Danis, *Aust. J. Chem.*, **25** (7), 1549 (1972).
237. T. Sasaki, K. Kanematsu, A. Kakehi, and G. Ito, *Tetrahedron*, **28** (19), 4947 (1972).
238. D. Moerler and F. Kroehnke, *Justus Liebigs Ann. Chem.*, **744,** 65 (1971).
239. W. Kiel, *Chem. Ber.*, **105** (11), 3709 (1972).
240. C. Leonte and I. Zugravescu, *Tetrahedron Lett.*, **20,** 2029 (1972).
241. Y. Tamura, Y. Sumida, S. Tamada, and M. Ikeda, *Chem. Pharm. Bull.*, **21** (5), 1139 (1973).
242. T. Sasaki, K. Kanematsu, Y. Yukimoto, and S. Ochiai, *J. Org. Chem.*, **36** (6), 813 (1971).
243. J. W. Lown and K. Matsumoto, *Can. J. Chem.*, **49** (8), 1165 (1971).
244. L. A. Walter, U.S. Patent 3,717,644 (Cl. 260/293.53; C 07d), February 20, 1973, Appl. 571,426, August 10, 1966; 5 pp.
245. N. Prostakov and O. B. Baklibaev, *Khim. Geterotsikl. Soedin*, **9,** 1220 (1972).
246. J. Froehlich and F. Kroehnke, *Chem. Ber.*, **104** (5), 1621 (1971).
247. N. S. Prostakov and O. B. Baktiaev, *Khim. Geterotsikl. Soedin.*, **7** (10), 1395 (1971).
248. W. Flitsch and E. Gerstmann, *Chem. Ber.*, **105** (7), 2344 (1972).
249. Y. Tamura, Y. Sumida, and M. Ikeda, *Chem. Pharm. Bull.*, **20** (20), 1058 (1972).
250. G. Caronna and S. Palazzo, *Atte Accad. Sci., Lett. Arte Palermo*, Part 1, **30,** 5–21; 1969–1970 (pub. 1971), *Chem. Abstr.*, **77,** 151797f (1972).
251. N. Saldabols and J. Popelis, *Khim. Geterotsikl. Soedin*, **5,** 691 (1972): *Chem. Abstr.*, **77,** 126503z, (1972).
252. W. Triebis, *Jerusalem Symp. Quantum Chem. Biochem.*, **3,** 91 (1971).
253. T. Okamoto, T. Irikura, S. Suzue, K. Ushiyama, Y. Matsui, Y. Nagatsu, and S. Sato, Japan Kokai 73 72, 193 (Cl. 16 E611), September 29, 1973, Appl. 7202,665, December 29, 1971; 3 pp.
254. J. A. Hickman and D. G. Wibberley, *J. Chem. Soc., Perkin Trans.*, **I** (23), 2954 (1972)
255. N. Saldabols, L. L. Zeligman, and S. Hillers, *Khim. Geterotsikl. Soedin*, **7** (6), 812 (1971).
256. W. W. Paudler, C. I. Chao, and L. S. Helmick, *J. Heterocycl. Chem.*, **9** (5), 1157 (1972).
257. A. V. El'tsov, M. L. Aleksandrova, and Yu. V. Lepp, *Zh. Org. Khim.*, **8** (2), 418 (1972).

258. G. Jones, G. R. Cliff, and J. Stanyer, *J. Chem. Soc. C.*, **20,** 3426 (1971).
259. J. A. Hickman and D. G. Wibberley, *J. Chem. Soc., Perkin Trans.*, **1** (23), 2958 (1972).
260. L. Greci, M. Collonna, P. Bruni, and G. Padovano, *Gazz. Chim. Ital.*, **101** (5), 396 (1971).
261. A. Bailey, B. R. Brown, and M. C. Churn, *J. Chem. Soc. C.*, **9,** 1590 (1971).
262. L. Greci and G. Padovano, *Atti. Accad. Sci. Ist Bologna, Cl. Sci. Fis., Rend.*, **7** (2), 84 (1970): *Chem. Abstr.*, **75,** 76510k (1971).
263. S. Huenig, G. Kiesslich, F. Linhart, and H. Schlaf, *Justus Liebigs Ann. Chem.* **752,** 182 (1971).
264. E. E. Glover and M. Yorke, *J. Chem. Soc. C.*, **19,** 3280 (1971).
265. S. Huenig and F. Linhart, *Tetrahedron Lett.*, **17,** 1273 (1971).
266. K. Borgdanowicz-Szwed and B. Kawalek, *Zesz. Nauk. Univ. Jagiellon., Pr. Chem.*, No. 18, 187 (1973): *Chem. Abstr.*, **80,** 47763t (1974).
267. D. J. Pointer, J. B. Wilford, and D. C. Bishop, *Nature (London)*, 239 (5371), 332 (1972).
268. D. J. Pointer and J. B. Wilford, *J. Chem. Soc., Perkin Trans.*, **2** (15), 2259 (1972).
269. G. Yu. Stepanova, *Visn. Kiiv. Politekh. Inst. Ser. Khim. Mashinobuduv, Tekhnol*, **7,** 107 (1970); *Chem. Abstr.*, **75,** 88448j (1971).
270. J. W. Carpenter, J. D. Mee and D. W. Heseltine, U.S. Patent 3,615,639 (Cl. 96/130; G 03c), October 26, 1971, Appl. 677,058, October 23, 1967; 16 pp.
271. L. Pentimalli, L. Greci, and G. Milani, *Ann. Chim. (Rome)*, **61** (4), 254 (1971).
272. J. W. Carpenter, J. D. Mee, and D. W. Haseltine, U.S. Patent 3,809,691 (Cl. 260–240E; C 09b), May 7, 1974, Appl. 677,058, October 23, 1967; 15 Division of U.S. 3,615,639; *Chem. Abstr.*, **81,** 51156k (1974).
273. C. Holstead, German Offen. 2,122,025 (Cl. G 03c) November 18, 1971, Brit. Appl. May 7, 1970; 37 pp; *Chem. Abstr.*, **76,** 40242f (1972).
274. C. Holstead, German Offen. 2,122,060 (Cl. G. 03c), November 18, 1971, Brit. Appl. May 4, 1970; 71 pp; *Chem. Abstr.*, **76,** 40240d (1972).
275. M. H. Fisher, U.S. Patent 3,701,780 (Cl. 260–294.8c; C 07d), October 31, 1972, Appl. 73,603, September 18, 1970; 8 pp; *Chem. Abstr.*, **78,** 43479r (1973).
276. M. H. Fisher and A. Lusi, *J. Med. Chem.*, **15** (9), 982 (1972).
277. A. V. De and B. P. Saha, *J. Pharm. Sci.*, **62** (11), 1897 (1973).
278. V. P. Arya, F. Fernandes, and V. Sudarsanam, *Indian J. Chem.*, **10** (6), 598 (1972).
279. J. N. Wells, J. N. Davisson, W. R. Campbell, S. Sangiah, and G. K. W. Yim, *J. Med. Chem.*, **16** (6), 700 (1973).
280. C. Casagrande, R. Ferrini, G. Miragoli, and G. Ferrari, *Farmaco. Ed. Sci.*, **27** (9), 715 (1972).
281. A. G. Brown and J. H. Nayler, German Offen. 2,046,904 (Cl. C. 07d), April 22, 1971, Brit. Appl. October 4, 1969.
282. David Jack and E. E. Glover, German Offen. 2,127,355 (Cl. C 07d, A61k) December 16, 1971, Brit. Appl. June 3, 1970; 30pp: *Chem. Abstr.*, **76,** 85818q (1972).
283. Lewis A. Walter, U.S. Patent 3,642,807 (Cl. 260–296B; C 07d), February 15, 1972, Appl. 571,426, August 10, 1966; 4pp: *Chem. Abstr.*, **76,** 140563e (1972).
284. K. R. Kallay and R. F. Doerge, *J. Pharm. Sci.*, **61** (6), 949 (1972).
285. N. Saldabols, S. Hillers, L. N. Alekseeva, and I. V. Dipan, *Khem.-Form. Zh.*, **6** (6), 16 (1972): *Chem. Abstr.*, **77,** 88402f (1972).
286. C. Casagrande, A. Invernizzi, R. Ferrini, and G. Miragoli, *Farmaco, Ed. Sci.*, **26** (12), 1059 (1971).
287. L. Bolger, R. T. Brittain, D. Jack, M. Jackson, L. E. Martin, J. Mills, D. Poynter, and M. B. Tyers, *Nature (London)*, 238 (5363), 354–535 (1972).

288. T. M. Savege, C. E. Blogg, L. M. Ross, M. Lang, and B. R. Simpson, *Anaesthesia*, **28** (3), 253 (1973).
289. A. J. Coleman, A. O'Brien, J. W. Downing, D. E. Jeal, D. G. Moyes, and W. P. Leary, *Anaesthesia*, **28** (3), 262 (1973).
290. L. Beregi, P. Hugon, J. Duhault, and P. Desnoyers, French Addn. 0300 (Cl A 61k, C 07d), March 16, 1970, Appl. 161,473, August 1, 1968; 3pp.
291. E. Camarri, L. Zaccherotti, and D. D'Alonzo, *Arzneim.-Forsch.*, **22** (4), 768 (1972): *Chem. Abstr.*, **77,** 83707z (1972).
292. G. Pallavicini, G. Cetta, and A. Calatroni, *Ball. Soc. Ital. Biol. Sper.*, **47** (20), 644 (1971).
293. A. Magnaini, L. Almirante, and W. Murmann, S. African 71 07, 020 (Cl. C 07d), January 25, 1973, Australian Appl. 22, 794/70, November 30, 1970; 7pp: *Chem. Abstr.*, **80,** 82984n (1974).
294. E. F. Rogers, J. Hannah, and R. A. Dybos, German Offen. 2,306,001 (Cl. C 07d, A 61k), August 23, 1973, U.S. Appl. 224,620, February 8, 1972; 39pp: *Chem. Abstr.*, **79,** 126518 (1973).
295. I. Irikura, K. Kasuga, T. Hashizume, M. Ohashi, and M. Yamada, Japan. Kobai 7230, 693 (Cl. 16 E 611), November 9, 1972, Appl. 71 16,744, March 23, 1971; 3pp.: *Chem. Abstr.*, **78,** 29772z (1973).
296. S. Suzue, M. Hirobe, and T. Okamoto, *Chem. Pharm. Bull.*, **21** (10), 2146 (1973).
297. T. Okamoto, M. Hirobe, S. Suzue, and Y. Nagatsu, German Offen. 2,118,917 (Cl. C 07d), March 30, 1972, Japan. Appl. 70,83,985, September 25, 1970; 11pp: *Chem. Abstr.*, **77,** 19641w (1972).
298. T. Irikwra, M. Hayashi, K. Koshirae, H. Urawa, and E. Hetsugi, German Offen. 2,315,801 (Cl. C 07d), October 11, 1973, Japan. Appl. 7332,033, March 30, 1972; 20pp: *Chem. Abstr.*, **80,** 14923s (1974).
299. Y. Tamura, Y. Mike, Y. Y. Sumida, and M. Ikeda, *J. Chem. Soc., Perkin Trans.*, **1** (21), 2580 (1973).
300. T. Sasaki, K. Kanematsu, A. Kahehi, and G. Ito, *J. Chem. Soc., Perkin Trans.*, **1** (19), 2089 (1973).
301. Y. Tamura, Y. Sumida, and M. Ikeda, *J. Chem. Soc., Perkin Trans.*, **1** (19), 2091 (1973).
302. J. Mirek and K. Bogdanowicz, *Zesz. Nauk Univ. Jagiellon., Pr. Chem.* (18), 179 (1973).
303. T. Afonia and F. Ya Perveeu, *Zh. Org. Khim*, **9** (10), 2006 (1973).
304. M. Sone, Y. Tominaga, R. Natsuki, Y. Matsuda, and G. Kobayashi, *Yakugaku Zasshi*, **94** (7), 839 (1974): *Chem. Abstr.*, **81,** 152106z (1974).
305. H. Duerr and R. Sergio, *Chem. Ber.*, **107** (6), 2027 (1974).
306. H. Duerr, H. Kober, R. Sergio, and V. Formacek, *Chem. Ber.*, **107** (6), 2037 (1974).
307. H. Duerr and W. Schmidt, *Justus Liebigs Ann. Chem.* (7), 1140 (1974).
308. T. Kato, Y. Yamamoto, and S. Takeda, *Yakugaku Zasshi*, **94** (5), 627 (1974).
309. B. M. Lynch and B. P. Lem. *J. Heterocycl. Chem.*, **11,** 223 (1974).
310. J. A. Hickman and D. G. Wibberley, *J. Chem. Soc., Perkin Trans.*, **1,** 1972 (23), 2954.
311. J. A. Hickman and D. G. Wibberley, *J. Chem. Soc., Perkin Trans.*, **1,** 1972 (23), 2958.
312. R. Jaxquier, H. Lopez, and G. Maury, *J. Heterocycl. Chem.*, **10** (5), 755 (1973).

CHAPTER III

Azaindolizine Systems Having More Than One Nitrogen Atom in the 6-Membered Ring

GEORGES MAURY

Laboratoire de synthèse et d'étude physicochimique d'hétérocycles azotés, Centre de Chimie organique. Université des Sciences et Techniques du Languedoc, Montpellier, France (and Département de Chimie. Faculté des Sciences. Rabat, Morocco)

I. Introduction

In recent years the azaindolizine system (**1**) has attracted much interest because of its aromaticity, as well as the usefulness of certain of its derivatives. The azaindolizines (**1**) formally derive from indolizine (**2**) by systematically replacing CH groups with nitrogen atoms. In this chapter we consider *azaindolizine systems having more than one nitrogen atom (N-4) in the 6-membered ring.* This represents most of the possibilities (exactly 120). Other members of the series are described in Chapter 2.

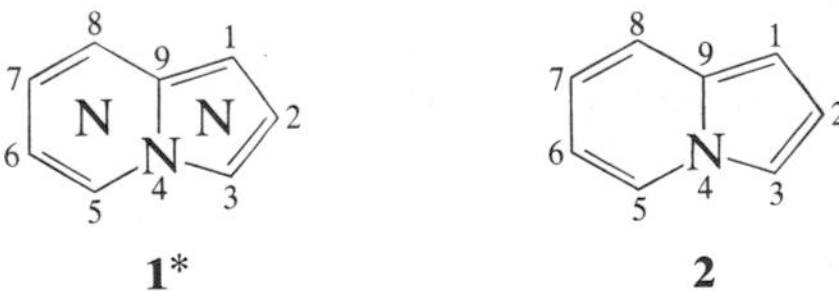

Considering the large number of heterocyclic systems taken into account, efforts have been made to organize the text as a comparative study rather than a detailed coverage of every series. It follows that, for the sake of clarity in comparing the various structures, *all individual systems* (**1**) *have been oriented as shown in structure* **1** although it is not always the

* The letter N in a ring means that one or several CH groups are replaced with nitrogen atoms. For the sake of simplicity the corresponding heterocycles have been referred as azaindolizines rather than polyazaindolizines.

numbering recommended by the *Ring Index* or by *Chemical Abstracts.* On the other hand, these heterocycles have been generally named according to the *Ring Index*, and current nomenclature rules. Because of the uncertainty in the position of tautomeric equilibria, hydroxy-, amino-, or mercaptoazaindolizine derivatives have been represented as such regardless of their actual structures (except of course in the section dealing with proton tautomerism of these compounds).

The azaindolizine systems known in 1958 have been individually reviewed by W. L. Mosby in an earlier volume of this series.[1] In the present article the aim has been to review the literature from 1959 to mid-1974. However references to papers published before 1959 are not necessarily excluded. Finally, no attempt has been made to review the literature dealing with reduced azaindolizines or benzo derivatives of these compounds.

II. Synthesis of Azaindolizine Ring Systems

Azaindolizine heterocycles are conveniently prepared by using two simple and general procedures: either the formation of the 5-membered ring from a suitably substituted azine or the formation of the 6-membered ring from an azole derivative.

A. Formation of the 5-Membered Ring from a Substituted Azine

1. *Pyrroloazines from N-(β-oxoalkyl)azinium derivatives*

CH_3 O Base → R

N N ⊕ CH_2 C R N N

3 4

This method is an extension of the Tschitschibabin procedure in the pyridine series[2]: cyclization of **3** occurs through proton abstraction with a base such as sodium bicarbonate. Consequently the success of the preparation depends on the acidity of the corresponding *N*-alkylazinium cation **3**. Thus the adducts of 2,4,6-trimethylpyrimidine and 4,6-dimethylpyrimidine with phenacyl bromide afford 2-phenyl-5,7-dimethylpyrrolo[1,2-*a*]pyrimidine[3] and 2-phenyl-7-methylpyrrolo[1,2-*c*]-pyrimidine,[4] respectively. 3,6-Dimethylpyridazine and ethyl bromopyruvate give ethyl 6-methylpyrrolo[1,2-*b*]pyridazine-2-carboxylate in low yield,[5] and attempts to cyclize 1-phenacyl-2,5-dimethylpyrazinium bromide have failed.[6] A recent study shows that cyclization does not occur if the 4-methyl group of 3-(β-oxoalkyl)-4-methylpyrimidinium halides is substituted with electron withdrawing groups.[7]

Another way of cyclizing **3** to a derivative of **4** uses 1,3-dipolar cycloaddition of a dialkyl acetylenecarboxylate on zwitterions of compounds **3** under dehydrogenating conditions. Thus **5** yields the pyrrolo-[1,2-*a*]pyrazines **6** and **7**.[6] Compound **7** was directly obtained from 1,3,6-trimethylpyrazinium bromide.[10] Similar, but more complex reactions from pyrazine[5] (or 2-methylpyrazine[8]) and 3-methylpyridazine[5] with dialkyl acetylenedicarboxylate afford trimethyl pyrrolo[1,2-*a*]pyrazine-1,2,3-tricarboxylate and trimethyl 6-methylpyrrolo[1,2-*b*]pyridazine-1,2,3-tricarboxylate, respectively.[9]

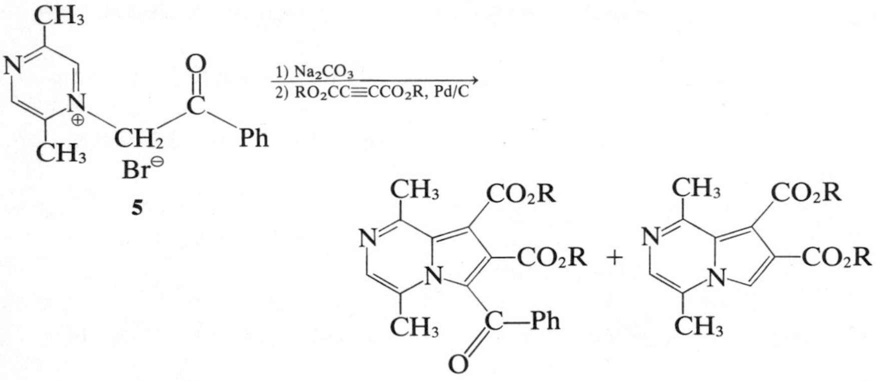

CH_3 N N$^\oplus$ CH_3 CH_2 C O Ph $Br^\ominus$

5

1) Na_2CO_3
2) $RO_2CC{\equiv}CCO_2R$, Pd/C

CH_3 CO_2R N N CO_2R CH_3 C—Ph O + CH_3 CO_2R N N CO_2R CH_3

6 **7**

Recently dipolar cycloaddition of azinium methylides and activated acetylene derivatives has been shown to be a general and efficient method to prepare pyrrolo[1,2]azines, for example, pyrrolo[1,2-*b*]pyridazines, pyrrolo[1,2-*a*]pyrazines, and pyrrolo[1,2-*a*]pyrimidines.[252–254] Similarly, pyridazine *N*-imides lead to pyrazolo[2,3-*b*]pyridazines:[252]

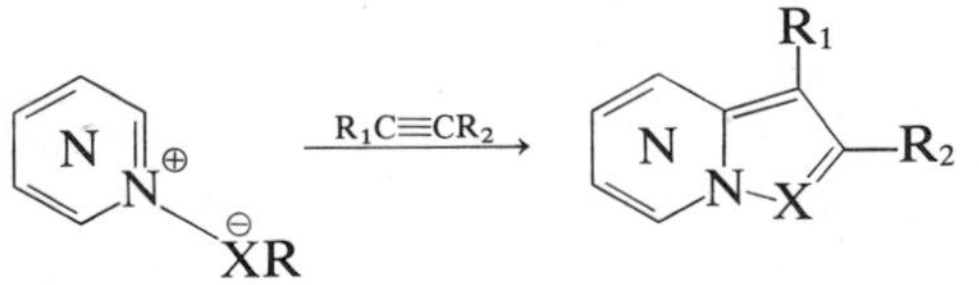

N N$^\oplus$ $\overset{\ominus}{X}R$ $\xrightarrow{R_1C{\equiv}CR_2}$ N N X R_1 R_2

(X = CR or N)

Pyrrolo[1,2-*b*]pyridazines, pyrrolo[1,2-*a*]pyrazines or -pyrimidines are also obtained in good yields in the reactions of the corresponding azines and cyclopropenones,[255] for example,

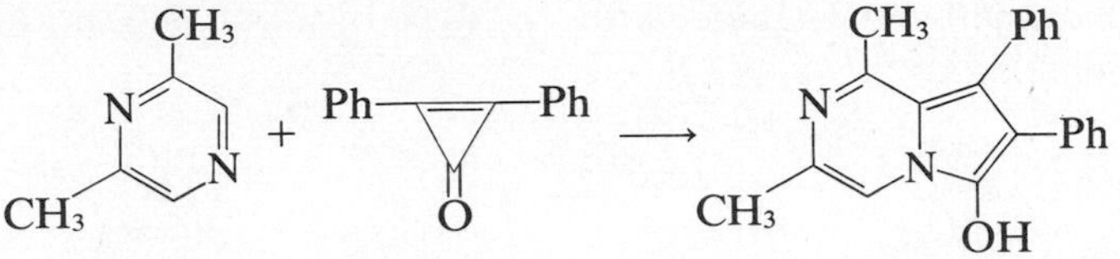

2. *Imidazoazines from Aminoazine Derivatives*

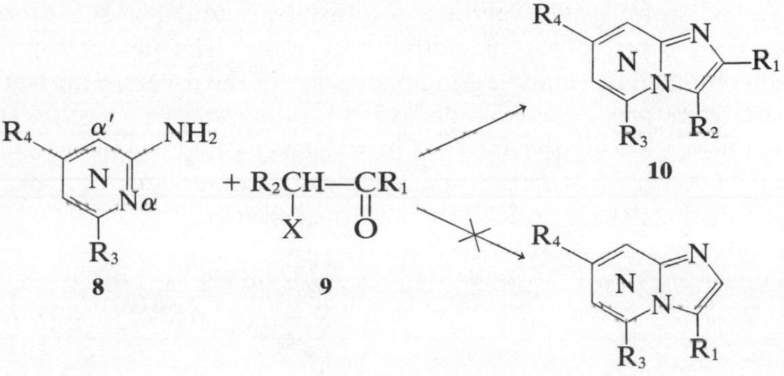

Condensations of this type could lead to either compound **10** or **11** assuming the formation of C–N bonds exclusively. However, only compound **10** is generally obtained unless the Dimroth rearrangement of this compound occurs. There is no simple explanation of the apparent specificity of the reaction especially since tautomerization of the aminoazine must be taken into account. The more favored intermediate presumably arises from the attack of the α-carbon in **9** by a ring nitrogen atom of **8**. This mechanism is currently accepted in the 2-aminopyridine series[11] and substantiated by the isolation of compound **13** in the condensation of 3-aminopyridazine **12** and phenacyl bromide.[12] If compound **9** is an α-halo acid halide, the same orientation prevails with the acylation of the exocyclic amino group as the initial step of the condensation, although *C*-acylation has also been reported.[13] The situation is further complicated if the α and α' positions of **8** are both occupied by a nitrogen atom: two isomeric products of type **10** can be formed in proportions depending on the relative bulks of R_3 and R_4. A general study of the applicability of the reaction suggests that its success depends largely on the nature of the substituents of the aminoazine nucleus.[14] In particular, nitro substituents are generally associated with low yields or absence of reaction.[14] Finally, the condensation is usually performed in refluxing methanol or ethanol,[14] although dimethylformamide with sodium bicarbonate has also been recommended.[15]

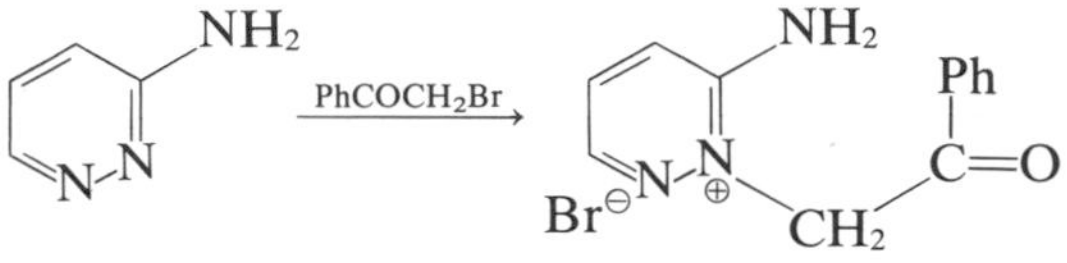

NH_2 PhCOCH$_2$Br → NH_2 Ph C=O $Br^{\ominus}$ $N^{\oplus}$ CH_2

12 **13**

Imidazo[1,2-*b*]pyridazines (**15**) are conveniently prepared (generally in good yields) from 3-aminopyridazines unsubstituted[12] or substituted at position 6 by chloro[12,16,17] or methoxy groups.[18] The parent compound has been obtained by catalytic dehalogenation of the 6-chloro derivative.[19] This procedure was later extended to the preparation of methyl substituted imidazo[1,2-*b*]pyridazines.[20] The existence of a long-range coupling between H-3 and H-8 in the nmr spectra of compounds of type **15** is useful to distinguish 2- and 3-substituted derivatives.[19]

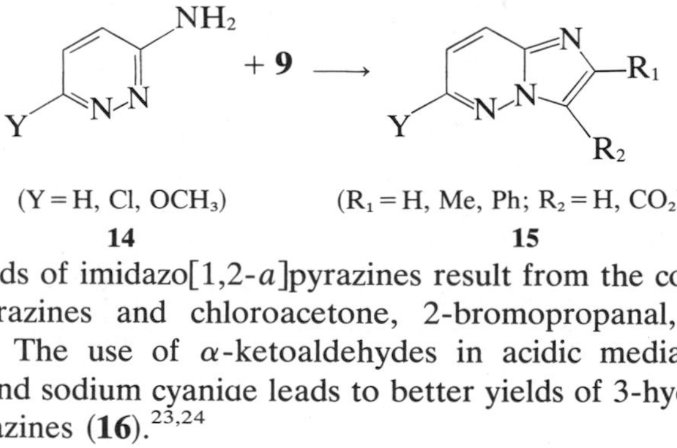

NH_2 + **9** → R_1 R_2

(Y = H, Cl, OCH_3) (R_1 = H, Me, Ph; R_2 = H, CO_2Et)[16,17,21]

14 **15**

Low yields of imidazo[1,2-*a*]pyrazines result from the condensation of 2-aminopyrazines and chloroacetone, 2-bromopropanal, or phenacyl bromide.[14] The use of α-ketoaldehydes in acidic media or of formaldehyde and sodium cyanide leads to better yields of 3-hydroxyimidazo-[1,2-*a*]pyrazines (**16**).[23,24]

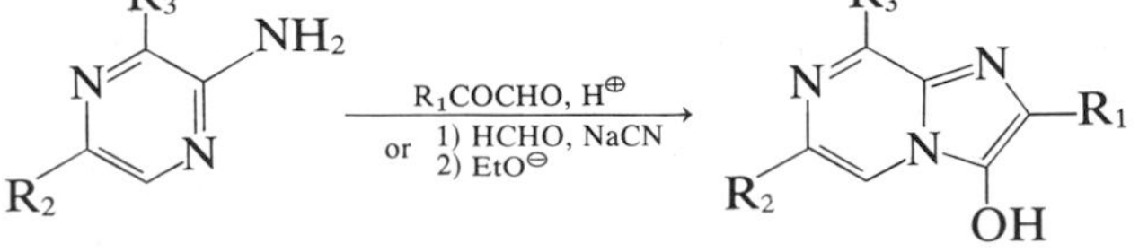

R_3 NH_2 R_2 —— R_1COCHO, $H^{\oplus}$ or 1) HCHO, NaCN 2) $EtO^{\ominus}$ → R_3 R_1 R_2 OH

16

A large variety of the easily available 2-aminopyrimidines have been condensed with α-halo carbonyl compounds, generally in good yields (some examples of this reaction are listed in Table I). However, the

TABLE I. PREPARATION OF IMIDAZO[1,2-*a*]PYRIMIDINES FROM 2-AMINOPYRIMIDINES

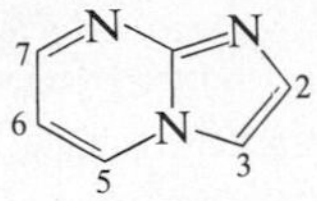

Substituents						
R_2	R_3	R_5	R_6	R_7	Yield	Ref.
H	H	H	H	H	69	25
Me	H	H	H	H	60	22
H	Me	H	H	H	50	22
H	H	Me	H	Me	78	25
Aryl	H	H	H	H	—	26
Ph	H	OH	H	Me	61	15
Me	H	H	Cl	H	—	27
Ph	H	H	Br	Me	63	28
Ph	H	OH	NO_2	Me	61	28
Aryl	H	H	OMe	H	33.5	29

influence of the nature and position of the substituents needs to be further investigated, particularly in view of the deactivating effect of the nitro group on the 2-aminopyrimidine nucleus.[14]

Two intermediates analogous to **13** can be formed in the condensation of dissymmetrically substituted 2-aminopyrimidines and α-halo carbonyl compounds. Recent experimental results suggest that the formation and cyclization of these intermediates are controlled by steric factors if the 4(6) position of the 2-aminopyrimidine nucleus is substituted. Thus the condensation of **17** ($R_3 = H$) and α-haloaldehydes or α-haloketones gives predominantly the 7-methyl isomer **18** regardless of the nature of the other substituents.[25,28,30b] Similarly, 2-amino-4-hydroxy-6-methylpyrimidines lead only to the corresponding 5-hydroxy-7-methylimidazo-[1,2-*a*]pyrimidines; however electronic factors probably play an important part in this reaction because the 5-hydroxy isomer is the only product of the reaction between 2-amino-4-hydroxypyrimidine and phenacyl

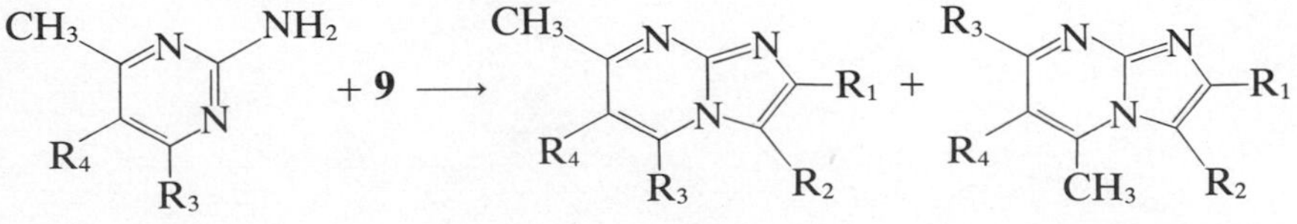

17 + **9** ⟶ **18** + **19**

bromide.[15] Intramolecular *N*-alkylation of 2-(β-phenoxyethylamino)-4-hydroxypyrimidines yields two isomeric dihydroimidazo[1,2-*a*]-pyrimidines.[256,257]

Less satisfactory results are obtained from 4-aminopyrimidines because of their apparent low reactivity and lack of specificity of the condensation in this case. Thus 4-aminopyrimidine and 2-bromopropanal give low yields of both 2- and 3-methylimidazo[1,2-*c*]pyrimidines; the former isomer probably arises from the rearrangement of the latter.[22] 2-Methylthio-4-hydroxy-6-aminopyrimidine (**20**) and chloracetaldehyde lead to compound **21**, rather than the expected 5-methylthio-7-hydroxy-imidazo[1,2-*c*]primidine.[13] However 1*N*-methylcytosine and chloracet-aldehyde give 6-methyl-5-oxo-5,6-dihydroimidazo[1,2-*c*]pyrimidine.[258] Condensation of compound **20** with chloracetyl chloride involves first the acylation of the exocyclic amino group, and then cyclization to a 2-hydroxyimidazo[1,2-*c*]pyrimidine system, although 4-aminouracil is initially *C*-acylated at position 5.[13] Phenacyl bromide and derivatives lead normally to 2-arylimidazo[1,2-*c*]pyrimidines.[27]

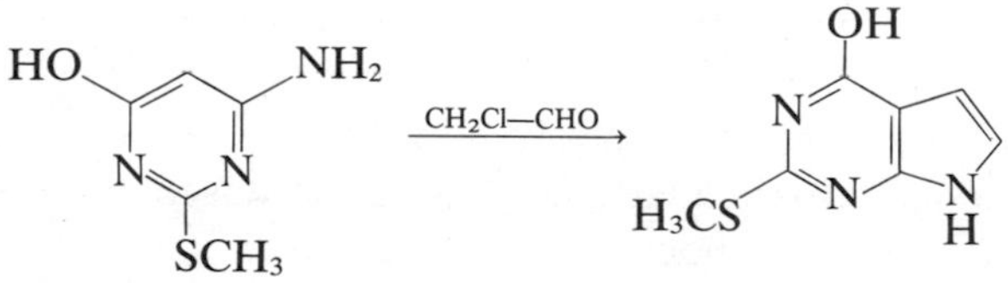

HO, NH_2, N, N, SCH_3 —(CH_2Cl—CHO)→ OH, N, H_3CS, N, N, H

20 **21**

The interaction of 3-amino-*as*-triazines and α-halo carbonyl compounds may lead to systems **22** or **23**.[31] Initially structure **22** had been chosen over **23** without justification,[32,33] or by analogy with the cyclization of 3-hydrazino-*as*-triazines known to involve N-2 rather than N-4.[14,31] The nmr spectrum of the condensation product of 5,6-dimethyl-*as*-triazine and 3-bromo-2-butanone shows no abnormal deshielding of the methyl groups, thus ruling out structure **23** in which the *peri*-situated 3- and 5-methyl groups are expected to interact sterically.[34] Similarly several imidazo[1,2-*a*]-*s*-triazines have also been prepared from 2-amino-*s*-triazine derivatives using the same method.[35,36]

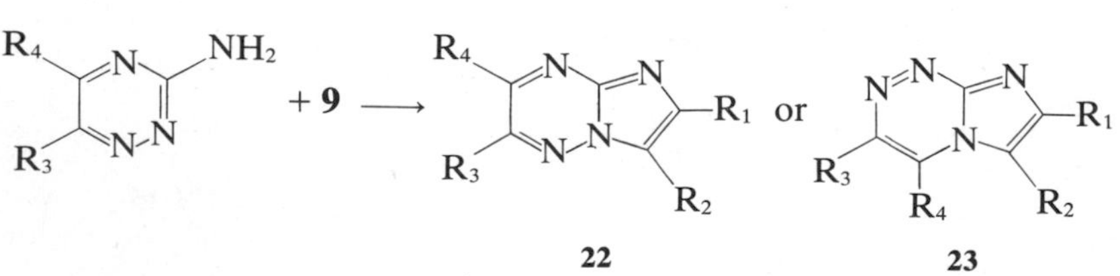

R_4, N, NH_2, R_3, N, N + **9** ⟶ R_4, N, N, R_1, R_3, N, N, R_2 or N, N, N, R_1, R_3, N, R_4, R_2

22 **23**

Few other preparations of imidazoazines from 6-membered heterocycles have been reported. Armarego prepared, in low yields, three parent imidazodiazines from the corresponding chloroazines and α-aminoacetaldehyde diethylacetal.[37] Cyclization of 1-(2-propynyl)-2-iminopyrimidines (**24**) gives **25** in ethanolic sodium ethoxide but also the rearranged compounds **26** in aqueous sodium hydroxide.[38–40]

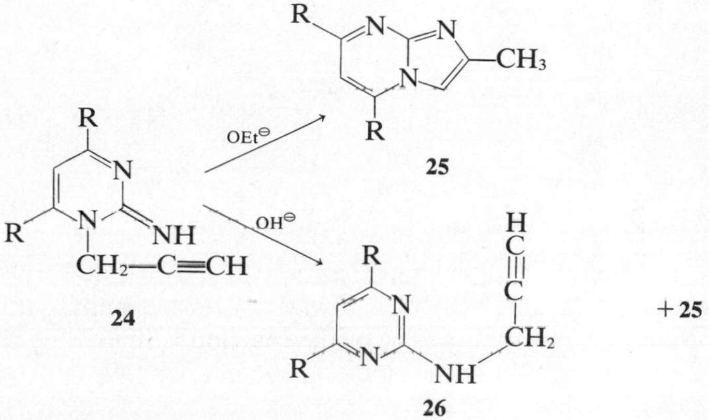

Imidazo[1,5]azines can be prepared from 2-aminomethylazines, but the usefulness of the method is limited by the instability of the amine. Thus the unsubstituted imidazo[1,5-*a*]pyrazine was prepared in moderate yield from 2-aminomethylpyrazine and formic acid.[259] The condensation of azadicarboxamidine and β-dicarbonyl compounds leads to the previously unknown imidazo[5,1-*f*]-*as*-triazine system,[260] for example,

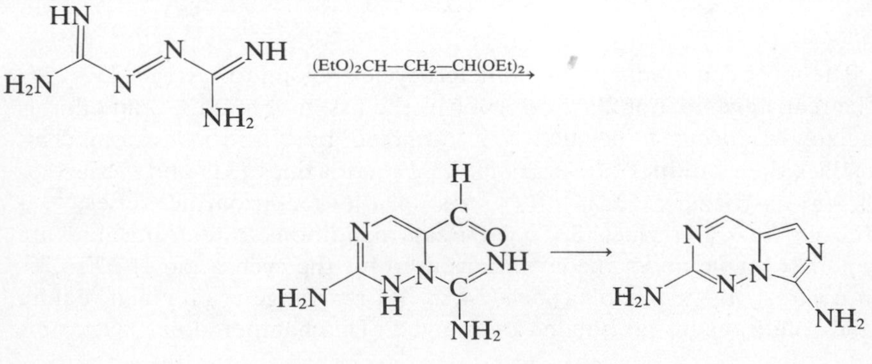

3. *s-Triazoloazines from Hydrazinoazine Derivatives*

a. *s*-TRIAZOLO[4,3] OR [3,4]AZINES

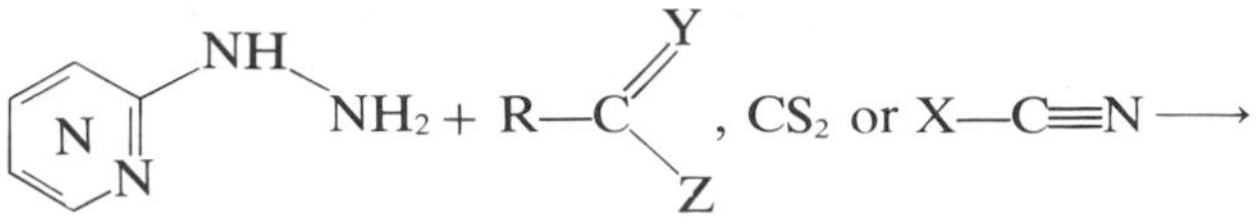

NH—NH$_2$ + R—C(=Y)—Z , CS_2 or X—C≡N ⟶

27 **28**

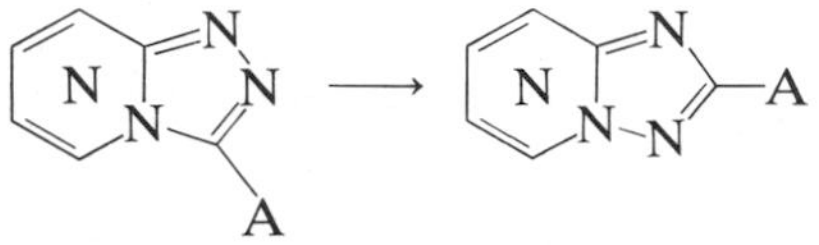

29 ⟶ **30**

The condensation of hydrazinoazines (**27**) with one-carbon cyclizing agents (**28**) is a simple and efficient way to prepare many *s*-triazoloazines (**29**). However, the applicability of the reaction is limited by the eventual rearrangement of (**29**) to the isomeric compounds (**30**) depending on the conditions.

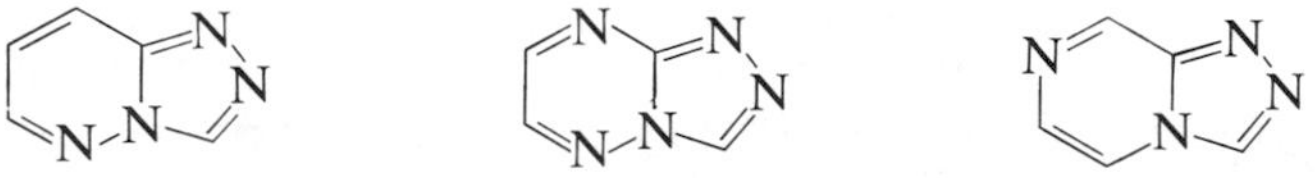

31 **32** **33**

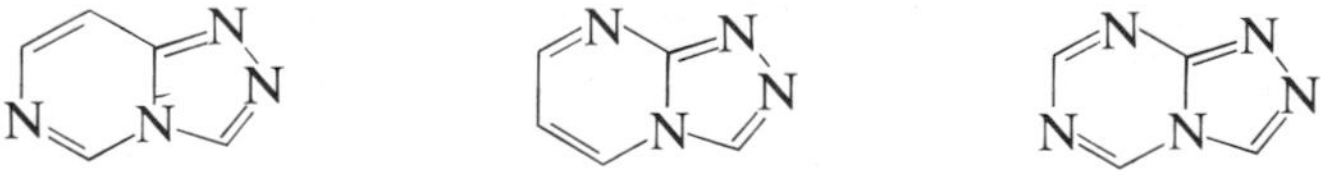

34 **35** **36**

The most commonly prepared heterocycles belong to systems **31** to **36**. The rearrangement of **29** to **30** involves the fisson of bond 4–5 and thus is unlikely to occur if position 5 is occupied by a nitrogen atom. This justifies the stabilities of *s*-triazolo[4,3-*b*]pyridazines (**31**) and *s*-triazolo-[4,3-*b*]-*as*-triazines (**32**).[41,261,262] As in the imidazoazine series,[22] *s*-triazolo[4,3-*a*]pyrazines (**33**) need drastic conditions to rearrange and are generally stable under the conditions used for the cyclization of **27** to **29**. In contrast, the *s*-triazoloazines (**34** to **36**) rearrange readily and require mild conditions to be isolated unchanged. This hampered early attempts to elucidate the product structures especially in the *s*-triazolo[4,3-*a*]pyrimidine series.[42]

Two possibilities of cyclizing 2-hydrazinopyrimidine,[43] 2-hydrazino-*s*-triazine, and 3-hydrazino-*as*-triazine derivatives exist (e.g., compounds **37** and **38** are obtained from 2-hydrazino-4-hydroxy-6-methylpyrimidine).[44a,45] The orientation of cyclization is influenced by the nature of substituents and the experimental conditions. For example, cyclization of 3-hydrazino-5-hydroxy-6-methyl-*as*-triazine with formic acid involves only N-2, whereas under neutral conditions (ethyl orthoformate) a 50% yield of 5-hydroxy-6-methyl-*s*-triazolo[4,3-*c*]-*as*-triazine is obtained.[46] Ring closure of 2-benzylidenehydrazino-4-methoxy-*s*-triazine is controlled by steric and electronic factors associated with substituents on the 6-membered ring and affords a 2.5 : 1 ratio of 7-methoxy- and of 5-methoxy-3-phenyl-*s*-triazolo[4,3-*a*]-*s*-triazine, respectively.[54]

$\xrightarrow{HC(OEt)_3}$

37 **38**

The substituent introduced at position 3 of **29** depends on the nature of the cyclizing agent (**28**).[263] Formic acid or acetic anhydride lead to **29** (A = H or Me, respectively).[31,47] Ethyl orthoformate, ethyl orthoacetate, or diethoxymethyl acetate react under milder conditions and are more appropriate to prepare derivatives of systems **34** to **36**,[44a,48,49] but they have also been used in the series **31** and **33**.[50,51] 3-Phenyl-*s*-triazoloazines (**29**) are

TABLE II. INTRODUCTION OF VARIOUS SUBSTITUENTS R AT POSITION 3 OF *s*-TRIAZOLOAZINES

Series	R = H, alkyl, or aryl (Ref.)	R = OH (Ref.)	R = NH_2 (Ref.)	R = SH (Ref.)
s-Triazolo[4,3-*b*]pyridazines (**31**)	41, 47, 50, 65–67.	56.	50, 68.	—
s-Triazolo[4,3-*b*]-*as*-triazines (**32**)	31, 46, 69.	—	—	70, 71.
s-Triazolo[4,3-*a*]pyrazines (**33**)	51,59,72.	57, 59, 72.	57, 59	59, 72.
s-Triazolo[4,3-*c*]pyrimidines (**34**)	48, 73.	48, 58, 74.	48, 59.	75.
s-Triazolo[4,3-*a*]pyrimidines (**35**)	44, 45, 49, 53, 77.	42, 76.	43, 78.	44b, 45, 76, 79, 80.
s-Triazolo[4,3-*a*]-*s*-triazines (**36**)	54, 55.	—	—	264.

obtained either in cyclizing *N*-benzoyl derivatives of **27**[51,52] or by dehydrogenative cyclizing its benzylidene derivatives with lead tetraacetate or bromine. The latter method has been especially useful in the preparation of *s*-triazolo[4,3-*b*]pyridazines,[50,52] *s*-triazolo[4,3-*a*]pyrimidines,[45,53] and *s*-triazolo[4,3-*a*]-*s*-triazines.[54,55]

Ethyl chlorocarbonate,[48,56] phosgene,[57,58] or urea[59] lead to intermediates of type **39**, which undergo cyclization to **29** (A = OH) generally without rearrangement (Table II). 3-Amino-*s*-triazoloazines **29** (A = NH_2) are conveniently obtained by condensing compounds **27** and cyanogen halides (Table II) at low temperature and in neutral media.[59] Similarly the interaction of **27** and carbon disulfide, phenylisothiocyanate or 1,1′-thiocarbonyldiimidazole yields 3-mercapto-*s*-triazolozines (Table II).

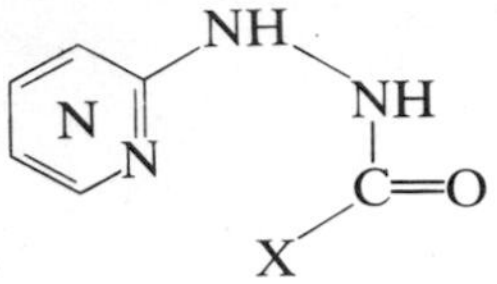

39 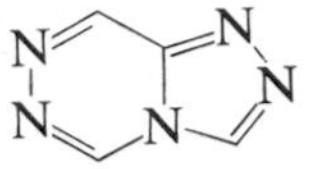 **40** 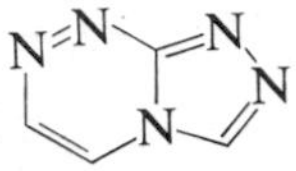 **41** 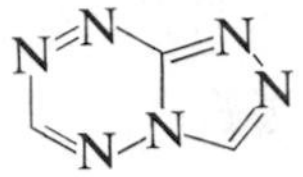 **42**

This procedure has also been useful (although less frequently applied) in the synthesis of derivatives of systems **40** to **42**. Thus 3-alkyl- and 3-mercapto-*s*-triazolo[4,3-*d*]-*as*-triazines have been obtained from 5-hydrazino-*as*-triazines[60,62,265] and derivatives of **41** from 3-hydrazino-*as*-triazines.[46,63] Finally, the condensation of 2-hydrazino-6-phenyl-*s*-tetrazine and ethyl orthoformate affords a 54% yield of 6-phenyl-*s*-triazolo[4,3-*b*]-*s*-tetrazine.[64]

b. *s*-Triazolo[2,3] or [3,2]azines (**30**)[9]. Most of the known *s*-triazoloazines (**30**) have been obtained via the Dimroth rearrangement of the appropriate isomer (**29**). The critical parameters influencing the rearrangement are the acidity or the basicity of the medium and the temperature. Depending on the series, the rearrangement is carried out under basic (NaOH, Na_2CO_3) or acidic (HCl, HCO_2H) conditions or through the application of heat alone[43] by refluxing in either nitrobenzene or in dichlorobenzene. The mechanisms of these rearrangements are discussed in another section.

The interaction of 2- and 4-hydrazinopyrimidines with cyanogen halides yields the rearranged isomer (**30**) under any conditions other than at low temperature and in neutral media.[43,59,78] However drastic conditions may induce the hydrolytic decomposition of the heterocycle as in the case of *s*-triazolo[2,3-*c*]pyrimidine derivatives in acidic media,[59] or the base-catalyzed rearrangements of 5,7-dimethoxy-*s*-triazolo[2,3-*a*]-pyrimidines.[43] The introduction of a 5-hydroxy substituent in the *s*-triazolo[4,3-*a*]pyrimidine ring seems to increase its resistance to rearrangement in acidic and basic media.[43,77] In contrast, a carbethoxy group at position 6 has been claimed to favor the rearrangement of 7-hydroxy-8-alkyl-*s*-triazolo[4,3-*a*]pyrimidines.[53] Attempted base- or acid-catalyzed rearrangements of *s*-triazolo[4,3-*a*]pyrazine derivatives under vigorous conditions gave low yields of rearranged products or led to an unexpected cleavage of the 6-membered ring.[81] Finally, a number of *s*-triazolo[2,3-*a*]-*s*-triazines,[54] and *s*-triazolo[2,3-*d*] or [3,2-*c*]-*as*-triazines[61,82,83] have been obtained in good yields from the appropriate isomers (**29**).

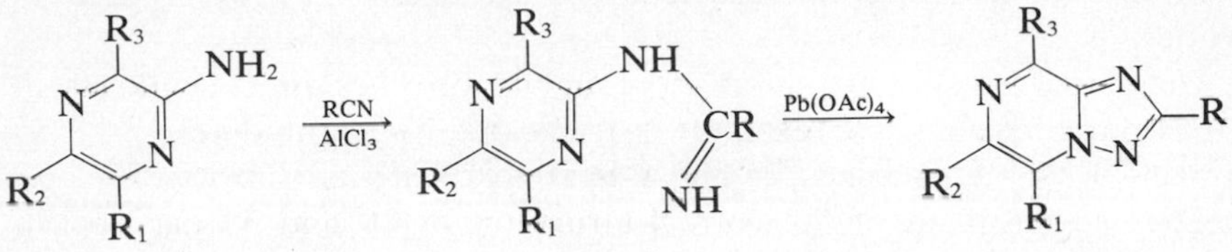

R_3, NH_2, N, N, R_2, R_1 —RCN / $AlCl_3$→ R_3, NH, N, CR, N, NH, R_2, R_1 —$Pb(OAc)_4$→ R_3, N, N, R, N–N, R_2, R_1

43 **44**

Direct syntheses of the *s*-triazoloazines (**30**) have also been developed. These are especially useful when the Dimroth rearrangement of the corresponding isomer (**29**) is difficult or impossible. Thus **44** was obtained through oxidative cyclization of the pyrazinylamidine **43**.[84,85] This procedure had previously been used in the preparation of *s*-triazolo[2,3-*a*]-pyrimidines.[86] Other related syntheses involve the cyclization of **45** to the *s*-triazolo[2,3-*a*]pyrimidine **46** in excellent yield,[43] and the condensation of hydrazine with 4-ureidopyrimidines.[74]

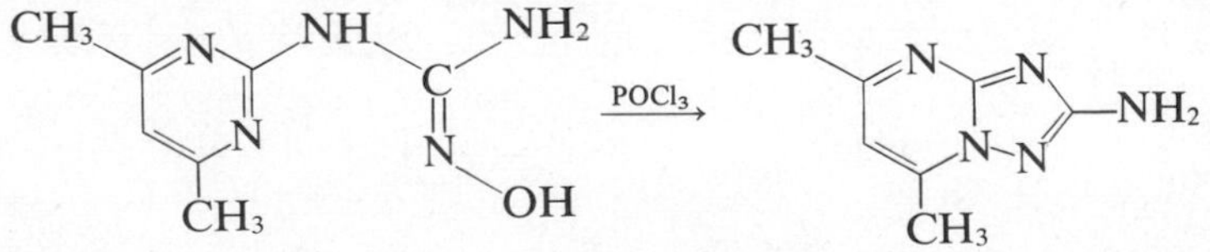

CH_3, N, NH, NH_2, C, N, N, OH, CH_3 —$POCl_3$→ CH_3, N, N, NH_2, N–N, CH_3

45 **46**

Recently the preceding methods have been extended to the synthesis of several *s*-triazoloazine systems. Thus the oxidative cyclization of 3-pyridazinylamidines leads to *s*-triazolo[2,3-*b*]pyridazines,[266] whereas the cyclization of hydroxyiminomethyleneaminoazines in polyphosphoric acid gives *s*-triazolo[2,3-*b*]pyridazine, *s*-triazolo[2,3-*a*]pyrazine, -pyrimidine or -*s*-triazine derivatives:[267,268]

$$\text{2-aminoazine (NH}_2\text{)} \longrightarrow \text{N=CH—N(CH}_3)_2 \longrightarrow \text{N=CH—NHOH} \longrightarrow \text{s-triazolo[2,3]azine}$$

A related reaction involving aromatic nitrile-oxides and sulfimides derived from 2-aminopyrimidines leads to *s*-triazolo[2,3-*a*]pyrimidine 3-oxides.[269]

Diamino-1,2-azines are suitable precursors for other direct syntheses of *s*-triazolo[2,3]azines, for example *s*-triazolo[2,3-*a*]pyrimidines,[270,271] *s*-triazolo[3,2-*c*]-*as*-triazines,[272] and *s*-triazolo[2,3-*d*]-*as*-triazines.[273]

The establishment of the correct structures of **29** and **30** has been a very important step in all of the preceding preparations of these compounds. The methods of characterization involve the synthesis of **29** or **30** by independent routes (often from a substituted azole), the rearrangement of **29** to **30** and the spectrometric study of these isomers (mainly nmr,[49] uv,[44a,59,80] mass spectrometry,[57] and even X-ray crystallographic examinations[59]). The characterization of the four possible isomers expected from the cyclization of 2-hydrazino-4-methyl-6-hydroxypyrimidine illustrates the necessity and the difficulty of determining unambiguously the structures of the products in these reactions.[77,80]

4. *Tetrazoloazines from Hydrazinoazine or Haloazine Derivatives*

$$\underset{\mathbf{27}}{\text{NH—NH}_2\ (\alpha', N\alpha)} \xrightarrow{HNO_2} \underset{\mathbf{47a}}{N_3} \rightleftharpoons \underset{\mathbf{47b}}{\text{tetrazolo}} \xleftarrow{NaN_3} \underset{\mathbf{48}}{X}$$

Potentially useful in the study of valence tautomeric processes, the tetrazoloazines (**47**) are prepared according to the general methods shown above: (1) treatment of hydrazinoazines (**27**) with nitrous acid, (2) nucleophilic substitution of substituent X (e.g., a halogen atom) in structure **48** by azide ion in DMF, ethylene glycol, or acidic ethanol.

As in the synthesis of *s*-triazoloazines, the eventual rearrangement of compounds of type **47** complicates the structure problem. It has been shown by ir and nmr spectroscopy that the nature of the products (**47a**, **47b**, or both) depends on the solvent[87,88] and on the physical state of the compounds; in particular, structure **47b** is favored in the solid state since the characteristic azide band is absent from the 2150-cm^{-1} infrared region[87–89] (however, exceptions have been reported in the case of tetrazolo[1,5-*a*]-*s*-triazines).[36] If both the α and α' positions of **27** or **48** are occupied by nitrogen atoms and the aromatic nucleus is dissymmetrically substituted, two alternative modes of ring closure exist and the equilibrium may involve two different tetrazolo isomers (**47b**).[90] The tetrazolo–azido interconversion is discussed in more detail in a forthcoming section.

The previously reported syntheses (**27** or **48** → **47**) have been mostly utilized to prepare derivatives of tetrazolo[1,5-*b*]pyridazine,[66,91–94] tetrazolo[1,5-*c*]pyrimidine,[87,95–97] and tetrazolo[1,5-*a*]pyrimidine.[44a,97–99,263] In particular, the diazotization of **49**, the condensation of 3-chloropyridazine-1-oxide and sodium azide,[94] or direct *N*-oxidation of tetrazolo[1,5-*b*]pyridazine[100] lead to 3-azidopyridazine-1-oxide (**50**). Valence isomerization to the azido form is also induced by the formation of a second azole ring fused to the 6-membered ring of the tetrazolo[1,5-*b*]pyridazine nucleus.[89,93,101] 8-Methyl-5-hydroxytetrazolo[1,5-*c*]pyrimidine was recently obtained in the reaction of ethyl *cis*-3-cyanocrotonate with aluminium azide in THF.[274]

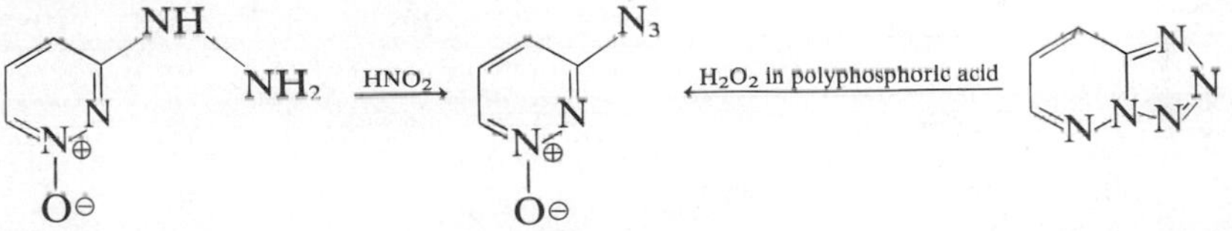

NH–NH_2 —HNO_2→ N_3 ←H_2O_2 in polyphosphoric acid—

49 **50**

Substitution of the chlorine atom at position 4 of the nitroazine **51** by azide ion in hot tetrahydrofuran afforded the oxadiazolopyrimidine *N*-oxide **52** and not the expected tetrazolo[1,5-*c*]pyrimidine derivative.[96] However, in aqueous hydrochloric acid the tetrazolopyrimidine was obtained and degraded.[96]

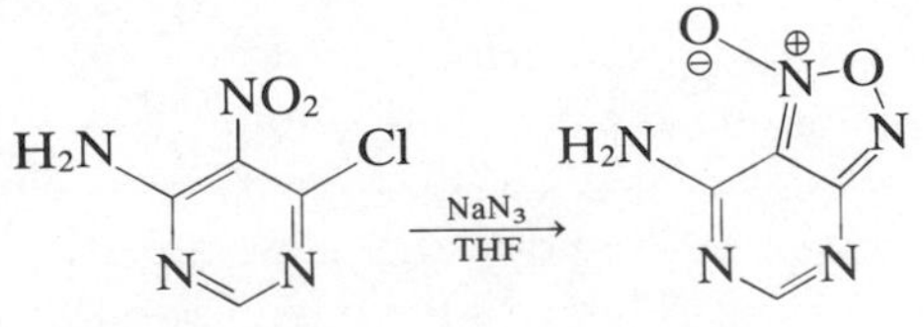

NaN_3, THF

51 **52**

The other tetrazoloazine series are less known, although a few tetrazolo[1,5-*a*]pyrazines,[89,97,102] tetrazolo-*s*-triazines,[36] tetrazolo-*as*-triazines,[60,63,89,103,265] and tetrazolo[1,5-*b*]-*s*-tetrazines[248] have been prepared from the corresponding azines **27** or **48**.

B. Formation of the 6-Membered Ring From a Substituted Azole

Reactions of this type are generally less specific and less widely applicable than the preceding syntheses from substituted azines. However, this is an important and efficient way to prepare azaindolizines especially when the substituted azoles used as starting materials are easily accessible.

1. *Azolo[b]pyridazines*

A number of pyrrolo[1,2-*b*]pyridazines have been recently prepared in good yields from 1-aminopyrrole and β-dicarbonyl compounds[104,105,275] or diketene.[246] Dissymmetric β-ketoaldehydes or β-diketones lead to two positional isomers (e.g., **54** and **55**). These compounds are characterized by uv spectroscopy or by an unequivocal synthesis. The proposed assignments[104] are consistent with the nmr spectra. Thus the condensation of **53** and benzoylacetone affords 6-methyl-8-phenylpyrrolo[1,2-*b*]pyridazine since there is no coupling between H-7 and the methyl group. With β-ketoesters *N*-(1-pyrrolyl)-β-enaminoesters are obtained, but cyclization does not give the expected pyrrolo[1,2-*b*]pyridazines.[105]

Ph–C(=O)–CH_2–C(=O)–H + **53** (H_2N–N pyrrole) ⟶ **54** (Ph) + **55** (Ph)

Similarly imidazo[1,5-*b*]pyridazine derivatives have been prepared by condensing 1-aminoimidazole and β-diketones.[276]

β-Ketoesters as well as β-diketones react with 4-amino-*s*-triazole (**56**) leading to *s*-triazolo[4,3-*b*]pyridazine derivatives.[41,66,106–108,277,278] In general, the structure of **57** was not unambiguously ascertained and 8-hydroxy-*s*-triazolo[4,3-*b*]pyridazines were assumed to result from the condensation of **56** with β-ketoesters. This was proved to be correct for the condensation of **56** and ethyl acetoacetate, a reaction that afforded 6-methyl-8-hydroxy-*s*-triazolo[4,3-*b*]pyridazine,[66] contrary to an earlier report.[109]

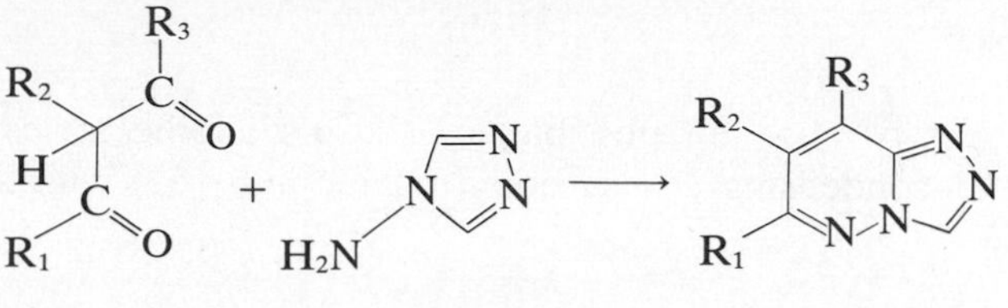

56 **57**

8-Hydroxy-6-methyl-*v*-triazolo[3,4-*b*]pyridazine has been claimed to be obtained from 1-amino-*v*-triazole.[110]

2. *Azolo[a]pyrazines*

Cyclization of the intermediates **59** obtained from 2-acylpyrroles (**58**) and 2-aminoacetaldehyde diethylacetal affords two isomeric products, **60** and **61**, as shown by uv[111] and nmr[112] spectral evidence. The yields of pyrrolo[1,2-*a*]pyrazines (**60**) are considerably greater[111] (and have been further improved[112]) than those of the corresponding compounds of structure **61**, indicating a marked preference for the attack on the nitrogen under these conditions.[111] Recently a similar synthesis has been reported involving the condensation of the sodium derivative of **58** and α-halocarbonyl compounds or their acetals and subsequent cyclization of the corresponding dicarbonyl intermediates with ammonium acetate.[113]

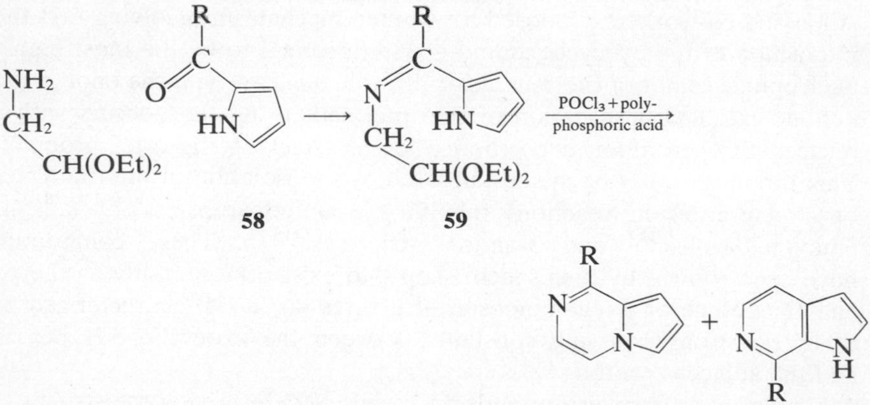

$POCl_3$ + polyphosphoric acid

58 **59**

60 **61**

Attempts to prepare *s*-triazolo[4,3-*a*]- or -[2,3-*a*]pyrazines from 3-aminomethyl-*s*-triazole and α-dicarbonyl compounds failed.[57]

3. *Azolo[a]pyrimidines*[9]

Derivatives of all eight possible azolo[a]pyrimidine series have been prepared by condensing β-dicarbonyl compounds (**62**) with aminoazoles of type **63**.

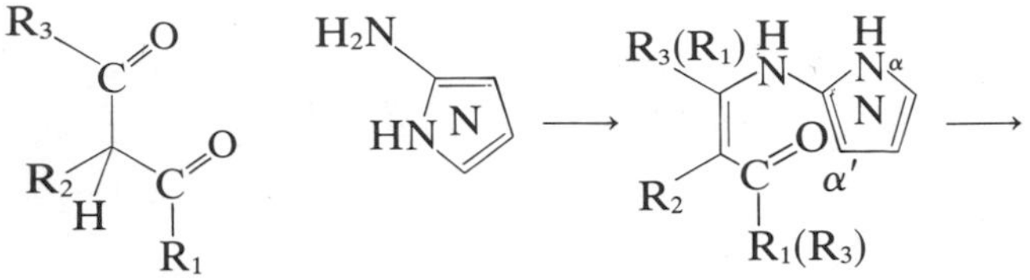

62 **63** **64**

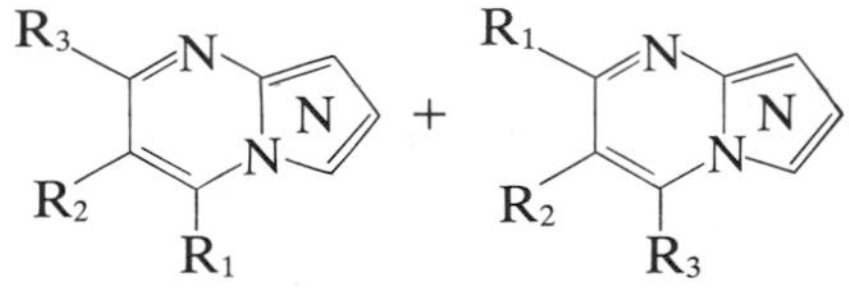

65 **66**

Mechanistic studies are hampered by the multiple possible orientations of the condensation and the potential tautomerism of both **62** and **63**. In discussing the condensation of 3-amino-*s*-triazole with ethyl acetoacetate, Allen and co-workers proposed a two-step mechanism involving first the interaction of the exocyclic amino group (assumed to be the most highly nucleophilic center of the 3-amino-*s*-triazole nucleus) with the enol of the ketonic carbonyl. Cyclization of the intermediate **64** then occurs with a nuclear nitrogen atom at position α or α' attacking the ester group.[44b] This proposition has been substantiated by the isolation of intermediates of structure **64** in reactions involving 3-aminopyrazoles,[114–116,281] 4-aminoimidazoles,[117] and 3-amino-*s*-triazoles.[118,119] These compounds have been shown, by nmr spectroscopy, to exist preferentially as the β-enaminoketone or β-enaminoester structures (as in **64**) characterized by relatively strong coupling constants between the exocyclic NH proton and the adjacent proton.[115,117]

Nmr spectrometry also provided a means of determining the structures of the positional isomers (**65** or **66**)[25] and prompted new investigations into the direction of cyclization of the reaction. Thus the nature and the amount of catalyst have been shown to influence the cyclization of intermediates (**67**) leading to **69** (R = H) in acidic medium and of both **68** and **69** under basic catalysis.[115] The selectivity of the reaction of **62** and **63** can be improved by condensing monoketals of compounds of type **62** in suitable media.[30b,117] If both the α and α' positions of **64** are occupied by nitrogen atoms, the occurrence of base- or acid-catalyzed Dimroth rearrangements of the products is possible, and this possibility must always be kept in mind.[22] A general study of the condensation (using 2-aminoimidazole and its 5-methyl derivative) suggests that the proportions of isomers **65** and **66** are partly controlled by steric effects in the ring closure of the corresponding intermediates (**64**).[30b] Finally, regardless of the nature of **63**, most of the studies show that its exocyclic amino group reacts preferentially with an aldehydic rather than a ketonic group and with the latter rather than an ester or nitrile function.

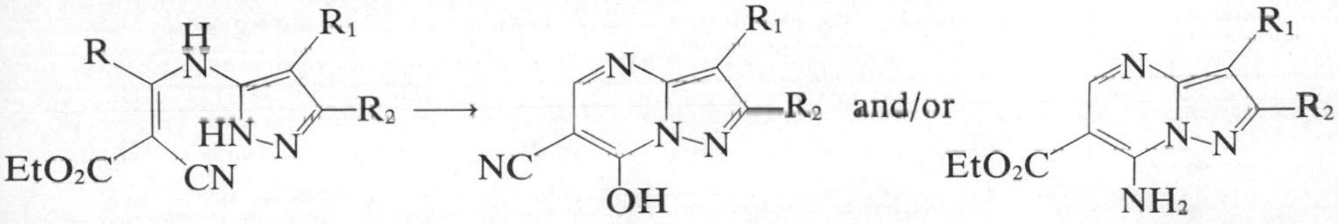

R, H, N, R_1, R_2, HN–N, EtO_2C, CN → N, R_1, R_2, N–N, NC, OH and/or N, R_1, R_2, N–N, EtO_2C, NH_2

67 **68** **69**

The preceding method has been successfully applied to the preparation of series of pyrrolo[1,2-*a*]pyrimidine derivatives from 2-amino-3-cyano-pyrroles and β-diketones, β-ketoaldehydes, or β-ketoesters.[120] Pyrrolo[1,2-*a*]pyrimidines can also be obtained in high yields by self-condensation of 2-(phenacylamino)-3-cyanopyrroles in the presence of potassium *t*-butoxide.[279]

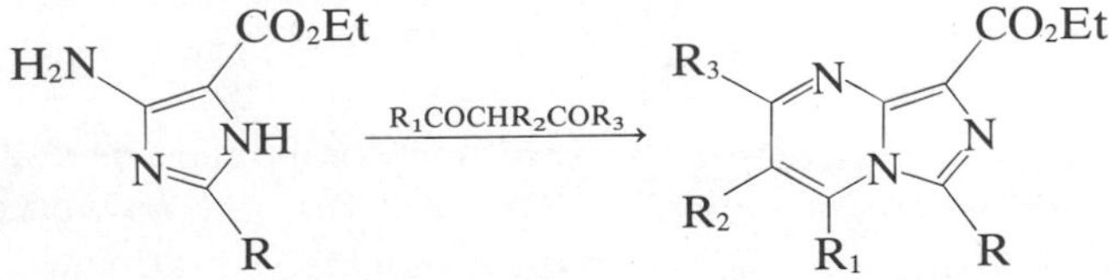

H_2N, CO_2Et, NH, N, R —$R_1COCHR_2COR_3$→ R_3, N, CO_2Et, N, N, R_2, R_1, R

70

4-Amino-5-carbethoxy (or 5-methyl) imidazoles and various β-dicarbonyl compounds lead to imidazo[1,5-*a*]pyrimidines (**70**).[117,121] The parent imidazo[1,5-*a*]pyrimidine was also prepared in low yield from the unstable 4-aminoimidazole.[30a] Similarly 2-aminoimidazoles are convenient starting materials in the synthesis of imidazo[1,2-*a*]pyrimidines. 5-Hydroxy derivatives are normally obtained from β-aldehydo- or β-ketoesters,[30b,44a,122,123] whereas 2-amino-4(5)-hydroxyimidazoles afford 2-hydroxyimidazo[1,2-*a*]pyrimidines.[124]

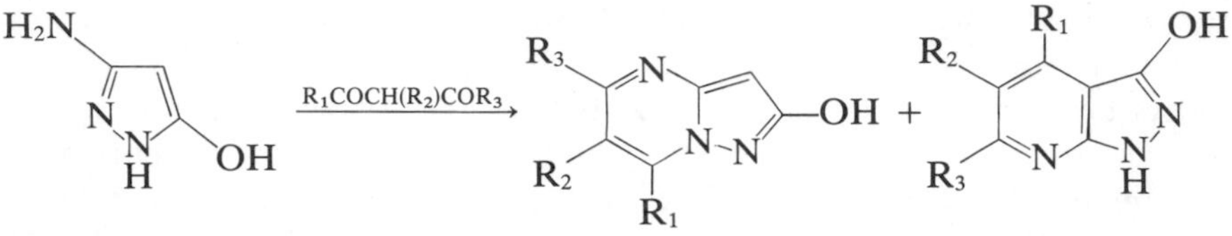

$R_1COCH(R_2)COR_3$

71 **72** **73**

Substituents on the pyrazole nucleus markedly influence the course of the condensation of 3-aminopyrazoles and β-dicarbonyl compounds. Thus the activation of position 4 of **71** towards electrophilic attack induces the formation of the pyrazolo[3,4-*b*]pyridines (**73**) along with the expected pyrazolo[2,3-*a*]pyrimidines (**72**). This observation caused a controversy over the actual structures of the products.[125] The influence of the substituents is also illustrated in this series by the condensation of the aminopyrazole **74** and acetylacetone under basic conditions.[126]

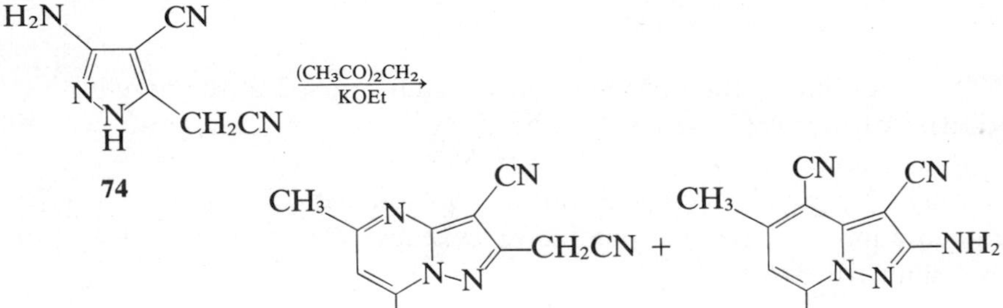

$(CH_3CO)_2CH_2$, KOEt

74 **75** **76**

Precursors or derivatives of β-dicarbonyl compounds have also been condensed with 3-aminopyrazoles. Takamizawa and co-workers have studied the condensations of β-ketonitriles or their ketals and various 3-aminopyrazoles leading to 5-aminopyrazolo[2,3-*a*]pyrimidines of pharmaceutical importance.[115,127–131] Low yields of 5-hydroxy- and 7-hydroxypyrazolo[2,3-*a*]pyrimidines result from the reaction of 3-aminopyrazole and acetylenic esters.[132] Diketene and cyanoacethydrazide yield the same pyrazolo[2,3-*a*]pyrimidine derivative as **71** and ethyl acetoacetate.[133] 5-Aminopyrazolo[2,3-*a*]pyrimidine is formed in the reaction of β-aminocrotonitrile and hydrazine.[282] Finally, the cyclization of intermediate **79** obtained from **77** and **78** has been used as a convenient route to 7-hydroxypyrazolo[2,3-*a*]pyrimidines (**80**)[116] or to 5-aminopyrazolo[2,3-*a*]pyrimidines[280] depending on the nature of R.

R_1O_2C, OR_2, NC, R: C=C (**77**) + R, CN, CO_2R_1, NH–NH_2: C=C (**78**) ⟶

77 **78**

R_1O_2C, NC, R: C=C–N(H)–N(H)–C(R)=C(CN)CO_2R_1 (**79**) ⟶ HO, NC, R, CO_2R_1, R, N, N–N (**80**)

79 **80**

Numerous investigations of the preparation of *s*-triazolo[2,3-*a*]-pyrimidines from 3-amino-*s*-triazoles, often motivated by the practical importance of these compounds, have been carried out. Ring closure of the corresponding intermediate (**64**) (seldom isolated) always involves N-2 of the 3-amino-*s*-triazole nucleus. *s*-Triazolo[4,3-*a*]pyrimidines have never been obtained in this reaction.[134] In attempts to justify this experimental fact, Williams qualitatively studied the charge distribution in the 3-amino-*s*-triazole nucleus and proposed that the exocyclic nitrogen atom or a nuclear nitrogen atom is the more nucleophilic site under acidic or basic conditions, respectively.[135] The condensation of **81** and **82** indeed results in the formation of different orientation products in acidic and basic media.[135] On the other hand, the exocyclic amino group is normally more reactive toward an aldehyde than a ketone, an ester, or a nitrile.[43,44b,80,135,179] Thus 5-hydroxy-7-alkyl-*s*-triazolo[2,3-*a*]pyrimidines are obtained from **81** and β-ketoesters[136] under strongly basic conditions. However acetalization of the aldehydic or ketonic carbonyl can reverse the trend and induce the formation of the 7-hydroxy isomers.[137] 7-Hydroxy-*s*-triazolo[2,3-*a*]- or -[4,3-*a*]pyrimidines have also been obtained in the condensation of **81** with ethyl 2,3-dibromopropionate,[138] with methyl phenylpropiolate,[283] or with ethyl tetrolate.[137] Ring closure of an intermediate of type **67** obtained from **81** affords a 5-amino-*s*-triazolo[2,3-*a*]pyrimidine derivative in acidic medium. Both the 5-hydroxy and 5-amino isomers (analogous to **68** and **69**) are formed under basic conditions,[118] as is also observed in the pyrazolo[2,3-*a*]pyrimidine series. Cyanacetic ester and **81** lead to **84**.[119,139] The latter is also prepared from **85** by successive replacements of the chlorine atoms with amino and hydroxy groups.[179] 2-Amino isomers result from the condensation of 3,5-diamino-*s*-triazole and β-dicarbonyl compounds.[43,44d] Finally, the condensations of ethyl acetoacetate and 3-amino-*s*-triazoles alkylated at N-4 or at the exocyclic amino group lead to 1-alkyl-5-hydroxy-*s*-triazolo[2,3-*a*]pyrimidines or 8-alkyl-5- or -7-hydroxy-*s*-triazolo-[2,3-*a*]pyrimidines, respectively.[140]

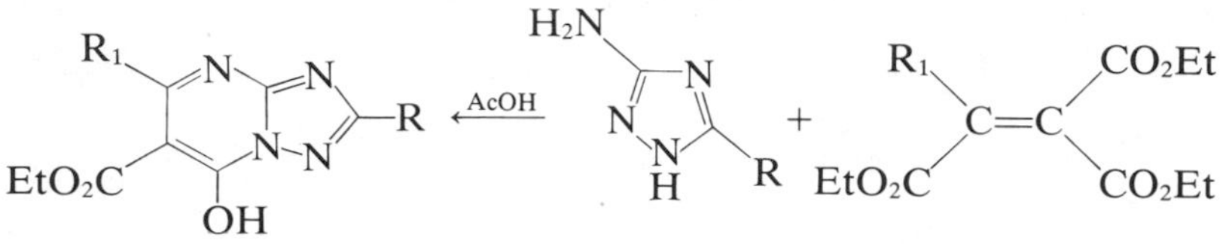

AcOH ← **81** + **82** → NaOEt → **83**

81 —NC—CH₂—CO₂Et→ **84** ← **85**

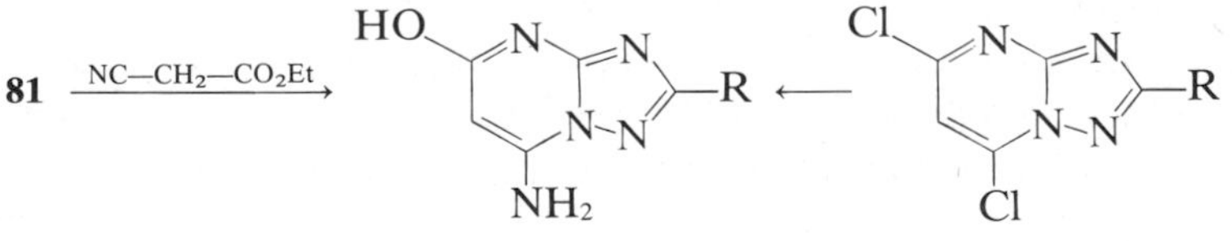

2-Amino-*v*-triazoles and β-dicarbonyl compounds similarly yield *v*-triazolo[1,5-*a*]pyrimidine derivatives in the presence of piperidine,[141,284,285] for example,

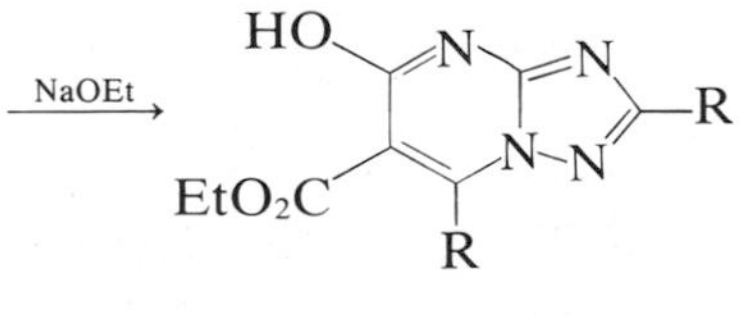

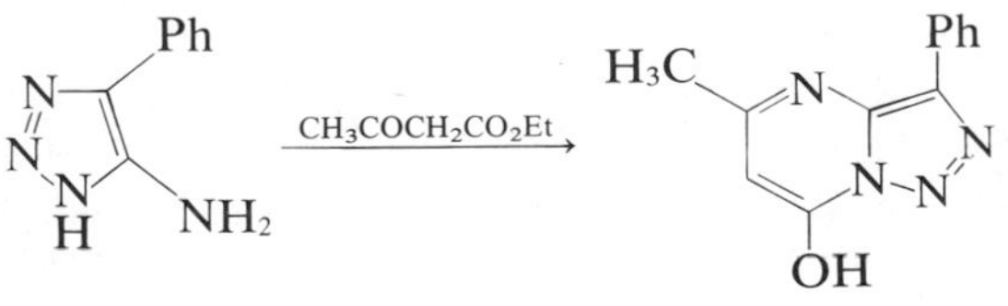

—$CH_3COCH_2CO_2Et$→

The interaction of 5-aminotetrazole and β-dicarbonyl derivatives[44d,142] or acetylenic esters[286] affords tetrazolo[1,5-*a*]pyrimidines, although this route has been much less utilized than the synthesis from pyrimidine derivatives.

4. *Azolotriazines and Azolotetrazines*

Azolo[*a*]-*s*-triazines are prepared from aminoazoles either directly (e.g, using dicyandiamide) or in two steps: (1) addition to isocyanates, isothiocyanates, or imidates; (2) ring closure with one-carbon cyclizing agents. Thus were obtained pyrazolo[2,3-*a*]-*s*-triazine,[288–290] *s*-triazolo-[4,3-*a*]-*s*-triazine,[143] and *s*-triazolo[2,3-*a*]-*s*-triazine[144–146,289,291–293] derivatives. In addition, the following routes have been proposed for the formation of 5-amino- or 5-hydroxy-*s*-triazolo[2,3-*a*]-*s*-triazines:[144]

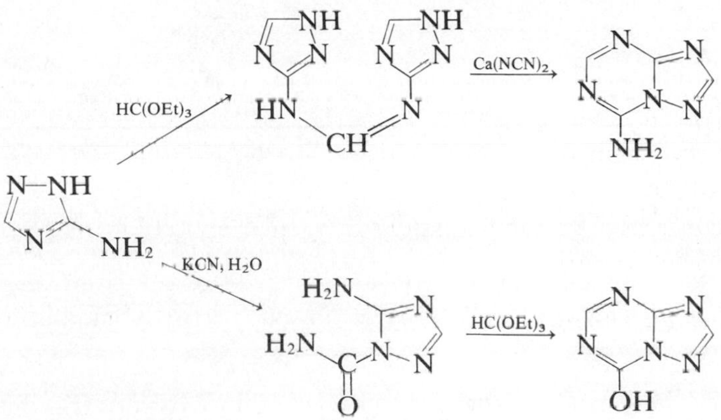

Pyrrolo[1,2-*b*]-*as*-triazines are formed by condensing 1,2,5-triamino-3,4-dicyanopyrrole and α-dicarbonyl compounds,[147] or alkylidenemalononitrile and diazoketones or -esters.[294] Similarly imidazo[1,2-*b*]-*as*-triazines[295] or imidazo[1,5-*b*]-*as*-triazines[296] are obtained in the reaction of α-dicarbonyl derivatives and 1,2-diaminoimidazoles or 1,5-diaminoimidazoles, respectively. 3,4-Diamino-*s*-triazoles afford *s*-triazolo[4,3-*b*]-*as*-triazine derivatives;[148–151,261] for example, the condensation of **86** and pyruvic acid.[70]

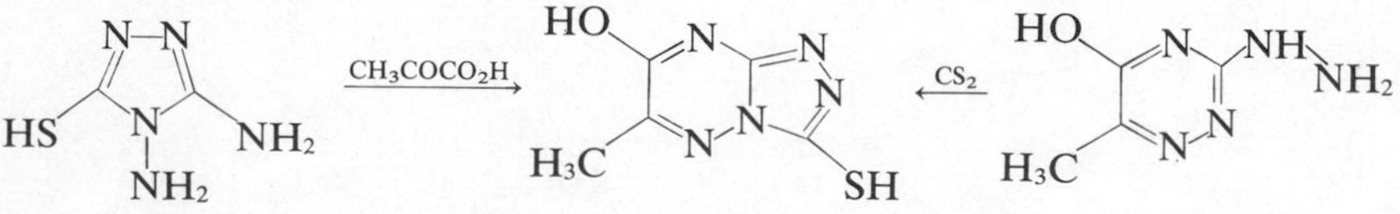

Derivatives of almost every azolo[*c*]-*as*-triazine have been prepared. Two syntheses of the imidazo[2,1-*c*]-*as*-triazine system are known: the condensation of 2-hydrazinoimidazolines and α-dicarbonyl compounds,[297] and the cyclization of 1-(β-oxoalkyl)-2-bromoimidazoles with hydrazines.[298] Diazotized 3-aminopyrazoles (**87**; X = CH) or 4-amino-*v*-triazoles (**87**; X = N) react with β-dicarbonyl compounds to give pyrazolo[3,2-*c*]-*as*-triazines (**88**; X = CH)[153] or *v*-triazolo[5,1-*c*]-*as*-triazines (**88**; X = N),[299] respectively. *s*-Triazolo[3,4-*c*]-*as*-triazines are obtained from 3-hydrazino-*s*-triazole and α-dicarbonyl derivatives.[154] The same method, applied to 5-hydrazinotetrazole, yields tetrazolo[1,5-*c*]-*as*-triazines.[63,103]

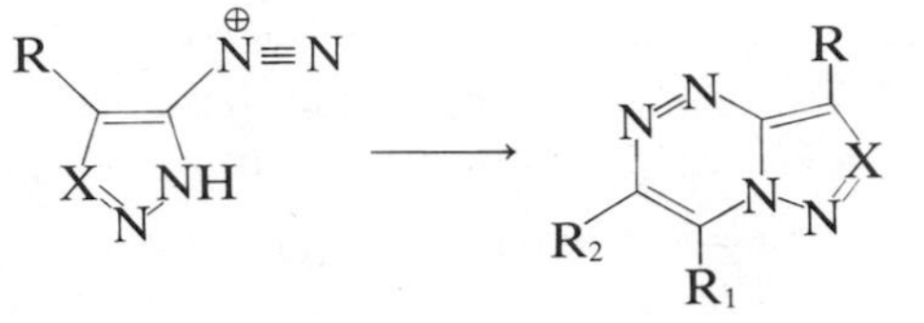

87 **88**

Recently the new pyrrolo[1,2-*d*]-*as*-triazine was synthesized by base-catalyzed cyclodehydration of pyrrole-2-carboxaldehyde formylhydrazone,[300] whereas the reaction of hydrazines and 1-carbethoxypyrrole-2-carboxaldehyde led to 5-oxo derivatives.[301] Treated with acetic anhydride or ethyl orthoformate, 5-methyl-3-pyrazolcarboxylic acid hydrazide yields a 5-hydroxypyrazolo[2,3-*d*]-*as*-triazine.[158,302] Finally, *s*-triazolo[3,4-*f*]-*as*-triazines are obtained in one step from *N*-(*s*-triazolyl)amidines:[303]

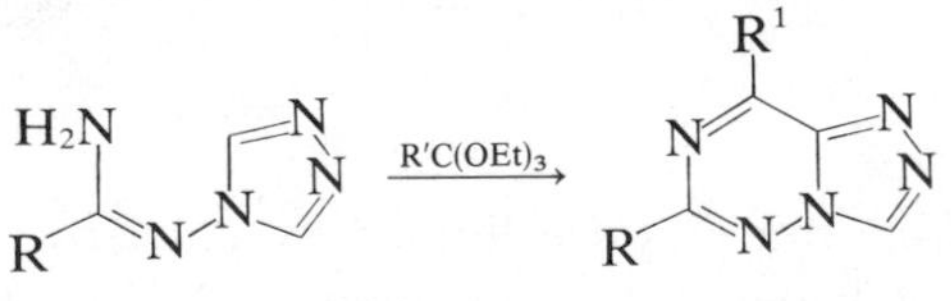

$R'C(OEt)_3$

To date only one azolo-*v*-triazine derivative is known that arises from ring-chain tautomerism of methyl 3-diazoacetyl-3-methyl-1-pyrazoline-4-carboxylate:[344]

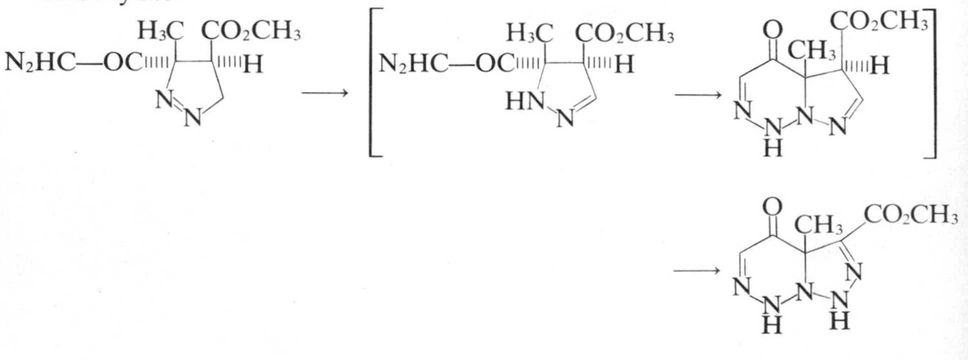

The formation of *s*-triazolo[4,3-*b*]-*s*-tetrazine derivatives (**90**) from **89** (R = SH) and aldehydes can be used as a specific test for the detection of aldehydes.[155] An analogous preparation involves the oxidative cyclization of 3-(benzylidenehydrazino)-4-amino-*s*-triazole.[154]

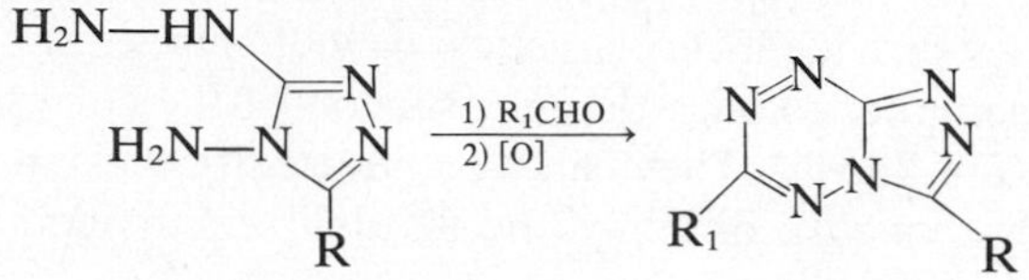

$$\xrightarrow[\text{2) [O]}]{\text{1) } R_1CHO}$$

89 **90**

Some of the preparations of azaindolizines cannot be included in the last two sections. They deal mostly with the rearrangements of various heterocyclic systems to azaindolizine derivatives,[152,287,304–307] for example:

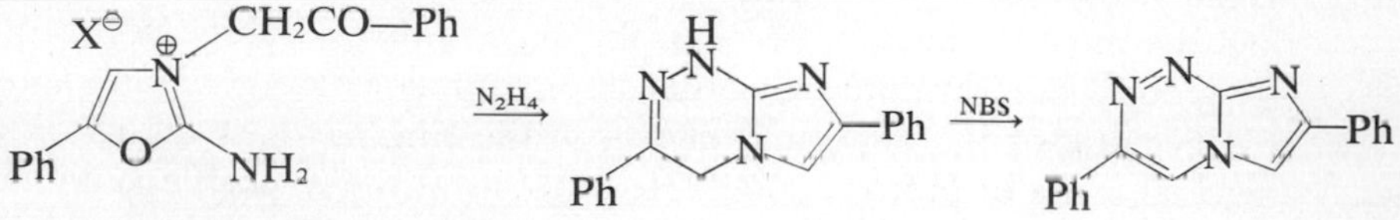

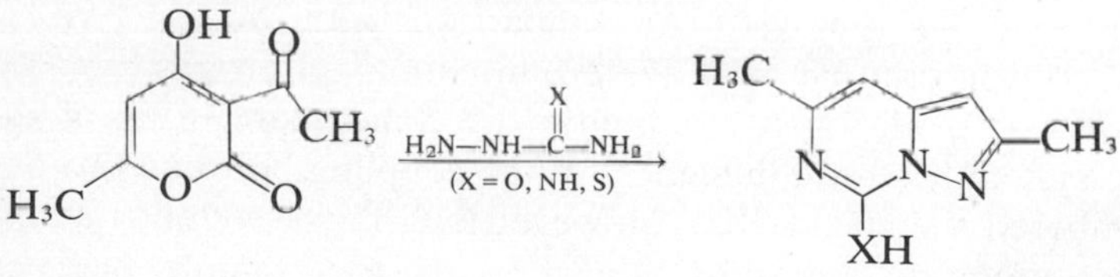

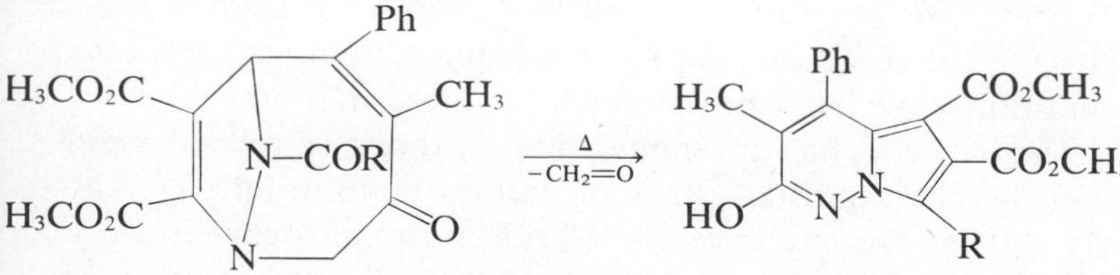

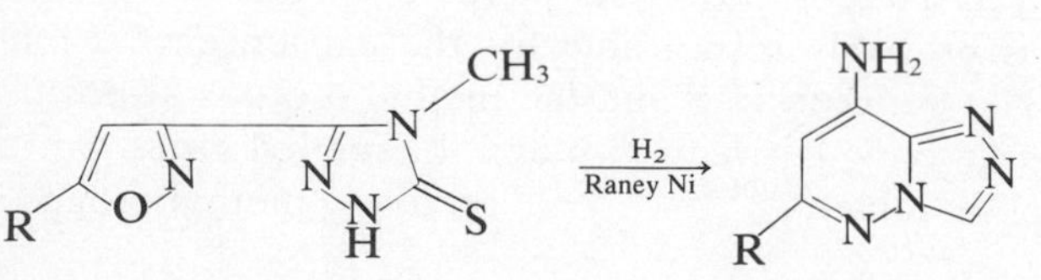

$$\xrightarrow{N_2H_4} \quad \xrightarrow{NBS}$$

$$\xrightarrow[(X = O, NH, S)]{H_2N{-}NH{-}\overset{X}{\overset{\|}{C}}{-}NH_2}$$

$$\xrightarrow[-CH_2{=}O]{\Delta}$$

$$\xrightarrow[\text{Raney Ni}]{H_2}$$

III. Physical Properties

A. Nuclear Magnetic Resonance Spectra

Studies of the nmr spectra of azaindolizines have served two main purposes. They provide a very practical way of characterizing homologous or isomeric compounds once the nmr spectra of the corresponding parent heterocycles are known. They can also be used to estimate the aromaticity of these systems. Accordingly, the discussion is divided into two parts.

1. *Assignments of Absorption Peaks*

Relatively few proton magnetic resonance spectra of unsubstituted azaindolizines have been reported in the literature; those that have been reported are collected in Table III (the spectrum of the fundamental heterocycle, indolizine, is described in ref. 159). Since the studied azaindolizines have more than one nitrogen atom in the 6-membered ring, the most complex patterns in the spectra involve three consecutive CH groups. Furthermore first-order approximations are generally applicable because of the important effects of the ring nitrogen atoms on the chemical shifts of protons bonded to carbons at the α-positions and also because of the relatively low values of the coupling constants. Ambiguities in the identification of neighboring protons within AB, AMX, or ABX systems can be removed by comparison with the spectra of homologs suitably substituted with deuterium or methyl groups.[25]

The introduction of nitrogen atoms into the 5-membered ring of the azaindolizine system has little influence on the coupling between protons of the 6-membered ring (Table III). However long-range coupling across the bridgehead atoms at N-4 and C-9 also occurs, for example between H-1 and H-5[112,117] or between H-3 and H-8,[22,112,301] or even between H-1 and H-6.[259] The observed coupling constants of some unsubstituted and alkyl-substituted azaindolizine systems are listed in Table IV.

The invariance of coupling constants within a series of related azaindolizines and the characteristic values of various bond orders[161,168] are undoubtedly due to the presence of a bridgehead nitrogen atom. In particular, the relatively large bond orders of the 5-6 bond in azolo[*a*]-pyrimidine systems is probably responsible for the existence of a long-range coupling ($\sim$1 Hz) between H-6 and the methyl protons at position 5.[164] In contrast, the coupling between H-6 and the methyl group proton at position 7 is not observable.[25,30b,49,88,117,163] Similarly methyl groups at

TABLE III. PMR SPECTRAL PARAMETERS OF SOME UNSUBSTITUTED AZAINDOLIZINES

Compounds	H-1[a]	H-2	H-3	H-5	H-6	H-7	H-8	Coupling constants (Hz)	Solvent	Ref.
Pyrrolo[1,2-*b*]pyridazine	6.31	6.67	7.52	—	7.79	6.32	7.51	$J_{12}=4, J_{13}=1, J_{23}=2.5, J_{67}=4.5, J_{68}=2, J_{78}=9$	CS_2	104
Pyrrolo[1,2-*c*]pyrimidine	6.5	6.94	7.37[b]	8.84	—	7.37[b]	7.37[b]	[c]	$CDCl_3$	157
Pyrrolo[1,2-*a*]pyrazine	6.85	6.97	7.46	7.89	7.58	—	8.91	$J_{12}=4.5, J_{13}=1.5, J_{15}=0.9, J_{23}=2.6, J_{38}=1, J_{56}=5.5, J_{58}=1.6$	$CDCl_3$	112
Pyrrolo[1,2-*d*]-*as*-triazine	6.8–7.1	7.81	9.45	—	—	9.15		$J_{58}=2.4$	DMSO-d_6	300
Imidazo[1,2-*b*]pyridazine	—	7.79	7.99	—	8.30	7.0	7.95	$J_{23}=1, J_{38}=0.8, J_{67}=4.5, J_{68}=2, J_{78}=10$	$CDCl_3$	19
Imidazo[1,2-*a*]pyrimidine	—	7.77	7.66	8.72	6.91	8.60	—	$J_{23}=1.4, J_{56}=6.9, J_{57}=2, J_{67}=4.1$	$CDCl_3$	25
Imidazo[1,2-*b*]-*as*-triazine	—	8.06	8.03	—	8.52	8.45	—	$J_{23}=1.6, J_{67}=1.8$	$CDCl_3$	34
Imidazo[1,5-*a*]pyrimidine	7.62	—	8.05	8.22	6.54	8.14	—	$J_{15}=1, J_{56}=7.2, J_{57}=1.7, J_{67}=3.8$	$CDCl_3$	117
Imidazo[1,5-*a*]pyrazine	7.83	—	8.28	7.58	7.91	—	9.03	$J_{56}=5$	—	259
s-Triazolo[4,3-*b*]pyridazine	—	—	9.15	—	8.40	7.17	8.20	$J_{67}=4.5, J_{68}=2, J_{78}=9.5$	$CDCl_3$	47
s-Triazolo[4,3-*a*]pyrimidine	—	—	9.28	9.02	7.15	8.77	—	$J_{56}=6.8, J_{57}=2, J_{67}=4$	DMSO-d_6	49
s-Triazolo[2,3-*b*]pyridazine	—	8.38	—	—	8.44	7.42	8.08	$J_{67}=4.2, J_{58}=8.9, J_{68}=1.8$	$CDCl_3$	267
s-Triazolo[2,3-*a*]pyrazine	—	8.50	—	8.58	8.19	—	9.37	$J_{56}=4.5, J_{58}=1.5$	$CDCl_3$	267
s-Triazolo[2,3-*a*]pyrimidine	—	8.52	—	9.00	7.19	8.87	—	$J_{56}=6.5, J_{57}=2, J_{67}=4.4$	$CDCl_3$	49
Tetrazolo[1,5-*b*]pyridazine	—	—	—	—	8.85	7.75	8.58	$J_{67}=4.5, J_{68}=2, J_{78}=9$	$CDCl_3$	97
Tetrazolo[1,5-*a*]pyrazine	—	—	—	8.80	8.32	—	9.63	$J_{56}=4.7, J_{58}=1.6, J_{68}\leqslant 0.5$	$CDCl_3$	97
Tetrazolo[1,5-*a*]pyrimidine	—	—	—	9.85	7.77	9.30	—	$J_{56}=7, J_{57}=1.8, J_{67}=4.1$	DMSO-d_6	88

[a] Chemical shifts δ(ppm)
[b] Unresolved multiplet
[c] Not specified.

TABLE IV. COUPLING CONSTANTS OF SOME UNSUBSTITUTED OR ALKYL-SUBSTITUTED AZAINDOLIZINES

Compounds	Coupling constants (Hz)	Ref.
Pyrrolo[1,2]azines	$J_{12}=4–5$, $J_{23}=2.5–3$, $J_{13}=1–1.6$	104, 112, 157
Imidazo[1,2]azines	$J_{23}=1–2$	18, 19, 25, 30*b*, 34
Azolo[*b*]pyridazines	$J_{67}=4.5$, $J_{78}=2$, $J_{68}=9–10$	18, 19, 20, 47, 97, 104
Azolo[*a*]pyrimidines	$J_{56}=6.5–7.2$, $J_{67}=3.6–4.4$, $J_{57}=1.6–2.4$	25, 30b, 49, 88, 117
Azolo[*a*]pyrazines	$J_{56}=4.5–5.5$, $J_{58}=1.2–2$	22, 85, 112

position 8 of azolo[*b*]pyridazine systems are coupled with H-7. Coupling between H-7 and the methyl group protons at position 6 has not been reported.[20,47] These observations are particularly useful in establishing the structures of isomeric methylazaindolizines in these series.[25,30b]

The chemical shifts of the methyl protons in 3,5-dimethylazaindolizine derivatives are considerably higher (in δ) than expected, based on observed chemical shifts of the methyl groups in the corresponding 3- and 5-methyl compounds. This deshielding effect has been reported in the imidazo[1,2-*a*]- and -[1,5-*a*]pyridine,[34] imidazo[1,2-*a*]pyrimidine,[30b] and *s*-triazolo[4,3-*a*]pyrimidine[49] series. If we assume that the azaindolizines are planar or nearly planar,[162] the origin of the *peri*-interaction between substituents at positions 3 and 5 is probably of a steric nature. This is consistent with the recent report that the Dimroth rearrangement of 3,5-dimethylimidazo[1,2-*a*]pyrimidines is faster than the rearrangement of the corresponding 3- or 5-methyl compounds.[22] Such a *peri*-effect may also exist, to a certain extent, in the pmr spectra of any azaindolizines disubstituted at positions 3 and 5 by bulky substituents and could eventually be used to assign the signals and characterize the structures, as in the *s*-triazolopyrimidine[49] or imidazo-*as*-triazine[34] series. Many other important applications of pmr spectroscopy have been developed in the azaindolizine series and are mentioned elsewhere in this chapter. Recently ^{13}C nmr spectroscopy has been used with success as an aid in molecular structure determination of azaindolizines.[308–311]

2. *Theoretical Studies*

Attempts have been made to correlate the observed proton chemical shifts with the electron densities obtained from molecular orbital calculations. In particular, the aromaticity of this class of compounds is established by the importance of the ring current contribution to the calculation of the proton chemical shifts in these compounds.

In their study of *s*-triazolo[2,3-*a*]pyrimidine, Makisumi and co-workers assume that ring current values can be approximated with those obtained for the isoelectronic indenyl carbanion.[163] Because of the presence of an excess unit charge in the indenyl ion, Paudler and Kuder prefer to use the calculated ring current of indolizine to correlate the chemical shifts of imidazoazine derivatives.[160] An SCF calculation of ring currents shows that aza-substitution at positions 1 and 3 of the indolizine nucleus enhances the current in the 6-membered ring.[165] Finally, a recent study of carbon-13 chemical shifts in this series confirms the high degree of aromaticity of azaindolizines through considerable delocalization of the bridgehead nitrogen lone pair.[161]

Contributions to the shielding of the ring protons also arise from the anisotropy effect of the ring nitrogen atoms and the field effect associated with their unshared electrons. However it has been assumed that the effect of N-4 is negligible and equivalent to the effect of a neighboring sp^2 carbon caused by the delocalization of the lone pair.[163,165] The contribution of other nitrogen atoms to the chemical shifts of protons attached to α-carbons has been estimated rather arbitrarily to be 0.25 ppm in the case of *s*-triazolo[2,3-*a*]pyrimidine,[163] whereas a value of 0.5 to 0.9 ppm is suggested in another investigation concerning imidazoazines.[160] Using indolizine as a reference, a linear relationship has been found between the HMO π-electron densities at nuclear carbon atoms of several azaindolizines and the chemical shifts of the corresponding adjacent protons.[362]

In conclusion, the semiempirical calculations of the chemical shifts of *s*-triazolo[2,3-*a*]pyrimidine[163] and imidazoazines (or their methyl derivatives)[160] give satisfactory results considering the approximations involved. More elaborate calculations of carbon-13 chemical shifts of various azaindolizines reproduce the experimental values, except for carbon atoms at bridgehead positions.[161]

B. Ultraviolet Spectra

In general, the uv spectra of azaindolizines (Table V) consist of three main bands, as does the spectrum of indolizine itself. These bands are found at about 210 to 240 nm ($\epsilon_{max} > 10^4$), 250 to 280 nm ($\epsilon_{max} < 10^4$), and 290 to 360 nm ($\epsilon_{max} < 10^4$).

Attempts to associate the absorption bands with particular electronic transitions have led to different conclusions. Using the indenyl anion as a model, Mason studied the perturbations caused by successive aza substitutions on the energy levels of the Hückel molecular orbitals. Numbering the molecular orbitals from the lowest to the highest, this author

TABLE V. ULTRAVIOLET SPECTRA OF UNSUBSTITUTED AZAINDOLIZINES

Compounds	λ_{max} (nm)	log ϵ_{max}	Solvent	Ref.
Pyrrolo[1,2-*a*]pyrazine	224.5, 235(i); 274(i), 283, 292.5; 388.5	4.39, 4.30; 3.33, 3.49, 3.48; 3.43	H_2O (PH=10)	37
Pyrrolo[1,2-*c*]pyrimidine	229; 272, 283; 345	4.44; 3.74, 3.77; 3.03	EtOH	157
Pyrrolo[1,2-*d*]-*as*-triazine	223, 232(i), 262, 268, 278, 322	4.57, 4.50, 3.80, 3.82, 3.74, 3.46	EtOH-95	300
Imidazo[1,2-*b*]pyridazine	221; 254, 260, 271; 325	4.20; 3.24, 3.20, 3.10, 3.59	H_2O (PH=7)	37
Imidazo[1,2-*c*]pyrimidine	214.5; 259.5, 269.5; 294, 298	4.29; 3.73, 3.73; 3.39, 3.39	H_2O (PH=7)	37
Imidazo[1,2-*a*]pyrazine	221.5; 276, 282; 302	4.21; 3.61, 3.67; 3.69	H_2O (PH=7)	37
Imidazo[1,2-*a*]pyrimidine	211, 223; 275, 283; 317	4.15, 4.30; 3.45, 3.47; 3.46	H_2O (PH=7)	37
Imidazo[1,5-*a*]pyrazine	260(i), 263(i), 269, 280, 331	3.41, 3.43, 3.52, 3.50, 3.33	EtOH-95	259
Imidazo[1,5-*a*]pyrimidine	223, 227, 254, 263, 268; 280; 358	4.55, 4.48, 3.24, 3.28, 3.27, 3.16; 3.20	EtOH-95	30a
s-Triazolo[4,3-*b*]pyridazine	~260, ~268(i); 300, 310, 317, 323	2.7, 2.75; 3.23, 3.27, 3.27, 3.27	Cyclohexane	166
s-Triazolo[4,3-*a*]pyrazine	206; 253(i), 262(i), 269(i); 292	4.46; 3.31, 3.37, 3.40; 3.57	EtOH-95	84
s-Triazolo[4,3-*a*]pyrimidine	209, 213(i); 250, 266(i); 298	4.04, 3.98; 3.20, 3.08; 3.31	EtOH	84
s-Triazolo[2,3-*a*]pyrimidine	208, 273	4.50, 3.57	EtOH	84
Tetrazolo[1,5-*b*]pyridazine	295, 312	3.13, 2.88	Cyclohexane	166
Tetrazolo[1,5-*a*]pyrazine	203, 270	4.08, 3.65	H_2O	102

assigned the band at lower wavelengths to the transition $4 \rightarrow 6$, the second band to a combination of $5 \rightarrow 6$ and $4 \rightarrow 7$, and the last band to $5 \rightarrow 7$.[167] This interpretation differs from the conclusions of a more recent and extensive study using the SCF MO method in which it is suggested that the first band is composed of two major bands close enough to overlap. The second band is caused by a combination of transitions $4 \rightarrow 6$ and $5 \rightarrow 7$, and the third corresponds essentially to $5 \rightarrow 6$.[168] Recently the electronic spectra of *s*-triazolo[4,3-*a*]pyrimidine, *s*-triazolo[2,3-*a*]-pyrimidine, and tetrazolo[1,5-*a*]pyrimidine have also been calculated using the PPP method.[314]

Inspection of the uv spectra of azaindolizines shows that the general effect of introducing nitrogen atoms into the indolizine ring is to induce hypsochromic shifts of the major bands, and that this effect tends to increase with the number of aza substituents.[168] The effect is more pronounced when nitrogen atoms are introduced into the 5-membered ring of indolizine (in particular at positions 1 and 3) than in the 6-membered ring. Thus aza-substitution at position 7 of indolizine results in almost no change in the ultraviolet spectrum.[37] The introduction of a nitrogen atom at any single position of the 6-membered ring of imidazo-[1,2-*a*]pyridine also has little influence on the spectrum compared to that of the reference compound.[37] In contrast, aza-substitution in the 6-membered ring of *s*-triazolo[4,3-*a*]pyridine induces a bathochromic shift of the long wavelength band and a hypsochromic shift of the intermediate band, especially in the case of *s*-triazolo[4,3-*b*]pyridazine.[84] Similar but more important shifts are observed in the spectrum of *s*-triazolo[2,3-*a*]-azines, suggesting a possible method for the characterization of isomeric *s*-triazolo[4,3]- and -[2,3]azines.[80,84] However this view has been contested in the case of methyl-substituted *s*-triazolo[4,3-*a*]- and -[2,3-*a*]-pyrimidines and pmr spectroscopy has been found to be a more appropriate method for characterization.[42,49] The existence of a *peri*-interaction between the 3-phenyl and the 5-methyl groups of 3-phenyl-5,8-dimethyl-*s*-triazolo[4,3-*a*]pyrazine has been proposed to explain the relatively low extinction coefficients of the major bands in the spectrum of this compound.[175]

Uv spectra have also been utilized to establish the tautomeric structures of azaindolizines substituted with OH, SH, or NH_2 groups. The corresponding spectra are discussed in the section devoted to the study of tautomerism. The determination of the sites of protonation of imidazo-[1,2]azines has been carried out using uv spectrometry.[37] The spectral changes observed upon protonation are probably due to the importance of resonance-contributing structures such as **91a** and **b**.

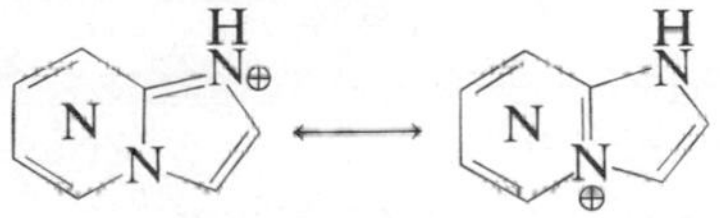

91a ⟷ **91b**

C. Mass Spectra

The presence of a bridgehead nitrogen atom causes important differences in the bond orders and thus leads to interesting fragmentation patterns in the mass spectra of azaindolizines. These spectra frequently involve loss of HCN from either the 5- or the 6-membered rings. Imidazo[1,2-*a*]pyrimidines[169] and *s*-triazolo[2,3-*a*]pyrimidines[42] undergo cleavage of the 5-membered ring (with loss of HCN) giving the bicyclic fragment **92,** which may ring expand to **93** prior to another loss of HCN. On the other hand the simultaneous cleavage of bonds 1-9 and 3-4 leads to a hetaryne (**94**) that can ring expand to **95** if suitably substituted. Imidazo[1,5-*a*]pyrimidine[30a] and imidazo[5,1-*f*]-*as*-triazines[260] behave similarly. The methyl derivatives of these heterocyclic systems have a tendency to lose initially a hydrogen atom instead of a methyl radical. The resulting ion either first ring expands and secondly loses HCN or the reverse.

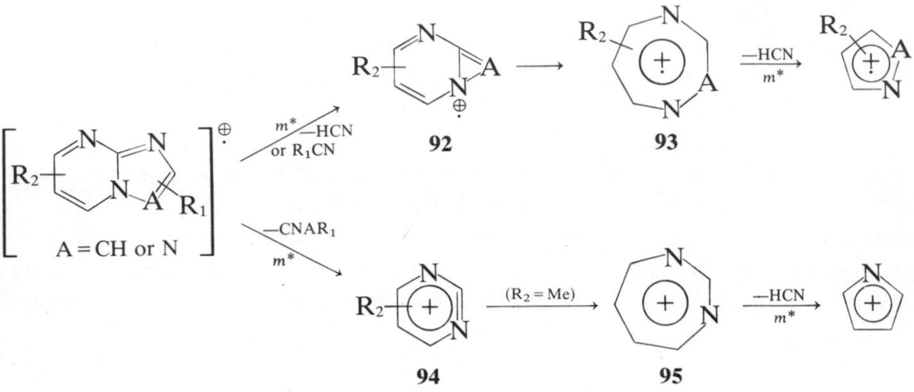

The fragmentation of pyrrolo[1,2-*a*]pyrazines occurs through loss of HCN or $C_2H_2N^{\oplus}$ from the 6-membered ring, whereas the primary cleavages of the 1- and 3-nitro derivatives probably yield diazaindenone ions.[10] The presence of a N–N bond in imidazo[1,2-*b*]pyridazines does not change the fragmentation patterns compared to those of the preceding series. The major cleavage involves loss of a HCN molecule from the 5-membered ring if position 2 is unsubstituted, and a loss of RCN from the 6-membered ring if this position is substituted.[170] In contrast, no loss of HCN was observed with various *s*-triazolo[4,3-*b*]pyridazines.[249] Spectra of pyrazolo[2,3-*c*]pyrimidines[287] and of pyrazolo[3,2-*c*]-*as*-triazines[250] primarily depend on the nature of the 6-membered ring substituents. One of the main features of the mass spectra of 7-diazo-8-oxo-*s*-triazolo[4,3-*b*]pyridazines is the loss of the fragment CN.[313]

Tetrazolodiazines (in potential equilibrium with the corresponding azidoazine forms) undergo fragmentation in a different way.[97] These compounds first lose N_2 and then HCN from the $M^{\oplus}–N_2$ fragment. This is probably a ring-expanded ion by analogy with the results in other series. If the equilibrium conditions favor the tetrazolodiazine form, the spectrum exhibits an intense $M^{\oplus}–N_2$ peak. If the azido form is favored, a major fragment of the spectrum is the $M^{\oplus}–26$ ion, which can be assumed to be the result of a thermal process involving a transient nitrene and

* Denotes metastable peaks.

hydrogen capture.[97] One of the most abundant species in spectra of tetrazolo[1,5-*b*]pyridazine derivatives corresponds to the loss of four nitrogen atoms (probably at positions 2, 3, 4, and 5).[249] 6-Azidotetrazolo[1,5-*b*]-pyridazines similarly present peaks corresponding to the loss of six nitrogens.[249]

D. Quantum Mechanical Calculations

Calculations of π-electron densities based on the HMO theory are useful in predicting the reactivity of azaindolizines. Streitweiser's parameters[171] are generally used, although attempts have been made to take the inductive effect of N-4 into account.[10] The calculated total π-electron densities are in good agreement with the experimentally observed sites of electrophilic substitution (generally C-3) in the pyrrolo[1,2-*a*]pyrazine,[10] imidazo[1,2-*b*]pyridazine,[172,173] and imidazo[1,2-*a*]pyrimidine[25] series. However frontier electron densities generally predict not only the sites of electrophilic substitution at nuclear carbon atoms but also the sites of protonation and *N*-alkylation.[25,173] Attempts have been made to take into account the nature of the electrophilic or nucleophilic reagent in these calculations.[316]

Application of the SCF method also indicates that carbon-3 of pyrrolodiazines and of imidazo[1,2]diazines has the highest π-electron density[168] in accord with the properties of these compounds. Calculations of bond orders suggests that the resonance contributing structure **1** contributes largely to the ground state of the molecules and emphasizes the high double bond characters of bonds 5-6 and 7-8.[168] Transition energies and relative intensities of the experimental uv spectra are satisfactorily reproduced.[168,314] Finally, the PPP procedure has been used to study an unknown class of mesoionic compounds derived from 5-oxoazolo[*a*]-pyrimidine and 5,7-dioxoazolo[*c*]pyrimidine.[315].

Recently the CNDO/2 method has been applied to calculate the carbon-13 chemical shifts[161] and dipole moments[312] of azaindolizines using idealized or adjusted geometries. In particular, the calculated dipole moments of pyrrolo[1,2-*a*]pyrazine and imidazo[1,2-*a*]pyrimidine are in good agreement with the experimental values (2.54 and 5.45 D, respectively).[174]

IV. Chemical Properties

A. Tautomerism

In contrast to purines, azaindolizines cannot exist as "NH-tautomers"[144] unless substituted at suitable positions with groups bearing delocalizable protons such as OH, NH_2, or SH[176] (Cf. **96**). To date no attempt has been made to study the relative stabilities of the possible tautomers in any of the azaindolizine systems discussed in this review.

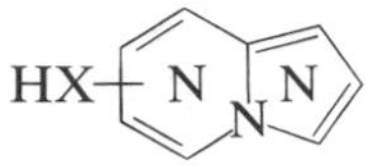

(X=NH, O, S)

96

Because of the difficulty of characterization, all of the possible tautomers have seldom been detected or even considered in any particular study. However the application of ir, uv, or nmr spectroscopy and, eventually, the use of model compounds, have generally been successful in determining the structure of the major tautomers.

As shown by the ir spectra, hydroxyazaindolizines exist largely in one or several oxo forms[42,60,74,77,116,132,184,301] with rare exceptions.[116] The frequencies[54,80,177] and intensities[15] of the carbonyl stretching bands seem to depend on the possibilities of resonance delocalization and on the inductive effects, in relation with the positions of the carbonyl group and the substituents on the nucleus. Uv spectroscopy has been especially utilized to study the tautomeric structures of 5- and 7-hydroxyazaindolizines.[314] Their spectra present, in general, the usual three major bands[46,54,77,132,287] assigned to oxo tautomers analogous to **97** and **98**. These quasi *ortho*- and *para*-quinonoid structures have characteristic uv absorptions used to identify the corresponding 7- and 5-oxo isomers.[44a,45,132] In particular, the ratio of the intensities of the band at intermediate wavelengths to the band at longer wavelengths, has been found to be greater than 1 for **97** and less than 1 for **98**.[44a,45,77,132] Oxo-tautomers and their *N*-alkyl derivatives do not have necessarily the same uv spectra. Therefore the latter compounds cannot always be used as models. The application of nmr spectroscopy leads to less ambiguous results than does ir or uv spectroscopy. Thus **97** and **98** are identified as the most stable tautomeric structures on the basis of characteristic coupling constants between the protons of the 6-membered ring.[30b,77,117,132] Uv

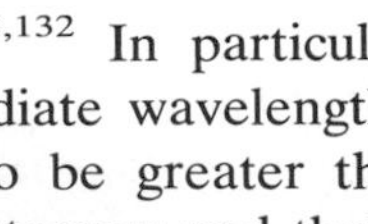

97 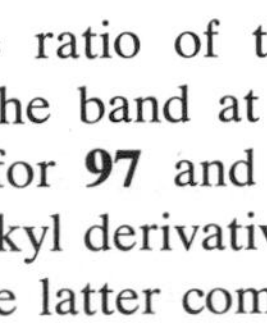 **98** 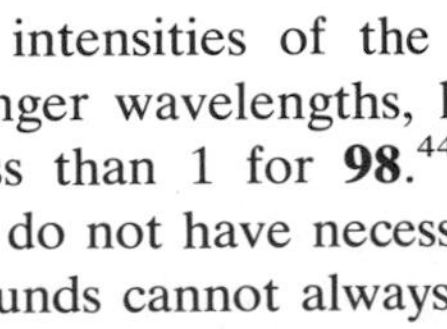 **99**

and nmr spectroscopy can be conveniently used to distinguish 5-oxo-5,8-dihydro-*s*-triazolo[4,3] and [2,3]pyrimidines from the isomeric 7-oxo-*s*-triazolopyrimidines.[77,80] This is particularly useful in the study of the Dimroth rearrangement of hydroxy-*s*-triazolo[4,3]pyrimidines. A quasi *para*-quinonoid form (**99**) is the favored tautomer of 2,7-dihydroxy-5-thiomethylimidazo[1,2-*c*]pyrimidine, as shown by nmr spectroscopy.[13] Both forms can exist in the case of 5-hydroxy-*s*-triazolo[4,3-*a*]-*s*-triazines, and the quasi *para*-quinonoid structure is favored.[54] Finally, the 2- or 3-oxo forms are the most stable tautomers of 2- or 3-hydroxy-imidazo[1,2]azines, respectively.[174] Recently the product of the base-catalyzed condensation of β-ketoesters and 1-alkyl-4-amino-*s*-triazolium ions has been shown spectroscopically to exist in the mesoionic form derived from the 1-alkyl-8-hydroxy-*s*-triazolo[4,3-*b*]pyridazinium cation:[278]

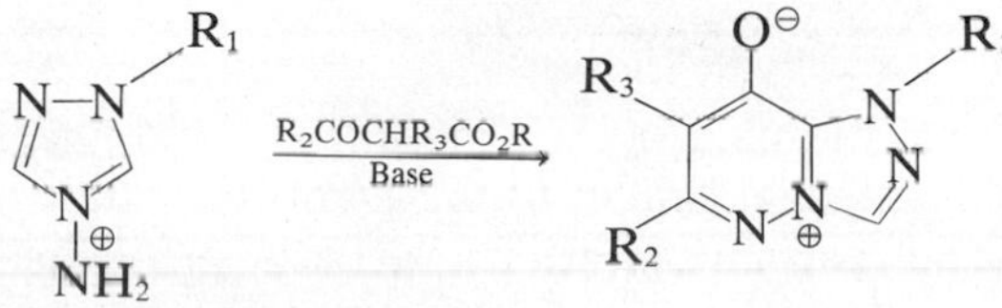

$$\xrightarrow[\text{Base}]{R_2COCHR_3CO_2R}$$

The amino- and imino-tautomers deriving from aminoazaindolizines are more difficult to characterize than the corresponding oxo- or hydroxy-tautomers. Amino derivatives of azaindolizines are assumed to exist in the amino form on the basis of their ir and uv spectra.[43,67,68,91,116,179,287] Although rarely used, nmr spectroscopy leads to more definite conclusions. For example, this method shows (through detection of NH_2 signals[58,182,183]) that the amino forms of 5-aminoimidazo[1,5-*a*]pyrazine,[178] 3-amino-*s*-triazolo[4,3-*a*]pyrazine,[57] and 5-aminopyrazolo[2,3-*a*]pyrimidine[180] are favored. Iminoazaindolizines may result from *N*-alkylation[180] or *N*-acylation[72] at suitably located ring nitrogen atoms. Finally, a few studies of the tautomerism of aminohydroxyazaindolizines or mercapto-azaindolizines[75,175,184] have been reported; in particular, 5-mercaptopyrazolo[2,3-*c*]pyrimidines[287] and 5- or 7-mercapto-*s*-triazolo[2,3-*a*]-*s*-triazines[317] exist in the thioxo form.

B. Rearrangements

1. *Dimroth-Type Rearrangements*

$$\xrightarrow[\text{or } \Delta]{\overset{\ominus}{O}H,\ \overset{\oplus}{H}}$$

(A,B = CH or N)

100 **101**

TABLE VI. AZAINDOLIZINE SYSTEMS THAT UNDERGO DIMROTH-TYPE REARRANGEMENTS

Azaindolizines	Conditions	Ref.
Imidazo[1,2-*a*]pyridines	Basic	363
Imidazo[1,2-*c*]pyrimidines	Basic	22, 188
Imidazo[1,2-*a*]pyrimidines	Basic	22
s-Triazolo[4,3-*a*]pyridines	Basic	185
s-Triazolo[4,3-*c*]pyrimidines	Basic, acid, or thermal	59, 74, 186
s-Triazolo[4,3-*a*]pyrimidines	Basic, acid, or thermal	44b, 77, 78, 134, 283
s-Triazolo[4,3-*a*]pyrazines	Basic	175
s-Triazolo[4,3-*a*]-*s*-triazines	Neutral or basic	54, 264
s-Triazolo[4,3-*d*]-*as*-triazines	Acid	61
s-Triazolo[3,4-*c*]-*as*-triazines	Acid	46
Tetrazolo[1,5-*c*]pyrimidines	Acid	96, 97
Tetrazolo[1,5-*a*]pyrimidines	Basic	88

Aza derivatives of imidazo[1,2-*a*]pyridines (**100**) undergo a rearrangement to **101** in a manner similar to the Dimroth rearrangement of iminopyrimidines. The rearrangement can be base- or acid-catalyzed or thermally induced under neutral conditions (Table VI).

The base-catalyzed rearrangements of 3-methylimidazo[1,2-*a*]pyridine and its monoaza derivatives have been investigated. The rearrangement of 3-methylimidazo[1,2-*a*]pyridine is not observed, but the introduction of a nitrogen atom at positions 6 or 8 induces the reaction. This rearrangement is faster for 3-methylimidazo[1,2-*c*]pyrimidine than for imidazo[1,2-*a*]pyrimidine.[22] In contrast, 3-methylimidazo[1,2-*a*]pyrazine does not rearrange under fairly basic conditions,[22] and *s*-triazolo[4,3-*a*]pyrazines need more vigorous conditions than do *s*-triazolopyrimidines.[175] That the ease of the rearrangement increases with the number of aza substitutions is suggested by a survey of the experimental conditions of the studies reported. *s*-Triazolopyrimidines, (**102**), for example, rearrange very readily under basic, acidic, or neutral conditions (Table VI). Since positions 2 and 3 of tetrazoloazines are both occupied by a nitrogen atom, it is difficult to demonstrate the occurrence of the rearrangement. However in this case, covalently hydrated species corresponding to the intermediates of the eventual rearrangement have been isolated.[87,96,97]

It is generally accepted that the initial step of the base- and acid-catalyzed rearrangements involve an equilibrium-controlled covalent hydration process with the initial addition of $OH^{\ominus}$ at position 5 of the azaindolizine nucleus.[22,44*b*,53,80,185] This step is followed by a tautomeric ring opening of the 6-membered ring to a carbonyl intermediate analogous to

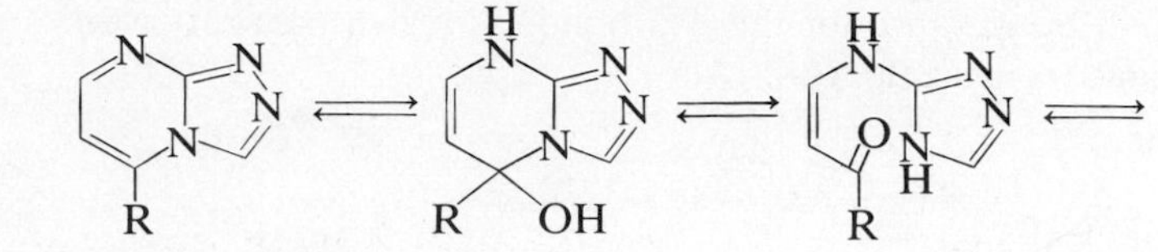

(R = H, alkyl, or OH)

102

103

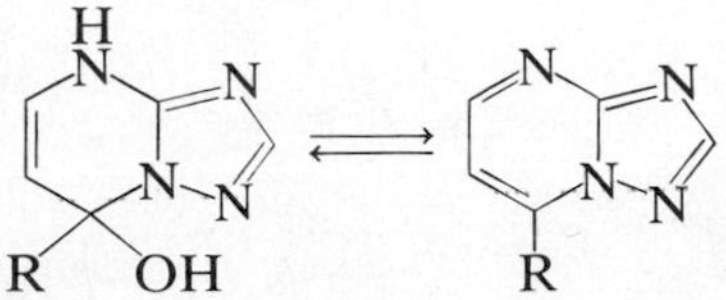

103. This species then cyclizes to the rearranged azaindolizine. It has been proposed that thermally induced rearrangements involve zwitterionic intermediates of type **104,** but no proof has been advanced to support this view.[59,187]

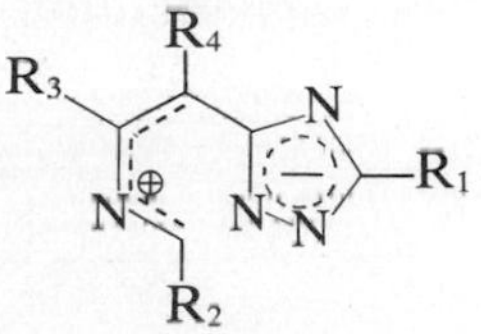

104

According to the mechanism above, increasing the electron deficiency at position 5 or stabilizing the intermediate **103** should facilitate the rearrangement. Indeed, 8-nitro-*s*-triazolo[4,3-*a*]pyridines rearrange more readily and in higher yields[189] than in the absence of the nitro substituent.[185] The electron-donating amino groups have the reverse effect.[189] Similarly it has been suggested that the carbethoxy substituent favors the rearrangement of 6-carbethoxy-*s*-triazolo[4,3-*a*]pyrimidines by decreasing the electron density at position 5 and by stabilizing the intermediate **103**.[53] Steric effects also influence the rate of the Dimroth rearrangement. Thus 3,5-dimethylimidazo[1,2-*a*]pyrimidines rearrange under basic conditions faster than do 3-methylimidazo[1,2-*a*]-pyrimidines.[22]

Related to the Dimroth rearrangement are the new syntheses of pyrazolo[3,4-*b*]pyridine (**106**) through thermally induced decarboxylation of **105** (X = CO_2H or CN),[190,318] and the acid catalyzed rearrangement of the amino-*s*-triazolo[4,3-*a*]pyrazine derivative **107**[81] involving the fission of the 7-8 bond (and not the 4-5 bond, as previously reported[57]) to give the imidazo-*s*-triazole **108**.

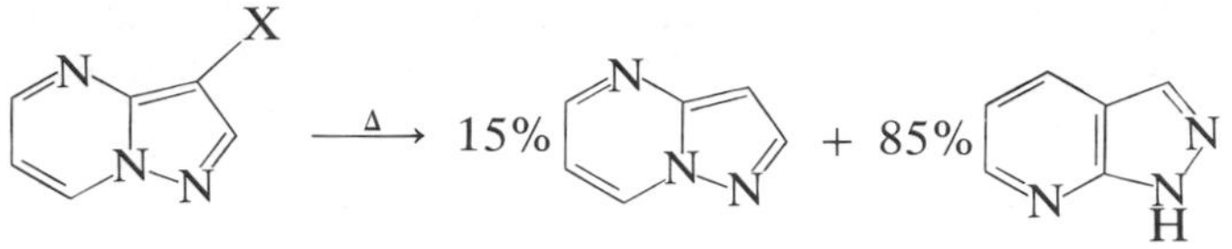

105 **106**

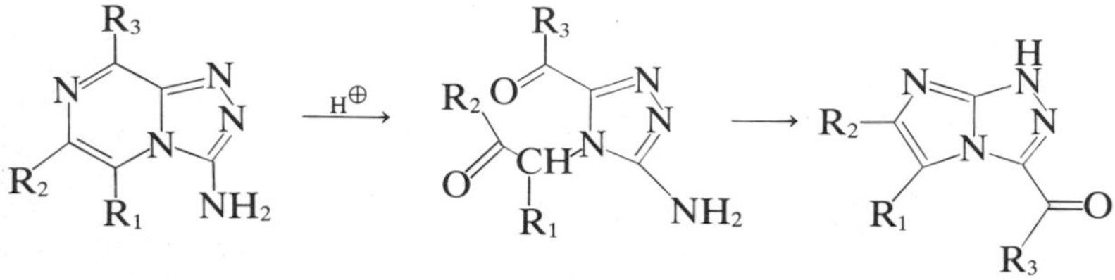

107 **108**

2. *Tetrazolo–Azidoazine Isomerization and Related Processes*

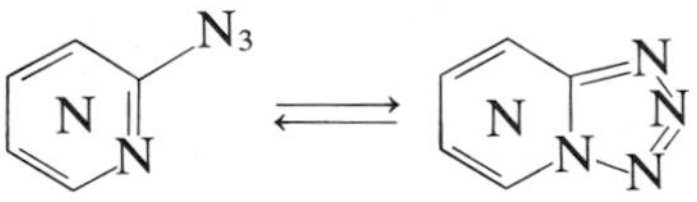

47a **47b**

Rearrangements of this type have been recently demonstrated in several tetrazoloazine series through the simultaneous applications of ir and nmr spectroscopy.[97] The influences of the substituents, the solvent, and the temperature on the equilibrium **47a** ⇌ **47b** have been particularly investigated.

In general, electron-attracting substituents stabilize the electron-donating azido group and thus **47a,** and destabilize the electron-withdrawing tetrazole group and tautomer **47b**. Electron-donating substituents have the reverse effect. Thus 6-azido derivatives of imidazo[1,2-*b*]pyridazine,[100,101] *s*-triazolo[4,3-*b*]pyridazine,[100,101] tetrazolo[1,5-*b*]-pyridazine,[89,94] and azido derivatives of tetrazolo[1,5-*a*]- or -[1,5-*c*]pyrimidines[97] exist in the corresponding azidoazaindolizine form in the solid state. However the ditetrazolo form derived from 6-methyl-7-azido-*s*-triazolo[4,3-*b*]-*as*-triazine is favored in chloroform solution.[89] Substitution at positions 5 and 7 of tetrazolo[1,5-*a*]pyrimidine, with methyl or methoxy groups, results in the stabilization of the tetrazolo form (**47b**).[88,97]

In contrast, a chlorine at position 7 of 5-methyltetrazolo[1,5-*a*]pyrimidine stabilizes the azido tautomer sufficiently for it to be isolated, along with the tetrazolo tautomer.[97] The same electron-donating substituents at positions 5 and 7 of tetrazolo[1,5-*c*]pyrimidine have been reported to favor the azido tautomer[97]; an amino group at position 8 has the reverse effect unless it is acetylated.[87] Finally, nitrogen atoms introduced in the 6-membered ring of **47** may be considered as electron-withdrawing substituents stabilizing **47a** or **47b** depending on their position.

Regardless of the tetrazoloazine (**47**), the tetrazolo tautomer is generally favored in the solid state or in dimethyl sulfoxide solution.[86,87,97] On the other hand, the azido form is stabilized in trifluoroacetic acid, presumably through protonation of an annular nitrogen and destabilization of the tetrazoloazine tautomer.[88,97] In other solvents both **47a** and **47b** are usually detected, except for the tetrazolo[1,5-*b*]pyridazines, which are stable, as the tetrazolo forms, in solution.[89,97]

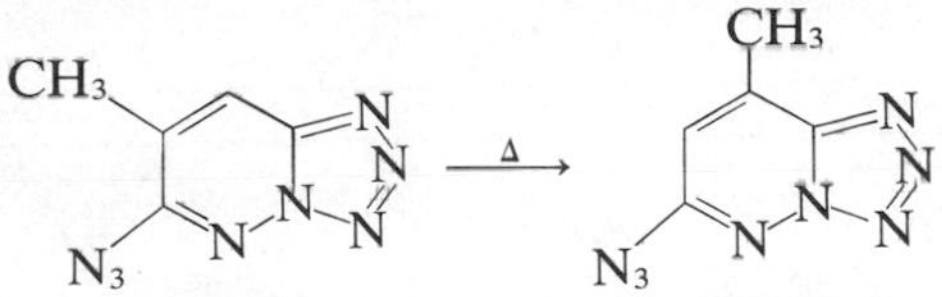

Increasing the temperature induces the isomerization of **109** to **110**, both compounds being stable at room temperature.[93] The tautomerization of 5,7-dimethyltetrazolo[1,5-*a*]pyrimidine is endothermic in deuterochloroform.[88] Rates of interconversion of **111a** to **111c** are first-order, and the values of the activation parameters (in particular the low value of the entropy of activation) suggest that the rate-determining step is the fission of the 3-4 bond to give **111b**.[90]

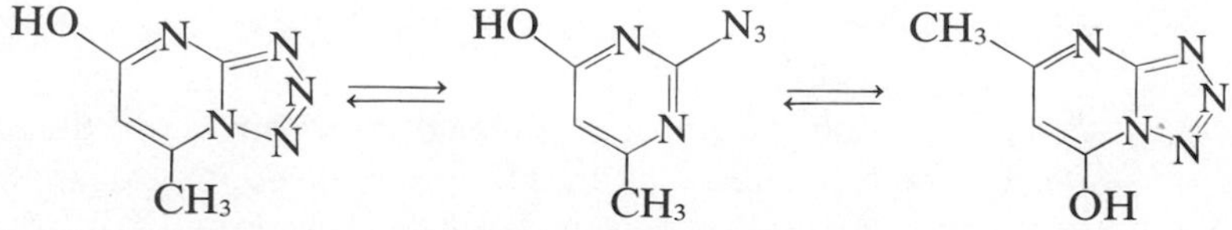

An analogous ring-chain tautomerism has been observed in the *v*-triazolo[1,5-*a*]pyrimidine series.[285,319] These compounds exist entirely in the fused triazole form at room temperature as shown by ir and nmr spectroscopy. However at higher temperatures the nmr peaks progressively coalesce and give rise to averaged signals, thus demonstrating the occurrence of the following equilibrium:

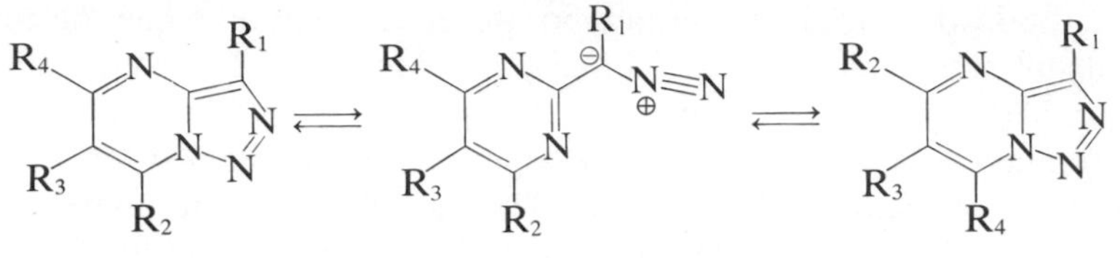

Electron-withdrawing substituents facilitate the tautomeric process and stabilize the diazo tautomer, which is then directly detected by ir spectroscopy.[319]

Very recently 5,7-dimethyl-2-*p*-tolyl-*s*-triazolo[2,3-*a*]pyrimidine 3-oxide has been shown by nmr to undergo a similar thermally induced ring-chain tautomerism:[269]

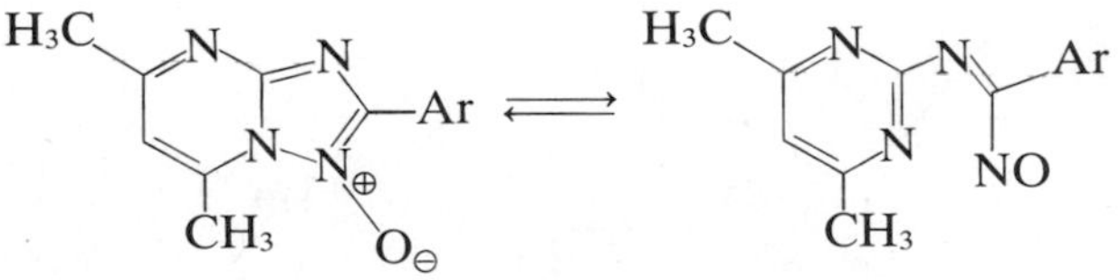

C. Electrophilic Reactions

The reactions discussed in this section include electrophilic attacks at an annular nitrogen atom (protonation and *N*-alkylation) and electrophilic substitutions at a ring carbon atom of the azaindolizine nucleus.

1. *Protonation*

Most of the protonation studies of azaindolizines having more than one nitrogen atom in the 6-membered ring refer to a basic investigation by Armarego dealing with mono or diazaindolizines.[37] The methods used to determine the protonation sites involve the comparison of uv or pmr spectra of the free bases and the protonated azaindolizines. As already stated, the change in the uv spectrum on protonation depends on how different the contributing resonance structures of the protonated molecule are compared to the free base.[37] It is generally admitted that protonation and *N*-alkylation occur at the same position if the pmr spectra of the corresponding protonated and *N*-alkylated azaindolizine derivatives present only minor differences.[25,191,192]

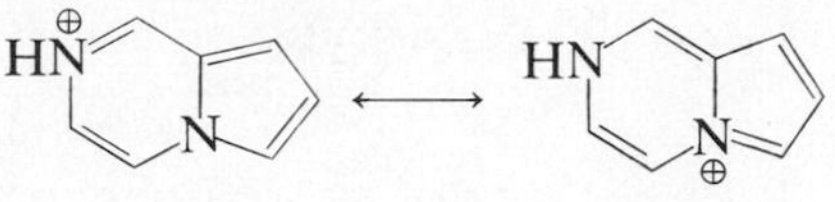

112a **112b**

In general the azaindolizines examined are protonated at a nonbridgehead nitrogen atom. Thus the protonation site of pyrrolo[1,2-*a*]-pyrazine is N-7, as suggested by the presence of spin coupling between H-1 and H-3 on protonation.[10] In contrast, pmr spectroscopy shows that protonation of 6,8-dimethyl (or-diphenyl) pyrrolo[1,2-*b*]pyridazine occurs at carbon-1 and carbon-3[246]; however another study of the protonation of substituted pyrrolo[1,2-*b*]pyridazines suggests that protonation at N-5 is kinetically favored whereas the corresponding 3*H*-cation is thermodynamically favored.[320,321] Imidazo[1,2]diazines have similar pK_a values regardless of the position of the nonbridgehead nitrogen in the 6-membered ring, in agreement with protonation at N-1.[37] This has been confirmed by nmr spectroscopy in the case of imidazo[1,2-*a*]pyrimidine[25] and imidazo[1,2-*b*]pyridazine[173] derivatives. *s*-Triazolo[4,3-*b*]-pyridazines[173] and *s*-triazolo[2,3-*b*]pyrimidine[191] are also protonated at position 1 while imidazo[1,5-*a*]pyrimidine[30a] is protonated at position 2.

Studies of the basicities of azaindolizines have seldom been carried out, and only a few pK_a values are mentioned in the literature (Table VII). In particular, these values show that the introduction of a nitrogen atom in the 6-membered ring of imidazo[1,2-*a*]pyridine ($pK_a = 6.79$) lowers the

TABLE VII. pK_a VALUES OF VARIOUS AZAINDOLIZINES

Compounds	pK_a	Ref.
Pyrrolo[1,2-*a*]pyrazine	6.28	37
Imidazo[1,2-*b*]pyridazine	4.57	37
Imidazo[1,2-*c*]pyrimidine	4.41	37
Imidazo[1,2-*a*]pyrazine	3.59	37
Imidazo[1,2-*a*]pyrimidine	4.81	37
2-Methyl derivative	5.15	40
5,7-Dimethyl derivative	5.76	37
2,5,7-Trimethyl derivative	6.12	40

basic strength, with a maximum effect observed in imidazo[1,2-*a*]-pyrazine.[37] In addition, some hydroxy azaindolizines show acidic properties[69,133,135] and give metal salts.[144,177,215,218]

2. *N-Alkylation*

N-Alkylation studies have been of interest to determine the position of protonation in azaindolizines and to provide models for studying tautomerism in these systems. Accordingly, the investigations deal either with unsubstituted and alkyl-substituted azaindolizines yielding quaternary salts or with hydroxy and amino-derivatives, leading to uncharged species under basic conditions.

The *N*-alkylation site of imidazo[1,2-*a*]pyrimidine and *s*-triazolo[2,3-*a*]pyrimidine has been shown to be N-1 in both cases, by means of hydrolytic degradation studies, which afford 1-methyl-2-aminoimidazole and 3-amino-4-methyl-*s*-triazole, respectively.[191] Methyl groups on carbons adjacent to a nonbridgehead nitrogen atom are normally more reactive (e.g., toward carbonyl compounds) when the nitrogen is quaternized. This can be tentatively used to locate the *N*-alkylation sites of pyrazolo[2,3-*a*]pyrimidine,[194] *s*-triazolo[4,3-*b*]pyridazine,[243] and imidazo[1,2-*b*]-*as*-triazine[193] derivatives. However evidence based on nmr spectroscopy is more often used. Thus quaternization of imidazo[1,2-*a*]pyrimidines occurs at N-1 because a methyl substituent present at position 2 affects the proton chemical shift of the *N*-methyl group markedly compared to the parent methiodide.[25] The steric *peri*-interaction between an N-1 methyl and a C-8 methyl group induces a deshielding of the *N*-methyl group compared to homologous *N*-alkyl derivatives with no substituent at the 8-position. This establishes the *N*-alkylation site and, according to this method, *N*-methylation of imidazo[1,2-*b*]pyridazines occurs at N-1.[173] HMO calculations predict the position of *N*-alkylation of *s*-triazolo[4,3-*b*]pyridazine to be N-1,[173] contrary to a report which shows that in nitromethane quaternization occurs at N-2.[108b] In methanol, however, both N-1 and N-2 methiodides were obtained. It has been suggested that the proportion of isomers is controlled by a combination of steric and electronic effects due to substituents at C-3 and C-8.[173]

Hydroxy- and aminoazaindolizines are *N*-alkylated at ring nitrogen atoms and neutral species are formed in basic medium if the *N*-alkyl derivatives arise from possible NH-tautomers. Thus *N*-alkylation of 5-hydroxy-*s*-triazolo[2,3-*a*]pyrimidines (or their silver salts) generally occurs at positions 1 or 8.[45,138,140] Evidence for this is based on independent syntheses of the products from *N*-alkyl derivatives of 3-amino-*s*-triazole and β-ketoesters.[140] 7-Hydroxy-*s*-triazolo[4,3-*a*] or [2,3-*a*]pyrimidines have been reported to be methylated exclusively at position 8[45,215] or on the hydroxy substituent. The *N*-alkylation of 5-hydroxy-*s*-triazolo[2,3-*a*]-*s*-triazine silver salt was shown to occur at position 6 by means of alkaline hydrolysis of the product.[144] 5-Aminopyrazolo[2,3-*a*]pyrimidines are alkylated at position 8,[180] whereas *N*-methylation of 3-phenyl-5-hydroxypyrazolo[2,3-*a*]pyrimidine has been claimed to yield the 3-methyl or 8-methyl derivatives, depending on the reaction conditions.[132,195] Methylation of 2-phenyl-6-hydroxyimidazo[1,2-*b*]pyridazine occurs at position 5.[12] Finally, the *N*-glycosylation sites in nucleosides derived from 5-hydroxypyrazolo[2,3-*a*]pyrimidine, 5-hydroxy-*s*-triazolo[2,3-*a*]-pyrimidine, and 5-hydroxypyrazolo[2,3-*a*]-*s*-triazine derivatives are conveniently determined from the ^{13}C nmr chemical shifts of neighboring carbons.[309]

N-Alkyl- and *O*-alkylazaindolizines can undergo thermal rearrangements involving the migration of the alkyl groups to other nucleophilic centers of the molecule. Thus both N-1 and N-2 methiodides of 8-methyl-*s*-triazolo[4,3-*b*]pyridazine interconvert at high temperature.[173] In the *s*-triazolo[4,3-*a*]pyridine series N-2 methiodides are totally rearranged to the corresponding N-1 isomers unless a methyl substituent is present at position 8.[196] 6-Methoxyimidazo[1,2-*b*]pyridazines can be transposed at high temperatures to the 5-methyl-6-oxo- and 1,5-dimethyl-6-oxo- derivatives.[322] The rearrangement of 5-alkoxy-7-methyl-*s*-triazolo[2,3-*a*]pyrimidines affords very good yields of 1-alkyl-5-oxo and 8-alkyl-5-oxo derivatives.[181] Similarly both 5-methoxy- and 6-methyl-*s*-triazolo[4,3-*a*]-*s*-triazines rearrange to the 8-methyl-5-oxo isomer.[54] The orientation of the rearrangement of 5-allyloxy-*s*-triazolo[2,3-*a*]-pyrimidines depends on the substitution at positions 6 and 7. Thermal treatment of **113** leads to **114a–c** in addition to the products of normal intermolecular alkyl migration.[197] Compounds **114a–c** are assumed to arise from an intramolecular Claisen-type rearrangement.[197] Similarly the formation of compounds **116a–c** from **115** has been explained by two successive allyl rearrangements.[198]

113 ($R = CH_2{-}CH{=}CH_2$) $\xrightarrow{150°}$ **114a** + **114b** + **114c**

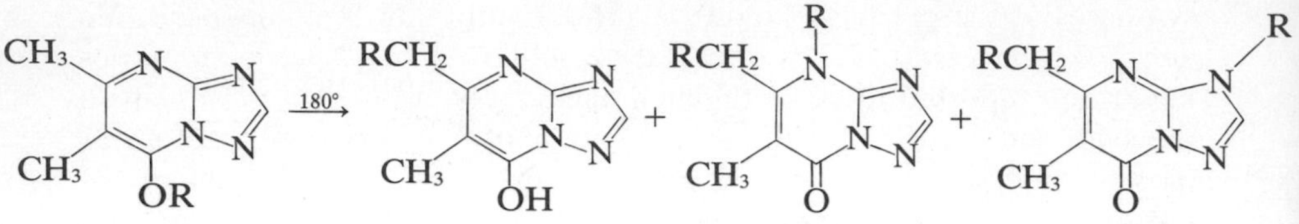

180°

115 **116a** **116b** **116c**

3. *Halogenation*

This is the most frequently encountered electrophilic substitution in the azaindolizine series. The site of electrophilic substitution of unsubstituted or alkyl- and aryl-substituted azaindolizines depends primarily on the number and locations of nitrogen atoms in the 5-membered ring. The following tentative generalizations are consistent with published data:

a. Pyrrolo[1,2]azines lead, in general, to 1- or 3-halo derivatives or 1,3-dihalo derivatives (e.g., pyrrolo[1,2-*b*]pyridazines,[105] pyrrolo[1,2-*c*]-pyrimidines,[245] pyrrolo[1,2-*a*]pyrazines,[112] and pyrrolo[1,2-*d*]-*as*-triazine[300]). However the bromination of 6,8-dimethylpyrrolo[1,2-*b*]-pyridazine with bromine has been reported to afford the 1,2,3-tribromo-derivative in low yield.[246]

b. If a nitrogen occupies position 1, halogenation occurs at position 3 (e.g., imidazo[1,2-*b*]pyridazine,[19] imidazo[1,2-*a*]pyrimidines,[25] and imidazo[1,2-*b*]-*as*-triazines[34,200]).

c. Aza-substitution at position 2 induces the halogenation of carbons 1 and/or 3 (e.g., imidazo[1,5-*a*]pyrimidine,[30a] and imidazo[1,5-*b*]-*as*-triazines[296]).

d. Pyrazolo[2,3]azines are halogenated at position 1 (e.g., pyrazolo-[2,3-*c*]pyrimidines,[287] pyrazolo[2,3-*a*]-*s*-triazines,[288] and pyrazolo[3,2-*c*]-*as*-triazines[201]).

e. Halogenation of *s*-triazolo[4,3]azines occurs at position 3 (e.g., *s*-triazolo[4,3-*a*]pyrazines[175] and *s*-triazolo[4,3-*b*]pyridazines[323]).

f. Finally, if nitrogen atoms occupy positions 1 and 3, halogenation occurs on the 6-membered ring (e.g., *s*-triazolo[2,3-*a*]pyrimidines[42,199]).

Alkyl or aryl substituents at unreactive sites, or aza-substitution in the 6-membered ring have generally little influence on the orientation of halogenation. In contrast, hydroxy or amino substituents have a tendency to activate the carbon atoms at the α-positions. This is probably related to the potential tautomeric changes of the structures.[44b,133,199,202,204,262] Chlorination of 2-hydroxy-5,7-dimethylpyrazolo[2,3-*a*]pyrimidine has been claimed to occur at position 3, as well as on the 7-methyl substituent.[133] The treatment of 1,3-dibromo-2,7-diphenylpyrrolo[1,2-*c*]-pyrimidine perbromide hydrobromide with alcohols leads to 8-alkoxy-1,3-dibromo and 1,3,8-tribromo derivatives, presumably through a covalent dibromo adduct.[245]

Although no mechanistic study has been carried out in the azaindolizine series, it is assumed that the mechanism of halogenation (using halogen or *N*-bromosuccinimide) is ionic[205] and involves nonprotonated species in water, by analogy with acid-catalyzed deuterium exchanges.[42,221] In general predictions based on frontier electron density calculations (or sometimes total π-electron density calculations) agree reasonably well with the actual nucleophilicities of annular carbon atoms.[25,35,42,112,172]

4. *Other Electrophilic Substitutions*

Nitration, nitrosation, and azo-coupling of azaindolizines generally occur at the same positions as halogenation (Table VIII) except if other positions are activated by α-hydroxy or α-amino groups.[277] Strong electrophiles may lead to polysubstituted products (nitration), and weaker nucleophiles effect only monosubstitutions (nitrosation, azo coupling), or fail to react, as in the pyrrolo[1,2-*a*]pyrazine[10] and *s*-triazolo[2,3-*a*]-pyrimidine[206] series. A phenyl substituent is generally nitrated at the

TABLE VIII. SITES OF ELECTROPHILIC SUBSTITUTIONS IN SOME AZAINDOLIZINES

Azaindolizines	Nitration	Nitrosation	Azo coupling	Ref.
Pyrrolo[1,2-*b*]-pyridazines	1,3-Dinitro derivative	3-Nitroso derivative	3-Arylazo derivative	105
Pyrrolo[1,2-*c*]-pyrimidines	—	3-Nitroso derivative	3-Arylazo derivative	7
Pyrrolo[1,2-*a*]-pyrazines	1-Nitro, 3-nitro, and 1,3-Dinitro derivatives	[a]	—	10
Imidazo[1,2-*b*]-pyridazines[b]	3-Nitro derivative	—	—	19, 172
Imidazo[1,2-*a*]-pyrimidines	3-Nitro derivative	3-Nitroso derivative	3-Arylazo derivative	28, 207
Pyrazolo[2,3-*a*]-pyrimidines	1-Nitro derivative	1-Nitroso derivative	1-Arylazo derivative	133, 194, 195, 209–211, 214
s-Triazolo[2,3-*a*]-pyrimidines	6-Nitro derivative	6-Nitroso derivative	6-Arylazo derivative	119, 206, 212, 213

[a] The parent compound fails to react.[10]

[b] Sulfonation with chlorosulfonic acid leads to a 3-sulfonic acid.[19]

para-position.[105,172,207] This process may be favored over substitution in the azaindolizine nucleus.[207] In general, the structures of the substitution products have been determined either by unequivocal syntheses,[28,194,207,208] as in the case of **117** and **118,** or, more conveniently, by pmr spectroscopy.[10,19,25,105,112,172] Nitration of 2-hydroxy-6-carbethoxy-7-methylpyrazolo[2,3-*a*]pyrimidine has been reported to involve not only nitration at position 1, but also the replacement of the 7-methyl substituent by a hydroxy group.[209] Azo-coupling in pyrazolo[2,3-*a*]-pyrimidines[194,120] and imidazo[1,2-*a*]pyrimidines[28] leads to colored compounds used as dyes.

Other related reactions include the acylation of pyrrolo[1,2-*b*]-pyridazine[105] or pyrazolo[2,3-*a*]pyrimidine[324] derivatives with trifluoroacetic anhydride, acylation of pyrrolo[1,2-*c*]pyrimidines with acetic anhydride, ethyl benzoylacetate, or phosgene.[7] The Reimer–Tiemann reaction on **119**[7] and Mannich base syntheses from imidazo[1,2-*b*]pyridazines,[18] imidazo[1,2-*a*]pyrimidines,[29] and pyrazolo[2,3-*a*]pyrimidines[325] have also been examined.

CH_3 N OH N N CH_3 $\xrightarrow{ArN_2^{\oplus}}$ Ar N=N CH_3 N OH N N CH_3 $\xleftarrow[\text{2) } \beta\text{-naphtol}]{\text{1) } HNO_2}$ CH_3 N NH_2 OH N N CH_3

117

N N Ar N $\xrightarrow{HNO_2}$ N N Ar N NO $\longleftarrow$ N NH_2 N + O C Ar C HON Cl

118

Ph N N CH_3 $\xrightarrow[KOH]{CHCl_3}$ Ph N N CH_3 CHO

119

D. *Nucleophilic Reactions*

1. *Covalent Hydration*

As already mentioned, covalent hydration of the 6-membered ring of azaindolizines very probably precedes the Dimroth rearrangement of these systems. The covalently hydrated species are usually not isolated, except in the case of **120** and **121,** where they are stabilized by the electron-withdrawing properties of the tetrazole ring.[87,97] The attack occurs, in general, at position 5, but positions 7[144,291] or 8[81] have also been implicated.

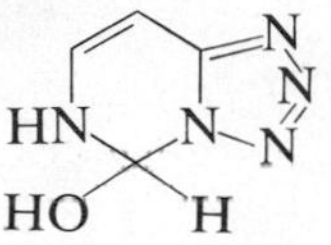

120 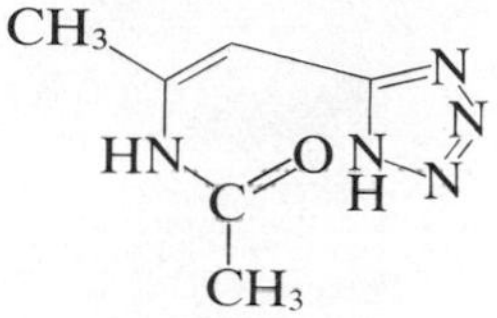 **121**

Intermediates were detected in the base-catalysed hydrolysis of *s*-triazolo[2,3-*a*]pyrazine and -pyrimidine derivatives; data from pK_a measurements, kinetic studies, and uv spectroscopy suggest that the corresponding intermediates have structures analogous to **121**.[326,327]

2. *Replacement of Halogen Atoms*

Nucleophilic attack of haloazaindolizines serves to introduce various substituents, namely OR, SR, NR_2, N_2H_2R, or N_3 by the use of NaOR, KSR (or thiourea), HNR_2, N_2H_3R, or NaN_3, respectively. These reactions have been applied to numerous azaindolizine systems, including azolo[*b*]-pyridazines,[16,19,52,56,94,268,323,328,329] azolo[*c*]pyrimidines,[74,183] azolo[*a*]-pyrazines,[85] azolo[*a*]pyrimidines,[44b,203,215–219,330] and azolo[*a*]-*s*-triazines.[288] The introduction of amino-, hydrazino, or azido groups is a key step in the attempted syntheses of the tricyclic heterocycles **122**–**125**,[52,89,101,215,267] or homologous systems.[335]

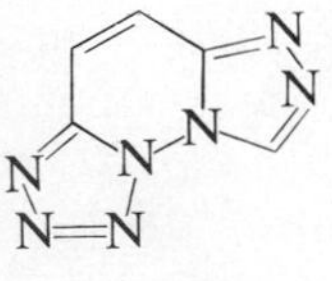

122 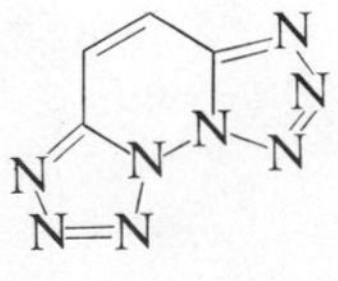 **123** 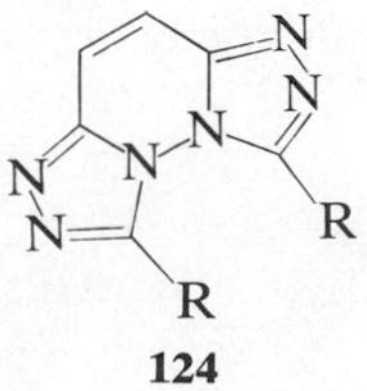 **124** 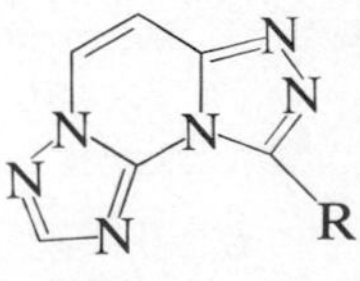 **125**

3. *Reactions with Metal Hydrides and Organometallic Compounds*

The reduction of azaindolizines with $LiAlH_4$ or $NaBH_4$ affords, in general, a dihydro derivative. Imidazo[1,2-*a*]pyrimidine,[42] *s*-triazolo[2,3-*a*]pyrimidine,[42] and imidazo[2,1-*c*]-*as*-triazine[152] derivatives lead to products analogous to **126**. In contrast, 2,6-diphenylimidazo[1,2-*b*]-*as*-triazine has been claimed to yield the 1,5-dihydro compound,[32] whereas other imidazo[1,2-*b*]-*as*-triazines afford the corresponding 5,6,7,8-tetrahydro derivatives.[332]

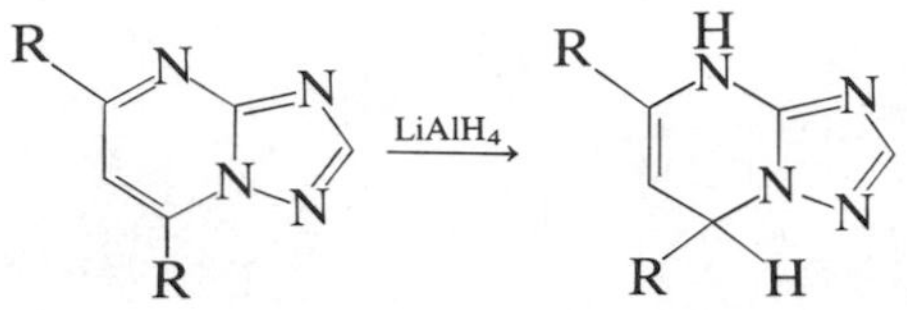

$LiAlH_4$

126

Recently the mechanism of the reduction of azolo[*b*]pyridazines with $NaBH_4$ was determined through selective deuteration experiments.[333]

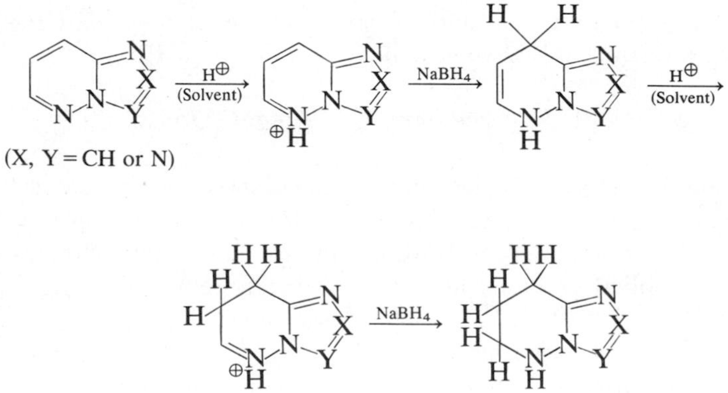

(X, Y = CH or N)

Attack of imidazo[1,2-*a*]pyrimidine or *s*-triazolo[2,3-*a*]pyrimidine with methyllithium occurs at position 7, probably via a 1,2-addition process, whereas, with phenyllithium, both 5-phenyl- and 7-phenyl-*s*-triazolo[2,3,-*a*]pyrimidines are obtained.[42] Pyrrolo[1,2-*a*]pyrazine appears to be resistant to *n*-butyllithium;[10] however 8-phenylpyrrolo[1,2-*a*]pyrazine is obtained on treatment with phenyllithium.[331] Imidazo[1,2-*b*]-*as*-triazine also reacts with phenyllithium and gives 7-phenylimidazo[1,2-*b*]-*as*-triazine.[331]

E. Miscellaneous Reactions

1. *Deuterium–Hydrogen Exchange Processes*

Deuterium–hydrogen exchanges of imidazo[1,2-*a*]pyrimidine and *s*-triazolo[2,3-*a*]pyrimidine derivatives have been investigated under acidic, as well as basic conditions.[220,221] In acidic medium, electrophilic substitution by $D^{\oplus}$ occurs at position 3 of imidazo[1,2-*a*]pyrimidine and at position 6 of *s*-triazolo[2,3-*a*]pyrimidine.[220] The exchange rates of the corresponding N-1 methiodides are slower under the same conditions, thus showing that the free bases and not the protonated species are involved in these reactions.[221] Similarly H-3 in imidazo[1,2-*b*]- and *s*-triazolo[4,3-*b*]pyridazines is exchanged more rapidly than H-8 under neutral, acid, or base-catalyzed conditions.[334]

The kinetics of base-catalyzed exchange reactions have a first-order dependency on both the heterocycle and the base. In addition, the existence of a primary deuterium isotope effect in the case of 3-deutero-imidazo[1,2-*a*]pyridine suggests that the proton transfer occurs in the rate-determining step.[221] Under base catalysis H-5 of imidazo[1,2-*a*]pyrimidine exchanges faster than does H-3, and H-3 exchanges faster than H-6 and H-2. On the other hand, H-5 of *s*-triazolo[2,3-*a*]pyrimidine is more reactive than H-6 and H-2 or H-7.[221] The H-3 and H-2 protons of the N-1 methiodides of various *s*-triazolo[2,3-*a*]pyrimidines exchange about 10^3 times faster than those in the free base.[221]

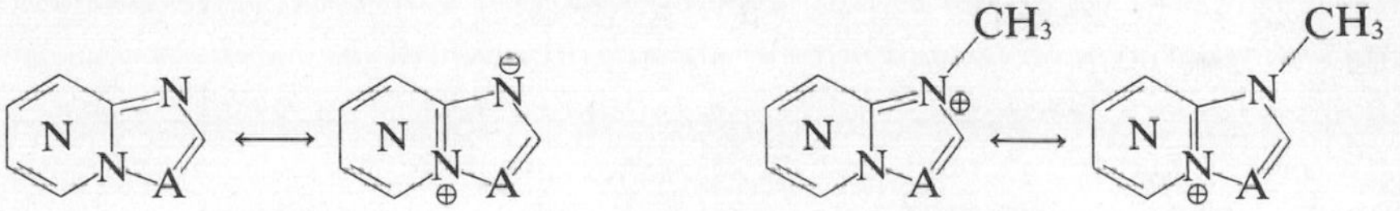

127 (A = CH or N) **128**

The preceding data are consistent with high contributions of the two resonance structures **127** or **128** in the ground state of the molecules. In particular, the partial positive charge at N-1 or N-4 stabilizes the developing negative charges at adjacent positions, thus favoring the exchange at these sites.[221] Comparison of overall reactivities of *s*-triazolo[2,3-*a*]pyrimidines, imidazo[1,2-*a*]pyridines or -pyrimidines, and pyrrolo[1,2-*a*]pyrazine suggests that the relative rates of deuterium exchange increase with aza substitution.[10,42,220,221]

2. *Catalytic Reduction or Hydrogenation*

Catalytic dechlorination (with palladium) or desulfurization (with Raney nickel) is currently used to prepare some substituted azaindolizines (e.g., *s*-triazolo[4,3-*b*]pyridazines and tetrazolo[1,5-*b*]pyridazines[56,67,91]). Replacement of 2- or 3-mercapto groups with a hydrogen atom has been particularly useful.[44b,75,134,178]

The 6-membered ring of *s*-triazolo[2,3-*a*]pyrimidine or of pyrrolo[1,2-*a*]pyrazine has been selectively hydrogenated over a palladium or platinum catalyst, leading to 5,6,7,8-tetrahydro derivatives[10,42] although other *s*-triazolo[2,3-*a*]pyrimidines have been reported to be resistant to catalytic hydrogenation.[177] In contrast, hydrogenation of 5,6-dimethylpyrazolo[3,2-*c*]-*as*-triazine over Adams catalyst affords the 3,5-dihydro derivative,[153] and the reduction of *s*-triazolo[4,3-*b*]pyridazine using palladium yields the 7,8-dihydro derivative.[333]

3. *Reactivity of Methyl Substituents*

Methyl groups substituted at carbon atoms adjacent to a nitrogen of the azaindolizine nucleus (N-4 excluded) have sufficient anionic character to react with aromatic carbonyl compounds or with orthoesters, leading to cyanine dye analogs (e.g., **129** or **130**).[194,222] The reactivity of the methyl groups is normally enhanced if the neighboring nitrogen is quaternized.[193,194,223]

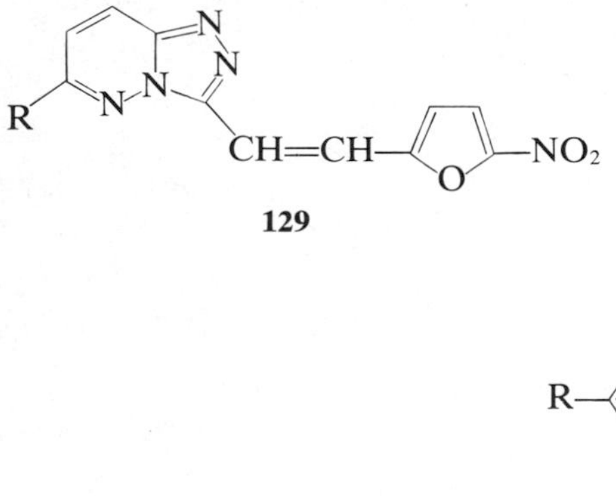

R, N, N, N, N, CH=CH—, O, —NO_2

129

$X^{\ominus}$ C_2H_5, C_2H_5, ⊕N, N, H, H, C, C, C, H, R—, —R, N-N, N-N, CH_3, CH_3

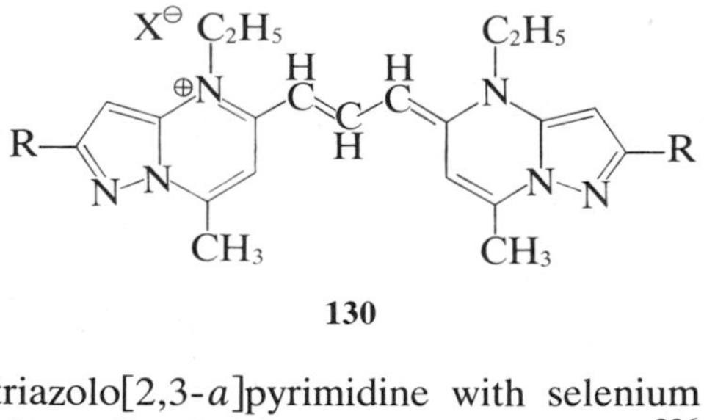

130

Oxidation of 5,7-dimethyl-*s*-triazolo[2,3-*a*]pyrimidine with selenium dioxide yields 5-methyl-*s*-triazolo[2,3-*a*]pyrimidine-7-carboxaldehyde:[336]

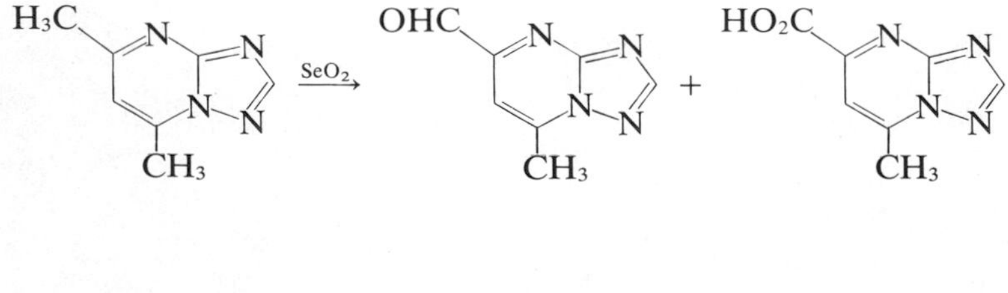

H_3C, N, N, N, N, CH_3 $\xrightarrow{SeO_2}$ OHC, N, N, N, N, CH_3 + HO_2C, N, N, N, N, CH_3

The Tschitschibabin reaction on compound **131** affords the dipyrrolo-[*ac*]pyrazine derivative **132**.[10]

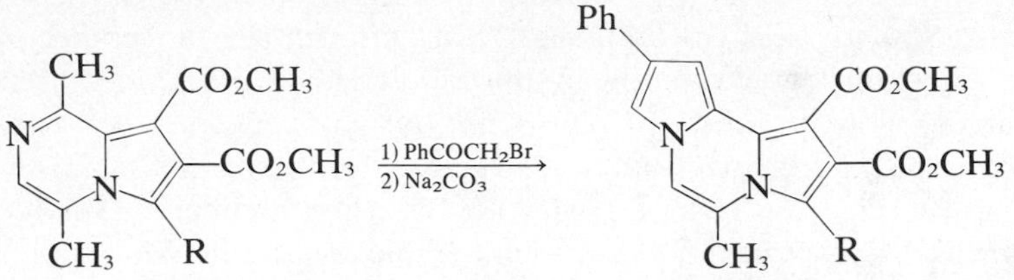

1) $PhCOCH_2Br$ 2) Na_2CO_3

131 **132**

4. *Ring Cleavage Reactions*

Depending on the conditions and the heterocycle, ring cleavage reactions affect the 5-membered ring or the 6-membered ring of the azaindolizine nucleus.

Alkaline cleavage has often been used to determine the structure of substitution products. As expected, the initial attack occurs at carbons adjacent to the bridgehead nitrogen, as in N-1 methiodides of imidazo-[1,2-*a*]pyrimidine, and *s*-triazolo[2,3-*a*]pyrimidine[191] or as in the case of compounds **133** leading to **134**.[108b,303] 6-Aryl-*s*-triazolo[4,3-*b*]-*s*-tetrazines[247] and 5-mercapto-*s*-triazolo[2,3-*a*]-*s*-triazines[291] are also easily cleaved in alkaline medium.

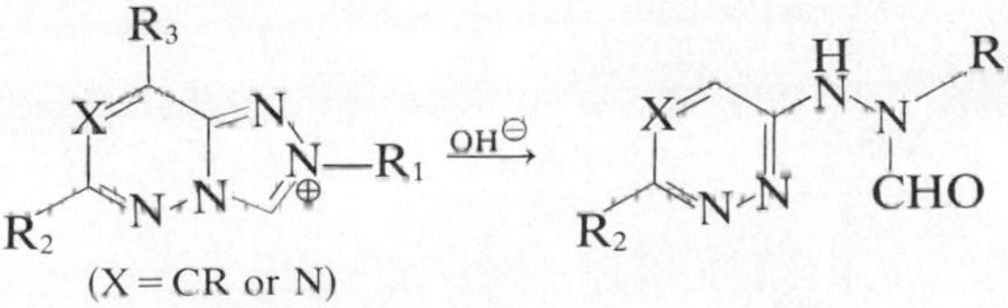

$OH^{\ominus}$

(X = CR or N)

133 **134**

Treatment of imidazo[1,2-*b*]pyridazine or *s*-triazolo[4,3-*b*]pyridazine derivatives with potassium *t*-butoxide or isopropoxide leads to the fission of the N4–N5 bond and affords mixtures of geometrical isomers of **135**.[337]

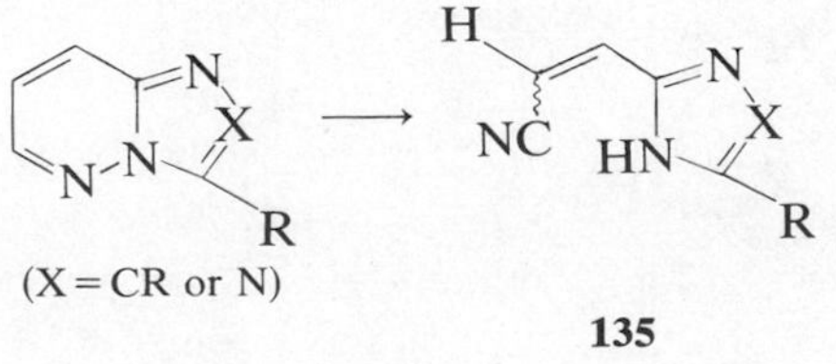

(X = CR or N)

135

It is likely that hydrazinolysis of imidazo[1,2-*a*]pyrimidine,[338] of *s*-triazolo[4,3-*a*]pyrimidine,[224] of *s*-triazolo[2,3-*a*]pyrimidine,[181,271,339] and of pyrazolo[3,2-*c*]-*as*-triazine[225] derivatives initially involves the fission of the N4–C5 bond (as in the base-catalyzed Dimroth rearrangement), since the products of degradation are 2-aminoimidazoles, 3-amino-*s*-triazoles, or 3-aminopyrazoles, respectively. Similarly hydrazinolysis of *s*-triazolo-[4,3-*b*]pyridazines yields 4-amino-*s*-triazole derivatives.[343]

Selective ring cleavages of *v*-triazolo[1,5-*a*]pyrimidines[141,284] or *v*-triazolo[5,1-*c*]-*as*-triazines[299] occur under acidic conditions. This represents a convenient synthesis of pyrimidine or *as*-triazine derivatives such as **136**.

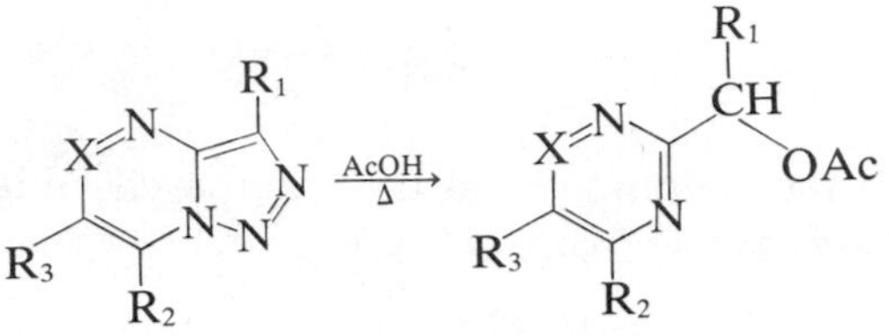

(X = CR or N)

136

Catalytic reduction of nitrosopyrrolo[1,2-*c*]pyrimidines or oxidation of 3-amino derivatives leads to the cleavage of the 5-membered ring supposedly via a nitrene intermediate.[226] Similarly hydroxy- or aminopyridazines (**138**) are prepared from hydroxy- or amino-*s*-triazolo[4,3-*b*]pyridazines (**137**).[277] A few oxidative degradations of azaindolizines have been carried out; in the reported reactions cleavage of the 6-membered ring occurs and gives azolecarboxylic acids.[266,287,304]

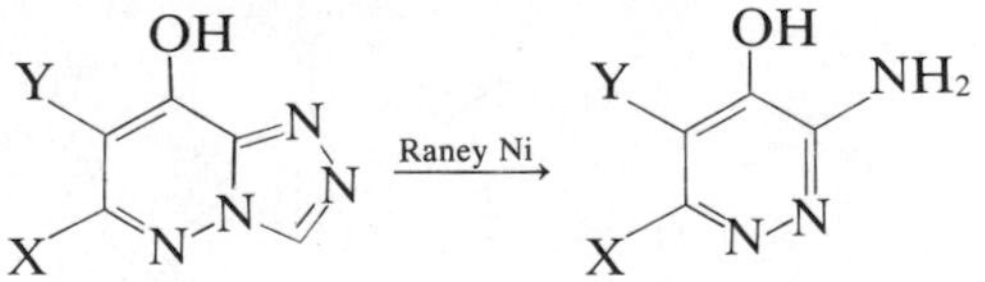

(X, Y = OH or NH_2)

137

138

The preceding degradation reactions may be regarded as "relay syntheses" of suitably substituted monocyclic heterocycles: for example, the preparations of substituted azines from readily available aminoazoles.[277,284,299,303]

5. *Photolysis and Thermolysis Reactions*

Except in the tetrazolo[1,5]azine series no general study of the photolysis or thermolysis of azaindolizines has been reported.

Photolysis of diazo- or mesoionic derivatives of 8-hydroxy-*s*-triazolo-[4,3-*b*]pyridazine involves a ring contraction process and yields pyrazolo-[3,2-*c*]-*s*-triazole (e.g., **139**).[340,361] Irradiation of *s*-triazolo[4,3-*b*]-pyridazines and *s*-triazolo[2,3-*b*]pyridazines in the presence of alkenes leads to mixtures of cycloaddition products such as **140** and **141**, respectively.[341]

140 (X = N, Y = CH)
141 (X = CH, Y = N)

Recently pyrolysis in the gas phase or in solution of tetrazolodiazines has been shown to yield 1-cyanoimidazole derivatives (from tetrazolo-[1,5-*c*]pyrimidines and tetrazolo[1,5-*a*]pyrazine), 1-cyanopyrazoles and 2-aminopyrimidines (from tetrazolo[1,5-*a*]pyrimidines), whereas the degradation of tetrazolo[1,5-*b*]pyridazine affords only nonheterocyclic compounds.[227,342] Photolysis or thermolysis of 5,6 diphenyltetrazolo[1,5-*a*]-pyrazine in acetic acid also yields imidazole derivatives.[89] A ring contraction of transient nitrene intermediates derived from the corresponding azidodiazines has been postulated to explain these results.[89,227,342]

Photolysis of tetrazolo[1,5-*b*]pyridazines gives 3-cyanocyclopropenes (**142**) and traces of 3-cyanopyrazoles (**143**), presumably via cyano-diazo intermediates and carbenes.[345]

Photolysis of the azido tautomer corresponding to 8-phenyltetrazolo-[1,5-*c*]pyrimidine in trifluoroacetic acid affords the pyrimidoindole **144** in excellent yield.[346]

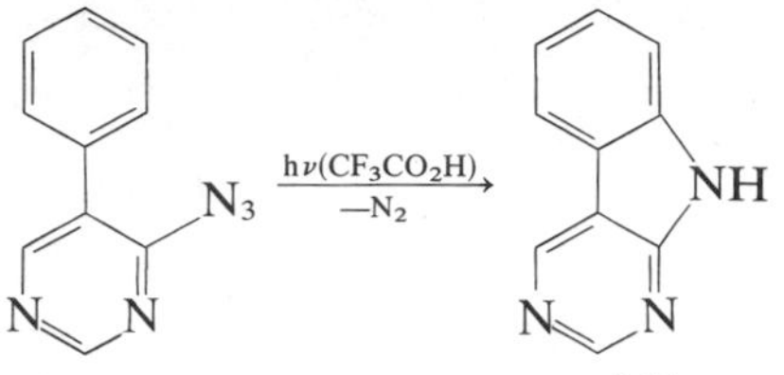

N_3 $\xrightarrow[-N_2]{h\nu(CF_3CO_2H)}$ NH

N N N N

144

Finally, irradiation or thermolysis of 6-azidoazolo[*b*]pyridazine derivatives results in the formation of nitrenes and either hydrogen abstraction,[347,348] dimerization,[347] or reaction with diethylamine[348] or alkyl acrylates,[349] depending on the conditions.

6. *Cycloaddition Reactions*

Attempts to prepare aza analogs of cycl[3.2.2.]azine (**145**) from azaindolizines having more than one nitrogen in the 6-membered ring have been limited to the case of the pyrroloazine series. The condensation of

N

145

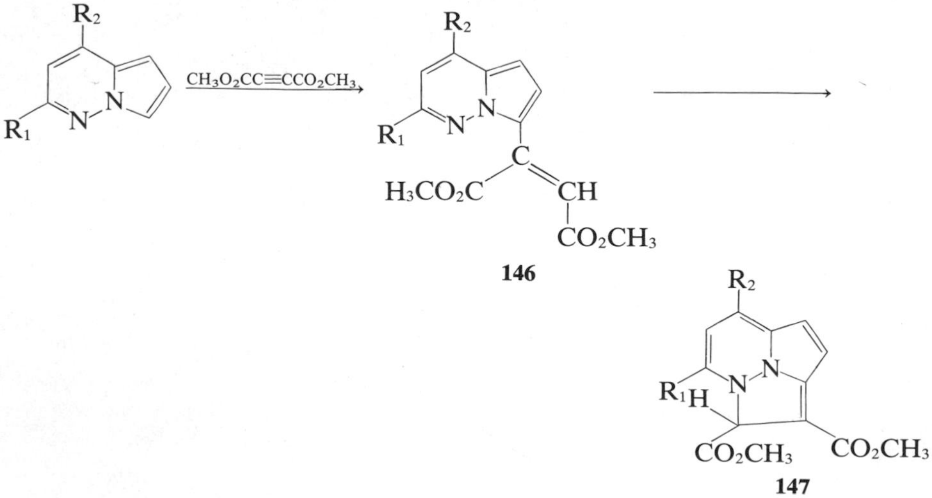

R_2 R_1 N N $\xrightarrow{CH_3O_2CC{\equiv}CCO_2CH_3}$ R_2 R_1 N N C H_3CO_2C CH CO_2CH_3 $\longrightarrow$

146

R_2 R_1 H N N CO_2CH_3 CO_2CH_3

147

pyrrolo[1,2-*b*]pyridazines with dimethyl acetylenedicarboxylate is a two-step reaction leading successively to **146** and **147**. The reaction fails with tetracyanoethylene.[104,105,246] Similarly 2-phenyl-7-methylpyrrolo[1,2-*c*]-pyrimidine affords 2-phenyl-6-methyl-5-azacycl[3.2.2]azine after decarboxylation.[4] Cycloaddition of pyrrolo[1,2-*a*]pyrazine with various dienophiles failed,[10] whereas cycloaddition of 5,6-diphenyltetrazolo[1,5-*a*]-pyrazine and dimethyl acetylenedicarboxylate occurs on the azido tautomer.[89]

V. Biological Properties and Uses

Few azindolizine derivatives are naturally occurring products. Hexahydroimidazo[1,2-*a*]pyrimidine alkaloids (alchornine, alchornidine and alchorneine)[228,242,350] and tetrahydropyrrolo[1,2-*c*]pyrimidine alkaloids (arenaine)[310] have been isolated from plants and their structures elucidated. Degradation of saxitonin, a substance found in the shellfish poison, led to the isolation of 5-oxo-7-methyl-1,2,3,5-tetrahydropyrrolo[1,2-*c*]-pyrimidine.[229] The antibiotic viomycin includes a dihydropyrrolo[1,2-*a*]-*s*-triazine nucleus in its molecule.[251] Cypridina luciferin, a bioluminescent substance, was recently shown to have structure **148,** and its total synthesis was achieved;[230] a possible biosynthesis from three amino acids was also proposed.[353] That the 3-oxo-3,7-dihydroimidazo[1,2-*a*]pyrazine

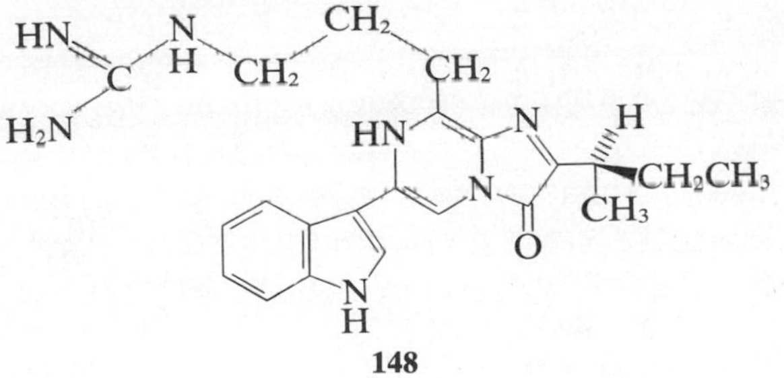

148

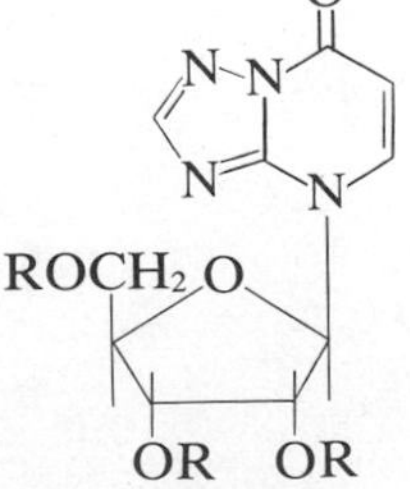

149

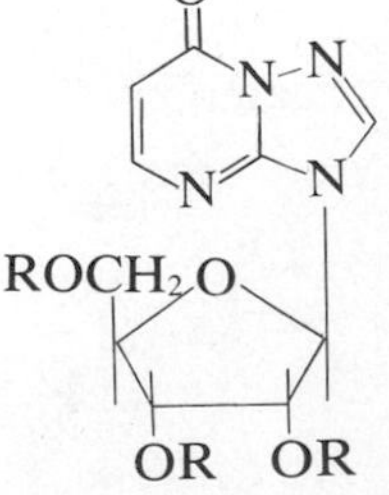

150

moiety is the essential part of the molecule of luciferin is suggested by the occurrence of light emission when 3-oxo-3,7-dihydroimidazo[1,2-*a*]pyrazine or its 2-methyl derivative is dissolved in aprotic solvents in the presence of oxygen.[24,351,352] The light-emission process has been shown to involve proton abstraction with a base, oxidative decarbonylation of the 5-membered ring, and release of the 2-acetamidopyrazine anion.[24] Recently the chemiluminescence of compounds related to the luciferins from certain coelenterates has been suggested to arise from an analogous reaction pathway.[355]

Many azaindolizine derivatives have been tested as potential growth inhibitors of microorganisms, since they are structurally related to purines. Alkyl- or aryl-substituted azaindolizines do not exist as NH-tautomers and cannot be converted to nucleoside analogs.[144] This is in contrast to hydroxy-, amino-, or mercaptoazaindolizines. Thus the condensation of the *O*-trimethylsilyl derivative of 7-hydroxy-*s*-triazolo[2,3-*a*]-pyrimidine and a ribofuranosyl bromide leads to nucleoside analogs **149** and **150**.[176] Other nucleoside analogs from imidazo[2,3-*a*]pyrimidine,[309,339] imidazo[1,2-*c*]pyrimidine,[244] pyrazolo[2,3-*a*]pyrimidine[309] and pyrazolo[2,3-*a*]-*s*-triazine[309] derivatives have recently been prepared and studied. Although many reports exist on the anti-cancer activity of azaindolizines (particularly in the pyrazolo[2,3-*a*]pyrimidine,[231,232,356] *s*-triazolo[2,3-*a*] and [4,3-*a*]pyrimidine,[233–238,356] pyrazolo[3,2-*c*]-*as*-triazine,[153] or *s*-triazolo[2,3-*a*]-*s*-triazine[357] series), it is not yet possible to relate structure and activity. Toxicity or insolubility may also limit the application.[148]

In addition, various azaindolizine derivatives present useful pharmacological properties. 5-Aminopyrazolo[2,3-*a*]pyrimidines,[128] 5- or 7-hydroxy-*s*-triazolo[2,3-*a*]pyrimidines, or 2-arylimidazo[1,2-*a*]pyrimidines[29] are antipyretic, antiinflammatory, and analgesic compounds. Other azaindolizines are useful as hypocholesteric agents,[239] vasodilators,[240] bronchodilators,[241] hypotensive agents,[358] insect sterilisants,[264] herbicides,[273,359] or antimalarial agents.[360]

Finally, properly substituted azaindolizines (mainly *s*-triazolo[2,3-*a*]-pyrimidine derivatives) have been widely used as stabilizers and antifogging agents for photographic emulsions.[354]

VI. References

1. W. L. Mosby, in "The Chemistry of Heterocyclic Compounds," A. Weissberger, Ed., Heterocyclic Systems with Bridgehead Nitrogen Atoms, Interscience, New York, 1961.
2. (a) A. E. Tschitschibabin, *Ber.*, **58,** 1704 (1925); (b) A. E. Tschitschibabin, *Ber.*, **59,** 2048 (1926).

3. E. Ochiai and M. Karii, *J. Pharm. Soc. Jap.*, **59,** 18 (1939); *Chem. Abstr.*, **33,** 3791 (1939).
4. V. Boekelheide and S. S. Kertelj, *J. Org. Chem.*, **28,** 3212 (1963).
5. R. L. Letsinger and R. Lasco, *J. Org. Chem.*, **21,** 764 (1956).
6. B. Boekelheide and K. Fahrenholz, *J. Am. Chem. Soc.*, **83,** 458 (1961).
7. J. Taylor and D. G. Wibberley, *J. Chem. Soc.* (*C*), 2693 (1968).
8. R. M. Acheson and M. W. Foxton, *J. Chem. Soc.* (*C*), 2218 (1966).
9. The numbering of ring positions of the azaindolizine nucleus has been set as in indolizine; cf., the introductory Section.
10. D. E. Dunham, Ph.D. Thesis. University of Ohio, Athens, Ohio, 1967.
11. E. T. Borrows and D. O. Holland, *Chem. Rev.*, **42,** 611 (1948).
12. F. Yoneda, T. Ohtaka, and Y. Nitta, *Chem. Pharm. Bull.* (*Tokyo*), **12,** 135 (1964).
13. C. W. Noell and R. K. Robins, *J. Heterocycl. Chem.*, **1,** 34 (1964).
14. L. M. Werbel and M. L. Zamora, *J. Heterocycl. Chem.*, **2,** 287 (1965).
15. S. C. Bell and W. T. Caldwell, *J. Am. Chem. Soc.*, **82,** 1469 (1960).
16. B. Stanovnik and M. Tisler, *Tetrahedron*, **23,** 387 (1967).
17. B. Stanovnik and M. Tisler, *Tetrahedron*, **23,** 2739 (1967).
18. J. G. Lombardino, *J. Heterocycl. Chem.*, **5,** 35 (1968).
19. J. Kobe, B. Stanovnik, and M. Tisler, *Tetrahedron*, **24,** 239 (1968).
20. A. Pollak, B. Stanovnik, and M. Tisler, *Tetrahedron*, **24,** 2623 (1968).
21. B. Stanovnik and M. Tisler, *Tetrahedron Lett.*, 2403 (1966).
22. P. Guerret, R. Jacquier, and G. Maury, *J. Heterocycl. Chem.*, **8,** 643 (1971).
23. S. Inoue, S. Sugiura, H. Kakoi, and T. Goto, *Tetrahedron Lett.*, 1609 (1969).
24. T. Goto, S. Inoue, and S. Sugiura, *Tetrahedron Lett.*, 3873 (1968).
25. W. W. Paudler and J. E. Kuder, *J. Org. Chem.*, **31,** 809 (1966).
26. N. P. Buu-Hoï and N. Dat Xuong, *C.R. Acad. Sci.* (*Paris*), **243,** 2090 (1956).
27. N. P. Buu-Hoï, L. Petit, and N. Dat Xuong, *C.R. Acad. Sci.* (*Paris*), **248,** 1832 (1959).
28. T. Pyl and W. Baufeld, *Ann. Chem.*, **699,** 112 (1966).
29. L. Almirante, L. Polo, A. Mugnaini, E. Provinciali, P. Rugarli, A. Gamba, A. Olivi, and W. Murmann, *J. Med. Chem.*, **9,** 29 (1966).
30. (a) P. Guerret, R. Jacquier, and G. Maury, *Bull. Soc. Chim.* (*Fr.*), 2481 (1972); (b) *Idem*, p. 3503.
31. R. Fusco and S. Rossi, *Rend. Ist. Lombardo Sci. Pt. I.*, **88,** 173 (1955); *Chem. Abstr.*, **50,** 10742 (1956).
32. B. Loev and M. M. Goodman, *Tetrahedron Lett.*, 789 (1968).
33. British Patent 1,001,064; *Chem. Abstr.*, **65,** 20152 (1966).
34. J. M. Barton, Ph.D. Thesis, University of Ohio, Athens, Ohio, 1966.
35. J. Kobe, B. Stanovnik, and M. Tisler, *Chem. Commun.*, 1456 (1968).
36. J. Kobe, B. Stanovnik, and M. Tisler, *Monatsh. Chem.*, **101,** 724 (1970).
37. W. L. F. Armarego, *J. Chem. Soc.*, 2778 (1965).
38. I. Iwai and T. Hiraoka, *Chem. Pharm. Bull.* (*Tokyo*), **12,** 813 (1964).
39. H. A. Wagner, U.S. Patent 3,244,717; *Chem. Abstr.*, **65,** 10598 (1966).
40. D. J. Brown and B. T. England, *J. Chem. Soc.* (*C*), 1922 (1967).
41. D. Libermann and R. Jacquier, *Bull. Soc. Chim.* (*Fr.*), 355 (1962).
42. L. S. Helmick, Ph.D. Thesis, University of Ohio, Athens, Ohio, 1968.
43. J. A. Bee and F. L. Rose, *J. Chem. Soc.* (*C*), 2031 (1966).
44. (a) C. F. H. Allen, H. R. Beilfuss, D. M. Burness, G. A. Reynolds, J. F. Tinker, and J. A. Van Allan, *J. Org. Chem.*, **24,** 779 (1959); (b) *Idem.* p. 787; (c) *Idem.* p. 793; (d) *Idem.* p. 796.
45. C. F. H. Allen, G. A. Reynolds, J. F. Tinker, and L. A. Williams, *J. Org. Chem.*, **25,** 361 (1960).

46. J. Daunis, R. Jacquier, and P. Viallefont, *Bull. Soc. Chim. (Fr.)*, 2492 (1969).
47. M. Japelj, B. Stanovnik, and M. Tisler, *Monatsh. Chem.*, **100,** 671 (1969).
48. C. Temple, R. L. Mac Kee, and J. A. Montgomery, *J. Org. Chem.*, **28,** 2257 (1963).
49. W. W. Paudler and L. S. Helmick, *J. Heterocycl. Chem.*, **3,** 269 (1966).
50. A. Pollak, B. Stanovnik, and M. Tisler, *J. Heterocycl. Chem.*, **5,** 513 (1968).
51. P. J. Nelson and K. T. Potts, *J. Org. Chem.*, **27,** 3243 (1962).
52. A. Pollak and M. Tisler, *Tetrahedron*, **22,** 2073 (1966).
53. R. G. W. Spickett and S. H. B. Wright, *J. Chem. Soc. (C)*, 498 (1967), and references therein.
54. J. Kobe, B. Stanovnik, and M. Tisler, *Tetrahedron*, **26,** 3357 (1970).
55. M. Jelenc, J. Kobe, B. Stanovnik, and M. Tisler, *Monatsh. Chem.*, **97,** 1713 (1966).
56. P. Francavilla and F. Lauria, *J. Heterocycl. Chem.*, **8,** 415 (1971).
57. S. E. Mallett and F. L. Rose, *J. Chem. Soc. (C)*, 2038 (1966).
58. C. Temple, B. H. Smith, and J. A. Montgomery, *J. Org. Chem.*, **33,** 530 (1968).
59. G. W. Miller and F. L. Rose, *J. Chem. Soc.*, 5642 (1963).
60. T. Sasaki and K. Minamoto, *Chem. Ber.*, **100,** 3467 (1967).
61. J. Daunis, R. Jacquier, and P. Viallefont, *Bull. Soc. Chem. (Fr.)*, 3670 (1969).
62. T. Sasaki, K. Minamoto, and S. Fukuda, *Chem. Ber.*, **101,** 2747 (1968).
63. A. Dornow, H. Menzel, and P. Marx, *Chem. Ber.*, **97,** 2185 (1964).
64. V. A. Ershov and I. Y. Postovskii, *Khim. Geterotsikl. Soedin.*, 1134 (1968); *Chem. Abstr.*, **70,** 68327 (1969).
65. N. Takahayashi, *J. Pharm. Soc. Jap.*, **75,** 1242 (1955); *Chem. Abstr.*, **50,** 8655 (1956).
66. S. Linholter and R. Rosenørn, *Acta Chem. Scand.*, **16,** 2389 (1962).
67. T. Kuraishi and R. N. Castle, *J. Heterocycl. Chem.*, **3,** 218 (1966).
68. N. K. Basu and F. L. Rose, *J. Chem. Soc.*, 5660 (1963).
69. K. Kalfus, *Coll. Czech. Chem. Commun.*, **33,** 2962 (1968).
70. A. Dornow, W. Abele, and H. Menzel, *Chem. Ber.*, **97,** 2179 (1964).
71. Belg. Patent 642,615. *Chem. Abstr.*, **63,** 18127 (1965).
72. J. Maguire, D. Paton, and F. L. Rose, *J. Chem. Soc. (C)*, 1593 (1969).
73. D. Shiho, S. Tagami, N. Takahayashi, and R. Honda, *J. Pharm. Soc. Jap.*, **76,** 804 (1956); *Chem. Abstr.*, **51,** 1196 (1957).
74. G. W. Miller and F. L. Rose, *J. Chem. Soc.*, 3357 (1965).
75. W. Broadbent, G. W. Miller, and F. L. Rose, *J. Chem. Soc.*, 3369 (1965).
76. K. Shirakawa, *Yakugaku Zasshi*, **80,** 1542 (1960) [*Chem. Abstr.*, **55,** 10450 (1961)] and earlier references in this paper.
77. H. Reimlinger and M. A. Peiren, *Chem. Ber.*, **103,** 3266 (1970).
78. A. Kreutzberger, *Chem. Ber.*, **99,** 2237 (1966).
79. D. J. Fry and B. R. D. Whitear, British Patent 874,204; *Chem. Abstr.*, **56,** 8213 (1962).
80. A. H. Beckett, R. G. W. Spickett, and S. H. B. Wright, *Tetrahedron*, **24,** 2839 (1968).
81. F. L. Rose, G. J. Stacey, P. J. Taylor, and T. W. Thompson, *Chem. Commun.*, 1524 (1970).
82. A. Dornow, H. Menzel, and P. Marx, *Chem. Ber.*, **97,** 2173 (1964).
83. A. Dornow and H. Pietsch, *Chem. Ber.*, **100,** 2585 (1967).
84. G. M. Badger, P. J. Nelson, and K. T. Potts, *J. Org. Chem.*, **29,** 2542 (1964).
85. T. Okamoto, Y. Torigoe, and M. Sato, *Chem. Pharm. Bull. (Tokyo)*, **16,** 1154 (1968).
86. K. Shirakawa, *Yakugaku Zasshi*, **80,** 956 (1960); *Chem. Abstr.*, **54,** 24762 (1960).
87. C. Temple, R. L. Mac Kee, and J. A. Montgomery, *J. Org. Chem.*, **30,** 829 (1965).
88. C. Temple and J. A. Montgomery, *J. Org. Chem.*, **30,** 826 (1965).
89. T. Sasaki, K. Kanematsu, and M. Murata, *J. Org. Chem.*, **36,** 446 (1971).

90. C. Temple, W. C. Coburn, M. C. Thorpe, and J. A. Montgomery, *J. Org. Chem.*, **30,** 2395 (1965).
91. T. Kuraishi and R. N. Castle, *J. Heterocycl. Chem.*, **1,** 42 (1964).
92. W. D. Guither, D. G. Clark, and R. N. Castle, *J. Heterocycl. Chem.*, **2,** 67 (1965).
93. B. Stanovnik and M. Tisler, *Tetrahedron*, **25,** 3313 (1969).
94. T. Itai and S. Kamiya, *Chem. Pharm. Bull.* (*Tokyo*), **11,** 348 (1963).
95. C. Temple, R. L. Mac Kee, and J. A. Montgomery, *J. Org. Chem.*, **27,** 1671 (1962).
96. C. Temple, C. L. Kussner, and J. A. Montgomery, *J. Org. Chem.*, **33,** 2086 (1968).
97. C. Wentrup, *Tetrahedron*, **26,** 4969 (1970).
98. K. Shirakawa, Japanese Patent 777 (157); *Chem. Abstr.*, **52,** 4699 (1958).
99. A. W. Spassov, E. V. Golovinsky, and G. C. Russev, *Z. Chem.*, **8,** 421 (1968).
100. B. Stanovnik, M. Tisler, M. Ceglar, and V. Bah, *J. Org. Chem.*, **35,** 1138 (1970).
101. A. Kovacic, B. Stanovnik, and M. Tisler, *J. Heterocycl. Chem.*, **5,** 351 (1968).
102. H. Rutner and P. E. Spoerri, *J. Heterocycl. Chem.*, **3,** 435 (1966).
103. A. Dornow, H. Pietsch, and P. Marx, *Chem. Ber.*, **97,** 2647 (1964).
104. W. Flitsch and U. Krämer, *Tetrahedron Lett.*, 1479 (1968).
105. M. Zupan, B. Stanovnik, and M. Tisler, *J. Heterocycl. Chem.*, **8,** 1 (1971).
106. E. A. Steck and R. P. Brundage, *J. Am. Chem. Soc.*, **81,** 6290 (1959), and references therein.
107. M. Pesson, French Patent 1,248,409; *Chem. Abstr.*, **56,** 10160 (1962).
108. (a) H. G. O. Becker and H. Boettcher, *Wiss. Z. Tech. Hochsch. Chem. Leuna-Merseburg*, **8,** 122 (1966). *Chem. Abstr.*, **69,** 27360 (1968); (b) H. G. O. Becker and H. Boettcher, *Tetrahedron*, **24,** 2687 (1968).
109. A. N. Kost and F. Gents, *J. Gen. Chem. USSR*, **28,** 2796 (1958).
110. E. J. Birr, *Z. Wiss. Phot.*, **47,** 2 (1952); *Chem. Abstr.*, **47,** 2617 (1953).
111. W. Hertz and S. Tocker, *J. Am. Chem. Soc.*, **77,** 6355 (1955).
112. W. W. Paudler and D. E. Dunham, *J. Heterocycl. Chem.*, **2,** 410 (1965).
113. (a) V. I. Shvedov, L. B. Altukhova, A. V. Bocharnikova, and A. N. Grinev, USSR Patent 237,153; from *Otkrytiya, Izobret., Prom. Obraztsy, Tovarnye Znaki*, **46,** 22 (1969); *Chem. Abstr.*, **71,** 13142 (1969). (b) V. I. Shvedov, L. B. Altukhova, and A. N. Grinev, *Khim. Geterotsikl. Soedin.*, 1048 (1970); *Chem. Abstr.*, **74,** 125628 (1971).
114. H. Dorn and A. Zubek, *Chem. Ber.*, **101,** 3265 (1968).
115. A. Takamizawa, Y. Hamashima, S. Sakai, and S. Nagakura, *Bull. Chem. Soc. Jap.*, **41,** 2141 (1968).
116. I. Hori, K. Saito, and H. Midorikawa, *Bull. Chem. Soc. Jap.*, **43,** 849 (1970).
117. P. Guerret, J.-L. Imbach, R. Jacquier, and G. Maury, *Bull. Soc. Chem.* (*Fr.*), 1031 (1971).
118. L. A. Williams, *J. Chem. Soc.*, 2222 (1962).
119. E. Tenor and C.-F. Kröger, *Chem. Ber.*, **97,** 1373 (1964).
120. V. I. Shvedov, I. A. Kharizomenova, L. B. Altukhova, and A. N. Grinev, *Khim. Geterotsikl. Soedin.*, 428 (1970). *Chem. Abstr.*, **73,** 25403 (1970).
121. E. Ochiai and M. Sibata, *J. Pharm. Soc. Jap.*, **59,** 185 (1939). *Chem. Abstr.*, **33,** 4988 (1939).
122. S. C. Bell and W. T. Caldwell, *J. Am. Chem. Soc.*, **82,** 1469 (1960), and references therein.
123. T. Pyl, S. Melde, and H. Beyer, *Ann. Chem.*, **663,** 108 (1963).
124. M. A. Prokof'ev and Y. P. Shvachkin, *Zhur. Obshch. Khim*, **25,** 975 (1955). *Chem. Abstr.*, **50,** 3458 (1956).
125. J.-L. Imbach, R. Jacquier, and J.-L. Vidal, *Bull. Soc. Chim.* (*Fr.*), 1929 (1970), and references therein.

126. E. C. Taylor and K. S. Hartke, *J. Am. Chem. Soc.*, **81,** 2452 (1959).
127. A. Takamizawa and S. Hayashi, Japanese Patent 2346 and 2347 (1962); *Chem. Abstr.*, **58,** 7952 and 10213 respectively (1963).
128. A. Takamizawa and Y. Hamashima, *Yakugaku Zasshi*, **84,** 1113 (1964); *Chem. Abstr.*, **62,** 5276 (1965).
129. A. Takamizawa and S. Hayashi, Japanese Patent 3173 (1967); *Chem. Abstr.*, **67,** 11500 (1967).
130. A. Takamizawa and Y. Hamashima, Japanese Patent 15583 (1966); *Chem. Abst.* **66,** 10957 (1967).
131. A. Takamizawa, Japanese Patent 7,030,334; *Chem. Abstr.*, **74,** 22867 (1971).
132. H. Reimlinger, M. A. Peiren, and R. Merenyi, *Chem. Ber.*, **103,** 3252 (1970).
133. W. Ried and E.-U. Köcher, *Ann. Chem.*, **647,** 116 (1961).
134. L. A. Williams, *J. Chem. Soc.*, 1829 (1960).
135. L. A. Williams, *J. Chem. Soc.*, 3046 (1961).
136. Y. A. Levin, N. A. Gulkina, and V. A. Kukhtin, *J. Gen. Chem. USSR*, **33,** 2603 (1963).
137. Y. A. Levin, A. P. Fedotova, and V. A. Kukhtin, *J. Gen. Chem. USSR*, **34,** 501 (1964).
138. R. G. W. Spickett and S. H. B. Wright, *J. Chem. Soc.* (*C*), 503 (1967).
139. Y. A. Levin, R. N. Platonova, and V. A. Kukhtin, *Izv. Akad. Nauk SSSR, Ser. Khim.*,. 1475 (1964); *Chem. Abstr.*, **64,** 15878 (1966).
140. Y. Makisumi, *Chem. Pharm. Bull.* (*Tokyo*), **11,** 129 (1963).
141. D. R. Sutherland and G. Tennant, *Chem. Commun.*, 1070 (1969).
142. L. E. Brady and R. M. Herbst, *J. Org. Chem.*, **24,** 922 (1959).
143. D. W. Kaiser, G. A. Peters, and V. P. Wystrach, *J. Org. Chem.*, **18,** 1610 (1953).
144. E. C. Taylor and R. W. Hendess, *J. Am. Chem. Soc.*, **87,** 1980 (1965).
145. G. Cipens, R. Bokaldere, and V. Grinsteins, USSR Patent 213,888; from *Izobret., Prom. Obraztsy, Tovarnye Znaki*, **45,** 35 (1968); *Chem. Abstr.*, **69,** 67437 (1968).
146. R. Bokaldere and V. Grinsteins, *Khim. Geterotsikl. Soedin.*, 563 (1970); *Chem. Abstr.*, **73,** 87897 (1970).
147. C. L. Dickinson, W. J. Middleton, and V. A. Engelhardt, *J. Org. Chem.*, **27,** 2740 (1962).
148. E. C. Taylor, W. H. Gumprecht, and R. F. Vance, *J. Am. Chem. Soc.*, **76,** 619 (1959), and references therein.
149. French Patent 1,379,480; *Chem. Abstr.*, **62,** 11838 (1965).
150. Belgian Patent 642,615. *Chem. Abstr.*, **63,** 18127 (1965).
151. H. Gehlen, R. Drohla, *Arch. Pharm.* (*Weinheim*), **303,** 650 (1970); *Chem. Abstr.*, **73,** 87862 (1970).
152. A. Hetzheim and H. Push, *Chimia*, **23,** 303 (1969).
153. M. W. Partridge and M. F. G. Stevens, *J. Chem. Soc.* (*C*), 1127 (1966).
154. French Patent. 1,379,479. *Chem. Abstr.*, **62,** 16278 (1965).
155. R. G. Dickinson and N. W. Jacobsen, *Chem. Commun.*, 1719 (1970).
156. V. A. Ershov and I. Y. Postovskii, *Khim. Geterotsikl. Soedin.*, 1134 (1968); *Chem. Abstr.*, **70,** 68327 (1969).
157. J. L. Wong, M. S. Brown, and H. Rapoport, *J. Org. Chem.*, **30,** 2398 (1965).
158. J. D. Kendall, G. F. Duffin, and H. R. J. Waddington, British Patent 862,825; *Chem. Abstr.*, **56,** 15076 (1962).
159. P. J. Black, M. L. Heffernan, L. M. Jackman, Q. N. Porter, and G. R. Understood, *Austr. J. Chem.*, **17,** 1128 (1964).
160. W. W. Paudler and J. E. Kuder, *J. Heterocycl. Chem.*, **3,** 33 (1966).

161. R. J. Pugmire, M. J. Robins, D. M. Grant, and R. K. Robins, *J. Am. Chem. Soc.*, **93,** 1887 (1971).
162. An X-ray study of a *s*-triazolo[2,3-*c*]pyrimidine derivative indicates that each of the fused rings is planar and that the two rings are inclined at an angle of only 6 deg. (cf., Ref. 59).
163. Y. Makisumi, H. Watanabe, and K. Tori, *Chem. Pharm. Bull.* (*Tokyo*), **12,** 204 (1964).
164. M. Ohtsuru, K. Tori, and H. Watanabe, *Chem. Pharm. Bull.* (*Tokyo*), 1015 (1967).
165. P. J. Black, R. D. Brown, and M. L. Heffernan, *Austr. J. Chem.*, **20,** 1305 (1967).
166. N. Takahayashi, *J. Pharm. Soc. Jap.*, **76,** 765 and 1296 (1956); *Chem. Abstr.*, **51,** 1192 and 6645 respectively (1957).
167. S. F. Mason, *J. Chem. Soc.*, 3999 (1963).
168. V. Galasso, G. De Alti, and A. Bigotto, *Theoret. Chim. Acta*, **9,** 222 (1968).
169. W. W. Paudler, J. E. Kuder, and L. S. Helmick, *J. Org. Chem.*, **33,** 1379 (1968).
170. M. Tisler and W. W. Paudler, *J. Heterocycl. Chem.*, **5,** 695 (1968).
171. A. Streitweiser, "Molecular Orbital Theory for Organic Chemists," Wiley, New York. 1961, p. 135.
172. B. Stanovnik and M. Tisler, *Croat. Chem. Acta*, **40,** 1 (1968).
173. M. Japelj, B. Stanovnik, and M. Tisler, *J. Heterocycl. Chem.*, **6,** 559 (1969).
174. W. W. Paudler, private communication.
175. K. T. Potts and S. W. Schneller, *J. Heterocycl. Chem.*, **5,** 485 (1968).
176. M. W. Winkley, G. F. Judd, and R. K. Robins, *J. Heterocycl. Chem.*, **8,** 237 (1971).
177. V. C. Chambers, *J. Am. Chem. Soc.*, **82,** 605 (1960).
178. A. Albert and K. Ohta, *J. Chem. Soc.* (*C*), 1540 (1970).
179. Y. Makisumi, *Chem. Pharm. Bull.* (*Tokyo*), **9,** 801 (1961).
180. A. Takamizawa and Y. Hamashima, *Chem. Pharm. Bull.* (*Tokyo*), **13,** 142 (1965).
181. Y. Makisumi and H. Kano, *Chem. Pharm. Bull.* (*Tokyo*), **11,** 67 (1963).
182. A. Takamizawa and Y. Hamashima, *Chem. Pharm. Bull.* (*Tokyo*), **13,** 1207 (1965).
183. C. L. Schmidt and L. B. Townsend, *J. Heterocycl. Chem.*, **7,** 715 (1970).
184. T. Sasaki, K. Minamoto, and M. Murata, *Chem. Ber.*, **101,** 3969 (1968).
185. K. T. Potts, H. R. Burton, and S. K. Roy, *J. Org. Chem.*, **31,** 265 (1966).
186. G. W. Miller and F. L. Rose, British Patent 898,409; *Chem. Abstr.*, **57,** 11210 (1962).
187. K. Shirakawa, *J. Pharm. Soc. Jap.*, **78,** 1395 (1958); *Chem. Abstr.*, **53,** 8150 (1959).
188. T. Ueda and J. J. Fox, *J. Org. Chem.*, **29,** 1762 (1964).
189. K. T. Potts, C. R. Surapaneni, *J. Heterocycl. Chem.*, **7,** 1019 (1970).
190. M. Ain Khan and B. M. Lynch, *J. Heterocycl. Chem.*, **7,** 247 (1970).
191. W. W. Paudler and L. S. Helmick, *J. Heterocycl. Chem.*, **5,** 691 (1968).
192. W. W. Paudler and H. L. Blewitt, *J. Org. Chem.*, **31,** 1295 (1966).
193. B. Mariani and R. Sgarbi, *Chim. Ind.* (*Milan*), **46,** 630 (1964); *Chem. Abstr.*, **61,** 12117 (1964).
194. W. Ried and E.-U. Köcher, *Ann. Chem.*, **647,** 144 (1961).
195. S. Checchi, P. Papini, and M. Ridi, *Gazz. Chim. Ital.*, **85,** 1160 (1955); *Chem. Abstr.*, **50,** 10098 (1956).
196. W. W. Paudler and R. J. Brumbaugh, *J. Heterocycl. Chem.*, **5,** 29 (1968).
197. Y. Makisumi, *Chem. Pharm. Bull.* (*Tokyo*), **11,** 851 (1963).
198. Y. Makisumi, *Chem. Pharm. Bull.* (*Tokyo*), **11,** 859 (1963).
199. Y. Makisumi, *Chem. Pharm. Bull.* (*Tokyo*), **9,** 808 (1961).
200. It is probable that bromination of 2,6-diphenylimidazo[1,2-*b*]-*as*-triazine occurs at position 3 rather than 7 as reported (cf., Ref. 32).
201. G. R. Bedford, M. W. Partridge, and M. F. G. Stevens, *J. Chem. Soc.* (*C*), 1214 (1966).

202. H. Kano, Y. Makisumi, S. Takahashi, and M. Ogata, *Chem. Pharm. Bull.* (*Tokyo*), **7,** 903 (1959).
203. Y. Makisumi and H. Kano, *Chem. Pharm. Bull.* (*Tokyo*), **7,** 907 (1959).
204. M. Yanai, T. Kuraishi, T. Kinoshita, and M. Nishimura, *J. Heterocycl. Chem.*, **7,** 465 (1970).
205. Bromine in water or NBS yield the same products, and addition of peroxide has no effect on the reaction rate (W. W. Paudler, private communication).
206. Y. Makisumi, *Chem. Pharm. Bull.* (*Tokyo*), **9,** 878 (1961).
207. (a) L. Pentimalli and V. Passalacqua, *Gazz. Chim. Ital.*, **100,** 110 (1970). (*b*) L. Pentimalli and G. Milano, *Gazz. Chim. Ital.*, **100,** 1106 (1970).
208. J. P. La Rocca, C. A. Gibson, and B. B. Thompson, *J. Pharm. Sci.*, **60,** 74 (1971).
209. M. Ridi, P. Papini, and S. Checchi, *Gazz. Chim. Ital.*, **92,** 209 (1962)
210. W. Ried and K.-P. Peuchert, *Ann. Chem.*, **660,** 104 (1962).
211. M. Ridi, P. Papini, and S. Checchi, *Gazz. Chim. Ital.*, **91,** 973 (1961).
212. Y. Makisumi, *Chem. Pharm. Bull.* (*Tokyo*), **9,** 873 (1961).
213. H. Kano and Y. Makisumi, Japanese Patent 6280 (1962); *Chem. Abstr.*, **60,** 9389 (1964).
214. A. Takamizawa and S. Hayashi, Japanese Patent 24377 (1965); *Chem. Abstr.*, **64,** 5110 (1966).
215. H. Reimlinger, *Chem. Ber.*, **103,** 3278 (1970).
216. T. Pyl and W. Baufeld, *Ann. Chem.*, **699,** 127 (1966).
217. W. Ried and K.-P. Peuchert, *Ann. Chem.*, **682,** 136 (1965).
218. Y. A. Levin and V. A. Kukhtin, *J. Gen. Chem. USSR*, **34,** 504 (1964).
219. Y. Makisumi, *Chem. Pharm. Bull.* (*Tokyo*), **9,** 814 (1961).
220. W. W. Paudler and L. S. Helmick, *Chem. Commun.*, 377 (1967).
221. W. W. Paudler and L. S. Helmick, *J. Org. Chem.*, **33,** 1087 (1968).
222. Neth. Appl. 6,609,264. *Chem. Abstr.*, **67,** 54151 (1967).
223. B. Mariani and R. Sgarbi, Italian Patent 752,234; *Chem. Abstr.*, **69,** 11415 (1968).
224. F. Baumbach, H.-G. Henning, and G. Hilgetag, *Z. Chem.*, **2,** 369 (1962)
225. M. W. Partridge and M. F. G. Stevens, *J. Chem. Soc.* (*C*), 1828 (1967).
226. (a) W. J. Irwin and D. G. Wibberley, *Tetrahedron Lett.*, 878 (1968). (b) W. J. Irwin and D. G. Wibberley, *J. Chem. Soc.* (*C*), 3237 (1971).
227. C. Wentrup and W. D. Crow, *Tetrahedron*, **26,** 4915 (1970).
228. (a) N. K. Hart, S. R. Johns, and J. A. Lamberton, *Chem. Commun.*, 1484 (1969); (b) N. K. Hart, S. R. Johns, and J. A. Lamberton, and R. I. Willing, *Austr. J. Chem.*, **23,** 1679 (1970).
229. W. Schuett and H. Rapoport, *J. Am. Chem. Soc.*, **84,** 2266 (1962).
230. Y. Kishi, T. Goto, S. Inoue, S. Sugiura, and H. Kishimoto, *Tetrahedron Lett.*, 3445 (1966).
231. Y. Makisumi, Japanese Patent 13640 (1963); *Chem. Abstr.*, **60,** 531 (1964).
232. A. Takamizawa and S. Hayashi, Japanese Patent 2346 (1962); *Chem. Abstr.*, **58,** 7952 (1963).
233. Y. Makisumi, Japanese Patent 5947 (1962); *Chem. Abstr.*, **59,** 1659 (1963).
234. Y. Makisumi, Japanese Patent 7984 (1962); *Chem. Abstr.*, **59,** 8762 (1963).
235. B. Camerino and T. La Noce, German Patent 1,071,088 (1959); *Chem. Abstr.*, **55,** 16576 (1961).
236. T. Okabayashi, *Chem. Pharm. Bull.* (*Tokyo*), **8,** 162 (1960); *Chem. Abstr.*, **55,** 16576 (1961).
237. Y. Makisumi, H. Kano, and S. Takahashi, Japanese Patent 9498 (1962); *Chem. Abstr.*, **59,** 5178 (1963).

238. T. Iwaki, *Kobe Ika Daigaku Kiyo,* **20,** 217 (1961); *Chem. Abstr.,* **61,** 3584 (1964).
239. A. Takamizawa, Y. Hamashima, S. Hayashi, and S. Sakai, *Chem. Pharm. Bull.* (*Tokyo*), **16,** 2195 (1968).
240. E. Tenor, H. Fueller, and F. Hausschild, East German Patent 55,956 (1967); *Chem. Abstr.,* **67,** 90830 (1967).
241. G. W. Miller and F. L. Rose, British Patent 898,414; *Chem. Abstr.,* **57,** 13775 (1962).
242. F. Khuong-Huu, J.-P. Leforestier, G. Maillard, and R. Goutarel, *C.R. Acad. Sci. Paris,* **270 (C),** 2070 (1970).
243 G. F. Duffin in "Advances in Heterocyclic Chemistry," Vol. 3, A. R. Katritzky, Ed., The Quaternization of Heterocyclic Compounds, Academic Press, New York, 1964, p. 45.
244. E. Zbiral and E. Hugl, *Tetrahedron Lett.,* 439 (1972).
245. W. J. Irwin, D. G. Wibberley, and G. Cooper, *J. Chem. Soc.* (*C*), 3870 (1971).
246. W. Flitsch and U. Krämer, *Ann. Chem.,* **735,** 35 (1970).
247. Y. I. Postovskii and V. A. Ershov, *Khim. Geterotsikl. Soedin.,* 708 (1971); *Chem. Abstr.,* **76,** 126,949 (1972).
248. V. A. Ershov and I. Y. Postovskii, *Khim. Geterotsikl. Soedin.,* 711 (1971); *Chem. Abstr.,* **76,** 126,947 (1972).
249. V. Pirc, B. Stanovnik, M. Tisler, J. Marsel, and W. W. Paudler, *J. Heterocycl. Chem.,* **7,** 639 (1970).
250. M. H. Palmer, P. N. Preston, and M. F. G. Stevens, *Org. Mass Spectrom.,* **5,** 1085 (1971).
251. J. R. Dyer, C. K. Kellogg, R. F. Nassar, and W. E. Streetman, *Tetrahedron Lett.,* 585 (1965).
252. Y. Kobayashi, T. Kutsuma, and K. Morinaga, *Chem. Pharm. Bull.* (*Tokyo*), **19,** 2106 (1971).
253. T. Sasaki, K. Kanematsu, Y. Yukimoto, and S. Ochiai, *J. Org. Chem.,* **36,** 813 (1971).
254. Y. Masaki, H. Otsuka, Y. Nakayama, and M. Hioki, *Chem. Pharm. Bull.* (*Tokyo*), **21,** 2780 (1973).
255. J. W. Lown and K. Matsumoto, *Can. J. Chem.,* **49,** 1165 and 3119 (1971).
256. J. Reiter, P. Sohar, and L. Toldy, *Tetrahedron Lett.,* 1411 (1970).
257. J. Reiter, P. Sohar, J. Liptak, and L. Toldy, *Tetrahedron Lett.,* 1417 (1970).
258. N. K. Kochetkov, V. N. Shibaev, and A. A. Kost, *Tetrahedron Lett.,* 1993 (1971).
259. E. Abushanab, A. P. Bindra, L. Goodman, and H. Peterson, *J. Org. Chem.,* **38,** 2049 (1973).
260. (a) A. Kreutzberger and R. Schücker, *Arch. Pharm.,* **306,** 561, 697 and 801 (1973); (b) A. Kreutzberger and P. Schücker, *Tetrahedron,* **29,** 1413 (1973).
261. M. F. G. Stevens, *J. Chem. Soc. Perkin Trans. I,* 1221 (1972).
262. M. Yanai, T. Kinoshita, S. Takeda, M. Nishimura, and T. Kuraishi, *Chem. Pharm. Bull.* (*Tokyo*), **20,** 1617 (1972).
263. A. W. Spassov and Z. Raikov, *Z. Chem.,* **11,** 422 (1971).
264. A. B. De Milo, J. E. Oliver, and R. D. Gilardi, *J. Heterocycl. Chem.,* **10,** 231 (1973).
265. C. Cristescu, *Rev. Roum. Chim.,* **16,** 311 (1971); *Chem. Abstr.,* **75**, 20358 (1971).
266. M. Zupan, B. Stanovnik, and M. Tisler, *Tetrahedron Lett.,* 4179 (1972).
267. S. Polanc, B. Vercek, B. Stanovnik, and M. Tisler, *Tetrahedron Lett.,* 1677 (1973).
268. S. Polanc, B. Vercek, B. Sek, B. Stanovnik, and M. Tisler, *J. Org. Chem.,* **39,** 2143 (1974).
269. T. L. Gilchrist, C. J. Harris, C. J. Moody, and C. W. Rees, *Chem. Commun.,* 486 (1974).
270. T. Tsuji and T. Ueda, *Chem. Pharm. Bull.* (*Tokyo*), **19,** 2530 (1971).

271. B. Simon and K.-H. Uteg, *Z. Chem.*, **12,** 20 (1972).
272. K. Dickore and L. Eue, German Patent 1965739; *Chem. Abstr.*, **75,** 129842 (1971).
273. M. Jautelat, K. Ley, and L. Eue, German Patent 2236340; *Chem. Abstr.*, **80,** 108583 (1974).
274. E. R. Wagner, *J. Org. Chem.*, **38,** 2976 (1973).
275. G. M. Coppola, G. E. Hardtmann, and B. S. Huegi, *J. Heterocycl. Chem.*, **11,** 51 (1974).
276. V. A. Chuiguk and G. M. Golubshina, USSR. 407903; from *Otkrytiya, Izobret., Prom. Obraztsy Znaki*, **50,** 83 (1973); *Chem. Abstr.*, **80,** 96001 (1974).
277. H. G. O. Becker, H. Böttcher, R. Ebisch, and G. Schmoz, *J. Prakt. Chem.*, **312,** 780 (1970).
278. H. G. O. Becker, D. Nagel, and H. J. Timpe, *J. Prakt. Chem.*, **315,** 97 (1973).
279. J. W. Sowell and C. D. Blanton, *J. Heterocycl. Chem.*, **10,** 287 (1973).
280. K. Saito, I. Hori, M. Igarashi, and H. Midorikawa, *Bull. Chem. Soc. Jap.*, **47,** 476 (1974).
281. H. Dorn and A. Zubek, *J. Prakt. Chem.*, **313,** 969 (1971).
282. E. Alcade, J. De Mendoza, J.-M. Garcia-Marquina, C. Almera, and J. Elguero, *J. Heterocycl. Chem.*, **11,** 423 (1974).
283. H. Reimlinger, R. Jacquier, and J. Daunis, *Chem. Ber.*, **104,** 2702 (1971).
284. D. R. Sutherland and G. Tennant, *J. Chem. Soc. (C)*, 2156 (1971).
285. D. R. Sutherland, G. Tennant, and R. J. S. Vevers. *J. Chem. Soc. Perkin, Trans. I*, 943 (1973).
286. H. Reimlinger, M. A. Peiren, and R. Merenyi, *Chem. Ber.*, **105,** 103 (1972).
287. E. Kranz, J. Kurz, and W. Donner, *Chem. Ber.*, **105,** 388 (1972).
288. J. Kobe, R. K. Robins, and D. E. O'Brien, *J. Heterocycl. Chem.*, **11,** 199 (1974).
289. L. Capuano and H. J. Schrepfer, *Chem. Ber.*, **104,** 3039 (1971).
290. A. H. Albert, R. K. Robins, and D. E. O'Brien, *J. Heterocycl. Chem.*, **10,** 885 (1973).
291. T. Hirata, L. Twanmoh, L. B. Wood, A. Goldin, and J. S. Driscoll, *J. Heterocycl. Chem.*, **9,** 99 (1972).
292. R. Bokaldere and A. Liepina, *Khim. Geterosikl. Soedin.*, 276 (1973); *Chem. Abstr.*, **78,** 136237 (1973).
293. R. Bokaldere, A. Liepina, I. Mazeika, I. S. Yankovskaya, and E. Liepins, *Khim. Geterosikl. Soedin.*, 419 (1973); *Chem. Abstr.*, **78,** 147921 (1973).
294. E. Fanghänel, K. Gewald, K. Pütsch, and K. Wagner, *J. Prakt. Chem.*, **311,** 388 (1969).
295. I. Lalezari and Y. Levy, *J. Heterocycl. Chem.*, **11,** 327 (1974).
296. P. Guerret, R. Jacquier, H. Lopez, and G. Maury, *Bull. Soc. Chem. (Fr.)*, 1453 (1974).
297. M. Brugger and F. Korte, *Ann. Chem.*, **764,** 112 (1972).
298. M. V. Povstyanoi and P. M. Kochergin, *Ukr. Khim. Zh.*, **40,** 99 (1974); *Chem. Abstr.*, **80,** 82902 (1974).
299. H. Mackie and G. Tennant, *Tetrahedron Lett.*, 4719 (1972).
300. J. P. Cress and D. M. Fokkey, *Chem. Commun.*, 35 (1973).
301. C. Jaureguiberry and B. Roques, *C.R. Acad. Sci. (Paris)*, **274** (**C**), 1703 (1972).
302. E. Ajello and C. Arnone, *J. Heterocycl. Chem.*, **10,** 103 (1973).
303. H. G. O. Becker, D. Beyer, G. Israel, R. Müller, W. Riediger, and H.-J. Timpe, *J. Prakt. Chem.*, **312,** 669 (1970).
304. J. A. Moore, R. C. Gearhart, O. S. Rothenberger, P. C. Thorstenson, and R. H. Wood, *J. Org. Chem.*, **37,** 3774 (1972).
305. V. Sprio and S. Plescia, *J. Heterocycl. Chem.*, **9,** 951 (1972).
306. E. Ajello, O. Migliara, and V. Sprio, *J. Heterocycl. Chem.*, **9,** 1169 (1972).

307. K. Tserng and L. Bauer, *J. Heterocycl. Chem.*, **11,** 163 (1974).
308. C. B. Chapleo and A. S. Dreiding, *Helv. Chim. Acta,* **57,** 873 (1974).
309. P. Dea, G. R. Revankar, R. L. Tolman, R. K. Robins, and M. P. Schweizer, *J. Org. Chem.*, **39,** 3226 (1974).
310. A. Rabaron, M. Koch, M. Plat, J. Peyroux, E. Wenkert, and D. W. Cochran, *J. Am. Chem. Soc.*, **93,** 6270 (1971).
311. H. Dürr, H. Kober, R. Sergio, and V. Formacek, *Chem. Ber.*, **107,** 2037 (1974).
312. W. W. Paudler and J. N. Chasman, *J. Heterocycl. Chem.*, **10,** 499 (1973).
313. B. Adler, H. G. O. Becker, and H. Böttcher, *J. Prakt. Chem.*, **314,** 36 (1972).
314. C. Glier, F. Dietz, M. Scholz, and G. Fischer, *Tetrahedron,* **28,** 5779 (1972).
315. R. A. Coburn, R. A. Carapellotti, and R. A. Glennon, *J. Heterocycl. Chem.*, **10,** 479 (1973).
316. (a) J. Arriau, O. Chalvet, A. Dargelos, and G. Maury, *J. Heterocycl. Chem.*, **10,** 659 (1973); (b) J. Arriau, O. Chalvet, A. Dargelos, and G. Maury, *J. Heterocycl. Chem.*, **11,** 1013 (1974).
317. L. Twanmoh, H. B. Wood, and J. S. Driscoll, *J. Heterocycl. Chem.*, **10,** 187 (1973).
318. T. Novinson, R. K. Robins, and D. O'Brien, *J. Heterocycl. Chem.*, **10,** 887 (1973).
319. G. Tennant and R. J. S. Vevers, *Chem. Commun.*, 671 (1974).
320. M. Fraser, *J. Org. Chem.*, **36,** 3087 (1971).
321. M. Fraser, *J. Org. Chem.*, **37,** 3027 (1972).
322. M. Furlan, B. Stanovnik, and M. Tisler, *J. Org. Chem.*, **37,** 2689 (1972).
323. S. Yurugi, A. Miyake, and M. Tomimoto, *Takeda Kenkyusko Ho,* **32,** 111 (1973); *Chem. Abstr.*, **80,** 37073 (1974).
324. T. Novinson, R. K. Robins, and D. E. O'Brien, *Tetrahedron Lett.*, 3149 (1973).
325. H. Dorn and A. Zubek, *J. Prakt. Chem.*, **313,** 211 (1971).
326. S. Nicholson, G. J. Stacey, and P. J. Taylor, *J. Chem. Soc. Perkin Trans. II*, 4 (1972).
327. M. Dukes, S. Nicholson, and P. J. Taylor, *J. Chem. Soc. Perkin Trans. II*, 1695 (1972).
328. B. Stanovnik, *Synthesis*, 424 (1971).
329. B. Stanovnik, M. Tisler, and V. Zigon, *Monatsh. Chem.*, **103,** 1624 (1972),
330. E. Tenor, H. Fueller, and E. Thomas, *German (East) Patent* 76677; *Chem. Abstr.*, **75,** 88633 (1971).
331. W. W. Paudler, C. I. Patsy Chao, and L. S. Helmick, *J. Heterocycl. Chem.*, **9,** 1157 (1972).
332. B. Loev, U.S. Patent 3719677; *Chem. Abstr.*, **78,** 148007 (1973).
333. P. K. Kadaba, B. Stanovnik, and M. Tisler, *Tetrahedron Lett.*, 3715 (1974).
334. V. Pirc, B. Stanovnik, and M. Tisler, *Monatsh. Chem.*, **102,** 837 (1971).
335. H. Reimlinger and M. A. Peiren, *Chem. Ber.*, **104,** 2237 (1971).
336. E. Lippmann and V. Becker, *Z. Chem.*, **13,** 293 (1973).
337. A. Pollak, B. Stanovnik, and M. Tisler, *Synthetic Commun.*, **1,** 289 (1971).
338. J. Clark and M. Curphey, *Chem. Commun.*, 184 (1974).
339. G. R. Revankar, R. K. Robins, and R. L. Tolman, *J. Org. Chem.*, **39,** 1256 (1974).
340. H. G. O. Becker and H. Böttcher, *J. Prakt. Chem.*, **314,** 55 (1972).
341. J. S. Bradshaw, B. Stanovnik, and M. Tisler, *J. Heterocycl., Chem.*, **10,** 801 (1973).
342. C. Wentrup, *Helv. Chim. Acta,* **55,** 565 (1972).
343. A. Bezeg, B. Stanovnik, B. Sket, and M. Tisler, *J. Heterocycl. Chem.*, **9,** 1171 (1972).
344. F. B. Culp, A. Nabeya, and J. A. Moore, *J. Org. Chem.*, **38,** 2949 (1973).
345. (a) T. Tsuchiya, H. Arai, and H. Igeta, *Chem. Commun.*, 1059 (1972); (b) T. Tsuchiya, H. Arai, and H. Igeta, *Chem. Pharm. Bull. (Tokyo)*, **21,** 2517 (1973).
346. J. A. Hyatt and J. S. Swenton, *J. Heterocycl. Chem.*, **9,** 409 (1972).
347. B. Stanovnik, *Tetrahedron Lett.*, 3211 (1971).

348. S. Polanc, B. Stanovnik, and M. Tisler, *J. Heterocycl. Chem.*, **10,** 565 (1973).
349. B. Stanovnik, *J. Heterocycl. Chem.*, **8,** 1055 (1971).
350. M. Caserio and J. Guilhem, *Acta Cryst.*, **B28,** 151 (1972).
351. S. Sugiura, S. Inoue, H. Fakatsu, and T. Goto, *Yakugaku Zasshi*, **90,** 1475 (1970); *Chem. Abstr.*, **74,** 141691 (1971).
352. K. Hori, J. E. Wampler, and M. J. Cormier, *Chem. Commun.*, 492 (1973).
353. F. McCapra and M. Roth, *Chem. Commun.*, 467 (1973).
354. E. J. Birr, *Chimia*, **24,** 125 (1970).
355. F. McCapra and M. J. Manning, *Chem. Commun.*, 467 (1973).
356. Communications 52–56, abstracts of the A. C. S. National Meeting. Division of Medicinal Chemistry, New York, 1972.
357. Communication 21, abstracts of the A. C. S. National Meeting. Division of Medicinal Chemistry, Washington, 1971.
358. P. L. Anderson, W. J. Houlihan, and R. E. Manning, U.S. Patent 3637690; *Chem. Abstr.*, **76,** 113238 (1972).
359. T. Jojima, *Sankyo Kenkyusha Nempo*, **24,** 121 (1972); *Chem. Abstr.*, **78,** 159538 (1973).
360. L. M. Werbel, E. F. Elslager, and V. P. Chu, *J. Heterocycl. Chem.*, **10,** 631 (1973).
361. H. G. O. Becker, D. Nagel, and H.-J. Timpe, *Z. Chem.*, **13,** 213 (1973).
362. E. Kleinpeter, R. Borsdorf, G. Fischer, and H.-J. Hofmann, *J. Prakt. Chem.*, **314,** 515 (1972).
363. R. Jacquier, H. Lopez, and G. Maury, *J. Heterocycl. Chem.*, **10,** 755 (1973).

CHAPTER IV

The Chemistry of Cyclazines

ALFRED TAURINS

Department of Chemistry, McGill University, Montreal, Quebec, Canada

I. Introduction: Structure and Nomenclature

In 1951, V. Boekelheide and co-workers[1,2] started the synthesis and investigation of a new class of nitrogen heterocycles for which they proposed a name cyclazines. According to their suggestion, the word cyclazine refers to a fused tricyclic conjugated ring system held planar by the three covalent bonds to an internal nitrogen atom. The rings involved

TABLE I. PARENT STRUCTURES OF CYCLAZINES

Structure	Name
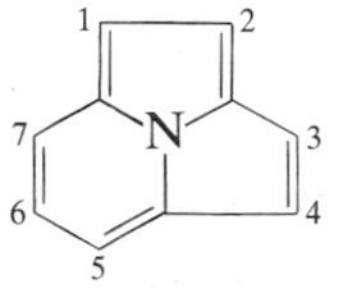 1	Cycl[3.2.2]azine[1,2] Pyrrolo[2,1,5-*cd*]indolizine
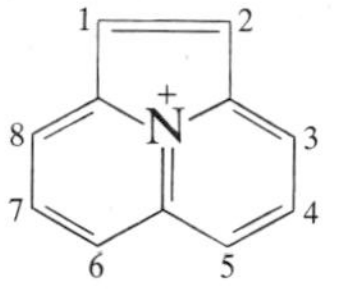 2	Cycl[3.3.2]azinium[1,2] Pyrrolo[2,1,5-*de*]quinolizinium
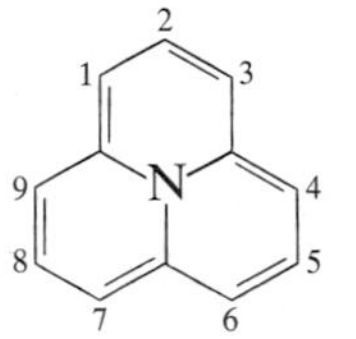 3	Cycl[3.3.3]azine[1,2] Pyrido[2,1,6-*de*]quinolizine
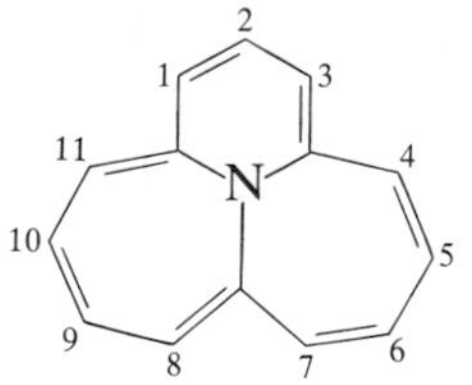 4	Cycl[4.4.3]azine[1,2]
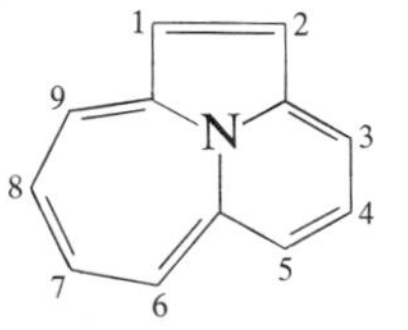 5	Cycl[4.3.2]azine

in the cyclazine formation can be 5-, 6-, and 7-membered. Various members of this class are distinguished by placing in brackets numbers indicating the number of atoms of the peripheral cycle between the atoms bonded to the internal nitrogen (Table I, structures **1** to **4**).

In naming cyclazines, *Chemical Abstracts* follows its own general rules of nomenclature of fused heteroaromatic systems. Some of these names are given in Table I. To the list of the four cyclazines (**1–4**) proposed initially[2] the structure **5** is added.

Boekelheide et al.[3,4] have shown that it is also possible to synthesize azacyclazines, derived from cyclazines by replacing CH groups by nitrogen atoms in the periphery of the system. If cycl[3.2.2]azine (**1**)

TABLE II. PARENT STRUCTURES OF THE KNOWN AZACYCLO[3.2.2]AZINES

Structure	Name
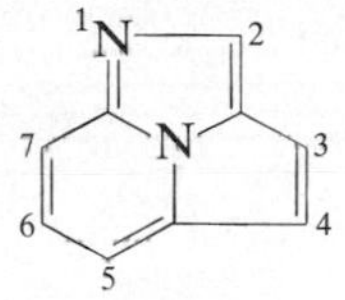 **6**	1-Azacycl[3.2.2]azine[3] Imidazo[5,1,2-*cd*]indolizine
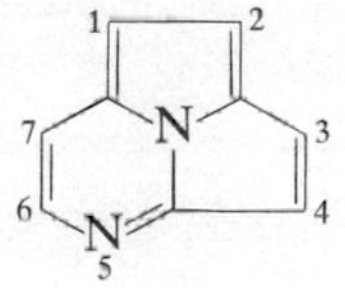 **7**	5-Azacycl[3.2.2]azine[4] Pyrimido[2,3,4-*cd*]pyrrolizine
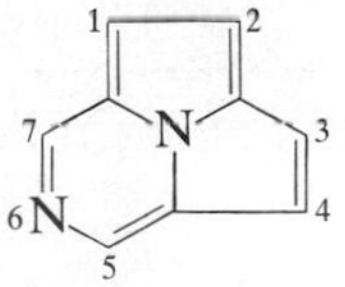 **8**	6-Azacycl[3.2.2]azine[5] Pyrazino[3,4,5-*cd*]pyrrolizine
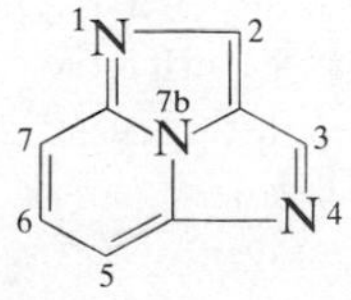 **9**	1,4-Diazacycl[3.2.2]azine[6,7] 1,4,7*b*-Triazacyclopent[*cd*]indene

TABLE III. PARENT STRUCTURES OF THE KNOWN AZACYCLAZINES

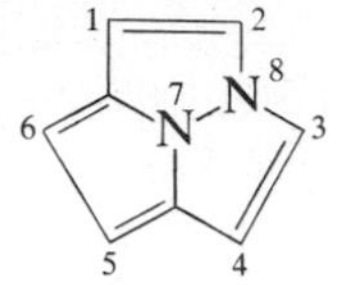

Structure	Name
10	8-Azacycl[2.2.2]azine[8] 2a,7-Diazacyclopenta[cd]pentalene
11	1,3,6-Triazacycl[3.3.3]azine[9]
12	1,3,6,7-Tetrazacycl[3.3.3]azine[10]
12a	1,3,4,7-Tetrazacycl[3.3.3]azine[10a]

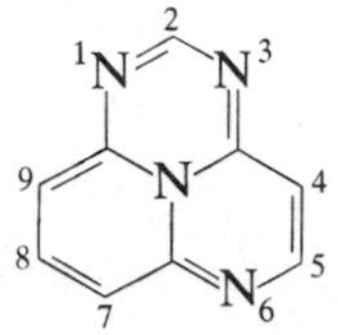

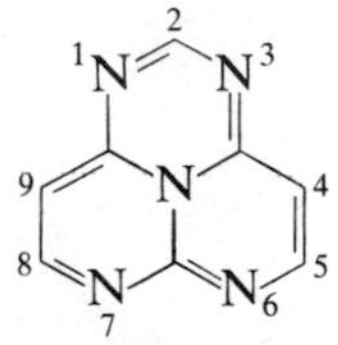

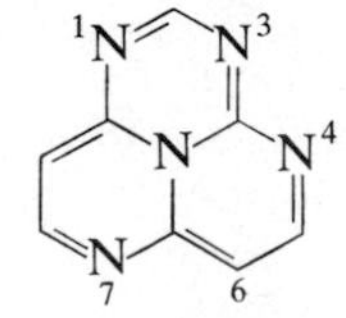

contains an additional nitrogen atom, four monoazacycl[3.2.2]azines are possible. There can be 12 diaza- and 11 triazacycl[3.2.2]azines. Since the cycl[3.3.3]azine (**3**) molecule possesses a threefold axis of symmetry, the number of isomeric azacycl[3.3.3]azines is relatively small; for example, only two monoaza- and six diazacycl[3.3.3]azines are possible.

The parent structures of the known azacyclazines are listed in Tables II and III.[5–10,10a]

II. Syntheses of Cyclazines

A. Cycl[3.2.2]azines from 5-Methylindolizines and *N,N*-Dimethylformamide or -Benzamide

The starting material, 5-methylindolizine (**13**), was prepared from 2-formyl-6-methylpyridine via vinyl-(6-methylpyridyl-2)methanol.[11] 5-Methylindolizine (**13**) was treated with *n*-butyllithium to form a deep red 5-indazolylmethyl anion followed by dimethylformamide. On hydrolysis the reaction product gave the aldehyde **14,** which was cyclized in boiling acetic acid to yield cycl[3.2.2]azine.[1.2.12]

In a similar manner, 2-phenylcycl[3.2.2]azine (**15**) was prepared from **13** and *N,N*-dimethylbenzamide, instead of dimethylformamide. 2,3-Diphenylcycl[3.2.2]azine (**17**) was obtained from 5-methyl-2-phenylindolizine (**16**) and *N,N*-dimethylbenzamide.[1,2,12]

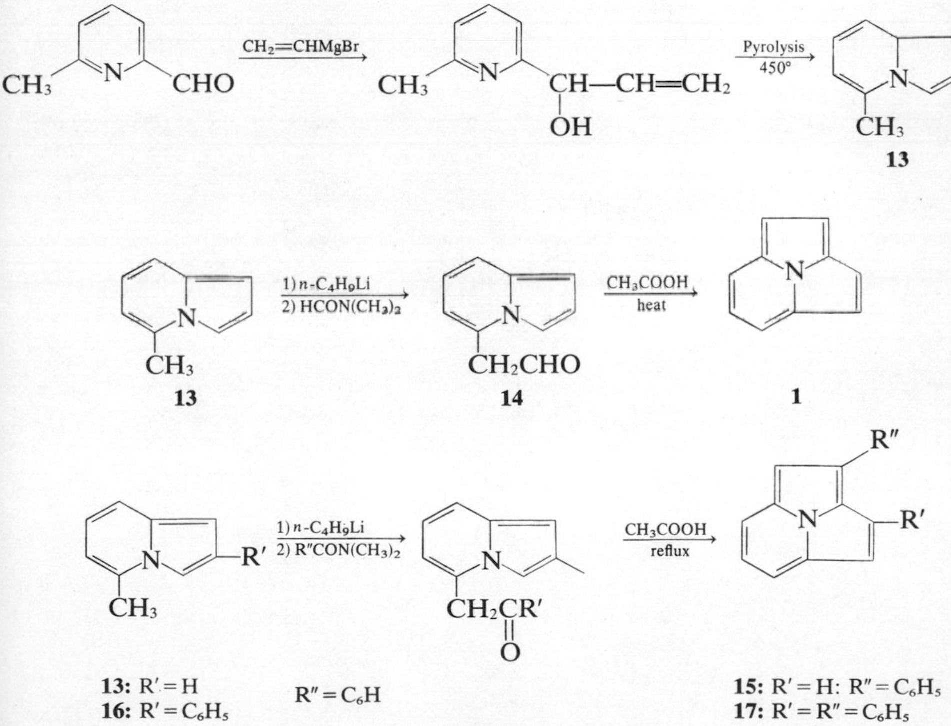

CH_3 N CHO $\xrightarrow{CH_2{=}CHMgBr}$ CH_3 N CH—CH=CH_2 OH $\xrightarrow[450°]{Pyrolysis}$ N CH_3 **13**

N CH_3 **13** $\xrightarrow[2)\ HCON(CH_3)_2]{1)\ n\text{-}C_4H_9Li}$ N CH_2CHO **14** $\xrightarrow[heat]{CH_3COOH}$ N **1**

N —R′ CH_3 $\xrightarrow[2)\ R''CON(CH_3)_2]{1)\ n\text{-}C_4H_9Li}$ N CH_2CR' ‖ O $\xrightarrow[reflux]{CH_3COOH}$ R″ N —R′

13: R′ = H
16: R′ = C_6H_5

R″ = C_6H

15: R′ = H: R″ = C_6H_5
17: R′ = R″ = C_6H_5

Scheme 1

B. Cycl[3.2.2]azines from Indolizines and Dimethyl acetylenedicarboxylate

Indolizine (**18**) and dimethyl acetylenedicarboxylate in toluene solution in the presence of 5% palladium-on-charcoal catalyst formed 1,2-dicarbomethoxycycl[3.2.2]azine (**19**) in a good yield.[13,14]

In addition to **19,** 1,2-dicarbomethoxy-3,4-dihydrocycl[3.2.2]azine (**20**) was isolated as a by-product in a 10 to 15% yield. Transformations of **19** into cycl[3.2.2]azine (**1**), the anhydride of 1,2-dicarboxycycl[3.2.2]azine (**23**), and 1,2-dicarbomethoxy-3,4,4*a*,5,6,7-hexahydrocycl[3.2.2]azine (**22**) are shown in Scheme 2.

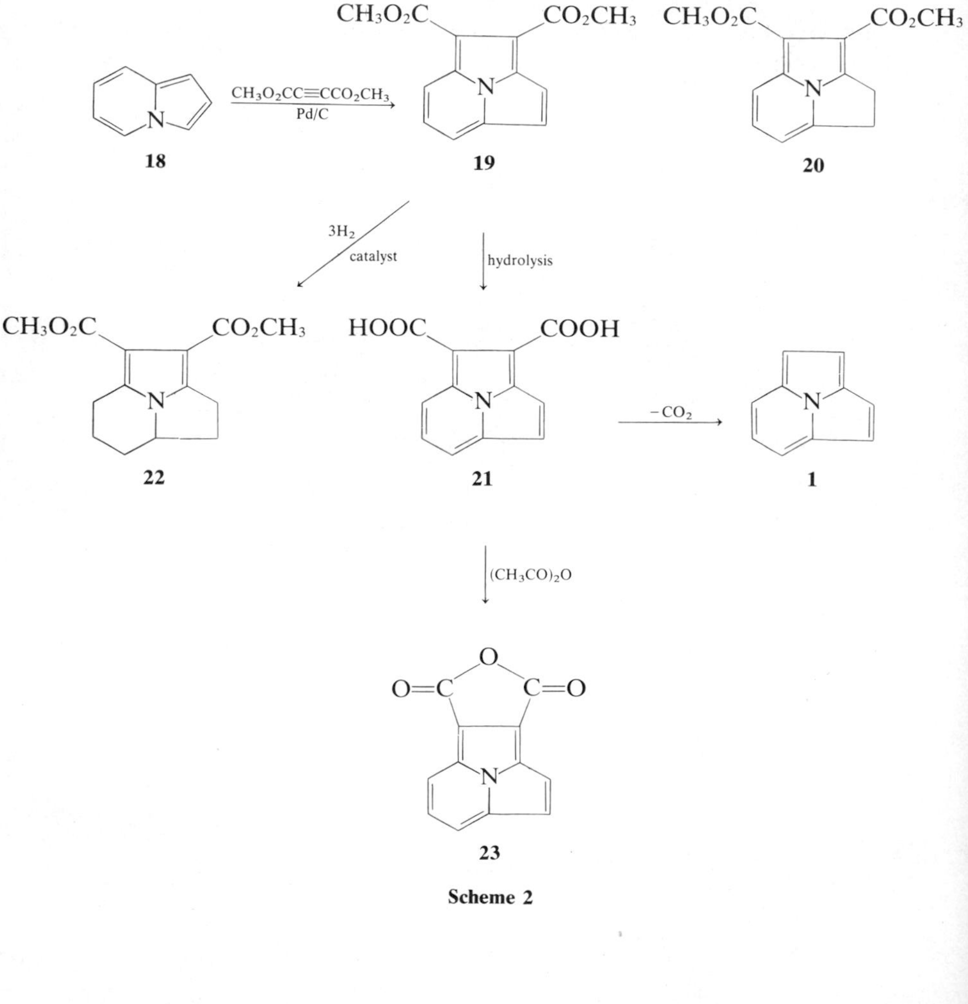

Scheme 2

Substitution of methyl phenylpropiolate for dimethyl acetylenedicarboxylate in the reaction with indolizine (**18**) led to the formation of 1-carbomethoxy-2-phenylcycl[3.2.2]azine, which was transformed into 2-phenylcycl[3.2.2]azine (**15**).[14]

2-Methylcycl[3.2.2]azine (**26**) was prepared[15] in a series of reactions starting from 2-methylindolizine (**24**) and dimethyl acetylenedicarboxylate. The first product, 1,2-dicarbomethoxy-3-methylcycl[3.2.2]azine (**25**), was hydrolyzed to a dicarboxylic acid which was decarboxylated to give **26**.

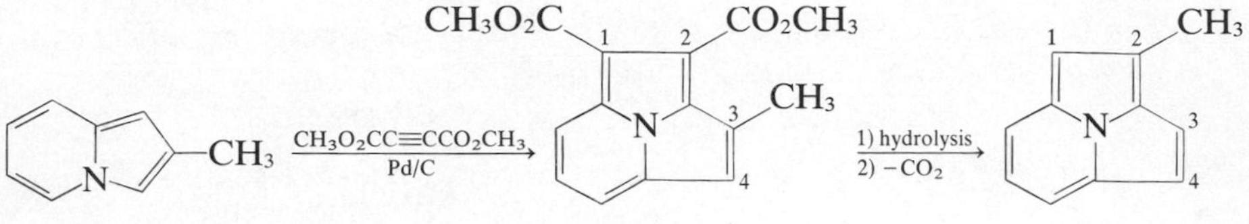

$CH_3O_2CC\equiv CCO_2CH_3$, Pd/C; 1) hydrolysis, 2) $-CO_2$

24 **25** **26**

Scheme 3

Dibenz[*b,g*]indolizine (**27**) and diethyl acetylenedicarboxylate were refluxed in toluene solution for 14 hr under nitrogen in the presence of 10% palladium-on-charcoal and a trace of hydroquinone to obtain 1,2-dicarbethoxydibenz[*a,g*]cycl[3.2.2]azine (**28**)[16] or benzo[6,7]pyrrolizino-[3,4,5-*ab*]isoquinoline.[17,18]

The compound **28** has no basic properties. It shows a strong yellow-green fluorescence in ether or benzene solution.

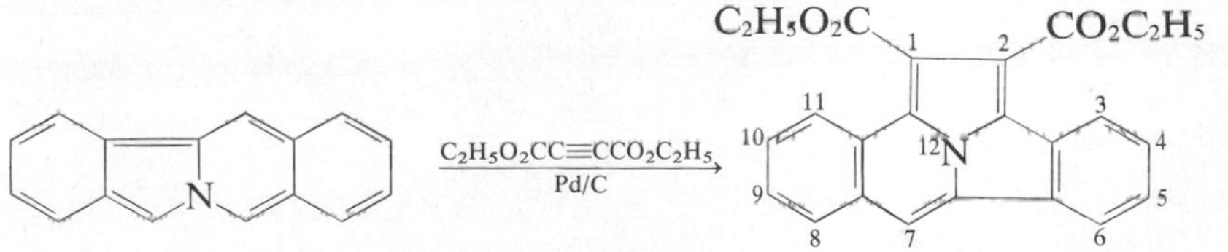

$C_2H_5O_2CC\equiv CCO_2C_2H_5$, Pd/C

27 **28**

Scheme 4

C. Cycl[3.2.2]azines from 3*H*-Pyrrolizine and Dimethyl acetylenedicarboxylate

3*H*-Pyrrolizine (**29**) was treated with dimethylformamide–phosphoryl chloride (Vilsmeier reagent) to form a salt which was isolated as the perchlorate (**30**).[19]

The salt **30** was reacted with dimethyl acetylenedicarboxylate in the presence of sodium hydride and the product, 5,6-dicarbomethoxycycl[3.2.2]azine (**31**), was isolated. The latter was transformed into cycl[3.2.2]azine (**1**) by the standard procedures involving hydrolysis and decarboxylation.

The treatment of the salt **30** with dimethylformamide and acetic anhydride gave bis(dimethylaminomethylene)pyrrolizinium salt (**32**), which was transformed into 6-azacycl[3.2.2]azine (**8**), 6-nitrocycl[3.2.2]azine (**33**), and cyclopenta[*h*]cycl[4.2.2]azine (**34**).

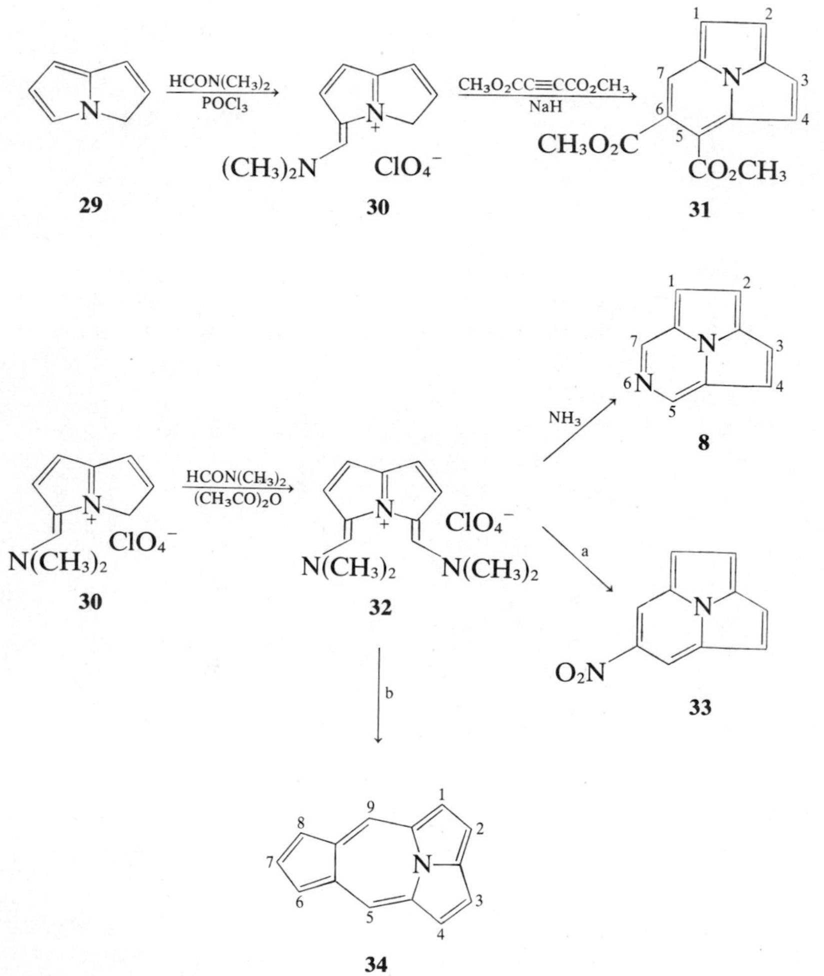

HCON(CH$_3$)$_2$, POCl$_3$; CH$_3$O$_2$CC≡CCO$_2$CH$_3$, NaH

(CH$_3$)$_2$N ClO$_4^-$ CH$_3$O$_2$C CO$_2$CH$_3$

29 **30** **31**

HCON(CH$_3$)$_2$, (CH$_3$CO)$_2$O; NH$_3$; a; b

N(CH$_3$)$_2$ ClO$_4^-$ N(CH$_3$)$_2$ N(CH$_3$)$_2$ ClO$_4^-$ O$_2$N

30 **32** **8** **33**

34

Scheme 5

Reagents: (*a*) CH$_3$NO$_2$ + *t*-C$_4$H$_9$OK; (*b*) Cyclopentadiene + NaH + HCON(CH$_3$)$_2$.

D. Cycl[3.2.2]azines from Pyridines and Methyl Propiolate

Pyridine, 3-methyl-, 4-methyl-, and 3,5-dimethylpyridine with methyl propiolate gave indolizines and cycl[3.2.2]azines in poor yields.[20,21] The products were identified mainly from their mass, nmr and uv spectra. 1,2-Dihydropyridines, reported previously,[22] were not the final products when the reagents were specially dried and purified. The reaction of pyridine with methyl propiolate was carried out in dry acetonitrile for 135 hr at room temperature to give 1-carbomethoxymethyl-2,4-dicarbomethoxycycl[3.2.2]azine (**35**). 3-Methyl-, 4-methyl-, and 3,5-dimethylpyridine required less time to produce similar triesters **36, 37,** and **38,** respectively.

The compound **38** was also obtained by Boekelheide procedure[14] from indolizine **39**.

The triester **38** was hydrolyzed to an acid that was decarboxylated to give 1-methylcycl[3.2.2]azine. The mass spectrum of 1-methylcycl[3.2.2]-azine showed a large molecular ion peak and a base peak m/e M–1 which was probably due to species **40** of ring-expanded [3.3.2] system.[23]

$H{-}C{\equiv}C{-}CO_2CH_3$

$CH_3O_2CCH_2$, CO_2CH_3, R^3, R^2, R^1, CO_2CH_3

35 $R^1, R^2, R^3 = H$
36 $R^1, R^2 = H$; $R^3 = CH_3$
37 $R^1, R^3 = H$; $R^2 = CH_3$
38 $R^2 = H$; $R^1, R^3 = CH_3$

CH_3, $CH_2CO_2CH_3$, CO_2CH_3, CH_3

39

40

Scheme 6

E. Cycl[3.3.2]azine Derivatives

The parent ring system of cycl[3.3.2]azine does not exist because there can only be structures with one, or more, extra hydrogens, or the cycl-[3.3.2]azinium (**2**) type derivatives.

A 1,2-dihydro derivative of cycl[3.3.2]azine was recognized[24] in the "second stable adduct" (**42**) of trans-stilbazole (**41**) and dimethyl acetylenedicarboxylate.[25] The three hydrogens at C-1 and C-2 in **42** displayed nmr bands of the ABX type in the high field. Strong acids protonated the compound **42** mainly at the position 6 yielding the cation **43** (R = H).

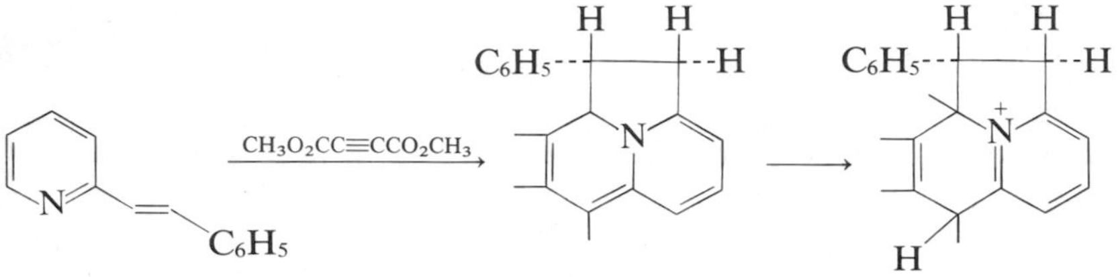

C_6H_5 $CH_3O_2CC{\equiv}CCO_2CH_3$ H H C_6H_5 H N   H H C_6H_5 H $\overset{+}{N}$ H

41   **42**   **43**

Scheme 7

An oxo derivative of 3*H*-cycl[3.3.2]azine (**44**) has been synthesized.[26] 5-Methyl-2-phenylindolizine (**16**) reacted with ethoxalyl chloride to produce 1,3-diethoxalyl-5-methyl-2-phenylindolizine (**45**). Heating this compound with ethanolic sodium ethoxide converted it into two products: (a) 1-ethoxalyl-3-carbethoxy-2-phenylcycl[3.2.2]azine (**46**), and (b) a sodium salt of hydroxycycl[3.3.2]azinone which was converted by the usual procedures of hydrolysis and decarboxylation into a bright yellow crystalline 4(or 3)-hydroxy-2-phenyl-3(or 4)*H*-cycl[3.3.2]azin-3(or 4)-one (**47** or **48**).

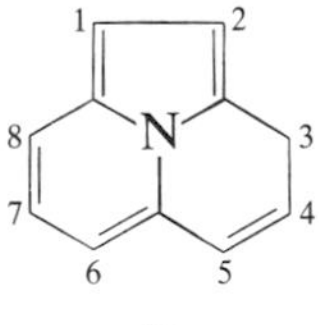

1 2 3 4 5 6 7 8 N

44

$C_6H_5 + 3ClCOCO_2C_2H_5 \xrightarrow{CH_2Cl_2}$ $CO{\cdot}CO_2C_2H_5$ C_6H_5 $CO{\cdot}CO_2C_2H_5$ $\xrightarrow[C_2H_5OH]{NaOC_2H_5}$

N CH_3 N CH_3

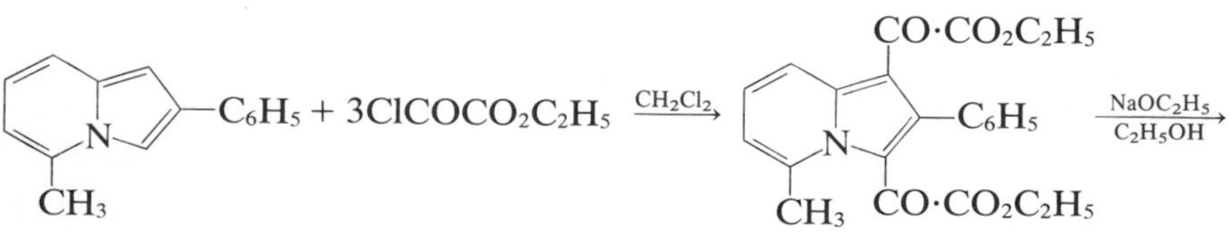

16   **45**

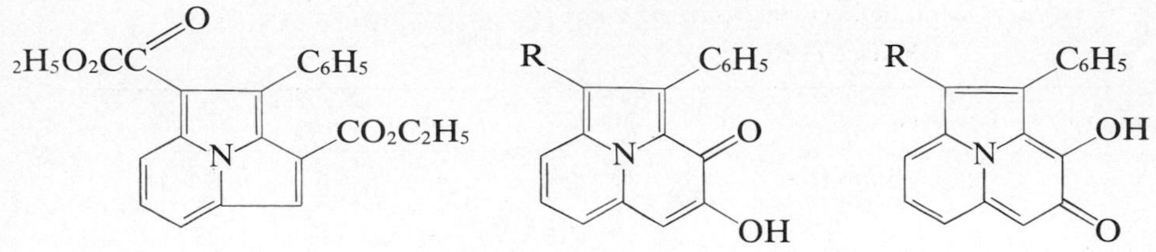

46 **47** ($R = CO{\cdot}CO_2C_2H_5$) **48**

Scheme 8

F. Cycl[3.3.3]azines

There have several unsuccessful efforts to synthesize cycl[3.3.3]azine (**3**),[27–31] but only recently its synthesis has been accomplished starting from 4-chloroquinolizinium perchlorate (**49**),[32] as outlined in Scheme 9.

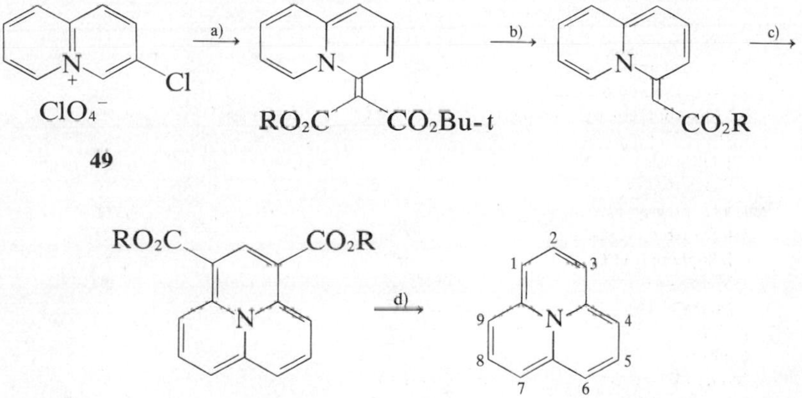

49 **50** **3**

Reagents: (*a*) $NaCH(CO_2R){\cdot}CO_2Bu\text{-}t$ + tetrahydrofuran; (*b*) $HCl + C_6H_6$ (for $R = C_2H_5$), or $C_6H_5SO_3H + CH_3COOH$ (for $R = t\text{-}C_4H_9$), NaOH; (*c*) $HC{\equiv}CCO_2R + C_6H_5NO_2$ at 210°; (*d*) $R = t\text{-}C_4H_9$; heat in a sealed, evacuated tube at 250 to 300° for 5 min.

Scheme 9

The diesters **50** ($R = C_2H_5$, or $t\text{-}C_4H_9$) were crystalline, stable compounds. However their solutions decomposed rapidly. The parent cycl[3.3.3]azine (**3**) was stable only in nitrogen atmosphere but decomposed in air or in solutions. Its nmr spectrum and structure are discussed further in Section V.A.3.

TABLE IV. PHYSICAL PROPERTIES OF CYCL[3.2.2]AZINE AND ITS DERIVATIVES

Number	Compound	M.p. (°C)	Spectra	Ref.
1	Cycl[3.2.2]azine (**1**)	63.5–64.5	uv[2,21,43], ir[2,21], nmr[2,15,43]	1, 2, 12, 13, 14, 15, 21, 43, 47
2	1-Acetamido-	143–145	uv[37]	37
3	1-Acetyl-	Oil	—	2
4	Anhydride of 1,2-dicarboxy-	267	uv[14]	14
5	1-Carbomethoxy	61–62	uv[37]	37
6	1-Carbomethoxymethyl-2,4-dicarbomethoxy- (**35**)	181.5–182	uv[21], ir[21]	20, 21
7	1-Carbomethoxymethyl-2,4-dicarbomethoxy-6-methyl-	204–206	uv[21], ir[21]	20, 21
8	1-Carbomethoxymethyl-2,4-dicarbomethoxy-7-methyl-	150–157	uv[21], ir[21]	20, 21
9	1-Carbomethoxymethyl-2,4-dicarbomethoxy-5,7-dimethyl-	178–179.5	uv[21], ir[21]	20, 21
10	1-Carbomethoxy-2-phenyl-	118–120	—	14
11	2-Carboxy-	231–233	uv[14]	14
12	1-Carboxymethyl-2-carboxy-[a]	>180° dec.	uv[21]	20, 21
13	1,4-Diacetyl-	183–184	uv[2]	2
14	1,4-Diacetyl-2,3-diphenyl-	189–190	—	2
15	1,4-Dibromo-	121–122	uv[2]	2
16	1,2-Dicarbomethoxy- (**19**)	91–92	uv[14]	13, 14
17	5,6-Dicarbomethoxy- (**31**)	—	—	19
18	1,2-Dicarbomethoxy-3-methyl- (**25**)	115–115.5	uv[15], nmr[15]	15
19	1,2-Dicarboxy-	>310	uv[14]	13, 14
20	5,6-Dicarboxy-	—	—	19
21	1,2-Dicarboxy-3-methyl-	>320	nmr[15]	15
22	2,4-Dicarboxy-1-methyl-[b]	>180° dec.	uv[21]	21
23	2,3-Diphenyl- (**17**)	141–141.5	uv[2]	1, 2
24	1-Methyl-	Oil	uv[21], nmr[20]	20, 21
25	2-Methyl- (**26**)	10–12	nmr[15], uv[14,21]	15
26	1-Nitro-	193.5–195	uv[2]	2
27	6-Nitro-	—	nmr[19]	19
28	Oxime of 1-acetyl-	139–140	—	37
29	2-Phenyl- (**16**)	93.5–94.5	uv[2]	1, 2
	Deuterium derivatives			
30	Cycl[3.2.2]azine-1,4-d_2	—	nmr[15]	15
31	2-Methylcycl[3.2.2]azine-1,4-d_2	—	nmr[15]	15
32	1,2-Dicarbomethoxycycl[3.2.2]-azine-4-*d*	—	nmr[15]	15
	Hydrogenated derivatives			
33	2-Carbomethoxy-1-carboxy-3,4-dihydroxycycl[3.2.2]azine	292	uv[14]	14
34	1,2-Dicarbomethoxy-3,4-dihydro- (**20**)	178–180	uv[14]	14
35	1,2-Dicarbomethoxy-3,4,4*a*,5,6,7-hexahydro- (**22**)	111–112	uv[14]	14
	Dibenz[a,g]*cycl*[3.2.2]*azines*			
36	1,2-Carbethoxydibenz[*a*,*g*] cycl[3.2.2]azine (**28**)	125–126	ir[16], uv[16]	16

[a] Or No. 22. [b] Or No. 12.

A series of substituted cyclopenta[*cd*]cycl[3.3.3]azines (**53**) have been prepared by a general scheme from 4-methyl- or 4-phenylcyclopenta[*c*]-quinolizines (**51**) and acetylenic esters (**52**).[33] They are stable crystalline compounds, and their nmr spectra indicate they are essentially aromatic in character.

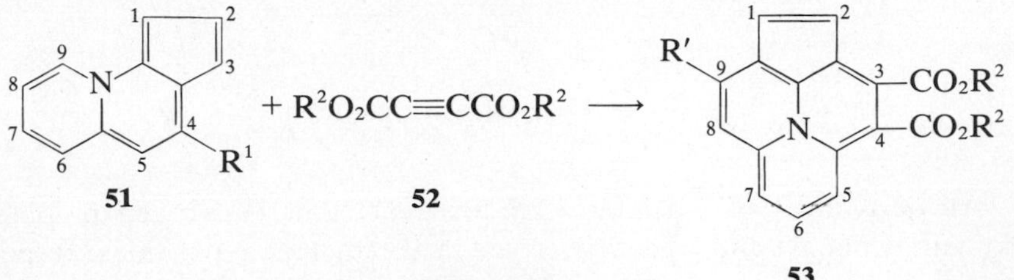

$$\textbf{51} + R^2O_2CC{\equiv}CCO_2R^2\ (\textbf{52}) \longrightarrow \textbf{53}$$

$R^1 = CH_3, C_6H_5$
$R^2 = CH_3, C_2H_5$

Scheme 10

III. Syntheses of Azacyclazines

A. 1-Azacycl[3.2.2.]azines

The parent compound, 1-azacycl[3.2.2]azine (**6**), has not been prepared despite all efforts. The known derivatives of **6** contain a phenyl group in 2-position, and their synthesis starts with the reaction of 2-phenyl-imidazo[1,2-*a*]pyridine (**54**)[34] in boiling toluene in the presence of

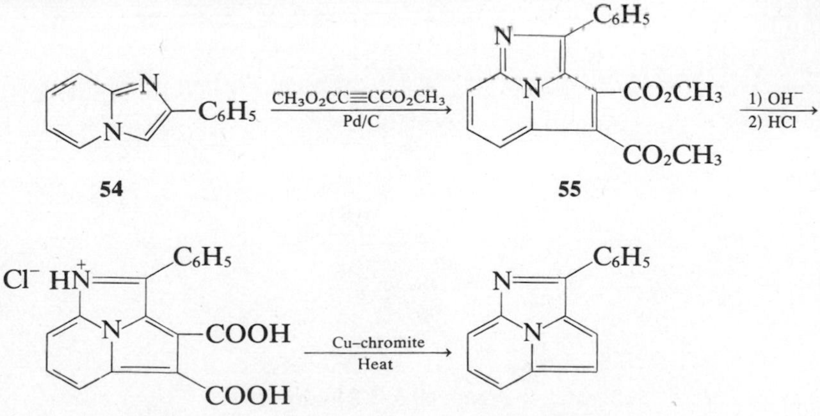

$$\textbf{54} \xrightarrow[\text{Pd/C}]{CH_3O_2CC{\equiv}CCO_2CH_3} \textbf{55} \xrightarrow[\text{2) HCl}]{\text{1) OH}^-} \textbf{56} \xrightarrow[\text{Heat}]{\text{Cu-chromite}} \textbf{57}$$

Scheme 11

palladium-on-charcoal.[3] The product formed, 3,4-dicarbomethoxy-2-phenyl-1-azacycl[3.2.2]azine (**55**), was hydrolyzed in alkaline methanol to give the potassium salt of a dicarboxylic acid which was transformed into the hydrochloride **56**. The later was mixed with copper chromite catalyst and refluxed in diphenyl ether to achieve the formation of 2-phenyl-1-azacycl[3.2.2]azine (**57**).

B. 5-Azacycl[3.2.2]azines

All derivatives of 5-azacycl[3.2.2]azine (**7**) that have been prepared contain a phenyl and a methyl group in the 6- and 2-positions, respectively, but the parent compound is unknown.[4] 4,6-Dimethylpyrimidine was converted with phenacyl bromide into the quaternary bromide **58,** and the latter was cyclized with aqueous bicarbonate solution to give 7-methyl-2-phenylpyrrolo[1,2-*c*]pyrimidine (**59**) (7-methyl-2-phenyl-6-azaindolizine). The reaction of **59** with dimethyl acetylenedicarboxylate and Pd/C produced 3,4-dicarbomethoxy-6-methyl-2-phenyl-5-azacycl-[3.2.2]azine (**60**) in a moderate yield. Hydrolysis of **60** followed by decarboxylation gave 6-methyl-2-phenyl-5-azacycl[3.2.2]azine (**61**).[4]

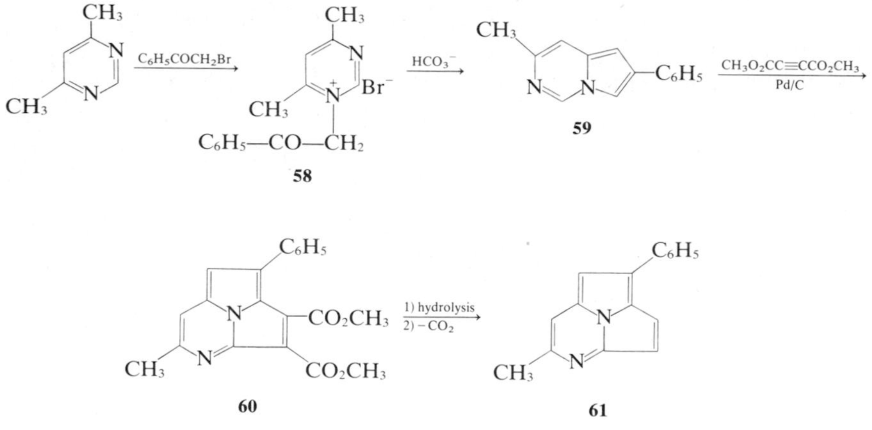

Scheme 12

C. 6-Azacycl[3.2.2]azine (8)

See Section II,C., Ref. 19

D. 1,4-Diazacycl[3.2.2]azines

1,4-Diazacycl[3.2.2]azine (**9**) has been synthesized starting from 2,6-diaminopyridine and α-chloroformylacetate.[6] The resulting 5-amino-3-carbethoxyimidazo[1,2-*a*]pyridine (**62**) was treated with sodium hydride in dimethylformamide to obtain 2-oxo-1,2-dihydro-1,4-diazacycl-[3.2.2]azine (**63**). The compound **63** was reduced with $LiAlH_4$ to give

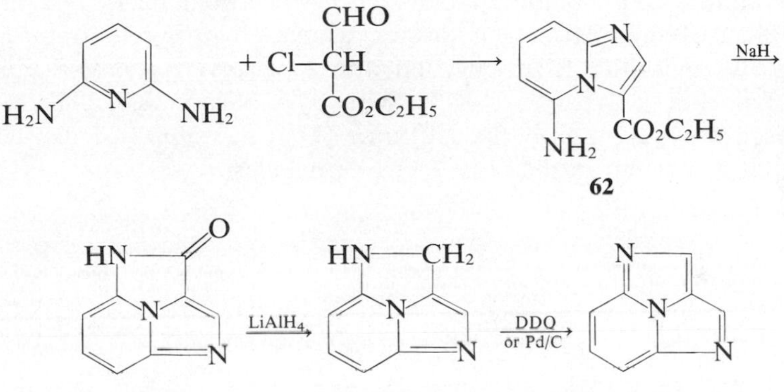

Scheme 13

1,2-dihydro-1,4-diazacycl[3.2.2]azine (**64**), which was treated with 2,3-dichloro-5,6-dicyanobenzoquinone (DDQ). The final product, 1,4-diazacycl[3.2.2]azine (**9**), formed yellow crystals.

2,3-Dimethyl-1,4-diazacycl[3.2.2]azine (**67**) was prepared in the following way. 5-Amino-2-methylimidazo[1,2-*a*]pyridine (**65**)[35] was acetylated with acetic anhydride. Two products were obtained, the expected 5-acetamido-2-methylimidazo[1,2-*a*]pyridine and 5-acetamido-3-acetyl-2-methylimidazo[1,2-*a*]pyridine (**66**). Compound **66** was refluxed in 25%

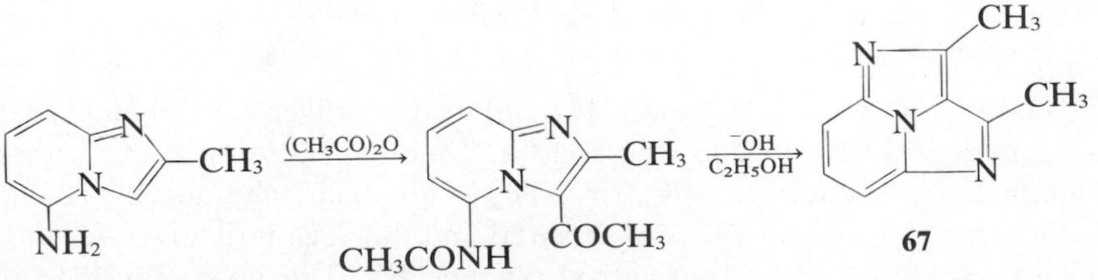

Scheme 14

sodium hydroxide solution for 1 hr, acidified with dilute hydrochloric acid, and made slightly alkaline with ammonium hydroxide. The final product was 2,3-dimethyl-1,4-diazacycl[3.2.2]azine (**67**).[7]

E. 8-Azacycl[2.2.2]azines

1-Benzoyl-2-phenyl-4,8-diazapentalene (**68**)[36] was treated with dimethyl acetylenedicarboxylate and Pd/C in toluene for 4 days at room temperature to obtain 2-benzoyl-5,6-dicarbomethoxy-1-phenyl-8-azacycl[2.2.2]azine (**69**).[8] The diester **69** was hydrolyzed to form the diacid **70** which was decarboxylated into 2-benzoyl-1-phenyl-8-azacycl-[2.2.2]azine (**71**). Treatment of **71** with ethanolic potassium hydroxide produced 1-phenyl-8-azacycl[2.2.2]azine (**72**) (1-phenyl-2*a*,7-diazacyclopenta[*cd*]pentalene) in the form of white needles.

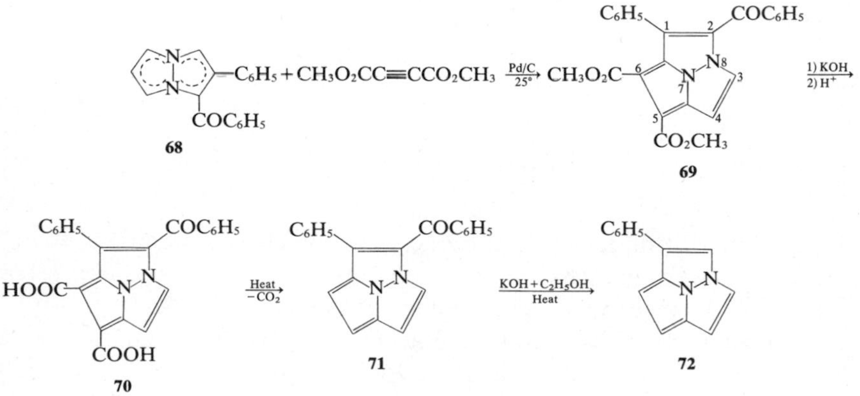

Scheme 15

F. 1,3,6-Triazacycl[3.3.3]azines

1,3,6-Triazacycl[3.3.3]azine (**11**) and its derivatives have been prepared by two general procedures[9] from a 2,6-diaminopyridine and ethoxymethylenecyanoacetate (**73**), or ethoxymethylenemalonoitrile (**78**). The intermediates **74** and **79** were acylated and the acyl derivatives **75** and **80** were cyclized to form derivatives **76** and **81**. The compound **76,** with R′ = H, served as the material for the preparation of the parent compound **11**.

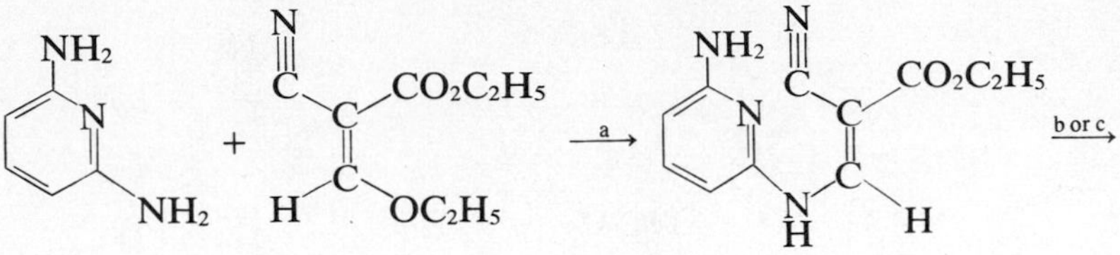

NH₂ N C CO₂C₂H₅ C C H OC₂H₅ NH₂ 73 74 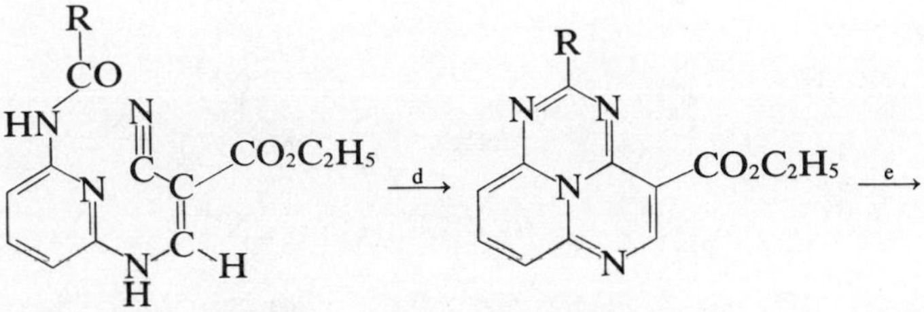 75 76 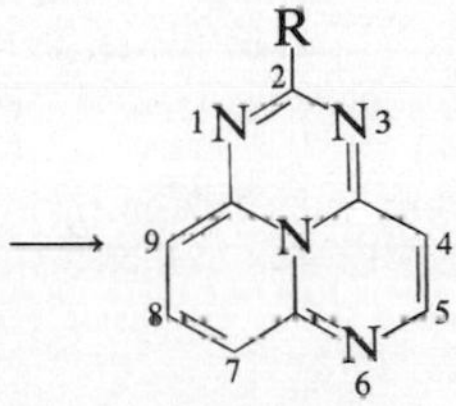 77

Reactions: (*a*) heat in benzene; (*b*) acetic-formic anhydride in pyridine (20°) (R = H); (*c*) acetic anhydride in pyridine (20°) (R = CH_3); (*d*) reflux in diphenyl ether (259°); (*e*) hydrolysis and decarboxylation.

Scheme 16

G. 1,3,6,7-Tetrazacycl[3,3,3]azines

The synthesis of 4-carbethoxy-2-methyl-1,3,6,7-tetrazacycl[3.3.3]azine (**82**) is outlined in Scheme 18.[10]

H. 1,3,4,7-Tetrazacycl[3.3.3]azines

The preparation of 9-carbethoxy-2-methyl-1,3,4,7-tetrazacycl[3.3.3]-azine (**12a**) has been described.[10a]

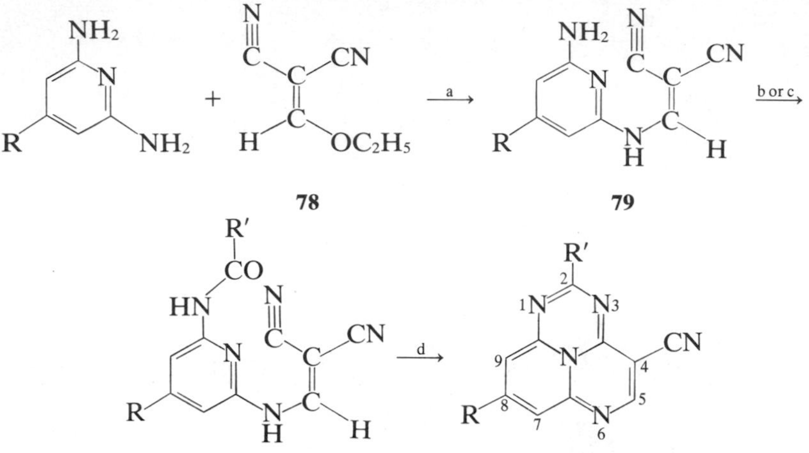

78 79

80 81

81a: R = H; R′ = H
81b: R = H; R′ = CH_3
81c: R = CH_3; R′ = CH_3

Reagents: (*a*) heat in benzene; (*b*) acetic–formic anhydride in pyridine (20°) (R′ = H); (*c*) acetic anhydride in pyridine (20°) (R′ = CH_3); (*d*) reflux in diphenyl ether (259°).

Scheme 17

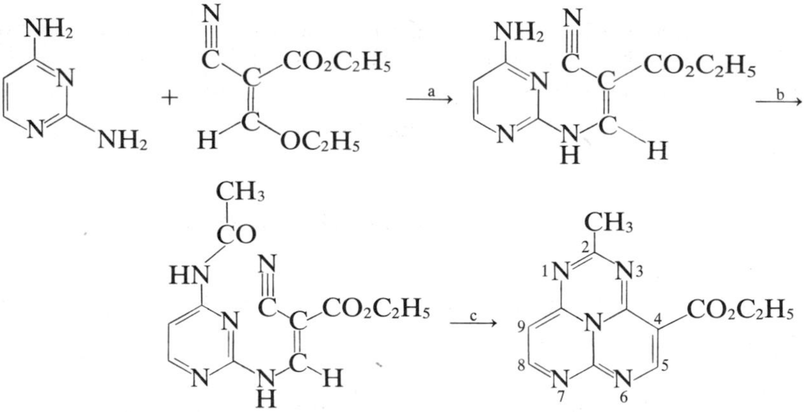

82

Reagents: (*a*) benzene, heat; (*b*) acetic anhydride and xylene, reflux; (*c*) *p*-toluene–sulfonic acid and diphenyl ether, heat.

Scheme 18

IV. Reactions of Cyclazines

Simple HMO calculations predicted[2] that the electrophilic substitution in cycl[3.2.2]azine (**1**) should occur most readily at C-1, and the following experiments confirmed these predictions.[2] Acetylation of **1** with acetic anhydride and stannic chloride at room temperature in *sym*-tetrachloroethane gave a mixture of 1-acetyl- and 1,4-diacetylcycl[3.2.2]azine, which were separated by chromotography. 2,3-Diphenylcycl[3.2.2]azine was acetylated with acetyl chloride and aluminum chloride at 60°, or with acetic anhydride and stannic chloride, and in both cases 1,4-diacetyl-2,3-diphenylcycl[3.2.2]azine was obtained.[2] Bromination of **1** with bromine at room temperature in dichloromethane yielded 1,4-dibromocycl[3.2.2]-azine.[2]

1-Nitrocycl[3.2.2]azine was formed when **1** was nitrated with cupric nitrate in acetic anhydride at room temperature.[2] The nitro compound could not be reduced to the corresponding amino derivative; however reductive acetylation of the 1-nitro compound produced 1-acetamidocycl-[3.2.2]azine.[37] The latter compound was also made from 1-acetyl derivative by transferring it into the oxime, followed by the Beckmann rearrangement.[37]

Cycl[3.2.2]azine, 2-phenyl-, and 2,3-diphenylcycl[3.2.2]azine form charge-transfer complexes with 1,3,5-trinitrobenzene in the ratio 1 : 1.[2]

1,4-Diazacycl[3.2.2]azine (**9**) and 2-methyl-1,4-diazacycl[3.2.2]azine are hydrolyzed very easily by dilute acid at room temperature.[6] The 1-2 bond in these compounds is broken with the formation of 5-amino-3-formylimidazo[1,2-*a*]pyridine from **9** and a mixture of 3-acetyl-5-amino- and 5-amino-3-formyl-2-methylimidazo[1,2-*a*]pyridine from the 2-methyl derivative.

2,3-Dimethyl-1,4-diazacycl[3.2.2]azine (**67**)[7] is less reactive toward hydrolysis, and the formation of the hydrolysis product, 3-acetyl-5-amino-2-methylimidazo[1,2-*a*]pyridine, occurs only when the solution is warmed.[6]

Bromination of 4-cyano-2-methyl-1,3,6-triazacycl[3.3.3]azine (**81b**) with *N*-bromosuccinimide, or with bromine in glacial acetic acid, gave 7-bromo-, 9-bromo-, and 7,9-dibromo derivatives.[38]

Photolytic bromination of **81b** with bromine occurred at the methyl group and the positions 7 and 9.[39]

V. Physical Properties

A. Structure and Quantum Chemical Calculations

Three essential factors must be considered in evaluating the stability and structure of cyclazines: (1) the number of π-electrons in the peripheral ring, (2) the participation of the unshared electrons of the nitrogen atom in the bonding with carbon atoms, and (3) the planarity of the cyclazine molecule. Since there could be a wide variations in these factors, each cyclazine type should be treated separately.

On the basis of the number of π-electrons in the peripheral ring, cycl[3.2.2]azine (**1**), cycl[3.3.2]azinium (**2**), and cycl[4.4.3]azine (**4**) should be aromatic $4n+2$ π-electron systems, containing 10, 10, and 14 π-electrons, respectively. On the other hand, cycl[3.3.3]azine (**3**) and cycl[4.3.2]azine (**5**) should be antiaromatic $4n$ π-electron systems with 12 π-electrons each. However the contribution of the nitrogen unshared electrons and planarity of the molecule could influence the antiaromaticity to a certain extent.

Of all the cyclazines, only cycl[3.2.2]azine (**1**) and some of its derivatives have been studied with respect to their stability and chemical behavior. They show a great stability toward air, light, solvents, and heat (sublime easily). Cycl[3.2.2]azine shows a complete lack of basicity.

Quantum chemical calculations on cyclazines **1, 3,** and **4** have been carried out by several groups of workers. Resonance energies, charge densities, free valencies, and bond orders have been estimated[2] by the Hückel (HMO) method, and from the values of resonance energies it was concluded that all three compounds should be aromatic. However in the light of more recent experimental work[32] and advanced calculations,[40] it appears that only **1** and **4** possess aromatic character but that compound **3** is antiaromatic.

The HMO calculations[2] have shown that in cycl[3.2.2]azine (**1**) the highest charge density is on C-1 and C-4, followed by C-5. The same gradation of charge densities for **1** was found by other workers.[41,42]

The correlation between the transition energies and electronic spectrum of cycl[3.2.2]azine has been investigated by the simplified Pople–Parisier–Parr method.[43]

HMO calculations have been also used in the interpretation of the electron paramagnetic resonance spectra of **1**.[44,45]

A semiempirical Streitwieser ω-type HMO method was applied to calculate the resonance energy, charge density, and free valencies of 1,3,6-triazacycl[3.3.3]azine (**11**)[9] and 1,3,6,7-tetrazacycl[3.3.3]azine (**12**).[10]

The charge densities and other parameters of 1,4-diazacycl[3.2.2]azine (**9**) have been estimated by the HMO method.[6,7]

B. Crystal Structure

The crystal structure determination for 1,4-dibromocycl[3.2.2]azine by the x-ray diffraction method revealed that the molecule is very nearly planar with the nitrogen configuration being also planar. The bond lengths and bond angles have been estimated.[46]

C. Spectra

1. *Ultraviolet*

The uv spectrum of cycl[3.2.2]azine (**1**) was measured in ethanol solution.[2] It displays the characteristic features of a heteroaromatic compound with three main bands which are classified as L_b, L_a, and B (Platt), or α, p, and β (Clar). In addition to that a number of other cyclazines were characterized by their uv spectra.[2,14,15,21] A more detailed study was carried out[43] by recording the ultraviolet absorption spectrum, the fluorescence polarization spectrum (Fig. 1). The characteristic features of the AP spectrum were used in assigning the directions of polarization of the molecule during electronic transitions.

2. *Infrared*

A single infrared spectrum of cycl[3.2.2]azine has been recorded, but no structure–vibrations correlations were offered.[2] Partial infrared data are available for a number of the cyclazine derivatives. No Raman spectroscopic studies have been carried out.

3. *Proton Magnetic Resonance Spectra*

The nmr spectrum of cycl[3.2.2]azine (**1**), recorded at 60 MHz in carbon tetrachloride solution,[47] consists of an A_2B multiplet arising from the protons 5, 6, and 7 of the 6-membered ring and an A^1B^1 quartet displayed by the protons of the two equivalent 5-membered rings. The higher field doublet of the A^1B^1 system, at δ7.20, was assigned to H-1 and H-4 because it was missing from the cycl[3.2.2]azine-1,4-d_2. The doublet at δ7.51 was assigned to H-2 and H-3, and $J_{A^1B^1} = 4.2$ Hz. The chemical shifts of H-5 and H-7 (A_2) and H-6 (B) were δ7.86 and 7.59, respectively; $J_{AB} = 8.0$ hz. These results show that cycl[3.2.2]azine absorbs in a rather narrow region of radio–frequencies characteristic of aromatic hydrocarbons. This is a strong indication that the peripheral

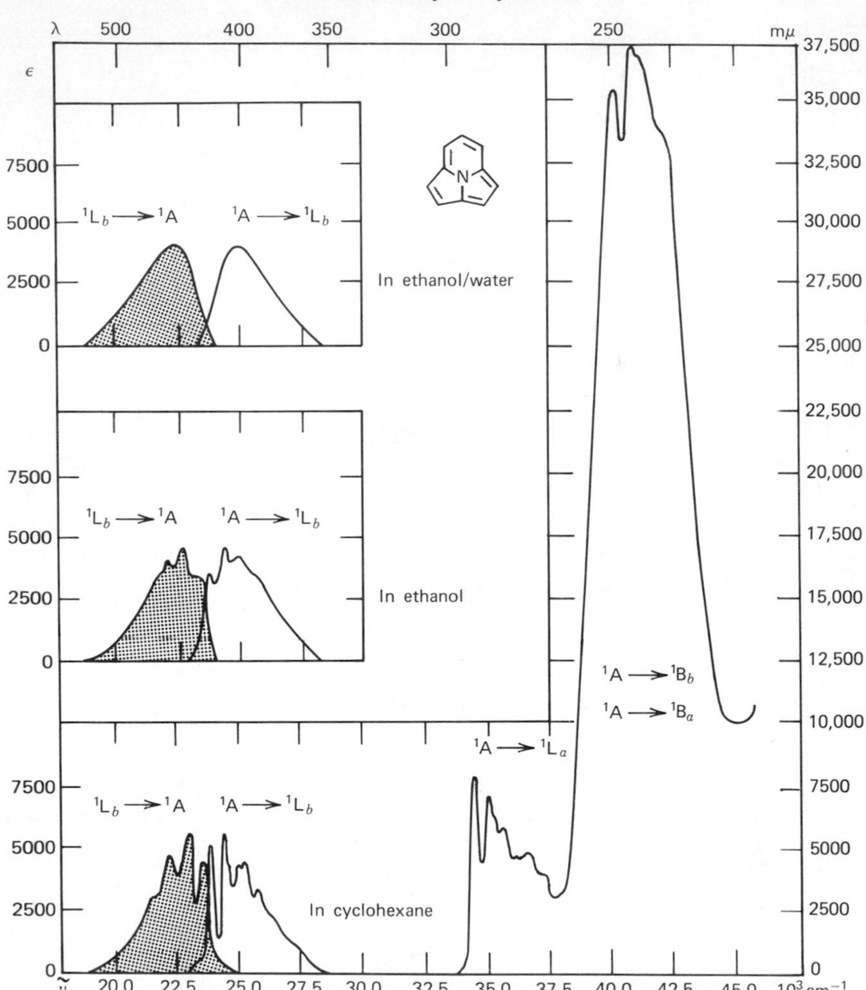

Fig. 1. Ultraviolet absorption and fluorescence spectra of cycl[3,2,2]azine.[43] Reprinted from *Helv. Chim. Acta.*, **46,** 1940 (1963).

carbons of **1** are sp^2 hydbridized, and that they form a planar system with the central nitrogen, which is also sp^2 hybridized (Fig. 2).

An independent study of the nmr spectra of cycl[3.2.2]azine and some derivatives confirmed[15] the previous results. A series of cycl[3.2.2]azines have been also studied by other workers.[19–21]

The nmr spectrum of cycl[3,3,3]azine (**3**), recorded in bis(trimethylsilyl)ether,[32] revealed a dramatic difference compared with **1**. The compound **3** displayed a triplet centered at δ3.65 (protons 2, 5, and 8) and a

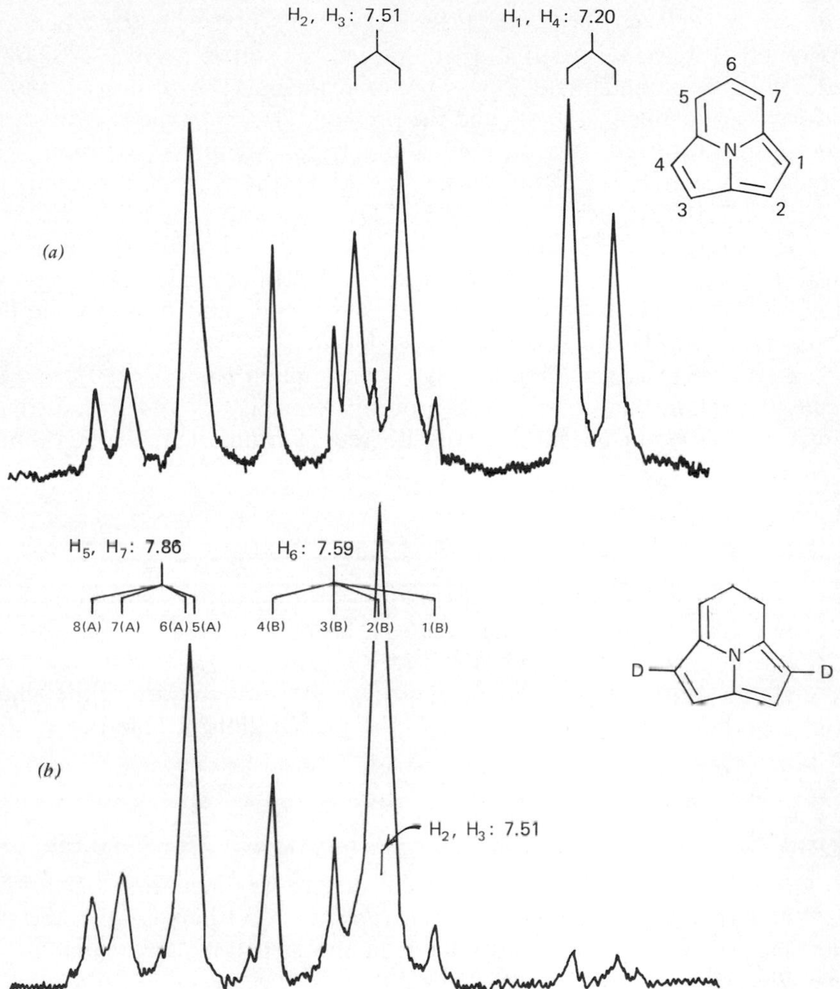

Fig. 2. Proton magnetic resonance spectra of (*a*) cycl[3,2,2]azine, (*b*) cycl[3,2,2]azine-d_2. Solvent: CCl_4.[47] Reprinted from *Helv. Chim. Acta.*, **46,** 1951 (1963).

doublet centered at $\delta 2.07$ (protons 1, 4, 6, 7, and 9). The nmr bands occur at the highest field yet reported for protons bonded to trigonal carbon. The authors of this work[32] explain this high degree of shielding as evidence of a paramagnetic ring current in the peripheral, nonaromatic system of 12 π-electrons[48–50] of the cycl[3.3.3]azine molecule.

The nmr spectrum of 1,4-diazacycl[3.2.2]azine (**9**), recorded at 100 MHz in $CDCl_3$ solution, was rather unusual because it showed only two singlets at δ 8.7 (H-2 and H-3) and 8.12 (H-5, H-6, H-7). When the spectrum of **9** was recorded after addition of the shift reagent Eu-2,2,6,6-tetramethylheptanedionale, it was better resolved. The protons 2 and 3 displayed a singlet at δ10.05, and the previous singlet of the protons 5, 6, and 7 was resolved into a doublet and triplet of an AX_2 system.[6] A similar AX_2 pattern of bands was also shown by the protons 5, 6, and 7 of 2-methyl-1,4-diazacycl[3.2.2]azine in $CDCl_3$ without the shift reagent.[6]

The spectrum of 2,3-dimethyl-1,4-diazacycl[3.2.2]azine (**67**), recorded in $CDCl_3$ solution at 60 MHz, displayed a band pattern characteristic of an A_2B system. The seven lines by the protons 5, 6, and 7 occurred in the aromatic region between δ8.16 and 7.70.[7]

The Biot–Savart shielding law has been applied to cycl[3.3.3]azine to study the relationship between its proton chemical shift and ring current, in order to obtain insight into the degree of aromaticity of this compound.[50a]

4. *Electron Paramagnetic Resonance*

The epr spectra of anions of cycl[3.2.2]azine and cycl[3.2.2]azine-1,4-d_2 have been measured and analyzed.[44,45] The values of the hyperfine interactions of two of the ring protons depend on the nature of the alkali metal, sodium, or potassium, used in the preparation of anions.

5. *Mass Spectra*

A detailed study of the mass spectrometric fragmentation of various 1,3,6-triazacycl[3.3.3]azines has been reported.[51] Otherwise this technique has been used only occasionally in the structure determination of cyclazines.[21]

D. Ionization

Protonation of cycl[3.2.2]azine (**1**) by strong acids has been followed by uv absorption (in ethanol–perchloric acid solution)[43] and by nmr spectroscopy methods (in concentrated sulfuric acid solution).[47] In both cases the protonation of **1** occurs at C-1. Otherwise cycl[3.2.2]azine is stable in acid solutions.

The difference in the spectra of **1** and its conjugate acid in the region 300 to 450 nm made it possible to determine the basicity of **1**. The measurements in ethanol–perchloric acid (1 : 4) system gave the value of $pK_a = -2.8$ on a Hammett H_0-scale.[43] The equilibrium protonation of **1** was also measured spectrophotometrically in aqueous perchloric acid solutions, and the value of -2.82 ± 0.10 for the pK_a was obtained.[52]

Rate coefficients for the acid-catalyzed detritiation of cycl[3.2.2]azine-1-*t* were measured and compared with those obtained for azulene-1-*t* ($pK_a = -1.8$). The determination of the thermodynamic functions E_a, ΔS^*, and ΔF^* for **1** and azulene indicated that in both cases the detritiation occurred by the same slow proton transfer process.[52]

VI. References

1. V. Boekelheide and R. J. Windgassen, *J. Am. Chem. Soc.*, **80,** 2020 (1958).
2. R. J. Windgassen, W. H. Saunders, and V. Boekelheide, *J. Am. Chem. Soc.*, **81,** 1459 (1959).
3. V. Boekelheide and A. Miller, *J. Org. Chem.*, **26,** 431 (1961).
4. V. Boekelheide and S. A. Kertelj, *J. Org. Chem.*, **28,** 3212 (1963).
5. M. A. Jessep and D. Leaver, *Chem. Comm.*, 790 (1970).
6. W. W. Paudler, R. A. VanDahm, and Y. N. Park, *J. Heterocycl. Chem.*, **9,** 81 (1972).
7. K. Valentin and A. Taurins, *Tetrahedron Lett.* 3621 (1966).
8. V. Boekelheide and N. A. Fedoruk, *Proc. Nat. Acad. Sci.*, **55,** 1385 (1966).
9. O. Ceder and J. E. Andersson, *Acta Chem. Scand.*, **26,** 596 (1972).
10. O. Ceder and J. F. Witte, *Acta Chem. Scand.*, **26,** 635 (1972).

10a. O. Ceder and K. Rosen, *Acta Chem. Scand.*, **27,** 2421 (1973).

11. V. Boekelheide and R. J. Windgassen, Jr., *J. Am. Chem. Soc.*, **81,** 1456 (1959).
12. R. J. Windgassen and V. Boekelheide, U.S. Patent 2,986,563 (1961); *Chem. Abstr.*, **56,** P3464i.
13. A. Gailbraith, T. Small, and V. Boekelheide, *J. Org. Chem.*, **24,** 582 (1959).
14. A. Gailbraith, T. Small, R. A. Barnes, and V. Boekelheide, *J. Am. Chem. Soc.*, **83,** 453 (1961).
15. L. M. Jackman, Q. N. Porter, and G. R. Underwood, *Austr. J. Chem.*, **18,** 1221 (1965).
16. J. C. Godfrey. *J. Org. Chem.*, **24,** 581 (1959).
17. Ring Index, No. 9207.
18. Chem. Abstr. Subject Index 1957–1961, p. 1663s.
19. M. A. Jessep and D. Leaver, *Chem. Comm.*, 790 (1970).
20. R. M. Acheson and D. A. Robinson, *Chem. Comm.*, 175 (1967).
21. R. M. Acheson and D. A. Robinson, *J. Chem. Soc.* (*C*), 1663 (1967).
22. A. Crabtree, A. W. Johnson, and J. C. Tebby, *J. Chem. Soc.*, 2497 (1961).
23. P. N. Rylander, S. Meyerson, and H. M. Grubb, *J. Am. Chem. Soc.*, **79,** 842 (1957).
24. R. M. Acheson and R. S. Feinberg, *Chem. Comm.*, 342 (1965).
25. O. Diels and F. Möller, *Lieb. Ann.*, **516,** 45 (1935).
26. W. K. Gibson, D. Leaver, J. E. Roff, and C. W. Cumming, *Chem. Comm.*, 214 (1967).
27. V. Boekelheide and W. G. Gall, *J. Org. Chem.*, **19,** 499 (1954).
28. H. V. Hansen and E. D. Amstutz, *J. Org. Chem.*, **28,** 393 (1963).
29. V. Boekelheide, H. Fritz, J. M. Ross, and H. X. Kaempfen, *Tetrahedron*, **20,** 33 (1964).

30. D. Leaver and J. D. R. Vass, *J. Chem. Soc.*, 1629 (1965).
31. G. R. Underwood, *J. Org. Chem.*, **33,** 1313 (1968).
32. D. Farquhar and D. Leaver, *Chem. Comm.*, 24 (1969).
33. R. P. Cunningham, D. Farquhar, W. K. Gibson, and D. Leaver, *J. Chem. Soc.* (*C*), 239 (1969).
34. A. E. Tschitschibabin, *Ber.*, **59,** 2048 (1926).
35. J. P. Paolini and R. K. Robins, *J. Heterocycl. Chem.* **2,** 53 (1965).
36. T. W. G. Solomons, F. W. Fowler, and J. Calderazzo, *J. Am. Chem. Soc.*, **87,** 528 (1965).
37. V. Boekelheide and T. Small, *J. Am. Chem. Soc.*, **83,** 462 (1961).
38. O. Ceder, J. E. Andersson, and L. E. Johansson, *Acta Chim. Scand.*, **26,** 624 (1972).
39. O. Ceder and M. L. Samuelsson, *Acta Chim. Scand.*, **27,** 2095 (1973).
40. B. A. Hess, Fr., L. J. Schaad, and C. W. Holyoke, Jr., *Tetrahedron*, **28,** 3657 (1972).
41. P. Carles and A. Julg, *J. Chim. Phys. Physico-Chim. Biol.*, **65,** 2030 (1968).
42. R. D. Brown and B. A. W., Coller, *Molec. Phys.*, **2,** 158 (1958).
43. F. Gerson, E. Heilbronner, N. Joop, and H. Zimmermann, *Helv. Chim. Acta.*, **46,** 1940 (1963).
44. N. M. Atherton, F. Gerson, and J. N. Murrell, *Mol. Phys.*, **6,** 265 (1963).
45. F. Gerson and J. D. W. van Voorst, *Helv. Chim. Acta.*, **46,** 2257 (1963).
46. A. W. Hanson, *Acta Cryst.*, **14,** 124 (1961).
47. V. Boekelheide, F. Gerson, E. Heilbronner, and D. Meuche, *Helv. Chim. Acta.*, **46,** 1951 (1963).
48. J. A. Pople and K. G. Untch, *J. Am. Chem. Soc.*, **88,** 4811 (1966).
49. K. G. Untch and D. C. Wysocki, *J. Am. Chem. Soc.*, **89,** 6386 (1971).
50. H. C. Longuet-Higgins, "Aromaticity," Special publication No. 21, The Chemical Society, London, 1976, p. 109.
50a R. C. Haddon, *Tetrahedron*, **28,** 3613, 3635 (1972).
51. O. Ceder and J. E. Andersson, *Acta Chim. Scand.*, **26,** 611 (1972).
52. R. J. Thomas and F. A. Long, *J. Org. Chem.*, **29,** 3411 (1964).

CHAPTER V

Dithiole and Dithiolium Systems

R. D. HAMILTON

Ft. Lewis College,
***Durango, Colorado,* 81301**

E. CAMPAIGNE

Indiana University,
***Bloomington, Indiana,* 47401**

I. Introduction

The 1,2- and 1,3-dithiolium ions (**1** and **2**) are unsaturated 5-membered rings containing two sulfur atoms. These ions may be formally

1

2

derived either from their corresponding dithioles by hydride abstraction or from the nonbenzenoid carbocyclic tropylium ion (**3**) by replacement of two carbon–carbon double bonds by two sulfur atoms. In either case the procedure would generate a cyclic $4n+2\pi$-electron system which, in accordance with the rule formulated by Hückel,[1] would be expected to

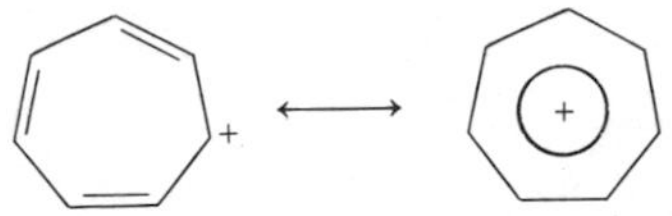

3

possess enhanced stability and to exhibit "aromatic" properties. The ability of sulfur to supply a pair of 3p π-electrons to a mesomeric bond system, and thereby to participate in conjugated systems, is well-documented.[2]

Much research interest has been shown in the dithiolium system over the past 15 years, owing largely to the realization of the close relationship between ions **1** and **2** and the tropylium ion (**3**). Among the first workers to take cognizance of this fact were Wizinger and Soder, who suggested as early as 1958 that "aryl-1,3-benzodithiol-1-sulfonium" salts could be considered as benzo-1,3-dithiolium derivatives.[3] Prior to that time only the pioneering work on the fused-ring benzo-1,3-dithiolium system by Hurtley and Smiles had been reported.[4–6] The parent, unsubstituted ions (**1** and **2**) were first synthesized by Klingsberg in 1960[7] and 1962.[8] Other work carried out during this period and subsequently has led to numerous research reports on synthetic, spectral, and mechanistic properties of this interesting class of compounds. Many of the data have been summarized and tabulated in previous review articles dealing with the 1,2- and 1,3-dithiolium ions,[9,10] and with electronic structure of heterocyclic sulfur compounds.[11]

II. Nature of Dithiolium Salts

A. Synthesis

Although a few preparative methods were known to produce substituted dithiolium salts prior to 1960, most of the synthetic work in this field has been conducted in the last 15 years. During this time procedures have been developed for the preparation of the parent, unsubstituted ions, **1** and **2**, as well as for many aryl, alkyl, methylthio, amino, and other substituted derivatives.

Many of the procedures available afford dithiolium salts in good to very good yields. Generally dithiolium salts are quite stable, crystalline compounds, which are readily purified by recrystallization. Perchlorate salts of these compounds, which sometimes decompose violently upon heating, would not generally be desirable for large-scale production, though they are often the salts of choice in research scale operations because of their high melting points and ease of purification.

1. *1,2-Dithiolium Salts*

The most important preparative scheme for 1,2-dithiolium systems—including the parent ion itself—involves a redox reaction first employed by Klingsberg.[7,12] In this method an appropriately substituted 1,2-dithiole-3-thione ("trithione") (**4a**) is treated with 40% peracetic acid in acetone solution. The extracyclic sulfur atom is eliminated as sulfate ion,

R^2 R^1 X S—S **4**

a: X = S
b: X = O

R^2 X$^-$ R^1 R^3 + S—S **5**

a: R^3 = H

and the dithiolium salt (**5a**) precipitates, a factor that probably accounts partially for the high yields.[12] Analogies to this reaction are found in the synthesis of thiazoles[13] and imidazoles[14,15] from appropriate 2-thione derivatives. Acid-catalyzed hydrolysis of an intermediate sulfonic acid might be involved.[16] Other reagents that have been used to effect similar oxidations are 30% H_2O_2 in glacial acetic acid,[17,18] methanol–sulfuric acid,[19] and chlorine in glacial acetic acid.[20] The effect of an alkaline oxidation medium on trithiones was previously found to lead to ring destruction.[21,22] Several procedures have been devised for the preparation of the precursor trithiones.[23–25] The 5-phenyl derivative is most readily prepared by a two-step sequence starting from ethyl cinnamate.[12]

A second procedure for the preparation of the parent 1,2-dithiolium ion (**1**) involves reaction of diacetal **6** with H_2S_2 in benzene solution saturated with hydrogen chloride.[26] A 67% yield of **1** is reported.

$$\underset{\mathbf{6}}{CH(OC_2H_5)_2-CH_2-CH(OC_2H_5)_2} + H_2S_2 \xrightarrow[C_6H_6]{HCl} \underset{\mathbf{1}}{\text{1,2-dithiolium}^{+}}\ Cl^-$$

Another major synthetic route to 1,2-dithiolium salts involves the reaction of higher hydrogen sulfides[17,26–28] or phosphorous pentasulfide[29,30] with 1,3-dicarbonyl compounds. Utilization of this procedure was reported for the first time by Leaver, Robertson, and McKinnon,[17] who prepared 3,5-diaryl and 3,5-arylalkyl-1,2-dithiolium perchlorates from β-diketones and hydrogen sulfide–hydrogen chloride in ether. Precursors used were dibenzoylmethane, benzoylacetone, and acetylacetone. Although no 1,2-dithiolium salts with purely aliphatic substituents were obtained in the work above, such compounds were synthesized by a similar procedure in later work.[28] Chloro- and amino-substituted 1,2-dithiolium salts may also be prepared under substantially the same conditions. Thus when dialdehyde **7** is reacted with higher hydrogen sulfides in ethanol–benzene solution saturated with hydrogen chloride, 4-chloro-1,2-dithiolium chloride (**5b**) is obtained in good yields.[26] Reaction of dicarbonyl compounds with P_2S_5 followed by acid yields 3-arylamino-1,2-dithiolium salts (5c).[29]

$$\underset{\mathbf{7}}{Cl-CH(CHO)_2} \xrightarrow[HCl]{H_2S_x} \underset{\mathbf{5}}{\text{3-}R^1\text{-4-}R^2\text{-5-}R^3\text{-1,2-dithiolium}^{+}\ X^-} \xleftarrow[HClO_4]{P_2S_5} \underset{\mathbf{8}}{R^1-C(=O)-CH_2-C(=O)-NHAr}$$

b: $R^1 = R^3 = H$; $R^2 = Cl$; $X = Cl$
c: $R^2 = H$, $R^3 = NHAr$, $X = ClO_4$

Mechanistic steps in the synthesis of 1,2-dithiolium salts form 1,3-dicarbonyl compounds and higher hydrogen sulfides (H_2S_x, $X \geqq 2$) may be represented formally as follows[28]:

$+H_2S_x(X \geq 2)$, $-H_2O$; $+HCl$; $+HCl$, $-Cl-S_{X-2}-H$; $HClO_4$, $-H_2O, Cl^-$; ClO_4^-; Cl^-

5

Another class of amino-1,2-dithiolium compounds—the 3,5-diamino-1,2-dithiolium salts (**10a**)—may be prepared in nearly quantitative yield by means of oxidation of dithiomalonamides (**9a**) with iodine in ethanol or hydrogen peroxide in acidic media.[31,32] *N,N′*-Diaryldithiomalonamides

RHN NHR → Oxid. → RHN NHR

9 **10**

a: R = H
b: R = Ar
c: R = $-COCH_3$

a: R = H
b: R = Ar
c: R = $-COCH_3$

(**9b**) are converted to 3,5-bis(arylamino)-1,2-dithiolium salts (**10b**) by this procedure, using iodine in ethanol as the oxidative medium.[33] 3,5-Diacetamido-1,2-dithiolium salts (**10c**) are also obtained by this procedure.[31] Derivatives of **10a** bearing substituents at the 4-position may also be obtained by oxidation of 2-substituted dithiomalonamides. However the precursor must have at least one free hydrogen on the C-2 atom to undergo ring closure; otherwise desulfurization occurs. This implies oxidation of the ene-thiol to a sulfenium ion as an intermediate step:

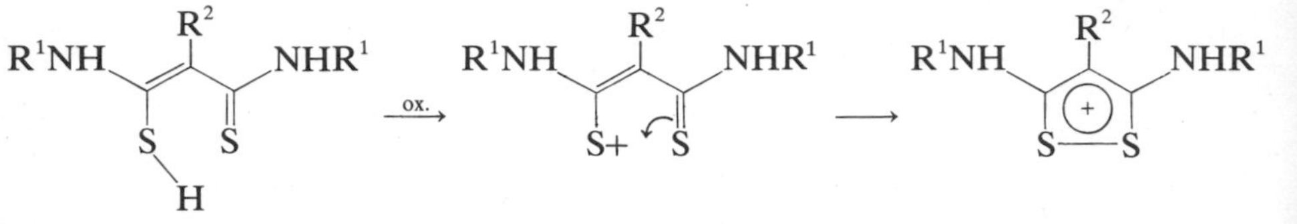

R^1NH–C(SH)=C(R^2)–C(=S)–NHR^1 (**9**) $\xrightarrow{\text{ox.}}$ R^1NH–C(S^+)=C(R^2)–C(=S)–NHR^1 ⟶ **10**

9 **10**

Highly labile 3-chloro-1,2-dithiolium compounds may be prepared from open-chain perchloro olefins or from cyclic trithiones. Thus when hexachloropropene was heated above 160° with sulfur, a 65% yield of 3,4,5-trichloro-1,2-dithiolium chloride was obtained,[34] first thought to be tetrachloro-1,2-dithiacyclopentene.[35] Compound **5d** has also been obtained in low yields by the reaction of heptachloropropane with sulfur and by the reaction of 1,1,1-trifluoro-2,3,3-trichloropropene with Cl_3CSH.[36] In addition, halogenation of trithiones (**4a**) with chlorine[37,38] or with

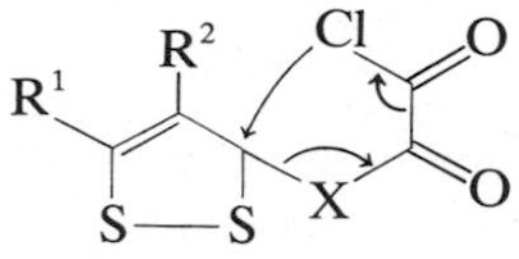

11: X = S or O

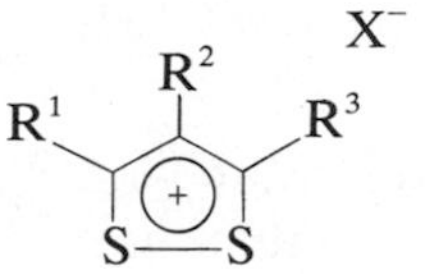

5d: $R_1 = R_2 = R_3 = Cl$; $X = Cl$
5e: $R_3 = Cl$; $X = Cl$
5f: $R_3 = SCH_3$

oxalyl halides[34] yields 3-halo (**5e**) derivatives instead of *gem*-dihalodithioles (**12**),[34,39] as was previously thought.[37,38] The latter reaction most likely proceeds via an intermediate of type **11** formed by nucleophilic attack of the 3-substituent of **4** on the carbonyl carbon atom of the oxalyl halide.[39] Hydrolysis-sensitive acetoxy derivatives of type **13** are

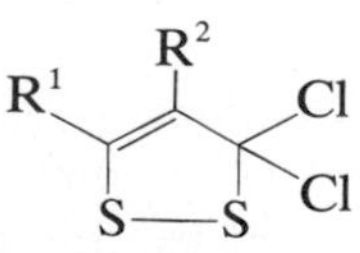

12

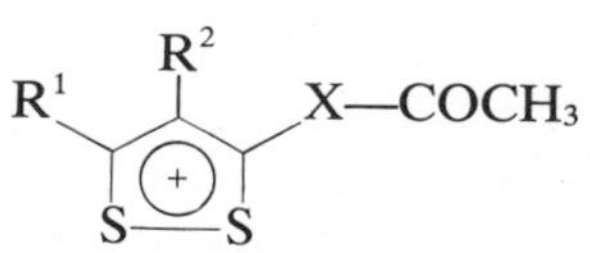

13: X = S or O

obtained when **4** is heated with acetic anhydride and then with perchloric acid.[34] Another class of highly reactive 1,2-dithiolium compounds—the 3-alkylthio-1,2-dithiolium salts (**5f**)—may be prepared by alkylation of corresponding trithiones using alkylating agents such as dimethyl sulfate, methyl iodide, or triethyloxonium borofluoride.[41–44]

Other reactions are also known to generate 1,2-dithiolium ions. For example, Knauer, Hemmerich, and Van Voorst[45] found that a violet complex of dithioacetylacetonatoiron (**14**) when dissolved in water at pH 3, gave a colorless solution containing the 3,5-dimethyl-1,2-dithiolium ion **5g**. The following equations describe the transformation:

$$2CH_3COCH_2COCH_3+4H_2S+4FeCl_4^- \rightarrow \underset{\mathbf{14}}{Fe(C_5H_7S_2)_2Cl_4}+3FeCl_4^{2-}+2H_3O^+ +2H_2O$$

$$Fe(C_5H_7S_2)_2Cl_4 \xrightarrow{H_2O} Fe^{2+}+4Cl^- +2(C_5H_7S_2)^+$$

$$\mathbf{5g}\text{: } R^1=R^3=CH_3,\ R^2=H$$

In the presence of halogen or sulfuryl chloride, dehydrogenation accompanied by ring contraction of 1,3-dithiacyclohexanes leads to a 4-substituted 1,2-dithiolium salt.[46] Addition products of trithiones with salts of heavy metals have been regarded by various authors as 1,2-dithiolium salts.[46–49]

2. *1,3-Dithiolium Salts*

Various routes are available for preparation of 1,3-dithiolium salts, with the method of choice depending largely on the type of nuclear substitution desired. Synthetic schemes can be roughly classified into three major groups: those producing 2-unsubstituted 1,3-dithiolium salts, those generating 2,4-disubstituted-1,3-dithiolium compounds, and finally those leading to 2-substituted benzo-1,3-dithiolium salts.

1,3-Dithiolium salts that are unsubstituted in the 2-position are conveniently accessible through oxidation of an appropriate 1,3-dithiole-2-thione ("isotrithione") intermediate according to the method of Klingsberg.[8] In this method, in analogy with the 1,2-series, oxidation of the isotrithione by means of 40% peracetic acid in acetone yields 1,3-dithiolium hydrogen sulfates. Even though the procedure has been employed only in the preparation of the parent 1,3-dithiolium ion,[8] it should be capable of extension to afford a wide variety of 1,3-dithiolium compounds.

Limitations in the procedure for the 1,3-dithiolium ion stem largely from the laborious procedures required for preparation of isotrithione precursors. To illustrate, 1,3-dithiole-2-thione (**17a**) may be prepared from 4,5-dicyano-1,3-dithiole-2-thione[8] (**16**) which in turn is obtainable by the action of thiophosgene on disodium dimercaptomaleonitrile **15**, itself accessible from sodium cyanide and carbon disulfide.[50] In an alternate procedure, bromination of ethylene dithiocarbonate **18** with *N*-bromosuccinimide to form **19**, followed by dehydrobromination with a

NC, NC, S^-Na^+, S^-Na^+ $\xrightarrow{CSCl_2}$ NC, NC, S, S, =S ⟶ S, S, =X

15 **16** **17**

a: X = S
b: X = O

tertiary amine in carbon tetrachloride yields 1,3-dithiole-2-one, **17b**.[51] Phosphorous pentasulfide converts **17b** into **17a** in 80% yield. Treatment of β-keto di- and trithiocarbonates with phosphorus pentachloride in refluxing tetralin furnishes another method of synthesizing 1,3-dithiole-2-thiones, though yields are generally low.[17,27] Compound **17a** has been

S, S, =O $\xrightarrow{NBS}$ Br, H, S, S, =O $\xrightarrow[CCl_4]{\text{tert-amine}}$ S, S, =O $\xrightarrow{P_2S_5}$ **17a**

18 **19** **17b**

obtained in low yield by the reaction of sodium acetylide, sulfur, and carbon disulfide at room temperature,[52] and has been obtained from the high boiling fraction (140–165°/15 mm) of a reaction mixture consisting of sulfur and acetylene.[40] An interesting new compound—a selenium analog of the isotrithiones—has been isolated from the reaction mixture of sodium phenylacetylide, powdered selenium, and carbon disulfide.[53] Whether this compound can be converted into a dithiolium analog has not yet been disclosed. Additional sequences leading to isotrithiones have been described.[4,17,27,51,52,54–56] Benzoisotrithiones, on the contrary, are readily obtainable via commercially available benzene-1,2-dithiols by the action of carbon disulfide in aqueous base.[54]

A second method by which 2-unsubstituted-1,3-dithiolium salts may be prepared was elucidated by Hurtley and Smiles in pioneering work in 1926.[4,5] In an acid-catalyzed condensation reaction between benzene-1,2-dithiol and ethyl orthoformate, a deeply colored solution containing benzo-1,3-dithiolium chloride was obtained. The chloride, too soluble to be isolated, was converted into the zinc chloride and chloroplatinate double salts, which precipitated from the reaction mixture. In a similar procedure, Soder and Wizinger found that benzene-1,2-dithiol, when heated first with formic acid and then with perchloric acid, yielded benzo-1,3-dithiolium salts **20a**.[54] A likely mechanism would involve formation of an intermediate thioester, followed by carbonyl protonation, cyclization, and elimination of water. In recent work Pazdro[57] utilized this reaction in the preparation of 4,5-diphenyl-1,3-dithiolium perchlorate (**22**) starting

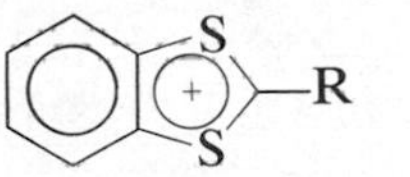

20 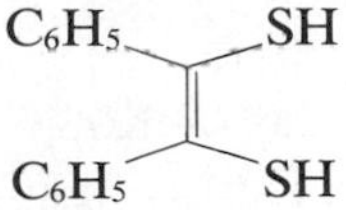 **21** 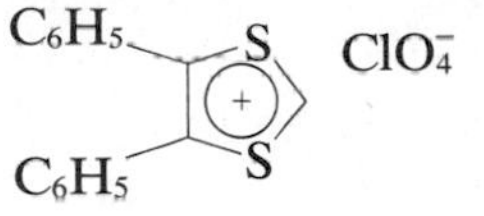 **22**

a: R = H
b: R = C_6H_5
c: R = CH_3

from *cis*-1,2-diphenylethylene-1,2-dithiol (**21**). A third method of preparation of 2-unsubstituted-1,3-dithiolium compounds involves reduction of 2-dialkylamino-1,3-dithiolium salts with sodium borohydride to give 2-amino-1,3-dithioles, which, on treatment with acids, afford 1,3-dithiolium salts instantly and in high yields.[58] Because of the nature of preparation of the 2-dialkylamino intermediates, the 1,3-dithiolium products are substituted in the 4-position with an aryl group.

An efficient method of obtaining 2,4-disubstituted-1,3-dithiolium salts involves acid-catalyzed cyclization of β-keto (**23g**), cyanomethyl (**23h**), or carboxymethyl (**23i**) dithioesters.[17,27,59,60] Campaigne, Jacobsen, and Hamilton found that the reactions in general proceeded rapidly (≤ 5 min) and in high yields in warm 70% perchloric acid or sulfates.[59,60] The 2-substituent of the products may consist of an alkyl, aryl, dialkylamino, or alkylthio group, whereas the 4-substituent may be an alkyl, aryl, or hydroxy group; not all combinations of these substituents are possible, however. The mechanism of this reaction under the above conditions was considered to be an acid-catalyzed cyclodehydration.[59] The perchlorate salts described in the above work were fairly stable in the crystalline state and in nonhydroxylic solvents, but in water and ethanol reverted to starting material.

23

24

a: $R = Ar, R^1 = CH_3$
b: $R = CH_3, R^1 = Ar$
c: $R = C_2H_5O, R^1 = Ar$
d: $R = CH_3S, R^1 = Ar$
e: $R = (CH_3)_2N, R^1 = Ar$
f: $R = (CH_3)_2N, R^1 = CH_3$
g: $R^1 = R$ or Ar
h: $R^1 = CH_2CN$
i: $R^1 = CH_2COOH$

Earlier work employing this general approach was reported in 1960 and 1962 by Leaver, Robertson, and McKinnon.[17,27] The authors synthesized several aryl-substituted 1,3-dithiolium salts using two sets of experimental conditions: the first involved treatment of an ethereal solution of **23** with dry gaseous hydrogen chloride and hydrogen sulfide for several days, whereas the second consisted of treating **23** with hydrogen sulfide and perchloric acid in boiling glacial acetic acid for several hours. Under these conditions, the conversion **23** → **24** failed unless either R or R_1 was aryl, and when phenacyl alkyl di- (**23c**) and tri- (**23d**) thiocarbonates were used as precursors. The yield of product was lower in the absence of hydrogen sulfide, and the authors attributed its function to conversion of the β-carbonyl group into a thiocarbonyl group, thus allowing the reaction to proceed as follows:

When hydrogen sulfide was not added, its formation was postulated by decomposition of half of the starting material. Dithiolium precursors used in the synthetic procedures above were prepared by allowing sodium or potassium salts of appropriate dithioacids or trithiocarbonates to react with phenacyl halides, chloroacetic acid, chloroacetone, or chloroacetonitrile.[17,27] Starting from closely related starting materials, S-α-oxoalkyl thioesters (**25**), alkyl, and aryl-substituted 1,3-dithiolium salts have been obtained generally in low yields, in a mixture of perchloric acid and glacial acetic acid.[17,61] Since yields never exceed 50%, it is likely that a large part of **25** is consumed in converting unreacted monothioester into dithioester intermediate. Addition of hydrogen sulfide to provide extra sulfur during the cyclization increases the yield substantially.

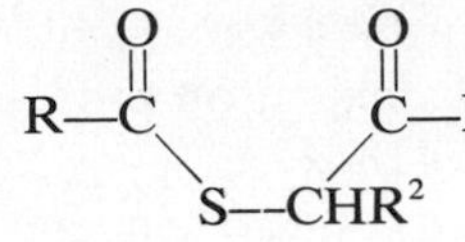

$$\mathrm{R{-}C({=}O){-}S{-}CHR^2{-}C({=}O){-}R^1}$$

25

Photolysis of aryl-substituted 1,2,3-thiadiazoles (**26**) gives rise to intermediate dithioles **27a**, which on protonation yield corresponding dithiolium derivatives (**28**).[62]

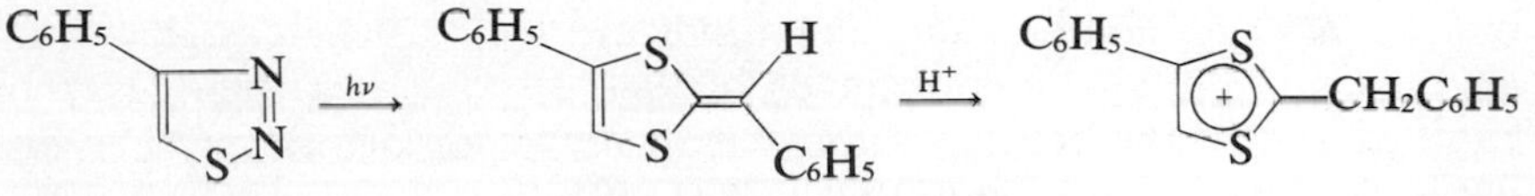

C_6H_5 N N S $\xrightarrow{h\nu}$ C_6H_5 S S H C_6H_5 $\xrightarrow{H^+}$ C_6H_5 S (+) S $-CH_2C_6H_5$

26 **27a** **28**

Closely related to the 1,3-dithiolium cation is the 1,3-oxathiolium ion (**31**). A synthetic route to the latter system has recently been reported which should be capable of wider application to produce a 2-alkylamino-1,3-dithiolium salt.[63] In this sequence *S*-chloroisothiocarbamoyl chlorides (**29**) are reacted with aliphatic and cycloaliphatic ketones (**30**):

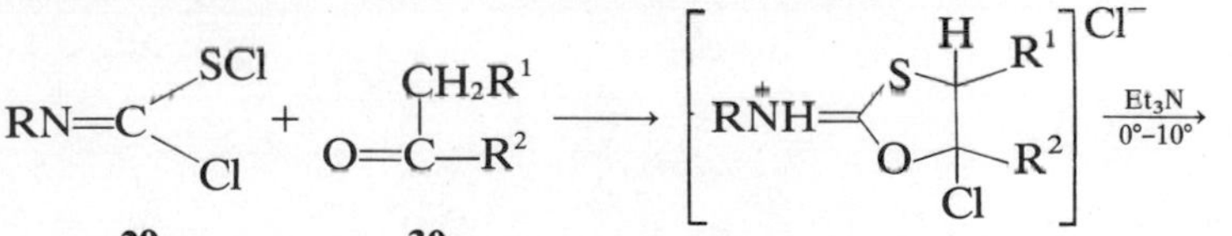

$\mathrm{RN{=}C(SCl)Cl}$ + $\mathrm{O{=}C(CH_2R^1){-}R^2}$ $\longrightarrow$ $\left[\mathrm{R\overset{+}{N}H{=}}\text{(S, O ring; H, }R^1\text{; Cl, }R^2)\right]Cl^-$ $\xrightarrow[0°-10°]{Et_3N}$

29 **30**

RN= (S, O ring; H, R^1; Cl, R^2) $\xrightarrow{>0°C}$ $\left[\text{RNH}-\text{(S, (+), O ring; } R^1, R^2) \longleftrightarrow \mathrm{RH\overset{+}{N}{=}}\text{(S, O ring; } R^1, R^2) \right] Cl^-$

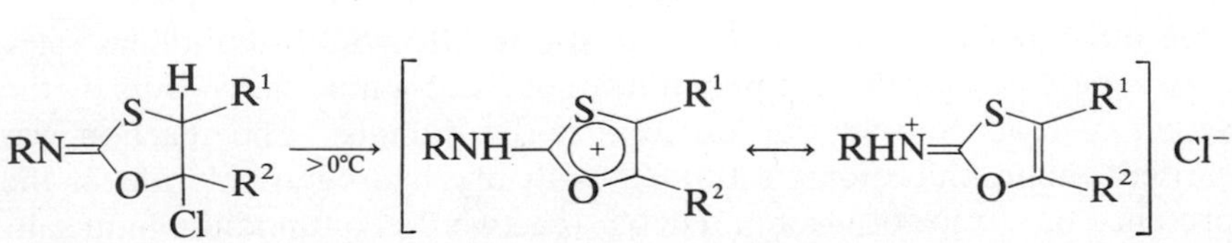

31

2-Substituted benzo-1,3-dithiolium salts constitute a third class of 1,3-dithiolium salts, the preparation of which has been researched in quite some detail. Indeed, the first 1,3-dithiolium salts isolated were the fused-ring benzo derivatives **20b**, which were obtained by Hurtley and Smiles upon treatment of benzene-1,2-dithiol with benzaldehyde in alcohol saturated with hydrogen chloride, followed by nitric acid.[4,5] In this acid-catalyzed condensation reaction an intermediate, 2-phenyl-1,3-benzodithiole, was isolated and oxidized to produce 2-phenylbenzo-1,3-dithiolium nitrate **20b**. When the aromatic aldehyde reactant was substituted in the *para* position, the oxidation process was favored by electron-releasing groups such as *p*-methoxy or *p*-hydroxy; a *p*-nitro substitutent had the opposite effect. A more facile sequence leading to the fused-ring 1,3-dithiolium salts is by condensation of benzene-1,2-dithiols with aliphatic and aromatic carboxylic acids.[3,54,64,65] The reaction is usually performed in refluxing phosphorus oxychloride, and proceeds rapidly and in good yield. The use of acetic acid as a reactant yields the 2-methyl derivative (**20c**), aromatic acids produce 2-aryl derivatives (**20d**), cinnamic acids afford 2-styryl compounds (**20e**),[3,64] and a dibasic acid such as malonic acid gives dithiolium compounds of type **32**, whereas glutaconic acid leads to dithiolium compounds of type **33**.[54,65] Aryl chlorides, too, can be used in the condensation reaction in place of carboxylic acids, often with an improvement in yield.[54,65]

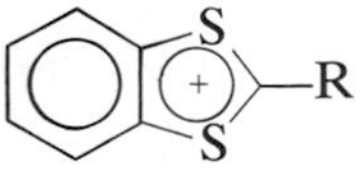

20

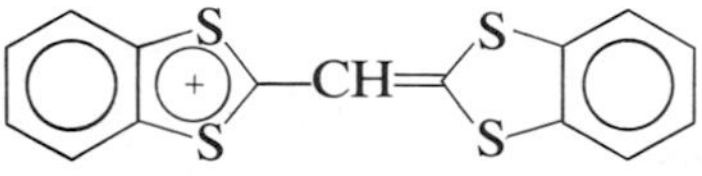

32

c: R = CH_3
d: R = Ar
e: R = CH = CHAr

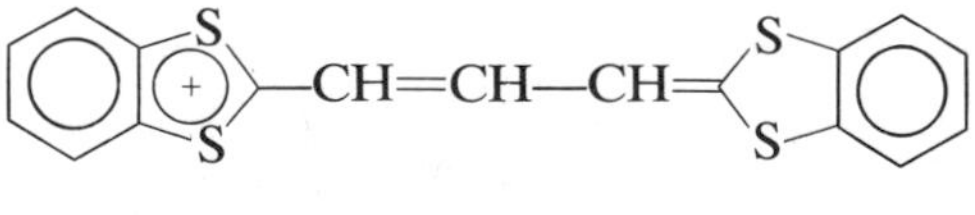

33

In other reactions leading to 2-substituted benzo-1,3-dithiolium salts, note should be made of the preparation of the 2-amino derivatives by the action of cyanogen chloride on toluene-3,4-dithiol.[66] The reaction was carried out in chloroform saturated with dry hydrogen chloride in the presence of a trace of alcohol. Highly reactive 2-alkylthiodithiolium salts have been obtained by alkylation of the extracyclic sulfur atom in 1,3-dithiole-2-thiones and benzo-1,3-dithiole-2-thiones.[40,54,65,67] Either dimethyl sulfate[54,65,67] or methyl iodide[40] in polar solvents was employed. Yields were nearly quantitative; the dithiolium ion was usually isolated as a perchlorate salt. Reaction between toluene-3,4-dithiol and methyl chlorothioformate in 70% perchloric acid produced 2-methylthio-5-methylbenzo-1,3-dithiolium perchlorate, but in low yield.[88] 2-Halo-1,3-dithiolium salts, analogous to 3-halo-1,2-dithiolium compounds, have not yet been reported.

B. Chemical Reactivity

1. *1,2-Dithiolium Salts*

As electron-deficient compounds, 1,2-dithiolium salts are highly reactive and unselective toward nucleophilic reagents. Theoretical studies have shown that the ion is best represented as a hybrid of carbonium (**34**) and sulfonium resonance structures (**35**), **34** being more important, and that the reactive site is the C-3 (or C-5) carbon atom.[69] Chemical experimentation, making use of a variety of nucleophiles, has been

34  **35**

carried out on the ring system and has borne out these conclusions. As a consequence of the highly electron-deficient C-3 atom, hydrogen atoms on attached methyl, methylene, or methine groups are labile and readily undergo condensation reactions with carbonyl reagents.

Owing to the presence of the disulfide linkage (–S–S–), certain basic reagents lead to ring-opening reactions, in which elemental sulfur is eliminated. In this connection probably the greatest distinctions between chemical reactivity of the 1,2- and 1,3-dithiolium series arise as a result of the disulfide bond. Electrophilic reagents (e.g., H_2SO_4) have no effect on the 1,2-dithiolium ion, even after long periods of storage. Most of the chemical reactions described in the following sections are based on aryl-substituted 1,2-dithiolium salts rather than the unsubstituted ion. Aryl substituents in the ring would be expected to stabilize certain transition states, thereby directing the course of the reaction along a preferred path. For this reason a particular reaction may not be capable of general application.

Carbon nucleophiles, which are most often generated from active methylene compounds, attack the 1,2-dithiolium nucleus at the C-3 position, as expected, and the products—3-ylidene compounds—may be considered as "1,2-dithiafulvalenes" or as "thiothiophthalene" derivatives. Accordingly, these chemical reactions are discussed in a later section

(Sections IIIA and B). A second type of carbon nucleophile that reacts with the 1,2-dithiolium nucleus is the electron-rich *para*-carbon atom of *N,N*-dialkylanilines,[34,70,71] phenols, or phenolic ethers.[68] In this reaction *N,N*-dialkylanilines react with 3-unsubstituted,[70] 3-methylthio-,[71] or 3-chloro-[34] 1,2-dithiolium salts with the elimination of hydrogen, methylmercaptan, or hydrogen chloride, respectively, to form highly conjugated dithiolium salts of type **36**. These products may be obtained directly from

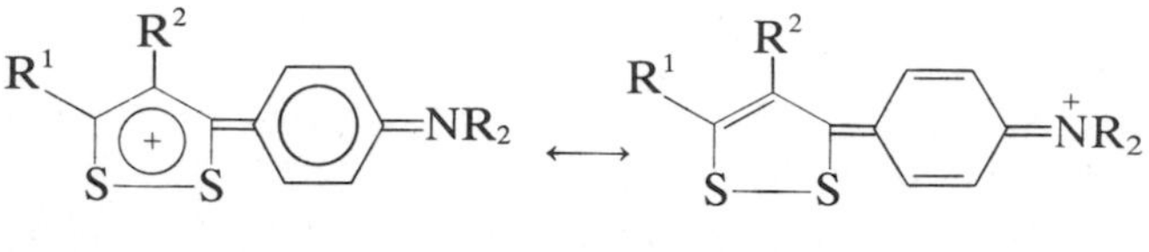

36

trithiones provided that catalytic amounts of trichloromethylsulfuryl chloride are present.[70] The effect of organometallic reagents such as methyl or phenyl lithium or the Grignard reagent on 1,2-dithiolium salts has not been discussed in the literature. 1,3-Dithiolium salts have been treated with Grignard reagent,[72] as discussed later (Section II.B.2.).

Nucleophilic nitrogen compounds that easily undergo reaction with the 1,2-dithiolium nucleus include ammonia, primary, and secondary amines, hydrazines, and *N,N′*-disubstituted hydrazines. There are no reports of the action of phosphorus nucleophiles on 1,2-dithiolium salts. Reaction of 1,2-dithiolium salts with ammonia takes several courses, depending on the number and nature of the dithiolium ring substituents, the solvents, and reaction conditions.

Conversion of 1,2-dithiolium compounds to isothiazoles by the action of ammonia in ether has been reported.[27,73–75] Thus for example, 4-phenylisothiazole (**38**) was obtained in 88% yield, starting from **37**, and 3,5-diphenylisothiazole was obtained in a yield of about 50% by this method.[74a] A saturated solution of ammonium acetate in acetic acid has

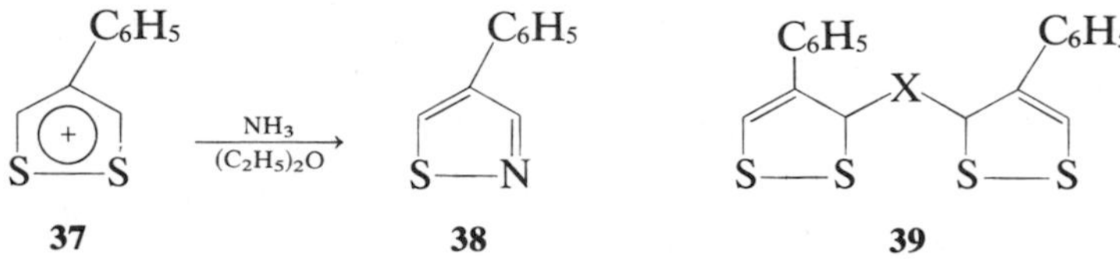

NH_3, $(C_2H_5)_2O$

37 **38** **39**

a: X = 0
b: X = S

also been used to effect these conversions.[74a] For example, 3,4-dimethyl-1,2-dithiolium ion yields 3,4-dimethylisothiazole in 15% yield, but 80% of

4,5-dimethylisothiazole.[74b] Analogies to this reaction are found in the reaction of pyrylium[76] and thiapyrylium[77] salts with ammonia. Of the possible mechanistic routes, a simple addition-elimination, followed by ring closure via displacement of sulfhydride ion, is most reasonable.[74a] Treatment of 1,2-dithiolium salts with hydroxylamine gave a mixture of isoxazoles and isothiazoles.[74a] Aqueous ammonia in the presence of ammonium chloride (pH of reaction mixture, ~8) effected a conversion of **37** into products **38**, trithione, and **39**.[78] Reaction of unsubstituted 1,2-dithiolium ion with ammonia did not yield the parent isothiazole system, however. A side reaction of 5-phenyl-3-methyl-1,2-dithiolium salts with ammonia was abstraction of methyl proton.[74a] When a nonpolar solvent such as benzene is used, the major product (60%) is the bisdithiolyl sulfide **39b** formed by the reaction of two dithiolium cations with one molecule of hydrogen sulfide, which is eliminated during formation of isothiazole.[73] In analogous fashion ditropyl sulfide is formed from tropylium cation and hydrogen sulfide.[79] In a different reaction ammonia abstracts a mole of perchloric acid from the *N*-arylamino-1,2-dithiolium perchlorate **40**, thus generating imine **41a**.[29] Primary and secondary amines such as aniline, *N*-methylaniline, or piperidine react like ammonia toward 3(5)-phenyl-1,2-dithiolium salts, leading to aminothione compounds such as **42**, formed by ring cleavage and elimination of elemental sulfur.[73,80]

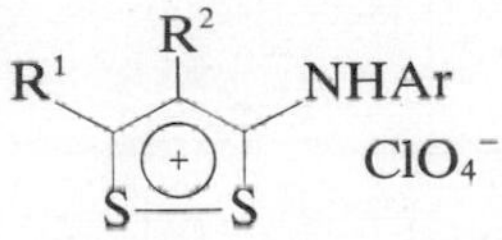

R^1, R^2, NHAr, ClO_4^-, S—S

40

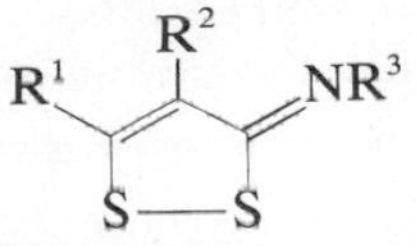

R^1, R^2, NR^3, S—S

41

C_6H_5, C_6H_5, S, NHR

42

a: $R^3 = Ar$
b: $R^3 = C_6H_5$

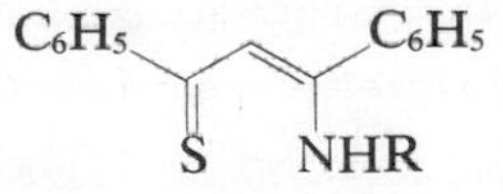

3-Methylthio-1,2-dithiolium salts react with primary amines and hydrazines to produce C=N bonded derivatives such as **41**.[47,81] These Schiff bases are obtained in boiling glacial acetic acid and in the presence of a tertiary amine in the cases of less basic amino acids and hydrazines.[81] Alkylation of **41** produces dialkylamino salts (**43**). The 1,2-dithiolium ion

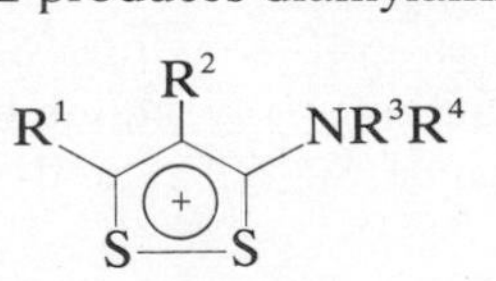

R^1, R^2, NR^3R^4, S—S

43

RNH, S, Ar, SCH_3

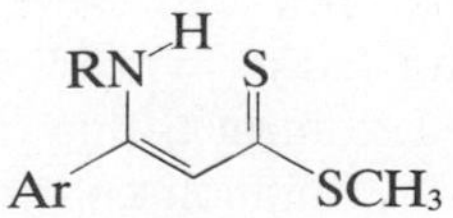

44

reacts with urea to form the expected imines.[83] 3-Methylthio-5-aryl-1,2-dithiolium salts have been found to form mixtures consisting of trithiones, methyl dithioacrylates (**44**), and imines (**41**) when treated with primary amines.[84] Reaction of 3-chloro-1,2-dithiolium ion with aniline yields imine **41b**.[34] Dithiolium derivative **45** reacts with aniline with ring cleavage to produce a cyclopentenedithioester (**46**) in 64% yield.[48] It appears

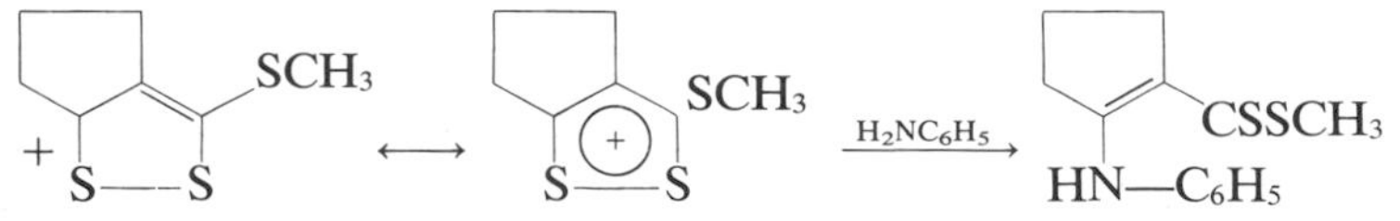

SCH_3 SCH_3 $\xrightarrow{H_2NC_6H_5}$ $CSSCH_3$ $HN—C_6H_5$

45a **45** **46**

that the reaction is unique with this particular fused-ring starting material in which the strain imposed by the 5-membered carbocyclic ring fused to the 4,5-position enhances the contribution of canonical form **45a,** with positive charge localized at the 5-position. Substituted hydrazines in the absence of water react exothermically with the parent, unsubstituted 1,2-dithiolium ion to afford substituted pyrazoles **47**.[7,12] A likely mechanism has been proposed:

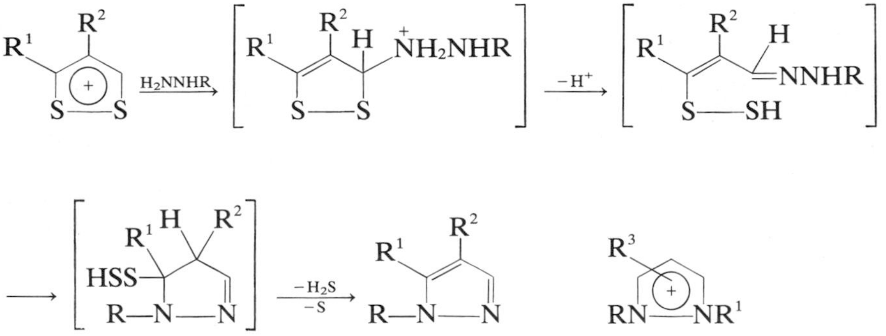

$\xrightarrow{H_2NNHR}$ $\overset{+}{N}H_2NHR$ $\xrightarrow{-H^+}$ $=NNHR$ $\xrightarrow[-S]{-H_2S}$

47 **48**

Other workers have proposed a similar sequence and have given evidence in its support.[86] *N,N′*-Disubstituted hydrazines react analogously with 4-phenyl-1,2-dithiolium salts at −20° to −40°, giving *N,N′*-disubstituted pyrazolium salts (**48**).[87] In these reactions only about half the starting material is converted to pyrazolium salts (50–65% yields), the remainder being sulfurated to form trithiones.

Reactions of 1,2-dithiolium compounds with secondary amines in tetrahydrofuran have been described.[88] Both the 1,3- and 1,2-dithiolium compounds exhibit similar behavior in this reaction, though yields are somewhat lower in the 1,2-series. This has been attributed to the formation of decomposition products along path 2 (below), which appears to be more important in the 1,2-series than in the 1,3-series.[88] Under slightly different conditions dimethylamine reacts with 3-aryl-1,2-dithiolium salts with ring-opening to produce mixtures consisting of 3-dimethylamino-1-aryl-2-propene-1-thiones and substituted ammonium perchlorates.[82]

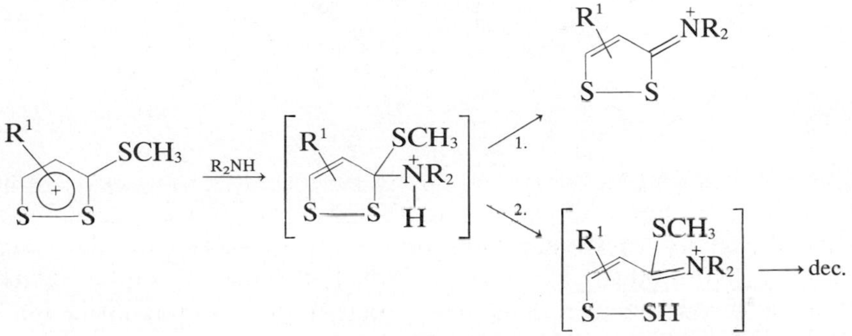

R^1 SCH_3 S S $\xrightarrow{R_2NH}$ [R^1 SCH_3 $\overset{+}{N}R_2$ S—S H] 1. → R^1 $\overset{+}{N}R_2$ S—S ; 2. → [R^1 SCH_3 $\overset{+}{N}R_2$ S—SH] ⟶ dec.

Reaction of 1,2-dithiolium salts with tertiary aromatic amines was discussed earlier in this section. Tertiary amines such as pyridine or acridine are methylated by 3-methylthio-1,2-dithiolium salts.[44,47] 3-Phenyl-1,2-dithiolium hydrogen sulfate, when treated with triethylamine in a water–ethanol mixture, affords a low yield (8%) of 2,5-diphenyl-1,6,6a,S^{IV}-trithiapentalene-3-carboxaldehyde (**49a**).[89] Similar reactions have been achieved by pyridine in water (see Section III.A).[90] Pyridine in ethanol affords a mixture of simple dithiole, 3-ylidine compound, and dimeric disulfide.[85]

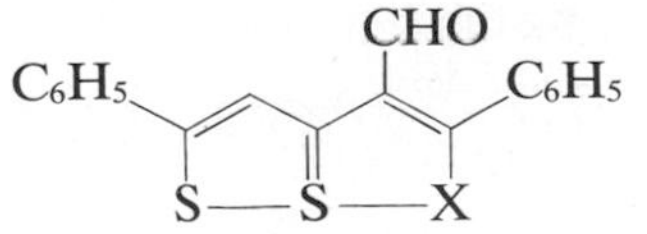

CHO C_6H_5 C_6H_5 S—S—X

49

a: X = S
b: X = 0

1,2-Dithiolium salts are extremely sensitive toward aqueous bases—that is, oxygen nucleophiles—and decompose rapidly with liberation of elemental sulfur. It appears likely that hydroxide ion attacks position 3 (or 5) to form an unstable hydroxy-1,2-dithole, which decomposes via ring cleavage to alkali-sensitive α,β-unsaturated ketones, which in turn lose elemental sulfur.[91] Direct attack of hydroxide ion on the disulfide linkage cannot be discounted;[92] however from studies of the acidic hydrolysis of 3-chloro-1,2-dithiolium salts, in which 1,2-dithiole-3-ones are produced, it appears likely that the C-3 carbon atom is first involved.[20,34–36,38,93] In this case the intermediate **50** is stabilized by expulsion of hydrogen chloride instead of ring opening. The interception of the

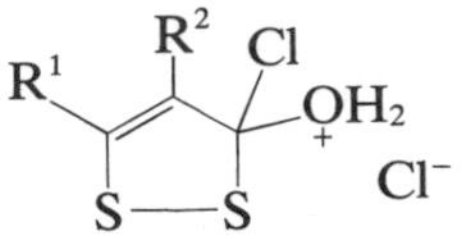

50

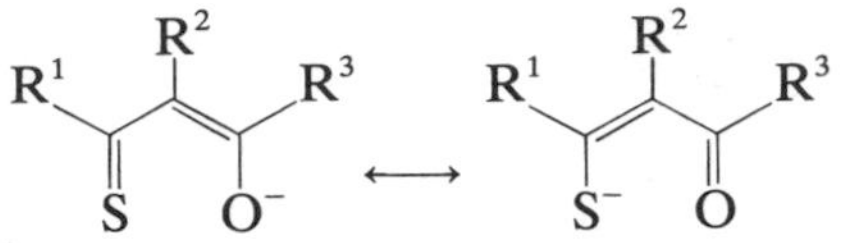

51

ambident anion **51** by benzyl chloride in the case of 3,5-diphenyl-1,2-dithiolium adds support to the proposed mechanism.[73] Like hydroxide ion, ethoxide ion also causes decomposition of the 1,2-dithiolium ion[73]; only the sulfurized product, 4-phenyl-1,2-dithiole-3-thione, could be isolated from the corresponding dithiolium ion upon reaction with ethoxide ion.[87] With the 3,5-diaryl substituted salt **5h**, it is possible to isolate the colorless ethoxydithiole **52**, which can be reconverted into the original salt by the action of perchloric acid.[73]

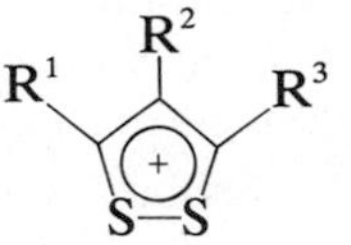

5

h: $R^1 = R^3 = C_6H_5$, $R^2 = H$
i: $R^1 = R^3 = H$, $R^2 = C_6H_5$, $X = HSO_4^-$
j: $R^1 = R^3 = H$, $R^2 = C_6H_5$, $X = ClO_4^-$
k: $R^1 = C_6H_5$, $R^2 = H$, $R^3 = CH_2C_6H_5$
m: $R^1 = C_6H_5$, $R^2 = H$, $R^3 = CH_3$

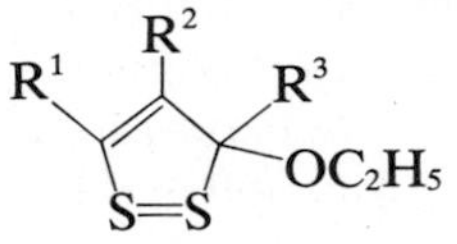

52

a: $R^1 = R^3 = C_6H_5$, $R^2 = H$

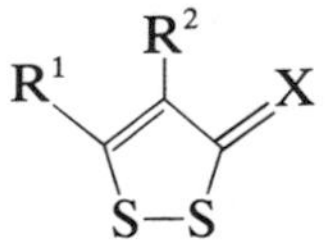

4

c: $R^1 = C_6H_5$, $R^2 = H$, $X = CHC_6H_5$
d: $R^1 = C_6H_5$, $R^2 = H$, $X = CH_2$

Exposure of **5i** to aqueous sodium bicarbonate for 30 min affords a yellow crystalline ether, **39a**, which gives **5j** in high yield upon treatment with perchloric acid.[18,78] This behavior is similar to that reported in the benzo-1,3-dithiolium series by Hurtley and Smiles.[5]

Sulfur nucleophiles have much the same effect as oxygen nucleophiles on 1,2-dithiolium salts. For example, 3,5-diamino-1,2-dithiolium ion is rapidly decomposed by hydrosulfide ion, the only reaction product isolated being dithiomalonamide.[31] Thioether **39b** was obtained when hydrogen sulfide was bubbled into an aqueous solution of **5i**.[78] Similar thioethers are obtained by the action of Na_2S_2 on substituted 1,2-dithiolium salts.[18] 3-Methylthio-4-substituted-1,2-dithiolium iodides react with sulfur to produce 5-methylthio 4-substituted trithiones.[94] 3-Halogenated dithiolium salts react with hydrogen sulfide to form trithiones.[34] Mercaptotrithiones can also be prepared.[95a] Detection of various anions (e.g., $Fe(CN)_6^{4-}$) is possible using 4-(*p*-tolyl)-1,2-dithiolium hydrosulfate.[95b]

The positively charged 1,2-dithiolium ring renders groups such as CH, OH, SH, and NH acidic if they are attached to the 3- or 5-position. For example, benzyl derivative (**5k**) is deprotonated to the stable, crystalline 1,2-dithiafulvene (**4c**) by aqueous soda solution. Depending on nuclear substitution, some 1,2-dithiafulvenes are unstable. Such is the case with **4d**, obtained from 3-methyl-5-phenyl-1,2-dithiolium ion (**5m**) by the action of ammonia. Reactivity of the 3-methyl group in **5m** finds parallels in the case of condensation of the methyl-substituted pyridinium,[96] pyrylium,[76] thiapyrylium,[77] and tropylium[97] salts. To illustrate, compound **51** reacts smoothly with *p*-dimethylaminobenzaldehyde to produce styryl-substituted **53**[27] and with methyl dithiobenzoate to form "thiothiophthene" **54**.[86] When compound **45** is treated with aldehydes in glacial acetic acid, formation of intensely colored compounds of type **55** ensues; these react further with primary amines to give cyanine dyes such as **56**.[85]

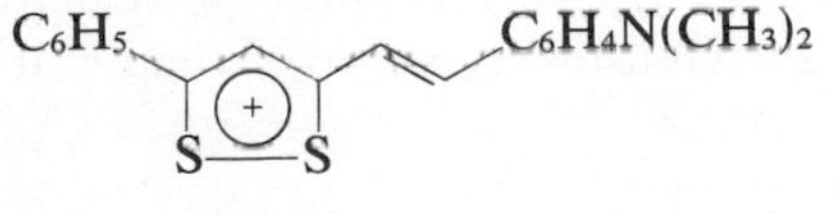

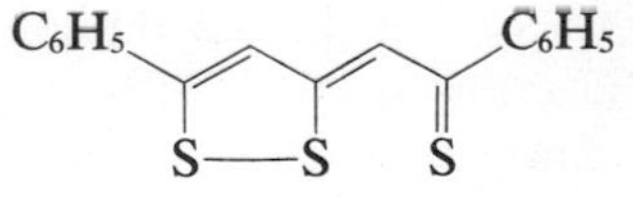

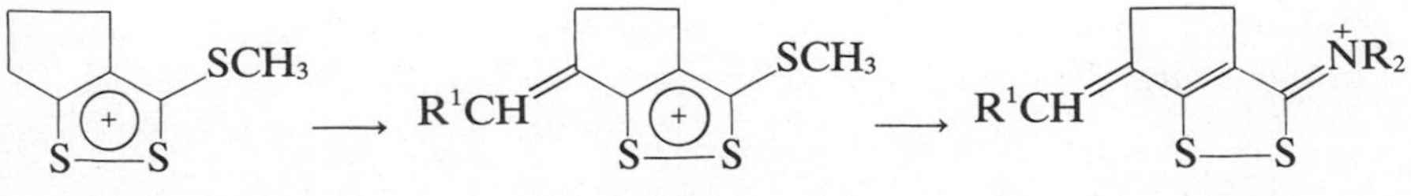

3,5-Diamino-1,2-dithiolium salts (**10a**) may be converted smoothly into *N,N′*-diacetyl salts (**10c**),[31] in which the N–H bonds are acidic. Reaction of **10c** with ammonia affords the stable base, **57**, for example,

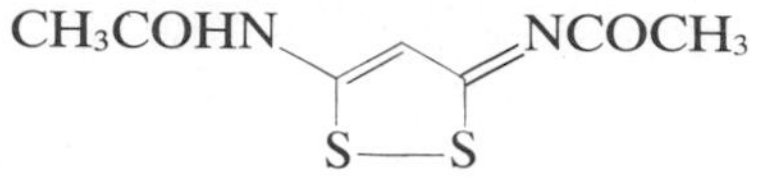

57

2. *1,3-Dithiolium Salts*

Chemical reactions undergone by 1,3-dithiolium salts are often analogous to those of the 1,2-series. One factor that distinguishes the two systems is the disulfide bond of the 1,2-series, as discussed earlier. Like the 1,2-series, 1,3-dithiolium salts partake of two main resonance hybrids, the sulfonium **58** and carbonium **59** structures, so that the positive charge is delocalized over only one carbon atom (i.e., the –S–C–S– grouping). The electron deficiency in the corresponding carbonium–sulfonium structures of the 1,2-dithiolium ion, on the other hand, is

58 **59**

distributed over two equivalent carbonium structures. It might be expected, then, that the electrophilic character of the equivalent C-3 and C-5 positions in the 1,2-series would be weaker than that of C-2 in **59**; chemical experimentation has borne out this expectation, for electrophilic character of the 1,3-dithiolium salts is distinctly stronger than that of the 1,2-series.

Several reactions undergone by the 1,3-dithiolium system are also observed with the saturated system, ethylenetrithiocarbonate. An attempt has been made to resolve the question as to whether **60**, in which the dipolar form **60b**, permits positive-charge delocalization over the entire ring, shows particular stability or "aromaticity" compared to **61**, in which positive charge delocalization occurs over only the –S–C(S)–S– linkage.[98]

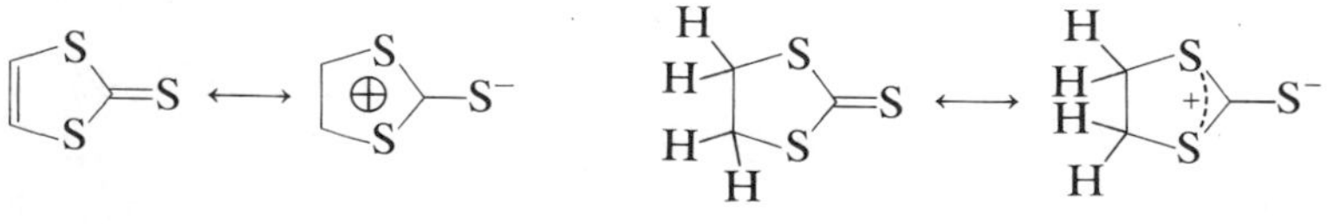

60a **60b** **61a** **61b**

Compound **61** could not be dehydrogenated to **60**, a somewhat surprising result in view of the expected energy loss in going to a more "aromatic" system. Apparently little difference in resonance stabilization exists between **60b** and **61b**. However **60** is thermally more stable than **61**. Compound **61** reacts with nucleophilic reagents with ring opening, while **60** usually does not. Both compounds can be methylated, and these intermediates behave similarly toward carbon, nitrogen, oxygen, and sulfur bases.[98] Dithiolium salts (**62**), undergo addition of nucleophiles at C-2. For example, **62a** (R = H) adds methanol to form the methyl ether, **63a**, which can be isolated as the neutral compound. When this is treated with Grignard reagents, the methoxy group is replaced by the Grignard carbanion (e.g., **62b**), and this dithiole is then oxidized by triphenylmethyl perchlorate to regenerate the 2-substituted dithiolium salt (e.g., **62f**).[72] This reaction opens a new route to 2-substituted-1,3-dithiolium systems.

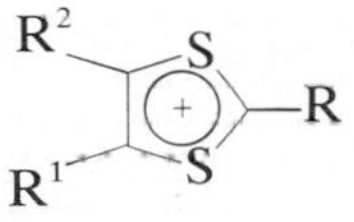

62

a: R = H or SCH_3
b: R = p-$C_6H_4NR_2$
c: R = NH_2
d: R = NR_2, R^1 = p-C_6H_4Cl
e: R = NR_2 or NRAr
f: R = C_6H_5

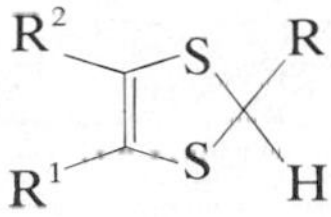

63

a: R = OCH_3
b: R = C_6H_5

The nature of the reactivity of **62** toward carbon bases has recently been studied.[99] Carbon nucleophiles attack the 1,3-dithiolium nucleus at C-2, leading in many cases to "1,4-dithiafulvalenes" (Section III.B.). In other reactions, the behavior of the fused-ring benzo-1,3-dithiolium salts toward strong C–H acids has been described in detail.[54,64,65] Thus reaction of **20c** with malonic ester leads, by double decarboxylation and dehydrogenation, to the formation of the monomethine salt, **33**.

The *para*-positions of compounds such as dimethylaniline, anisole, or phenol are electron rich, and these nucleophiles react with the positively charged 1,3-dithiolium nucleus attacking the electron-deficient C-2 carbon atom and yielding 2-p-substituted phenyl derivatives **62b**.[3,54,71] Diphenylamine also exhibits this type of reactivity toward 2.[68] Glacial acetic acid and acetic acid–acetic anhydride are useful solvents.[54] With dimethylaniline yields are generally above 90%; with anisole yields are lower, and stronger reaction conditions must be used.[54] Condensation proceeds in normal fashion with phenols, but subsequent oxidation occurs to produce quinone derivatives.[100] For example, when **2** is treated with anthrone in glacial acetic acid–pyridine, compound **64** is obtained. Hurtley and Smiles were the first to describe the course of this reaction.[4,5] 1,2-Dithiolium salts do not generally react with phenols or phenol ethers, owing to their decreased electrophilic character relative to 1,3-dithiolium salts. When 2,4,5-triphenyl-1,3-dithiolium perchlorate was treated with sodium cyanide in ethanol, 2-cyano-1,3-dithiole, **65a**, was obtained, along with some triphenyl-2-ethoxy-1,3-dithiole, **65c**.[74]

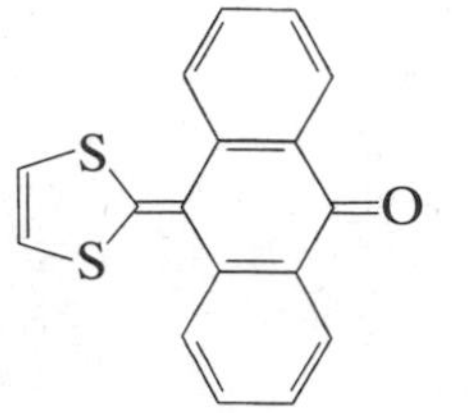

64

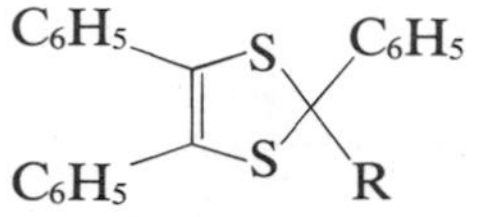

65

a: R = CN
b: R = NH_2
c: R = OC_2H_5
d: R = SH

In general, nitrogen nucleophiles react with 1,3-dithiolium salts to yield well-defined condensation products. The reaction with ammonia, however, remains irregular, depending on the number and type of substituents on the heterocyclic nucleus. For example, in attempts to prepare 2-amino-1,3-dithiolium compounds of type **62c** by the action of ammonia on 2-methylthio-1,3-dithiolium salts **62a**, only tars were obtained.[101] 2-Aminobenzo-1,3-dithiolium derivatives are, however, accessible as described earlier (Section II.A.2.). Treatment of 2-dimethylamino-4-*p*-chlorophenyl-1,3-dithiolium perchlorate **62d**, with ammonia in ethanol gives a 43% yield of 4-*p*-chlorophenyl-1,3-dithiole-2-thione **66**.[40] On material balance it was assumed that 1 mole of starting material decomposed during the reaction to provide the extra sulfur atom.

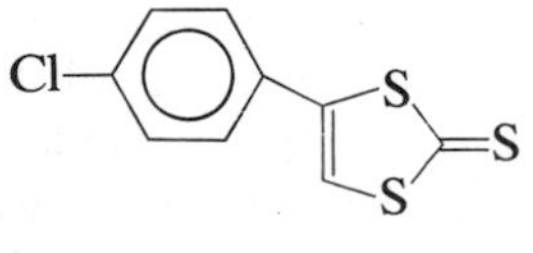

66

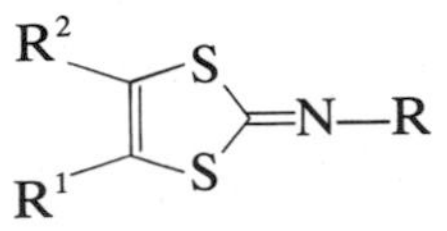

67

A crystalline aminodithiole adduct, **65b**, was isolated from the reaction of 2,4,5-triphenyl-1,3-dithiolium perchlorate and dry ammonia in benzene,[74] but when **65b** was treated with perchloric acid, the dithiolium salt was regenerated. Compound **65b** decomposed in boiling ethanol with liberation of hydrogen sulfide to yield elemental sulfur, desoxybenzoin, thiobenzamide, triphenyl-1,3-thiazole, and 2,4,5-triphenyl-1,3-dithiole. A 2-amino-1,3-dithiole was obtained from 4-phenyl-1,3-dithiolium ion in a similar manner.[72]

Aromatic and aliphatic primary amines react with 2-methylthio-1,3-dithiolium salts with evolution of methyl mercaptan to produce either neutral imino derivatives, **67**, or aminodithiolium salts, **62e**, depending on the number of moles of primary amine employed.[51,101] With excess base, compounds of type **67** are formed;[101] with 1 mole of primary amine, compounds of type **62e** are readily accessible.[51]

Aliphatic and arylalkyl secondary amines react with 2-methylthio-1,3-dithiolium salts **62a** to give good yields of the corresponding 2-dialkyl- or 2-arylalkylamino-1,3-dithiolium compounds, **62e**. The reaction probably involves attack of secondary amine at C-2 to afford **68**, which can then spontaneously expel methyl mercaptan to form the more stable dialkylaminodithiolium system **62e**. Reactions of **62a** with aliphatic secondary amines have been carried out in tetrahydrofuran or methanol.

R² S SCH₃ HNR₂ R¹ S **62a** → [R² S SCH₃ NR₂ R¹ S N] **68** →

[R² S =NR₂ R¹ S ⟷ R² S NR₂ R¹ S] **62e**

With less basic arylalkyl and diarylamines such as methylaniline and diphenylamine, a more polar solvent such as dimethyl formamide is necessary.[68] 2-Dialkylamino-1,3-dithiolium salts can be reduced in the presence of sodium borohydride to produce 2-amino-1,3-dithiole derivatives.[58] Transmethylation has been observed to occur in the reaction of a tertiary amine, such as pyridine, with 2-methylthio-1,3-dithiolium salts.[40,51] This reaction proceeds by attack on the alkylmercapto group instead of on the C-2 carbon atom. When 4,5-diphenyl-1,3-dithiolium perchlorate was reacted with triethylamine in acetone, *bis*-4,5-diphenyl-1,3-dithiol-2-ylidene, **69a**, was obtained.[57a] *N*-Ethyldiisopropylamine

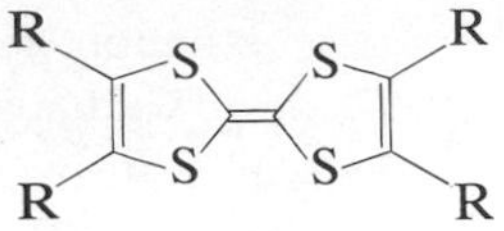

69 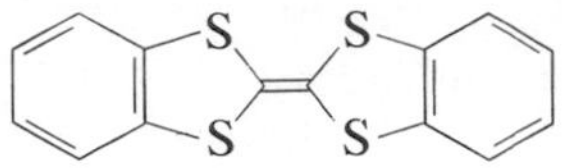 **70** 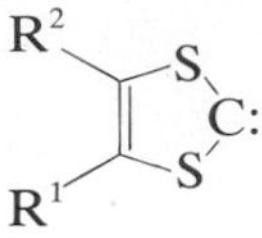 **71**

a: $R = C_6H_5$
b: $R = -CO_2CH_3$
c: $R = H$

effects a similar reaction with 4-substituted-1,3-dithiolium salts.[103] In earlier work by Hurtley and Smiles,[6] benzo-1,3-dithiolium tetrachlorozincate was converted to **70** when heated in acetic anhydride. Compound **70** is also obtained by the action of triethylamine on benzo-1,3-dithiolium salts.[57b] The mechanism of these reactions undoubtedly involves a dithiocarbene, **71,** obtained by proton abstraction from the dithiolium conjugate acid. Open-chain intermediates of this type have recently been postulated in the formation of tetrakisalkylmercaptoethylenes from *O*-thioformates and strong bases,[104,105] and in the thermal decomposition of the toluene sulfonylhydrazone salts of trithiocarbonates.[106,107] Moreover the cyclic nucleophilic carbene is implicated in the mechanism of the condensation of carbon disulfide with active acetylene[102a] and in the mechanism of the reaction of hexafluoro-2-butyne with active acetylene or with carbon disulfide and ketones.[102b] Two products of these reactions are **69b** and **72**.

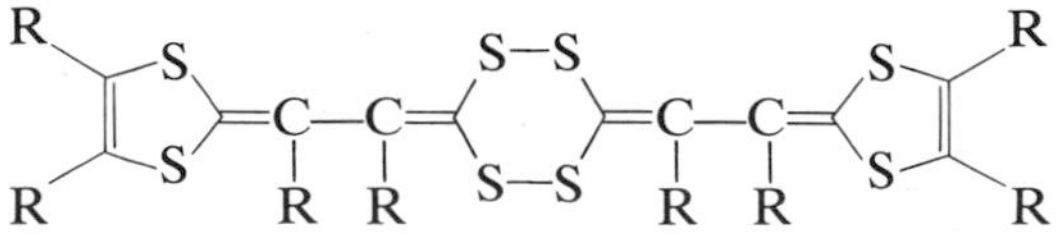

72 ($R = CO_2CH_3$)

Compounds of structure **69** and **70** have an electron-rich central double bond and react with electron-deficient olefins such as tetracyanoethylene.[164] Both **69c** and **70** form unusually stable radical cations on oxidation with lead tetraacetate or chlorine or photolytically in air.[163,164]

Reaction of azide with 2-methylthio-1,3-dithiolium ion provides a new route to some nitrogen–sulfur heterocycles.[108] Thus 1,4,2-dithiazines, **73**, are produced by the decomposition of the unstable azide intermediate **73a** at room temperature. Compounds of type **73** are themselves thermally unstable, and yield compounds of type **74** at 160 to 180°C.[108]

Towards an oxygen nucleophile such as water, benzo-1,3-dithiolium salts behave as normal Lewis acids, with the products existing essentially

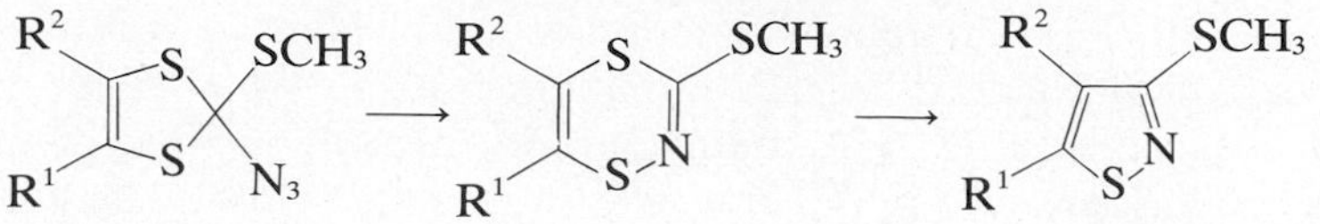

73a 73 74

in the form of nonionized carbinols **75**.[4,5] The equilibrium between **75** and **20a** lies predominantly on the side of the dimer ether, **76**. Ethers of this type have also been encountered in the 1,2-dithiolium series.[78] 2-Alkoxy-1,3-dithioles, **65c**, have been obtained by the action of sodium

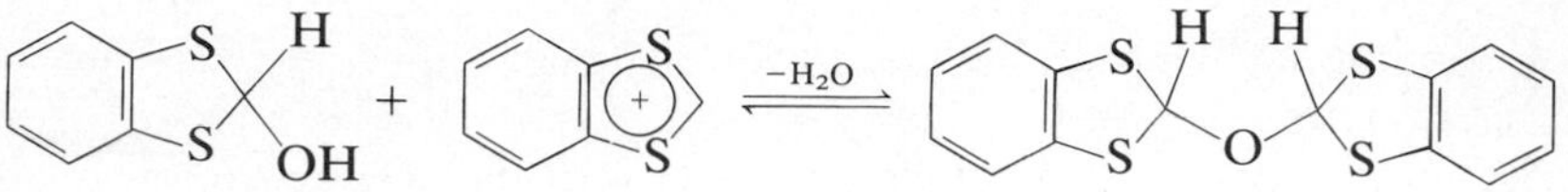

75 20 76

ethoxide in ethanol on 4-aryl-1,3-dithiolium salts.[72,73] These derivatives are easily reconvertible into the original dithiolium salts when treated with acid. Reduction of 2,4,5-triphenyl-1,3-dithiolium ion occurs in the presence of sodium hydrosulfide to afford the dithiole **65d**.[73] Reduction of 2-dialkylamino dithiolium salts to 2-aminodithioles takes place with lithium aluminum hydride.[58] 2-(Phenylthio)- and 2-(dialkylaminothlo-carbonylthio)-1,3-dithioles such as **77** have been obtained from dithiolium salts and corresponding nucleophiles.[72]

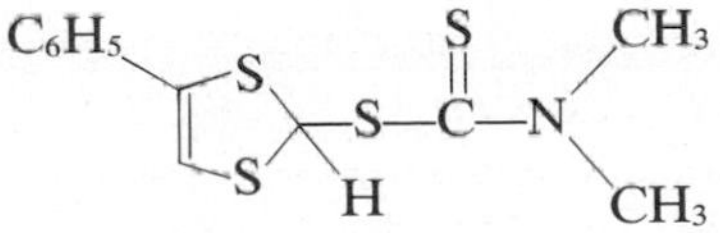

77

4-Phenyl-1,3-dithiole-2-one reacted with mercuric acetate to form the metalated derivative, 4-phenyl-5-acetoxymercuri-1,3-dithiol-2-one.[166] Reaction of the mercuri derivative with iodine in chloroform replaced the mercury with iodine, forming 4-phenyl-5-iodo-1,3-dithiol-2-one. This mercuration reaction may therefore prove useful in the formation of a variety of substituted 1,3-dithiolium salts.

As with 3- or 5-substituents in the 1,2-dithiolium series, 2-methyl or 2-methylene substituents on the 1,3-dithiolium ring undergo condensation reactions with aldehydes,[3,17,54] and under stronger conditions, even with some ketones.[3] With aromatic aldehydes styryl derivatives are obtained in good yields, the conversion taking place rapidly in hot acetic acid. 2-Methyl-1,3-dithiolium compounds also undergo condensation reactions with 2-methylthio-1,3-dithiolium salts to form highly conjugated dyes of type **32**.[54] A methyl group in position 4 is unreactive toward aldehydes, even in the presence of pyridine or triethylamine, thus indicating the lack of positive charge delocalization into the 4-position.[17]

C. Physical Properties

1. *Theoretical Calculations*

Quantum mechanical calculations were carried out on the 1,2- and 1,3-dithiolium ions as early as 1961 by Zahradník and Koutecký.[69] In a subsequent review, Zahradník has given a detailed account of the methods and problems associated with calculations on the electronic structures of sulfur-containing heterocyclic compounds.[91]

Zahradník's treatment was based on a Hückel LCAO MO approach—first including the sulfur *d* orbitals and then excluding participation of the *d* orbitals.[109] For the 1,2-dithiolium ion, it was found that the most electron-deficient carbon was C-3 and that charge distribution was essentially independent of the value of the resonance integral, β_{ss}. Moreover the C–C bond order proved to be much higher than the C–S bond order. An alternate approach was carried out at about the same time by Bergson,[110] who employed an SCF LCAO MO method. He found substantially greater double-bond character in the C–S bond relative to the C–C bond. A certain amount of cyclic delocalization, as in a typical "aromatic" ion, may be inferred from the high bond order of the S–S bond and from the high double-bond character in the C–S bond.

In the context of valence-bond theory the 1,2-dithiolium ion may be represented as a resonance hybrid between structures **78** to **81**, with the

78 ⟷ **79** ⟷ **80** ⟷ **81**

greatest importance being given to the sulfonium structure, **78**, and with **80** to **81** making relatively small contributions. Based on electron charge distribution, nucleophilic or free-radical attack is assumed to occur only at C-3.[69] The heterocyclic nucleus should be stable toward electrophilic reagents. In chemical reactivity studies, however, nucleophilic attack on the disulfide bridge has been observed.

Zahradník and Koutecký have also carried out quantum mechanical calculations on the 1,3-dithiolium ion using the LCAO MO method.[11,109] Results of the calculations show the theoretical centers of nucleophilic and free-radical reactivity to be the electron-deficient C-2 position. The C-4–C-5 bond order, being quite high, approximates the value for an isolated C=C bond. Delocalization energy was calculated to be 25 kcal/mole,[111] low compared to 49 kcal/mole for the thiopyrylium ion.[112]

2. *Molecular Structure*

Hordvik and co-workers[113–118] have obtained x-ray diagrams for a number of substituted 1,2- and 1,3-dithiolium salts by the oscillating crystal method and by the Weissenberg method, using Cu Kα radiation. Unit cell dimensions may be calculated from the x-ray diagrams.

A refined crystal and molecular structure for 3-phenyl-1,2-dithiolium iodide has been determined.[114] Bond lengths for the 1,2-dithiolium ring are significantly shorter than single bonds, and it therefore appears that π-orbital delocalization is present. The angle between the plane of the benzene ring and that of the 1,2-dithiolium ring is 27°. Good correlation exists between experimental bond lengths and calculated bond orders.

Based on S–S bond lengths determined from x-ray diagrams of 3,5-diamino-1,2-dithiolium iodide, conjugation does not extend over the S–S bond.[115] Thus the relative weight of resonance forms containing S–S double bonds should be smaller in the 3,5-diamino derivative than in the unsubstituted 1,2-dithiolium ion. The predominant iminium character of the former is further substantiated by nmr data (Section II.C.4). In 3,5-diacetamino-1,2-dithiolium iodide, partial bonding between sulfur and oxygen probably occurs, thus leading to a shortened S–S bond.[114] Although dimensions of the unit cells of 2,4-diphenyl- and 2,4,5-triphenyl-1,3-dithiolium perchlorate have been determined, accurate molecular data are not available.

Some interesting related data on the structure of *bis*(1,3-dithiol-2-ylidene), **69c**, have recently been reported,[125] comparing the crystal structure with the CNDO calculations. The central bridging double bond was longer than the ring C–C bonds, in agreement with heptafulvalene data. Also, the bonds between the S atoms and the bridging carbon atoms are longer than those between the same S atom and the ring double bonds. These are longer C–S bonds than have been reported in dithienyl systems. The overall molecule was not planar, but slightly distorted into a chair conformation, bent at the S atoms, with all four S atoms and the central C atoms in one plane.

3. *Infrared and Visible-Ultraviolet Data*

A systematic and reliable interpretative approach toward the ir spectra of even the simplest 1,2- and 1,3-dithiolium salts is not yet available. Only a few ir data for these compounds have been reported. The energy of the S–S stretching vibration in the 1,2-series would be of interest in view of the postulated "double" bond character of the S–S bond. Unfortunately difficulties arise in assignment of S–S stretching bands in unsaturated cyclic disulfides.[119] No systematic investigations have as yet been carried out on the ir spectra of isotrithiones.

Electronic absorption data for various 1,2- and 1,3-dithiolium systems have recently been tabulated and discussed.[9] The parent 1,2-dithiolium ion displays two intense bands, with the long wavelength maximum showing slight solvatochromism. 3-Phenyl-1,2-dithiolium and phenyltropylium ions show similar absorption behavior. The long wavelength band in these compounds is probably due to electron transfer from the phenyl groups to the pseudoaromatic nucleus. Electron-donating substituents in the 3- and 5-positions of the 1,2-dithiolium ring effect a greater bathochromic shift than in the 4-position, where the substituent is less likely to participate in the mesomerism of the heterocyclic ring system. Intense colors of compounds of type **36** are indicative of pronounced charge separation due to resonance between dithiolium and quaternary ammonium structures.[70] Consistent with this interpretation is

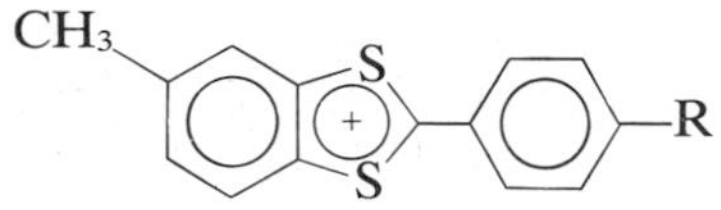

CH_3 S + S R

82

the fact that the band observed at 515 to 540 mμ in weakly acidic media undergoes a shift to 375 to 390 mμ in 70% perchloric acid, in which the dimethylamino group is fully protonated. Auxochromic substituents effect bathochromic shifts on the long wavelength feature with the fused-ring colored salts. For example, the dimethylamino group causes a bathochromic shift of 148 mμ in the case of the 2-phenyldithiolium system, and 198 mμ in the styryl system.[64]

4. *Nuclear Magnetic Resonance Data*

The magnetic equivalence of H-3 and H-5 in the unsubstituted ion and in the 4-phenyl derivative has been taken as evidence that the charge distribution in the 1,2-dithiolium ion is symmetrical.[12] According to x-ray diagrams, electron density at C-3 or C-5 (1,2-dithiolium ion) and C-2 (1,3-dithiolium ion) is appreciably less than at other carbon atoms in the heterorings.[114–118] Because of arguments put forth in a preceding section (Section II.B.2.), electron density at C-2 (1,3-dithiolium ion) is expected to be less than at either C-3 or C-5 (1,2-dithiolium ion). These observations have been confirmed by nuclear magnetic resonance studies.[120] In the 1,2-dithiolium ion, for example, H-4 and H-3 signals occur at $+1.12\tau$ and -0.57τ, respectively. The low chemical shift value of H-3 indicates very little diamagnetic shielding, which implies a high degree of "electron deficient" character about C-3. Comparing chemical shift value of H-5 (3-phenyl-1,2-dithiolium ion) with that of H-2 (4-phenyl-1,3-dithiolium ion), one finds the values -0.3τ and -1.31τ, respectively. As expected, these values indicate greater electron deficiency (less shielding) at C-2 of the 1,3-dithiolium ion than at C-5 of the 1,2-dithiolium ion. Protons attached to the positively charged dithiolium nucleus appear at lower fields than protons attached to simple dithioles or uncharged aromatic rings.

The high chemical shift value for H-4 (3.48τ) in the 3,5-diamino-1,2-dithiolium derivative (**10a**), in comparison with the H-4 value ($\tau = 1.08$) of 3-phenyl-1,2-dithiolium ion, may be interpreted to mean that **10a** exists primarily in the iminium form with the positive charge largely localized on the exocyclic nitrogen atoms. The same conclusions may be drawn from a study of the spectra of 2-dialkylamino-1,3-dithiolium salts. In this case spin–spin splitting patterns suggest that hindered rotation exists about the C–N bond, indicating its double bond character, as would be expected in the iminium form.[59] Studies of the effect of *p*-substituents on the position of the H-5 proton of 4-phenyl-2-dimethylamino-1,3-dithiolium salts have been reported.[59] Electron-donor groups increased shielding in the order of $OH > OCH_3 > CH_3 > H \geq Cl$, while electron-withdrawing groups (NO_2) reduced the shielding effect.

A greater paramagnetic shift of the H-4 signal was noted on passing from carbon disulfide to trifluoroacetic acid solvents for 1,2-dithiole-3-thiones (**4a**) than for the similarly substituted 1,2-dithiol-3-ones (**4b**).[120] An explanation for this would be that resonance structure **84** in relation to **83** is more important than that of **86** in relation to **85**, in accordance with observations that a carbonium center is more effectively stabilized by an adjacent oxygen than by sulfur.

Of the three 4-hydroxy-1,3-dithiolium derivatives (**87**, **88**, and **89**) the 2-phenyl derivative (**87**) gives the long-wave absorption typical of alkyl and aryl derivatives in the UV region; **88** and **89**, on the other hand, do not absorb in this region. In agreement with these findings, the H-5 NMR signal of **87** shows that **87** is in fact a 4-hydroxy-1,3-dithiolium derivative.[60] The olefinic signal is not present, however, in the nmr spectra of the perchlorates **88** and **89** in trifluoroacetic acid or 70% perchloric acid; these spectra contain instead a two-proton signal in the methylene region.

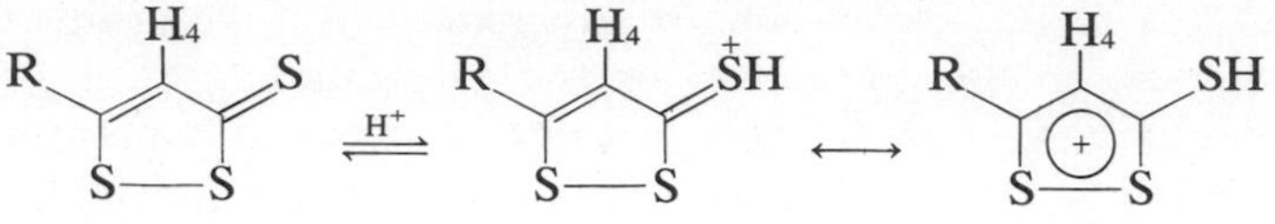

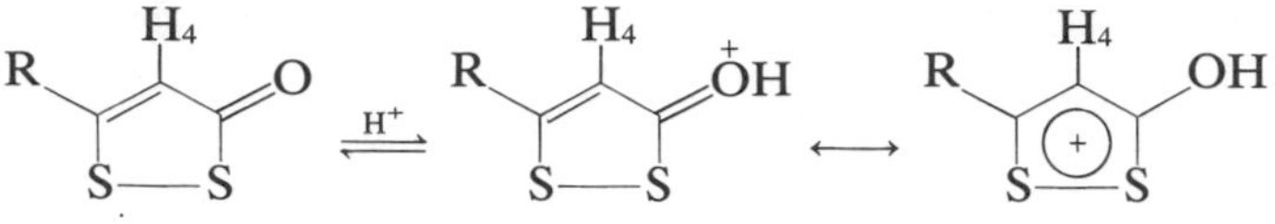

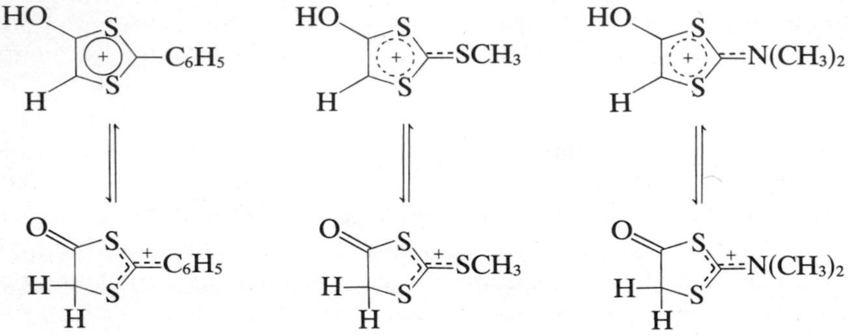

Thus these salts, both of which contain donor substituents in position 2, exist mainly in the keto form in these solvents.[60] It appears, therefore, that the salts **88** and **89** owe little stabilization energy to the formation of the π-electron sextet of the 1,3-dithiolium system by enolization and incorporation of the C_4–C_5 double bond into the mesomeric system.

III. Dithiolium Character in Dithioles

A. Thiothiophthene "No-Bond" Resonance System

An interesting class of compounds that is accessible via 1,2-dithiolium salts is the so-called thiothiophthene no-bond resonance system, otherwise named as 6*a*-thiathiophthens, meribicyclo-3,5-epidithiopenta-2,4-diene-1-thiones, and 1,6,6*a*-S^{IV}-trithiapentalenes.[121] Interest in compounds of this type focuses primarily on the nature of the bonding between the three sulfur atoms. For example, these compounds have been variously formulated as one-bond–no bond resonance stabilized systems **90**, as valence tautomers **91**, as bicyclic aromatic systems such as **92**, and as delocalized structures such as **93**.

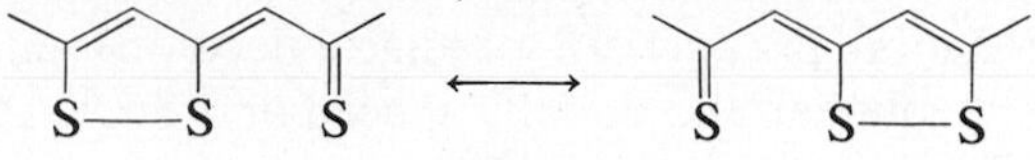

90

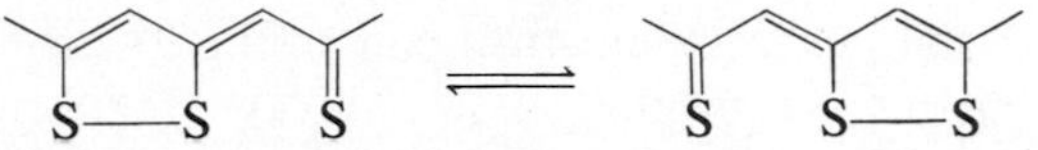

91

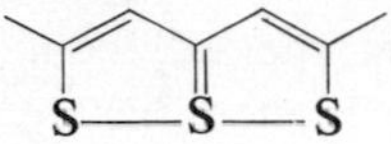

92 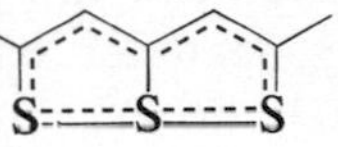 **93**

Although several synthetic routes lead to compounds of this class,[121] procedures involving 1,2-dithiolium salts are quite simple and give good yields.[122] Conversion of 1,2-dithiolium salts into the thiothiophthene system was first reported in 1963 by Klingsberg.[122a] The condensation apparently proceeds by attack of dithiolium electrophile on the enol or anionoid form of the carbonyl component; the adduct **94** is then dehydrogenated to the fully conjugated **95** at the expense of additional dithiolium salt, which is used in excess to give yields of 40 to 50%, based on ketone. Reaction of the oxygen isomer (**95**) with phosphorus pentasulfide gives the thiothiophthene derivative (**90**). In the work above Klingsberg, using previously unknown unsymmetrical, substituted thiothiophthenes, demonstrated the no-bond resonance phenomena. Thus on reaction with phosphorus pentasulfide, the oxygen isologs (**95a** and **b**),

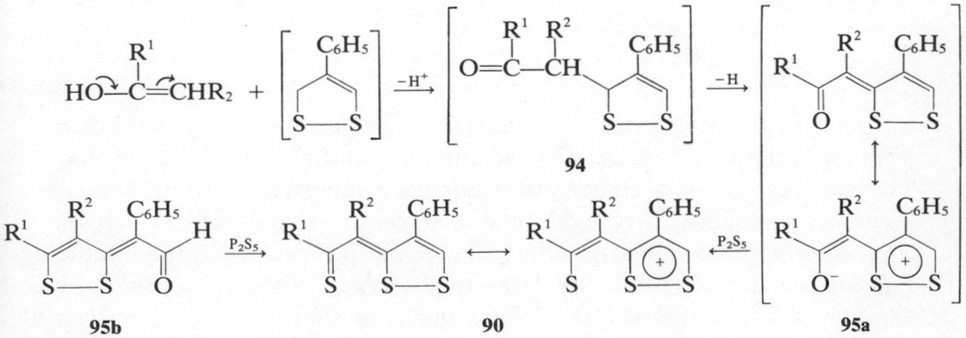

$HO-C(R^1)=CHR_2$ + [C_6H_5-dithiolium] $\xrightarrow{-H^+}$ [$O=C(R^1)-CH(R^2)$-dithiolyl-C_6H_5] **94** $\xrightarrow{-H}$ **95a**

95b $\xrightarrow{P_2S_5}$ **90** $\xleftarrow{P_2S_5}$ **95a**

gave the same product.[122a] Other chemical evidence to this end has been reported by French workers[123] and by Behringer and co-workers,[124] following the lead of data obtained by instrumental methods.

X-Ray diffraction analysis gave the first hint to the phenomena of thiothiophthene no-bond resonance.[122b] It was found that the three sulfur atoms in the reaction product **92** of diacetylacetone and phosphorous pentasulfide were colinear and equally spaced at 2.36 Å. A normal RSSR bond distance by comparison is 2.04 Å.

The bonding for the central sulfur atom in structure **92** might involve p^3 σ-bonds, with the unshared electron pair in the $3s$ orbital and through conjugation using the $3d$ orbitals. Alternatively, rehybridization of the sulfur bonding electrons might occur so that σ-bonds become sp^2, the unshared electron pair being promoted to $3d$ orbitals and the $3p_z$ orbital being open for conjugation with the electrons from the adjacent $2p_z$ orbitals on carbon.[126] Thiothiophthene no-bond resonance systems of type **95** exhibit neither normal carbonyl absorption nor normal chemical shift values for methyl ketone protons in the nmr spectrum.[127] The absence of ketonic character in **95** is attributed to a highly polarized dithiolium enol structure. Similar effects and explanations are already familiar from the cases of tropone[128] and 1,4-thiopyrone.[129] A detailed comparison of the nmr spectra of thiothiophthenes with those of related dithiolylidene ketones has led to the conclusion that the no-bond resonance system is capable of sustaining a ring current, a conclusion consistent with the view that it has aromatic character.[130]

3-Phenyl-1,2-dithiolium perchlorate condenses with cyclohexanone, and subsequent dehydrogenation gives the mesoionic species **96**.[131] 3-Methylthio-1,2-dithiolium salts may be converted to thiothiophthene

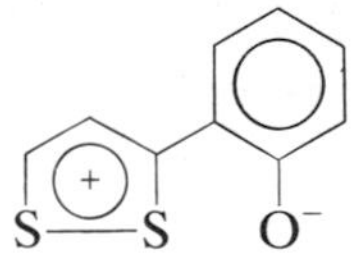

S S O⁻

96

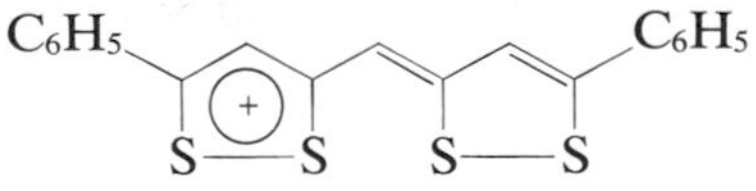

C_6H_5 C_6H_5

S S S S

97

no-bond resonance systems.[124] Tertiary amines convert 1,2-dithiolium salts into thiothiophthenes.[89,90] Recently no-bond resonance of the thiothiophthene type of a higher order has been reported by Klingsberg.[132] Thus in compounds of type **97**, it was shown by x-ray analysis that the four sulfur atoms are nearly collinear, with partial bonding of the internal pair, which are separated by 3.00 to 3.10 Å, a distance considerably shorter than twice the van der Waals radius of sulfur.

Reports of the preparation of α-cyano-5-phenyl-1,2-dithiole-$\Delta^{3,\alpha}$-thioacetamide (**98**) have been made.[133,134] Compound **98** probably represents a new class of amino-thiothiophthene no-bond resonance compound on the basis of spectral and chemical properties.[134]

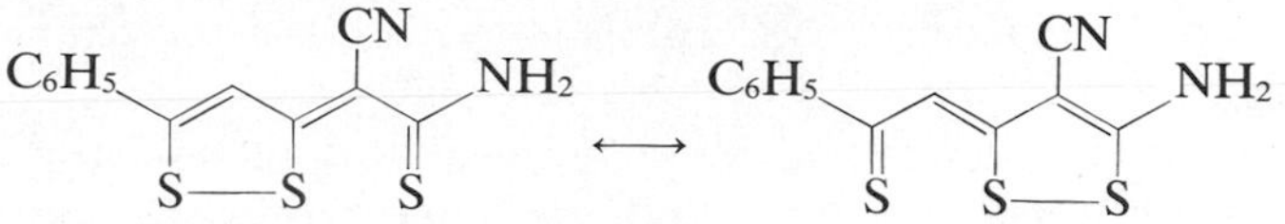

98

The vinylogous relationship between thiothiophthene no-bond resonance compounds and 1,2-dithiole-3-thiones has been demonstrated by methylation.[134] The methylated products, vinylogs of trithionium salts, were converted into the corresponding anils (**99**)—the first known thiothioisosters in which sulfur is replaced by a fifth group element.

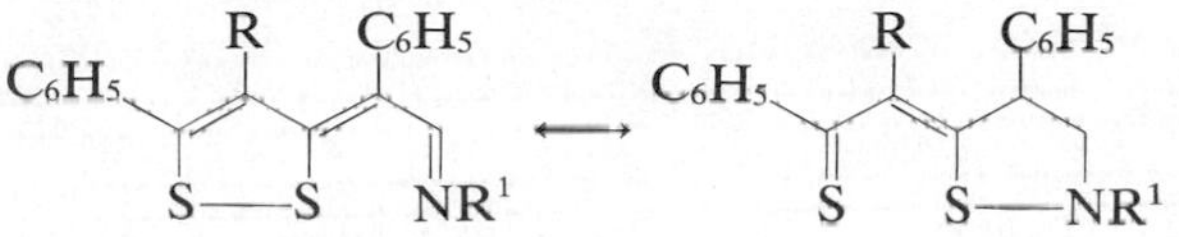

99

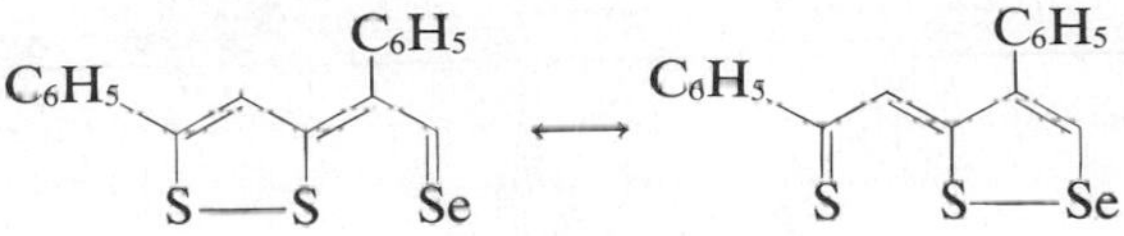

100

Incorporation of selenium in place of sulfur in the thiothiophthene no-bond resonance system (**100**) and x-ray crystallographic studies of this new material have been reported by Van den Hende and Klingsberg.[135] Investigations are underway to define a mechanism and to characterize the intermediates involved in a three-step reaction sequence that leads to

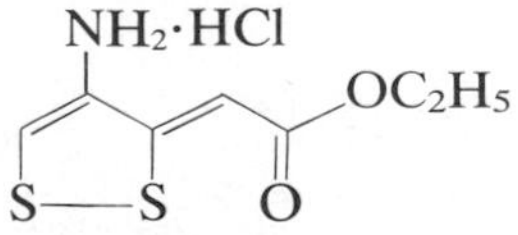

101

the no-bond resonance compound **101**.[136] The preparation and ir properties of an oxygen-18 enriched thiothiophthene derivative have been described.[137a] Electrophilic bromination, nitration, and nitrosation reactions have been carried out on the thiothiophthene nucleus.[137b]

B. Dithiafulvalenes

Dithiolium salts serve as useful intermediates in the preparation of 1,2- and 1,4-dithiafulvalenes and dithiafulvenes, a class of compounds that has been of interest in recent years. Efforts have been aimed chiefly toward answering the question of whether these compounds exist as in the covalent structural formula or in the dipolar formulation, for example, **102a** or **b**, respectively. Initial synthetic work was concerned with the preparation of highly substituted derivatives, whereas later work has shown that the parent dithiafulvalenes can be produced.

102a **102b**

The most general method of obtaining 1,2-dithiafulvalenes involves a condensation reaction between 3-methylthio-1,2-dithiolium salts (**5f**) and sodium salts of weak C–H acids, which may be cyclic (e.g., cycloalkanones,[138] pyrazolones, dimedone, indane-1,3-dione, phenalane-1,3-dione[139] and barbituric acids[44,139]) or open chain (e.g., malonates[44,139] and benzoylacetonitrile[124]) structures. Reaction of **5f** with such strong bases as the sodium salts of cyclopentadiene, indene, or fluorene, however, are often somewhat unpredictable, and the 1,2-dithiafulvalenes are generally obtained in low yields, along with a large number of by-products. Some of the by-products arise from the fact that 1,2-dithiafulvalenes react as fulvenes and add on strong bases.[140] A competing reaction that often interferes is carbanion attack on the S-methyl group with displacement of trithione, analogous to the reaction of **5f** and pyridine.[44,47] *C*-Methylation ability of **5f** has also been observed, for

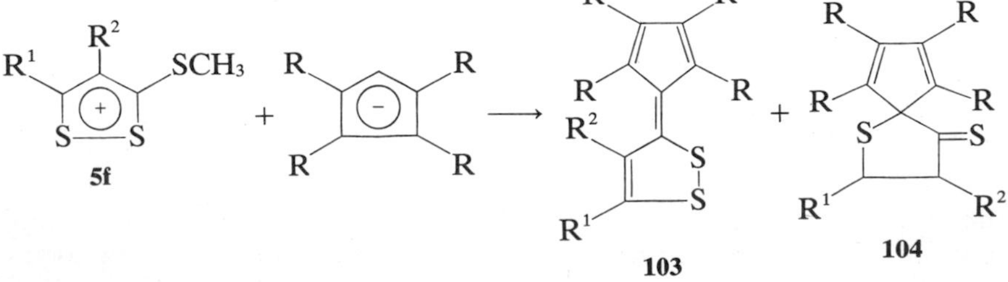

5f **103** **104**

a: $R = C_6H_5$
b: $R = Cl$

example, in its reaction with sodium acetoacetate or the sodium salt of fluorene.[44] Reaction of **5f** with highly stabilized tetraphenylcyclopentadiene affords only two products, namely **103a** and the spiro compound **104**.[67] 5-Aryl-3-methylthio-1,3-dithiolium salts react with tetrachlorocyclopentadiene to give high yields of a single tetrachlorodithiafulvalene derivative (**103b**). The reaction proceeds in methanol at 20° in the absence of base and is unaffected by side reactions. A 3-dialkylamino-1,2-dithiolium salt has been used as a substrate in place of the 3-methylthio derivative in the reaction above.[141] When 3-methylthio-1,2-dithiolium is replaced by a vinylogous substrate (**105**), conjugated structures of type **106** are obtained in the reaction with malononitrile.[142]

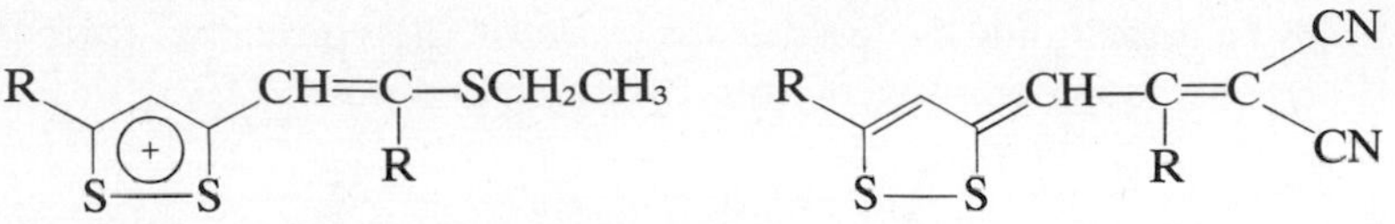

105 **106**

Among the first examples of a 1,4-dithiafulvene was **27a**, obtained by photolysis of aryl-substituted 1,2,3-thiadiazoles[62] (Section II.A.2.). Other examples of this type of compound have been prepared, usually by the reaction of 2-methylthio-1,3-dithiolium compounds with weak C–H acids.[51,67,71,143,144] Reaction conditions involve boiling glacial acetic acid containing a trace of pyridine[71,143] or tetrahydrofuran at room temperature.[51,67] The reaction may be considered to proceed *via* a nucleophilic attack of the active methylene carbanion on the electron-deficient C-2 site of **62a** in the cases where pyridine is used; subsequent elimination of methyl mercaptan gives the products—2-ylidene-1,3-dithioles (or 1,4-dithiafulvalenes).[71] Similar results are obtained when iminium salts were used in the place of the alkylthio derivatives (**62**). Both 1,4-dithiafulvenes

$$\textbf{62a}\ (\text{-SCH}_3) + {-CHR^1R^2} \xrightarrow{-CH_3SH} \textbf{27}$$

(**27**) and the mono oxygen analogs, 1,4-oxathiafulvenes, could be obtained in this way.[165] Active methylene compounds that have been employed include malonitrile, ethyl cyanoacetate, pentane-2,4-dione, ethyl acetoacetate, and ethyl benzoylacetate.[71,142] In general, weak C–H acids are not alkylated by 1,3-dithiolium salts without the help of strong base.[99] 4-Aryl-1,3-dithiolium salts react with sodium salts of tetraphenylcyclopentadiene or fluorene to give 10 to 20% yields of 1,4-dithiafulvalenes. 2-Methylthio-1,3-dithiolium salts react with tetrachlorocyclopentadiene in absolute methanol and under nitrogen to yield 90% of tetrachloro-1,4-dithiafulvalene and with cyclopentadienyl sodium to give unsubstituted 1,4-dithiafulvalene (**102**).[145–146] Several 3-alkyl and 3-alkenyl 1,3-dithiolium salts have been obtained by protonation of 1,4-dithiafulvalenes, which are comparable in basicity to the isomeric 1,2-dithiafulvalenes.

Similar to the iso-π-electronic sesquifulvalenes and their iso-π-electronic O-[147,148] and *N*-analogs,[149–152] 1,2-dithiafulvalenes are strong bases, being fully protonated even by moderately strong acids. The site of electrophilic attack, and hence the structure of the conjugate acids (**107** and **108**), has been determined from uv and nmr studies.[153] On the other

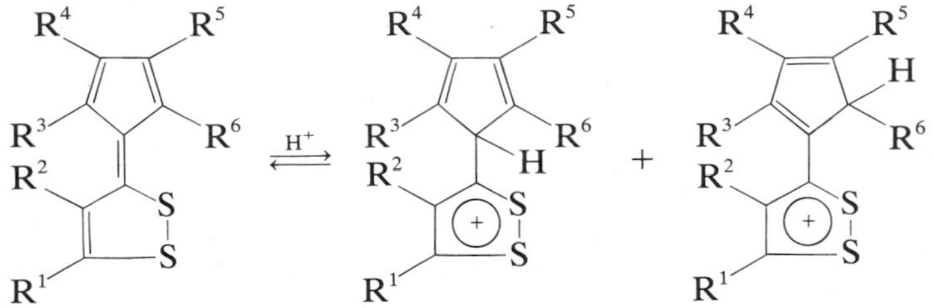

107 **108**

hand, protonation of dicyanodithiafulvalenes **109** and **27d**, **e**, **f** does not occur, as indicated by the failure of these compounds to undergo any particular alternation in trifluoroacetic acid, as manifested by spectral data.[67,71] Positions of absorption maxima of **109** and **27d**, **e**, **f** in the 220

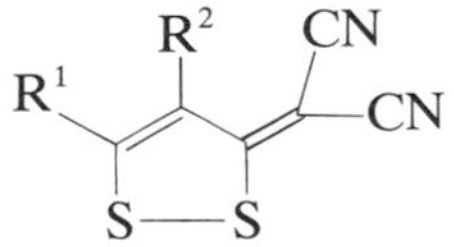

109

to 600 mμ region are roughly independent of the dielectric constant of the solvent. Therefore a covalent structural formula, **109**, best shows the actual electronic disposition in the ground state, in contrast to a dipolar formulation. 1,4-Dithiafulvalenes exhibit analogous properties toward acids, as manifested by uv and nmr studies. Thus in an acidic medium it is possible for these compounds to exist in the 1,3-dithiole form or in the protonated 1,3-dithiolium form, depending on their relative basicities. Comparison of chemical shift data of the C-5 proton of several of these compounds in trifluoroacetic acid has been made.[71] Results indicate that **27b** and **27e** exist as the protonated, charge-delocalized species **110b** and

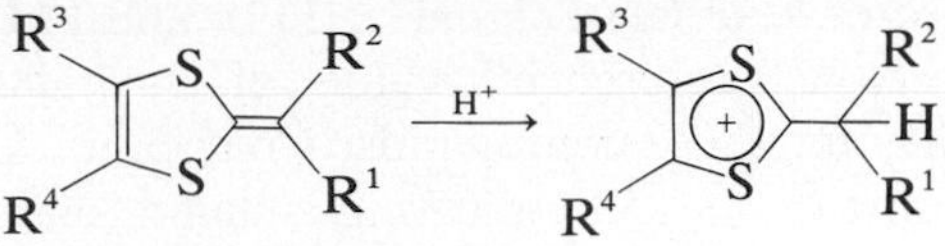

$$\xrightarrow{H^+}$$

27 **110**

a: $R^1 = C_6H_5$, $R^2 = H$, $R^3 = C_6H_5$
b: $R^1 = R^2 = COCH_3$, $R^3 = C_6H_5$
c: $R^1 = R^2 = COCH_3$, $R^3 = p\text{-}BrC_6H_4$
d: $R^1 = R^2 = CN$, $R^3 = C_6H_5$
e: $R^1 = CN$, $R^2 = CO_2Et$, $R^3 = C_6H_5$
f: $R^1 = CN$, $R^2 = CO_2Et$, $R^3 = p\text{-}BrC_6H_4$

110c, respectively, in which additional stabilization can result from hydrogen bonding with the carbonyl function. The derivatives **27d**, **e**, and **f**, being less basic, are not protonated and thus exhibit the C-5 proton resonance further upfield. That **110b** and **110c** are protonated in trifluoroacetic acid is also manifested in the uv region by a comparison of the spectra in acidic versus nonacidic media.[71]

Close resemblance of spectra of **112a** to spectra of similarly 2-substituted 1,3-dithiolium salts constitutes a proof of the site of protonation in **112a**. Protonation in the 5-position of **111b** to form **113b** is indicated by the relatively strong red shift of the long wavelength maxima of the conjugate acids of **111b**.[144] Absorption data for conjugate acids of tetrachloro-1,4-dithiafulvalenes also indicate that protonation occurs in the 5-position, with **113c** being generated.[144] Methoxy groups cause a

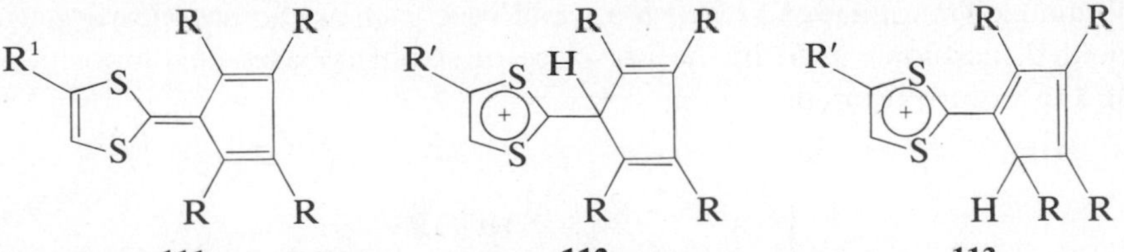

111 **112** **113**

a: R = —CH=CH—CH=CH— **a:** R = —CH=CH—CH=CH— **a:** R = —CH=CH—CH=CH—
b: $R = C_6H_5$ **b:** $R = C_6H_5$ **b:** $R = C_6H_5$
c: R = Cl **c:** R = Cl **c:** R = Cl

very slight bathochromic shift of the maximum at the longest wavelength in the spectra of the bases **111a**, **b** and **c**, as compared with a shift of 56 mμ in the spectra of the acid **112a**. This has been interpreted as an indication that π-bonds in this cross-conjugated system, which are analogous to the nonalternate bond types of sesquifulvalene, are also largely delocalized, thus permitting little polarization of the molecule toward the betaine structure having two aromatic sextets.[154] Comparison of the C-5 proton resonance signal of **111a** and **111b** ($\tau = 3.36$ and 3.35, respectively) to a "reference" value of 3.47 indicates fairly rigidly localized bonds for these compounds. Compound **111c** shows somewhat weaker shielding ($\tau = 3.15$), indicating a slight polarization that could not be quantitatively determined.[154] This effect is understandable, however, in view of the high acidity of tetrachlorocyclopentadiene. Structural assignment of the conjugate acids **112a**, **113b** and **113c** was based on the relationship between local π-electron distribution and chemical shift.[154] As 2-alkyl-4-aryl-1,3-dithiolium derivatives, the acids **112a**—and salts with similar substituents in the 4-position—have comparable electron densities on C-5 and hence also similar H-5 chemical shift values. Electron deficiency in **113b** must be effectively delocalized over the cyclopentadienyl residue in the conjugate acid, since a strong diamagnetic shift in the H-5 signals is observed. Evidence based on the chemical shift of H-5 indicates protonation of the tetrachloro compounds in the 9-position (as in **112**), whereas electronic absorption measurements strongly indicate attack by the proton in the 5-position to form **113c**.[154]

C. Mesoionic Compounds

Ohta and Sugiyama, in 1963, reported a sequence of reactions that apparently led to the first mesoionic dithiolium compound, a 4-*N*-acyl derivative.[155] Thus when cyanomethyl dithiobenzoate (**114**) was allowed to react with an acyl halide such as propionyl chloride, an acyl derivative (**115**) which, on the basis of ir studies, existed in tautomeric forms was obtained. Treatment of **115** with a weak base such as sodium bicarbonate yielded mesoionic **116**. In the presence of stronger bases decomposition of **116** occurred rapidly.

C_6H_5–C(=S)–S–CH_2CN (**114**) $\xrightarrow[C_6H_6]{RCOCl}$ [**115a** (NHCOR) ⇌ **115b** (N=C(OH)–R)] Cl^- $\xrightarrow{NaHCO_3}$ **116**

114 **115a** **115b** **116**

These authors also treated carboxymethyl dithiobenzoate with boron trifluoride in acetic anhydride at 60°C and, after aqueous workup, isolated a compound they believed to have the mesoionic structure **117a**.[155] However it was later shown by Potts and Singh[156] that the compound actually obtained by Ohta and Sugiyama was instead the anhydro base, anhydro-5-hydroxy-2-phenyl-4(thiobenzoylthiomethylcarbonyl)-1,3-dithiolium hydroxide (**118**). Preparation of **117a** in 75% yield was achieved by using triethylamine in acetic anhydride at 6 to 10°C and employing "dry" workup conditions.[156] Potts also prepared the 2-*p*-methoxyphenyl derivative **117b**.

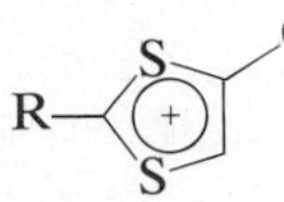

117 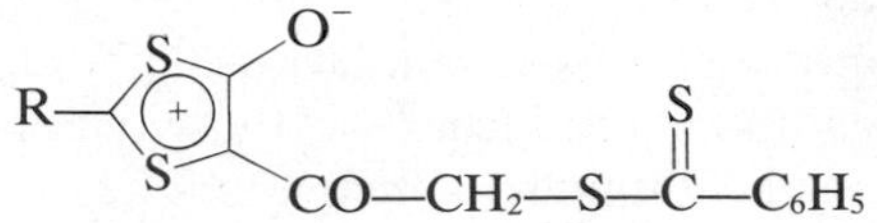 **118**

a: $R = C_6H_5$
b: $R = p\text{-}CH_3OC_6H_4$

A compound that would be expected to serve as a precursor to the anhydro bases would be **87**, first reported by Campaigne, Jacobsen, and Hamilton in 1964, by the action of 70% perchloric acid on carboxymethyl dithiobenzoate.[60] Although **87** is susceptible to decomposition in the presence of certain basic reagents, utilization of the conditions described

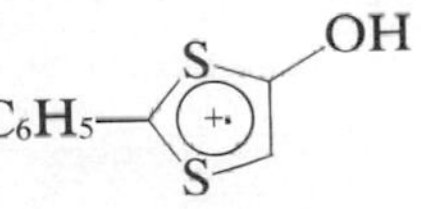

87

by Potts and Singh might be expected to afford **117a**. The first fully characterized members of the class of anhydro-1,3-dithiolium bases were reported by Gotthardt and Christl in 1968[157] These authors treated carboxyphenylmethyl dithiobenzoate (**119**) with triethylamine in acetic anhydride for 90 min at temperatures less than 10°C to obtain 90% yields of mesoionic compound **120**.

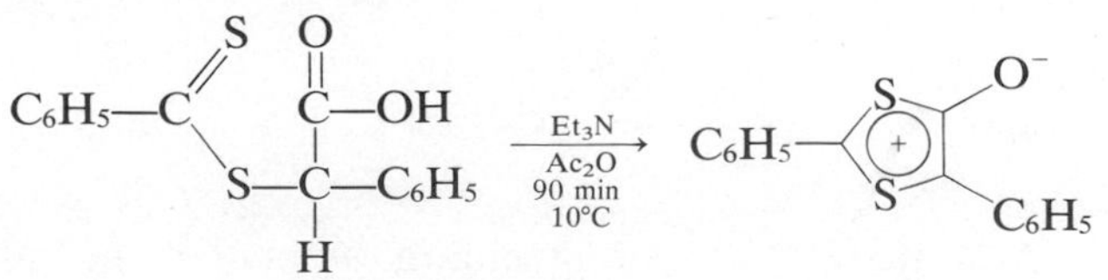

Et_3N, Ac_2O, 90 min, 10°C

119 **120**

A characteristic reaction of mesoionic anhydro bases is 1,3-dipolar cycloaddition with certain activated acetylenes to produce substituted thiophenes. The first report of this reaction was made in 1968 by Gotthardt and Christl,[158] who obtained high yields of tetrasubstituted thiophenes on reaction of certain acetylenic compounds (e.g., dimethyl acetylenedicarboxylate) with 2,4-diphenyl-anhydro-5-hydroxy-1,3-dithiolium hydroxide (**120**) in xylene at 130°C. The mechanism of the reaction, in which carbon oxysulfide is lost, is parallel to that of the formation of pyrazoles[159] and pyrroles[160] from sydnones and of mesoionic oxazolones with alkynes. Potts and Singh have also reported 1,3-dipolar cycloadditions of this type using acetylenic dipolarophiles.[156]

Anhydro bases also react with alkenes to form interesting dithiabicyclo-cycloaddition products.[161] For example, when 2,4-diarylanhydro-5-hydroxy-1,3-dithiolium hydroxides (**121**) were heated (120–130°) with various open-chain alkenes in xylene, cycloadducts of type **122** were obtained. Dehydrohalogenation of **122** with palladium/charcoal leads to substituted thiophenes (**123**). Analogous reactions occur with cyclic alkenes; cyclopentene yields compounds of type **124**, *N*-phenylmaleimide

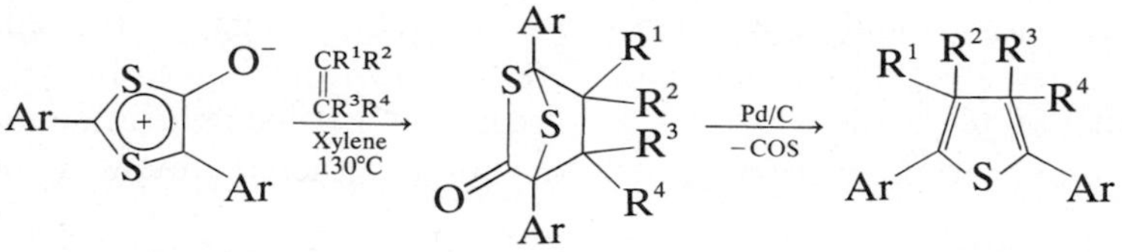

121 → ($CR^1R^2{=}CR^3R^4$, Xylene, 130°C) → **122** → (Pd/C, −COS) → **123**

and maleic anhydride produce structures **125**, and acenaphthalene yields compounds such as **126**. The acenaphthalene adduct (**126**) can be reacted further with alkenes such as norbornadiene, norbornene, and tetracyanoethylene. With norbornadiene, for example, structure **127** was obtained.

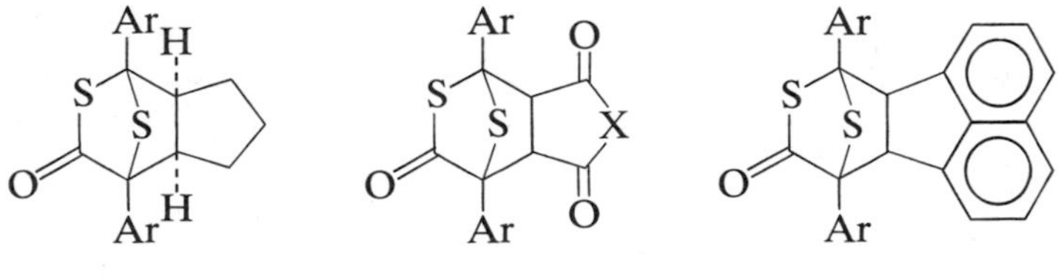

124 **125:** X = N—C_6H_5 or O **126**

Pyrolysis of **127** at 180° caused evolution of pentadiene and carbon oxysulfide, thus affording diaryl-substituted thiophenes (**128**).

127

128

In a recent communication Inouye, Sato, and Ohta[162] have described a remarkable preparation of a mesoionic 1,2-dithiolium derivative (**130**), using the following sequence:

C_6H_5—CH_2—CO—CH_2—C_6H_5 $\xrightarrow[CHCl_3\text{-}EtOH]{H_2S/HCl}$ [**129**] $\xrightarrow{base}$ **130**

The unstable hydroxydithiolium compound **129** was obtained in 40% yield. Treatment of dibenzyl ketone with P_4S_{10} also gave **130** in poor yield.

IV. References

1. E. Hückel, *Z. Phys*, **70,** 204 (1931).
2. G. Cilento, *Chem. Rev.*, **60,** 147 (1960).
3. R. Wizinger and L. Soder, *Chimia*, **12,** 79 (1958).
4. W. R. H. Hurtley and S. Smiles, *J. Chem. Soc.*, 1821 (1926).
5. W. R. H. Hurtley and S. Smiles, *ibid.*, 2263 (1926).
6. W. R. H. Hurtley and S. Smiles, *ibid.*, 534 (1927).
7. E. Klingsberg, *Chem. Ind.*, 1568 (1960).
8. E. Klingsberg, *J. Am. Chem. Soc.*, **84,** 3410 (1962).
9. H. Prinzbach and E. Futterer, in A. R. Katritzky, "Advances in Heterocyclic Chemistry," Vol. 7, Academic Press, New York, 1966, pp. 39–151.
10. E. Campaigne and R. D. Hamilton, *Quart. Rep. Sulfur Chem.*, **5,** 275 (1970).
11. R. Zahradník, in A. R. Katritzky, "Advances in Heterocyclic Chemistry," Vol. 5, Academic Press, New York, 1965, p. 1.
12. E. Klingsberg, *J. Am. Chem. Soc.*, **83,** 2934 (1961).
13. E. R. Buchman, A. O. Reims, and H. Sargent, *J. Org. Chem.*, **6,** 764 (1941).

14. I. E. Balaban and H. King,, *J. Chem. Soc.*, 1858 (1927).
15. T. O. Norris and R. L. McKee, *J. Am. Chem. Soc.*, **77,** 1056 (1955).
16. C. M. Suter, "The Organic Chemistry of Sulfur," J. Wiley, New York, 1944, p. 387.
17. D. Leaver, W. A. H. Robertson, and D. M. McKinnon, *J. Chem. Soc.*, 5104 (1962).
18. M. G. Voronkov and T. Lapina, *Khim. Geterotsikl. Soedin.*, **4,** 452 (1970); *Chem. Abstr.* **73,** 87820 (1970).
19. W. Walter and J. Curts, *Ann.*, **649,** 88 (1961).
20. H. Quiniou and N. Lozac'h, *bull. Soc. Chim. Fr.*, 1167 (1963).
21. B. Böttcher and A. Lüttringhaus, *Ann.*, **557,** 89 (1947).
22. F. Bauer, *Chem. Ztg.*, **75,** 3 (1951).
23. P. S. Landis, *Chem. Rev.*, **65,** 237 (1965).
24. J. Faust and R. Mayer, *J. Prakt. Chem.*, **26,** 340 (1964).
25. R. Mayer, P. Wittig, J. Fabian, and R. Heitmüller, *Chem. Ber.*, **97,** 654 (1964).
26. M. Schmidt and H. Schulz, *Z. Naturforsch*, 1540 (1968).
27. D. Leaver and W. A. H. Robertson, *Proc. Chem. Soc.*, 252 (1960).
28. M. Schmidt and H. Schulz, *Chem. Ber.*, **101,** 277 (1968).
29. J. P. Biton, G. Duguay, and H. Quiniou, *Acad. Sci., Paris, Ser. C.*, **267,** 586 (1968).
30. G. Purrello and A. LoVullo, *Boll. Sedute Acad. Gioenia Sci. Natur. Catania*, **4,** 20 (1967); *Chem. Abstr.*, **70,** 3885y (1969).
31. K. A. Jensen, H. R. Buccaro, and O. Buchardt, *Acta Chem. Scand.*, **17,** 163 (1963).
32. U. Schmidt, *Chem. Ber.*, **92,** 1171 (1959).
33. G. Barnikow, *ibid.*, **100,** 1389 (1967).
34. J. Faust and R. Mayer, *Angew. Chem.*, **75,** 573 (1963); *Ann.*, **688,** 150 (1965).
35. F. Boberg, *Angew. Chem.*, **72,** 629 (1960); *Ann.*, **679,** 109 (1964).
36. G. R. Schultze and F. Boberg, *German Patent* 1,102,174 (1959); *Chem. Abstr.*, **56,** 7326 (1962).
37. R. S. Spindt, D. R. Stevens, and W. E. Baldwin, *J. Am. Chem. Soc.*, **73,** 3693 (1951).
38. P. S. Landis and L. A. Hamilton, *J. Org. Chem.*, **25,** 1742 (1960).
39. J. Faust, H. Spies, and R. Mayer, *Naturwissenschaften*, **54,** 537 (1967).
40. E. Campaigne and R. D. Hamilton, Indiana University, unpublished results, 1964.
41. F. Challenger, E. A. Mason, E. C. Holdsworth, and R. Emmott, *Chem. Ind.*, 714 (1952); *J. Chem. Soc.*, 292 (1953).
42. J. Teste and N. Lozac'h, *Bull. Soc. Chim. Fr.*, 437 (1955).
43. B. Böttcher and F. Bauer, *Ann.*, **568,** 227 (1950).
44. A. Lüttringhaus and U. Schmidt, *Chem. Ztg.*, **77,** 135 (1953).
45. K. Knauer, P. Hemmerich, and J. D. W. Van Voorst, *Angew. Chem. Int. Edit.*, **6,** 262 (1967).
46. A. Lüttringhaus, M. Mohr, and N. Engelhard, *Ann.*, **661,** 84 (1963).
47. Y. Mollier and N. Lozac'h, *Bull. Soc. Chim. Fr.*, 614 (1961).
48. R. Mayer and H. Hartman, *Chem. Ber.*, **97,** 1886 (1964).
49. N. Loac'h, *Bull. Soc. Chim. Fr.*, 840 (1949).
50. G. Bahr and G. Schleitzer, *Chem. Ber.*, **90,** 438 (1957).
51. R. Mayer and B. Gebhardt, *ibid.*, **97,** 1298 (1964).
52. R. Mayer, B. Gebhardt, J. Fabian, and A. K. Müller, *Angew. Chem.*, **76,** 143 (1964).
53. R. Mayer and A. K. Müller, *Z. Chem.*, **4,** 384 (1964).
54. L. Soder and R. Wizinger, *Helv. Chim. Acta*, **42,** 1733 (1959).
55. R. Huisgen and V. Weberndörfer, *Experientia*, **17,** 566 (1961).
56. F. Range, Z. El-Hewehl, H. J. Renner, and E. Taeger, *J. Prakt. Chem.*, **11,** 284 (1960).

57. (a) K. M. Pazdro, *Rocz. Chem.*, **43,** 1089 (1969); *Chem. Abstr.*, **71,** 91357b (1969); *ibid.*, **44,** 1823 (1970); *Chem. Abstr.*, **75,** 5761g (1971). (b) D. Buza, A. Gryff-Keller, and S. Szymanski, *Rozc. Chem.*, **44,** 2319 (1970); *Chem. Abstr.*, **75,** 20252k (1971).
58. A. Takamizawa and K. Hirai, *Chem. Pharm. Bull.*, **17,** 1924 (1969).
59. E. Campaigne and N. W. Jacobsen, *J. Org. Chem.*, **29,** 1703 (1964).
60. E. Campaigne, N. W. Jacobsen, and R. D. Hamilton, *ibid.*, **29,** 1708 (1964).
61. D. B. Easton, D. Leaver, and D. M. McKinnon, *J. Chem. Soc.*, 642 (1968).
62. W. Kirmse and L. Horner, *Ann.*, **614,** 4 (1958).
63. G. Ottmann, G. D. Vickers, and H. Hooks, Jr., *J. Heterocycl. Chem.*, **4,** 527 (1967).
64. L. Soder and R. Wizinger, *Helv. Chim. Acta,* **42,** 1779 (1959).
65. R. Wizinger and D. Dürr, *ibid.*, **46,** 2167 (1963).
66. R. W. Addor, *J. Org. Chem.*, **29,** 738 (1963).
67. A. Lüttringhaus, E. Futterer, and H. Prinzbach, *Tetrahedron Lett.*, 1209 (1963).
68. N. Lozac'h and C. T. Pedersen, *Acta. Chem. Scand.*, **24,** 3189 (1970).
69. R. Zahradnik and J. Koutecky, *Tetrahedron Lett.*, 632 (1961).
70. E. Klingsberg and A. M. Schreiber, *J. Am. Chem. Soc.*, **84,** 2941 (1962).
71. E. Campaigne and R. D. Hamilton, *J. Org. Chem.*, **29,** 1711 (1964).
72. A. Takamizawa and K. Hirai, *Chem. Pharm. Bull.*, **17,** 1931 (1969).
73. D. Leaver, D. M. McKinnon, and W. A. H. Robertson, *J. Chem. Soc.*, **32,** (1965).
74. (a) R. A. Olofson, J. M. Landesberg, R. O. Berry, D. Leaver, W. A. H. Robertson, and D. M. McKinnon, *Tetrahedron,* **22,** 2119 (1966). (b) J. C. Poite, A. Perichaut, and J. Roggero, *CR Acad. Sci., Ser. C,* **270,** 1677 (1970).
75. F. Hübenett, F. H. Flock, W. Hansel, H. Heinze, and H. Hoffmann, *Angew. Chem.*, **75,** 1189 (1963).
76. K. Dimroth, *ibid.*, **72,** 331 (1960).
77. R. Wizinger and P. Ulrich, *Helv. Chim. Acta,* **39,** 207 (1956).
78. H. Newman and R. B. Angier, *Chem. Commun.*, 353 (1967).
79. W. von E. Doering and L. H. Knox, *J. Am. Chem. Soc.*, **79,** 352 (1957).
80. J. Bignebat, H. Quiniou, and N. Lozac'h, *Bull. Soc. Chim. Fr.*, 1699 (1966).
81. U. Schmidt, A. Luttringhaus, and F. Hubinger, *Ann.*, 631, 138 (1960).
82. F. Clesse, A. Reliquet, and H. Quiniou, *CR Acad. Sci., Ser. C,* **272,** 1049 (1971).
83. E. Klingsberg, *Chem. Ind.*, 1813 (1968).
84. G. LeCoustamer and Y. Mollier, *CR Acad. Sci., Paris, Ser. C.*, **267,** 1423 (1968).
85. G. Cailland and Y. Mollier, *Bull. Soc. Chim. Fr.*, 331 (1971).
86. D. Leaver and D. M. McKinnon, *Chem. Ind.*, 461 (1964).
87. E. Klingsberg, *J. Org. Chem.*, **28,** 529 (1963).
88. E. Campaigne and R. D. Hamilton, *ibid.*, **29,** 2877 (1964).
89. J. Bignebat and H. Quiniou, *CR Acad. Sci., Paris, Ser. C.*, **267,** 180 (1968); *ibid.*, **270,** 83 (1970).
90. C. Bouillon and J. Vialle, *Bull. Soc. Chim. Fr.*, 4560 (1968).
91. Ref. 9., p. 63.
92. W. A. Pryor, "Mechanism of Sulfur Reactions," McGraw-Hill, New York, 1962, p. 60.
93. F. Boberg and A. Marei, *Ann.*, **666,** 88 (1963); F. Boberg, **Ann., 681,** 169, 178 (1965).
94. A. Grandin, C. Bouillon, and J. Vialle, *Bull. Soc. Chim. Fr.*, 4555 (1968).
95. (a) L. E. Carosino, U.S. Patent 3,109,772; *Chem. Abstr.* **60,** 2933 (1964). (b) E. Luksa, T. Lapina, M. G. Voronkov, and J. Bankovskis, *Khim. Geterotsikl. Soedin,* 595 (1970); *Chem. Abstr.* **73,** 56003c (1970).

96. E. Shaw, in "Pyridine and its Derivatives," E. Klingsberg, Ed., Part 2, Interscience, New York, 1961, Chap. 3.
97. K. Hafner, H. W. Riedel, and M. Danielisz, *Angew. Chem.*, **75,** 344 (1963).
98. R. Mayer and K. Schäfer, *J. Prakt. Chem.*, **26,** 279 (1964).
99. R. Gompper and E. Kutter, *Angew. Chem.*, **75,** 919 (1963).
100. R. Gompper and E. Kutter, *Chem. Ber.*, **98,** 1365 (1965).
101. E. Campaigne, T. Bosin, and R. D. Hamilton, *J. Org. Chem.* **30,** 1677 (1965).
102. (a) D. L. Coffen, *Tetrahedron Lett.*, 2633 (1970). (b) H. D. Hartzler, *J. Am. Chem. Soc.*, **92,** 1412 (1970); *ibid.*, **92,** 1413 (1970).
103. H. Prinzbach, H. Berger, and A. Lüttringhaus, *Angew. Chem.*, **77,** 453 (1965).
104. J. Hine, R. P. Bayer, and G. G. Hammer, *J. Am. Chem. Soc.*, **84,** 1751 (1962).
105. A. Fröling and J. F. Arens, *Rec. Trav. Chim.*, **81,** 1009 (1962).
106. U. Schöllkopf and Wiskott, *Angew. Chem.*, **75,** 725 (1963).
107. D. M. Lemal and E. H. Banitt, *Tetrahedron Lett.*, **5,** 245 (1964).
108. E. Fanghänel, *Z. Chem.*, **5,** 386 (1965).
109. R. Zahradník and J. Koutecký, *Collect. Czech. Chem. Commun.*, **28,** 1117 (1963).
110. G. Bergson, *Arkiv Kemi*, **19,** 181 (1962).
111. J. Koutecký, *Collect. Czech. Chem. Commun.*, **24,** 1608 (1959).
112. J. Koutecký, J. Paldus, and R. Zahradník, *ibid.*, **25,** 617 (1960).
113. A. Hordvik, *Acta. Chem. Scand.*, **14,** 1218 (1960); **15,** 1186 (1961); **17,** 1809, 2575 (1963).
114. A. Hordvik and H. M. Kjoege, *ibid.*, **19,** 523, 935 (1965).
115. A. Hordvik, *ibid.*, **19,** 1039, 1253 (1965).
116. A. Hordvik, E. Sletten, and J. Sletten, *ibid.*, **20,** 1171 (1966).
117. A. Hordvik and J. Sundsfjord, *ibid.*, **19,** 753 (1965).
118. A. Hordvik and S. Joys, *ibid.*, **19,** 1539 (1965).
119. L. Schotte, *Arkiv Kemi*, **9,** 441 (1956).
120. See ref. 9, p. 97ff and p. 143ff.
121. (a) R. J. S. Beer, *Mech. React. Sulfur Compounds*, **2,** 121 (1967). (b) D. H. Reid, J. G. Dingwall, and J. D. Symon, *J. Chem. Soc.*, **C,** 2412 (1970).
122. (a) E. Klingsberg, *J. Am. Chem. Soc.*, **85,** 3244 (1963). (b) E. Klingsberg, *Synthesis*, 213 (1971). (c) E. I. G. Brown, D. Leaver, and D. M. McKinnon, *J. Chem. Soc.*, **C,** 1202 (1970).
123. C. G. Pfister-Guillouzo and N. Lozac'h, *Bull. Soc. Chim. Fr.*, 153 (1963).
124. H. Behringer, M. Ruff, and R. Wiedenmann, *Chem. Ber.*, **97,** 1732 (1964).
125. W. F. Cooper, N. C. Kenny, J. W. Edmonds, A. Nagel, F. Wudl, and P. Coppens, *Chem. Commun.*, 889 (1971).
126. C. A. Price and S. Oae, "Sulfur Bonding," Ronald Press, New York, 1962, p. 167.
127. H. G. Hertz, G. Traverso, and W. Walter, *Ann.*, **625,** 43 (1959).
128. T. Nozoe, in "Non-Benezenoid Aromatic Compounds," D. Ginsburg, Ed., Interscience, New York, 1959, Chap. 7.
129. L. J. Bellamy, in "Organic Sulfur Compounds," N. Kharasch, Ed., Vol. 1, Pergamon Press, Oxford, 1961, Chap. 6.
130. R. Pinel, Y. Mollier, and N. Lozac'h, *Bull. Soc. Chim. Fr.*, 856 (1967).
131. R. Pinel, Y. Mollier, and N. Lozac'h, *CR*, **260,** 5065 (1965).
132. E. Klingsberg, *J. Heterocycl. Chem.*, **3,** 243 (1966).
133. H. Behringer and R. Wiedenmann, *Tetrahedron Lett.*, 705 (1965).
134. E. Klingsberg, *J. Org. Chem.*, **31,** 3489 (1966); *ibid.*, **33,** 2915 (1968).
135. J. H. van den Hende and E. Klingsberg, *J. Am. Chem. Soc.*, **88,** 5045 (1966).
136. G. Claeson and J. Pedersen, *Tetrahedron Lett.*, 3283 (1967).

137. (a) D. Festal and Y. Mollier, *Tetrahedron Lett.*, 1259 (1970). (b) R. J. S. Beer, D. Cartwright, R. J. Gait, and D. Harris, *J. Chem. Soc.*, **C,** 963 (1971).
138. O. Coulibaly and Y. Mollier, *Bull. Soc. Chim., Fr.*, **9,** 3208 (1969).
139. Y. Mollier and N. Lozac'h, *Bull. Soc. Chim., Fr.*, 157 (1963).
140. K. Hafner, *Angew. Chem.*, **75,** 1041 (1963).
141. M. Piattelli and A. LoVullo, *Boll. Sedute Accad. Gioenia Sci. Natur. Catania*, **4,** 33 (1967).
142. H. Behringer and A. Falkenberg, *Chem. Ber.*, **102,** 1580 (1969).
143. E. Campaigne and F. Haaf, *J. Org. Chem.*, **30,** 732 (1965).
144. A. Lüttringhaus, H. Berger, and H. Prinzbach, *Tetrahedron Lett.*, 2121 (1965).
145. R. Gompper and E. Kutter, *Chem. Ber.*, **98,** 2825 (1965).
146. M. Brown, U.S. Patent 3,057,875 (1962); *Chem. Abstr.* **58,** 6837 (1963).
147. A. Schönberg and M. M. Sidky, *J. Am. Chem. Soc.*, **81,** 2259 (1959).
148. G. V. Boyd, *Proc. Chem. Soc.*, **93** (1959).
149. F. Kröhnke, K. Ellegast, and E. Bertram, *Ann.*, **600,** 176 (1956).
150. J. A. Berson, E. M. Evleth, and Z. Hamlet, *J. Am. Chem. Soc.*, **82,** 3793 (1960); **87,** 2887 (1965).
151. G. V. Boyd and L. M. Jackman, *J. Chem. Soc.*, 548 (1963).
152. H. Meerwein, W. Florian, W. Schön, and G. Stopp, *Ann.*, **641,** 1 (1961).
153. Ref. 9, p. 91ff.
154. Ref. 9, p. 137ff.
155. M. Ohta and M. Sugiyama, *Bull. Chem. Soc. Jap.*, **36,** 1437 (1963); **38,** 596 (1965).
156. K. T. Potts and U. P. Singh, *Chem. Commun.*, 569 (1969).
157. H. Gotthardt and B. Christl, *Tetrahedron Lett.*, 4743 (1968).
158. H. Gotthardt and B. Christl, *ibid.*, 4747 (1968).
159. R. Huisgen and H. Gotthardt, *Chem. Ber.*, **101,** 536, 1059 (1968).
160. R. Huisgen, H. Gotthardt, H. O. Bayer, and F. C. Schaefer, *Angew. Chem.*, **76,** 185 (1964).
161. H. Gotthardt and B. Christl, *Tetrahedron Lett.*, 4751 (1968).
162. K. Inouye, S. Sato, and M. Ohta, *Bull. Chem. Soc. Jap.*, **43,** 1911 (1970).
163. S. Hunig, H. Schlaf, G. Kiesslick, and D. Scheutzow, *Tetrahedron Lett.*, 2271 (1969).
164. F. Wudl, G. M. Smith, and E. J. Hufnagel, *Chem. Commun.*, 1453 (1970).
165. K. Hirai, *Tetrahedron Lett.*, 1137 (1971).
166. I. D. Rae, *Int. J. Sulfur Chem.*, **1,** 59 (1971).

CHAPTER VI

Heteropentalenes

K. T. POTTS

Department of Chemistry
Rensselaer Polytechnic Institute
Troy, New York

I. Introduction and Nomenclature

The term *heteropentalene* is a rather loose and unsystematic description applied to a variety of heterocycles derived by replacement of one or more carbon atoms in the bicyclic, 8π-electron hydrocarbon pentalene which has been of great interest to organic chemists since it was first postulated as a possible aromatic system by Armit and Robinson[1] in 1922. The pentalene problem has motivated numerous experimental and theoretical studies during the past 50 years. Several highly substituted pentalene derivatives have been reported in the literature.[2] However 1-methylpentalene, prepared by a flash vacuum pyrolysis technique, was found[3] to exist at −196° and had a pronounced tendency to dimerize even at temperatures below −140°. 1,3-Dimethylpentalene, 2-methylpentalene, and pentalene itself have all been found[4] to share this characteristic dimerization, the only simple 8π-electron system of this type not undergoing dimerization being 1,3,5-tri-*tert*-butylpentalene, in which the steric effects of the *tert*-butyl groups prevent dimerization.[5] This *tert*-butyl system was stable for several hours at room temperature in the absence of air, while a low concentration in *n*-hexane solution was stable for several days under N_2.

The heteropentalenes are of special interest because their properties should be governed to a large degree not only by the nature of the heteroatoms but also by their orientation. Thus an azapentalene with a trigonal nitrogen atom (pyridine type) at the nonfused positions (i.e., all positions other than 3*a*, 6*a*) would contain 8π-electrons and be expected to be nonaromatic. The simplest stable 2-azapentalene prepared[6] to date, 3-dimethylamino-1-ethoxy-2-azapentalene, formed bluish violet needles and is an 8π-electron heterocycle. The nmr spectrum showed no evidence for an induced ring current, and it was found to be stable in air for several days. It should be noted that this system contains formally a pentafulvene and an azapentafulvene system, which no doubt contribute to its stability.

However in the 10π-electron systems a completely different situation exists. These may be divided into four main types: those that can only be represented as charge-separated systems; those that have covalent representations; anionic systems; and the thiathiopthens and related systems containing other heteroatoms in addition to sulfur. Conforming to Hückel's $(4n+2)$ π-electron rule, they show varying degrees of aromatic character and are isoelectronic with the pentalene dianion.[7]

In Fig. 1 a classification of these different types of heteropentalenes is shown. This review discusses those derivatives that can be represented by charge-separated structures only, occasionally referred to incorrectly as mesoionic systems, from the view point of whether their points of fusion contain carbon atoms or heteroatoms, and are further subdivided according to the number and type of heteroatoms present in the nucleus.

There are two types of nomenclature currently in use. Pentalene is an acceptable hydrocarbon name, and accordingly the Stelzner and *Chemical Abstracts* methods have found considerable use. Thus as pentalene is

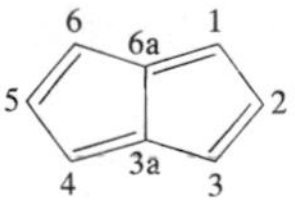

6 6a 1 5 2 4 3a 3

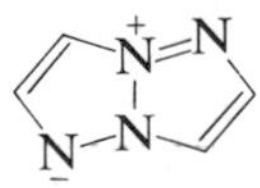

N N N N

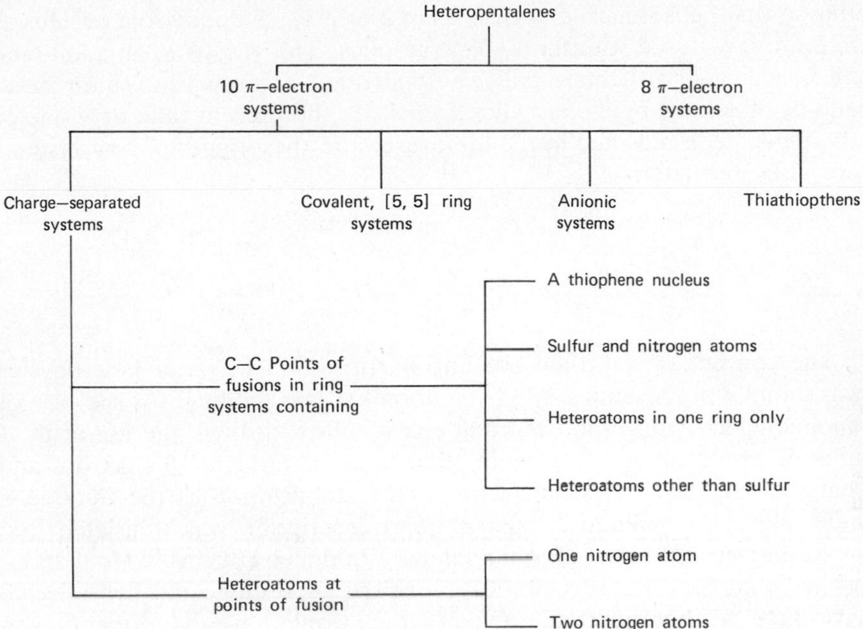

Fig. 1. Classification of heteropentalenes.

numbered as shown opposite, its tetraaza derivative would be called by both the Stelzner and *Chemical Abstracts* method 1,3*a*,4,6*a*-tetraazapentalene. Under the more systematic IUPAC and Ring Index method it is called 1*H*,5*H*-*v*-triazolo[2,1-*a*]-*v*-triazole (RIS 11776) or 1*H*,5*H*-1,2,3-triazolo[2,1-*a*][1.2.3]triazole. Both systems of nomenclature are used throughout this chapter.

II. Heteropentalenes with Ylidic Characteristics

A. Ring Systems with C–C Points of Fusion

1. *Containing a Thiophene Nucleus*

The nonclassical condensed thiophenes[8] may be considered the most important members of this group of heteropentalenes. Of the nine conceivable nonclassical condensed thiophenes represented by structure **1** (where X = Z = C, N; Y = O, S, NR), seven have been synthesized to date and details of their chemistry studied. The scope of this concept is increased considerably by introducing nitrogen atoms into the thiophene ring so that an isothiazole nucleus as in **2** or a 1,2,5-thiadiazole nucleus as in **3** are now fused with the X–Y–Z systems. This results in an additional 18 conceivable structures that would contain a tetravalent sulfur atom and be stabilized by dipolar ylide forms. At the present time only one of these has been reported in the literature, and these types of ring systems are discussed later.

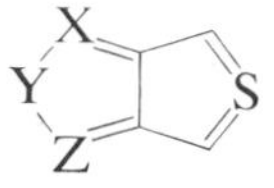

1 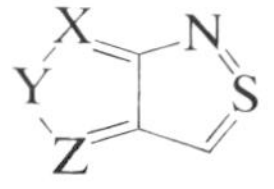 **2** 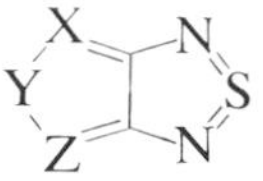 **3**

The concept of *d* orbital bonding participation in sulfur heterocycles was initially proposed in 1939 by Schomaker and Pauling[9] for the case of thiophene. The suggestion was made that sulfur, through the use of its *d* orbitals, could expand its valence-shell to accommodate 10 electrons and that this could be a special factor in the stabilization of the thiophene molecule. This qualitative valence-bond description was translated into molecular orbital terms 10 years later by Longuet-Higgins.[10] He demonstrated that the $3p_z$, $3d_{xz}$, and $3d_{yz}$ orbitals of sulfur could hybridize to give three pd^2 hybrid orbitals, two of which would have the proper energy and symmetry characteristics required for overlap with the $2p_z$ orbitals of adjacent ring carbon atoms. This led to sulfur being treated as equivalent to a vinyl group in Hückel-type calculations on thiophene and other sulfur-containing heterocycles.[11]

These proposals stimulated numerous investigations, both theoretical and experimental in nature, designed to evaluate the importance of *d* orbital participation in organosulfur bonding,[12–14] and this concept has also been covered recently in two review articles.[15,16] This issue is still not completely resolved. Some results tend to indicate that if the *d* orbitals are involved at all, it is to a minor extent. However the studies that lead to this conclusion involve almost exclusively systems that can be represented by classical valence bond structures not necessitating the use of *d* orbitals, and the study of these present tetravalent sulfur systems is thus of particular interest.

In 1967 Cava and Pollack[17a] attempted the synthesis of 1,3-dimethylthieno[3,4-*c*]thiophene (**5**; R = CH_3), a novel 10π-electron system for which the only uncharged resonance contributors are structures containing "tetravalent" sulfur. Since the valence shell of the "tetravalent" sulfur atom contains 10 electrons, bonding participation of the sulfur $3d$ orbitals is required in structural representations such as **5**. It was anticipated that **5** would offer new insights into the bonding capabilities of sulfur $3d$ orbitals.

Dehydration of the sulfoxide **4,** obtained by periodate oxidation of 1,3-dihydro-4,6-dimethylthieno[3,4-*c*]thiophene, which itself was formed by ring closure of 3,4-bis(chloromethyl)-2,5-dimethylthiophene with sodium sulfide, did not lead to an isolatable product, polymeric material being obtained when the sulfoxide was heated in the presence of neutral alumina. However when **4** and *N*-phenylmaleimide were refluxed in acetic anhydride for 4 hr, *exo*- and *endo*-adducts **7** and **8,** respectively, were isolated in the ratio 2.4 : 1. The authors interpreted these observations in terms of the transient existence of **5** ($R = CH_3$). Similarly 1,3-dicarbomethoxythieno[3,4-*c*]thiophene (**5**; $R = COOCH_3$), anticipated on heating **4** ($R = COOCH_3$) with acetic anhydride, proved too sensitive to

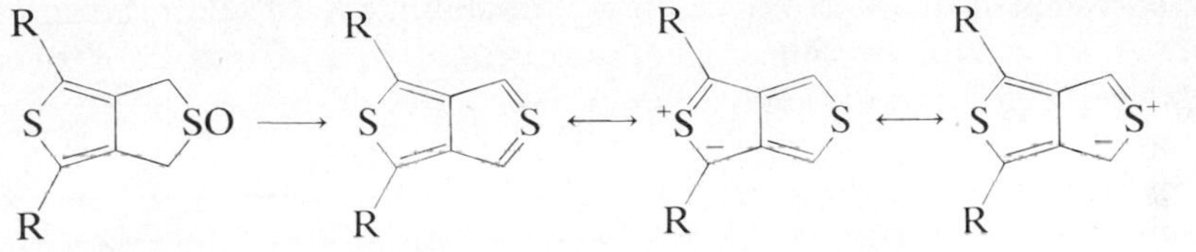

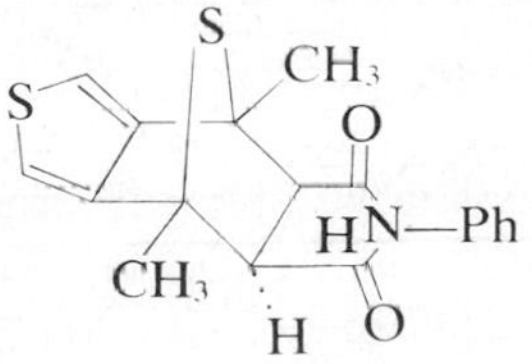

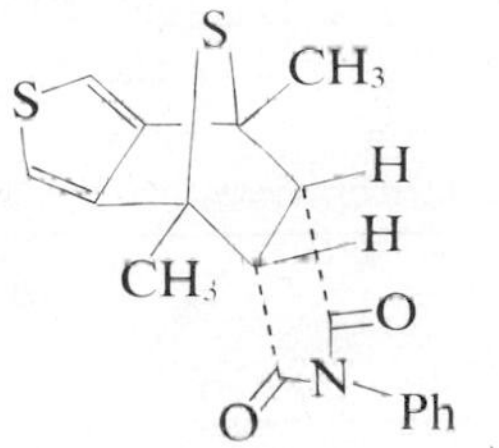

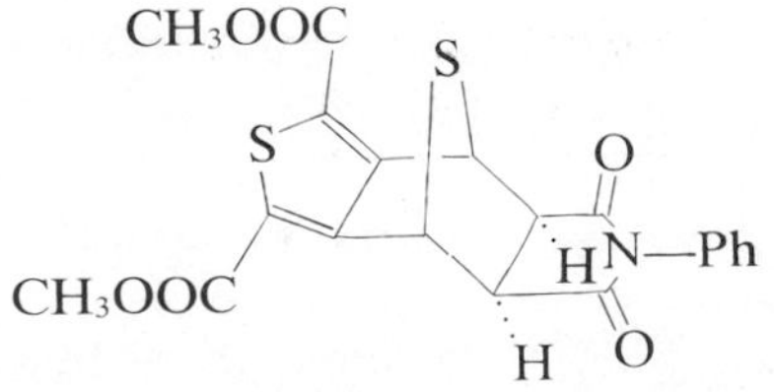

light and air to permit its isolation, but it was trapped with *N*-phenylmaleimide.[17b] A mixture of *exo*- and *endo*-adducts (**9** and **10**) was obtained, and it is interesting to note that addition in these cases occurred at the expected sites of highest electron density[17b] shown in **6**.

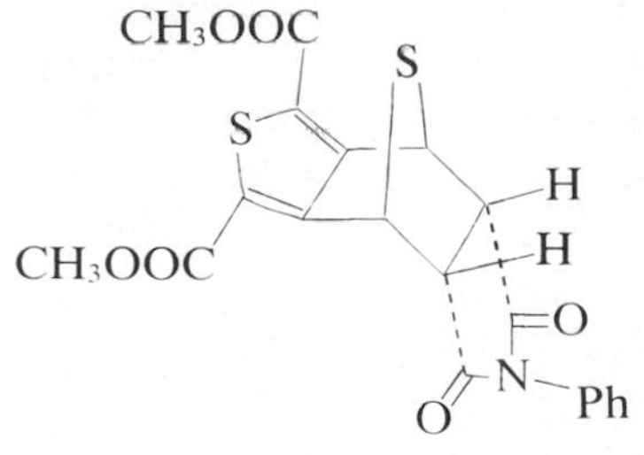

Arguments have been advanced against *d* orbital participation in **5,** and it has been demonstrated that the extreme reactivity of the thieno[3,4-*c*]-thiophene system could be adequately explained using a model incorporating only the 3*s* and 3*p* orbitals of sulfur. Dewar and Trinajstic,[18] employing this type of model, calculated a resonance energy of −33.9 kcal/mole for this system in comparison to a resonance energy of benzene of 22.6 kcal/mole. Based on similar calculations, a triplet ground state was proposed[19] for this ring system, and it was concluded that it would be too unstable for isolation.

More recent calculations, however, cast a different light on this question. Using the CNDO/2 method, it has been predicted[20] that **5** would have a singlet ground state and it was calculated to have substantial π-character in the C–S bonds, approximately 55% of which being attributed to p_π–d_π overlap. There was considered to be little overlap between the central bridging atoms.[20]

a. SYNTHESIS. Two general synthetic routes lead to these tetravalent sulfur systems. The first, involving dehydration of intermediate dihydrosulfoxides, has been utilized in the synthesis of several representatives and has the advantage over the alternative route in that the final step does not involve a relatively high-temperature ring closure with phosphorus pentasulfide when nuclear oxygen atoms are usually replaced with sulfur. Dehydration of an intermediate sulfoxide was utilized[21] in 1969 in the successful isolation of the first stable example of a nonclassical condensed thiophene system, tetraphenylthieno[3,4-*c*]thiophene (**14**). Treatment of tetrabenzoylethane (**11**) with phosphorus pentasulfide in refluxing xylene afforded 1,3-dihydro-1,3,4,6-tetraphenylthieno[3,4-*c*]-thiophene (**12**) in 46% yield. Oxidation to its corresponding sulfoxide **13** was accomplished with periodate ion, and refluxing **13** in acetic anhydride for 4 hr resulted in the formation of **14,** which was obtained as glistening, purple needles. A one-step process involving tetrabenzoylethylene and phosphorus pentasulfide in refluxing xylene also afforded **14,** but in only 3% yield.

This same approach was utilized[20] for the synthesis of tetraphenyl-thieno[3,4-*c*]furan (**21**), this being necessary as the reaction of 3,4-dibenzoyl-2,5-diphenylfuran (**15**) with P_4S_{10} in pyridine gave **14** directly.[20,22] Reduction of the diketone 15 to the corresponding diol **16,** followed by reaction with P_4S_{10} in carbon disulfide, gave a mixture of *cis*- and *trans*-1,3-dihydro-1,3,4,6-tetraphenylthieno[3,4-*c*]furans (**17** and **18**) and 1,3-dihydro-1,3,4,6-tetraphenylfurano[3,4-*c*]furan (**19**). When the *cis*-sulfoxide **20,** obtained by oxidation of **17** with periodate, was refluxed in acetic anhydride under nitrogen, a violet color appeared with λ_{max} 550 nm. Extreme sensitivity to light and air prevented direct characterization of **21** but when the dehydration was carried out in the presence of dimethyl acetylenedicarboxylate, the 1 : 1 adduct **22** was obtained. Deoxygenation of **22** with hot triethylphosphite afforded the isobenzothiophene **23,** indicating that **21** was involved in the reaction.

The second synthetic method also involves formation of the thiophene nucleus as the last step. A suitable heterocycle containing vicinal diketone substituents is treated with P_4S_{10} in either hot xylene or pyridine and, with the reservation that the heterocycle cannot contain groups sensitive to P_4S_{10}, this procedure offers an opportunity to prepare several further examples of this type of heteropentalene.

The second stable example of a nonclassical condensed thiophene was prepared by this route, diphenylthieno[3,4-*c*][1.2.5]thiadiazole (**25**) being obtained as brilliant, purple needles in 79% yield when 3,4-dibenzoyl-1,2,5-thiadiazole (**24**) was heated[23] in dioxane with P_4S_{10}. It has also been utilized in the synthesis of the other known examples of this class of compounds, the major limiting factor, that is, the availability of the precursor dibenzoyl heterocycles, being overcome by applying cycloaddition reactions with dibenzoylacetylene to prepare these intermediates.

Several mesoionic ring systems containing "masked" 1,3-dipoles have been shown to react with acetylenic dipolarophiles, giving a variety of 5-membered heterocycles after extrusion of a small, stable fragment from the initial cycloadduct. Thus *anhydro*-4-hydroxy-2,3,5-triphenylthiazolium hydroxide (**26**) and dibenzoylacetylene gave 3,4-dibenzoyl-2,5-diphenylthiophene (**28**) in 42% yield, the reaction presumably involving the initial addition of dibenzoylacetylene across the thiocarbonyl ylide dipole of **26** affording the primary cycloadduct **27,** which then decomposes to **28** by elimination of phenyl isocyanate.[24] When **28** was treated

15 16 17 18

21 20 19

DMAD

22 23

DMAD = $CH_3OOCC{\equiv}CCOOCH_3$

with P_4S_{10} in refluxing pyridine, tetraphenylthieno[3,4-*c*]thiophene (**14**) was isolated in 83% yield as glistening, purple needles.[25] The basic nature of the solvent in this reaction is not significant in that the same product is formed when refluxing xylene is used. It has been our experience that, in the workup of the pyridine reaction mixture by pouring into ice water and filtering the precipitated product, the water dissolves the phosphorus pentasulfide residues, and the tetraphenylthieno[3,4-*c*]thiophene is of relatively good purity. When the reaction was carried out in xylene, the thieno[3,4-*c*]thiophene along with phosphorus pentasulfide residues precipitate from solution forming a dark, viscous tar, and purification was accomplished by chromatography.

An alternative route to the thiophene diketone **28** allows variation of

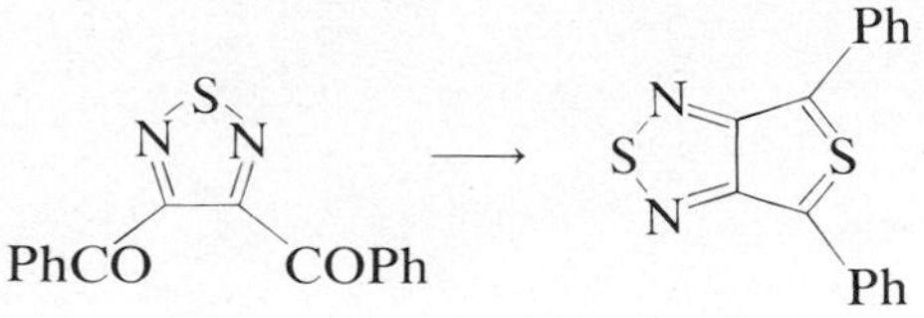

24 → 25

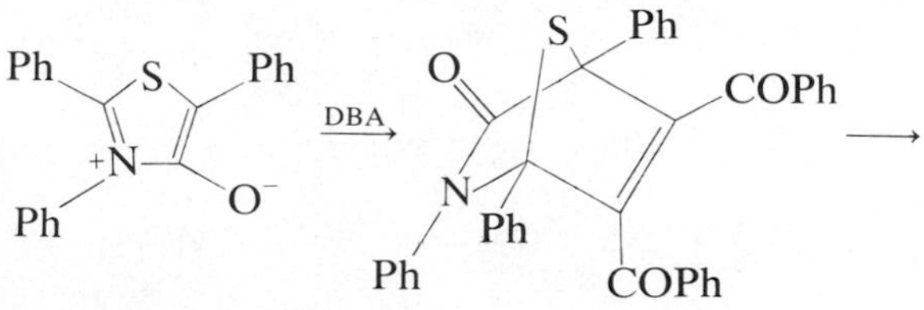

26 —DBA→ 27 → 28 → 14

DBA = PhCOC≡CCOPh

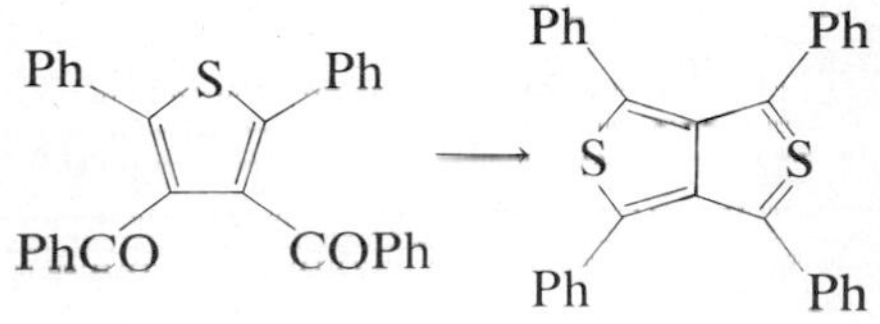

the substituents in the 2- and 5-positions, as well as the ketone groups in the 3- and 4-positions. *Anhydro*-4-hydroxy-2-*p*-methoxyphenyl-5-phenyl-1,3-dithiolium hydroxide (**29**) and dibenzoylacetylene afforded[26] 3,4-dibenzoyl-2-*p*-methoxyphenyl-5-phenylthiophene (**31**); in this case carbonyl sulfide was eliminated from the unstable initial cycloadduct **30**. Reaction of **31** with P_4S_{10} in refluxing pyridine afforded the corresponding 1-*p*-methoxyphenyl-3,4,6-triphenylthieno[3,4-*c*]thiophene (**32**) in excellent yield as deep purple needles.[27]

Variation of the 3,4-substituents comes of necessity by variation of the acetylenic dipolarophile. Replacement of the phenyl groups with *tert*-butyl groups was achieved by using 2,2,7,7-tetramethyl-4-octyne-3,6-dione (**34**) which, on reaction with *anhydro*-2,5-diphenyl-4-hydroxy-1,3-dithiolium hydroxide (**33**) in refluxing toluene, afforded 3,4-bis(2,2-dimethylpropionyl)-2,5-diphenylthiophene (**35**) in 95% yield.[28] This ketone is a particularly interesting intermediate for evaluation of the relative importance of electronic and steric effects in the stabilization of these nonclassical condensed thiophenes. Since the *tert*-butyl group should have minimum charge delocalizing ability and an inductive effect that, if anything, would slightly destabilize the ring by introducing excess negative charge, any stabilization of the ring system observed would best be explained as a steric inhibition of reaction leading to decomposition of the system (e.g., nucleophilic or electrophilic attack, concerted addition of singlet oxygen, etc.). Such a stabilizing effect has been observed to attend the introduction of *tert*-butyl substituents into a number of unrelated systems.[5,29,30]

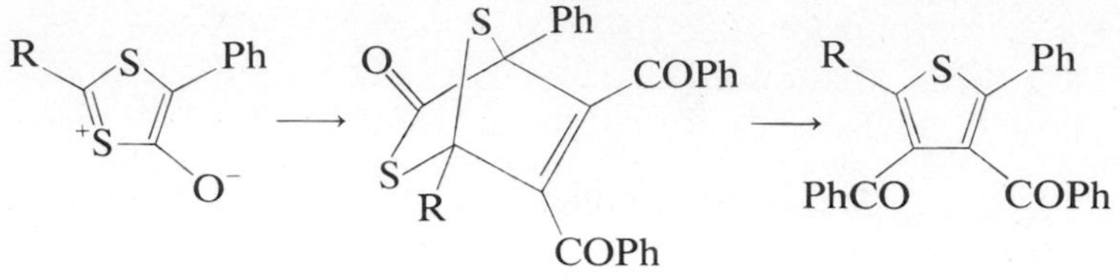

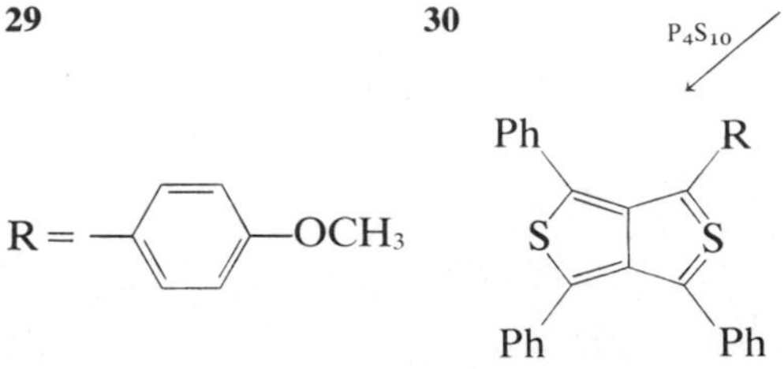

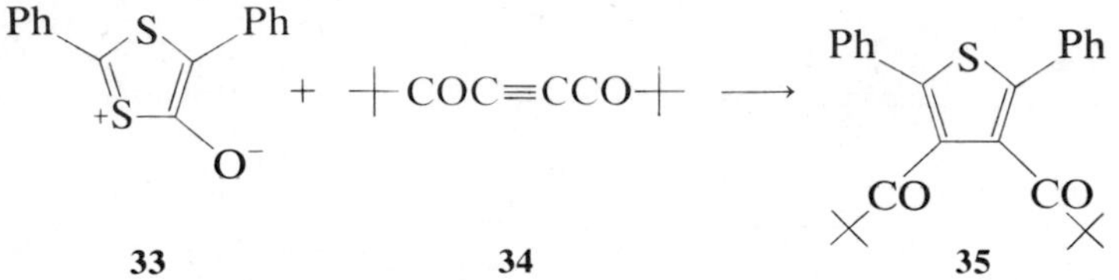

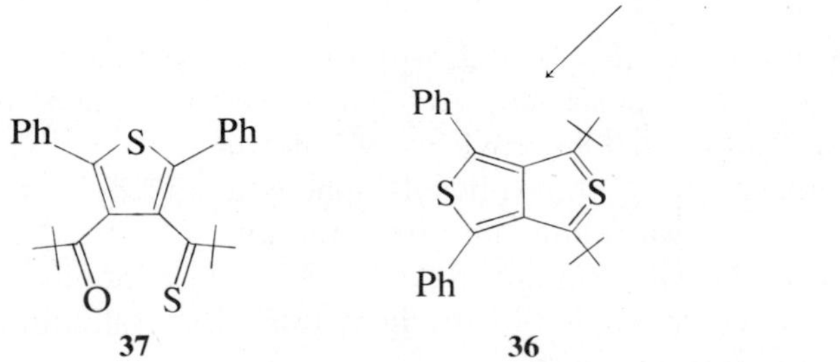

The diacylthiophene **35** quickly gave a reddish purple color on heating with phosphorus pentasulfide in pyridine. Reaction workup resulted in the isolation[28] of purple prisms of the thioketone **37** rather than the heteropentalene **36**. This type of thioketone has been obtained from **14** in benzene solution in air,[21b] which may indicate its mode of formation in this present instance.

Variation of the mesoionic ring system in the reaction with dibenzoylacetylene offers a convenient means of obtaining a variety of *vic*-dibenzoyl heterocycles. 3,4-Dibenzoylpyrrole derivatives were obtained in good yields from the reaction of the *anhydro*-5-hydroxyoxazolium hydroxide system with the acetylene. Because of the extreme reactivity of the mesoionic system, its members are often difficult to isolate in a pure state,[31] but they can be readily generated[32] *in situ* by acetic anhydride cyclization of the corresponding *N*-benzoylglycine derivatives **38,** and when this was carried out in the presence of dibenzoylacetylene, 3,4-dibenzoylpyrrole derivatives **41** were isolated in good yields. Compounds **39** and **40** are no doubt intermediates in the reaction, but the product

$R = CH_3$, Ph
$R^1 = H$, Ph

isolated from the reaction mixture is **41**. Phosphorus pentasulfide–pyridine treatment of those systems that contained a hydrogen atom at the 5-position resulted in formation of the corresponding dithiobenzoyl compound **42** only, irrespective of whether the *N*-substituent was methyl or phenyl. Although the actual mechanism of formation of the tetravalent sulfur system is unknown, dithiobenzoyl compounds have been proposed as intermediates in this process,[33] but why the reaction apparently stops at this stage in this instance is difficult to understand. It is possible that the fused thiophene systems that would be formed are unstable, high energy systems and not attainable under the reaction conditions or, if they were formed, are not sufficiently stable for isolation and revert to the dithiobenzoyl products. The presence of the highly reactive azomethine ylide dipole in **43** is consistent with this idea. This ylide is stabilized by substitution of a phenyl substituent (or acyl group), and it would be anticipated that 5-methyl-1,3,4,6-tetraphenylthieno[3,4-*c*]pyrrole (**43**; $R = CH_3$; $R^1 = Ph$) would be a stable system. This was found to be the case with the thienopyrrole being formed from **41** ($R = CH_3$; $R^1 = Ph$) and phosphorus pentasulfide–pyridine in 60% yield as small, brilliant red needles.[25]

The Paal–Knorr synthesis of pyrroles has also been utilized to obtain the intermediate dibenzoylpyrroles. Reaction of tetrabenzoylethane with methylamine or aniline readily afforded the *N*-methyl- or *N*-phenyl-3,4-dibenzoyl-2,5-diphenylpyrrole.[20]

Introduction of another nitrogen atom into **43** was readily achieved[34] by utilizing 3,4-dibenzoyl-1-phenylpyrazole (**46**; $R = Ph$; $R^1 = H$) which was obtained in 85% yield on treatment of dibenzoylacetylene with *anhydro*-5-hydroxy-3-phenyl-1,2,3-oxadiazolium hydroxide (**45**; $R = Ph$; $R^1 = H$), itself prepared by acetic anhydride cyclization of *N*-nitroso-*N*-phenylglycine (**44**; $R = Ph$; $R^1 = H$). It should be noted that **45** need not be isolated in this case also but may be generated *in situ* and condensed with dibenzoylacetylene with only minor reduction in yield.[32a] Phosphorus pentasulfide–pyridine treatment of **46** resulted in its conversion into 2,4,6-triphenylthieno[3,4-*c*]pyrazole (**47**; $R = Ph$; $R^1 = H$), obtained as brick red needles in 85% yield.

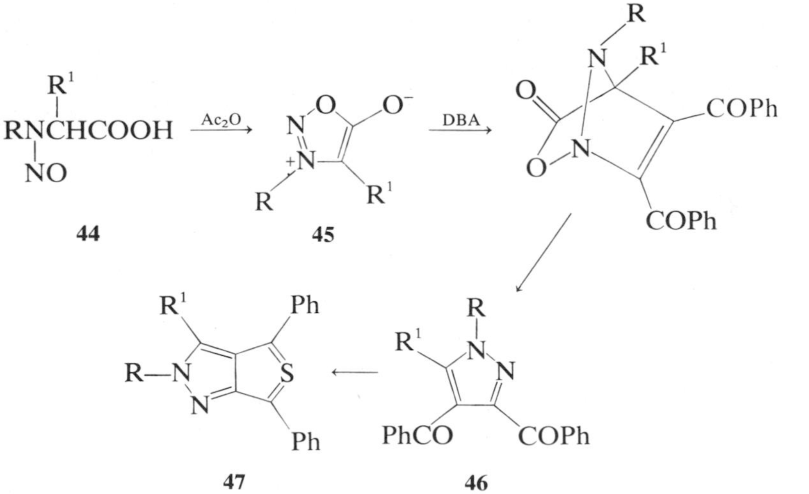

R[1]
RNCHCOOH $\xrightarrow{Ac_2O}$ **45** $\xrightarrow{DBA}$ [adduct: N–R, R[1], COPh, COPh] → **46** → **47**
NO

44 **45**

47 **46**

In contrast to the thieno[3,4-*c*]pyrrole system **43,** the thieno[3,4-*c*]-pyrazole system **47** was relatively unaffected by the nature of the substituents in the pyrazole ring, no doubt due to the stabilizing effect of the extra nuclear nitrogen atom. Thus utilization of the appropriately substituted **44** or **45** and dibenzoylacetylene gave 3,4-dibenzoyl-1-methylpyrazole (**46**; R = CH_3; R^1 = H) and 3,4-dibenzoyl-5-methyl-1-phenylpyrazole (**46**; R = Ph; R^1 = CH_3). These were readily converted into the corresponding fused thiophene systems 4,6-diphenyl-2-methylthieno[3,4-*c*]pyrazole (**47**; R = CH_3; R^1 = H) isolated as orange prisms in 76% yield and 3-methyl-2,4,6-triphenylthieno[3,4-*c*]pyrazole (**47**; R = Ph; R^1 = CH_3) obtained as lustrous brown plates in 64% yield.

The stabilizing influence of the second nitrogen atom in **47** compared to the thieno[3,4-*c*]pyrrole system **43** was quite dramatic. It was of interest then to introduce a nitrogen atom into the thieno[3,4-*c*]-thiophene system **14,** especially since the thieno[3,4-*c*][1,2,5]thiadiazole system **25** containing two nitrogen atoms had already been synthesized.[23] The requisite precursor to the required thieno[3,4-*c*]isothiazole system, 3,4-dibenzoyl-5-phenylisothiazole (**50**), was ultimately prepared by utilizing the cycloaddition of dibenzoylethylene to *anhydro*-5-hydroxy-4-phenyl-1,3,2-oxathiazolium hydroxide (**48**). The reactions of this mesoionic system illustrate the perversity of these systems and the difficulty in making generalizations regarding their behavior as "masked ylides." Although **48** undergoes reaction with dimethyl acetylenedicarboxylate,[35] reaction with dibenzoylacetylene was too sluggish at 90° so that both thermal and photochemical degradation of the mesoionic ring system occurred. However dibenzoylethylene reacted at 60°, presumably giving the intermediate **49** that, under reaction work-up conditions, underwent oxidation and loss of carbon dioxide in that or reverse order. Reaction of **50** with P_4S_{10} in pyridine readily afforded[36] 3,4,6-triphenylthieno[3,4-*c*]isothiazole (**51**) as reddish purple needles, stable in the solid state but rapidly bleached in solution, in 85% yield.

Our attempts to prepare the oxygen analog of **25** have been unsuccessful using the vicinal dibenzoyl heterocycle route. 3,4-Dibenzoyl-1,2,5-oxadiazole (**53**) was prepared from the corresponding 3,4-dibenzoyl-1,2,5-oxadizole-*N*-oxide (a furazan-*N*-oxide) (**52**), itself prepared from acetophenone and nitric acid, by treatment with stannous chloride and hydrochloric acid.[37] In this case reaction of **53** with P_4S_{10} was more effective in xylene than in pyridine, but in either solvent the product

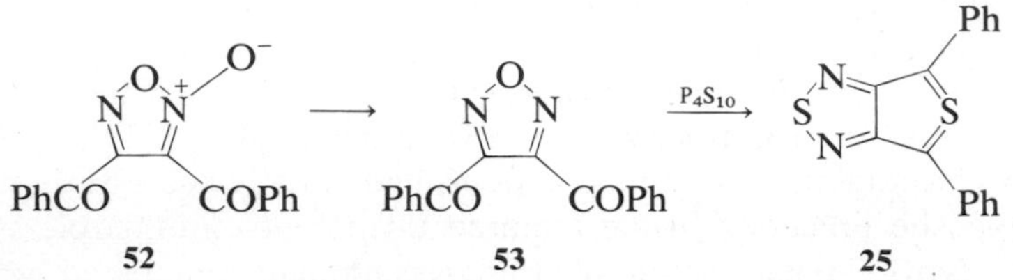

isolated was diphenylthieno[3,4-*c*][1.2.5]thiadiazole (**25**), which was obtained[36] in 65% yield. Reduction of **53** with sodium borohydride has given the corresponding dialcohol, and the synthesis of the oxygen analog of **25** is being attempted by the sulfoxide route.

The last of these nonclassical condensed thiophenes synthesized[38] to date is the thieno[3,4-*c*][1.2.3]triazole system. Reaction of hydrazoic acid with dibenzoylacetylene gave 3,4-dibenzoyl-1,2,3-triazole (**54**). Alkylation with methyl iodide is possible at two sites, and a mixture of 4,5-dibenzoyl-2-methyl- and 4,5-dibenzoyl-1-methyl-1,2,3-triazole, (**55**) and (**56**), respectively, was obtained. As this mixture was quite difficult to separate into its components, it was reacted with P_4S_{10}–pyridine, and 4,6-

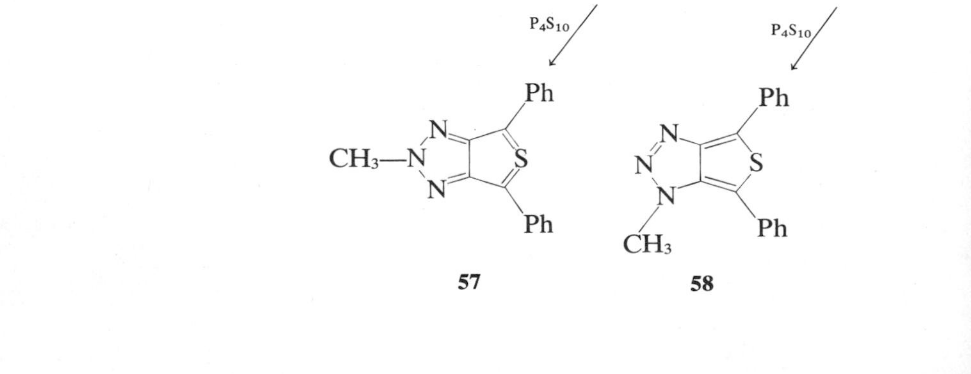

diphenyl-2-methylthieno[3,4-*d*][1.2.3]triazole (**57**) was separated from its classical isomer, 4,6-diphenyl-1-methylthieno[3,4-*d*][1.2.3]triazole (**58**), by chromatography on silica gel. The nonclassical thiophene 57 was obtained as orange, irregular prisms, m.p. 158–160° and was found to be quite stable while its isomer **58** formed orange-yellow needles, m.p. 185–186°.

This marked difference between the classical and nonclassical condensed thiophenes was also very apparent with **47** and its isomer. The latter was readily synthesized by the addition of the nitrile imine 1,3-dipole **59** to dibenzoylacetylene. The 4,5-dibenzoyl-1-(2,4-dibromophenyl)-3-phenylpyrazole (**60**) obtained[39] from this cycloaddition reacted with P_4S_{10} in pyridine to give 1-(2,4-dibromophenyl)-3,4,6-triphenylthieno[3,4-*d*]pyrazole (**61**) which was obtained as pale yellow needles.[40]

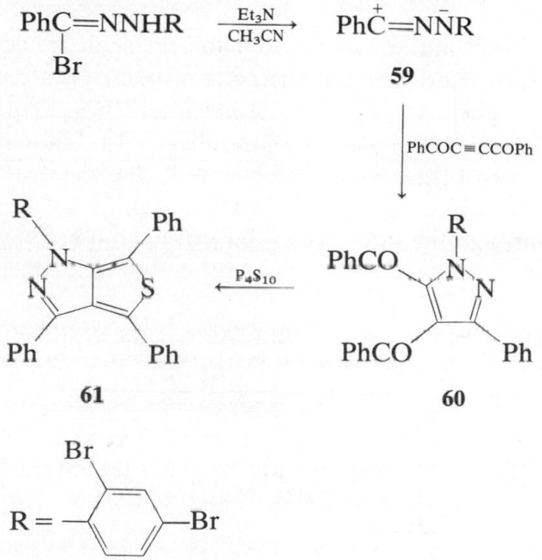

PhC(Br)=NNHR →(Et₃N, CH₃CN) Ph$\overset{+}{C}$=N$\overset{-}{N}$R (**59**) →(PhCOC≡CCOPh) **60** →(P_4S_{10}) **61**

3,4,6-Triphenylthieno[3,4-*d*]isothiazole (**65**) also formed pale yellow needles. It was prepared from 4,5-dibenzoyl-3-phenylisothiazole (**64**) and P_4S_{10}–pyridine, the necessary precursor (**64**) in this case being prepared by a cycloaddition reaction between benzonitrile sulfide (**63**) and di-. benzoylacetylene. This interesting 1,3-dipole was generated[41] *in situ* by thermolysis of 3-phenyloxathiazol-4-one (**62**) and trapping it with dibenzoylacetylene.

b. Physical Characteristics. Without exception all these nonclassical, condensed thiophenes are highly colored products with a long wavelength absorption in the 450 to 550 nm region. Their uv and visible absorption spectral data are shown in Table I. Their ir spectra show few distinguishing characteristics, but their mass spectra are extremely useful for identification purposes and show some interesting features. The mass spectrum of tetraphenylthieno[3,4-*c*]thiophene (**14**), shown in Scheme 1, illustrates the type of fragmentation observed. The spectrum is dominated by an intense molecular ion *a*, m/e 444, accounting for over 75% of the total ion current, and the intensity of the $[M+2]$ ion (15%) relative to the $M^{+\cdot}$ ion indicates the presence of two sulfur atoms in the molecule. The

TABLE I. ULTRAVIOLET AND VISIBLE ABSORPTION SPECTRA FOR SOME NONCLASSICAL CONDENSED THIOPHENES

Compound	λ_{max} nm (log ϵ)	Ref.
14	258 (4.30), 262 sh (4.27), 292 (4.15), 551 (3.92)[a]	25
	255 (4.23), 296 (4.30), 553 (4.11)[b]	21
43 ($R=CH_3$; $R^1=Ph$)	256 (4.41), 533 (3.15)[a]	25
43 ($R=R^1=Ph$)	247 (4.34), 260 (4.35), 345 sh (3.82), 526 (3.86)[b]	20
47 ($R=Ph$; $R^1=H$)	250 (4.12), 276 (4.21), 298 (4.18), 497 (4.06)[c]	34
47 ($R=CH_3$; $R^1=H$)	264 (4.40), 313 sh (3.70), 455 sh (4.41), 465 (4.44)[c]	34
47 ($R=Ph$; $R^1=CH_3$)	289 (4.33), 322 sh (3.79), 478 (4.33)[a]	34
25	275 (4.22), 312 (4.33), 330 (4.30), 558 (3.94)	23
51	237 sh (4.15), 256 sh (4.20), 275 (4.25), 297 sh (3.97), 329 sh (3.73), 526 (3.85)[c]	36
57	265 (4.29), 292 sh (4.06), 313 (4.00), 322 sh (3.98), 466 (4.24)[a]	38

[a] $CHCl_3$.
[b] $C_2H_4Cl_2$.
[c] CH_3OH.
[d] CH_2Cl_2.

Ph Ph]+· Ph Ph]++

$PhC\equiv\overset{+}{S}$ ⟵ S S $\xrightarrow{-1e}$ S S

Ph Ph Ph Ph

d, *m/e* 121(10) *a*, *m/e* 444(100) *b*, *m/e* 222(10)

* | −CS ↓ * | −Ph· ↓

Ph

Ph^+ S S +

Ph Ph

e, *m/e* 77(8) *c*, *m/e* 367(50)

*metastable transition.

Proposed fragmentation pathway for tetraphenylthieno[3,4-*c*]thiophene (**14**).

Scheme 1

doubly charged molecular ion *b*, m/e 222 is the next most abundant ion in the spectrum, and the presence of the M^{2+} ion in the mass spectra of many aromatic compounds has been such a common observation that it has led to the suggestion that their appearance be used as an experimental, empirical indication of aromatic character.[42] Other significant ions observed[21,25] include the thiobenzoyl ion *d*, m/e 121 and the phenyl ion *e*, m/e 77.

An analogous fragmentation (Scheme 2) was also observed[25] with 5-methyl-1,3,4,6-tetraphenylthieno[3,4-*c*]pyrrole (**43**; $R = CH_3$; R = Ph). The intense molecular ion *a*, m/e 441 (100), along with its doubly charged analog g, m/e 220.5 (**21**), dominated the spectrum. As would be anticipated, other significant ions in the spectrum were the thiobenzoyl ion *c*, m/e 121, the iminium ion *d*, m/e 118, and the methyl isocyanide radical ion *f*, m/e 41. Even though the molecular weights of the dibenzoyl precursor and the nonclassical condensed thiophene are the same, there is no difficulty in recognizing the latter from its characteristic, simple mass spectrum.

The incorporation of an additional heteroatom into the nucleus of **43** results in an even simpler mass spectrum. In 2,4,6-triphenylthieno[3,4-*c*]-pyrazole (**47**; R = Ph, $R^1 = H$), an intense molecular ion, m/e 352, again

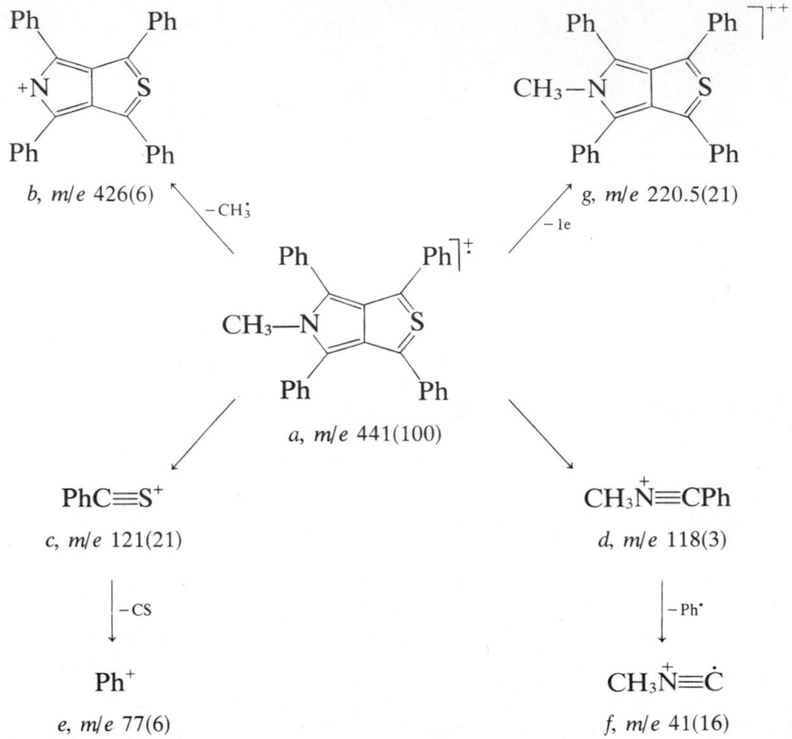

Proposed fragmentation pathway for 5-methyl-1,3,4,6-tetraphenylthieno[3,4-*c*]pyrrole (**43;** R = CH_3, R^1 = Ph)

Scheme 2

dominates the spectrum. The only other significant ions observed were the doubly charged molecular ion, m/e 176, and the thiobenzoyl, phenyl, and carbon monosulfide ions.

The nmr spectra of these ring systems, although useful for structural verification, show no unifying features. Particularly interesting is the nmr spectrum of tetraphenylthieno[3,4-*c*]thiophene (**14**) in which the phenyl protons absorb as a sharp singlet at $\delta 7.15$, suggesting that the phenyl groups are all twisted out of the plane of the fused ring system. A recent x-ray study has shown[43] this to be the case, establishing that the thienothiophene nucleus is planar and that the phenyl substituents are rotated out of the plane of the nucleus by 39.6° for those at C_1 and C_4, and by 58.4° for those at C_3 and C_6. The C-phenyl bond length was found to be 1.48 Å, corresponding approximately to an sp^2–sp^2 single bond, and these facts tend to preclude any major stabilization of the thienothiophene system by electronic interaction with the substituents, suggesting that their stabilizing effect might be steric in nature. On this basis it is difficult to understand why the introduction of *tert*-butyl groups as substituents described above did not result in an isolatable product.

This x-ray study showed some interesting features (Fig. 2). The C–S bond length is 1.706 Å, a relatively short bond while the 1-7 bond and the 7-8 bonds are 1.407 and 1.452 Å, these values being relatively long when compared to the corresponding bond lengths found for thiophene (1.714, 1.370, and 1.423 Å, respectively).

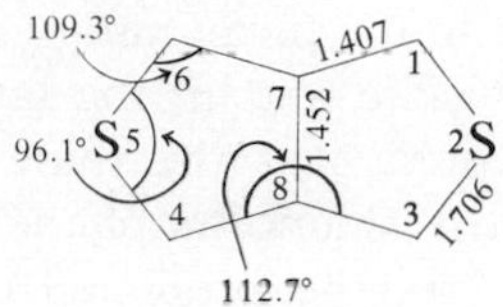

Fig. 2. Bond lengths and Bond angles for the thieno[3,4-*c*]thiophene system.

In contrast to the system in Fig. 2, the phenyl protons in 5-methyl-1,3,4,6-tetraphenylthieno[3,4-*c*]pyrrole (**43**; R = CH_3; R^1 = Ph) occurred as a multiplet at δ7.00 to 7.51. However the shape of this multiplet suggests that there is a good likelihood that the four phenyl substituents in **43** are also not coplanar with the nucleus. The chemical shift of the *N*-methyl substituent δ3.07 suggests that the nitrogen atom has very little positive charge associated with it, although this is predicted on the basis of recent molecular orbital calculations (see below). Although the nmr spectrum of diphenylthieno[3,4-*c*][1.2.5]thiadiazole (**25**) showed[23] a multiplet for the aromatic protons at ~δ8.2 with the ortho protons deshielded, this may not be due entirely to coplanarity of the phenyl substituents with the nucleus but could also be caused by the adjacent nuclear nitrogen atoms.

Earlier theoretical calculations suggested a triplet ground state for these nonclassical condensed thiophenes. However attempts to obtain an esr signal for **14** and **43** (R = CH_3; R^1 = Ph) were unsuccessful, showing that they have a singlet ground state.[21b] Recent theoretical calculations[20] using a semiempirical molecular orbital approach, CNDO/2, in which *d* orbital participation is included in contrast to the earlier calculations in which it was ignored, predict a singlet ground state for those systems studied. Figure 3 shows the relative π-overlap for the thieno[3,4-*c*]thiophene, thieno[3,4-*c*]pyrrole, and thieno[3,4-*c*]furan systems, and in Fig. 4 the π-charge densities are shown.

These data predict considerable π-bonding between carbon and sulfur of which over half was attributed to $d\pi - p\pi$ overlap. The unsymmetrical

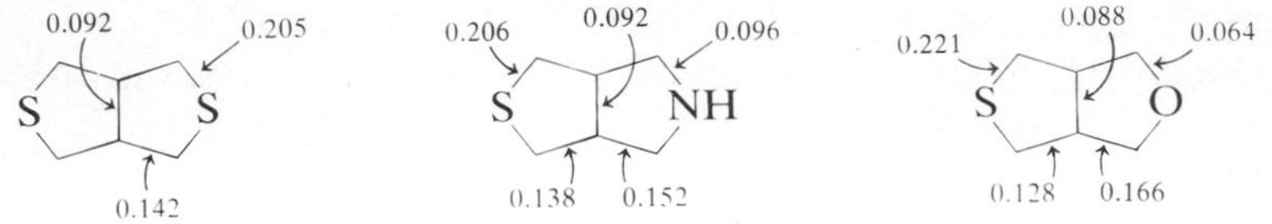

Fig. 3. Relative π-overlap for several tetravalent sulfur systems.

compounds exhibit π-charge separations of greater magnitude than that of the symmetrical thienothiophene, indicative of greater contributions to the electronic structure from dipolar canonical forms. The degree of charge separation is greater in the furan and pyrrole rings, indicating that in the former the carbonyl ylide and in the latter the azomethine ylide forms are energetically favored over the thiocarbonyl ylide. The calculated dipole moment for the thienothiophene was 0.0 D and for the thienofuran and thienopyrrole, the predicted dipole moments are 0.15 and 3.21 D, respectively. These theoretical models are particularly interesting in view of the known chemical reactions of these systems discussed below.

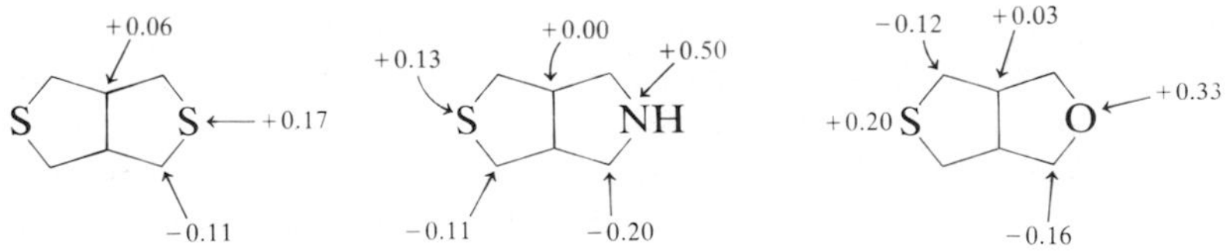

Fig. 4. π-Charge densities for several tetravalent sulfur systems.

c. Chemical Characteristics. Although tetraphenylthieno[3,4-*c*]-thiophene (**14**) is stable in the solid phase, solutions of it are rapidly bleached on exposure to light and air. This photooxidation probably involves an intermediate such as the thioozonide **66,** since the thioketone **67,** the thioketone *S*-oxide **68,** and the diketone **28** were all formed in the reaction.[21b] This ready photooxidation is probably self-sensitized and occurs in all members of this series, the ease of oxidation generally depending on the nature of the nuclear heteroatoms and substituents in the ring. 5-Methyl-1,3,4,6-tetraphenylthieno[3,4-*c*]pyrrole (**43**; $R=CH_3$; $R^1=Ph$) was readily oxidized[25] in solution in air to 3,4-dibenzoyl-2,5-diphenyl-1-methylpyrrole (**41**; $R=CH_3$; $R^1=Ph$). This oxidation occurred so readily that it necessitated working under conditions excluding light. This process also takes place in the solid state over prolonged periods of time. However the *N*-phenyl analog (**43**; $R=R^1=Ph$) is considerably more stable and could be recrystallized from acetic anhydride. Although additional heteroatoms imparted stability in the solid phase, the other

Ph Ph S S Ph Ph $\xrightarrow[h\nu]{O_2}$ Ph S S Ph O O Ph Ph ⟶ Ph S Ph PhCO CSPh

14 **66** **67**

Ph S Ph PhCO CPh ‖ SO   Ph S Ph PhCO COPh

68 **28**

known systems all underwent oxidation in solution to, principally, their diketone precursors.

The chemistry of these heteropentalenes is characterized by their tendency to undergo cycloadditions with a variety of electron-deficient dipolarophiles. Tetraphenylthieno[3,4-*c*]thiophene (**14**) underwent[21a] ready reaction with *N*-phenylmaleimide to give a 3 : 1 mixture of *endo* **69** and *exo* **70** primary 1 : 1 cycloadducts. Structural assignments were based

Ph Ph S S Ph Ph ⟶ Ph S S Ph H H Ph Ph O N—Ph O + Ph S S Ph O N—Ph O Ph Ph H H

14 **69** **70**

on the chemical shifts of the protons *α* to the imide carbonyl groups. It was originally thought that the bridgehead phenyl groups of **70** greatly deshielded the aliphatic protons *endo* to the sulfur bridge, but it was shown recently that this effect is insignificant.[21b] The difference in chemical shift between the aliphatic protons of **69** and **70** was interpreted in terms of the deshielding effect of the bridge sulfur atom in the case of the former. As a result, the resonance signal for the *exo* protons of adduct **69** was found at low field position relative to the *endo* protons of adduct **70**.

The former appeared as a singlet at δ5.16 while the latter were observed at δ4.27. This deshielding of adjacent hydrogens by a sulfur bridge has been used successfully to determine the stereochemistry of a variety of cycloadducts of this type.[44]

This cycloaddition of *N*-phenylmaleimide to **14** may be regarded as the addition of a 2π-electron system to a 4π-electron system and reactions of this type have been predicted to occur in a concerted manner based on the selection rules developed by Woodward and Hofmann.[45] The 4π-electron system in the present case has the π-electrons distributed over three atomic centers and may be regarded as a thiocarbonyl ylide[46] shown in Fig. 5. The molecular orbitals of the 4π-system can be represented as

Tetravalent Structure ⟷ Octet Structure ⟷ Sextet Structure

Fig. 5. The thiocarbonyl ylide 1,3-dipole.

combinations of *p* atomic orbitals, and it is assumed that these molecular orbitals take the form of those proposed for the allyl anion.[45b]

The cycloaddition reaction can be formulated as a parallel approach of the thiocarbonyl ylide and olefinic system, ultimately resulting in bond formation between the terminal carbon atoms.[47] A plane of symmetry perpendicular to the plane of the reactants and bisecting both the C–S–C angle of the thiocarbonyl ylide and the C–C bond of the olefin is maintained throughout this conversion. This is illustrated in Fig. 6.

A total of 10 molecular orbitals, shown in Fig. 7, are involved in this

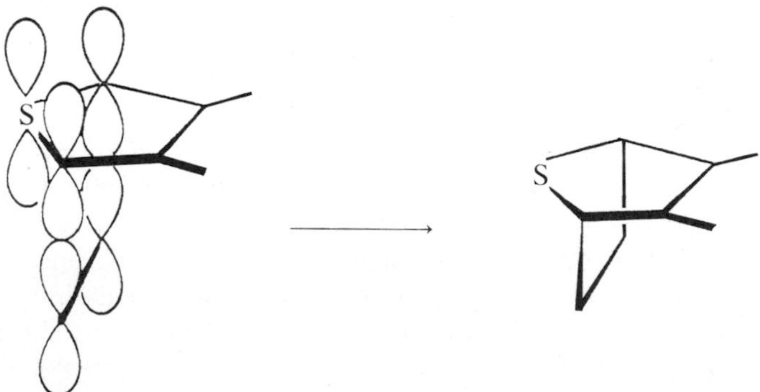

Fig. 6. Reaction of a thiocarbonyl ylide 1,3-dipole with an olefin.

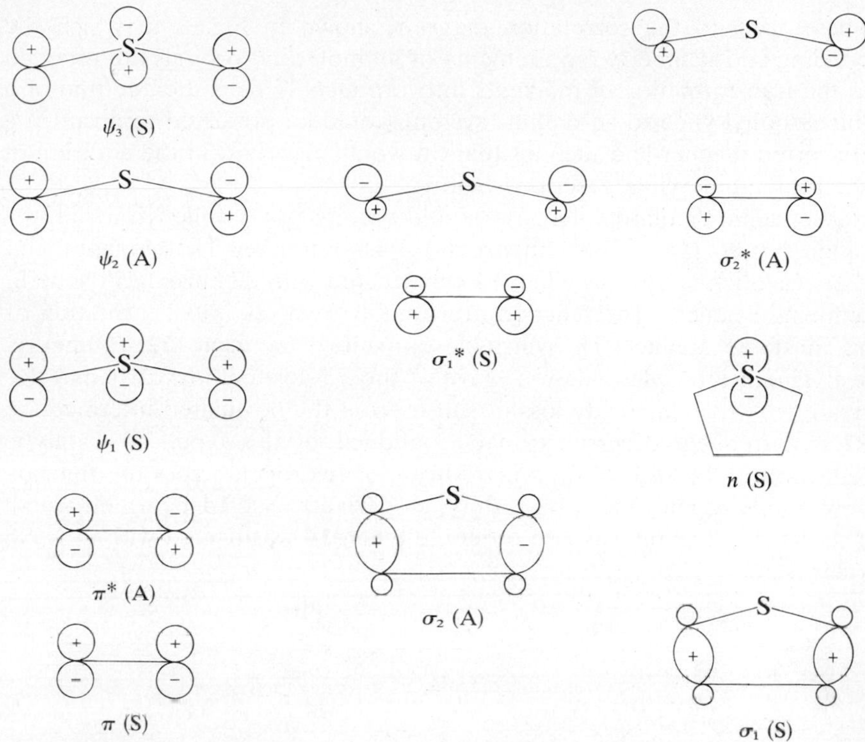

Fig. 7. Molecular orbitals involved in the cycloaddition reaction of a thiocarbonyl ylide and olefinic system.

cycloaddition: two from the olefin (π and π^*), three from the thiocarbonyl ylide (ψ_1, ψ_2 and ψ_3), and five from the product (σ_1, σ_2, n, σ_1^* and σ_2^*), and these are designated either symmetric (S) or antisymmetric (A) with respect to the plane of symmetry.

The various possible interactions among reactant orbitals and the product orbitals formed in these interactions are summarized in Table II.

TABLE II. THE VARIOUS POSSIBLE ORBITAL INTERACTIONS OF REACTANTS IN THE 1,3-DIPOLAR CYCLOADDITION REACTION OF A THIOCARBONYL YLIDE AND AN OLEFIN

Interacting reactant orbitals	Resulting product orbitals
ψ_1 and π	σ_1, σ_1^* or n
ψ_2 and π^*	σ_2, σ_2^*, or n
ψ_3 and π	σ_1, σ_1^*, or n

These lead to the correlation diagram shown in Fig. 8 in which the bonding and symmetry requirements of all molecular orbitals are retained in the transformation of reactants into products.[48] Thus the addition of a thiocarbonyl ylide to an olefinic system would be predicted to occur in a concerted manner and such a situation would also exist in the addition of the azomethine ylide described below.

Acetylenic dipolarophiles also added readily to tetraphenylthieno[3,4-*c*]thiophene (**14**). 5,6-Dibenzoyl-1,3,4,7-tetraphenylisothionaphthene (**72**) was obtained (61%) when **14** was treated with dibenzoylacetylene in refluxing xylene.[25] The reaction presumably involves initial formation of an unstable adduct **71** which loses sulfur to form **72**. Dimethyl acetylenedicarboxylate likewise gave[21b] the 5,6-bis(methoxycarbonyl) derivative of **72**. This ready loss of sulfur from the postulated intermediate **71** is a frequent occurrence in cycloadducts of this type.[24,49] Catalytic reduction of **14** also resulted in addition of hydrogen across the thiocarbonyl ylide dipole, the *cis*-1,3-dihydro derivative of **14** being obtained. This same reduction was also observed when **14** was heated with P_4S_{10} in xylene.[21b]

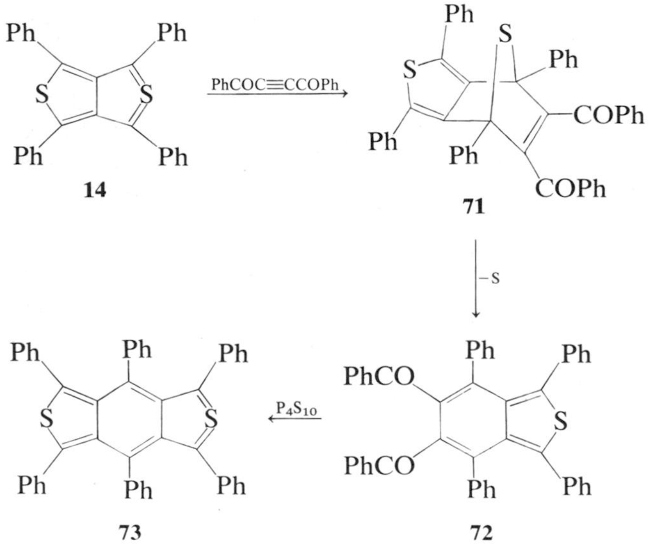

Ph Ph S S Ph Ph **14** —PhCOC≡CCOPh→ Ph S Ph S COPh Ph Ph COPh **71**

—S↓

Ph Ph Ph S S Ph Ph Ph **73** ←P_4S_{10}— PhCO Ph Ph S PhCO Ph Ph **72**

By an extension of the P_4S_{10} procedure that led to **14,** reaction of **72** gave a novel 14π-electron system containing a tetravalent sulfur atom in a 5-membered ring, hexaphenylthieno[3,4-*f*]isothionaphthene (**73**) which, in contrast to **14,** was obtained as blue needles. Although **73** added[25] olefinic dipolarophiles exclusively across the thiocarbonyl ylide dipole, its reaction with acetylenic dipolarophiles was less satisfactory, leading to

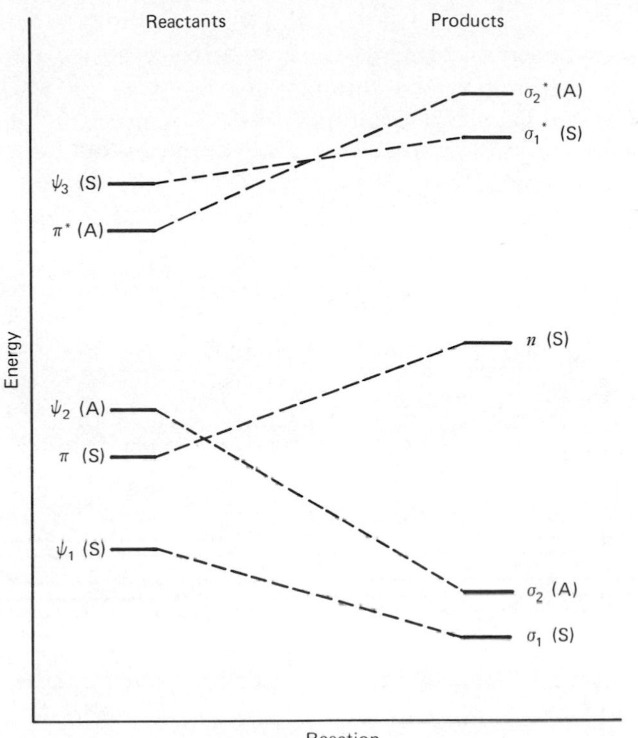

Fig. 8. The orbital correlation diagram for the cycloaddition reaction of a thiocarbonyl ylide and an olefin.

complex reaction mixtures probably due to a multiplicity of reaction sites.

The thieno[3,4-*c*]pyrrole system **43** offers more interesting possibilities compared to **14** for reaction with dipolarophiles as it contains both a thiocarbonyl ylide **43a** and an azomethine ylide **43b,** both reactive dipoles in their own right, and its reactions turned out to be of considerable interest.

Reaction of 5-methyl-1,3,4,6-tetraphenylthieno[3,4-*c*]pyrrole (**43**; R = CH_3; R^1 = Ph) with fumaronitrile in either refluxing toluene or xylene led to two products, only one of which was a primary 1 : 1 cycloadduct. There are two possible structures for this adduct: **74,** corresponding to addition across the thiocarbonyl ylide dipole of **43** (R = CH_3, R^1 = Ph), and **75,** in which addition has occurred across the azomethine ylide of **43**. The chemical shift of the N–CH_3 group in the adduct was δ3.13, not significantly different from the chemical shift of the N–CH_3 group in **43,** indicating structure **74** for the adduct. Resonance signals for the protons α to the nitrile groups were observed as doublets at δ4.12 and δ4.58

consistent with the shielding effect of the bridge sulfur atom. The observed coupling constant of 3.9 Hz indicated that the *trans* nature of the dipolarophile was maintained throughout the course of addition. The second product isolated was identified as 5,6-dicyano-2-methyl-1,3,4,7-tetraphenylisoindole (**76**), which was also obtained from **74** on treating with sodium methoxide.[25]

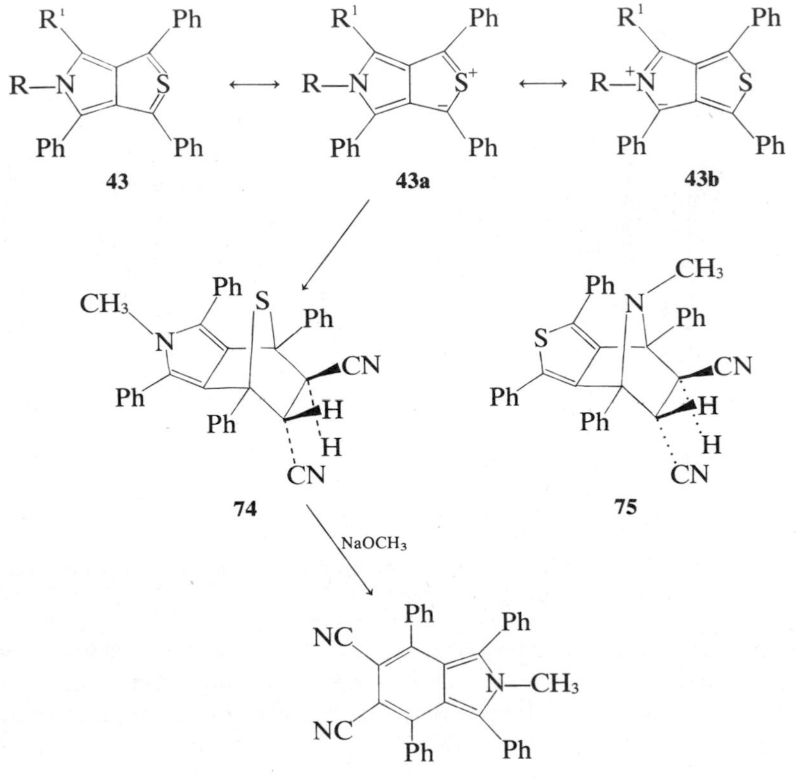

The yields of the primary cycloadduct and the isoindole were very dependent on reaction conditions. In refluxing toluene, the yield of primary cycloadduct **74** was 67%, with the yield of the isoindole **76** being 5%. In refluxing xylene the yield of the latter was increased to 53%, apparently at the expense of the former, whose yield was decreased to 10%.

This suggests that the isoindole is formed from the primary cycloadduct by thermal elimination of the elements of hydrogen sulfide. This may involve an initial loss of a sulfur atom from the primary adduct to give a diradical species from which the sulfur atom could then extract the hydrogen atoms α to the nitrile groups leading to the formation of **76** and hydrogen sulfide. This would be consistent with the dehydrogenating ability of sulfur. An alternative process involving an initial oxidation of **74,** followed by extrusion of sulfur, appears unlikely.

It is particularly interesting that in refluxing benzene, the reaction of fumaronitrile took another course. In this instance addition occurred exclusively across the azomethine ylide to give **75**.

The major difference between the adducts resulting from azomethine ylide addition and thiocarbonyl ylide addition was thermal stability. On attempted dissolution in warm solvents, the adduct **75** underwent a facile reversion to the initial reactants. On the other hand, adduct **74** was readily recrystallized from solvents such as ethanol and acetonitrile. A similar thermal instability was also observed in the solid state. When adduct **75** was heated in refluxing xylene for 14 hr, 5,6-dicyano-2-methyl-1,3,4,7-tetraphenylisoindole (**76**) was formed in 60% yield, and trace amounts of the adduct **74** were detected in the reaction mixture.

These results were rationalized in terms of an initial retro-1,3-dipolar cycloaddition of adduct **75** to the initial **43** and fumaronitrile, which then recombine to form adduct **74**. Under the reaction conditions **74** undergoes thermal loss of the elements of hydrogen sulfide to form **76**.

All of the observations encountered in the addition of fumaronitrile to the thienopyrrole **43** suggest that this is a reaction that displays both kinetic and thermodynamic product control, with the adduct **75** being the kinetically controlled product and the adduct **74** being the thermodynamically controlled product.

Similar modes of addition were observed with several other olefinic dipolarophiles. In refluxing benzene addition of *N*-phenylmaleimide to **43** occurred across the azomethine ylide to give adduct **77**, whereas in refluxing xylene addition occurred across the thiocarbonyl ylide to give adduct **78**. In this instance, however, no thermal loss of the elements of hydrogen sulfide occurred.

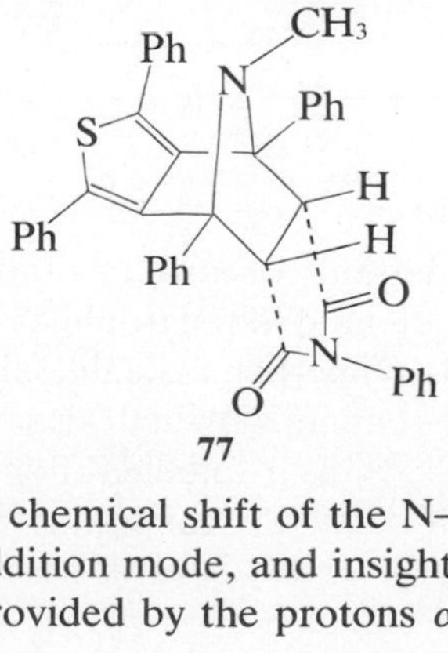

77 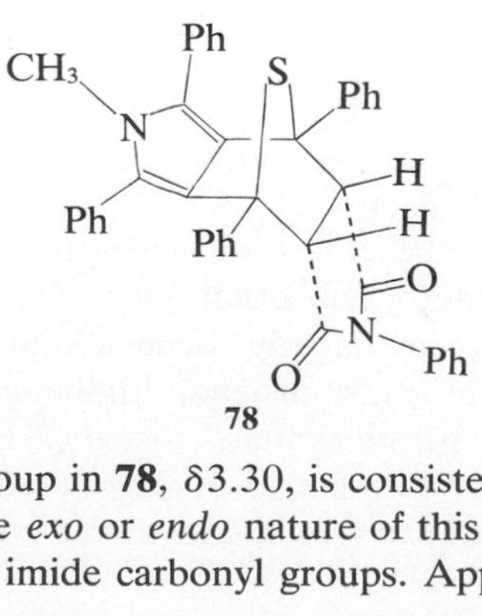 **78**

The chemical shift of the N–CH_3 group in **78**, $\delta 3.30$, is consistent with this addition mode, and insight into the *exo* or *endo* nature of this adduct was provided by the protons α to the imide carbonyl groups. Appearing as a singlet at δ5.15, they favor the latter configuration. However in the case of adduct **77** the question of *exo* or *endo* isomerism is not as readily answered due to a lack of suitable model systems. An *endo*-adduct was suggested, however, based on the Diels-Alder *endo*-rule.

Acrylonitrile and ethyl acrylate both undergo addition across the azomethine ylide of **43** in refluxing benzene, giving the adduct **79** (R = CN and COOEt, respectively). In toluene or xylene, isoindole derivatives **80** (R = CN and COOEt, respectively) were formed in addition to these azomethine ylide adducts. In these cases no thiocarbonyl ylide adducts were isolated, apparently undergoing facile conversion into the isoindoles **80**. The results above are summarized in Table III. It is interesting to note that these modes of addition are consistent with the theoretical model for this ring system described earlier.

TABLE III. PRODUCT DISTRIBUTION IN THE CYCLOADDITIONS OF 5-METHYL-1,3,4,6-TETRAPHENYLTHIENO[3,4-*c*]PYRROLE (**43**; R = CH_3, R^1 = Ph) WITH OLEFINIC DIPOLAROPHILES

			Yield %		
Dipolarophile	Solvent[a]	Time (hr)	Azo-methine ylide adduct	thio-carbonyl ylide adduct	Isoindole
Fumaronitrile	B	1	63		
Fumaronitrile	T	12		67	5
Fumaronitrile	X	12		10	53
N-Phenylmaleimide	B	0.5	68		
N-Phenylmaleimide	X	12		73	
Acrylonitrile	B	12	74		
Acrylonitrile	T	9	48		8
Acrylonitrile	X	24	4		55
Ethyl acrylate	B	15	68		
Ethyl acrylate	T	12	49		10
Ethyl acrylate	X	30	5		61

[a] All reactions carried out at the reflux temperature of the solvent: B = benzene; T = toluene; X = xylene.

From the ease with which these cycloadditions occurred, it would be anticipated that other electron-deficient olefinic dipolarophiles would undergo an equally facile addition to **43**. Dimethyl maleate, dimethyl fumarate, norbornene, diphenylcyclopropenone, phenyl isocyanate, phenyl isothiocyanate, benzoyl isocyanate, and *p*-toluenesulfonyl isocyanate all failed to undergo reaction with **43** under conditions ranging from refluxing benzene to refluxing xylene. While there is no obvious explanation for this lack of reaction, it could conceivably be due to unfavorable steric interactions in the cycloaddition process.

Acetylenic dipolarophiles added readily to **43** ($R=CH_3$; $R^1=Ph$). Dimethyl acetylenedicarboxylate reacted[21b] readily in the cold to form the nitrogen-bridged adduct **81** ($R=COOCH_3$); on the other hand dibenzoylacetylene required refluxing benzene for the reaction to occur,[25]

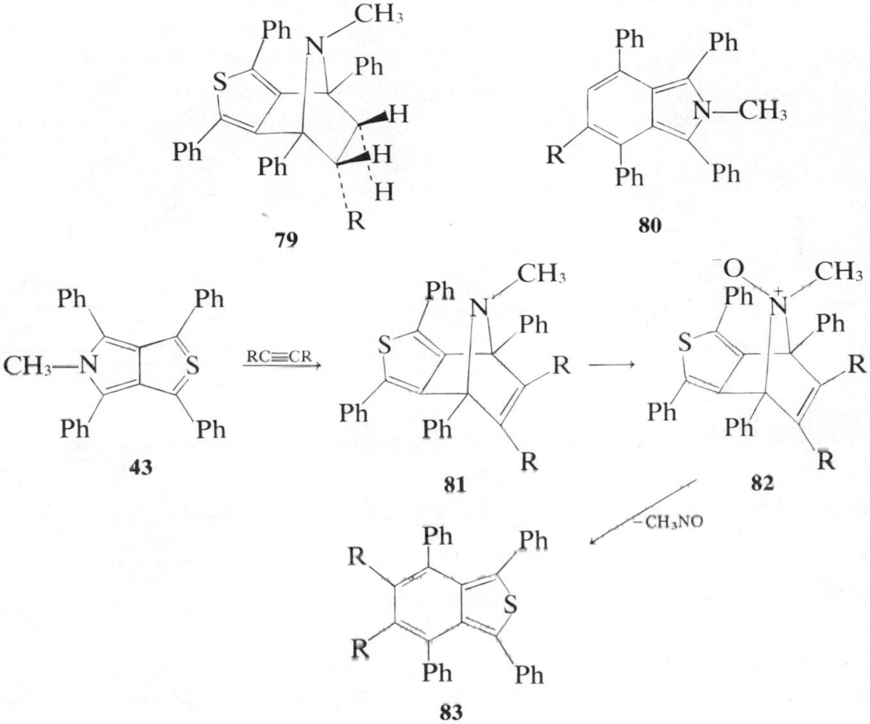

in this case giving **81** (R=COPh). As would be anticipated in a product like **81,** the chemical shift of the NCH_3 group was observed at $\delta 2.02$, considerably upfield relative to its position in **43**. Oxidation of **81** with *m*-chloroperoxybenzoic acid gave a 5,6-disubstituted-1,3,4,7-tetraphenylisothianaphthene (**83**; $R=COOCH_3$, COPh), presumably through the intermediacy of the *N*-oxide **82**. Under a variety of reaction conditions with dibenzoylacetylene, addition to **43** occurred exclusively across the azomethine ylide.

Introduction of an additional nitrogen atom into **43** not only stabilizes the ring system but also greatly influences the course of its reactions.

Considered to have contributions to its electronic structure from a thiocarbonyl ylide **47a** and an azomethine imine ylide **47b**, the thieno[3,4-*c*]-pyrazole system **47** was found to undergo cycloadditions exclusively at the thiophene ring.[34]

With *N*-phenylmaleimide in refluxing benzene, two primary 1 : 1 cycloadducts were obtained from **47** (R = Ph; R^1 = H). Distinction as to the site of addition was readily made from the nmr spectra, in which only two protons were observed in the aliphatic region. This is consistent with addition across the thiocarbonyl ylide of **47**, and the products represent *exo*- and *endo*-adducts **84** and **85**, respectively. Addition across the azomethine imine ylide to give **86** or its *endo*-analog would result in the observation of three protons in the aliphatic region of the spectrum.

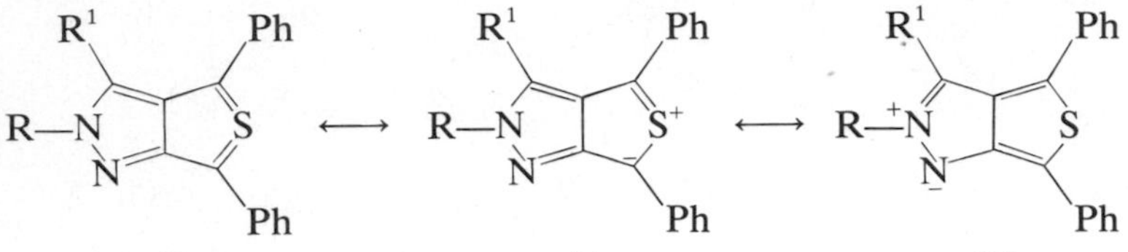

47 **47a** **47b**

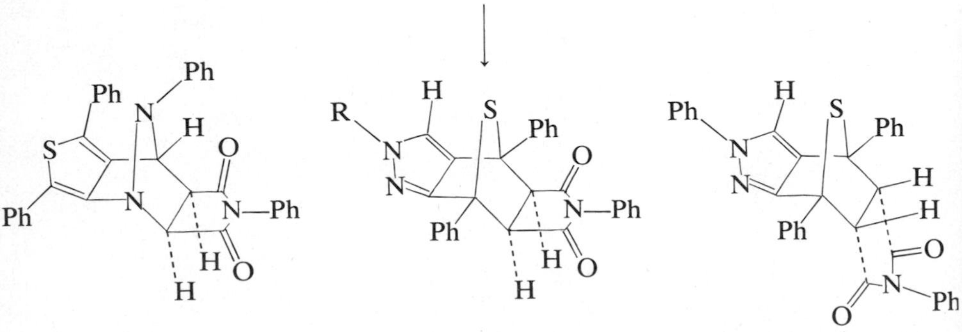

86 **84** **85**

Structural assignments for **84** and **85** were based on the chemical shifts of the protons α to the imide carbonyl groups considering the effect of the bridge sulfur atom. Thus these protons in **84** were observed at δ4.24 while the protons in **85** were deshielded and exhibited a signal at δ4.82.

The isomer ratio in this cycloaddition was found to be temperature dependent. A value of 18 : 1 favoring the *endo*-adduct was found when the reaction was carried out in refluxing benzene. This fell to 9 : 1 when refluxing toluene was used as solvent.

Formation of both stereoisomers was not observed consistently. In the thienopyrazoles **47**, when R = CH_3, R^1 = H, and R = Ph, R^1 = CH_3, only one primary 1 : 1 cycloadduct was obtained. Addition occurred across the thiocarbonyl ylide, and assignment of the *endo* structures was based on the chemical shift of the protons α to the imide carbonyl groups at $\delta 4.77$ and $\delta 4.92$, respectively.

The reaction of fumaronitrile and 2,4,6-triphenylthieno[3,4-*c*]pyrazole (**47**; R = Ph; R^1 = H) in refluxing benzene gave two products: a thermally unstable primary 1 : 1 cycloadduct of uncertain stereochemistry (**87a** or **87b**) and 5,6-dicyano-2,4,7-triphenyl-2*H*-indazole (**88**). Two protons in the aliphatic region of the nmr spectrum of **87a** (or **87b**) established that the addition had occurred across the thiocarbonyl ylide of **47,** and a coupling constant, J = 4.0 Hz for the protons α to the nitrile groups confirmed that the stereochemistry of the dipolarophile was maintained throughout the addition. In refluxing xylene the mode of addition was not altered; however the yield of the 2*H*-indazole system was increased apparently at the expense of the primary cycloadduct, which was also converted into **88** in a base-catalyzed process.

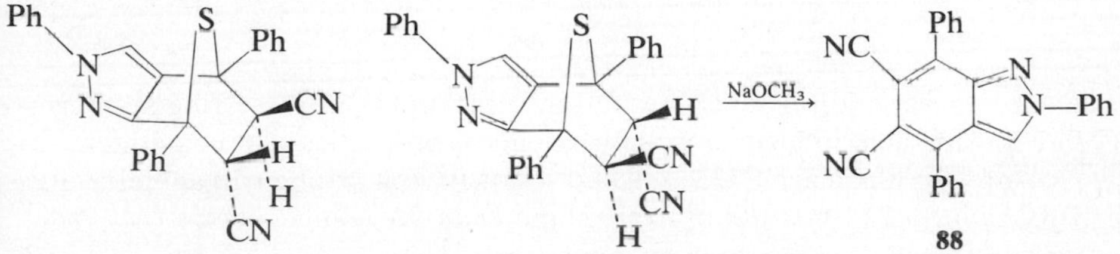

87a **87b** $\xrightarrow{NaOCH_3}$ **88**

As in the thienopyrrole system, olefinic dipolarophiles such as dimethyl fumarate, dimethyl maleate, *trans*-dibenzoylethylene, ethyl vinyl ether, ketene diethyl acetal, norbornene, and diphenylcyclopropenone did not form cycloadducts. A similar lack of reactivity was observed also with the heterocumulenes phenyl isocyanate, phenyl isothiocyanate, and *p*-toluenesulfonyl isocyanate.

Acetylenic dipolarophiles were found to add to **47** exclusively across the thiocarbonyl ylide dipole regardless of the reaction conditions. The initially formed cycloadducts were unstable, readily eliminating sulfur. Thus 5,6-bis(methoxycarbonyl)-2*H*-indazole derivatives (**90**; R^2 = $COOCH_3$) were formed from **47** (R = Ph, R^1 = H; R = CH_3, R^1 = H; R = Ph, R^1 = CH_3) and dimethyl acetylenedicarboxylate, apparently via the intermediate **89**. Although dibenzoylacetylene reacted with **47** (R = Ph, R^1 = H), giving 5,6-dibenzoyl-2,4,7-triphenyl-2*H*-indazole (**90**; R = Ph, R^1 = H, R^2 = COPh), diphenylacetylene failed to react.

The contiguous dibenzoyl groups in **90** (R = Ph, R^1 = H, R^2 = COPh), allowed further elaboration of the ring system to the thieno[3,4-*f*]-2*H*-indazole system **91**, a 14π-electron nonclassical condensed thiophene related to **73**.

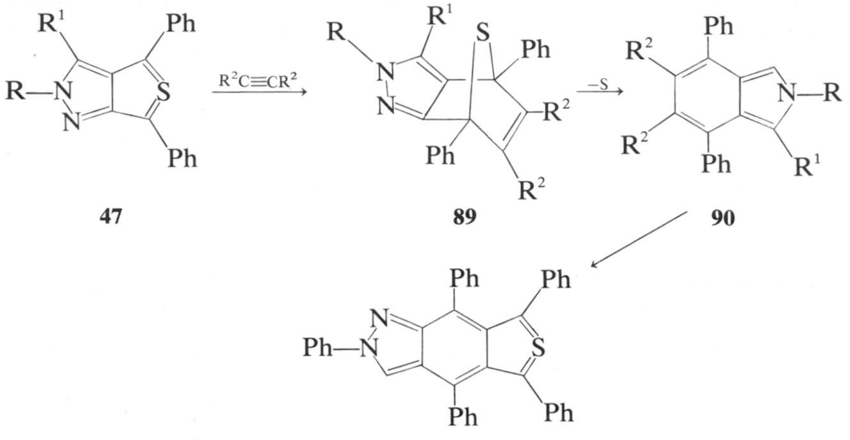

47 → ($R^2C{\equiv}CR^2$) **89** → (−S) **90** → **91**

Further variation of the heteroatom constituents in these ring systems has considerable influence on their chemical properties. Diphenylthieno-[3,4-*c*][1.2.5]thiadiazole (**25**) reacted[23] slowly with *N*-phenylmaleimide at 140° giving a 2:1 mixture of *exo* **92** and *endo* **93** isomers. The structures were assigned[23] on the basis of nmr data.

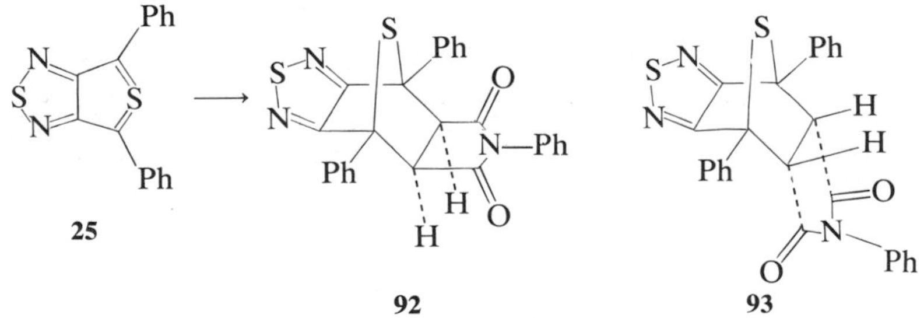

25 → **92** + **93**

In contrast to **14**, the thienothiadiazole[25] on irradiation of a rigorously degassed methylene chloride solution gave a colorless, crystalline dimer (30%) that reverted to its precursor on heating. This dimer may be represented as **94** or its isomer.

Reaction of **25** with dibenzoylacetylene in either toluene or xylene gave[36] 5,6-dibenzoyl-4,7-diphenyl-2,1,3-benzothiadiazole (**95**), a reaction analogous to those observed with other representatives of this class of compounds.

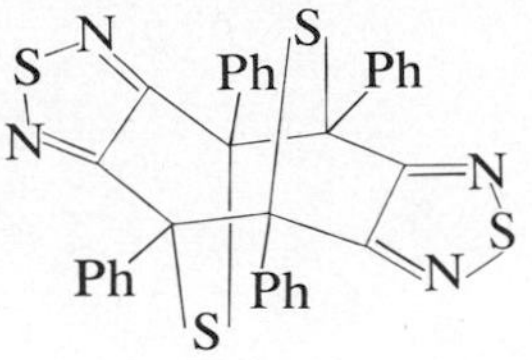

94

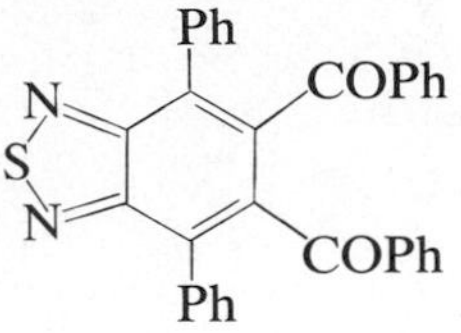

95

The reaction of *N*-phenylmaleimide with 3,4,6-triphenylthieno[3,4-*c*]-isothiazole (**65**) proved[36] to be so sluggish that, for most practical purposes, no reaction occurred. However fumaronitrile did react over 45 days in refluxing xylene. Instead of the initial primary 1:1 adduct, the product isolated was shown to be 5,6-dicyano-3,4,7-triphenyl-2,1-benzothiazole (**96**; R = CN). Loss of the elements of hydrogen sulfide from the primary adduct no doubt occurred due to the prolonged reaction time at elevated temperatures.

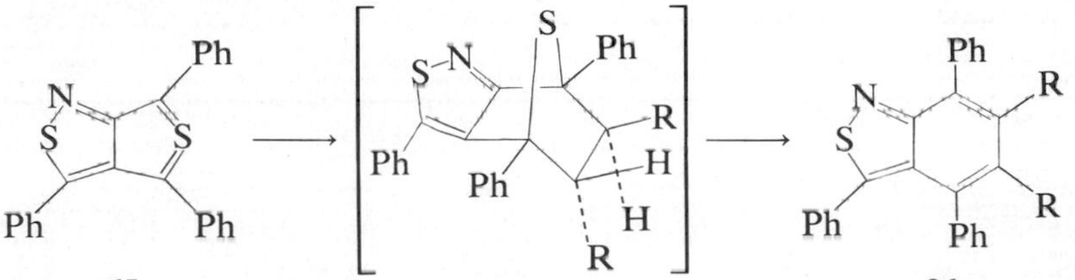

65

96

Dimethyl acetylenedicarboxylate and dibenzoylacetylene underwent reaction with **65** no more readily than fumaronitrile, approximately 45 days in refluxing xylene being required for the reaction to be completed. The 2,1-benzothiazole **96** (R = $COOCH_3$ and COPh) was formed in each case in 61% and 45% yield, respectively.

In contrast to the reactions observed with diphenylthieno[3,4-*c*]-[1.2.5.]thiadiazole (**25**), the thieno[3,4-*c*][1.2.3]triazole system **57** was not observed to undergo reaction with acetylenic and olefinic dipolarophiles within a reasonable time.[38]

2. *Containing Sulfur in a Thiadiazole Nucleus*

The only heteropentalene of this type prepared to date is 1,2,5-thiadiazolo[3,4-*c*][1.2.5]thiadiazole (**97**), prepared[50] by ring closure of 3,4-diamino-1,2,5-thiadiazole 1,1-dioxide with excess sulfur monochloride in DMF (59% yield) or excess sulfur dichloride in DMF (75% yield). It was also prepared by ring closure of 3,4-diamino-1,2,5-thiadiazole with 2 moles of $SOCl_2$ in pyridine solution (81.5% yield). A similar high yield

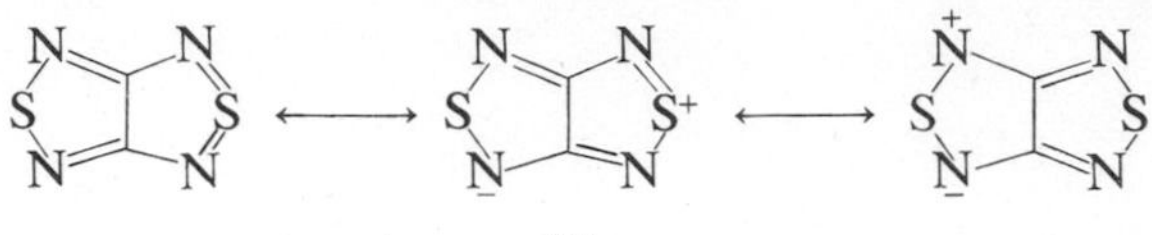

97

(83%) was obtained by using sulfur monochloride in DMF as the cyclization agent. The most convenient synthesis of **97** was the sulfur dichloride cyclization of oxamide dioxime in DMF at 55°, in this case a 66% yield being obtained.

The physical properties of **97** are in sharp contrast to those of the nonclassical, condensed thiophenes described above. It is a planar, nonpolar, volatile, weakly basic colorless solid, soluble in organic solvents and only very slightly soluble in water. It is hydrolyzed by water at 75° to 3,4-diamino-1,2,5-thiadiazole, oxamide and elemental sulfur.

An x-ray structure analysis of **97** showed it to be planar with the following characteristics: bond angles: NSN 103.2°, SNC 104.4° and NCC 114°; bond lengths: NS 1.62 Å, NC 1.35 Å and CC 1.44 Å.

This leaves seven members of this type of heteropentalene to be synthesized. The synthesis of furo[3,4-*c*][1.2.5]thiadiazole (**98**) and pyrrolo[3,4-*c*][1.2.5]thiadiazole (**99**) and a study of their reactions would be particularly interesting in view of the great reactivity of the analogous systems in the thiophene series.

Although no representatives have been reported to date, a series of heteropentalenes in which the tetravalent sulfur is in an isothiazole nucleus are also possible. Represented in general terms by structure **100**, where Y = 0, S, NR, and X = Z = N, C, it is possible to devise 11 such ring systems as, for example, the pyrrolo[3,4-*c*]isothiazole (**101**) and the pyrazolo[3,4-*c*]isothiazole (**102**).

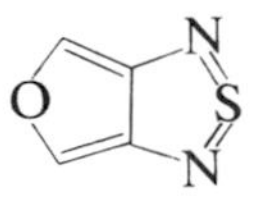

98

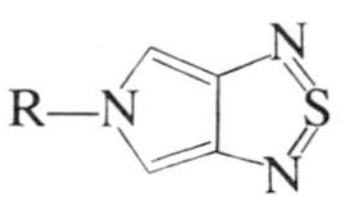

99

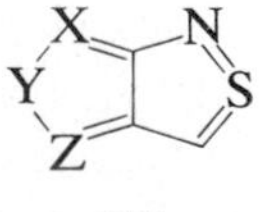

100

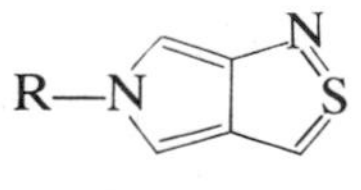

101

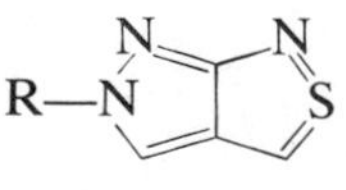

102

3. *Heteroatoms in One Ring Only*

The heteropentalenes above contain heteroatoms in both 5-membered rings, but it is also possible to have analogous systems with C–C points of fusion in which the heteroatoms are in one ring only. Represented by the general structure **103**, where X = Y = O, S, NR, and Z = C, N, there are 12 such systems that can be represented as internal ylides.

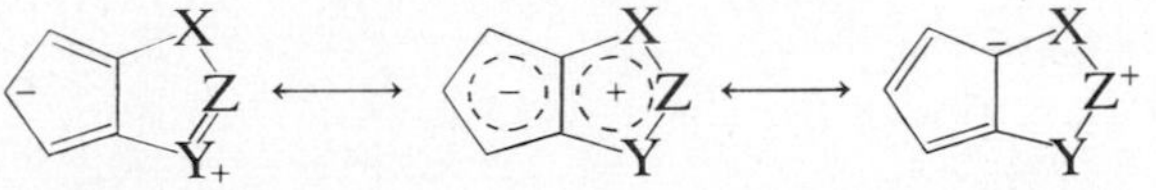

103 ⟷ **103a** ⟷ **103b**

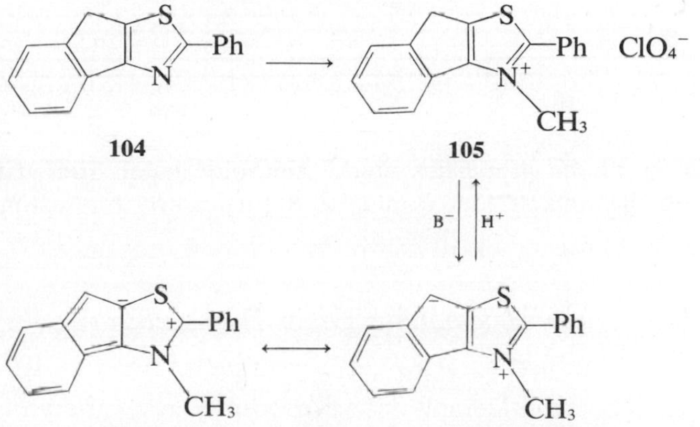

104 ⟶ **105** (ClO_4^-)

105 ⇌ (B^- / H^+) **106**

106a ⟷ **106**

A representative of this type of system has been described[51] and was prepared from 2-phenylindeno[2,1-*d*]thiazole (**104**) by methylation and subsequent treatment of the salt **105** with base. 3-Methyl-2-phenylindeno[2,1-*d*]thiazole (**106**) was obtained as an unstable, light-sensitive dark-purple solid that decomposes above 164° and reverts to its cation on treatment with acid. Molecular orbital calculations indicate an unequal distribution of charges in **106,** and the position of the longest wavelength band in the spectrum shifted to shorter wavelengths with increasing polarity of the solvent [λ_{max} (nm) 550 (benzene), ~540 sh (acetone), ~490 sh (ethanol), ~480 sh (methanol)] indicative of an intramolecular charge transfer.

A number of indeno[1,2-*b*]imidazoles have also been prepared[52] from the corresponding 2-amino-1-indanones (**107**). Formation of the thiol-substituted imidazole ring **108** with subsequent removal of the thiol group with Raney nickel gave **109**. Although it has not yet been reported, conversion of **109** into the internal ylide **110** should be readily possible

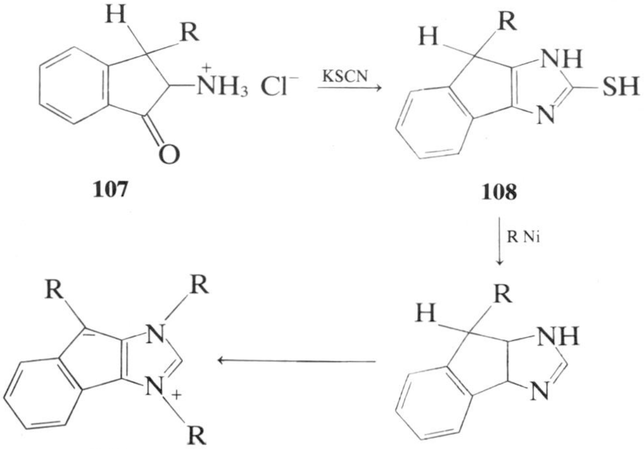

107 **108** **110** **109**

and, in view of the properties of the thiazole analog **106,** it would be anticipated that this system would be a particularly interesting one for study.

4. *Heteroatoms Other Than Sulfur*

There has been considerable interest recently in a variety of azapentalenes, and those of the type currently under discussion can be represented by the general formula **111** where $a = c = d = f =$ N, C and $b = e =$ O, NR. Consideration of these variants in the heteroatom composition leads to 17 possible systems, such as the 2*H*,5*H*-pyrrolo[3,4-*c*]-pyrrole system **112**, the 2*H*,5*H*-furo[3,4-*c*]pyrrole system **113**, and the 2*H*,5*H*-pyrazolo[4,3-*c*]pyrazole system **114**. Those most readily available, partly because of ease in synthesis and partly because of their stability, are those containing at least two heteroatoms in each ring of the fused system.

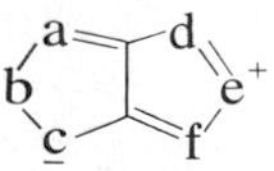

111 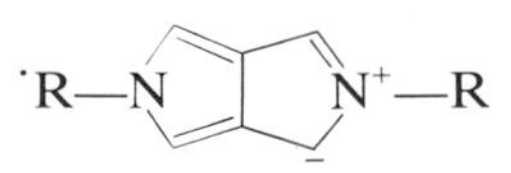 **112** 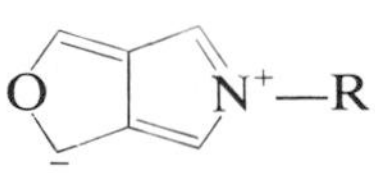 **113** 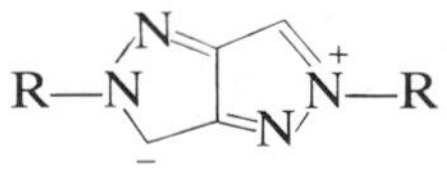 **114**

Tetraazapentalene derivatives, especially the 2*H*,5*H*-pyrazolo[4,3-*c*]-pyrazole system **114**, are readily available by several routes.[53] Dehydrohalogenation of 1,2-dichloroglyoxal-bis-phenylhydrazone (**115**) generated the oxalodinitrile-bis-phenylimine (**116**) which, in the absence of dipolarophiles, dimerized to the dark red 2,5-diphenyl-3,6-di(phenylazo)-2*H*,5*H*-pyrazolo[4,3-*c*]pyrazole (**118**), conceivably via an intermediate such as **117**. This dimerization occurred in the absence of suitable dipolarophiles. In the presence of olefinic dipolarophiles, however, normal 1,3-dipolar cycloaddition was observed, and from **116** and ethyl fumarate a 3,3′-bipyrazoline was obtained in 80% yield. The dimerization

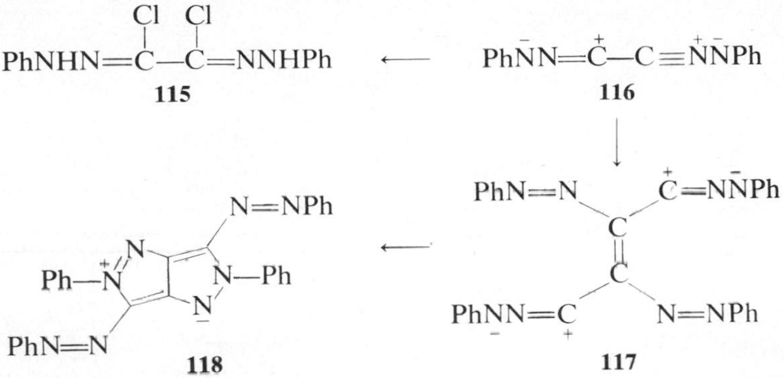

PhNHN=C(Cl)—C(Cl)=NNHPh ⟵ $Ph\bar{N}N=\overset{+}{C}—C\equiv\overset{+}{N}\bar{N}Ph$

115 **116**

118 ⟵ **117**

of **116** to the bicyclic system is sensitive to steric considerations. Although the phenyl and *o*-tolyl analog of **116** undergo the dimerization, the strongly sterically hindered 2,6-dimethylphenyl analog does not form the tetraazapentalene. Instead intramolecular rearrangement occurred to (7-methylindazolyl-2)-(2,6-dimethylphenylimino)acetonitrile (**119**). Oxidation of **118** with hydrogen peroxide gave the corresponding bis-*N*-oxide whereas treatment with nitric acid resulted in the formation of 2,5-diphenyl-3,6-dinitro-2*H*,5*H*-pyrazolo[4,3-*c*]pyrazole (**120**).

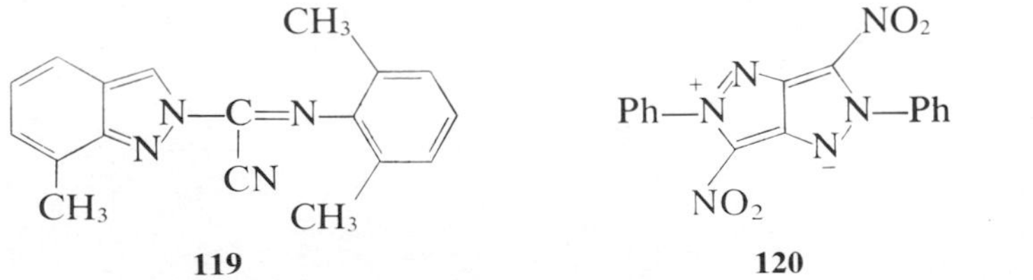

119 **120**

This same ring system has also been obtained from arylazoethynylarenes, which readily dimerize on heating.[53,54] Thus *p*-chlorophenylazoethynylbenzene (**121**), prepared from *p*-chlorobenzenediazonium chloride and silver phenylacetylide, on prolonged heating in boiling cyclohexane, gave 2,5-bis(*p*-chlorophenyl)-3,6-diphenyl-2*H*,5*H*-pyrazolo[4,3-*c*]-pyrazole (**122**) as pale yellow crystals. Oxidation of 122 with potassium permanganate gave[55] 3-benzoyl-1-*p*-chlorophenyl-4-*p*-chlorophenylazo-5-phenylpyrazole (**123**), whereas perbenzoic acid resulted in ring opening to **123** and further oxidation to the *N*-oxide **124**. Treatment of **124** with hot triethyl phosphite resulted in regeneration of **123,** and when the reaction period was extended to 21 hrs, the initial tetraazapentalene **122** was formed in 68% yield.

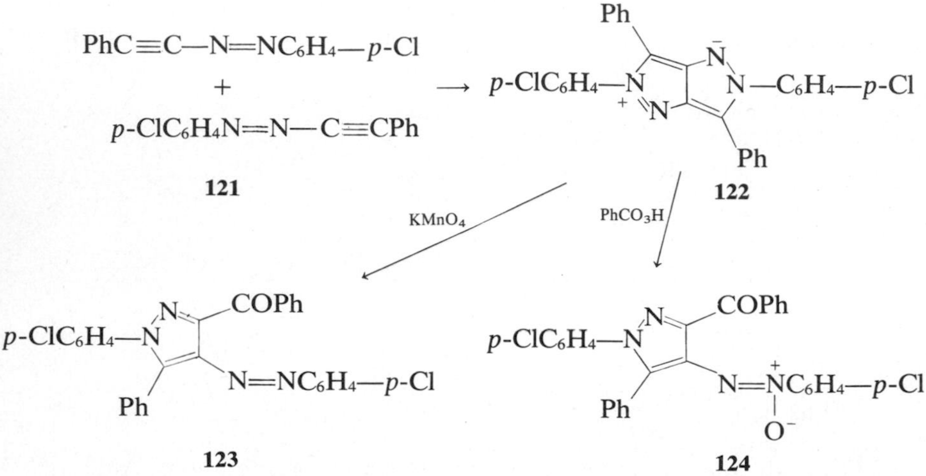

PhC≡C—N=NC$_6$H$_4$—*p*-Cl
+
p-ClC$_6$H$_4$N=N—C≡CPh

121

→ **122** (*p*-ClC$_6$H$_4$—N, N—C$_6$H$_4$—*p*-Cl, Ph, Ph)

KMnO$_4$ → **123**: *p*-ClC$_6$H$_4$—N, COPh, N=NC$_6$H$_4$—*p*-Cl, Ph

PhCO$_3$H → **124**: *p*-ClC$_6$H$_4$—N, COPh, N=N$^+$(O$^-$)C$_6$H$_4$—*p*-Cl, Ph

This reductive cyclization route to azapentalenes shows considerable promise as a means of preparing diverse structures.[56] Phenyl and substituted phenyl groups have been introduced into the 2- and 5-positions of **122**. Though there is no direct evidence, it has been suggested[56] that either an oxyphosphonium betaine **125** or a carbene **126** are intermediates in this reaction.

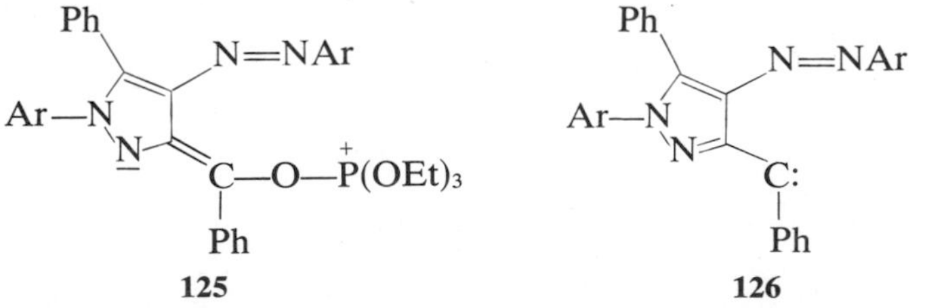

125: Ph, Ar—N, N=NAr, C—O—P$^+$(OEt)$_3$, Ph

126: Ph, Ar—N, N=NAr, C:, Ph

This reductive process led the same authors to consider an oxidative cyclization as a route to azapentalenes. This was realized by utilizing selenium dioxide as an oxidizing agent for a strategically placed methylene group. 5-Benzyl-1-methyl-3-phenyl-4-phenylazopyrazole

(**127**), on treatment with selenium dioxide in acetic acid at 80 to 90°, afforded 1-methyl-3,5,6-triphenyl-1*H*,5*H*-pyrazolo[3,4-*d*]pyrazole (**128**). Although **128** need not be represented as an ylide, variation of the nitrogen substitution in the precursor does enable the azapentalenes with ylidic characteristics to be formed. Thus reaction of 3-benzyl-1,5-diphenyl-4-phenylazopyrazole (**129**; R = Ph) and 3-benzyl-1-*p*-chlorophenyl-4-*p*-chlorophenylazo-5-phenylpyrazole (**129**; R = *p*-ClC_6H_4) on oxidation with selenium dioxide gave the phenyl analog of **122** in 12% yield and **122** itself in 14% yield, respectively.

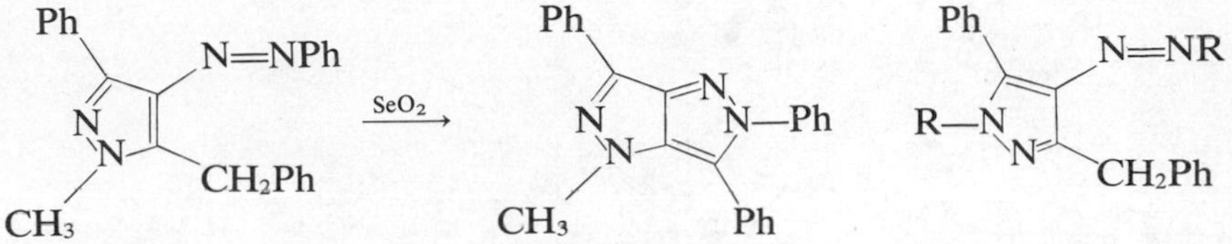

127 —SeO_2→ **128**; **129**

It has been possible to replace additional ring carbon atoms with nitrogen atoms and the representative of this system reported[57] is 2,5-diphenyl-2*H*,5*H*-1,2,3-triazolo[4,5-*d*][1.2.3]triazole (**133**). Condensation of 4,5-diamino-2-phenyl-1,2,3-triazole (**130**) with nitrosobenzene in aqueous potassium hydroxide afforded 4-amino-2-phenyl-5-phenylazo-1,2,3-triazole (**131**), which was readily converted into the corresponding azide (**132**). Gentle heating of the azide in decahydronaphthalene gave a quantitative yield of the hexaazaproduct **133** as colorless leaflets. A variety of aryl derivatives of **133** has been prepared in this way.

In a new approach to the synthesis of ring systems of this type that shows considerable promise of adaptation to the synthesis of other systems, a nitrene ring closure onto a basic nitrogen atom was used as the final step.[58] *C*-Phenyl-*N*-(ethyl oxalyl)sarcosine (**134**), prepared from *N*-methyl-*C*-phenylglycine hydrochloride and ethyl oxalyl chloride, and

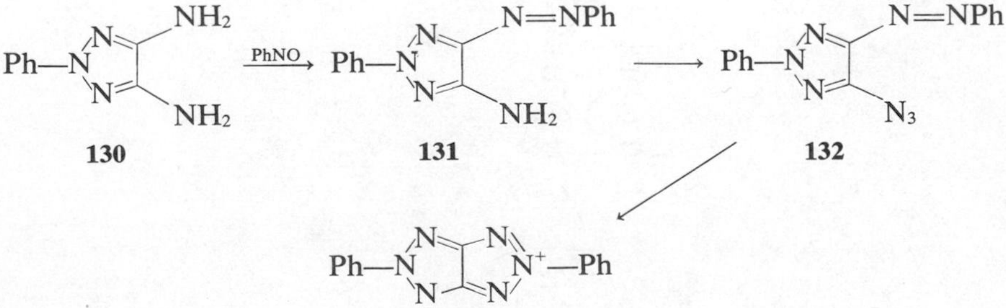

130 —PhNO→ **131** → **132** → **133**

2-ethynylpyridine in the presence of acetic anhydride gave ethyl 1-methyl-5-phenyl-4-(2-pyridyl)pyrrole-2-carboxylate (**137**), the mesoionic oxazolone **135,** and the 1 : 1 adduct **136** being intermediates in the reaction. Nitration of **137** with fumic nitric acid afforded the corresponding nitro compound **138**. Prolonged reflux (7 days) of **138** with triethyl phosphite in xylene in a nitrogen atmosphere afforded a triazapentalene derivative, ethyl 2-methyl-1-phenylpyrrolo[3′,4′ : 2,3]pyrazolo[1,5-*a*]pyridine-3-carboxylate (**139**), as golden yellow crystals. This azapentalene reacted with *N*-phenylmaleimide in refluxing xylene (7 days) giving, not the primary 1 : 1 cycloadduct **140** but compound **141,** clearly derived from **140** by extrusion of methylamine. The prolonged reaction period and elevated temperature were no doubt responsible for extrusion of methylamine from the primary cycloadduct. In contrast, the reaction of **139** with dimethyl acetylenedicarboxylate took place over 12 hr in refluxing benzene. In this instance the product isolated was the primary cycloadduct **142**.

B. Ring Systems with Heteroatoms at Points of Fusion

1. *One Nitrogen Atom*

The introduction of heteroatoms at the points of fusion greatly extends the possibilities for heteropentalenes with ylidic characteristics. With one nitrogen atom at a point of fusion, the heteropentalene can be represented by the general expression **143**. If these ring systems are restricted in the first instance to those containing nitrogen and sulfur atoms, then by varying $a = S$ and $b = c = d = e = f = C$, N, then there are 24 structures that fit into the category under discussion. Extending this concept to systems containing oxygen, that is by varying $a = O$ and $b = c = d = e = f = C$, N, an additional 24 structures are possible, and when $a = NR$, a further 24 structures need be considered. Thus the total number of possible structures is 72. Ring systems that illustrate this type of heteropentalene are the 5*H*-thiazolo[3,4-*b*]pyrazole system **144**,

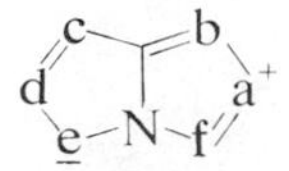

143 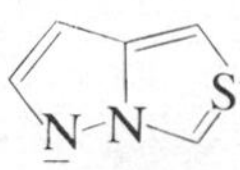 **144** 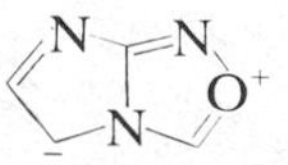 **145** 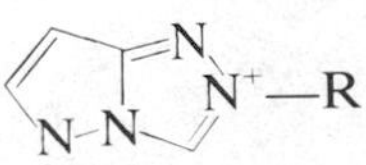 **146**

the 7*H*-imidazo[2,1-*c*][1.2.4]oxadiazole system **145**, and the 2*H*,5*H*-pyrazolo[3,2-*c*][1.2.4]triazole system **146**.

Only one such system has been synthesized to date.[59] Deoxygenation of ethyl 4-(2-nitrophenyl)-2-phenylthiazole-5-carboxylate (**148**; R= COOEt) with triethyl phosphite gave ethyl 3-phenylthiazolo[3,4-*b*]-indazole-1-carboxylate (**149**) as maroon needles, the precursor **148** (R= COOEt) being prepared by a reaction sequence from ethyl 2′-nitro-benzoylacetate **147** involving bromination and subsequent reaction with thiobenzamide.

The 5-substituent in the thiazole ring is essential to direct the attack of the nitrene intermediate to the trigonal nitrogen atom. In the absence of a substituent, for example, in **148** (R=H), C–H bond insertion is preferred over N–N bond formation, and the product isolated was identified as 2-phenylthiazolo[5,4-*b*]indole (**150**), a heteropentalene that does not have ylidic characteristics. Conversion of **148** (R=H) into the corresponding azide followed by thermolysis or photolysis also resulted in the formation of **150**. The preference for C–H bond insertion over N–N bond formation may be the result of thermodynamic control with the possibility that the N–N bond is made preferentially under kinetic control. The iminium imine thus formed could conceivably redissociate to the nitrene, which then reacts to form the observed product, possible nitrene formation by dissociation of $\overset{+}{N}-\bar{N}-$ being of interest for some time.[60]

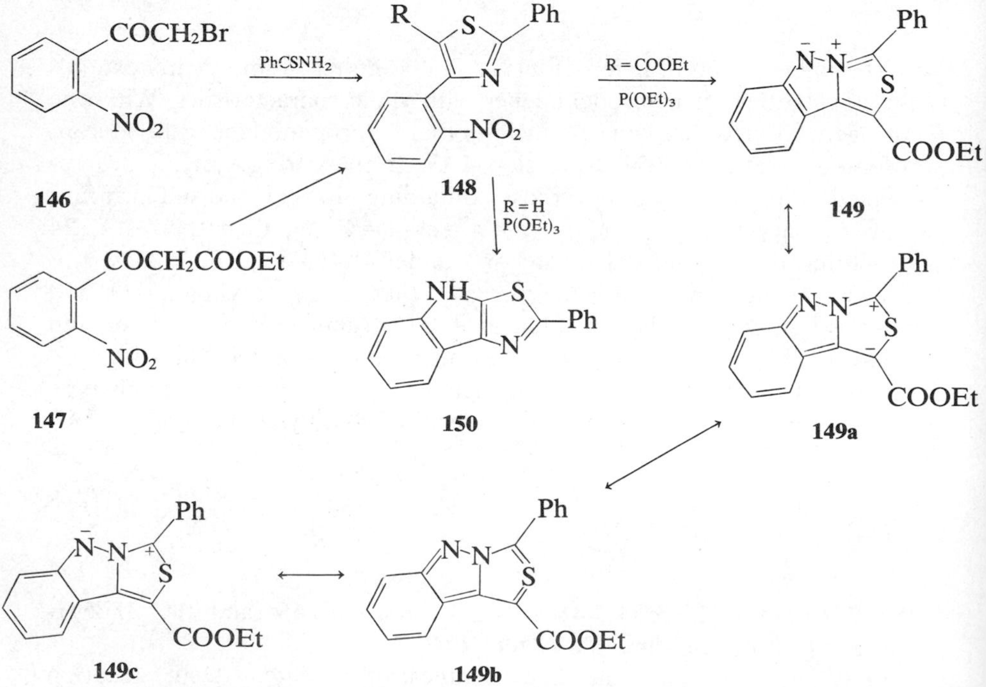

The heteropentalene **149** may be represented by several possible ylide structures and by one containing a "tetravalent sulfur" atom **149b**. *N*-Phenylmaleimide was found to add across the thiocarbonyl ylide dipole **149a** as was observed with the tetravalent sulfur systems discussed above. In this case reaction occurred in 17 hr in refluxing xylene. Both the primary 1 : 1 adduct **151** and its hydrogen sulfide elimination product **152** were isolated from the reaction. The latter was also formed from **151** by refluxing in xylene or by treatment with methanolic sodium methoxide. Elimination of H_2S from the primary 1 : 1 adduct removes from consideration addition occurring across the azomethine imine dipole **149c,** as the adduct **153** would not be expected to lose H_2S.

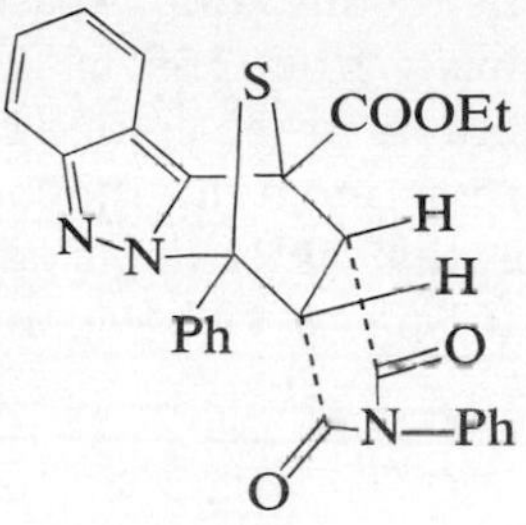

151

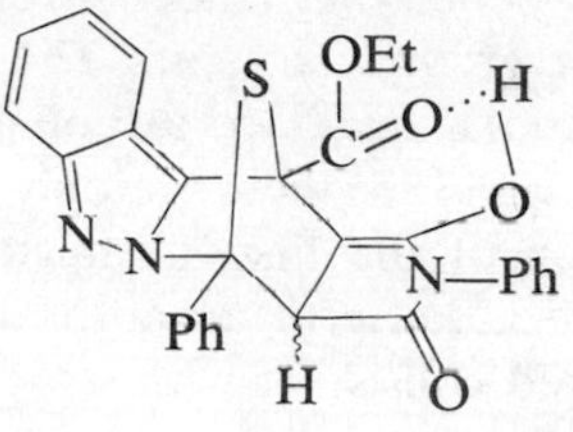

151a

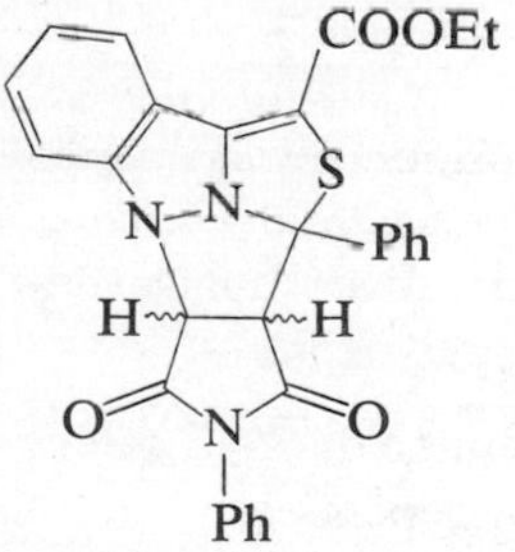

153

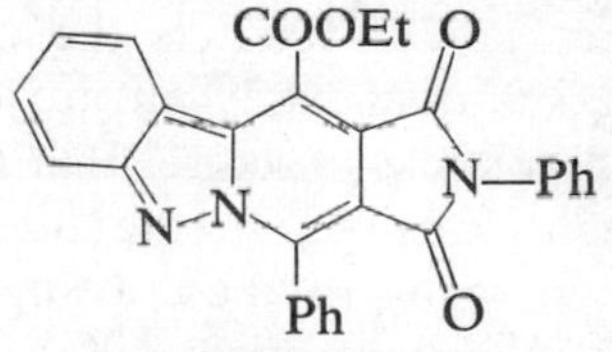

152

The nmr data for **151** indicated that this 1 : 1 adduct existed in the enol form **151a,** a feature not previously observed with imidic cycloadducts with epithio bridges. With dimethyl acetylenedicarboxylate the reaction took a different course, a 2 : 1 adduct being obtained as colorless prisms in 82% yield. Two structures were considered plausible for this cycloadduct, the epithiodihydroazocinoindazole **154** or the dihydrothiazolodihydrodiazepinoindazole **155,** representing addition of two molecules of dimethyl acetylenedicarboxylate across the thiocarbonyl ylide **149a** or

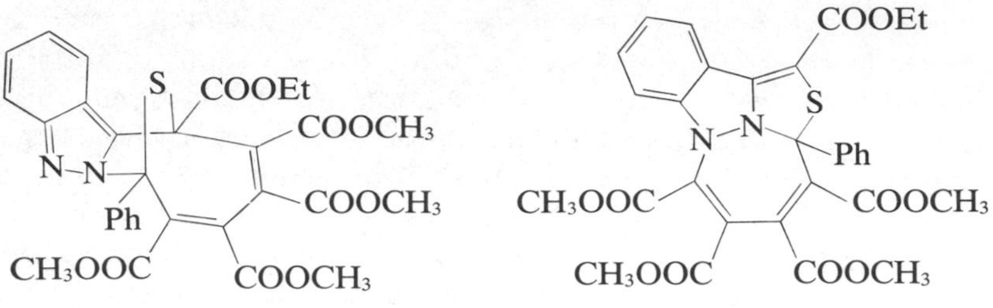

154 **155**

azomethine imine ylide **149c,** respectively. The ^{13}C PFT spectrum provided evidence that lended strong support to structure **155**. Apart from the ester groups, structure **154** differs principally from **155** in that the former contains two sp^3 carbon atoms at the termini of the epithio bridge, whereas in **155** there is only one sp^3 carbon atom at the original 2-position of the thiazole nucleus. An absorption at 114 ppm and the complete absence of an absorption between 61.62 to 127.49 ppm were considered consistent with structure **155**.

2. *Two Nitrogen Atoms*

Incorporation of two nitrogen atoms at the ring junction in heteropentalenes greatly restricts the number of possible systems as, in this instance, sulfur and oxygen are excluded from the fused system. This leaves 18 possible structures that need be considered and several of these have been described in the literature. Experimental convenience has resulted in many of these being benzo derivatives but several of the simpler systems are known.

a. Diazapentalenes. The parent unsubstituted diazapentalene was initially prepared in 1965 by two groups who independently used the same route.[61] 1-Allylpyrazole (**156**), on treatment with bromine, underwent cyclization to **157** which, with base, formed 1*H*-pyrazolo[1,2-*a*]pyrazole (**158**). Introduction of a variety of substituents is possible by this route, the allylation of pyrazoles proceeding readily regardless of the substituents in the ring. As would be anticipated, the ease of intramolecular

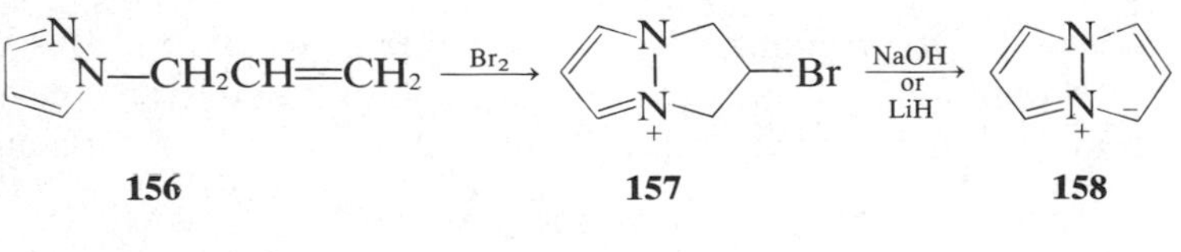

N—$CH_2CH{=}CH_2$ $\xrightarrow{Br_2}$ —Br $\xrightarrow[\text{LiH}]{\text{NaOH or}}$

156 **157** **158**

quaternization of the intermediate dibromo compound to **157** depended on the nature of the nuclear substituents. Alkyl-substituted pyrazoles quaternized readily at 100°, but longer heating was required for the dibromide derived from 1-allyl-4-bromopyrazole whereas a 4-cyano- or a 3,4,5-tribromopyrazole could not be quaternized.

The double dehydrohalogenation of these intermediate quaternary salts **157** can be carried out in several ways, the simplest being aqueous alkali. These 3*a*,6*a*-diazapentalenes **158** are colorless solids, soluble and stable in water but extremely sensitive to air giving highly colored products and requiring handling in an inert atmosphere in which their lifetime is still relatively short.

These heteropentalenes undergo ready electrophilic substitution giving, for example, stable 1,3-diacetyl, 1,3-dibenzoyl, and 1,3-dicyano-1*H*-pyrazolo[1,2-*a*]pyrazoles **159** (R = $COCH_3$, COPh, and CN, respectively) on treatment of **158** with acetic anhydride, benzoyl chloride, and cyanogen chloride, a similar substitution pattern being observed on acetylation and benzoylation of the 2-bromo derivative of **158**. Reaction of 1,2,3-trimethyl-1*H*-pyrazolo[1,2-*a*]pyrazole (**160**) with ethoxymethylenemalononitrile yielded an orange-colored 5-(2,2-dicyanovinyl)-1,2,3-trimethyl-4*H*-pyrazolo[1,2-*a*]pyrazole (**161**). All these derivatives of **158** with electron-withdrawing substituents were found to be stable in air over long periods, and some photosensitivity was observed with the diacetyl derivatives. However attempts to form a cycloadduct from **158** and methyl propiolate or dimethyl acetylenedicarboxylate were unsuccessful.

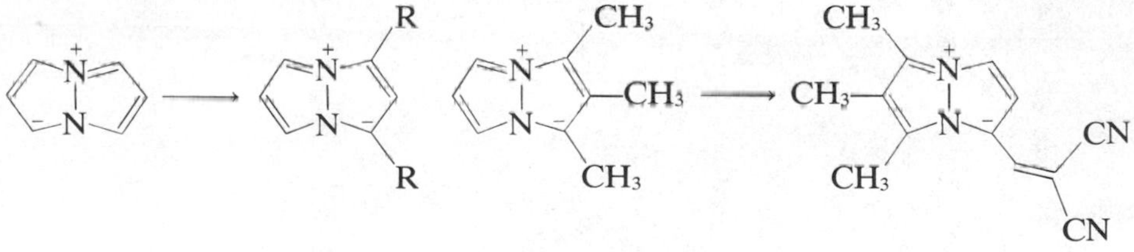

R R CH_3 CH_3 CH_3 CH_3 CH_3 CH_3 CN CN

158 **159** **160** **161**

The monoacyl derivatives of **158** have also been synthesized by an alternative route.[62] Pyrazole was converted into a diphenacylpyrazolium salt **162** by reaction with 2 moles of phenacyl bromide or the *p*-bromo-, *p*-chloro-, or *m*-nitrophenacyl bromide and, on treatment with aqueous sodium bicarbonate, these salts underwent cyclization to the 1-aroyl-2-aryl-1*H*-pyrazolo[1,2-*a*]pyrazoles (**163**; R = aryl). Heating of **163** (R = Ph) with concentrated hydrochloric acid resulted in removal of the 1-benzoyl substituent, the resulting 2-phenyl-1*H*-pyrazolo[1,2-*a*]-pyrazole being isolated as the hydropicrate. The pyrazolopyrazole ring

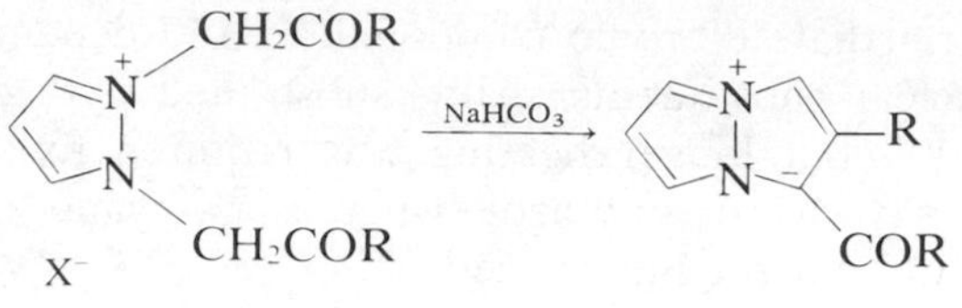

$\mathrm{R = Ph,\ \mathit{p}\text{-}BrC_6H_4,\ \mathit{p}\text{-}ClC_6H_4,\ \mathit{m}\text{-}NO_2C_6H_4}$.

162 163

system also has considerable thermal stability for **163** (R = Ph) can be sublimed in vacuo at 160°.

b. Triazapentalenes. Triaza-,tetraaza-, and hexaazapentalenes have all been described in the literature and found to have interesting characteristics. The reaction of an intermediate nitrene with a trigonal nitrogen atom was utilized[63] in the synthesis of the simplest 1,3*a*,6*a*-triazapentalene, 5*H*-pyrazolo[1,2-*a*]benzotriazole (**166**). Reaction of 1-(2′-nitrophenyl)pyrazole (**164**) with hot triethyl phosphite presumably generated the nitrene **165**, which formed **166**. However an alternative C–H insertion reaction to form **167** is also possible, this heteropentalene differing principally in its spectral characteristics from **166** by the presence of an infrared NH absorption. The data reported by the authors indicated the absence of **167**, but at the time this work was carried out preference for C–H insertion over N–N bond formation in nitrene reactions had not been reported. In recent years several examples of this

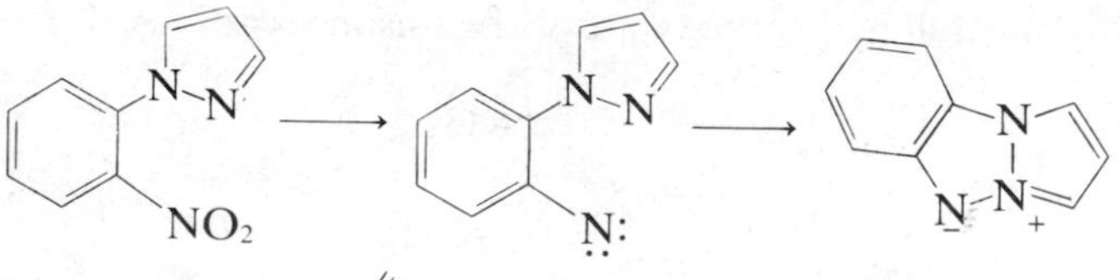

164 165 166

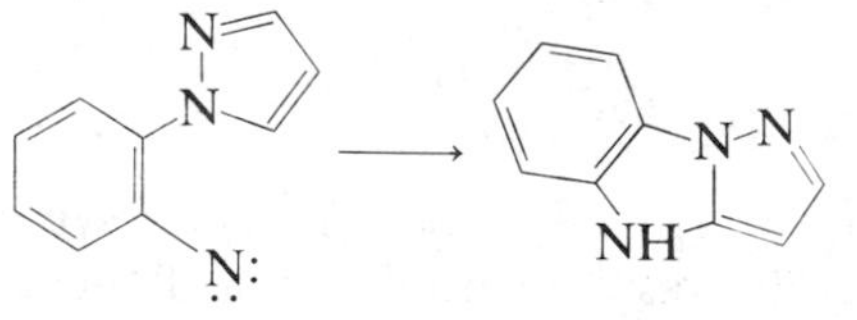

165 167

duality in reaction pathways have been described which make the reaction above worthy of reinvestigation.

This same general route has also been used for the synthesis of dibenzo analogs of the 1,3*a*,6*a*-triazapentalene system and is a recurrent theme in this general area. The photolysis of 1-(2′-azidophenyl)-1*H*-indazole

(**169**), prepared from 1-(2′-nitrophenyl)-1*H*-indazole (**168**), gave[64] a mixture of dibenzo[*b,e*]-1*H*-pyrazolo[1,2-*a*][1.2.3]triazole (**171**) (23%) and 2,2′-*bis*(1-indazolyl)azobenzene (**172**) (20%). The nitrene **170** is assumed

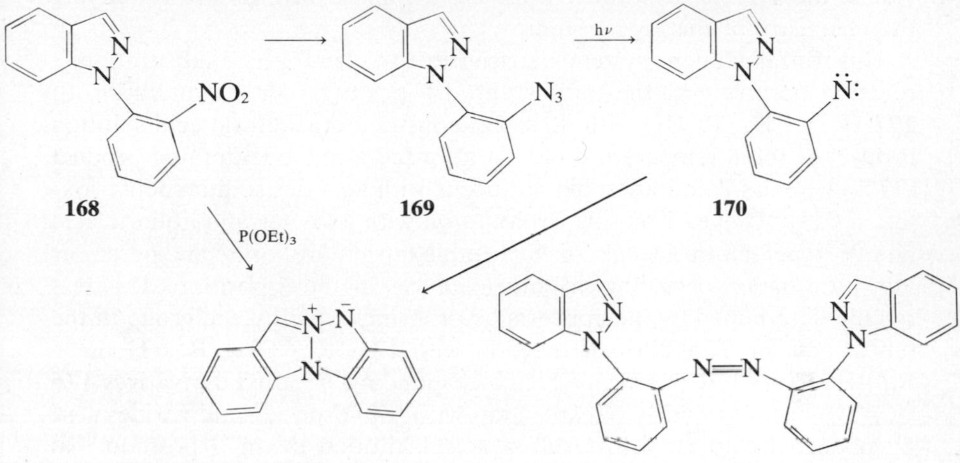

NO2 N3 hν N: P(OEt)3 N=N

168 **169** **170**

171 **172**

to be an intermediate in this reaction. Use of triethyl phosphite to generate the nitrene **170** directly from **168** gave a 30% yield of the triazapentalene **171** together with a 31% yield of another product that was tentatively assigned the dimeric structure **173** or **174**. Oxidation of this second product with sodium dichromate in acetic acid gave 1,1′-bis-(dibenzo[*b,e*]-1*H*-pyrazolo[1,2-*a*][1.2.3]triazol-7-one) (**175**).

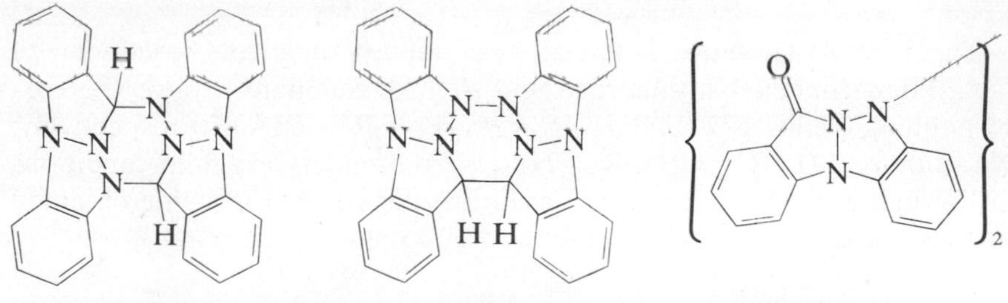

173 **174** **175**

This route to **171** has been shown[65] to be of considerable use for the introduction of substituents, the requisite starting materials being readily available by reaction of a substituted indazole and a substituted *o*-chloronitrobenzene in the presence of potassium acetate–cupric acetate. None of the 2-aryl isomers were isolated from this reaction. Treatment of **168** with hot triethyl phosphite[66] has given a variety of derivatives of **171** in yields ranging from 11 to 33%. These are yellow, crystalline products, all with relatively high melting points, and although the parent dibenzo derivative is relatively unstable and can be prepared only in moderate yields, the methyl-substituted products are more favorable for studying the chemistry of this ring system.

This triazapentalene system is susceptible to electrophilic substitution,[67] reaction occurring at the unsubstituted 7-position. Thus deuteration of **171** ($R^1 = CH_3$; $R = H$) with 40% aqueous deuteriosulfuric acid solution in D_2O at room temperature (12 hr) afforded a monodeuterated product (37%) whereas deuteration did not occur with aqueous sodium deuterioxide in CH_3OD over 1 month. Protonation with 47% tetrafluoroboric acid and 70% perchloric acid also gave a stable tetrafluoroborate and perchlorate, protonation occurring in both instances at the 7-position. This was readily determined by the appearance of a singlet methylene group in the nmr spectrum. The Vilsmeir reaction with **171** ($R^1 = CH_3$, $R^2 = H$; $R^1 = R^2 = H$; $R^1 = Cl$, $R^2 = H$) gave the corresponding 7-formyl derivatives **176** ($R = H$), and a Friedel–Crafts reaction with trifluoroacetic anhydride or *p*-nitrobenzoyl chloride also resulted in substitution in the 7-position. No catalyst was required in the Friedel–Crafts reaction for the formation of **176** ($R = CF_3$ or p-NO_2–C_6H_4).

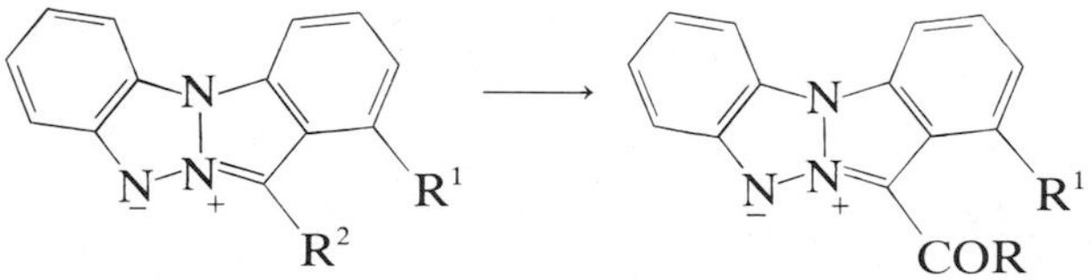

171 **176**

The formyl derivative behaved as a typical aldehyde reacting with benzylidenetriphenylphosphorane and diethyl malonate to give the corresponding olefins **177** ($R^1 = H$, $R^2 = Ph$; $R^1 = R^2 = COOEt$) respectively. Nitration of **171** ($R^1 = CH_3$; $R^2 = H$) occurred under very mild conditions (10% aqueous HNO_3, 20°, 3 hr), a dinitro derivative **178** being obtained in 59% yield.

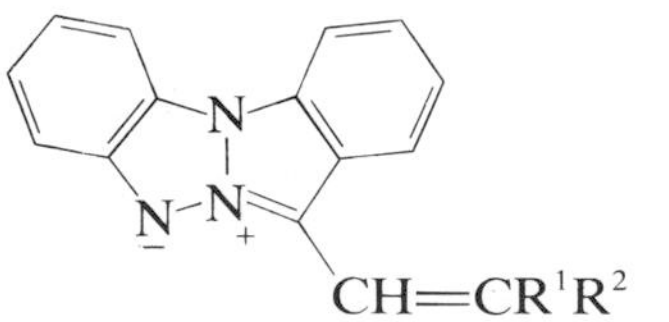

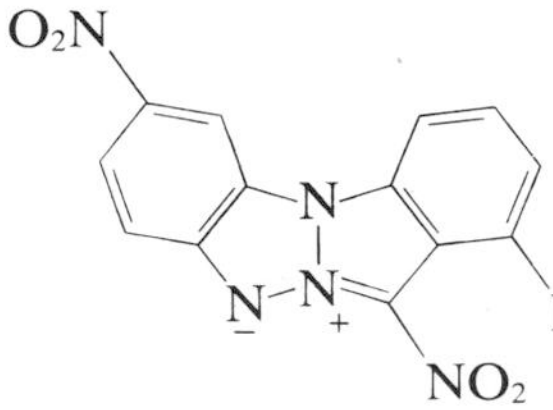

177 **178**

The 1,3*a*,6*a*-triazpentalene (**171**) can be represented by several mesomeric forms **171a, 171b,** and so on. Molecular orbital calculations[67] indicated that in the HOMO there is a significant contribution from the azomethine imine **171a,** and this ring system would, consequently, be anticipated to undergo cycloaddition with electron-deficient dipolarophiles.

Reaction of **171** ($R^1 = CH_3$; $R^2 = H$) with 1 equiv of dimethyl acetylenedicarboxylate in methylene chloride at 30° over 24 hrs afforded[68] two isomeric 1 : 1 adducts **179** and **180** in 32% and 1% yields,

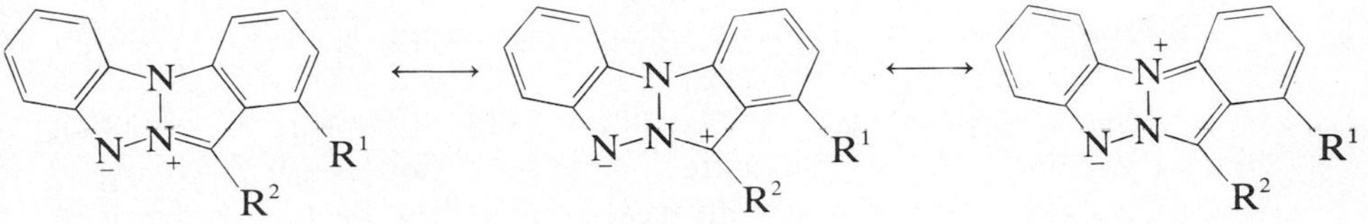

171 ⟷ **171a** ⟷ **171b**

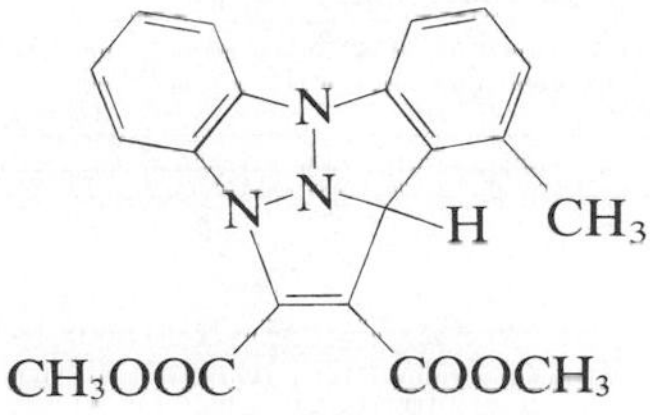

179 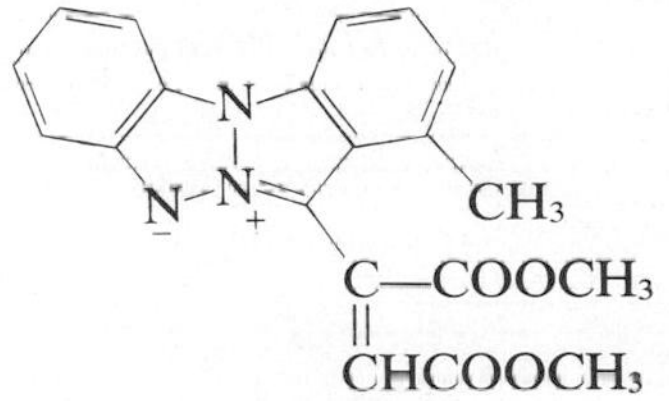 **180**

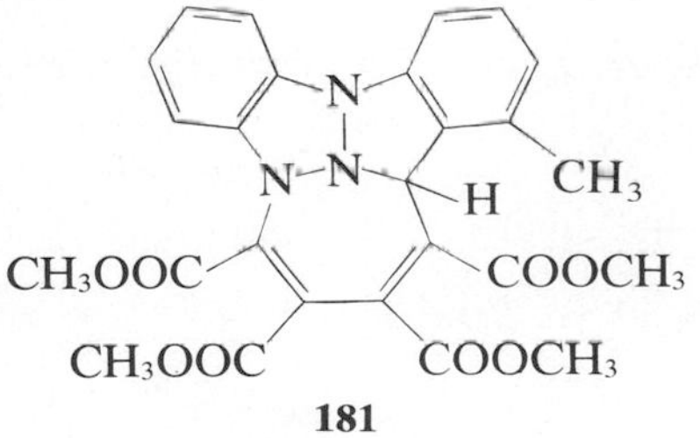

181

respectively. Use of refluxing ethanol as the reaction medium and a catalytic amount of sulfuric acid for 24 hr increased the yields of these two products dramatically, and they were then obtained in 60% and 32% yield, respectively. However when **171** ($R^1 = CH_3$; $R^2 = H$) and dimethyl acetylenedicarboxylate (2 moles) were reacted under similar acid catalytic conditions, the tricyclic adduct **181** was obtained in 90% yield, and no open-chain product corresponding to **180** was detected. In contrast to **179** the product 180 was obtained as dark violet needles suggesting retention of ylidic character. Its structure appears quite logical as it was not formed when $R^2=CH_3$ in **171,** and it formed a tetrabromo product on bromination.

Diethyl acetylenedicarboxylate also reacted with **171** ($R^1=CH_3$; $R^2=H$), giving the corresponding 1 : 1 adducts in 23% and 2% yields, respectively, and diphenylacetylene required more vigorous conditions for reaction to occur. In refluxing xylene reaction occurred with **171** ($R^1=CH_3$; $R^2=H$); however only the tricyclic adduct analogous to **179** was obtained in 18% yield.

Changing the relative amount of the dipolarophile in the reaction resulted in a different reaction product. Two equivalents of dimethyl acetylenedicarboxylate and **171** ($R^1=CH_3$; $R^2=H$) in refluxing methylene chloride for 12 hr resulted in the formation of a 1 : 2 adduct in 60% yield. This product was tentatively assigned the diazepine structure **181,** and its formation indicates that the reaction of **171** with these acetylenic dipolarophiles probably does not occur by a concerted process.

Asymmetrical acetylenes also readily added to **171** ($R^1=CH_3$; $R^2=H$), giving[69] the corresponding [3+2] cycloadducts as the major products. Their reactions are illustrated by the behavior of methyl propiolate with the 1,3*a*,6*a*-triazapentalene. A 1 : 1 adduct **182** was the major product, together with another 1 : 1 adduct **183** and a 1 : 2 adduct **184,** the yields

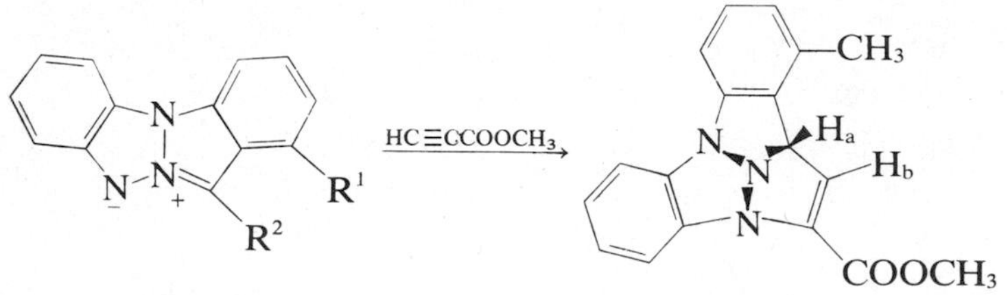

N N N + R¹ R² HC≡CCOOCH₃ → CH₃ N N N Hₐ H_b COOCH₃

171 **182**

N N N + CH₃ CH=CHCOOCH₃ CH₃ CH_c=CHₐCOOCH₃ N N N H_b COOCH₃

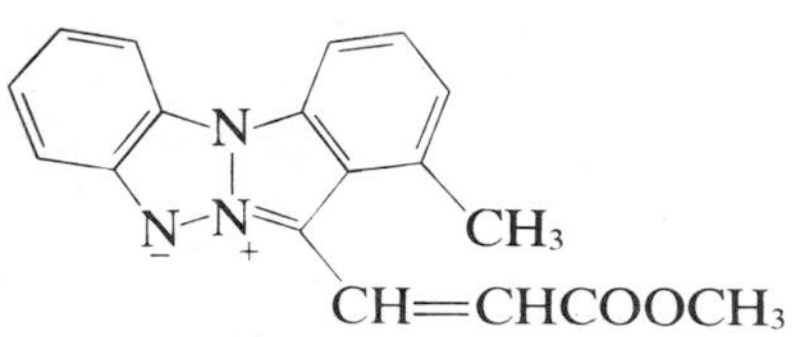

183 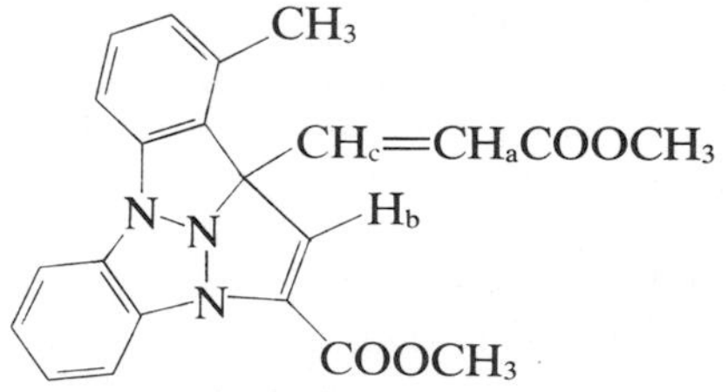 **184**

of these products depending on the nature of the solvent (Table IV). Nmr data were used to establish these structures and indicated that the configuration of **182** was more reasonable than that in which Ha was in a β orientation to the nitrogen bridge. The nmr spectrum of **182** showed

TABLE IV. PRODUCT DISTRIBUTION IN THE REACTION OF **171** ($R^1 = CH_3$; $R^2 = H$) WITH METHYL PROPIOLATE AT 80° FOR 24 HR

	% Yield of product		
Solvent	**182**	**183**	**184**
Benzene	53	0	1.5
Ethanol	25	10	3
Acetonitrile	27	13	0

two doublets (1*H*, J = 3 Hz) at δ5.67 (Ha) and 6.28 (Hb), besides the methyl and aromatic proton signals. Reaction of **171** ($R^1 = CH_3$; $R^2 = H$) with methyl monodeuteriopropiolate gave a product in which the doublet at δ6.28 was absent and a singlet was observed at δ5.67 in its nmr spectrum.

The structure of the product **183** was shown to be a Michael-type adduct by its preparation from the 7-formyl-1,3*a*,6*a*-triazapentalene and diethyl methoxycarbonylmethylphosphonate. When **171** ($R^1 = CH_3$; $R^2 = H$) was treated with 2 moles of ethyl propiolate, the yield of the third product was increased to greater than 90%. This same product was also obtained when **182** was treated with methyl propiolate but was not isolated from the reaction of **183** and methyl propiolate. This established the close relationship of **182** and **184**, and the structure of **184** was further evident from the nmr data. Three signals, at δ5.0 (Ha, *d*, J = 15 Hz), 7.2 (Hb, *s*), and 7.67 (Hc, *d*, J = 15 Hz), and collapse of the Hc doublet to a singlet on irradiation at δ5.0 are compatible with the assigned structure.

An interesting reversal in the mode of addition of phenylacetylene and *p*-chlorophenylacetylene to **171** ($R^1 = CH_3$; $R^2 = H$) was observed. Both dipolarophiles gave 1 : 1 adducts, and their formulation as Michael-type adducts **185** (R = Ph and *p*-ClC_6H_4) was readily discarded by the synthesis of **185** from the corresponding 1-formyl-1,3*a*,6*a*-triazapentalene and the appropriate Wittig reagent. The following considerations led to the assignment of structures **186** and **187** to these adducts.

The nmr spectrum of **186** showed one olefinic proton at δ5.65 together with multiplets of methine (1H) and aromatic protons (12H) at δ6.9 to 8.3. However the nmr spectrum of **187** showed two doublets at δ5.4 and 5.76 (each 1H, $J = 3$ Hz), besides aromatic and methyl protons. These data are best accounted for by addition occurring in the manner shown.

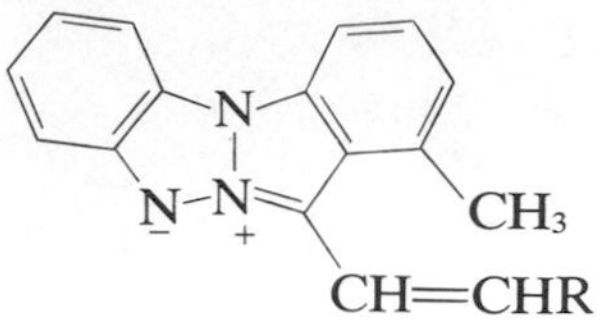

185 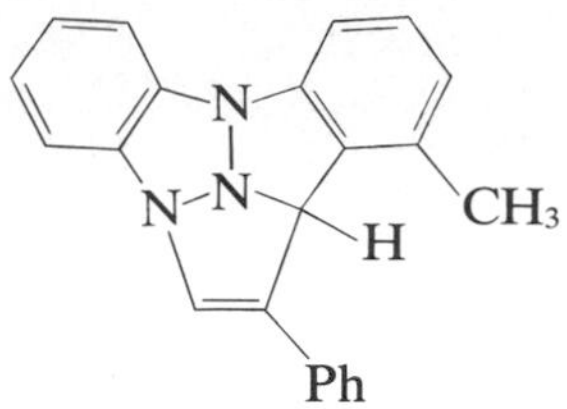 186 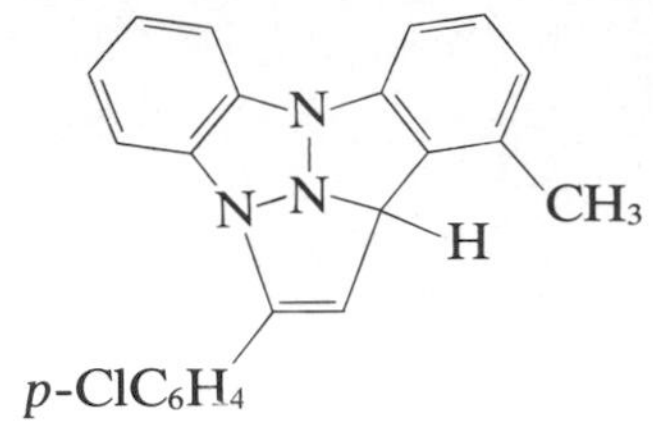 187

Similar nonspecificity was shown in the reaction of **171** ($R^1 = CH_3$; $R^2 = H$) with methyl phenylpropiolate, both modes of addition being observed with the relative amounts of the adducts **188** and **189** being in the ratio 1 : 3. The orientation of **188** was readily established by hydrolysis and decarboxylation to **186**.

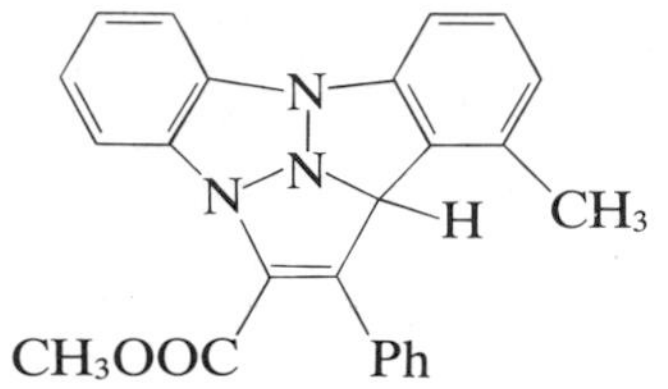

188 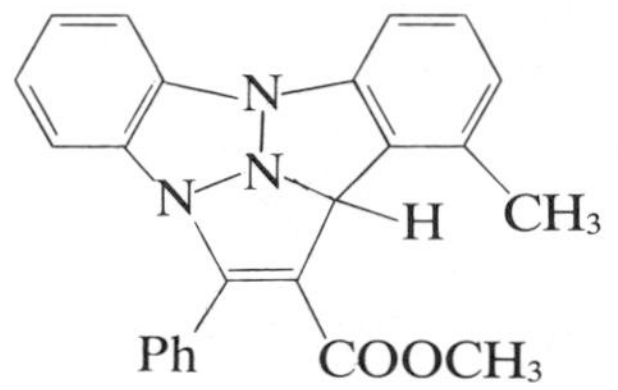 189

c. Tetraazapentalenes. The simplest tetraazapentalene with nitrogen at the ring junction was described[70] in 1963. This was readily prepared by acid catalyzed (e.g., 90% formic acid) dimerization of the α-monohydrazones of ethyl acetylglyoxylate (**190**). Thus reflux of **190** with 90% formic acid for 1 hr resulted in the separation (73% yield) of 2,6-dimethyl-3,7-bis(ethoxycarbonyl)-1*H*-1,2,3-triazolo [2,1-*a*][1,2,3]triazole (**191**; R = COOEt). This ring system is quite stable. Hydrolysis of **191** gives the corresponding acid that can be decarboxylated. The acid can be converted into its acid chloride which, in turn, forms an amide and a hydrazide. A variety of other reactions served to emphasize the stability of the ring system, but oxidation of **191** (R = COOH) with potassium permanganate resulted in degradation of the nucleus to give 4-methyl-1,2,3-triazole-5-carboxylic acid (**192**).

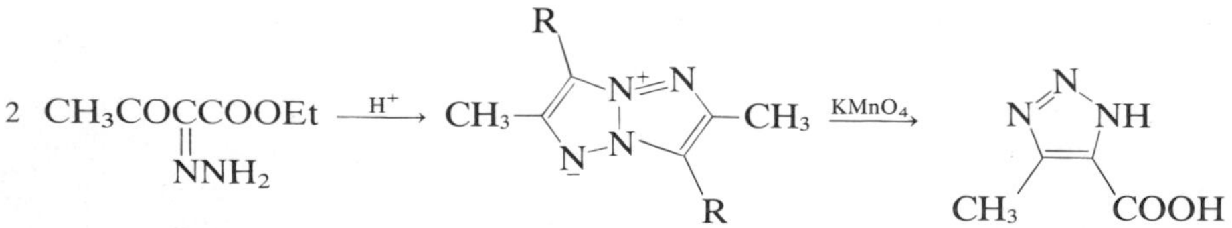

190 191 192

These data are strong evidence for the dimerization of **190** to **191,** rather than to a 1,2,5,6-tetrazacyclooctatetraene. This was confirmed[71] by x-ray analyses of two of these products, which established their planar nature and indicated their aromatic character. The observed bond lengths are shown in Fig. 9.

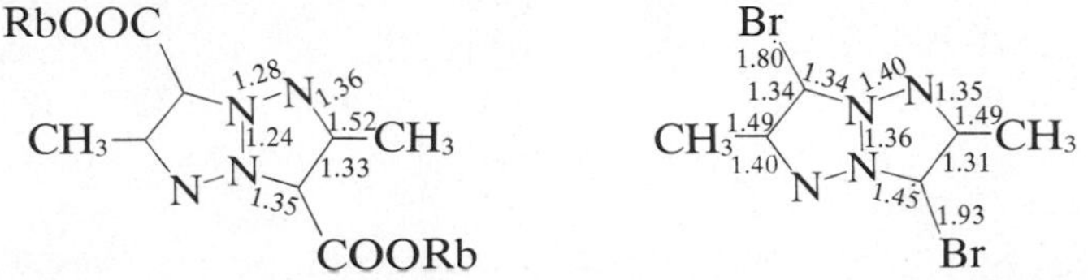

Fig. 9. Bond lengths in angstroms for some 1,3*a*,4,6*a*-tetraazapentalene derivatives.[71]

Other tetraazapentalenes described in the literature are either benzo analogs of the above system or members of the isomeric 1,3*a*,6,6*a*-system. In the majority of cases the most convenient synthesis is via a nitrene cyclization reaction, and the thermal or photochemical decomposition of the *o*-azidophenyl derivatives of 1*H*- and 2*H*-benzotriazoles gave[72] the isomeric dibenzo-1,3*a*,6,6*a*-tetraazapentalene and the 1,3*a*,4,6*a*-tetraazapentalene, respectively.

Oxidation of *o*-diaminobenzene (**193a**) with lead dioxide gave 2,2′-diaminoazobenzene (**193**), which was converted into the corresponding azide **194**. Thermolysis of the azide in a high-boiling solvent such as decahydronaphthalene or *o*-dichlorobenzene resulted in a smooth loss of 2 moles of nitrogen, and the tetraazapentalene **195** separated from the cooled concentrate.

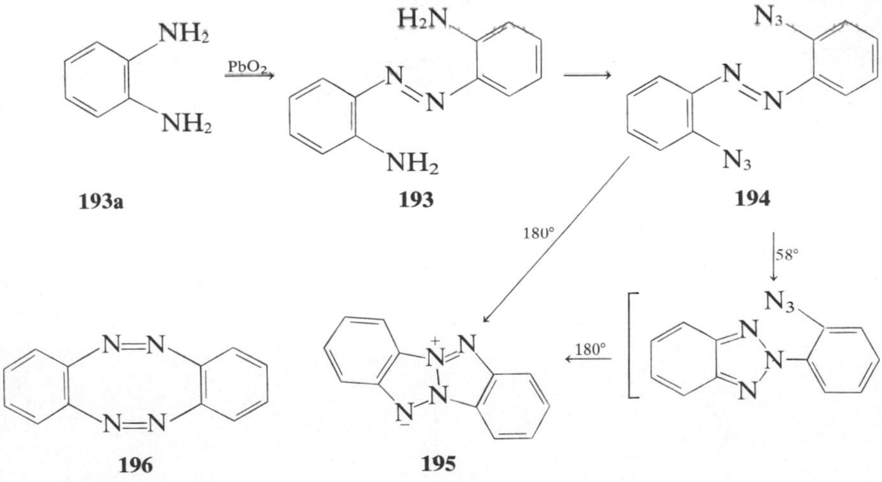

193a **193** **194** **196** **195**

The evolution of N_2 was observed to occur in two stages; the first mole was eliminated at 58° and the second mole at 170°. Although there is no direct evidence for nitrene intermediates, it is not unreasonable to assume that they were involved in the formation of **195**.

An alternative possibility for the product of this reaction is the dibenzo-1,2,5,6-tetraazacyclooctatetraene (**196**). However, x-ray data showed[73] that **195** is indeed the correct structure, and this was confirmed by chemical and nmr evidence. The proton spectrum of **195** was invariant over a wide temperature range, indicating no isomerism between structures **195** and **196**.

Additional evidence in support of this contention was obtained by the thermolysis of **197**. The developing nitrene intermediate derived from **197**, 2-(2′-azidophenyl)-2*H*-naphtho[1,2-*d*][1,2,3]triazole, could close at either of the 1- or 3-positions of the triazole ring. If the tetraazapentalene structures were correct, and if there were a large barrier for the valence tautomeric change, two isomeric products **198** and **199** should be formed. However formation of a tetraazacyclooctatetraene should result in only one isolatable product. Thermal decomposition of **197** gave two high melting products, separable by chromatography, and represented by structures **198** and **199**.

The isomeric system, dibenzo[*b,e*]-1,3*a*,6,6*a*-tetraazapentalene (**201**), was readily prepared by the thermal decomposition of 1-(2′-azidophenyl)-1*H*-benzotriazole (**200**). A colorless, crystalline product, this heteropentalene closely resembled its isomer **195** in physical and chemical properties. It has a large dipole moment (4.36 D) consistent with the unsymmetrically distributed charge implicit in its formulation as **201**; as anticipated, **195** had no detectable dipole moment in solution.

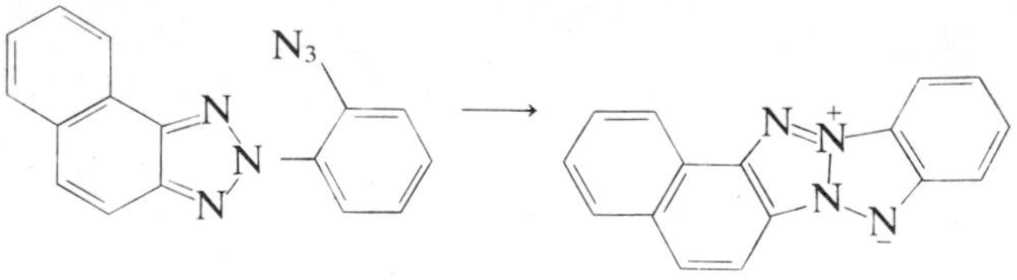

N_3 →

197 **198**

N_3 →

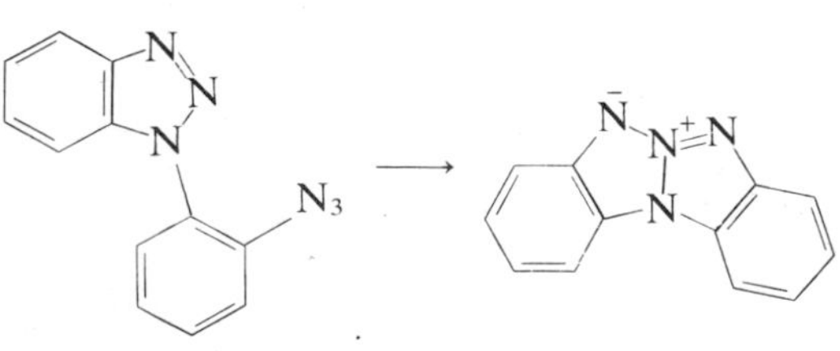

200 **201** 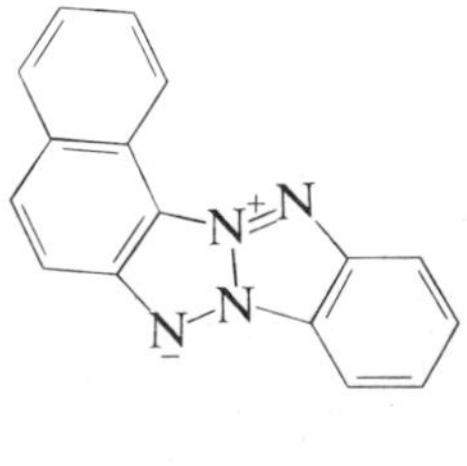 **199**

This synthetic approach has since been extended[74] to include benzo-substituted derivatives of **195**, and the use of triethyl phosphite to generate the nitrene functions from appropriate nitro precursors, for example, the conversion of 2,2′-dinitroazobenzene into **195**, has also been studied.[75,76,77]

Monobenzo-1,3*a*,4,6*a*-tetraazapentalenes are also available by application of this general synthetic route. 2-(2′-Nitrophenyl)-2*H*-1,2,3-triazole (**202**), prepared from 1,2,3-triazole and *o*-fluoronitrobenzene, on treatment with trimethyl phosphite, gave[76] benzo[*b*]-1*H*-1,2,3-triazolo[2,1-*a*][1,2,3]triazole (**203**). Similarly, reductive cyclization of 1-(2′-nitrophenyl)-1*H*-1,2,3-triazole (**204**) gave[76] the isomeric benzo[*b*]-1*H*-1,2,3-triazolo[1,2-*a*][1,2,3]triazole (**205**). This same system has also been synthesized by thermolysis of 1-(2′-azidophenyl)-1*H*-1,2,3-triazole.

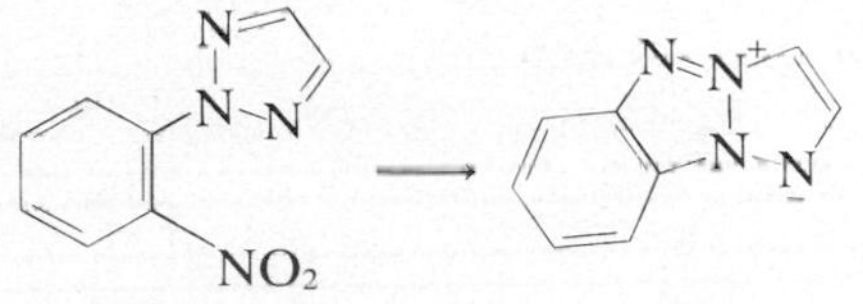

NO_2 **202** → **203**

NO_2 **204** → **205**

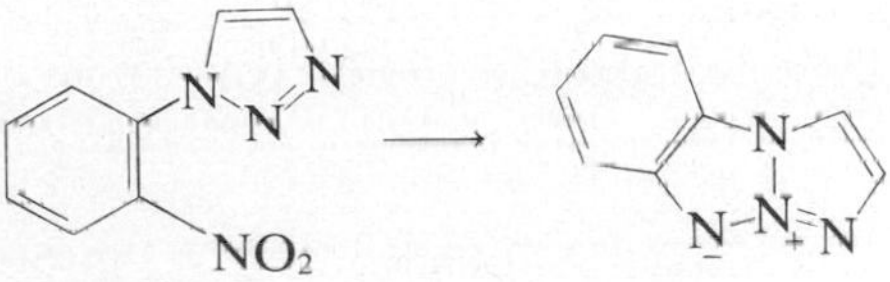

Fair to good yields of the respective tetraazapentalenes were obtained by this reductive cyclization procedure, being governed to some extent by the solubility of the by-products and the nature of the phosphorus compound. Thus the reaction of 1- or 2-(2′-nitrophenyl)-1*H*-1,2,3-triazole (**204** or **202**) with tributylphosphine was more rapid than with the less nucleophilic triethyl phosphite. However with the latter reagent the yields and product purities were much higher, but occasionally the reaction may follow a different course. Thus reaction of the triazole derivative **206** with triethyl phosphite afforded the anticipated tetraazapentalene **207** but with tributylphosphine, a product tentatively assigned the triazoloquinoxaline structure **208** was obtained.

The dibenzo-1,3*a*,4,6*a*-tetraazapentalene **195** undergoes relatively facile electrophilic substitution reactions with or without disruption of the nucleus. Chlorine or bromine in acetic acid or chloroform gave the dihalogenated derivative **209a,b** in good yields. With *N*-bromosuccinimide in acetonitrile, however, a monobromo derivative **210b** was obtained in

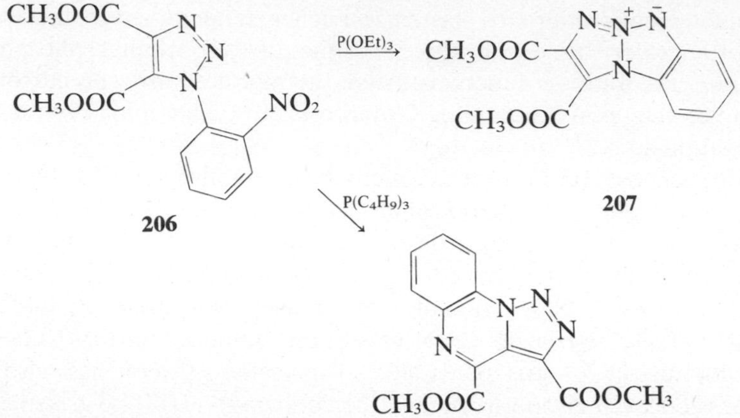

206 → ($P(OEt)_3$) → 207

206 → ($P(C_4H_9)_3$) → 208

70% yield, together with a small amount (5 to 10%) of the dibromo derivative. However, chlorination of **195** with *N*-chlorosuccinimide yielded a mixture of the monochloro derivative **210a,** the dichloro derivative **209a,** and unreacted **195**. Nitration at 5° with 70% nitric acid gave a mixture of dinitro derivatives from which a predominant isomer **209c** was isolated. However treatment of **195** with 90% nitric acid resulted in a tetranitro product **209e** in excellent yield. This same tetranitro product was also obtained when **195** was heated at 60° with 25% aqueous nitric acid but at 0° only a mononitro product was formed. Chlorosulfonic acid at 90° also produced a bis(sulfonyl chloride) **209f** in 50% yield. Reduction of the above nitro groups could not be effected cleanly with stannous chloride but 10% palladium-on-carbon in dimethylformamide gave the desired amino compounds.

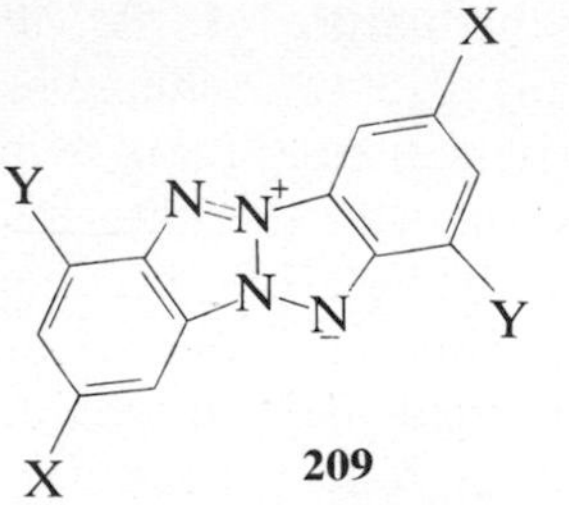

209

a: X = Cl, Y = H
b: X = Br, Y = H
c: X = NO_2, Y = H
d: X = NH_2, Y = H
e: X = Y = NO_2
f: X = SO_2Cl, Y = H

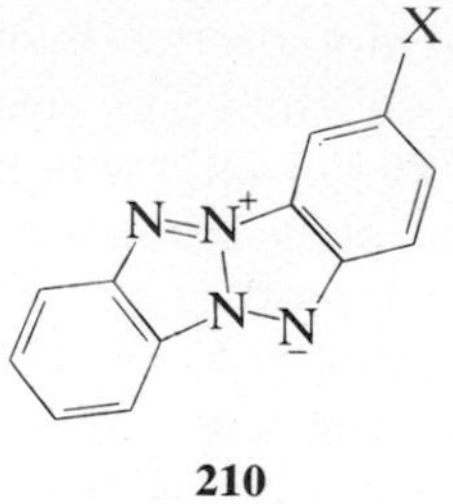

210

a: X = Cl
b: X = Br
c: X = NO_2
d: X = NH_2

The halogenation of **195** occurred to a very large extent in the benzene rings *para* to the nonfused nitrogen atoms. Electrophilic substitution at this position results in a transition state intermediate that should be more stable than the one where substitution occurs at an *ortho* position to the nonfused nitrogen atoms.

Tetracyanoethylene did not give a tetracyanoethyl or tricyanovinyl derivative of **195** but rather resulted in π-complex formation. This is in contrast to the monobenzo derivative **203** which in tetrahydrofuran solution formed a deep blue complex whose solution slowly liberated hydrogen cyanide. Dilution of this solution with water resulted in precipitation of the tricyanovinyl derivative **211**.

Reaction of **195** with propylene in concentrated sulfuric acid resulted in mainly reaction at a nitrogen atom, the *N*-isopropyl salt **212** ($R = i\text{-}C_3H_7$) being obtained. No substitution was observed in the benzene rings. The

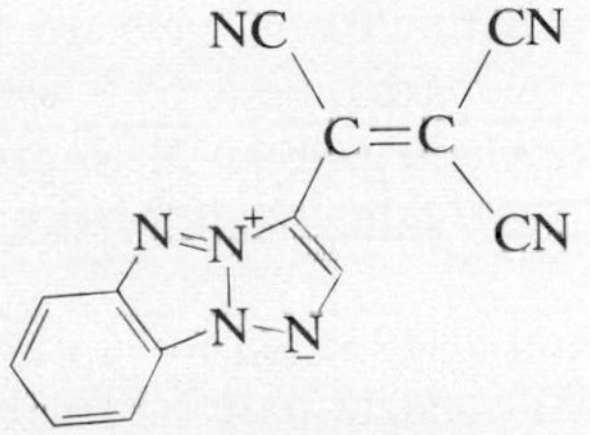

NC CN C=C CN N=N+ N-N X− R

211 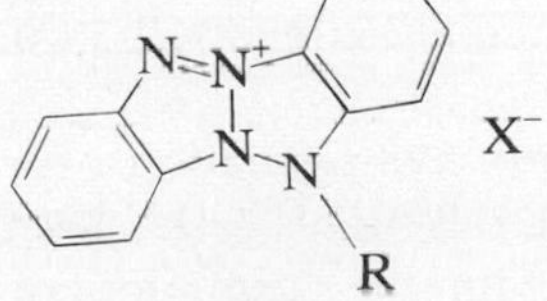 **212**

N-methyl derivative **212** ($R = CH_3$) was prepared by prolonged treatment of **195** with methyl iodide, or more readily with methyl sulfate at 150°. These crystalline salts were light sensitive.

Several methods are available for opening the ring system in **195**. Although stable to potassium permanganate in acetone, it was readily attacked by peracetic acid under mild conditions with cleavage of the N-N bond. This oxidative cleavage resulted in the formation of 2-(2′-nitrosophenyl)-2*H*-benzotriazole (**214**; R = NO), probably via the intermediate **213**, as a nonfused nitrogen atom with its relative high electron density would be a favorable site of attack for an electrophilic peracid.

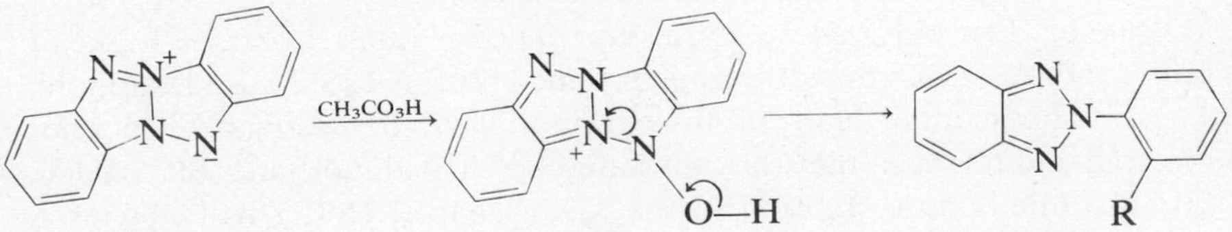

N=N+ N-N $\xrightarrow{CH_3CO_3H}$ N-N N+ N O-H ⟶ N N-N R

195 **213** **214**

Lithium aluminum hydride is also an effective ring-opening reagent. In this case reductive ring opening occurred with the formation of 2-(2′-aminophenyl)-2*H*-benzotriazole **214** ($R=NH_2$). Hydrogen in the presence of palladium-on-carbon in acetic acid solution similarly effected ring opening, the acetyl derivative of **214** ($R=NH_2$) being obtained.

The chemical reactivity of the dibenzo[*b*,*e*]-1,3*a*,6,6*a*-tetraazapentalene (**201**) paralleled closely[76] that reported for the isomeric system **195**. Mononitration occurred in 25% nitric acid solution at room temperature. Dinitro and tetranitro derivatives were formed at 0 to 5° in 70% and 90% nitric acid, respectively. Treatment of **201** with chlorine or chlorosulfonic acid yielded the dichloro or bis(chlorosulfonyl) derivatives, respectively. Although the positions of substitution in these electrophilic reactions have not been rigorously proved, charge density and localization energy considerations favor positions *ortho* and *para* to the nonfused nitrogen atoms as sites for attack.

Methyl iodide at 100° reacted slowly with **201** to give an unstable adduct isolated in an impure state. However treatment of a solution of **201** in concentrated sulfuric acid with propylene gave an *N*-isopropyl salt that was readily converted into the corresponding iodide with sodium iodide.

As before, hydrogenation over palladium at 125° led to rupture of the N–N bond, but in this case subsequent hydrogenation of the benzene ring occurred with the formation of 1-(2′-aminophenyl)-4,5,6,7-tetrahydro-1*H*-benzotriazole (**215**). This could be converted into the azide **216** and, subsequently, into the new tetraazapentalene derivative **217,** which was similar to a monobenzotetraazapentalene.

In contrast to the isomeric dibenzo derivatives **195** and **201**, the monobenzotetraazapentalene **205** reacts readily with tetracyanoethylene in dimethylformamide. A deep blue complex formed initially in concentrated solution, followed by elimination of hydrogen cyanide and the development of the purple color of the tricyanovinyl derivative **218**.

Utilizing experimental heats of combustion and sublimation for derivatives of the 1,3*a*,6,6*a*- and 1,3*a*,4,6*a*-tetraazapentalene ring systems, heats of formation (referred to the gas phase in standard states) were determined to be dibenzo-1,3*a*,4,6*a*-tetraazapentalene (**195**), 142.8 ± 1.3 kcal/mole; dibenzo-1,3*a*,6,6*a*-tetraazapentalene (**201**), 132.1 ± 1.5 kcal/mole; benzo-1,3*a*,4,6*a*-tetraazapentalene (**203**), 136.4 ± 1.2 kcal/mole; benzo-1,3*a*,6,6*a*-tetraazapentalene (**205**), 128.2 ± 1.3 kcal/mole. These experimental heats of formation reflect the resonance energies directly, and it is seen that the more angular shaped molecule **201** is some 10.7 kcal/mole more stable than the linear-shaped **159**. This behavior is typically that of pure benzenoid hydrocarbons; for example, the angular

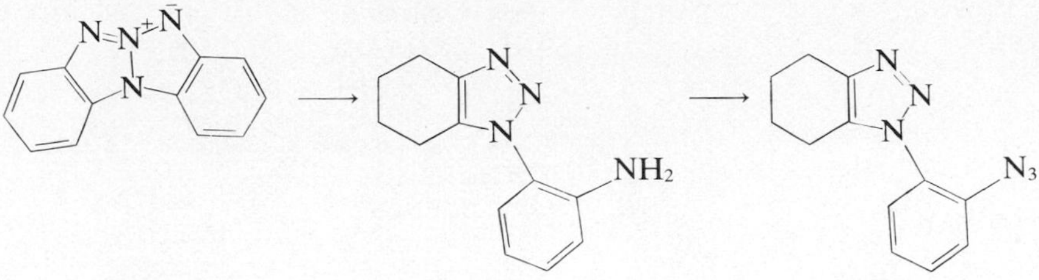

201 215 216

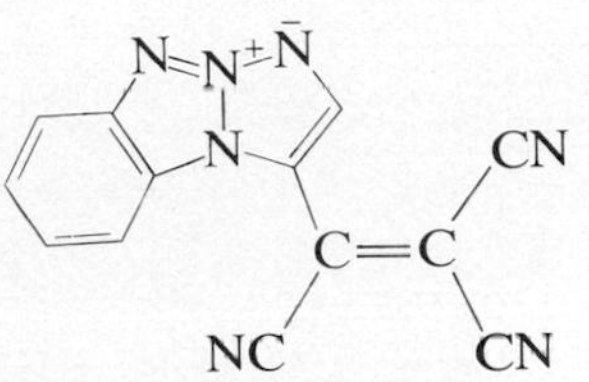

218 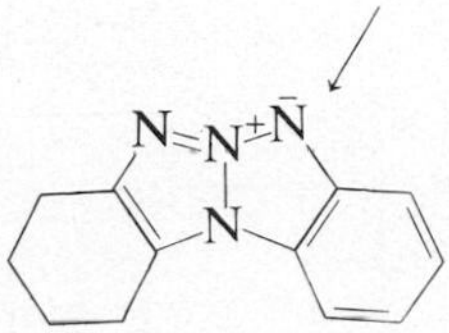 217

chrysene (116.5 kcal/mole) is more stable than the liner naphthacene (110 kcal/mole) by 6.5 kcal/mole. It is interesting that the resonance energies of **159** and **201** are larger than those of their isoelectronic hydrocarbon counterpart.

If each benzene ring in **195** and **201** were considered to contribute roughly 36.0 kcal/mole to the resonance energy, then the contributions of the 1,3*a*,4,6*a*- and 1,3*a*,6,6*a*-tetraazapentalene nuclei in **195** and **201**, respectively, are approximately 49.7 and 60.4 kcal/mole. A similar study for the isomeric monobenzo derivatives **203** and **205** gave contributions for the tetraazapentalene nuclei of 48.7 kcal/mole and 56.8 kcal/mole, respectively, in reasonable agreement with the values derived from the dibenzo derivatives.[78]

Hückel calculations for the tetraazapentalene systems indicate that the structure is highly stabilized by electron delocalization on the basis of the delocalization energy per π-electron compared to its valence tautomer tetraazacyclooctatetraene. On calculation of molecular orbitals, it was found that all bonding levels were displaced to lower energies relative to the pentalene dianion on introduction of nitrogen. These calculations showed that the positions of the nitrogen atoms had very little effect on the energy levels or on the total energy, a behavior anticipated if the tetraazapentalene nucleus is aromatic and its π-electrons highly delocalized. Bond orders and charge densities for the two isomeric systems are shown in Table V, and the closeness of these values should be noted.

Charge distributions for the isomeric dibenzo- and benzotetraazapentalenes are shown in Fig. 10. If electrophilic substitution were to

TABLE V. BOND ORDERS AND CHARGE DENSITIES FOR THE 1,3*a*,4,6*a*- AND 1,3*a*,6,6*a*-TETRAAZAPENTALENES[78]

1,3a,4,6a		1,3a,6,6a	
Bond order			
12	0.421	12	0.443
23	0.514	34	0.581
34	0.739	45	0.751
45	0.588	56	0.491
26	0.433	26	0.435
Charge densities			
1,5	1.497	1,3	1.490
2,6	1.447	2	1.436
		6	1.454
3,7	1.080	5,7	1.082
4,8	0.976	4,8	0.983

parallel electron density in the ground state, then the 2- and 4-positions in the benzene rings of **195** and **201** should be preferentially nitrated, chlorinated, and so on, and, as described above, such was found to be the case. Electrophilic attack on the tetraazapentalene nucleus would be predicted to occur at the positions marked with an asterisk, again in agreement with experimental observation.

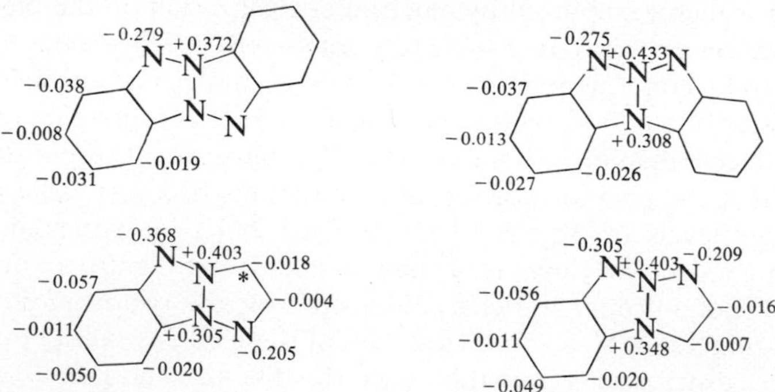

Fig. 10. Charge distribution for isomeric tetraazapentalenes.

This area of chemistry is not without surprises for in a forerun of more recent chemistry of aminobenzotriazole derivatives, it was found that thermolysis of 1,1′-bibenzotriazole (**219**) and 1,2′-bibenzotriazole (**220**)

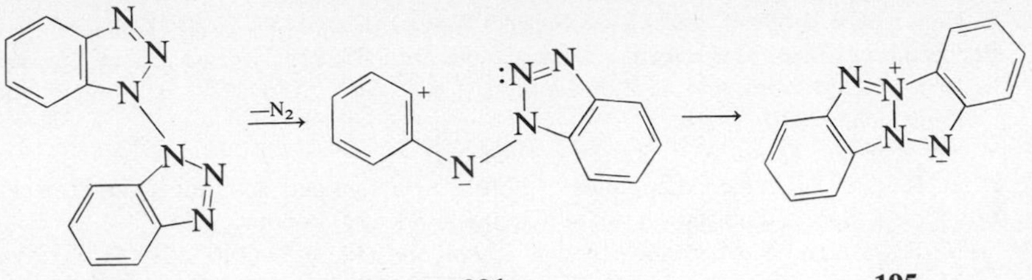

219 **221** **195**

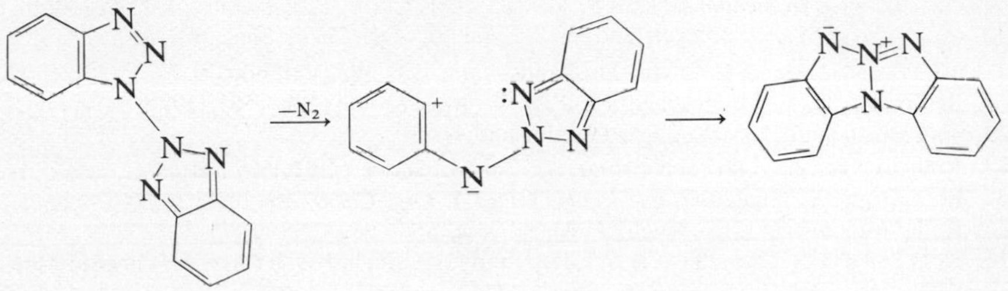

220 **222** **201**

leads to loss of nitrogen at 300° and the formation of dibenzo-1,3*a*,4,6*a*- and 1,3*a*,6,6*a*-tetraazapentalene (**159**) and (**201**), respectively. It has been suggested[79] that intermediates such as **221** and **222** were involved in this thermolysis which is reminiscent of the Graebe–Ullmann synthesis of carbazoles from 1-phenylbenzotriazoles.

III. References

1. J. W. Armit and R. Robinson, *J. Chem. Soc.*, **121,** 827 (1922); **127,** 1604 (1925).
2. E. Le Goff, *J. Am. Chem. Soc.*, **84,** 3975 (1962); K. Hafner, K. F. Bangert, and V. Orfanos, *Angew. Chem. Int. Ed.*, **6,** 451 (1967); K. Hartke and R. Matusch, *ibid.*, **11,** 50 (1972); C. T. Blood and B. P. Linstead, *J. Chem. Soc.*, 2255, 2263 (1952); C. C. Chuen and S. W. Fenton, *J. Org. Chem.*, **23,** 1538 (1958).
3. R. Block, R. A. Marty, and P. de Mayo, *J. Am. Chem. Soc.*, **93,** 3071 (1971).
4. K. Hafner, R. Dönges, E. Goedecke, and R. Kaiser, *Angew. Chem. Int. Ed.*, **12,** 337 (1973).
5. K. Hafner and H. U. Süss, *ibid.*, **12,** 575 (1973).
6. K. Hafner and F. Schmidt, *ibid.*, **12,** 418 (1973).
7. T. J. Katz and M. Rosenberger, *J. Am. Chem. Soc.*, **84,** 865 (1962); T. J. Katz, M. Rosenberger, and R. K. O'Hara, *ibid.*, **86,** 249 (1964).
8. M. P. Cava and M. V. Lakshmikantham, *Accounts Chem. Res.*, **8,** 139 (1975).

9. V. Schomaker and L. Pauling, *J. Am. Chem. Soc.*, **61,** 1769 (1939).
10. H. C. Longuet-Higgins, *Trans. Faraday Soc.*, **45,** 173 (1949).
11. A. Streitwieser, "Molecular Orbital Theory for Organic Chemists," Wiley, New York, 1961.
12. C. Kissel, R. J. Holland, and M. C. Caserio, *J. Org. Chem.*, **37,** 2720 (1972).
13. R. Gerdil and E. A. C. Lucken, *J. Am. Chem. Soc.*, **87,** 213 (1965).
14. K. Bocek, A. Mangini, and R. Zahradnik, *J. Chem. Soc.*, 255 (1963); A. Mangini and C. Zauli, *ibid.*, 2210 (1960).
15. W. G. Salmond, *Quart. Rev.*, **22,** 253 (1968).
16. D. H. Reid, in "Organic Compounds of Sulfur, Selenium and Tellurium," D. H. Reid, Ed., Vol. I, Special Publication of the Chemical Society, London, 1970.
17. (a) M. P. Cava and N. M. Pollack, *J. Am. Chem. Soc.*, **89,** 3639 (1967); (b) M. P. Cava, N. M. Pollack, and G. A. Dieterle, *ibid.*, **95,** 2558 (1973).
18. M. J. S. Dewar and N. Trinajstic, *ibid.*, **92,** 1453 (1970).
19. D. T. Clark, *Tetrahedron Lett.*, 5257 (1967); D. T. Clark, *Tetrahedron*, **24,** 2567 (1968).
20. M. P. Cava, M. A. Sprecker, and W. R. Hall, *J. Am. Chem. Soc.*, **96,** 1817 (1974).
21. (a) M. P. Cava and G. E. M. Husbands, *ibid.*, **91,** 3952 (1969); (b) M. P. Cava, M. Behforouz, G. E. M. Husbands, and M. Srinwasan, *ibid.*, **95,** 2561 (1973).
22. K. T. Potts and D. McKeough, unpublished results.
23. J. D. Bower and R. H. Schlessinger, *J. Am. Chem. Soc.*, **91,** 6891 (1969).
24. K. T. Potts, E. Houghton, and U. P. Singh, *J. Org. Chem.*, **39,** 3627 (1974).
25. K. T. Potts and D. McKeough, *J. Am. Chem. Soc.*, **96,** 4268 (1974).
26. K. T. Potts, D. R. Choudhury, A. J. Elliott, and U. P. Singh, *J. Org. Chem.*, **41,** 1724 (1976).
27. K. T. Potts and G. H. Williams, unpublished results.
28. K. T. Potts and J. L. Marshall, unpublished results.
29. K. E. Wilzbach and L. Kaplan, *J. Am. Chem. Soc.*, **87,** 4005 (1965); F. D. Greene and S. S. Hecht, *J. Org. Chem.*, **35,** 2482 (1970).
30. V. Jäger and H. Viehe, *Angew. Chem. Int. Ed.*, **8,** 273 (1969).
31. C. V. Greco, R. P. Gray, and V. G. Grosso, *J. Org. Chem.*, **32,** 4101 (1967).
32. (a) K. T. Potts, J. Baum, E. Houghton, D. N. Roy, and U. P. Singh, *J. Org. Chem.*, **39,** 3619 (1974); (b) H. O. Bayer, R. Huisgen, R. Knorr, and F. C. Schaefer, *Chem. Ber.*, **103,** 2581 (1970).
33. J. M. Hoffmann, Jr. and R. H. Schlessinger, *J. Am. Chem. Soc.*, **91,** 3953 (1969).
34. K. T. Potts and D. McKeough, *ibid.*, **96,** 4276 (1974).
35. H. Gotthardt, *Chem. Ber.*, **105,** 196 (1972).
36. K. T. Potts and J. C. Connors, unpublished observations.
37. H. R. Snyder and N. E. Boyer, *J. Am. Chem. Soc.*, **77,** 4233 (1955); E. R. Alexander, M. R. Kinter, and J. D. McCollum, *ibid.*, **72,** 801 (1950); E. Durio, *Gazz. Chim. Ital.*, **61,** 589 (1931).
38. K. T. Potts and J. Considine, unpublished results.
39. K. T. Potts and A. J. Elliott, *J. Org. Chem.*, **38,** 1769 (1973).
40. K. T. Potts and A. J. Elliott, unpublished results.
41. J. E. Franz and L. L. Black, *Tetrahedron Lett.*, 1381 (1970); R. K. Howe and J. E. Franz, *J. C. S. Chem. Comm.*, 524 (1973).
42. R. Engel, D. Halpern, and B. A. Funk, *Org. Mass Spectrom.*, **7,** 177 (1973).
43. M. D. Glick and R. E. Cook, *Acta Crystallogr., Sect. B*, **28,** 1336 (1972).
44. K. T. Potts, J. Baum, and E. Houghton, *J. Org. Chem.*, **39,** 3631 (1974).
45. (a) R. Hofmann and R. B. Woodward, *J. Am. Chem. Soc.*, **87,** 2046 (1965); (b) R. B. Woodward and R. Hofmann, "The Conservation of Orbital Symmetry," Verlag Chemie, GmbH., Weinheim/Bergstr., Germany, 1970.

46. R. Huisgen, *Angew. Chem. Int. Ed.*, **2,** 565 (1963).
47. R. Huisgen, *J. Org. Chem.*, **33,** 2291 (1968).
48. T. L. Gilchrist and R. C. Storr, "Organic Reactions and Orbital Symmetry," Cambridge University Press, London, 1972, p. 112.
49. G. Wittig, E. Knauss, and K. Niethammer, *Justus Liebigs Annalen der Chemie* **630,** 10 (1960); R. Helder and H. Wynberg, *Tetrahedron Lett.*, 605 (1972).
50. A. P. Komin, R. W. Street and M. Carmack, *J. Org. Chem.*, **40,** 2749 (1975).
51. G. V. Boyd, *Tetrahedron Lett.*, 1421 (1965).
52. H. Paul and K. Walter, *J. Prakt. Chem.*, **28,** 297 (1965).
53. C. Grundmann, S. K. Datta, and R. F. Sprecher, *Justus Liebig's Annalen der Chemie*, **744,** 88 (1971); S. J. Huang, V. Paneccasio, F. Di Battista, D. Picker, and G. Wilson, *J. Org. Chem.*, **40,** 124 (1975).
54. J. H. Lee, A. Matsumoto, O. Simamura, and M. Yoshida, *Chem. Commun.*, 1393 (1969).
55. A. Matsumoto, J. H. Lee, M. Yoshida, and O. Simamura, *Chem. Lett.*, 455 (1973).
56. J. H. Lee, A. Matsumoto, M. Yoshida, and O. Simamura, *Chem. Lett.*, 951 (1974).
57. M. Yoshida, A. Matsumoto, and O. Simamura, *Bull. Chem. Soc. Jap.*, **43,** 3587 (1970).
58. K. T. Potts and S. K. Datta, unpublished results.
59. K. T. Potts and J. L. Marshall, *J. Org. Chem.*, **41,** 129 (1976).
60. W. Lwowski, *Nitrenes*, Interscience, New York, 1970; R. A. Abramovitch, T. D. Bailey, T. Takaya, and V. Uma, *J. Org. Chem.*, **39,** 340 (1974) and references listed therein; R. A. Abramovitch and T. Takaya, *ibid.*, **38,** 3311 (1973).
61. S. Trofimenko, *J. Am. Chem. Soc.*, **87,** 4393 (1965); T. W. G. Solomons and C. F. Voight, *ibid.*, **87,** 5256 (1965); **88,** 1992 (1966).
62. T. W. G. Solomons, F. W. Fowler, and J. Calderazzo, *ibid.*, **87,** 528 (1965).
63. B. M. Lynch and Y.-Y. Hung, *J. Heterocycl. Chem.*, **2,** 218 (1965).
64. O. Tsuge and H. Samura, *ibid.*, **8,** 707 (1971).
65. O. Tsuge and H. Samura, *Org. Prep. Proc. Int.*, **4,** 273 (1972).
66. For a review of the use of triethyl phosphite in reactions of this kind see: J. I. G. Cadogan, *Quart. Rev.*, 222 (1968); J. I. G. Cadogan and R. K. Makie, *ibid.*, 87 (1974).
67. O. Tsuge and H. Samura, *Chem. Lett.*, 175 (1973).
68. O. Tsuge and H. Samura, *Tetrahedron Lett.*, 597 (1973).
69. O. Tsuge and H. Samura, *Heterocycles*, 27 (1974).
70. R. Pfleger, E. Garthe, and K. Rauer, *Chem. Ber.*, **96,** 1827 (1963).
71. M. Brufani, W. Fedeli, G. Giacomello, and A. Vaciago, *ibid.*, **96,** 1840 (1963).
72. R. A. Carboni, J. C. Kauer, J. E. Castle, and H. E. Simmons, *J. Am. Chem. Soc.*, **89,** 2618 (1967).
73. M. E. Burke, R. A. Sparks, and K. N. Trueblood, *Acta Cryst.*, **16,** A64 (1963).
74. J. H. Hall, J. G. Stephanie, and D. K. Nordstrom, *J. Org. Chem.*, **33,** 2951 (1968).
75. J. I. G. Cadogan, M. Cameron-Wood, R. K. Makie, and R. J. G. Searle, *J. Chem. Soc.*, 4831 (1965).
76. J. C. Kauer and R. A. Carboni, *J. Am. Chem. Soc.*, **89,** 2633 (1967).
77. R. A. Carboni, J. C. Kauer, W. R. Hatchard, and R. J. Harder, *ibid.*, **89,** 2626 (1967).
78. Y. T. Chia and H. E. Simmons, *ibid.*, **89,** 2638 (1967).
79. R. J. Harder, R. A. Carboni, and J. E. Castle, *ibid.*, **89,** 2643 (1967).

CHAPTER VII

Borazaromatic Compounds

ALBERT J. FRITSCH

Co-Director
Center for Science in the Public Interest
Washington, D.C.

I. Introduction

In recent years it has been recognized that the boron atom can participate in conjugated systems in as great a degree as can the common electronegative heteroatoms nitrogen, sulfur, and oxygen. The novel "borazaromatic" compounds described here are derived from normal aromatic hydrocarbons by the replacement of adjacent carbon atoms by boron and nitrogen. These compounds are similar in physical and chemical properties to the better known aromatic hydrocarbons to which they are related. I first review the synthesis and general properties of a great variety of these borazaromatic compounds and in the last two sections treat their common properties in the light of their aromaticity.

A neutral carbon atom in an aromatic hydrocarbon can be replaced by an isoelectronic positive nitrogen ion resulting in the pyridinium ion ($C_5NH_6^+$). This relatively stable ion is *isoconjugate* with benzene in much the same way that pyridine is isoconjugate with phenyl anion. The replacement of a carbon atom by an isoelectronic boron ion should result in the hypothetical isoconjugate negative ion ($C_5BH_6^-$). Thus one can expect that the replacement of a pair of carbon atoms by one nitrogen and one boron atom should yield a neutral borazaromatic system. Three molecules should be capable of existence which are isoconjugates of benzene: 2,1-borazarobenzene (**1**), 3,1-borazarobenzene (**2**), and 4,1-borazarobenzene (**3**)[1,2] (see Appendix concerning nomenclature).

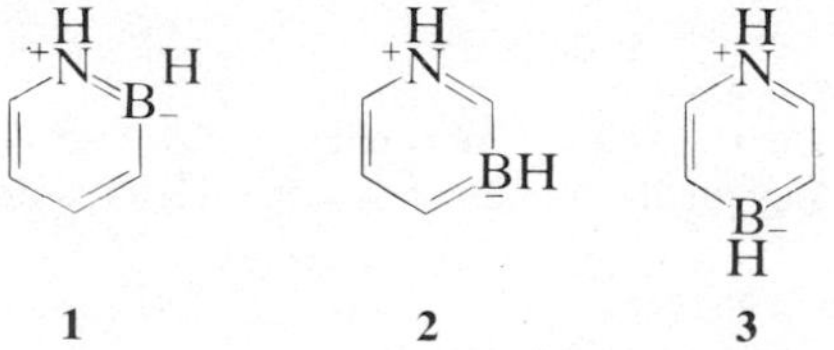

1 **2** **3**

It has long been known that borazine (**4**) "inorganic benzene," and one form of boron nitride are analogous to benzene and graphite, respectively. The synthesis and reactivity of borazine and its derivatives have been reviewed extensively elsewhere, and for the sake of brevity, I limit discussion to those borazaromatic compounds containing skeleton carbon atoms. Though generally inorganic synthetic techniques have been used for the borazine series and organic methods for these borazaromatic compounds, the division is arbitrary and in several instances there can be conversion from one system to the other.

Likewise I speak principally of boraza- compounds and only insert mention of the less stable analogous boroxa- and borathia-compounds for the sake of comparison. The aromaticity of these last two systems

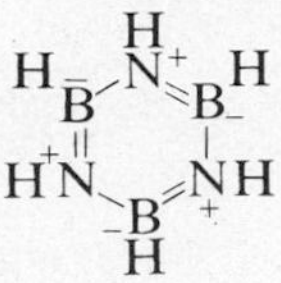

H
H B N⁺ B H
H N B N H
H

4

involves the less important contributions of dipolar resonance structures containing =O–$^{+}$ and the =S–$^{+}$.[2] The more important –N^{+}=B–$^{-}$ contribution is written as such to emphasize the aromatic character of the borazaromatic compounds. However, unless this notation is properly understood, it could be very misleading. The net charge on the boron atom is almost certainly quite small. In fact, CNDO-2 (complete neglect of differential overlap) calculations indicate that the net formal charge on boron is almost equal to that of nitrogen. In borazine the values are nitrogen = 1.0651 and boron = 0.9264 and in 2,1-borazarobenzene nitrogen = 1.1290 and boron = 0.9536.[3] These results indicate that the opposing σ- and π-electron polarizations are somewhat neutralized in these compounds.

II. The 2,1-Borazarene System

A. 2,1-Borazarobenzene

Since the 2,1-borazarene system appears to be most stable, synthetic efforts have been concentrated on this class of borazaromatic compounds. Several successful routes have been developed for making the simplest member of this series (**1**). Dewar and Marr[4] condensed methyl-2-styryl-3-aminothiophene-5-carboxylate with phenylboron dichloride to give 2-carbomethoxy-5,6-diphenyl-5,4-borazarobenzothiophene (**5**) the first sulfur-containing borazaromatic compound. Desulfurization of **5** with Raney nickel gives methyl β-(2,3-diphenyl-5-borazaryl) propionate, **6**.

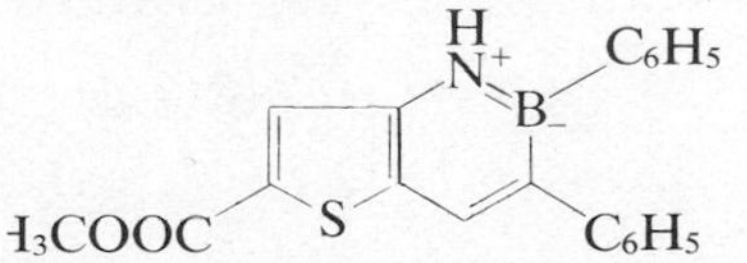

H N⁺ B C_6H_5 S H_3COOC C_6H_5

5

H_3COOC—$(CH_2)_2$ H N⁺ B C_6H_5 C_6H_5

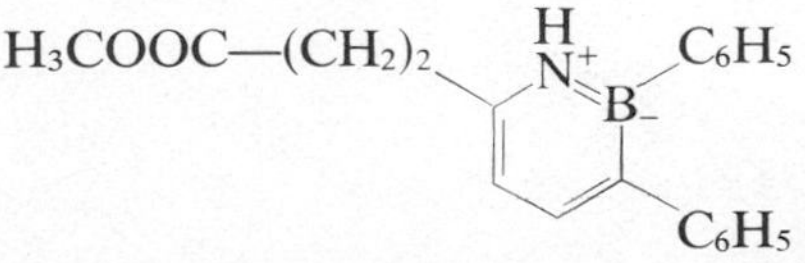

6

White has prepared 2-phenyl-2,1-borazarobenzene, **7**, by condensing 4-amino-1-butene with diborane followed by dehydrogenation. He obtained a fairly stable compound, although it can be easily destroyed by treatment with acid or alkali.[5]

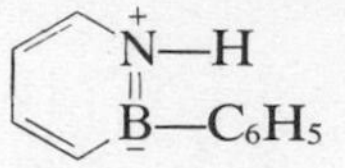

7

A third route to the simple 2,1-borazarene system is through decomposition of 14,16,18-tribora-13,15,17-triazarotriphenylene (**8**) an analog of triphenylene.[6] This intermediate was made by first reacting 4-amino-1-butene with lithium borohydride under pressure and dehydrogenating the product **9** over palladized charcoal at 300°C. The triphenylene analog underwent rapid hydrolysis in cold dilute ethanolic alkali yielding three molecules of 2-hydroxy-2,1-borazarobenzene, **10**. Since this route had several low-yield steps and the results were very erratic, a new route was developed starting with the hydroboration of 4-benzylamino-1-butene, **11**. The 1-benzyltetrahydroborazarobenzene, **12**, was reacted with phenylmagnesium bromide to give the 2-phenyl derivative, **13**, which was dehydrogenated to give 1-benzyl-2-phenyl-2,1-borazarobenzene, **14**, and a small amount of **7**.

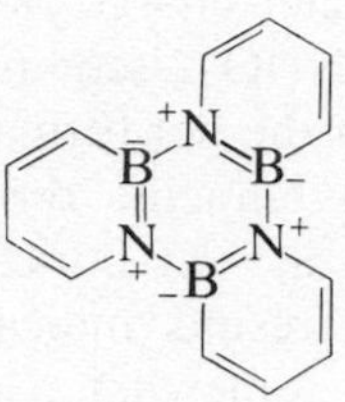

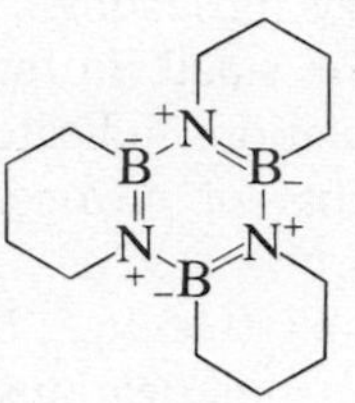

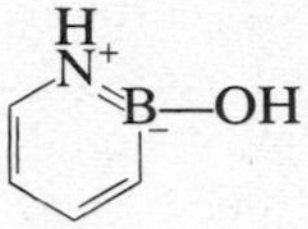

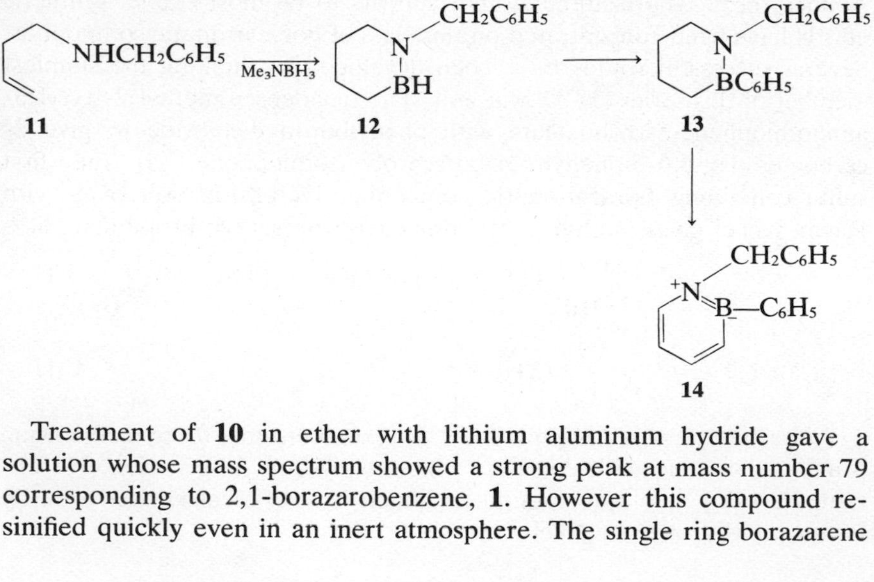

8 **9** **10**

Me_3NBH_3

11 **12** **13** **14**

Treatment of **10** in ether with lithium aluminum hydride gave a solution whose mass spectrum showed a strong peak at mass number 79 corresponding to 2,1-borazarobenzene, **1**. However this compound resinified quickly even in an inert atmosphere. The single ring borazarene system appears to be very reactive and quite prone to polymerization. However the stability of this ring is enhanced by the presence of phenyl groups on either the nitrogen or the boron atom; thus both **7** and **14** are less reactive than 2,1-borazarobenzene or its alkyl or hydroxy derivatives. This extra stabilization is presumably due to the conjugation between the borazarene and the benzene rings.[7] In **14** the *ortho*-benzyl group forces the rings out of coplanarity and thus results in lower stability.

Dehydrogenation of the tetrahydroborazene **4a** gave the *N*-methyl-2,1-borazarene, **4b**, which because of its reactivity has only been identified by mass and IR spectroscopy.[7a]

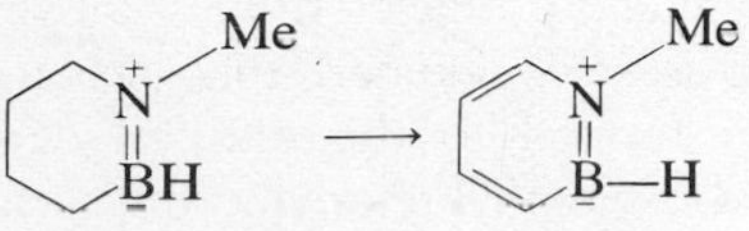

4a **4b**

B. 2,1-Borazaronaphthalene

The first borazaromatic compound isoconjugate with naphthalene, 2,1-borazaronaphthalene, **15**, was prepared by Dewar and Dietz.[8] They reacted *o*-aminostyrene with boron trichloride and formed the chloro derivative, **16a**, which was reacted with phenylmagnesium bromide

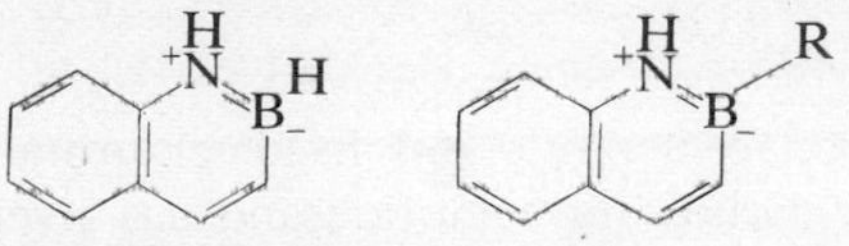

15 **16**

16a: R = Cl
16b: R = C_6H_5
16c: R = OH
16d: R = CH_3
16e: R = OHC_3
16f: R = OC_2H_5

to yield **16b**, hydrolyzed to give **17** the anhydride of **16c**. Likewise this 2-chloro derivative has been reacted with methylmagnesium iodide to give **16d** and with methyl and ethyl alcohol to produce **16e** and **16f**. Likewise **16b** has been directly prepared by reacting the starting styrene with phenylboron dichloride, and by reacting **16e** with an excess of the proper Grignard reagent.[9] The same product has been formed by reacting bisborazaronaphthyl ether, **17**, with excess Grignard reagent. The parent

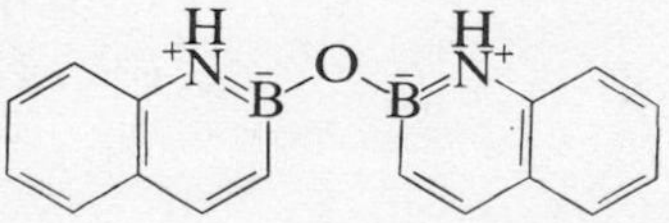

17

compound, **15**, has been obtained in good yield by treating **16e** with lithium aluminum hydride.[9]

The 2,1-borazaronaphthalene derivatives resemble naphthalene in their spectral features (see Section VI.A) and their general stability. With the exception of **15** these compounds show extraordinary stability to hydrolysis by hot concentrated alkali or acid. The 2-phenyl derivative can be recovered unchanged after boiling 3 hr with concentrated hydrochloric acid or after 2 hr with 40% aqueous potassium hydroxide. Prolonged boiling of **17** with base failed to produce an ultraviolet spectrum different from that of the anion, which resulted instantly and reversibly by the addition of basic ethanolic solution. This stability is most certainly not due to the cyclic nature of the ring containing heteroatoms. When one reduces **17** with sodium and produces the 3,4-dihydro derivative, **18**, the boron–nitrogen bond is so weakened that, when warmed with acetic anhydride, **18** is transformed to the diacetyl derivative.[10] The stable internal ester of *o*-hydroxymethyl phenylboronic acid, **19**, has been found to be resistant to hydrolysis and capable of being nitrated.[11,12] Though this stability had been attributed to the cyclic nature of the compound rather than to aromatic resonance, the same explanation cannot be extended to 2,1-borazaronaphthalenes (see Section VII.C).

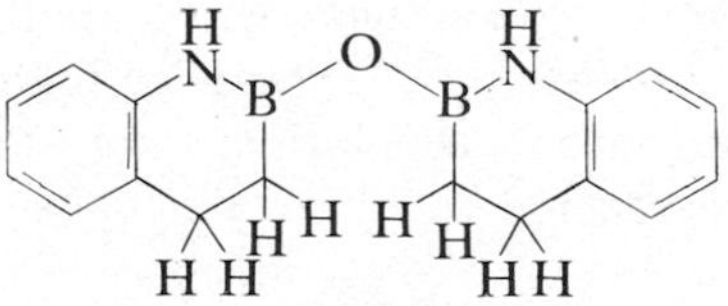

18 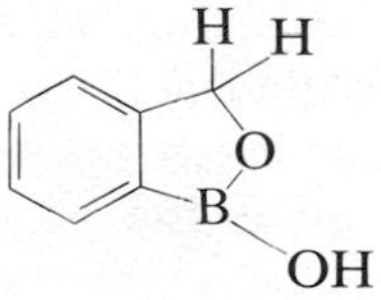 **19**

The 2-methyl-2,1-borazaronaphthalene, **16d**, has been brominated and chlorinated in acetic acid or carbon tetrachloride and has given a quantitative yield of the 3-bromo (**20a**) and the 3-chloro (**20b**) product together with some deboronated product.[13] The structure of 3-chloro compound was proven by hydrolyzing the one:one adduct of boron trichloride and ω-chloro-2-aminostyrene, and the resulting **21** was crystallized from methanol to give **20c** which, in turn, when treated with excess methylmagnesium iodide gave **20b**. One explanation of the de-

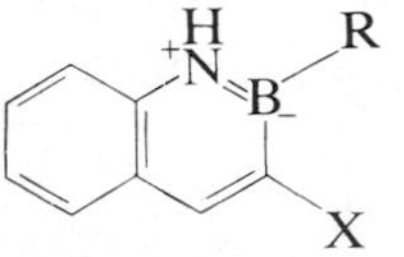

20 **21**

20a: X = Br, R = CH_3
20b: X = Cl, R = CH_3
20c: X = Cl, R = OCH_3

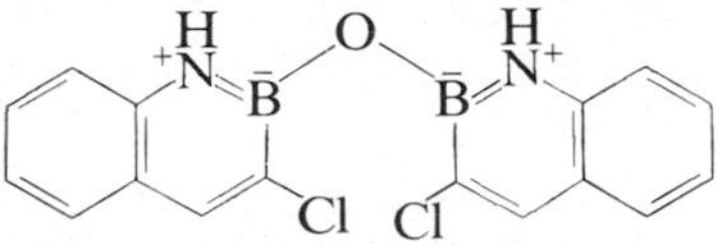

boronated by-product might be the primary attack of the 3,4-bond with production of the π-complex. If one assumes that the parent compound **15** is less aromatic than naphthalene (see Section VII.C), then the 3,4-bond must have a high bond order. Recent calculations show that this bond has the highest bond order (Section VI.C).[3] The lack of any detectable 4-halo compound from the π-complex precurson may be ascribed to steric hindrance. The π-complex could rearrange perhaps to form the arenonium ion, **22**, which either loses a proton to form the 3-halo compound or reacts with the acetic acid solvent to form the deboronated product **23**.[2]

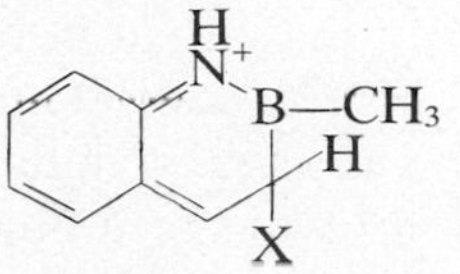

H N+ B—CH₃ H X

22

NH₂ X

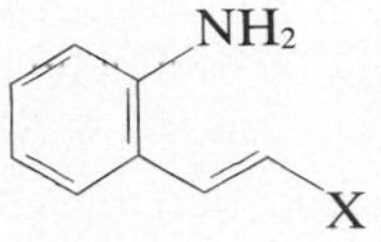

23

Methyllithium was reacted with **16d** to give the *N*-lithio derivative, **24a**, and this intermediate was used to make a whole series of *N*-alkyl derivatives as possible agents for neutron capture therapy of cancer.[14] This **24a** reacted with dimethyl sulfate, allyl bromide, 1-bromo-3-chloropropane, and ethyl chloroformate to give **24b**, **24c**, **24d**, and **24e**, respectively. The reaction of **24d** with the appropriate secondary amines gave the dialkylamino derivatives **25a–d**.

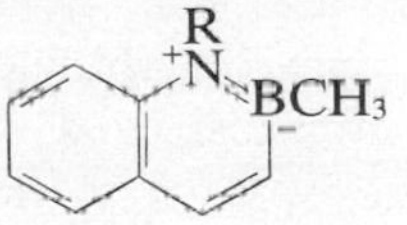

R +N BCH₃ −

24

$CH_2—CH_2—CH_2NR_2$ +N B—CH₃ −

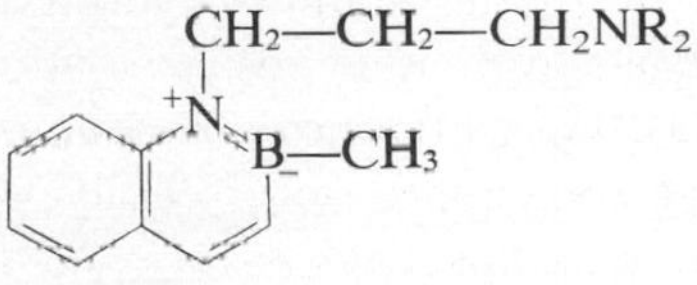

25

24a: R = Li
24b: $R = CH_3$
24c: $R = -CH_2CH{=}CH_2$
24d: $R = —(CH_2)_3Cl$
24e: $R = —COOC_2H_5$

25a: $R = CH_3$
25b: $R = C_2H_5$
25c: NR_2 = piperidino
25d: NR_2 = morphino

C. 10,9-Borazaronaphthalene

Since it was most desirable to synthesize a member of the 2,1-borazarene series which contained a boron atom free from substituents, much effort has been expended for making 10,9-borazaronaphthalene, **26**. This very interesting compound has a boron atom at the bridgehead of two fused aromatic rings and linked to three atoms other than hydrogen. Thus the influence of the nitrogen and boron atoms on charge density, reactivity, and stability become far more pronounced than in the case of 2,1-borazaronaphthalene.

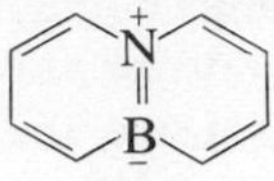

26

Compound **26** was first prepared by dehydration of 1,3-dihydroxybutane to 4-hydroxy-1-butene, conversion of this through its chloride into di-3-butenylamine, **27**, hydroboration to 9-aza-10-boradecalin, **28**, and dehydrogenation to the parent borazaromatic compound **26**.[15] Because of the poor overall yield, several modifications were made including the synthesis of **27** *via* the bromide made from 4-hydroxy-1-butene and dehydrogenation of the resulting **28** in an autoclave at 300° with palladized charcoal as catalyst and hexene-1 or octene-1

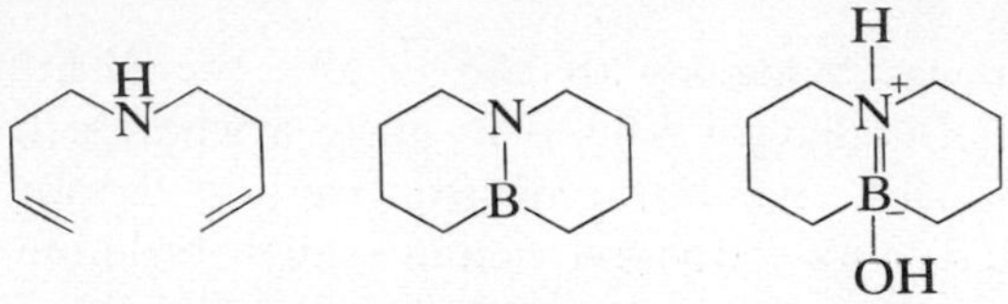

27 **28** **29**

as hydrogen acceptor.[16] Compound **28** is very unstable in air and is easily converted to the crystalline hydrated **29**. The parent compound **26** undergoes substitution at positions adjacent to the boron, yielding upon bromination in acetic acid, 4-bromo-10,9-borazaronaphthalene, **30a**. Deuteration took place very readily when **26** was dissolved in dioxane containing deuteriosulfuric acid, and displacing protons at the 4- and 5-position (**30b**). Acetylation with one mole of acetyl chloride gave the 4-acetyl derivative **30c**.[17]

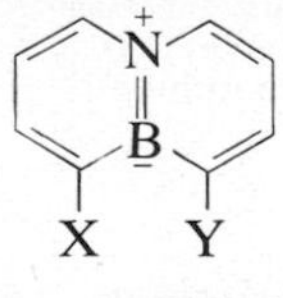

30a: X = Br, Y = H
30b: X = D, Y = D
30c: X = —$COCH_3$, Y = H

The 10,9-borazaronaphthalene resembles naphthalene in many physical properties including odor, melting point, violatility, crystalline structure, uv spectrum, and the tetracyanoethylene π-complex absorption. The phase diagram of the binary system of **26** and naphthalene indicates that the solid solutions form over a whole range of composition.[16] Likewise **26** resembles naphthalene in its remarkable stability being resistant to aerial oxidation or hydrolysis even in the presence of alkali or acid and it forms a stable crystalline picrate.

Some of these properties are quite unexpected especially the low melting point and the solubility of 10,9-borazaronaphthalene in nonpolar solvents. One would expect **26** to be highly polar in view of the resonance structure drawn. But the dipole moment must be quite low due to the above-mentioned physical properties and yet would seem at first glance to be quite high due to indicated polarity. One can only conclude from the observed physical properties that the B–N σ-bond must be highly polar with its moment in a direction opposite to that due the π-electron polarization; the polarity of the σ-bond will increase with the increase in negative charge on the boron atom and positive charge on the nitrogen atom. This "reciprocal polarization" of the σ- and π-electrons causes **26** to be nonpolar in actuality and yet, granting the limitations of present chemical notation, the resonance structure drawn is a reasonable expression of π-electron distribution in the borazarene ring.

The properties of naphthalene, 10,9-borazaronaphthalene, and 9-aza-10-boradecalin have been further studied using ^{11}Bnmr (see Section VI.B), mass spectral data, and photoelectron spectroscopy.[18] The parent peak in the mass spectra of **26** and naphthalene is the base peak and the fragmentation patterns of both are similar. However **26** gives higher yields of fragmentation products, and this would be expected, since **26** is less aromatic than naphthalene. Since **28** is not aromatic, it behaves like other amines and so loses a hydride next to the nitrogen atom giving as the most intense peak that of mass number of 135. The photoelectron spectra of **26** and naphthalene are quite similar with a first ionization potential of the former being 8.24 eV and of the latter being 8.11 eV. The first ionization potential of **28** (8.47 eV) corresponds to the loss of a nitrogen lone-pair electron. The values for the two aromatic compounds represent ionization from antisymmetric πMO's, and bear no relation to the 8.47 eV value for **28**. In 9-aza-10-boradecalin the nitrogen lone-pair electrons occupy a single molecular orbital centered on nitrogen, but these electrons in 10,9-borazaronaphthalene are dispersed over the three symmetric π-molecular orbitals. This study has confirmed the differences in electronic structure of **26** and **28** and has shown the similarities between **26** and naphthalene.

D. 10,9-Borazarophenanthrene

1. *Preparation*

The borazaromatic analog of phenanthrene, **31**, was the first 6-membered compound of this series synthesized.[19] It was prepared by the condensation of 2-aminobiphenyl with boron trichloride using aluminum chloride. This Friedel–Crafts type cyclization most probably proceeded through an intermediate arylaminoboron dichloride which then formed **32a**. Hydrolysis of this 10-chloro derivative yielded the free acid **32b**, and reaction with Grignard reagents gave the 10-methyl and 10-phenyl derivatives, **32c** and **32d**.[20] Likewise **32a** can be reduced to the parent compound, **31**, by means of lithium aluminum hydride and **32b** yielded

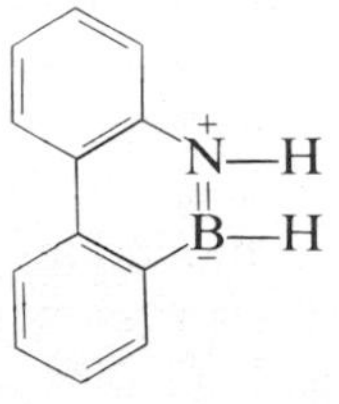

32a: R = Cl
32b: R = OH
32c: $R = CH_3$
32d: $R = C_6H_5$
32e: $R = OCH_3$
32f: $R = OC_2H_5$

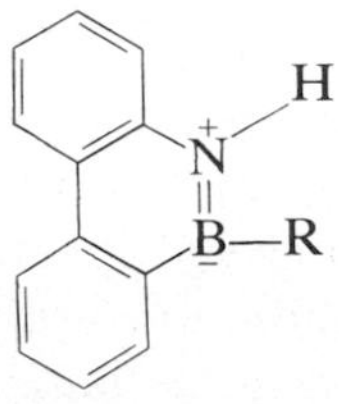

31 **32**

the esters **32e** and **32f** upon being dissolved in the proper alcohol. The stable 10-hydroxy derivative has been converted directly to **31** by reducing it with a slight deficiency of lithium aluminum hydride together with one-third molar equivalent of aluminum chloride.[9] Likewise the alkyl and aryl derivatives **32c** and **32d** have been obtained from **32b** in better yields than from the 10-chloro derivative. An alternate route to **32d** involved the treatment of 2-iodoaniline with chlorodiphenylborane giving an assumed intermediate (**33**) which yielded the 10-phenyl derivative upon irradiation.[21] The same method was extended to the synthesis of boroxaromatic and borathiaromatic compounds, **34a** and **35**.

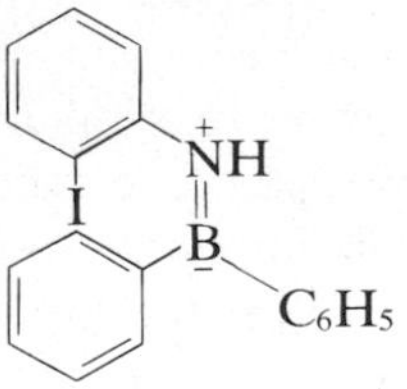

34a: $R = C_6H_5$
34b: R = Cl
34c: R = OH
34d: R = OMe
34e: R = OEt

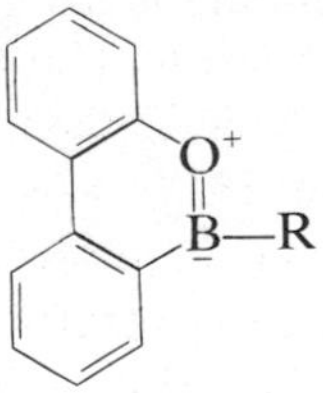

33 **34** 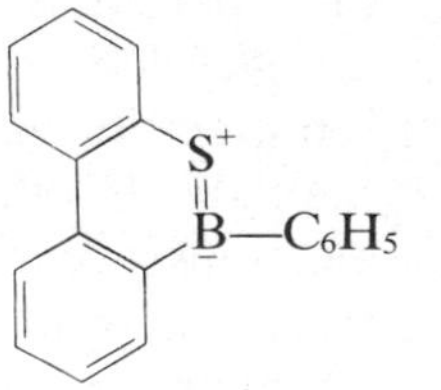 **35**

2. *Comparison with Boroxaro and Borathiaro Analogs*

The first 10,9-boroxarophenanthrene was made by the reaction of 2-hydroxybiphenyl with excess boron trichloride and aluminum chloride catalyst.[22] An alternate route has been found to be the hydrolysis of the diazotization product of **32c**.[23] Reactions similar to those for 10,9-borazarophenanthrene gave the derivatives **34a** to **34e**. Since oxygen is more electronegative than nitrogen, the contribution of the pseudoaromatic structure **34** should be less important. The 10,9-borazarophenanthrene derivatives have been shown to resist attack by boiling base or acid, whereas the **34a** was oxidized and hydrolyzed in water or ethanol to **34c**. However, the oxidation of **34c** with potassium permanganate and chromium trioxide was difficult, and it resisted attack in boiling hydrochloric acid or base but was deboronated in the presence of warm sulfuric acid.

Bis(10,9-borathiarophenanthryl)ether, **36**, which was made by reacting 2-mercaptobiphenyl with boron trichloride and aluminum chloride, proved very sensitive to hydrolytic solvents.[24] It was converted immediately to the boronic ester, **37**, upon treatment with cold ethanol. This is to be expected since sulfur is much less effective at forming π-bonds than nitrogen or oxygen. Further evidence of this difference in aromaticity can be found in ultraviolet and ^{11}B nmr data (see Section VI.A and B). Likewise the mass spectra showed prominent parent peaks as expected if aromatic. However, in the cases of the nitrogen and oxygen 10-hydroxy anhydrides, the base peaks corresponded to fragments formed by the B–O–B cleavage with the phenanthrene rings remaining intact. In the case of **36** the base peak corresponded to a dibenzothiophene fragment implying the cleavage of the central ring.

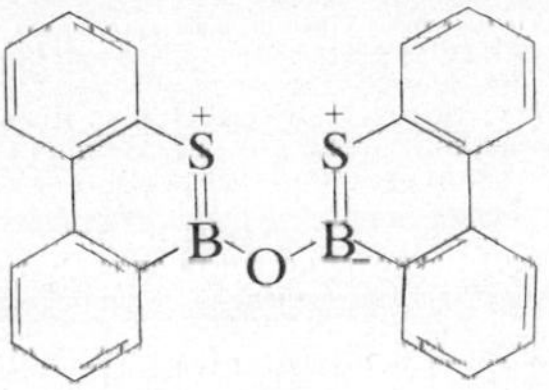

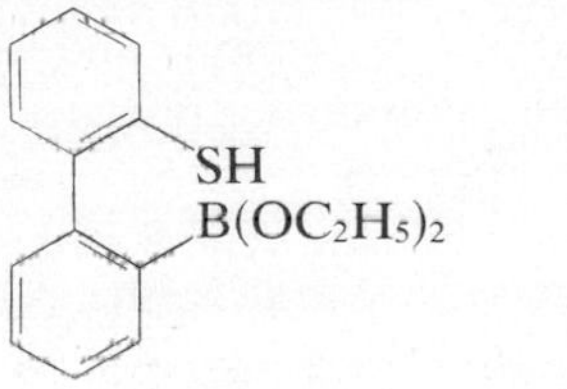

SH
$B(OC_2H_5)_2$

36 **37**

3. *Substitution Reactions*

Nitration of **32b** and **32c** using nitric acid in acetic anhydride gave about one-third of the 6-substitution products, **38a** and **39a**, and two-thirds of the 8-substitution products **38b** and **39b**. Reduction by means of hydrazine over palladized charcoal gave the corresponding 6- and 8-amino derivatives, **38c**, **38d**, **39c**, and **39d**.[25] Partial rate factors for formation of the 8- and 6-nitro derivatives have indicated a very reactive ring system.[26] Chlorination of **32c** with one molar equivalent of chlorine in acetic acid has given mainly the 8-chloro-10-methyl-10,9-borazarophenanthrene, **38e**, and a small amount of the 10-hydroxy derivative,

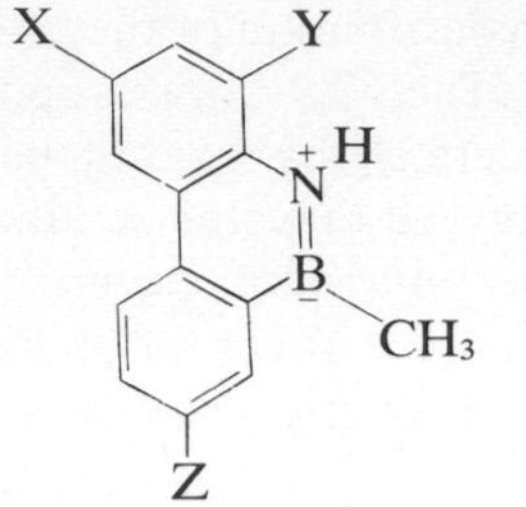

38a: $X = NO_2$, Y, Z = H
38b: $Y = NO_2$, X, Z = H
38c: $X = NH_2$, Y, Z = H
38d: $Y = NH_2$, X, Z = H
38e: Y = Cl, X, Z = H
38f: X = Cl, Y, Z = H
38g: Z = Cl, Y, X = H
38h: $X = COCH_3$, Y, Z = H
38i: X, $Y = COCH_3$, Z = H

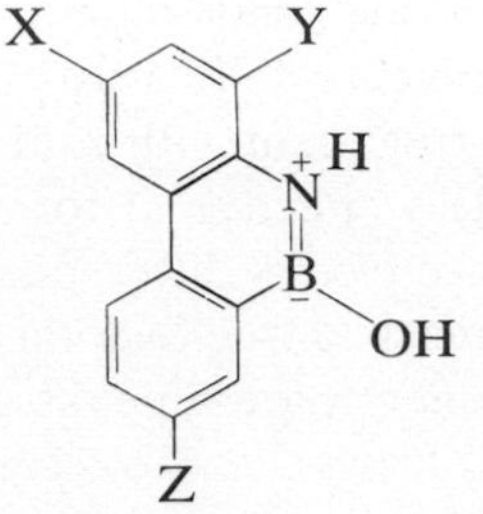

38

39a: $X = NO_2$, Y, Z = H
39b: $Y = NO_2$, X, Z = H
39c: $X = NH_2$, Y, Z = H
39d: $Y = NH_2$, X, Z = H
39e: Y = Cl, X, Z = H
39f: X, Y = Cl, Z = H
39g: X, Y = Br, Z = H
39h: X, Y, Z = Cl
39i: $X = COCH_3$, Y, Z = H
39j: X, $Y = COCH_3$, Z = H

39

39e. The 2-chloro, 6-chloro, and 8-chloro derivatives (**38e–g**) were synthesized separately by reacting the proper aminochlorodiphenyl with boron trichloride. Chlorination or bromination of the 10-hydroxy compound yielded the 6,8-dichloro and dibromo derivatives (**39f** and **39g**) and no detectable monohalo product.[27] Further chlorination gave the 2,6,8-trichloro product, **39h**. Several monofluoro derivatives have been made beginning with the appropriate fluoro-2-amino-biphenyls.[28]

Acetylation of the 10-methyl and 10-hydroxy-10,9-borazarophenanthrene with 1 mole of acetyl chloride in carbon disulfide and with aluminum chloride catalyst gave 6-monoacetyl and some 6,8-diacetyl derivatives, **38h**, **38i**, **39i**, and **39j**.[29] With 2 moles of acetyl chloride only the diacetyl derivative was formed and no monoacetyl product could be isolated. These unusual results have been ascribed to deactivation of the borazarophenanthrene through coordination with aluminum chloride. Since the monoacetyl derivative has a nitrogen with greatly reduced basicity, there is less tendency to coordinate with the catalyst. Thus further acetylation of the monoacetyl derivative competes with the initial acetylation of **32b** and **c**.[29]

4. Some *N*-Substituted Products

The *N*-lithio derivatives of **32c** and **32d** were obtained using methyllithium and these active products **40a** and **41a** were reacted with dimethyl

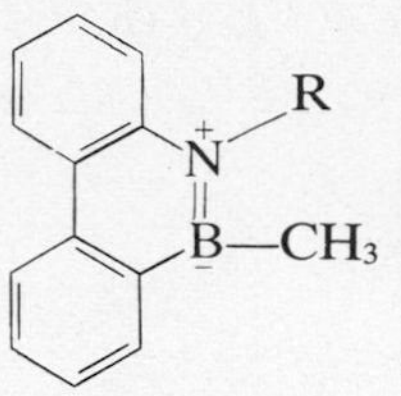

40a: R = Li
40b: R = CH_3
40c: R = COOEt
40d: R = $CH_2CH_2CH_2Cl$
40e: R = $CH_2CH{=}CH_2$

40

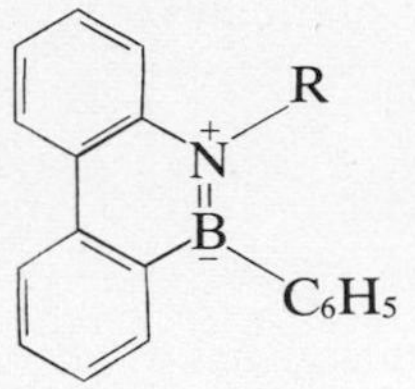

41a: R = Li
41b: R = CH_3
41c: R = COOEt

41

sulfate to give 9,10-dimethyl-19,9-borazarophenanthrene **40b** and 9-methyl-10-phenyl-10,9-borazarophenanthrene, **41b**.[30] Two moles of methyllithium reacted with **32c** to give the 9,9-dilithio-10,10-dimethyl derivative, **42**, which was found to be stable only in the absence of air and moisture. The reaction of **32d** with 2 moles of phenyllithium after hydrolysis gave 10,10-diphenyl-9,9-dihydro-10,9-borazarophenanthrene, **43a**. The boron in these compounds is internally chelated to the nitrogen, and this results in resistance to aerial oxidation and disproportionation reactions. The reaction of **42** with carbon dioxide followed by acid hydrolysis resulted in the formation of **43b**. Spectral data confirmed that the phenanthrene system had been replaced by one analogous to 9,10-dihydrophenanthrene, **44**. Thus conjugation in the borazarene ring had been lost and only a biphenyl conjugated system remained.

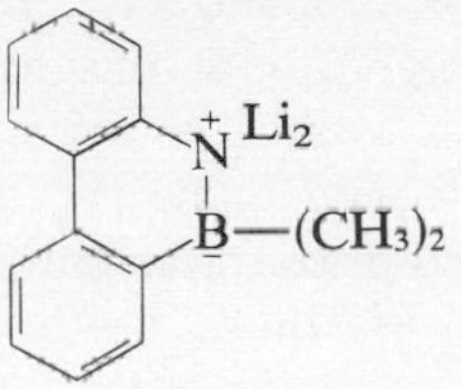

42 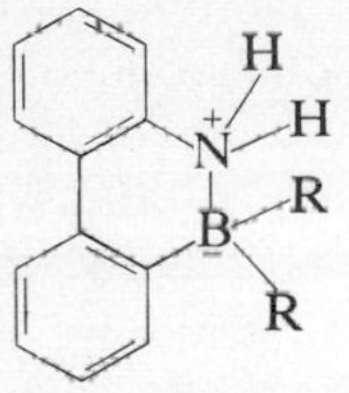 **43** **44**

a: R = C_6H_5
b: R = CH_3

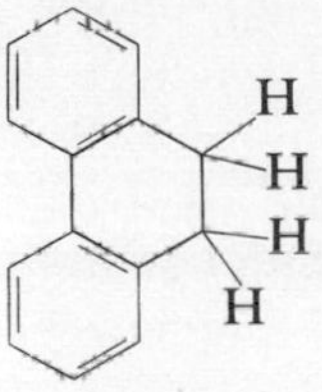

Reaction of the 9-lithio derivatives **40a** and **41a** with ethyl chloroformate gave the corresponding urethans, **40c** and **41c**[31]; however, unlike **40b** and **41b**, these urethans were rapidly oxidized in air and lost their methyl or phenyl substituent. It appeared that the attachment of this electron-withdrawing 9-ethoxycarbonyl group reduced the availability of the π-electrons of the nitrogen for bonding with boron. On the other hand, electron-donating groups such as the hydroxyl substituent on boron reduced the tendency to accept electrons from the nitrogen. The B–N bond has been known to remain intact to oxidative conditions in 10-hydroxy-10,9-borazarophenanthrene, **32b**, but breaks easily in the case of 10-hydroxy-9-ethoxycarbonyl-10,9-borazarophenanthrene, **45**. It appears that the B–N bond can survive one such substituent but not two simultaneously.

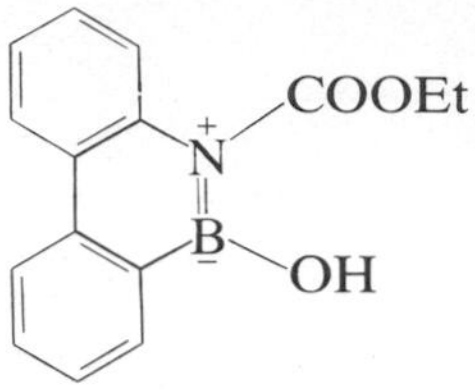

45

A similar effect has been observed where there was a strong electron-withdrawing group at the 8-position in the borazarophenanthrene molecule. Thus 8-nitro-10,9-borazarophenanthrene, **38b**, has been reported to be demethylated during nitration due to the strong interaction between the amino nitrogen and the nitro group. However this reaction did not occur in the case of the 6-nitro derivative, **38a**, or in the case of 10-methyl-10,9-borazarophenanthrene itself. In fact the strong resonance interactions of the borazarene central ring causes the 10-methyl compound to resist oxidation. These interactions would make one expect a very small dipole moment for **32c** (see Section II.C). Huisgen and co-workers[32] have measured the dipole moment of 10-methyl-10,9-borazarophenanthrene and found it to be 0.16 D.

The reaction of **40a** with 1-bromo-3-chloropropane gave 9-(3-chloro-1-propyl)-10-methyl-10,9-borazarophenanthrene, **40d**; this product was subsequently reacted with a series of amines to give products similar to **25a** to **25d**.[14] A second route to **40d** was found to be the addition of hydrogen bromide to 9-allyl-10-methyl-10,9-borazarophenanthrene, **40e**, which had previously been made by the reaction of **40a** with allyl bromide.

E. Other Borazaromatic Compounds

An anthracene analog containing two 2,1-borazarene rings has been formed by the reaction of 1,3-diamino-4,6-*bis*(8-styryl)benzene, **46**, with 2 molar equivalents of phenylboron dichloride. The resulting 2,3,6,7-tetraphenyl-2,7-diboro-1,8-diazaanthracene, **47a**, was quite stable to oxidation and hydrolysis conditions. When **46** was reacted with lithium aluminum hydride the parent 2,7-dibora-1,8-diazaroanthracene, **47b**, was prepared.[33]

A phenanthrene analog containing a boron atom in a bridgehead position was formed by condensing 1-lithio-2-methyl-2,1-borazaronaphthalene with 1,4-dibromobutane.[34] This bromide, **48**, was converted

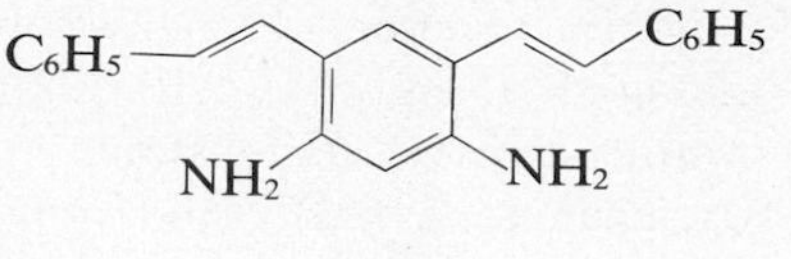

46

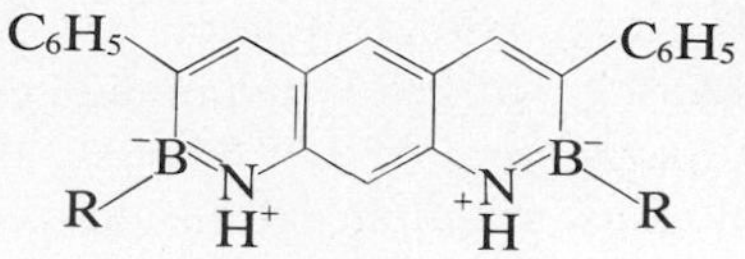

47

a: R = C_6H_5
b: R = H

to a Grignard reagent, **49**, which underwent cyclization with the elimination of methylmagnesium bromide. Dehydrogenation of **50** gave 11,12-borazarophenanthrene, **51**, in 30% yield. The aromatic stability of the product was demonstrated by the rigorous dehydrogenation conditions, and by the similarity of the uv spectrum to those of other borazarophenanthrene derivatives.

The same scheme used for making **51** has been used for the synthesis of 14,13-borazarotriphenylene, **52**.[6] 9-Lithio-10-methyl-10,9-borazarophenanthrene, **40a**, was condensed with 1,4-dibromobutane and the product converted to a Grignard reagent and cyclized to form the 1,2,3,4-tetrahydro compound **53**. This was dehydrogenated with sulfur to form **52** in good yield. This triphenylene analog has been found to be a remarkably stable aromatic system, being unaffected by prolonged exposure to air or treatment with strong acid or alkali and even resistant to attack by maleic anhydride.

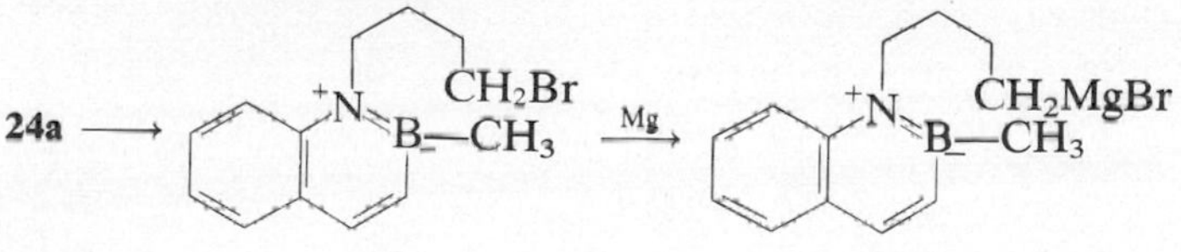

24a ⟶ **48** $\xrightarrow{Mg}$ **49**

$\xrightarrow{-CH_3MgBr}$ **50** $\xrightarrow[Hexene]{Pd/C}$ **51**

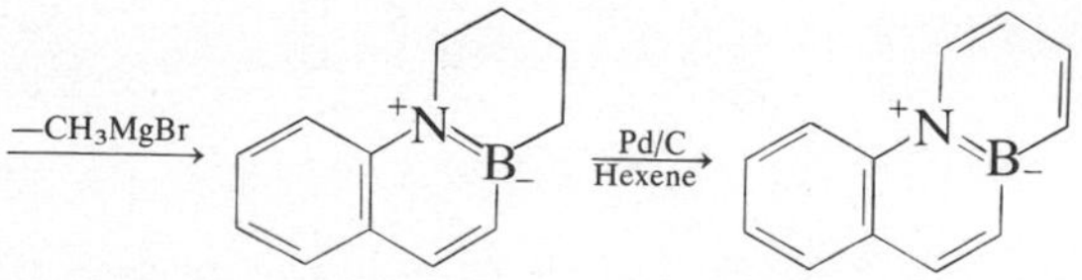

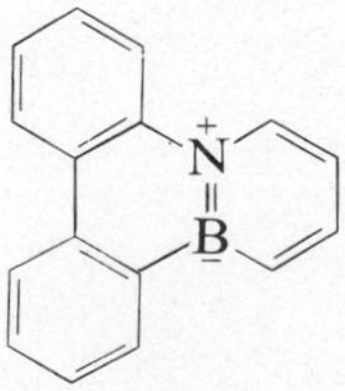

52

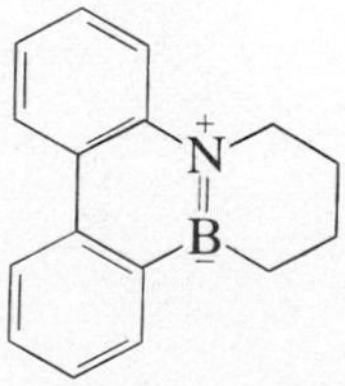

53

The first borazaropyrene was made by the reaction of phenylboron dichloride with 2,6-diaminodiphenyl giving 4,10-diphenyl-4,10-dibora-5,9-diazapyrene, **54**.[33] Though this product proved quite stable, the analogous 4,9-diphenyl-4,9-dibora-5,10-diazapyrene (**55**) could not be obtained by reacting 2,2′-diaminodiphenyl with phenylboron dichloride.

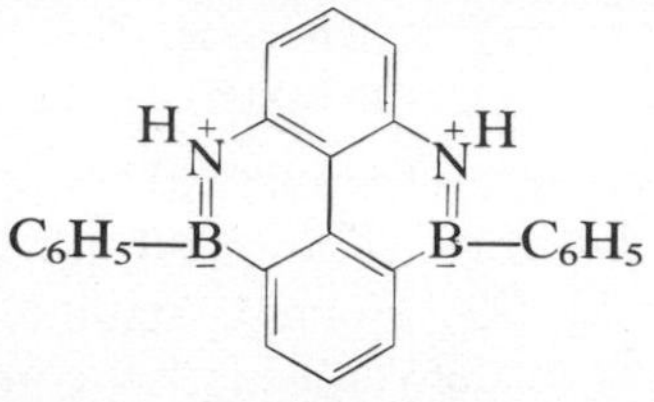

54 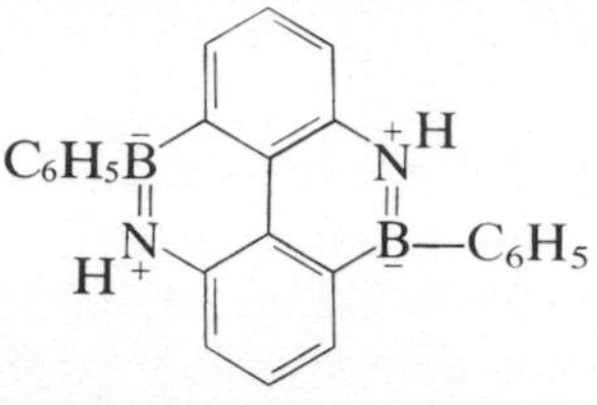 **55**

Another analog of pyrene, 5,4-borazaropyrene, **56a**, was synthesized with some difficulty by a Friedel–Crafts cyclization of the adduct from 4-aminophenanthrene and boron trichloride.[35] This reaction gave the 5-chloro derivative, **56b**, which was not isolated as such but reacted by methods previously mentioned to give the derivatives, **56c** to **e**. An alternative route involved the reduction of 4-nitro-9,10-dihydrophenanthrene and treating the product with boron trichloride and aluminum chloride; hydrolysis yielded 5-hydroxy-9,10-dihydro-5,4-borazaropyrene, **57**, which was dehydrogenated with palladized charcoal in hexene under pressure at 260°. The charge transfer spectra of complexes from derivatives of 5,4-borazaropyrene and tetracyanoethylene resembled those of the corresponding derivatives of 10,9-borazarophenanthrene.

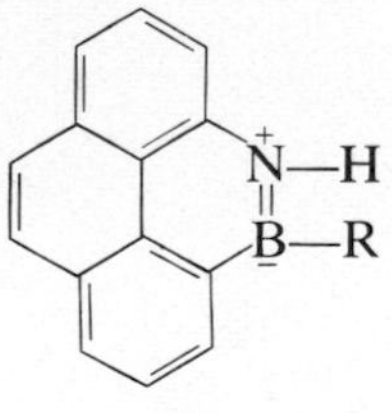

a: R = H
b: R = Cl
c: R = OH
d: R = CH_3
e: R = OCH_3
f: R = OEt

56 **57**

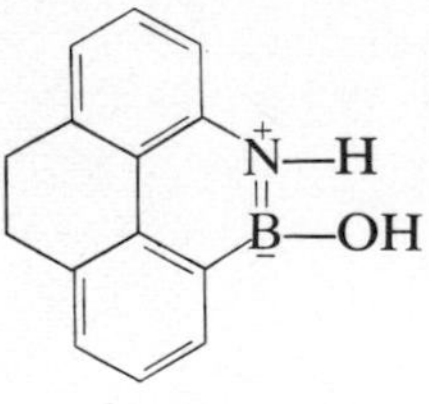

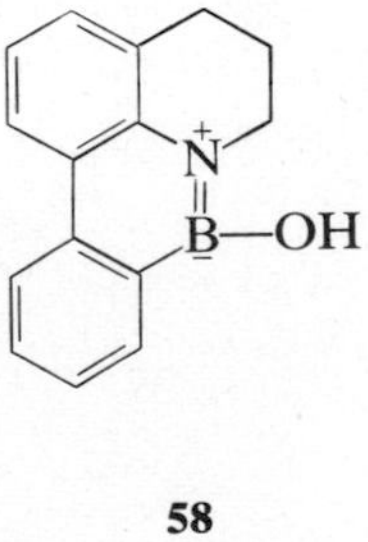

58

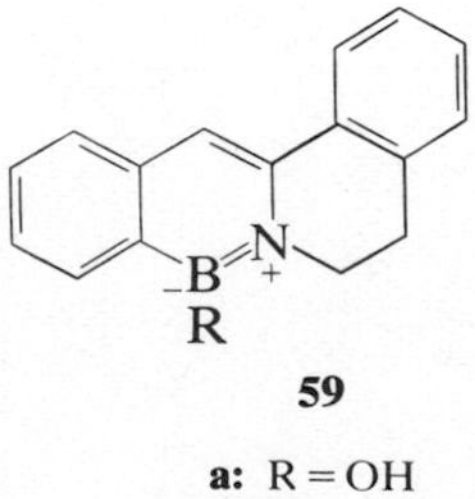

59

a: R = OH
b: R = C_6H_5

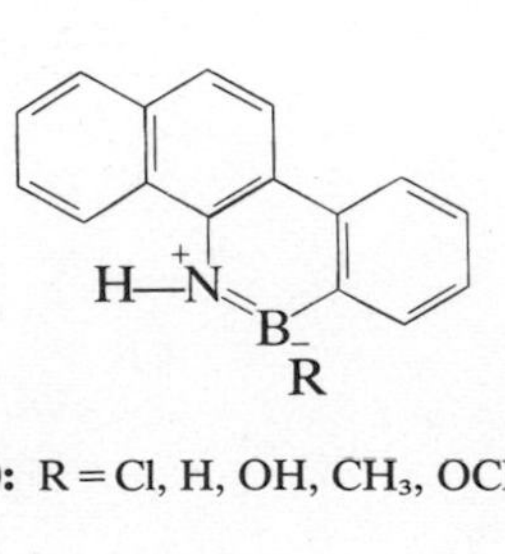

60: R = Cl, H, OH, CH_3, OCH_3

The hydroxy derivative of 8,9-trimethylene-10,9-borazarophenanthrene, **58**, was prepared by reducing 8-phenylquinoline to 8-phenyl-1,2,3,4-tetrahydroquinoline and reacting this product with boron trichloride followed by cyclization with aluminum chloride.[34] This compound proved to have a high degree of stability but the oxygen could not be removed by Grignard reagents perhaps due to the sterically hindered boron atom. In a similar manner a benz[*a*]anthracene derivative, **59a**, was obtained by heating 1-benzyl-3,4-dihydroisoquinoline with boron trichloride and aluminum chloride.[34] The 7-phenyl derivative, **59b**, was obtained using phenylboron dichloride. This ring system proved much less stable than most of the previously prepared borazaromatic compounds and was very susceptible to decomposition by hydrolysis and oxidation. The 6-chloro derivative of 6,5-borazarochrysene, **60**, was prepared in the manner of **58** and **59** and a number of stable derivatives were obtained.[35]

Two fully aromatic benzanthracenes have been prepared, namely, 6,5-borazarobenz[*a*]anthracene, **61a**, and 5,6-borazarobenz[*a*]anthracene, **62a**.[36] The first was prepared by reacting a mixture of 2-iodonaphthalene, 1-bromo-2-nitrobenzene, and copper bronze and forming 2-(*o*-nitrophenyl)naphthalene which was reduced and treated with boron trichloride and catalyst to give **61b**. The second was synthesized by diazotization of 3-nitro-2-naphthylamine to give 2-nitro-3-bromonaphthalene, treating with copper in excess iodobenzene, reducing the 2-nitro-3-phenylnaphthalene to an amine, and heating with boron trichloride to give **62b**. From the chloro derivatives the methoxy and methyl derivatives could be obtained, **61c**, **61d**, **62c**, and **62d**, which formed stable colorations with tetracyanoethylene. When the diazonium solution of **61c** or **62c** was poured in boiling water, nitrogen was eliminated and replaced by oxygen thus forming boroxarobenz[*a*]anthracenes **63** and **64**. The uv spectra of these compounds were similar to those of **61** and **62** but the boroxaro compounds were far less aromatic (see Section II.D.2).

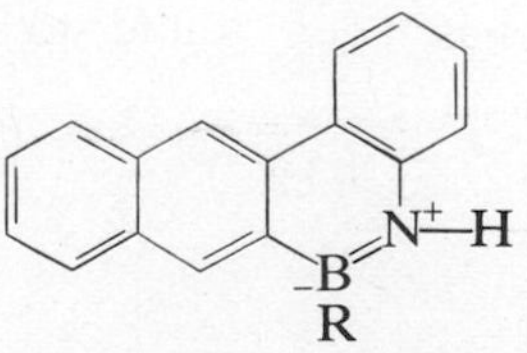

61

a: R = H
b: R = ½O
c: R = OCH_3
d: R = CH_3

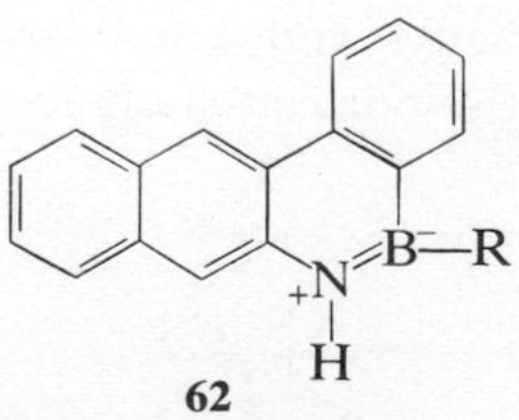

62

a: R = H
b: R = ½O
c: R = OCH_3
d: R = CH_3

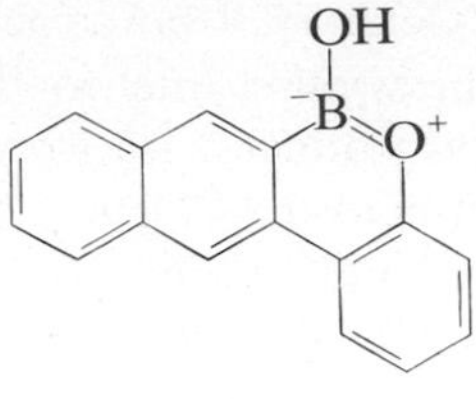

63

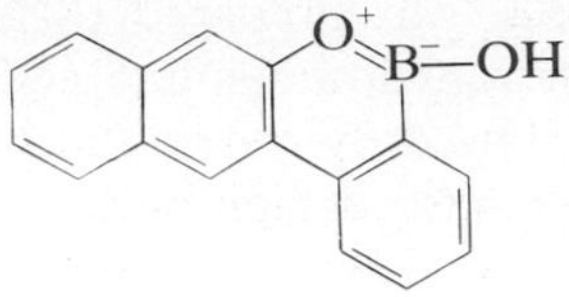

64

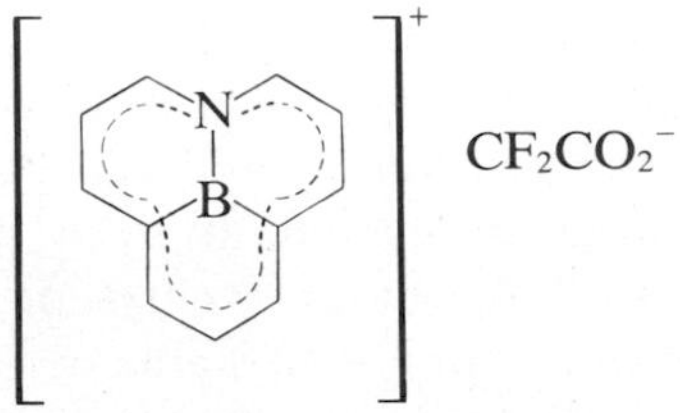

$CF_2CO_2^-$

65

It has been reported that the addition of malonaldehyde *bis*-diethylacetal to 10,9-borazaronaphthalene yielded the tricyclic 12,11-borazarophenalenium cation **65**.[37] However Michl[38] has studied a great number of ultraviolet spectra of borazaromatic compounds and found that the absorption of this salt is nearer what would be expected for the dimeric product of the starting materials.

III. The 4,1-Borazarene System

All the compounds discussed so far have been derived from or have contained the 2,1-borazarene ring. When the boron atom is directly bonded to nitrogen in a 6-membered conjugated ring, there is considerable stabilization of the ring due to the contributions of $-{}^{+}NH{=}BH^{-}-$. However one compound containing a 4,1-borazarene and one containing a 4,1-boroxarene system have been synthesized. The reaction of 2,2′-dilithiodiphenyl and boron trichloride has been found to yield 10-hydroxy-10,9-boroxaroanthracene, **66**.[39]

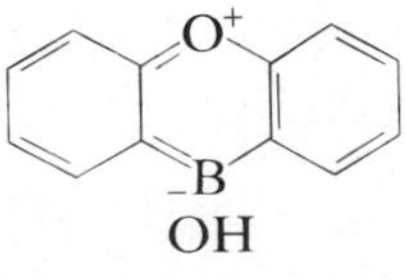

66

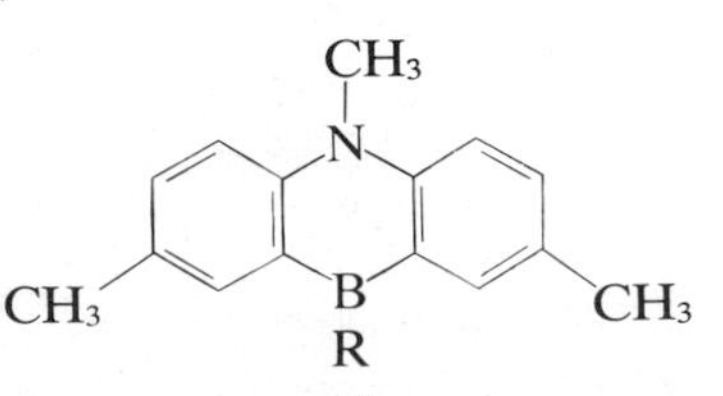

67

a: $R = C_6H_5$

b: $R = OH$

c: $R = \frac{1}{2}O$

Maitlis prepared several 10-substituted-3,6,9-trimethyl-10,9-borazaroanthracenes, **67**, from 2,2′-dilithio-4,4′,*N*-trimethyldiphenylamine with dibutyl phenylboronate and butyl orthoborate followed by hydrolysis.[40] The uv spectra resembled those of anthracene and acridine-type compounds (see Section VI.A). Besides this indication of aromaticity, the central 4,1-borazarene ring shows some stability toward aerial oxidation. However deboronation occurred by mild bromine reagent and the addition of small amounts of alkali to the ethanolic solution of **67a** and **67b** gave an immediate change to a diphenylamine-type ultraviolet spectra. Even the phenyl derivative, **67b**, which would be expected to be quite stable, has been found to be slowly oxidized in ethanol to the anhydride, **67c**. The stabilization of borazaro compounds is thus considerably reduced by separating the boron and nitrogen atoms by two carbons.

It has been pointed out that nuclear quadrupole spectra of α- and γ-chloropyridine indicates that there is a significant contribution by the dipolar resonance structure, **68**, where the nitrogen is next to the carbon linked to the chlorine atom, and there is virtually no contribution by **69** where the charge separation is obviously greater.[41,29]

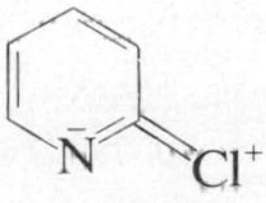

68 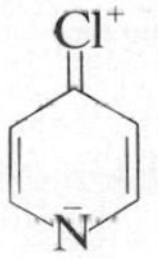 **69**

Though the uncharged forms may be written for the 2,1- and the 4,1-borazarene systems, the same is not possible in the case of the 3,1-isomer (**2**), where the hypothetical molecule would be a mesoionic compound for which no uncharged classical structure is able to be written.[29] Thus it is reasonable to expect that the 3,1-borazarene system is less stable than the 4,1-borazarene, which in turn has been found experimentally to be less stable than the 2,1-system. These predictions have been substantiated by recent molecular orbital calculations where the calculated heats of atomization for the three analogs of benzene follow the correct predicted order[3] (see Section VI.C).

IV. Five-Membered Borazaromatic Compounds

Though there have been a sizeable number of borazaro 5-membered rings fused to other aromatic rings, there are very few reports of monocyclic 5-membered borazaromatic rings.[42] One of these is 2,3,4,5-tetraphenyl-1,3,2-oxazaborole, **70**, which has been synthesized together with a small amount of 2,4,5-triphenyl-1,3,2-dioxaborole, **71**, by the reaction of phenylboronic acid, benzoin, and aniline.[43] This compound was autooxidized with ease and underwent hydrolysis when crystallization was attempted from aqueous ethanol, reverting to phenylboronic acid and benzoin monoanil. The uv spectrum of **71** indicated some aromatic character in the ring and also proved to be fairly resistant to hydrolysis. However **71** was oxidized easily to benzil. Neither compound showed much resonance stabilization.

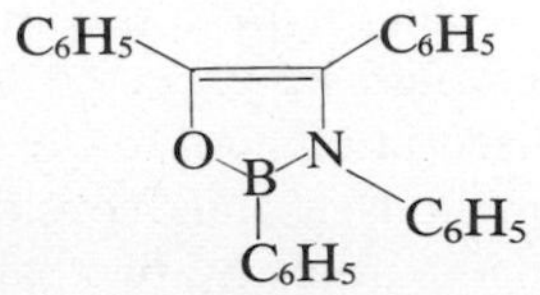

70 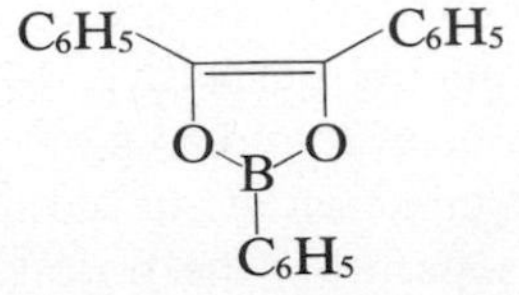 **71**

The first known heteroaromatic boron compound outside of the borazine series was 2-methylbenzo-1,3-diaza-2-borole, **72**, which was first reported by Goubeau and co-workers.[44,45] They did not pursue the theoretical ramifications of this new family of compounds. Shortly after this the 2-phenyl derivative, **73**, was synthesized from the phenylboron dichloride and *o*-phenylenediamine.[46] Likewise using *o*-aminophenol and *o*-aminothiophenol the corresponding oxaazaborole **74** and thiaazaborole **75** were obtained. These were hydrolyzed by acid to the

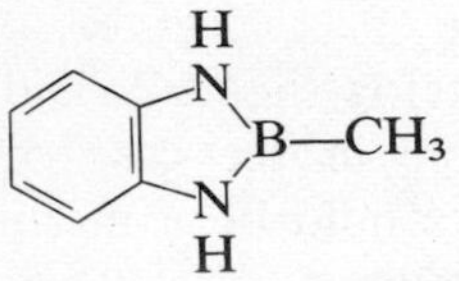

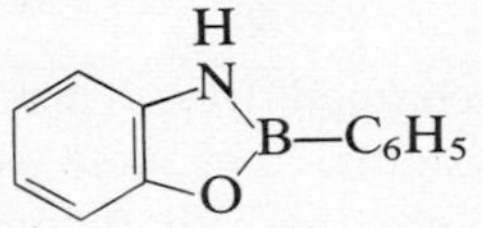

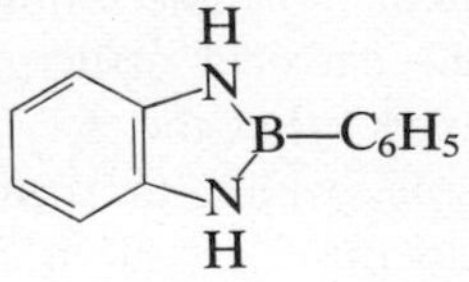

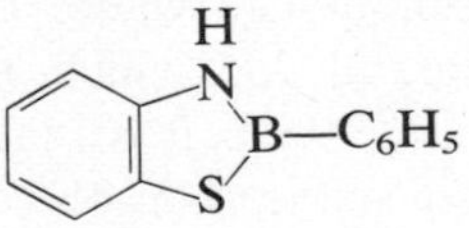

72 **73** **74** **75**

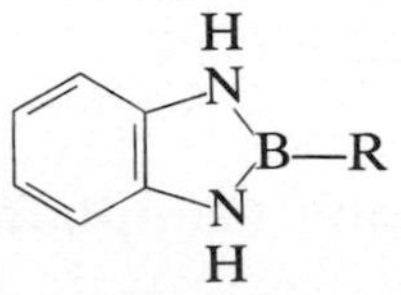

76

a: $R = n\text{-}C_3H_7$
b: $R = i\text{-}C_3H_7$
c: $R = n\text{-}C_4H_9$
d: $R = s\text{-}C_4H_9$
e: $R = i\text{-}C_4H_9$
f: $R = t\text{-}C_4H_9$
g: $R = n\text{-}C_5H_{11}$
h: $R = n\text{-}C_6H_{13}$
i: $R = CH_2C_6H_5$
j: $R = C_6H_{11}$

ortho-substituted aniline starting material. However the uv spectra of **62** and **63** resembled 2-phenylbenzoxazole and 2-phenylbenzathiazole respectively and differed markedly from the spectra of the starting aromatic compounds suggesting that these new borazaromatic compounds were isoconjugate with the conventional aromatic compounds of the indole type.

Hawthorne reacted a series of trimethylamine-alkylboranes with *o*-phenylenediamine in benzene at reflux temperature and obtained reasonably stable boron-substituted benzoboroles, **76**.[47,48] Nyilas and Soloway[49] have synthesized a series of boron-*para*-phenyl and 5-benzo substituents, **77a** to **i**, and Catlin and Snyder[50] prepared a fairly stable boron-*ortho*-phenyl compound **78** by reacting ethylamine-(*N*-*B*)-*o*-tolylborane with *o*-phenylenediamine.

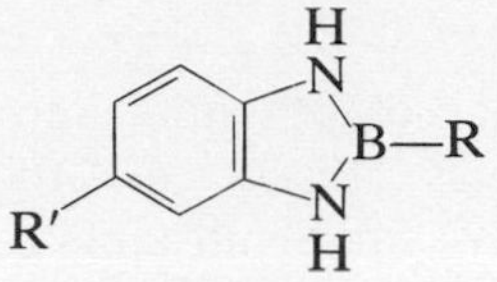

77

a: $R = C_6H_5$, $R' = CH_3$
b: $R = C_6H_5$, $R' = CH_3O$
c: $R = C_6H_5$, $R' = Cl$
d: $R = C_6H_5$, $R' = NO_2$
e: $R = C_6H_5$, $R' = COOH$
f: $R = p\text{-}ClC_6H_4$, $R' = H$
g: $R = p\text{-}NO_2C_6H_4$, $R' = H$
h: $R = p\text{-}COOH$, $R' = H$
i: $R = p\text{-}CH_3O$, $R' = H$

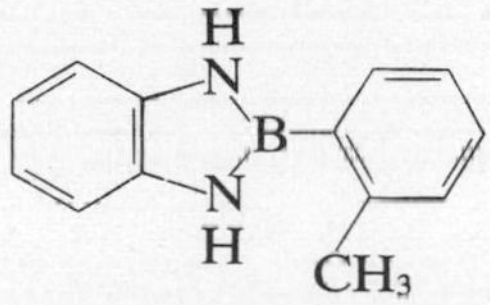

78

Hohnstedt and Pellicciotto[51] reacted boron trichloride with *o*-phenylenediamine and formed the highly reactive 2-chloro derivative, **79**, which was later found to decompose to form the borazine derivative, **80**.[52] This reaction was later extended to include other halogens, alkoxy, and dialkylamino groups substituted on the boron atom which underwent elimination of RH to give **80**.[53,54] It was found that this intermolecular condensation yielding the borazine derivative appears to predominate over normal substitution reactions at the boron atom in the 5-membered borazaromatic compound, thus indicating the borazine product to be a more stable ring system.[55]

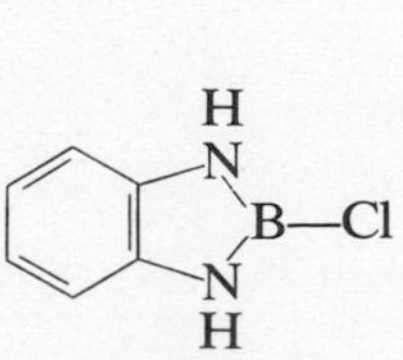

79

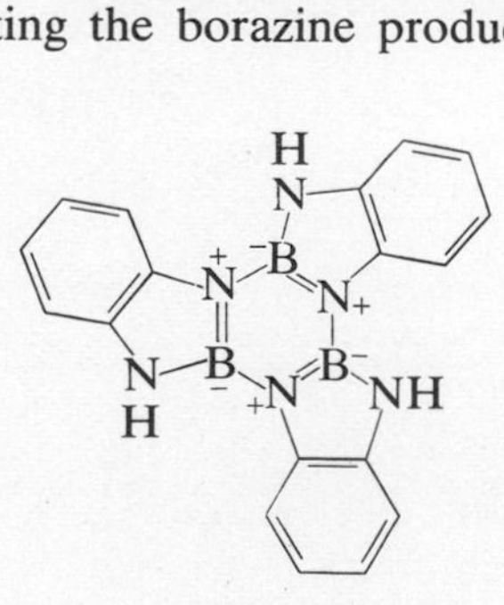

80

Some polymers of the 1,3,2-benzodiazaborolidine system have been prepared involving bonding of the phenyl rings and coupling by hydrocarbon bridges between the boron atoms. They have been found to be quite unstable and are readily solvolyzed and undergo autooxidation. There is little resonance stabilization in the system.[55,56]

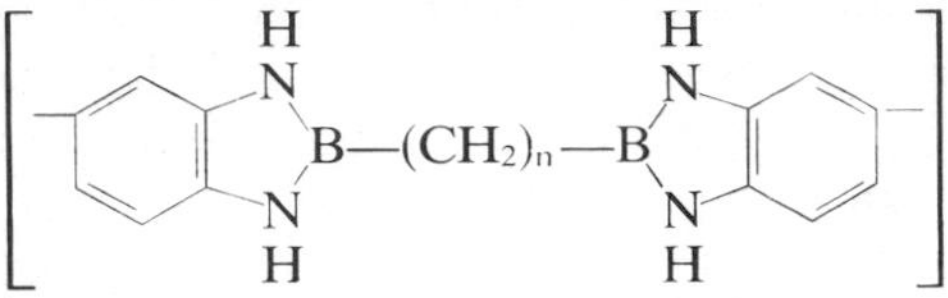

81

Zimmer and co-workers[57] have prepared a series of potential anticancer compounds containing heterocyclic rings fused to the diazaborolidine ring, **82** and **83**. The compounds were rather unstable to oxidation and hydrolysis.

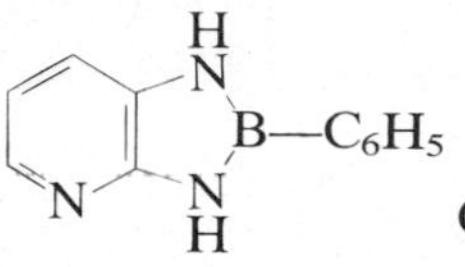

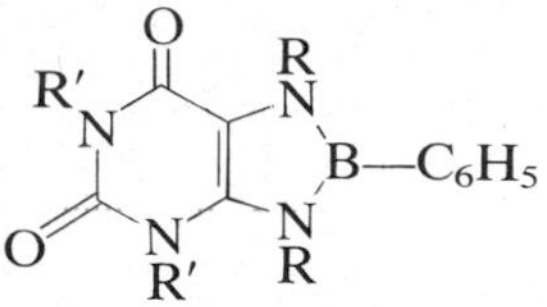

a: $R=H$, $R'=CH_3$
b: $R=H$, $R'=Bu$
c: $R=R'=CH_3$

82 **83**

The 1,3,2-benzodiazaborolidine boron has been made to form stable covalent metal bonds as in triphenylphosphine tetracarbonyl manganese–boron compound, **84**.[58] ^{11}B nmr studies have interpreted this stability in terms of back-donation of the metal *d*-electrons to the tervalent boron. The complex underwent hydrolysis with great ease.

Some recently synthesized 5-membered borazaromatic compounds include the following:

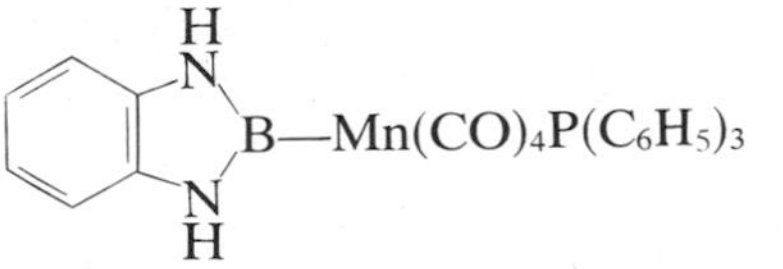

84

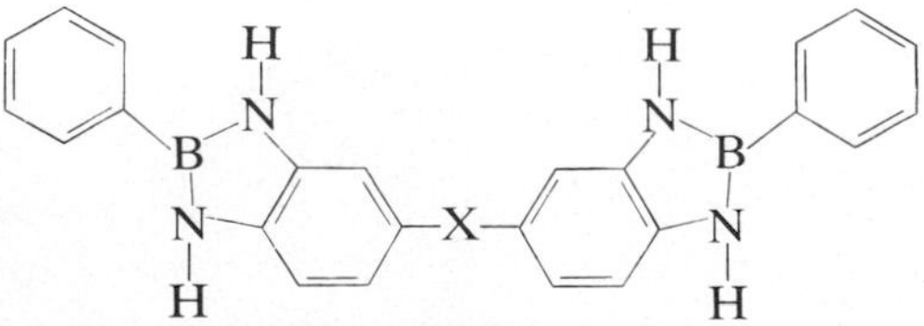

x = bond line, O, CO

84a **84b**

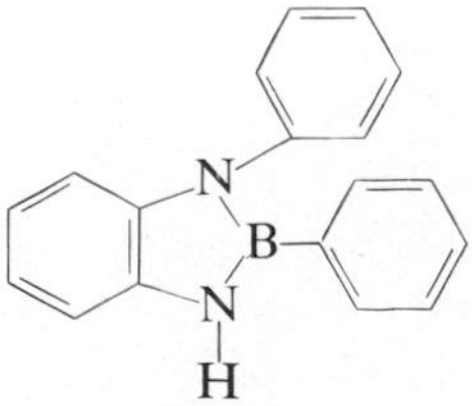

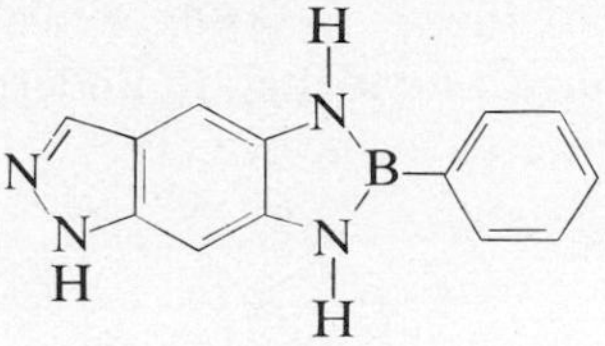

84c 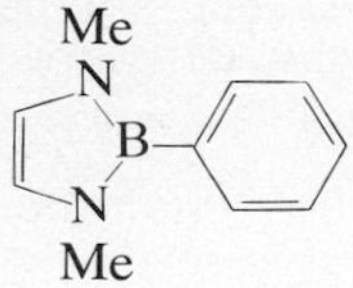 **84d**

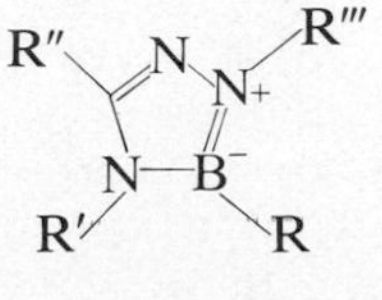

84e 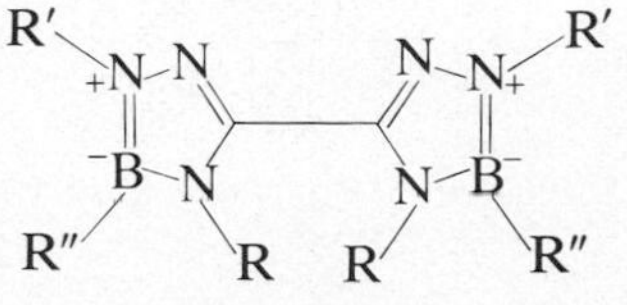 **84f**

R = phenyl, butyl
R′ = H, phenyl
R″ = CH_3, H, 2-pyridyl, phenyl
R‴ = CH_3, H, phenyl

R, R′ = H, phenyl
R″ = phenyl, butyl

V. Borazaromatic Compounds Containing Another Heteroatom in a 6-Membered Ring

A. Borazaroisoquinolines

A compound that is isoelectronic with isoquinoline, 4,3-borazaroisoquinoline, **85**, was first prepared by reacting *o*-formylphenylboronic acid with anhydrous hydrazine.[59,60] The first isolated derivative of this compound, **86a**, and many of the other derivatives **86b** to **86g**, possess unusual stability to hydrolysis and oxidation, even when compared with other borazaromatic compounds. The chemical properties of **86a** have been found to be analogous to those of 2-hydroxy-2,1-borazaronaphthalene, reacting with methylmagnesium iodide to form **86b** which, in turn, can be further methylated to **86c** with dimethyl sulfate in the presence of potassium hydroxide. Many previous mentioned borazaromatic compounds have been *N*-alkylated by treating the *N*-lithio intermediate with the proper alkylating agent. The 3,4-dimethyl-4,3-borazaroisoquinoline can likewise be obtained by this indirect method. Thus the imino proton appears quite acidic and the infrared spectrum shows an NH stretching vibration at 3280 cm^{-1} which would be expected if the acidity is due to resonance effects of the ring containing a B–CH_3

group. Upon standing, **86c** was autooxidized to **86d**; however, the monomethyl derivative, **86b**, does not have a tendency to undergo aerial oxidation. This autooxidation must be due to steric effects.

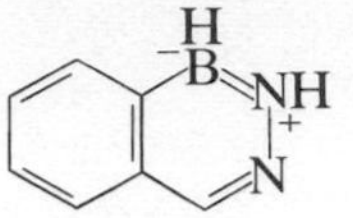

85

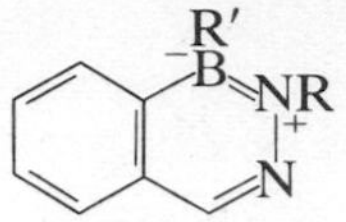

86

a: $R=H$, $R'=\frac{1}{2}O$
b: $R=H$, $R'=CH_3$
c: $R=CH_3$, $R'=CH_3$
d: $R=CH_3$, $R'=OH$
e: $R=C_6H_5$, $R'=OH$
f: $R=CH_3$, $R'=H$
g: $R=H$, $R'=OC_2H_5$

The *N*-substituted 4,3-borazaroisoquinolines have been prepared using substituted hydrazines. Phenylhydrazine readily reacted with *o*-formylphenylboronic acid to form the 3-phenyl-4-hydroxy derivative **86e**, isolated as an anhydride. However 3-methyl-4,3-borazaroisoquinoline, **86f**, can be made by the reaction of lithium aluminum hydride on **86d**, whereas using the same reducing agent on **86a** gave 1,2,3,4-tetrahydro-4,3-borazaroisoquinoline, **87**. This compound was found resistant to hydrolysis which could only be true if it is aromatic or the boron is quadricovalent; the infrared spectrum indicated the latter. The over-reduction of the ring may be due to salt formation with the metal giving an intermediate **88**, in which the boron-containing ring is no longer aromatic. With methyl iodide **86b** gave the methiodide **89** which was unaffected by boiling water and dilute acid. This is further evidence for the aromaticity of the borazaroisoquinoline system; otherwise this structure would be expected to show a great tendency toward hydrolysis.

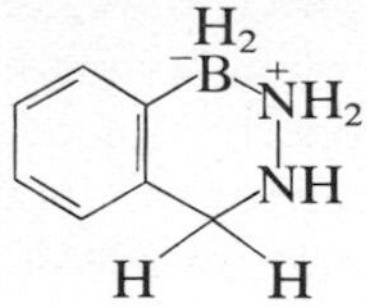

87

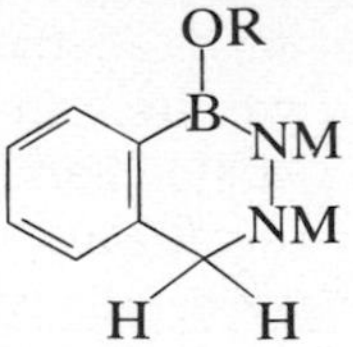

88

(M = metal)

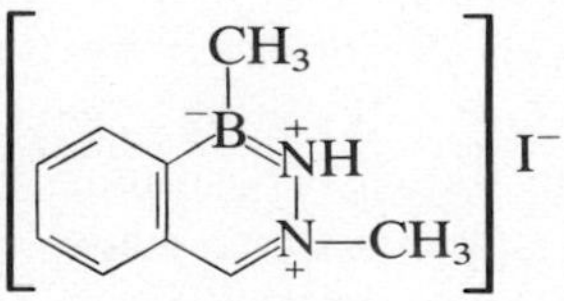

89

Substitution reactions with 4-methyl-4,3-borazaroisoquinoline, **86b**, yielded on reaction with nitric acid in sulfuric acid the 8-nitro-4-methyl-4,3-borazaroisoquinoline, **90**, accompanied by considerable oxidation by-products.[61] Upon heating, the 8-nitro compound was found to be quite unstable; perhaps this behavior is due to the mutual conjugation between the imino and nitro compounds (Section II.D) and can be illustrated in resonance terminology by the contribution of the dipolar structure, **91**. This enhanced polarity may account for both the instability of **90** and the fact that it dissolves only in polar solvents.

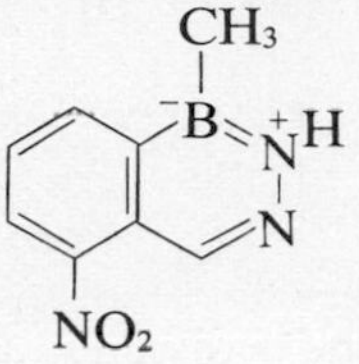

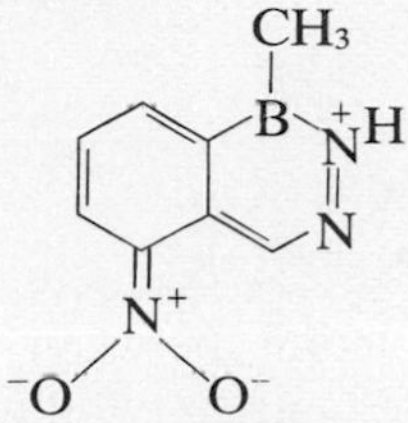

90 **91**

4-Methyl-4,3-borazaroisoquinoline, **86b**, has been found to resist acetylation by acetyl chloride in the presence of either stannic chloride or aluminum chloride. The bromination of **86b** with 1 molar equivalent of bromine in acetic acid gave a product that was oxidatively degradated in excess potassium permanganate to give *o*-bromobenzoic acid. Namtvedt and Gronowitz[62] have presented evidence that the brominated product was not the reported 8-bromo derivative but most likely 1-bromo-4-methyl-4,3-borazaroisoquinoline, **92**. They have argued that during acidic hydrolysis the excess permanganate and manganese dioxide oxidized the bromine ion formed to free bromine which then rapidly gave *o*-bromobenzoic acid in a bromodeboronation reaction with the intermediate oxidized product **93**. Indeed, when the excess permanganate was carefully removed, the completely deboronated benzoic acid was isolated in good yields, thus indicating that the brominated product was truly **92**.

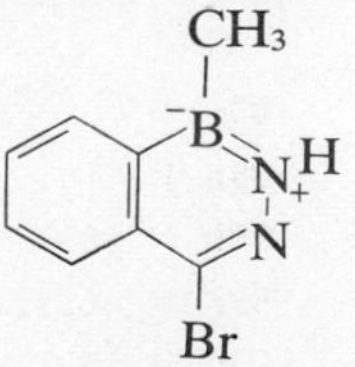

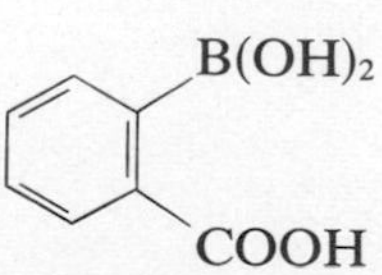

92 **93**

It is not unusual for electrophilic aromatic substitutions to occur in different positions; in isoquinoline the sulfonation and nitration occur at the 5-position along with a small amount of 8-nitroisoquinoline,[63] but the 4-position is the site of bromination and mercuration.

A number of other compounds possessing a B–N–N chain within a conjugated system have been synthesized and exhibit surprising stability. One of these is 4,5-borazarothieno[2,3-*c*]pyridine, **94a**, which was prepared as the *B*-hydroxyl derivative by reacting hydrazine with 2-formyl-3-thiopheneboronic acid, **95**.[64] This boronic acid was prepared through

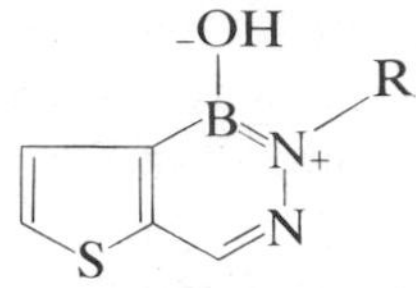

94

94a R = H

94b R = C_6H_5

94c R = CH_3

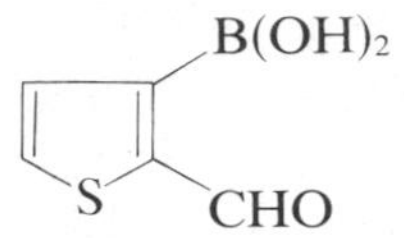

95

halogen–metal interaction between the ethylene acetal of 3-bromothiophene-2-aldehyde and butyllithium at −70° and treatment of the 3-lithio derivative with butyl borate followed by hydrolysis. By the reaction of phenyl and methylhydrazine with the boronic acid, **95**, the 3-phenyl-, **94b**, and 3-methyl-4,5-borazarothieno[2,3-*c*]pyridine, **94c**, have been prepared.

Likewise, the 7-hydroxy-7,6-borazarothieno[3,4-*c*]pyridine, **96**, was prepared by reacting hydrazine with 4-formyl-3-thiopheneboronic acid and 7-hydroxy-7,6-borazarothieno[3,2-*c*]pyridine, **97a**, by hydrazine with 3-formyl-2-thiopheneboronic acid. The last mentioned boronic acid reacted with methyl, phenyl, and 4-carboxyphenylhydrazine to yield the derivatives **97b**, **c**, and **d**.

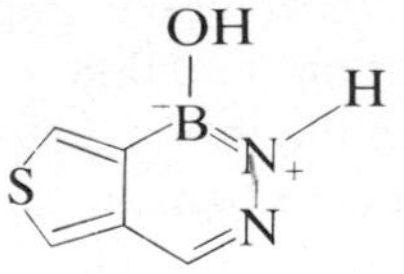

96

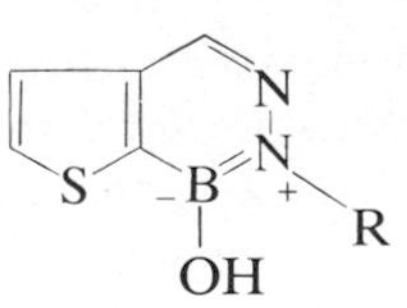

97

97a R = H

97b R = CH_3

97c R = C_6H_5

97d R = *p*-C_6H_5COOH

While it is relatively easy to vary the *N*-substituents on the boraza ring of these compounds by proper choice of hydrazines, it has proved considerably more difficult to synthesize a variety of boron-alkylated derivatives. However, in the case of **94a** and **97a** the boron-hydroxy group could be converted to the butyl ester and reacted with Grignard reagents to make **98** and **99**.[65] The less stable **96** decomposed under these reaction conditions.

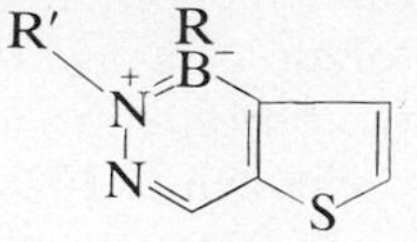

98

98a $R = CH_3$, $R' = H$

98b $R = R' = CH_3$

98c $R = C_4H_9$, $R' = H$

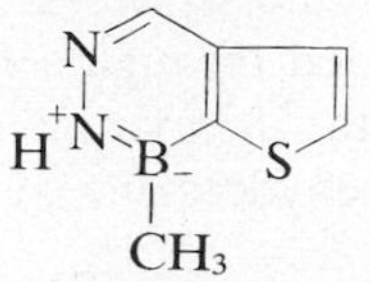

99

Derivatives of **94** and **97** are effective against bacteria.[65a]

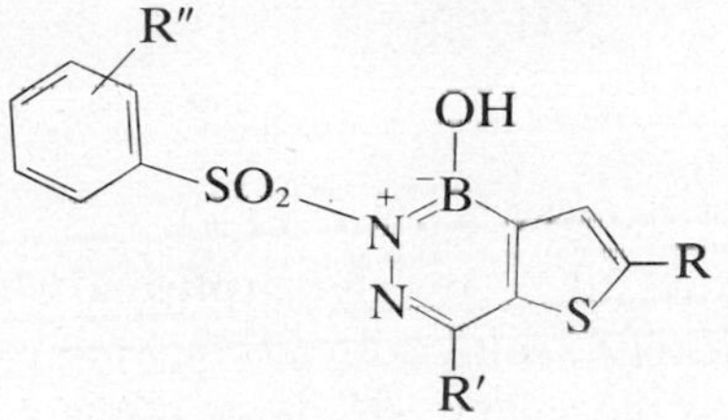

94d

$R = H, CH_3, NO_2$

$R' = H, CH_3$

$R'' = p\text{-}NO_2, p\text{-}NH_2$, p-NHAc, m-$NO_2$, H

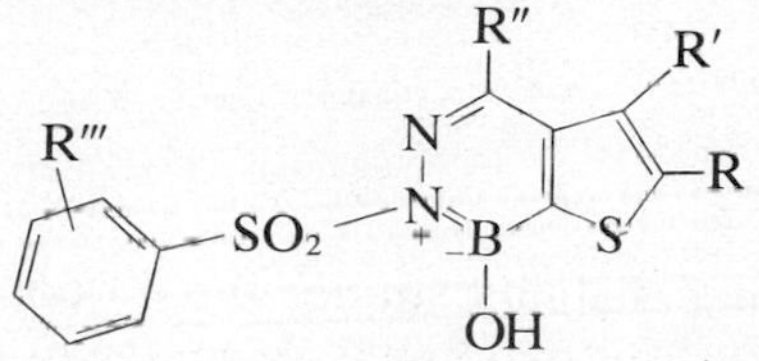

97e

$R = Cl, CH_3, NO_2$, Et, Pr

$R' = H, Cl$

$R'' = H, CH_3$

$R''' = H, m\text{-}NO_2, p\text{-}NO_2$, p-NHAc, p-$NH_2$

When 3-formyl-2-thiopheneboronic acid was reacted with o-carboxyphenylhydrazine the cyclic lactone **100** was formed; likewise the 2-formyl-3-thiopheneboronic acid reacted with the same hydrazine to yield **101**. Both of these products proved to be quite stable to hydrolysis and oxidation conditions.

Compound **94a** was not hydrolyzed by concentrated hydrochloric acid and was stable when refluxed with 2*N*-sodium hydroxide for 2 hr; likewise

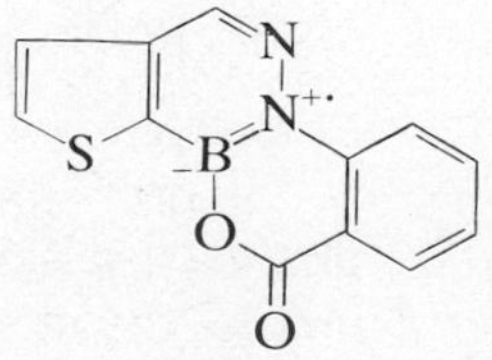

100

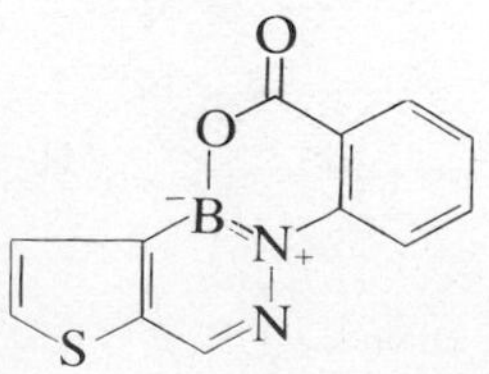

101

97a was not hydrolyzed by concentrated acid but was deboronated under the drastic basic conditions used for **94a**; however at room temperature with basic treatment 87% recovery of unreacted **97a** was effected. Compound **96** was the least stable of the borazarothienopyridines and was easily hydrolyzed by weak acid. Basic hydrolytic cleavage of **96** apparently proceeded through a ring-opening at the boron–nitrogen bond and resulted in the formation of the boronic acid azine **102** and the 3-thiophenealdehyde azine **103**.

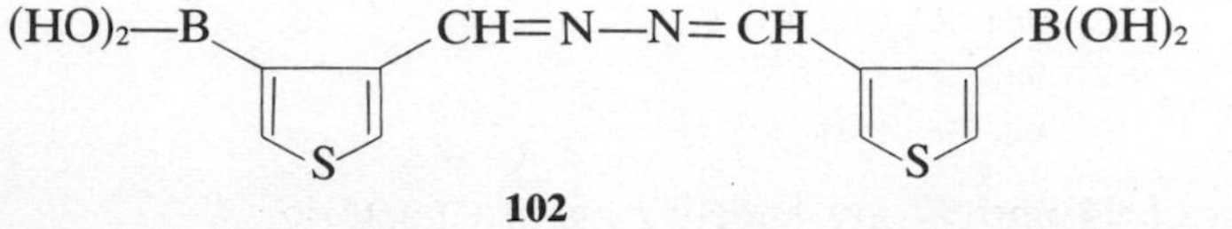

(HO)$_2$—B CH=N—N=CH B(OH)$_2$

S S

102

CH=N—N=CH

S S

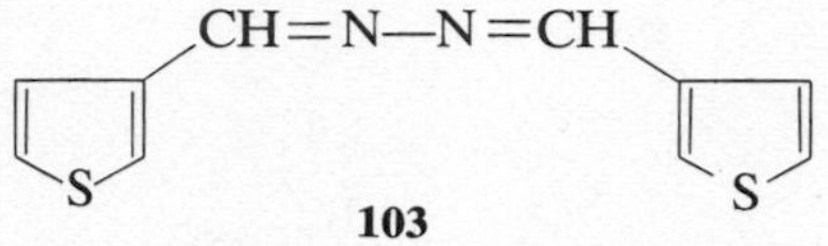

103

These stability studies should not prove to be overly surprising when one realizes that **94a** is an analog of thieno[2,3-*c*]pyridine, **104**, which is a fairly stable aromatic system. On the other hand, **96** is an analog of isothionaphthene, **105**, which has been found to be very unstable[66] and decomposes even at −30° after a few days.

Some electrophilic substitution reactions have been attempted with derivatives of 4,5-borazarothieno[2,3-*c*]pyridine and with 7,6-borazarothieno[3,2-*c*]pyridine.[67] Nitration of **97a**, **97b**, and **99** with nitric acid-sulfuric acid yielded the 3-nitro derivatives, **106a**, **b**, and **c**, as the only

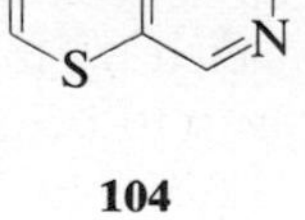

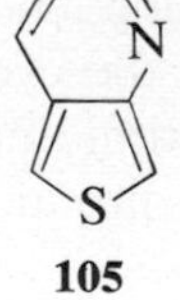

S N N S

104 **105**

NO$_2$ N N S B R R′

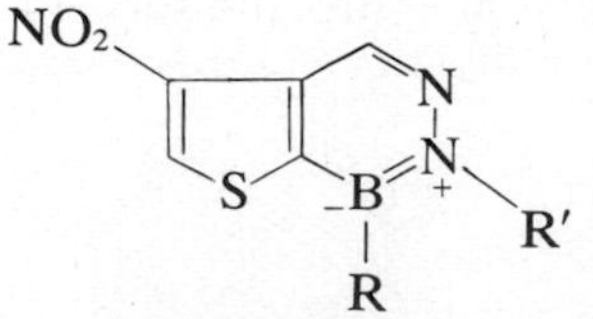

106

106a R = OH, R′ = H

106b R = OH, R′ = CH$_3$

106c R = CH$_3$, R′ = H

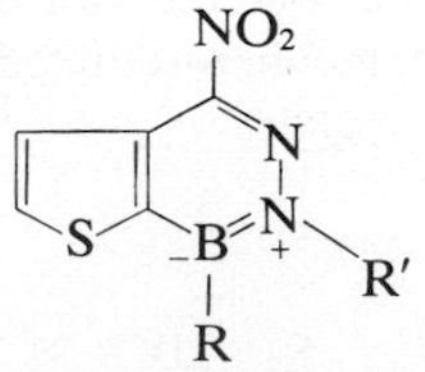

NO$_2$ N N S B R R′

107

107a R = H, R′ = OH

107b R = CH$_3$, R′ = OH

107c R = H, R′ = CH$_3$

detectable isomers but with fuming nitric acid in acetic anhydride the 4-nitro products, **107a**, **b**, and **c**, were obtained. On the other hand, nitration of **94a** using nitric acid–sulfuric acid mixture gave the 3- and 2-mononitro derivatives in an 8:1 ratio and using the same conditions with **94c** gave the same isomers in a 4:1 ratio (**108** and **109**). The same substrates yielded the 7-nitro derivative, **110**, when nitration was carried out in acetic anhydride.

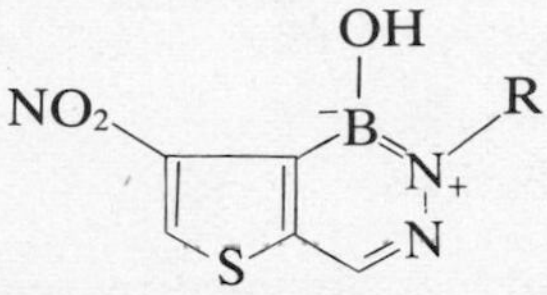

108 R = H, CH_3 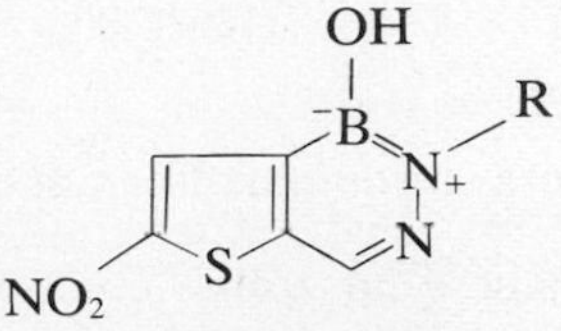 **109** R = H, CH_3 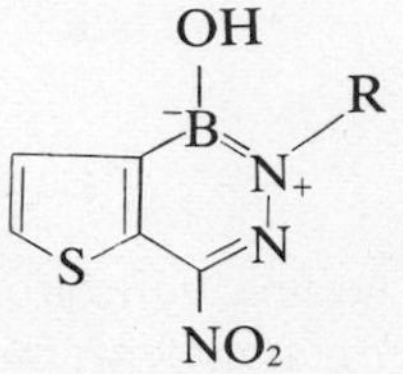 **110** R = H, CH_3

Similar differences in substitution have been found in the nitration of quinoline, which yields the 5-nitro and the 8-nitro derivatives when sulfuric acid is used and the 3-, 6-, and 8-nitro isomers using nitric acid–acetic anhydride.[68] Perhaps, the orientation here resembles the addition of acetyl nitrate across the 1,2-bond of quinoline, therein forming a substituted aniline which undergoes further substitution. Bromination of **98a** in acetic acid led to substitution in the 7-position of the borazaro ring yielding **111**[69] (see Section VII.B).

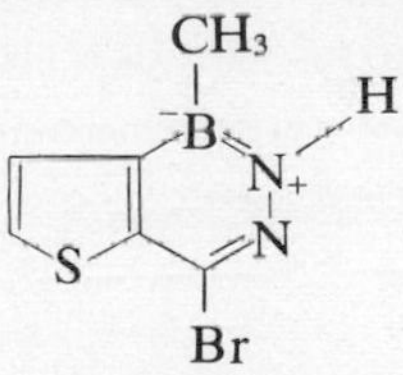

Raney Ni

111 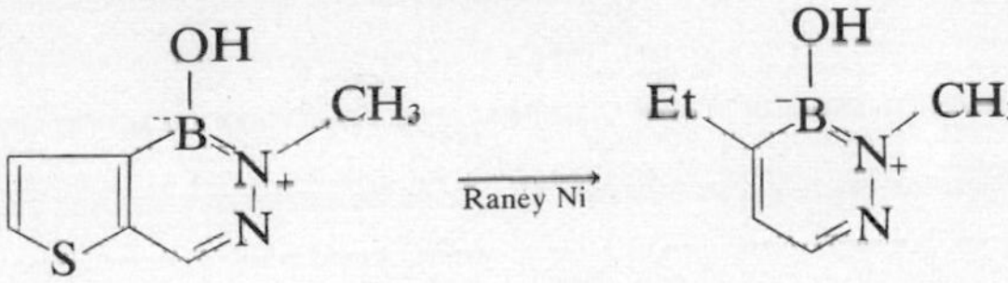 **112**

Desulfurization of **94c** led to preparation of 4-ethyl-3-hydroxy-2-methyl-3,2-borazaropyridine, **112**.[70] The ultraviolet spectrum resembled that of 3-hydroxypyridine and no change was observed upon standing in acid or base. This together with the fairly drastic conditions required for desulfurization indicated considerable aromaticity. Other more stable 3,2-borazaropyridine compounds have been prepared, **112a** and **112b**,[70a] by desulfurization of various borazarothienopyridines. The 3,2-borazaropyridines are very resistant to hydrolytic ring opening. Compound **112a** brominates with ease in the 6-position with bromine in pyridine-CCl_4; in **112b** the 4- and 6-positions show similar reactivities toward the brominating agent, and the dibromo derivative is also obtained.[70b]

Pyrolysis of **112c** yields 1,4,5-trimethylpyrazole and 1,3,4-trimethylpyrazole by an apparently novel photochemical boron extrusion.[70c]

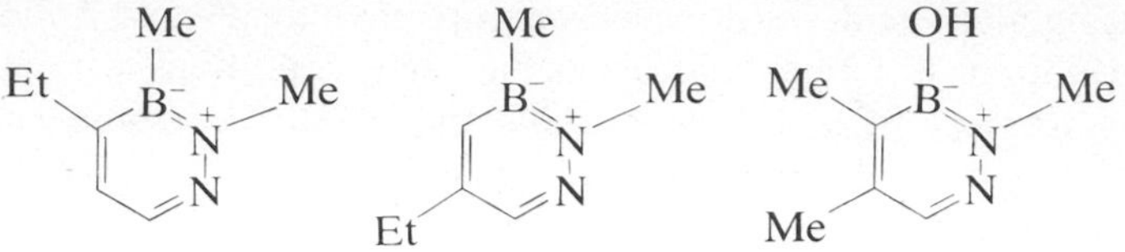

112a **112b** **112c**

B. Borazaroquinazolines and Purines

The first borazaromatic quinazoline compound was obtained by reacting *o*-aminobenzamide with phenylboron dichloride, with dibutylphenylboronate or with phenylboronic acid.[71,72] This product, 2-phenyl-2-boradihydro-4-quinazolone, **113**, was a colorless crystalline solid which was stable to air or moisture. However the uv spectrum was found to change and resemble a mixture of the starting materials upon standing in ethanol for 24 hr. This tendency to solvolyze was also observed in acidic and basic media. It has been assumed that this behavior indicated less aromaticity in this series of compounds than in many of the previous reported borazaromatics. This conclusion was supported by a large bathochromic shift in the uv spectrum upon addition of alkali due to the deprotonation of **113** to the conjugate ion, **114**.

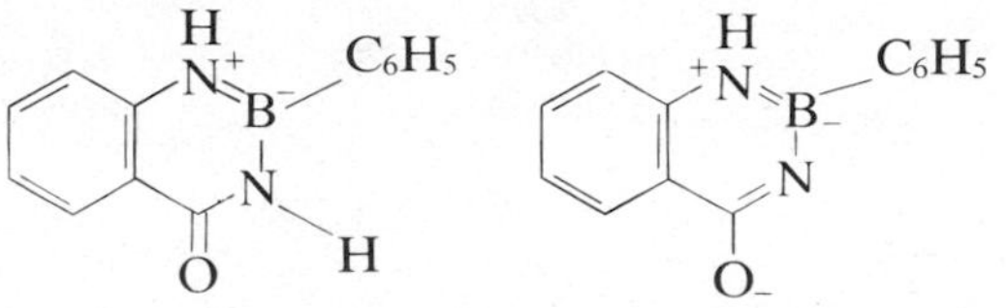

113 **114**

Yale and co-workers[73] have prepared a series of *N*- and *B*-substituted quinazolines by the condensation of a variety of *N*-substituted anthranilamides with aryleneboronic acids, and Fried and co-workers[74] have prepared a series of benz-substituted derivatives by treating an appropriately substituted benzamide with a boronic acid in a nonaqueous medium. The latter group of compounds showed central nervous system depressant activity. The *N*-substituted derivatives appeared to have greater stability than **113** itself.

An analog of purine, 2-phenyl-6-hydroxy-7-methyl-2-boradihydropurine, **116**, has been prepared by the action of 4-amino-1-methyl-5-imidazolecarboxamide, **117**, with dibutyl phenylboronate.[72] Likewise

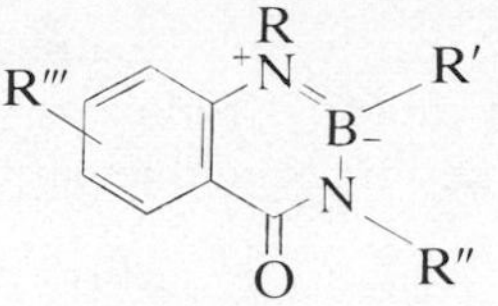

115

	R	R′	R″	R‴
a:	H	Mesityl	H	H[73]
b:	CH_3	C_6H_5	CH_3	H
c:	CH_3	C_6H_5	H	H
d:	H	1-Naphthyl	CH_3	H
e:	H	C_6H_5	2,6-$Me_2C_6H_3$	H[74]
f:	H	C_6H_5	2,6-$Me_2C_6H_3$	6-NH_2
g:	H	C_6H_5	2,6-$Me_2C_6H_3$	6-AcNH
h:	H	C_6H_5	2,6-$Me_2C_6H_3$	$Me_2N(CH_2)_3$
i:	H	C_6H_5	$C_6H_5CH_2$	H
j:	H	1-Naphthyl	$C_6H_5CH_2$	H
k:	H	OH	$C_6H_5CH_2$	H
l:	H	OH	C_6H_5	H

an 8-azapurine, **118**, has been prepared but both this and **116** proved quite unstable. The behavior of these purine analogs may be due to the introduction of hydroxyl groups and because both ring atoms adjacent the boron are nitrogen atoms; in the former case the addition reduces the aromaticity through cross-conjugation and in the latter the boron is more susceptible to nucleophilic attack. The purines are less stable than the borazaroquinazolines and **118** is less than **116**.

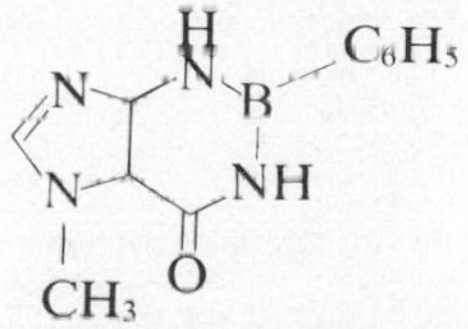

116 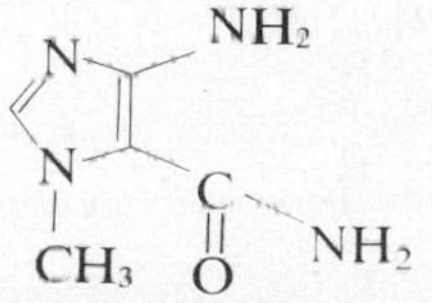 **117** 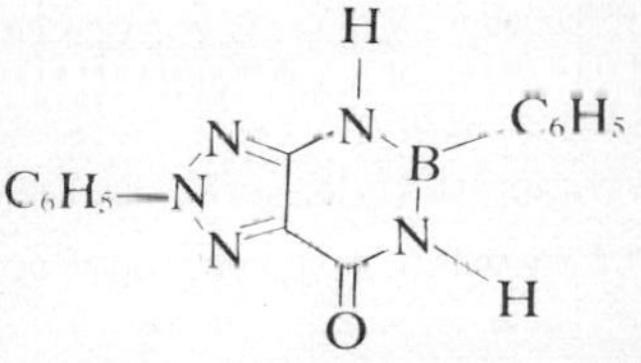 **118**

One other N–B–N bonded compound has been prepared, 2-phenyl-2-boradihydroperimidine, **119**.[72] It was obtained in low yield from 1,8-diaminonaphthalene and phenylboron dichloride. Its uv spectrum was different from 2-phenylperimidine, **120**, but **119** was quite stable to solvolysis even though it decomposed with aerial oxidation. At most **119** proved to be only slightly aromatic.

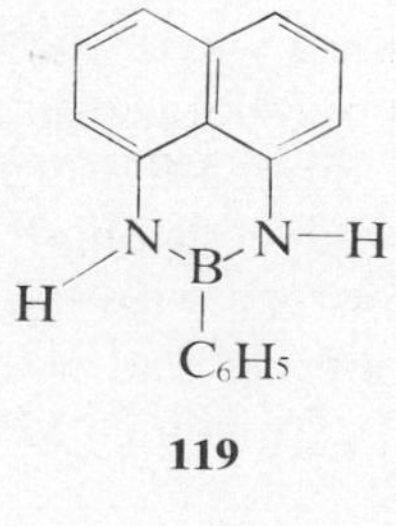

119 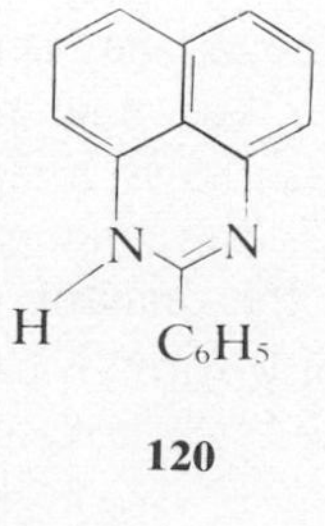 **120**

C. Diboradiazanaphthalenes

Koester and co-workers[75,76] have reported a number of interesting boron compounds obtained by thermal cyclizations. One of these involved a dimerization of 2 moles of **121** to form **122** which upon heating at 180 to 200° formed 2,4-diethyl-3-phenyl-2,4-diboro-1,3-diazaronaphthalene, **123**. Likewise the treatment of 2 moles of toluidine and

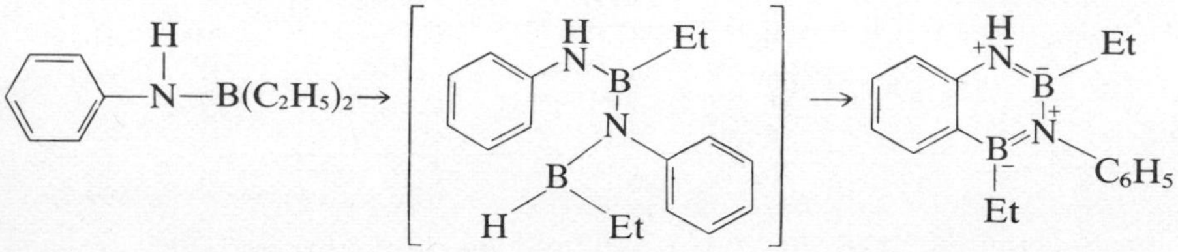

121 **122** **123**

boron trichloride yielded **124** as one of the products.[77] These diboradiazanaphthalenes have been found to be easily split by hydrolysis but have been shown to be remarkably resistant to aerial oxidation.

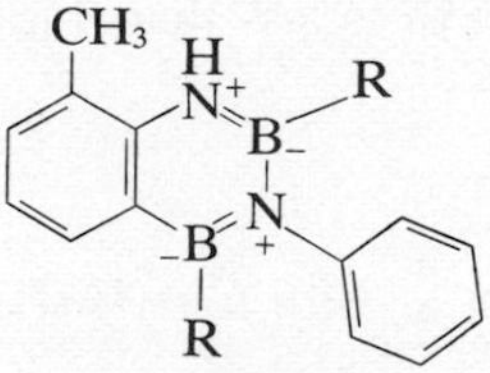

124 R = Cl
R = $N(Me)_2$

Among other cyclic reactions which have been found by Koester[76] though not directly related to the class treated here is the cyclization of the *N*-aryl borazine, **125**, to form **126**.

Treating 4-chloro- or 4-bromo-1-naphthylamine with BCl_3 gave **126a** (R = Cl, Br). Similarly, 1-aminoanthracene gave **126b**.[76a]

D. Other Compounds

Only one compound has been synthesized where the boron is directly linked to another boron in a 6-membered borazaromatic ring. Noeth and co-workers[78] have prepared 2,3-dibutyl-2,3-diboro-1,4-diazaronaphthalene, **127**, by condensing *o*-phenylenediamine with **128** which was prepared by the reaction of tributylborane with $B_2[N(Me)_2]_4$.

The reaction of 2-formyl-3-thiopheneboronic acid with hydroxylamine led to 4-hydroxy-4,5-boroxarothieno[2,3-*c*]pyridine, **129**.[64] It has been

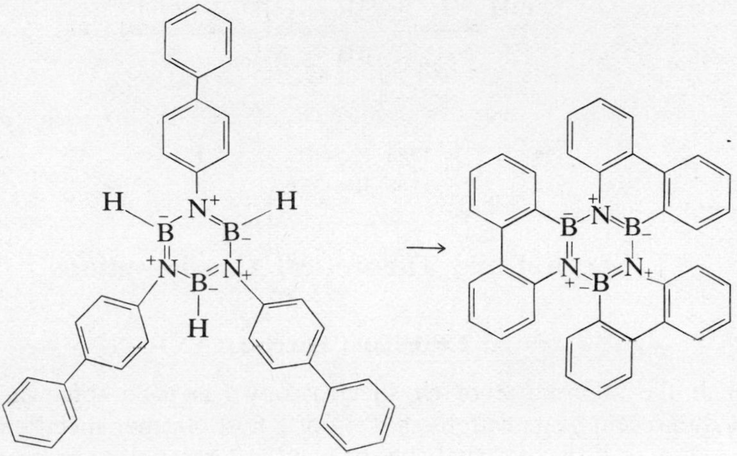

125 **126**

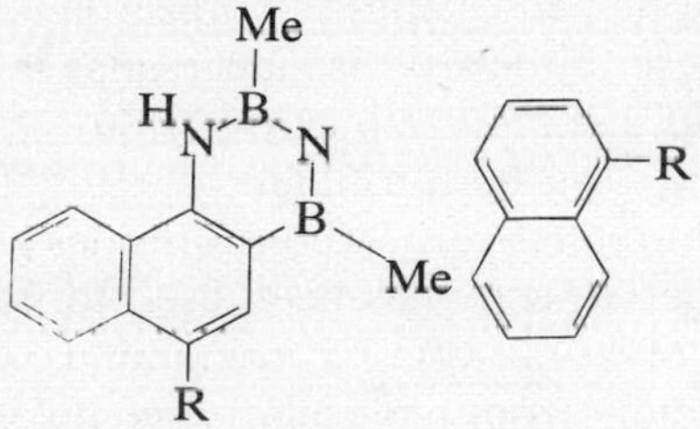

126a

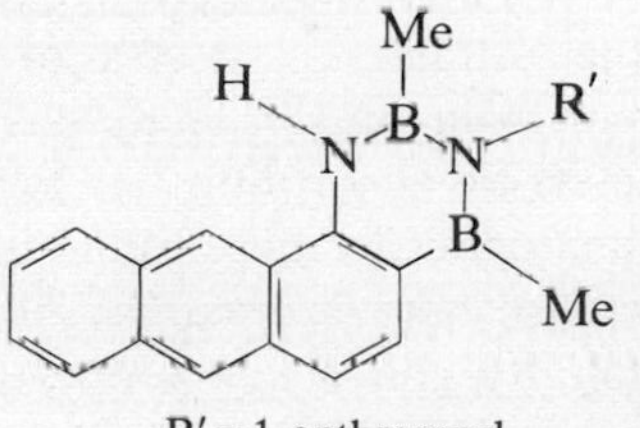

R′ = 1-anthracenyl
126b

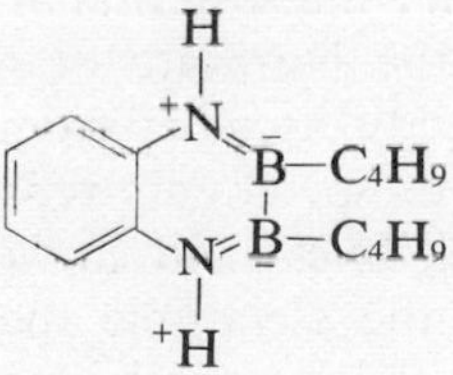

127 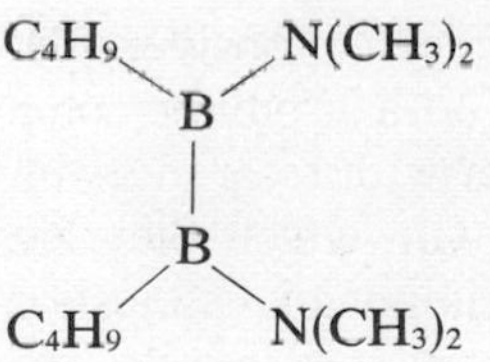 **128**

shown that this compound is somewhat stable but not as much as is 4-hydroxy-4,3-boroxaroisoquinoline, **130a**. This latter compound was first prepared by Snyder and co-workers[11] who did not comment on its properties. It was found to be resistant to hydrolysis and to act as a Lewis acid (see Section VII.A) but attempts to replace the hydroxyl group by a methyl group have failed. Certainly it has proven to be less aromatic than its nitrogen analog, **86a**.[59] The ethyl ether has been easily formed by dissolving **130a** in ethanol. Soloway[79] reported a 1,4-borazarene with oxygen present, **131**, but the physical properties were not noted.

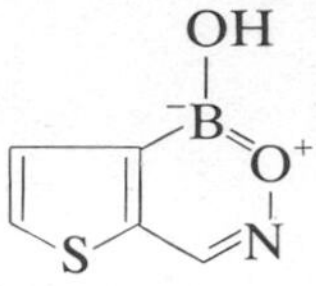

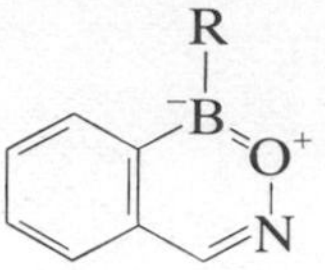

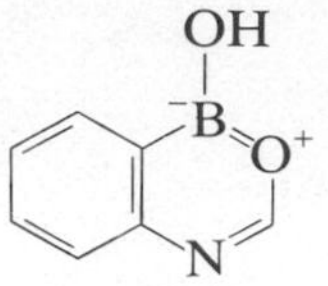

129 **130a** R = OH **131**

130b R = OEt

VI. Spectral and Theoretical Considerations

A. Ultraviolet Spectra

Though the importance of uv spectroscopy has been somewhat diminished in recent years with the advent of a host of other spectral tools, it still continues to have a vital role to play in determining the aromatic properties of compounds. Historically speaking, the uv spectra of many of the compounds just mentioned have helped establish "borazaromaticity." For the ultraviolet spectra of these borazaro products appear quite different from starting materials and a comparison of them show certain characteristic absorption bands which can be attributed to π–π^* transitions analogous to those found in alternate hydrocarbons. Besides helping to establish borazaromaticity, uv spectroscopy has been used to monitor the stability and acidic properties of borazaromatic compounds and to determine molecular orbital parameters from molecular π-complexes.

The similarity of the uv spectra of borazaromatic compounds with the parent hydrocarbon has been considered a leading criterion of aromaticity and has occupied considerable space in the literature of this subject. Table I lists borazaromatic compounds already discussed from Sections II to V together with references in which the uv spectra are reproduced and compounds with which they are compared. Michl and co-workers[80] are currently making an exhaustive study of the electronic spectra of the major borazaro compounds, which will allow for extended predictions of other properties such as ionization potentials and reactivity.

The introduction of heteroatoms into an alternate hydrocarbon does not alter appreciably the positions of the main absorption bands but the intensity of the α-band is greatly increased.[81] These new heteroatoms can also influence the energies of the π-molecular orbitals and the π–π^* transitions in several ways.[2] For example, 2,1-borazaronaphthalene, **15**, does resemble naphthalene in uv absorption bands, but the intensity of the α-band (320-nm region) is much greater.[2,8] From simple resonance considerations one would predict that the spectrum of **15** would resemble more closely that of isoquinoline, **132**,[2] than that of quinoline, **133**.

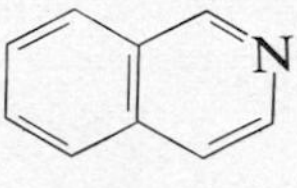

132 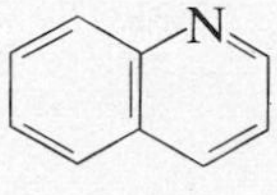 133

10,9-Borazaronaphthalene, **26**, has a uv spectrum with maxima at nm 236, 242, 248 (log ϵ = 4.509, 4.657, 4.487); there is a superficial resemblance with the first three $\pi-\pi^*$ transitions of naphthalene[16] but perhaps a closer similarity to that of quinoline, **133**. However attention has been called to bands that have been overlooked in the weaker part of the spectrum. The uv spectra of the derivatives of 4,3-borazaroisoquinoline, **85**, also have a more intense α-band than naphthalene and resemble isoquinoline and even more so the uv spectrum of quinazoline, **134**.

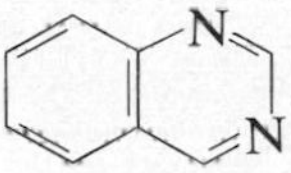

134 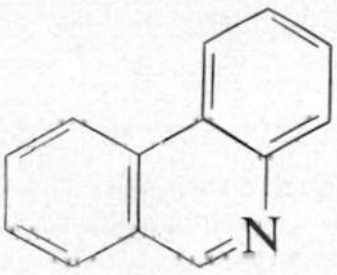 135

Likewise the spectra of 10,9-borazarophenanthrene, **31**, and its derivatives **32** resemble the spectrum of phenanthridine, **135**, and also that of phenanthrene.[2,19] But the uv spectrum of 10,9-borazarophenanthrene differs considerably from that of the wavelength of the absorption bands of the 11,12-borazaro isomer, **51**, perhaps due to the great difference between the bond orders of the corresponding bonds of the two isomers (compare Fig. 1). The uv spectra of 10-phenyl-10,9-borazarophenanthrene, **32d**, and its boroxaro and borathiaro analogs are quite similar and indicate considerable aromaticity in these compounds.[22] The spectra of 10-phenyl-10,9-borazaro, 10-phenyl-10,9-boroxaro, **34a**, and 10-phenyl-10,9-borathiarophenanthrene, **35**, show the characteristic aromatic bands at 329 [ϵ = 10 (900), 323 (6300), and 338 (7850), respectively] in cyclohexane.[21]

Acetylation of 10,9-borazarophenanthrene derivatives at the 2-, 6-, and 8-positions produces marked bathochromic shifts.[29] The wavelengths of these acetylated products are much less than those for the nitro-substituted products which would be expected since the acetyl groups have weaker electron-withdrawing properties.[25]

The analog of triphenylene, 14,16,18-triboro-13,15,17-triazarotriphenylene, **8**, has an uv spectrum that resembles that of the parent triphenylene far more than does that of the other isomer, 14,13-borazarotriphenylene, **52**.[6] The spectrum of **8** in ethanol has maxima at nm (log ϵ) 277, 287, 304, and 316 (4.69, 4.71, 4.03, and 3.80) with the

TABLE I. ULTRAVIOLET SPECTRA OF BORAZAROMATIC COMPOUNDS

Compound (solvent)	Compound to which compared (solvent)	Ref.
2,1-Borazaronaphthalene, **15** (cyclohexane)	Naphthalene (cyclohexane)	[a]
	Isoquinoline (cyclohexane)	[b]
2-Methyl, **16d** (ethanol)	2-Methylnaphthalene (ethanol)	[a]
2-Phenyl, **16b** (cyclohexane)	2-Phenylnaphthalene (cyclohexane)	[a]
2-Hydroxy anhydride, **17** (cyclohexane)	2-Naphthol (ethanol)	[a]
3-Bromo-2-methyl, **20a**	2-Methyl-2,1-borazaronaphthalene	[c]
3-Chloro-2-methyl, **20b**	2-Methyl-2,1-borazaronaphthalene	[c]
10,9-Borazarophenanthrene, **31** (cyclohexane)	Phenanthrene (cyclohexane)	[d]
10-Hydroxy, **32b** (cyclohexane)	10-Hydroxy, **32b** (10% aq. alkali)	[d]
2-, 6-, *and* 8-Chloro-10-methyl, **38e**, **f**, **g**		[e]
6- and 8-Nitro-10-methyl, **38a**, **b**		[e]
6- and 8-Nitro-10-hydroxy, **39a**, **b**		[e]
6- and 8-Amino-10-methyl, **38c**, **d**		[e]
6- and 8-Amino-10-hydroxy, **39c**, **d**		[e]
6- and 6,8-Diacetyl-10-hydroxy, **39i**, **j**	**32b**	[f]
6- and 6,8-Diacetyl-10-methyl, **38h**, **i**	**32c**	[f]
10-Methyl-9-ethoxycarbonyl, **40c**	**32c**	[g]
10-Phenyl-9-ethoxycarbonyl, **41c**	**32d**	[g]
10,9-Boroxarophenanthrene		
10-Hydroxy, **34c** (ethanol)	2-Hydroxybiphenyl (cyclohexane)	[h]
10-Phenyl, **34a** (cyclohexane)	**32d** (cyclohexane)	[h,i]
11,12-Borazarophenanthrene, **51** (ethanol)	**31**	[j,b]
Borazarotriphenylene		
14,16,18-Triboro-13,15,17-triazaro, **8** (ethanol)	Triphenylene (ethanol)	[k]
14,13-Borazarotriphenylene, **52** (ethanol)	Triphenylene (ethanol)	[k]
2-Phenylbenzo-2-borole		
1,3-Diaza, **73**	2-Phenylbenziminazole and *o*-Phenylenediamine	[l]
1,3-Oxaza, **74**	2-Phenylbenzoxazole	[l]
1,3-Thiaza, **75**	2-Phenylbenzothiazole	[l]
10,9-Borazaroanthracene		
10-Hydroxy, **67a** (ethanol)	*o*-Phenoxyphenylboronic acid (cyclohexane)	[m]
10-Phenyl, **67b** (ethanol)	Acridine (ethanol)	[m]
4-Methyl-4,3-borazaroisoquinoline, **86b** (ethanol)	Isoquinoline (cyclohexane)	[n]
2-Phenyl-2-boradihydro-4-quinazolone, **113** (ethanol)	2-Phenylquinazolone	[o]

[a] M. J. S. Dewar and R. Dietz, *J. Chem. Soc.*, 2728 (1959).
[b] M. J. S. Dewar, *Adv. Chem. Ser.*, **42,** 227 (1964).
[c] M. J. S. Dewar and R. Dietz, *J. Org. Chem.*, **26,** 3252 (1961).
[d] M. J. S. Dewar, V. P. Kubba, and R. Pettit, *J. Chem. Soc.*, 3073 (1958).
[e] M. J. S. Dewar and V. P. Kubba, *Tetrahedron*, **7,** 213 (1959).
[f] M. J. S. Dewar and V. P. Kubba, *J. Am. Chem. Soc.*, **83,** 1757 (1961).

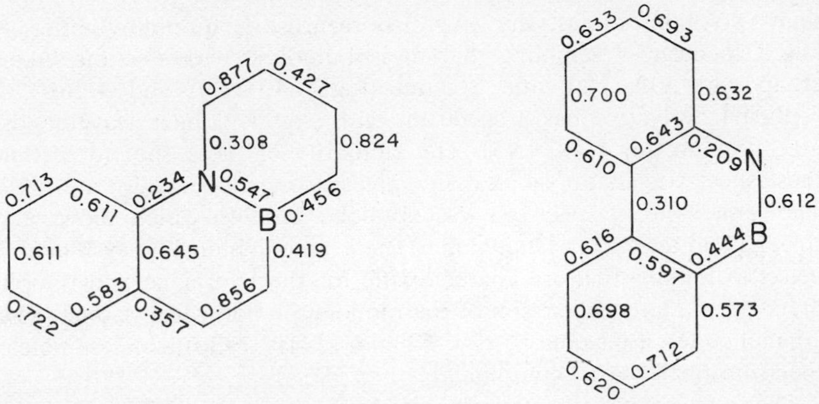

Fig. 1. Comparison of bond orders of 11,12-borazarophenanthrene with those of 10,9-borazarophenanthrene.

characteristic absorption disappearing in cold dilute ethanolic acid or basic solutions.[7] In the case of 14,13-borazarotriphenylene, the uv spectrum in ethanol has maxima at nm (log ϵ) 258, 269, 295, 316, and 330 (4.71, 4.69, 3.62, 3.93, and 3.97).[6]

Since the borazarobenz[*a*]anthracenes, **61a** and **62a**, are isoconjugate with benz[*a*]anthracene and differ by 2 of 18 conjugated atoms, one would expect the uv spectra to be similar except for an increase in the extinction coefficient of the weak α-band, due to the removal of degeneracy between two transitions which is found in alternate hydrocarbons but not in the isoconjugate borazaromatic systems.[2,36] The assigned α-band maxima of benz[*a*]anthracene, **61a** and **62a** are 384, 365, and 369 nm (log $\epsilon = 2.95$, 3.48, and 3.77, respectively) in methylcyclohexane.[36]

The 10,9-borazaroanthracene derivative, **67b**, has a uv spectrum which resembles those of anthracene and acridine and has a strong peak at 382 nm (log 3.86) in ethanol; this peak is not found in the uncyclized diphenylamine spectrum, but the addition of small amounts of alkali to a solution of **67b** results in the disappearance of the long waveband and a spectrum resembling that of diphenylamine.[40]

Among the 6-membered borazaromatic compounds containing more

---

[g] M. J. S. Dewar and P. M. Maitlis, *Tetrahedron*, **15,** 35 (1961).
[h] M. J. S. Dewar and R. Dietz, *J. Chem. Soc.*, 1344 (1960).
[i] M. J. S. Dewar, *Prog. Boron Chem.*, **1,** 235 (1964).
[j] M. J. S. Dewar, C. Kaneko, and M. K. Bhattacharjee, *J. Am. Chem. Soc.*, **84,** 4884 (1962).
[k] G. C. Culling, M. J. S. Dewar, and P. A. Marr, *J. Am. Chem. Soc.*, **86,** 1125 (1964).
[l] M. J. S. Dewar, V. P. Kubba, and R. Pettit, *J. Chem. Soc.*, 3076 (1958).
[m] P. M. Maitlis, *J. Chem. Soc.*, 425 (1961).
[n] M. J. S. Dewar and R. C. Dougherty, *J. Am. Chem. Soc.*, **86,** 433 (1964).
[o] S. S. Chissick, M. J. S. Dewar, and P. M. Maitlis, *J. Am. Chem. Soc.*, **83,** 2708 (1961).

than two heteroatoms, the *bis*(4,3-borazaro-4-isoquinolinyl)ether, **86a**, has a spectrum resembling that of isoquinoline with maxima taken in ethanol at 300, 280, and 269 nm (log 3.64, 3.85, and 4.26).[59] The 3-phenyl derivative has a spectrum with slightly longer wavelengths at 303, 284 nm (log 4.38, 4.45). The similarity of these spectra to that of these spectra to that of isoquinoline gives strong evidence for aromaticity. The same type of spectrum was obtained for **94b** which shows a very strong band at 306 nm belonging to the aromatic binuclear system and no peaks in regions that are characteristic for the hydrazine and thiophene precursors. The uv spectrum of the monocyclic borazaropyridine, **112**, in ethanol shows a maxima at 228, 273 nm (2410, 7930) and resembles the spectrum of 3-hydroxypyridine.[70]

The use of uv spectroscopy for monitoring stability of various borazaromatic compounds have been mentioned in the discussion of reactivity; its use in the study of the acidity of these compounds will be treated in a later section (VII.A).

Since the borazaromatic compounds can form π-complexes with tetracyanoethylene and *s*-trinitrobenzene, the uv spectra of these complexes have been studied by simple Hückel treatment[82,83] (see Section VI.C). The heats of formation of these complexes are in the range of 2 to 6 kcal/mole and thus the orbitals of the complex have been assumed not to be perturbed by the complex formation.[1] The charge transfer spectra of the molecular complexes have been used to deduce ionization potentials to which the necessary parameters were fixed for calculating borazaromatic compounds.

B. Nuclear Magnetic Resonance Spectra

Proton and ^{11}boron nuclear magnetic resonance spectroscopy have proved to be very valuable tools for both structure determination and the elucidation of the aromatic properties of borazaromatic compounds. Proton nmr has been used to check different molecular orbital methods for predictions of charge densities on boron-nitrogen compounds[16,84]; ^{11}B nmr has been used to find additional information about the bonding of borazaromatic compounds and has also led to a better understanding of the factors that determine ^{11}B resonance frequencies.[85]

One compound in this series which has been extensively studied by proton nmr is 10,9-borazaronaphthalene, **26**.[15,16] Since this compound has heteroatoms at the ring junctions (positions 9 and 10), there is a plane of symmetry passing through the B–N bond and perpendicular to the rings and the nmr spectrum can be treated as a four proton system (Fig. 2). Since coupling constants for *ortho* protons are much larger than for

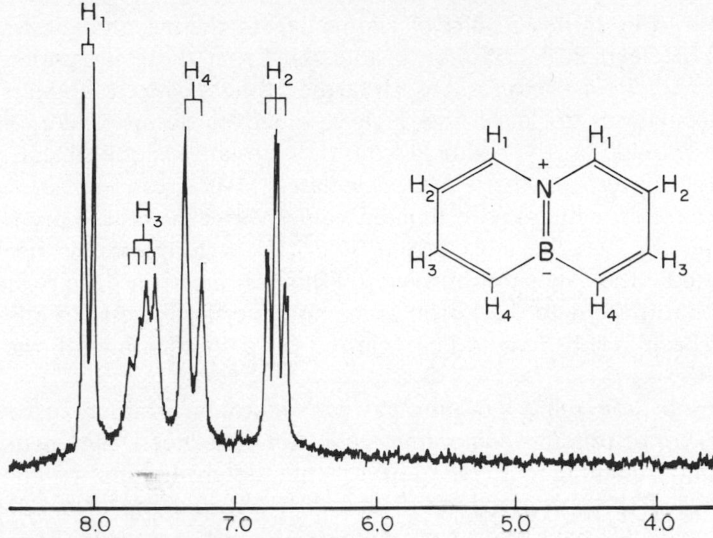

Fig. 2. Proton nmr spectrum of 10.9-borazaronaphthalene.

nonadjacent protons, the two doublets correspond to the 1- and 4-position protons and the two multiplets to the 2- and 3-positions. The spectrum has been further analyzed and the specific assignments made. One can estimate from inductive effects of the boron and nitrogen atoms and the assumed similarity of chemical shift differences between α- and β-protons as in naphthalene what the relative chemical shifts in the spectrum of **26** would be for uniform π-electron distributions (see Table II). The difference from the actual values in Fig. 2 and the estimated values should be a good indicator of charge distribution on the carbon atoms in the 1,2-borazarene ring.

TABLE II. CHEMICAL SHIFTS OF PROTONS AND π-ELECTRON DISTRIBUTIONS OF 10,9-BORAZARONAPHTHALENE

Position	Estimated chem. shift	Observed	Difference	HMO calculations	Pople calculations
1	9.11[a]	8.03[a]	−1.08[a]	0.907[b]	1.0396[c]
2	7.56	6.72	−0.84	1.056	0.9510
3	7.56	7.66	+0.10	0.944	0.9427
4	7.01	7.31	−0.30	1.093	0.9037

[a] M. J. S. Dewar and R. Jones, *J. Am. Chem. Soc.*, **90,** 2137 (1968).
[b] M. J. S. Dewar, "The Molecular Orbital Theory of Organic Chemistry," McGraw-Hill, New York, 1969, p. 383.
[c] M. J. S. Dewar, A. J. Fritsch, and M. J. Garcia, unpublished work, 1970.

However the earlier results of simple HMO theory predicted a very different π-electron distribution (Table II). From these predictions one would expect the 4-proton to absorb farther upfield, since it appears from these calculations to have the highest electron density; likewise the 1-proton should absorb downfield from the estimated chemical shift, since it has the lowest of the HMO calculated π-electron densities. The differences from actual and estimated values show that these predictions are wrong. In fact, the most recent SCF MO calculations by the Pople method show the 1-position to have the highest electron density and the 4-position, adjacent to the boron atom, to have the lowest[3] (see Section VI.C). These results might be regarded as a vindication of the SCF method.[84]

Besides proton nmr, ^{11}B nmr has proved to be a valuable tool for studying borazaromatic compounds. ^{11}B nmr has been shown to be a simple and unambiguous criterion for the determination of the acid properties of these compounds (Section VII.A) and the aromaticity of these compounds have been confirmed by this tool.

Oxygen is more effective than nitrogen in shifting the ^{11}B resonance upfield in the aliphatic series,[86] but this is no longer true when boron is part of a heteroaromatic ring. In the 6-membered rings the oxygen and nitrogen are equally effective (compare **86g** with **130b** and **32f** with **34e** in Table III).[85,87] However, compounds with 5-membered rings as 2-phenylbenzo-1,3-diaza-2-borole, **73**, absorb at a higher field than the boroxaromatic analog, **136**. Thus the π-bonding between boron and the adjacent heteroatom in the compounds listed on Table III is very important especially when contrasted with the open chain analogs. Since the aromaticity of these heteroaromatic compounds depends on extensive bonding to boron,[1] these results are to be expected.

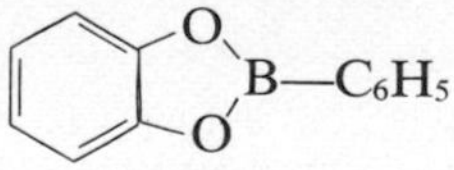

O, B—C_6H_5, O

136

^{11}B nmr spectra can be used to determine the influence of substitution at different positions of the borazaromatic compounds. The introduction of a nitro group at the 8-position of 10-methyl-10,9-borazarophenanthrene to form **38b** leads to a downfield shift and thus a reduction in aromaticity due to mutual conjugation between the nitro and the imino groups. The 6-nitro derivative, **38a**, shows a slight up upfield shift from that of the parent **32c**. Replacement of chloro by ethoxy groups leads to a greater upfield shift for the boroxarophenanthrene series than for the borazarophenanthrene series and thus verifies the earlier predictions that the boroxaro compounds are less aromatic than the borazaro compounds.[85]

TABLE III. ^{11}B NMR SPECTRAL DATA OF BORAZAROMATIC COMPOUNDS

Compound (solvent)	Chemical shift	Linewidth[e]
2,1-Borazaronaphthalene		
2-Methyl, **16d** (carbon disulfide)	−37.5	97[a]
2-Ethoxy, **16f** (ethanol)	−29.7	223[a]
10,9-Borazaronaphthalene, **26** (*n*-hexane)	−28.4	85[a]
4-Bromo, **30a** (acetic acid)	−27.5	175[a]
9-Aza-10-boradecalin, **28** (chloroform)	−42.9±0.5	111[d]
Hydrate, **29** (acetone)	−1.0±0.5	80[d]
10,9-Borazarophenanthrene		
10-Methyl, **32c** (benzene)	−39.0	250[a]
10-Ethoxy, **32f** (ethanol)	−29.3	221[a]
10-Phenyl, **32d** (tetrahydrofuran)	−36.5	226[a]
10-Chloro, **32a** (benzene)	−33.7	224[a]
6-Nitro-10-methyl, **38a** (tetrahydrofuran)	−38.2	443[c]
6-Nitro-10-hydroxy, **39a** (dimethyl sulfoxide)	−36.9	1250[c]
8-Nitro-10-methyl, **38b** (carbon disulfide)	−41.9	350[a]
10,9-Boroxarophenanthrene		
10-Ethoxy, **34e** (ethanol)	−28.8	233[a]
10-Chloro, **34b** (carbon disulfide)	−36.9	127[a]
14,13-Borazarotriphenylene, **52** (benzene)	−30.4	150[a]
1,2,3,4-Tetrahydro, **53** (carbon disulfide)	−38.1	181[a]
2-Phenylbenzo-1,3-diaza-2-borole, **73** (acetonitrile)	−27.7	115[a]
2-Phenylbenzo-1,3-dioxa-2-borole, **136** (carbon disulfide)	−31.9	111[a]
4,3-Borazarolsoquinoline		
4-Methyl, **86b** (acetic acid)	−37.5	305[b]
4-Ethoxy, **86g** (acetic acid)	−30.0	290[b]
4-Ethoxy-4,3-boroxaroisoquinoline, **130b** (ethanol)	−30.0	290[a]

[a] F. A. Davis, M. J. S. Dewar, and R. Jones, *J. Am. Chem. Soc.*, **90,** 706 (1968).
[b] M. J. S. Dewar and R. Jones, *J. Am. Chem. Soc.*, **89,** 2408 (1967).
[c] M. J. S. Dewar, R. Jones, and R. H. Logan, Jr., *J. Org. Chem.*, **33,** 1353 (1968).
[d] F. A. Davis, M. J. S. Dewar, R. Jones, and S. D. Worley, *J. Am. Chem. Soc.*, **91,** 2094 (1969).
[e] At half height in hertz.

The ^{11}B nmr spectrum of 10,9-boraxaronaphthalene has the smallest linewidth of all the compounds measured and follows the hypothesis that charge symmetry around boron produces sharper lines. When the spectrum of this compound is compared with that of its decalin precursor, **28**, one finds a very notable upfield shift which might be partly ascribed to the aromaticity of the former. However, based on results in the literature for vinylboron derivatives, one may conclude that this shift is due to local interactions due to hydridization of the carbon atoms adjacent to boron.[18] Most certainly the large difference between the decalin **28** and its hydrate **29** is due to this change of hydridization at the boron atoms.

C. Molecular Orbital Calculations

The unusual stability and structure of borazine and its derivatives has stimulated numerous molecular orbital calculations using a variety of methods.[88–103] A number of these studies use parameters calculated from ultraviolet, ^{11}B nmr, and ^{14}N nmr spectra of borazine and derivatives. Conclusions and discussions have centered on the stabilization of the ring by σ- or π-bonding or a combination of both, on the bond order of the B–N bond, and on charge distributions at the different atoms. The use of the CNDO method gives calculated ionization potentials for the methyl-substituted borazines in reasonable agreement with experimental values.[99] Baird and Whitehead[104] investigated a great variety of B–N chains and rings with π-electron networks by LCAO MO methods. The results for the chains predict a very weak alternation of bond order and for the B–N rings predict a lessening of the difference in π-energy stabilization between monocyclic systems with $4n+2$ electrons relative to $4n$ electrons than that found in carbon–carbon systems.

By comparison with the borazine series, the calculations for borazaromatic compounds are quite sparse. Some results using simple Hückel molecular orbital methods for calculating molecular orbital parameters for boron and nitrogen from charge transfer bands have already been mentioned, and the limitations of this method have been emphasized.[1,2,83] The LCAO technique with Hückel, Pariser–Parr, and Pople approximations was used on the π-systems of the borazarobenzenes, 10,9-borazarophenanthrene, and 2,1-borazaronaphthalene using several theoretical models for choice of input parameters.[105] The most reasonable model for calculation was found to correspond to the $B^-–N^+$ model for the boron–nitrogen bonds in the borazaromatic system. The order of stabilities for the borazarobenzenes agreed with the predicted order (see Section III) using the Pople method and the localization energies generally agreed with experimental results. Some LCAO MO calculations for σ- and π-electron systems of benzene, the borazarobenzenes, the diboradiazarobenzenes, and three triboratriazarobenzenes have been performed.[96,97] Among the last group borazine was found to be most stable and the order of the borazarobenzenes was as predicted.[2] Likewise the qualitative features of the charge distributions for 10,9-borazarophenanthrene and 2,1-borazaronaphthalene were as those calculated by Dewar.[2]

Carbo and co-workers[106] have studied the three borazarobenzenes with a number of common methods and concluded that the molecules should have fairly high dipole moments. The all-valence-electron methods did not predict any preferential stability. The more reliable CNDO assigned practically the same energy to the three molecules. Most of the methods indicated that the B–N polarity is B^{-}–N^{+} in reference to the π-electrons and is inverted in the σ-skeleton, but giving a total polarity of a positive boron and a negative nitrogen. In a previous study of borazine these workers used the B^{-}–N^{+} π-polarity assumption but chose neutral B and N for the study of the borazarobenzenes, since it was confusing how conjugation occurs in the presence of C atoms.

Recently borazine and a large number of borazaromatic compounds have been calculated by the Pople method using the SCF LCAO MO approximation[3] in which differential overlap is neglected and in which the σ-electrons are regarded as localized and using the modifications introduced for one-electron core resonance integrals.[107] The stabilities of the

TABLE IV. COMPARISON OF HEATS OF ATOMIZATION OF BORAZAROMATIC COMPOUNDS WITH PARENT HYDROCARBONS' OBSERVED VALUES

Borazaromatic compounds[a]	eV	Hydrocarbons	eV
2,1-Borazarobenzene	−54.94	Benzene	−57.16[b]
3,2-Borazarobenzene	−53.55		
4,1-Borazarobenzene	−53.60		
Borazine	−54.42		
2,1-Borazaronaphthalene	−88.68	Naphthalene	−90.61[c]
10,9-Borazaronaphthalene	−88.40		
10,9-Borazarophenanthrene	−122.47	Phenanthrene	−124.20[d]
12,11-Borazarophenanthrene	−122.13		
11,12-Borazarophenanthrene	−122.15		
12,11-Borazaroanthracene	−121.58	Anthracene	−123.93[d]
16,15-Borazaropyrene	−136.09	Pyrene	−138.88[e]
14,16,18-Tribora-13,15,17-triazarotriphenylene	−154.96	Triphenylene	−157.76[d]
4,3-Borazaroisoquinoline	−83.12	Phthalazine	(−79.22)[f]
4,5-Borazarothieno[2,3-*c*] pyridine	−66.46	Benzo[*b*]thiophene	(−74.08)[g]
7,6-Borazarothieno[3,2-*c*] pyridine	−66.55	Benzo[*c*]thiophene	(−73.40)[g]
7,6-Borazarothieno[3,4-*c*] pyridine	−66.23	Pyridazine	−45.59[h]
3,2-Borazaropyridine	−49.38	Pyridine	−51.79[i]

[a] M. J. S. Dewar, A. J. Fritsch, and M. J. Garcia (Calc. by SCF MO method).
[b] American Petroleum Institute Project 44, Carnegie Press, Pittsburgh, Pa., 1955.
[c] D. M. Speros and F. D. Rossini, *J. Phys. Chem.*, **6a,** 1723 (1960).
[d] A. Magnus, H. Hartmann, and F. Becker, *Z. Phys. Chem.* (*Leipzig*), **197,** 75 (1951).
[e] F. Klages, *Chem. Ber.*, **82,** 358 (1949).
[f] Calculated by M. J. S. Dewar and T. Morita, *J. Am. Chem. Soc.*, **91,** 796 (1969).
[g] Calculated by M. J. S. Dewar and N. Trinajstić, *J. Am. Chem. Soc.*, **92,** 1453 (1970).
[h] J. Tjebbes, *Acta Chem. Scand.*, **16,** 916 (1962).
[i] J. D. Cox, A. R. Challoner, and A. R. Meetham, *J. Chem. Soc.*, 265 (1954).

borazarobenzenes have been found to be according to the predicted order with the calculated heats of atomization of the 2,1-, 3,1-, and 4,1-borazarobenzenes being 54.94, 53.55, and 53.60 eV, respectively. The heats of atomization for a number of other borazaromatic compounds have been calculated and compared with the observed values of the parent hydrocarbons (see Table IV). These calculations indicate a high degree of aromaticity among this series of compounds.

The charge densities of the various borazaromatic compounds have been found to be in very good agreement with previous predicted and experimental results (Fig. 3). We have already noted that these results are

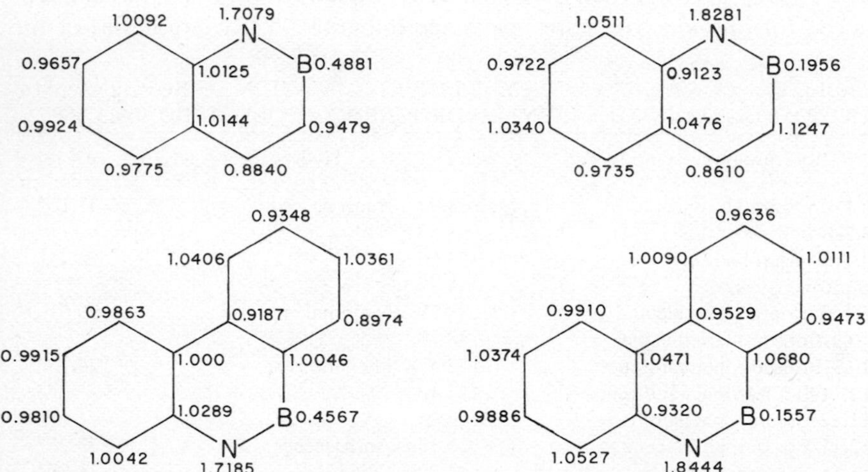

Fig. 3. Comparison of charge densities of 10,9-borazarophenanthrene and 2,1-borazaronaphthalene by the SCF-HMO and HMO methods.

considerably better than those obtained for 10,9-borazaronaphthalene by the HMO method (see Section VI.B). The charge densities and bond orders of borazaroisoquinoline and the borazaropyridines have been likewise calculated and are dscussed in conjunction with the localization energies of these compounds (Section VII.B).

VII. Comparison of Properties of Borazaromatic Compounds

A. Acid Properties

Trivalent boron compounds generally behave as Lewis acids and form coordination compounds with the appropriate Lewis bases. But when boron contains a hydroxyl group, it can either behave as a Lewis acid or a protic acid, in the former case deprotonating to give the planar R_2BO^- and in the latter forming the quadricovalent $R_2B^-(OH)_2$. Experimental evidence seems to indicate that boric acid and the arylboronic acids behave like Lewis acids. However, an interesting feature of borazaromatic compounds containing the BOH group is their behavior as protic acid; this difference has been attributed to aromaticity of the borazaromatic ring where the boron is reluctant to adopt a tetrahedral geometry.[2,87]

Several methods of proof of this behavior have been used. The uv spectra of 10-hydroxy-10,9-borazarophenanthrene, **32b**, and 2-hydroxy-2,1-borazaronaphthalene, **16c**, in either ethanol or base are almost identical, thus implying the loss of a proton from the hydroxyl group rather than coordination of a base to the boron. This proton loss leaves an ion –B–O⁻ with the negative charge residing mainly on the oxygen atom. The ultraviolet spectra of 4-hydroxy-4,3-boroxaroisoquinoline, **130a**, show almost no change in going from neutral to basic solution; however ^{11}B nmr shows that this apparently protic acid is really a Lewis acid which is not surprising due to the low aromaticity. The borazaro analog, **86a**, gave similar uv results that suggested a protic acid and in this case the ^{11}B nmr confirmed these results. Though 10-hydroxy-10,9-boroxarophenanthrene, **34c**, was also thought from the ultraviolet spectra to be a protic acid, the ^{11}B nmr and a reexamination of the uv data showed that this is really a Lewis acid.[108]

The ^{11}B nmr spectra of many of the B-hydroxy borazaro and boroxaro compounds have been measured in both neutral and alkaline solution. Salt formation tended to shift the boric acid and arylboronic acid resonances far upfield and at the same time the linewidths remained unchanged or became narrower, indicating that they behave as Lewis acids. However, in the cases of **32b**, **16c**, **86a**, and **130a** the salt formation resulted in a small downfield shift in the ^{11}B resonances and also in considerable broadening of the lines. Here again the salt formation implies proton transfer in agreement with the earlier uv evidence. Had there been a change in symmetry the linewidths would have decreased.

It has also been found that 6- and 8-nitro-10,9-borazarophenanthrene, **39a** and **39b**, also give pronounced colors on treatment with alkali. However one cannot attribute this to loss of a proton from the imino group for that would require it to be more acidic than the hydroxyl hydrogen. No ^{11}B nmr spectra could be obtained for the 6-nitro derivative, **39a**, but the signal for **38a**, **38b**, and **39b** showed both very large upfield shifts on addition of alkali and decreases in linewidth.[109] The salt formation involved addition of base to boron rather than loss of a proton from oxygen or nitrogen, thus resulting in compounds with structures of the type **137**.

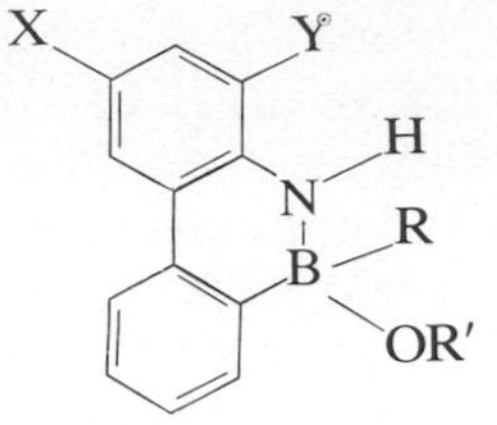

X = Y = NO_2 or H

R = OH or CH_3

R′ = H or CH_3

137

Ir spectra and proton nmr confirmed these results, especially the magnitude of the shifts in the latter.

The addition of base to these compounds (**137**) led to large bathochromic shifts indicating the compounds are quite acidic in contrast to 2-hydroxy-2,1-borazaronaphthalene, **16c**, and 10-hydroxy-10,9-borazarophenanthrene, **32b**. In the latter cases the ultraviolet spectra were unaltered by the additon of small amounts of acid, indicating that neither compound is appreciably basic and that both possess strong B–N bonds. It seems that introduction of a nitro group leads to cross conjugation, which consequently decreases the π-electron density on the boron. Addition of base removes it from conjugation with the adjacent imino nitrogen, thus greatly increasing the interaction of the nitrogen with the nitro substituent.[109]

B. Localization Energies

The borazaromatic compounds have been known to undergo electrophilic substitution reactions, and this has been used as a criterion for aromaticity. A mere comparison of calculated electron densities is not sufficient for predicting sites of substitution even though in some cases there is a relation. In fact the direct correlation of electron densities with chemical reactivity is invalid[110]; for any treatment that neglects the transition state is incorrect in principle, and the success of such treatments is due solely to fortuitous correspondence between the quantities in question and the energy differences which really determine the reaction rates.

It is necessary to obtain the localization energy for these compounds or, in other words, the difference in the energy of the activated complex, **138**, and the energy of the unreacted molecule. The π-energy of the activated complex is assumed to be equivalent to the computed energy of the molecule, **139**, which lacks the skeleton carbon at the site of sigma complexification. This difference in energy is the best indicator of sites of preferential reactivity.

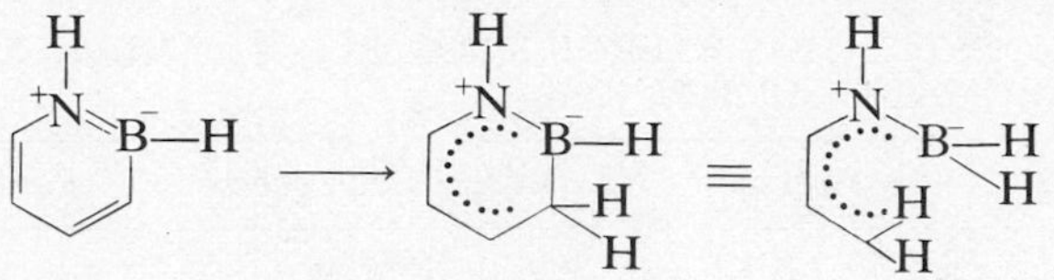

138 139

In calculating localization energies of borazaromatic compounds it appears that sigma polarization factors become quite important and cannot be ignored (Sections I and II.C). For molecules having formal charges at the site of complex formation there might very well be a tendency to distribute the charge over the whole molecular framework and the excited molecule would therefore behave more like a pure carbon–carbon system than a borazaromatic system with nitrogen and boron core charges of +2 and 0, respectively. It has been found that using atomic constants for a model containing nitrogen with a core charge of 1 and boron with a core charge of 1, gives results in excellent agreement with experimental data for reactivity.[3]

Nitrogen, being more electronegative than carbon, deactivates the position of opposite parity to itself, and boron, being less electronegative, activates the position of opposite parity to itself. For 2,1-borazarobenzene, **1**, the 3- and 5-positions should be activated and the 2- and 4-positions deactivated. For the 3,1-borazarobenzene, **2**, one would expect the 2-, 4-, and 6-positions to be deactivated by nitrogen and the 2- and 4-positions to be activated by boron. Likewise for **3** one expects the 3- or the equivalent 5-position to be more reactive to electrophilic substitution. Recent SCF MO calculations have been found consistent with these predictions, as can be seen from the data in Table V (also see Fig. 4).[3]

The calculated localization energies for 2,1-borazaronaphthalene are in agreement with experimental results and indicate that the 3-position next to the boron is the most reactive. Likewise the experimentally confirmed 4-position of 10,9-borazaronaphthalene is the site calculated for electrophilic attack. Here the use of charge density data would erroneously predict the 1-position to be the site of attack (see Section VI.B). The order of reactivity for 10,9-borazarophenanthrene is in general agreement with experimental evidence showing the 6- and 8-positions to be the most reactive sites.

This study has been extended to several interesting hypothetical or partially investigated compounds. In 12,11-borazarophenanthrene, **140**, the expected 4-position was found to be most favored with the 9-position second. In 12,11-borazaroanthracene, **141**, the anticipated 9-position is calculated to be the most likely for electrophilic attack with a somewhat

TABLE V. LOCALIZATION ENERGIES FOR ELECTROPHILIC SUBSTITUTION OF BORAZAROMATIC COMPOUNDS

Compound	Position	eV	Compound	Position	eV
2,1-Borazarobenzene	3	−3.379	10,9-Borazaro-		
			phenanthrene	1	−2.669
	4	−1.546		2	−3.166
	5	−3.137		3	−2.491
	6	−2.206		4	−3.258
3,1-Borazarobenzene	2	−2.148		5	−3.196
	4	−1.758		6	−3.461
	5	−2.066		7	−3.135
	6	−2.046		8	−3.502
4,1-Borazarobenzene	2	−1.021	12,11-Borazaro-		
			phenanthrene	1	−2.997
	3	−2.710		2	−3.497
2,1-Borazaro-				3	−2.492
naphthalene	3	−3.405		4	−3.784
	4	−2.016		5	−3.124
	5	−3.088		6	−3.642
	6	−3.248		7	−2.956
	7	−2.906		8	−3.461
	8	−3.376		9	−3.792
10,9-Borazaro-				10	−3.392
naphthalene	1	−3.052			
	2	−3.399			
	3	−2.383	14,16,18-Triboro-	1	−12.127
	4	−3.784	13,15,17-Triazaro-	2	−10.321
11,12-Borazaro-	1	−3.853	triphenylene	3	−11.772
phenanthrene	2	−2.337		4	−10.994
	3	−3.557			
	4	−2.907	4,3-Borazaroiso-	1	−2.305
	5	−3.547	quinoline	5	−2.371
	6	−3.288		6	−2.285
	7	−3.480		7	−2.076
	8	−3.384		8	−2.565
	9	−2.662			
	10	−3.782	7,6-Borazarothieno-	2	−2.386
12,11-Borazaro-	1	−4.222	[3,2-*c*]pyridine	3	−2.050
anthracene	2	−3.196		4	−2.283
	3	−3.829			
	4	−3.771	4,5-Borazarothieno-	2	−2.729
	5	−3.485	[2,3-*c*]pyridine	3	−2.112
	6	−3.604		7	−2.239
	7	−2.942			
	8	−4.005	7,6-Borazarothieno-	1	−2.938
	9	−4.822	[3,4-*c*]pyridine	3	−3.154
	10	−4.038		4	−2.532
16,15-Borazaropyrene	2	−3.374			
	3	−4.720	3,2-Borazaropyridine	4	−2.095
	4	−4.418		5	−1.402
	5	−3.939		6	−1.849
	6	−4.306			
	7	−4.181			

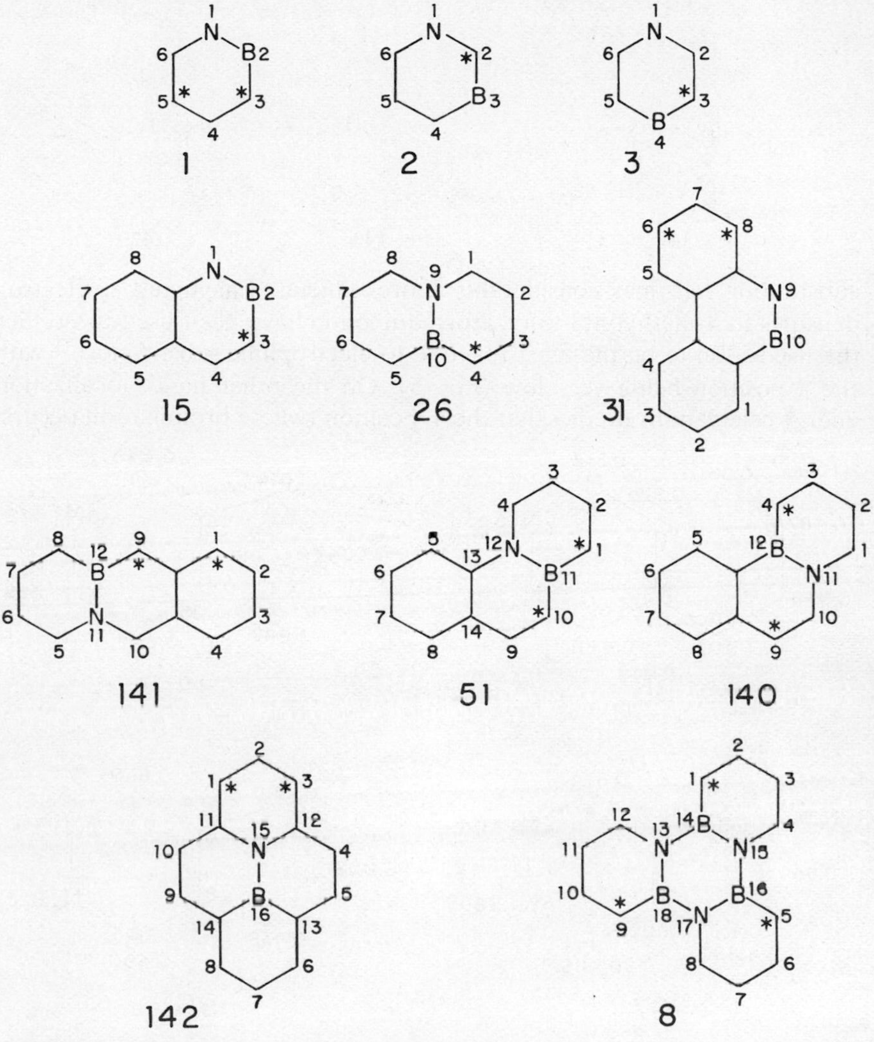

Fig. 4. Sites of reactivity of some borazaromatic compounds predicted by SCF-HMO calculations.

less favored 1-position as second. The 3- and equivalent 1-position of 16,15-borazaropyrene, **142**, has been calculated to be most active; this would be a very interesting result to attempt to verify by experimental evidence. The calculated sites for **8** and **51** are quite reasonable being adjacent to the boron atom.

With borazaromatic compounds containing other heteroatoms either in the borazarene or in the fused ring, the prediction of sites for electrophilic

141 **140** **142**

substitution becomes considerably more difficult. Calculated π-electron densities in 4-methyl-4,3-borazaroisoquinoline have been used to predict the 8-position to be the most reactive to electrophilic substitution,[61] with the 1-position being very low (Fig. 5). On the other hand, localization energy calculations predict that the 1-position (where bromination occurs)

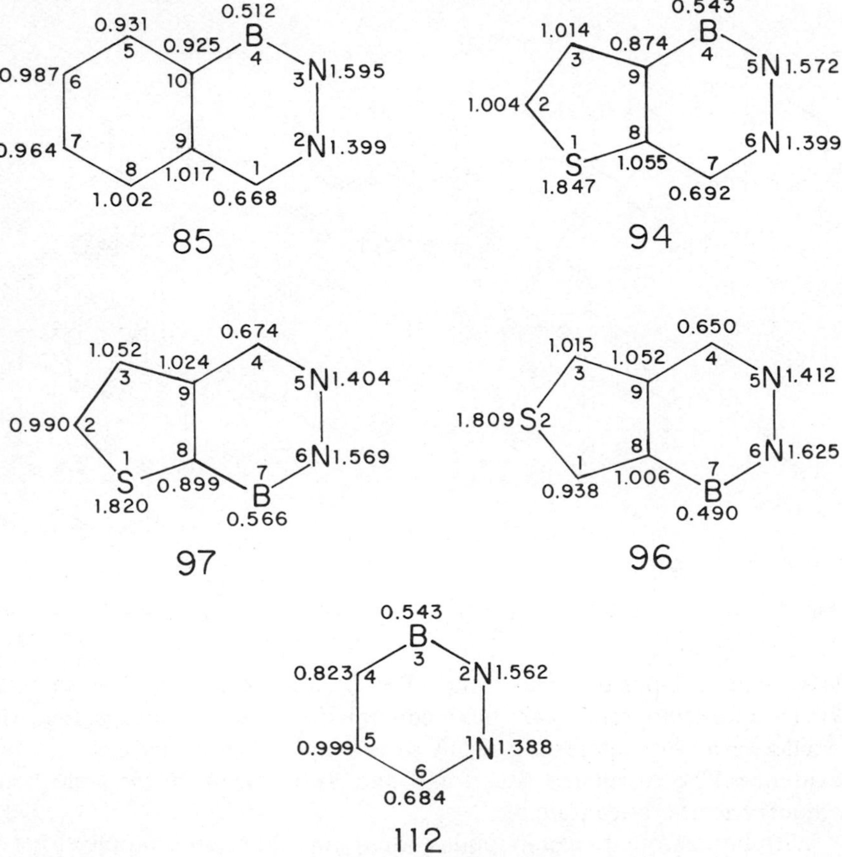

Fig. 5. Charge densities of borazaroisoquinolines and pyridines.

is the most susceptible and the 8-position (where nitration occurs) to be second.[3] The bond order for the carbon–nitrogen bond has been calculated to be the highest (Fig. 6).

As stated before (Section V.A), these results can be profitably compared with those of the isoelectronic "normal" compound, isoquinoline. From recent calculations of charge densities and localization energies, it appears that the 4-position should be most favored for electrophilic attack (where bromination and mercuration occur) and that the 5-position should be second (where nitration and sulfonation occur). The resulting calculations are assembled in Fig. 7. Using the Wheland–Pauling molecular orbital approximation, Longuet–Higgins and Coulson had previously calculated the 5-position to have the highest and the 4-position to possess the second lowest charge density.[111]

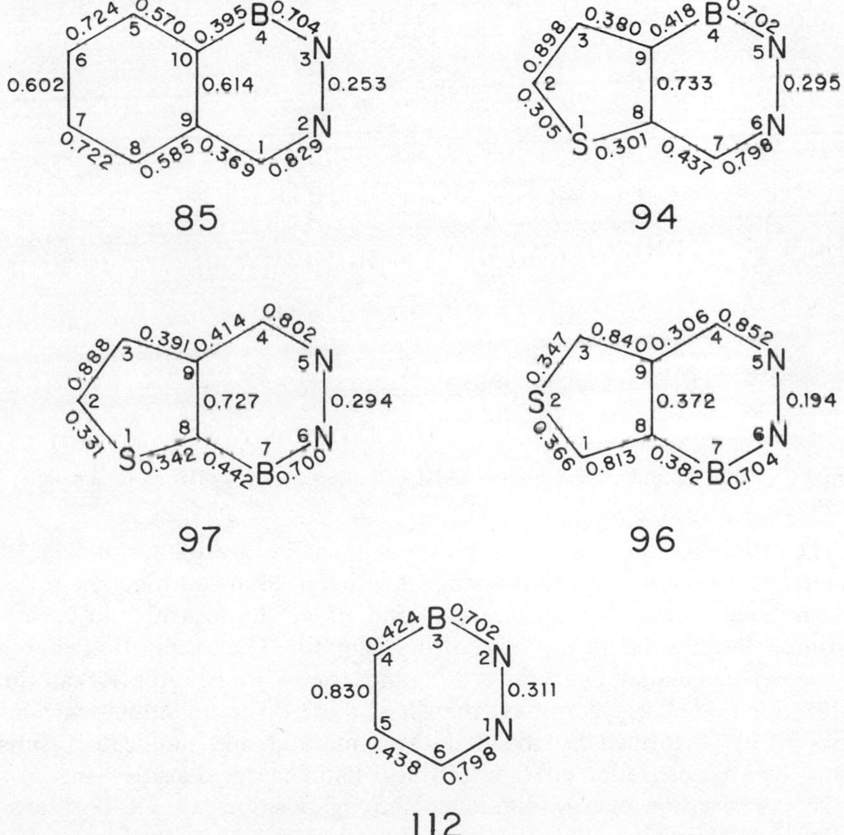

Fig. 6. Bond orders of borazaroisoquinolines and pyridines.

CHARGE DENSITIES

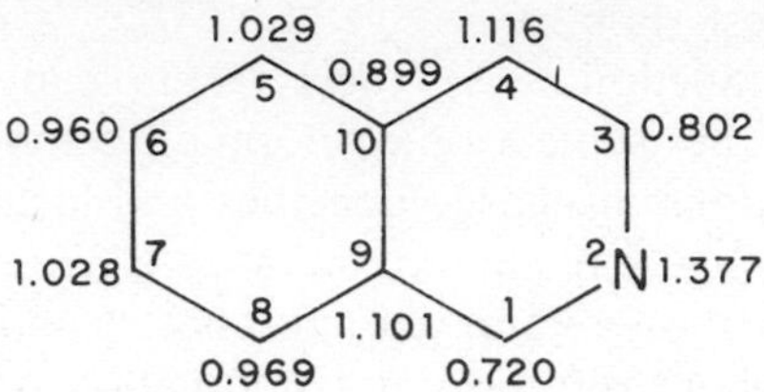

BOND ORDERS

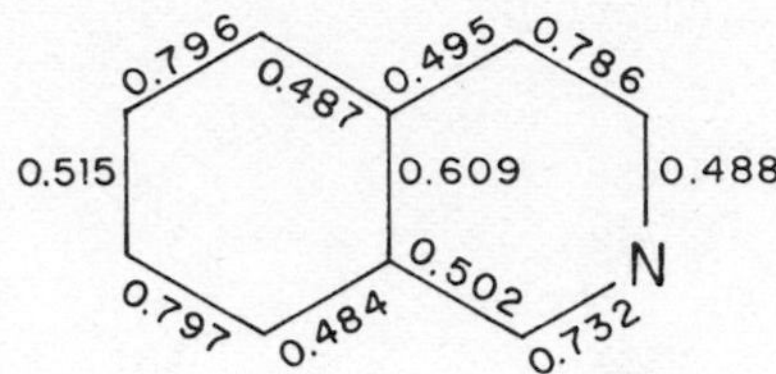

LOCALIZATION ENERGIES

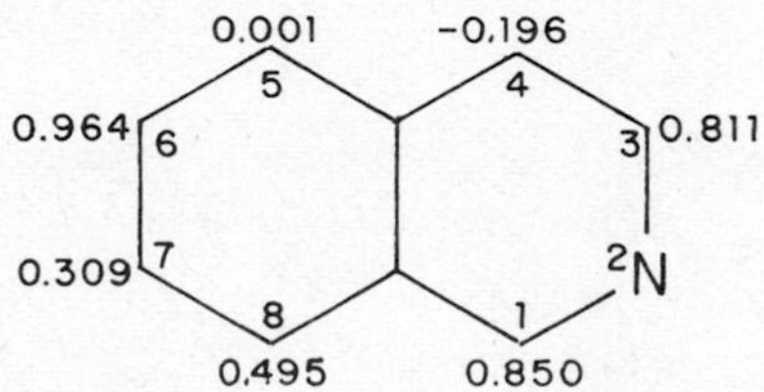

Fig. 7. Charge densities, bond orders, and localization energies (eV) of isoquinoline.

The differences in substitution patterns of the borazaromatic analogs of thianaphthene and isothianaphthene have been discussed (Section V.A). It has been found that the C_2–C_3 bond of 4,5-borazarothieno[2,3-*c*]-pyridine, **94**, has the highest bond order (Fig. 6).[3] The localization energy has been calculated by Namtvedt[67] and found to be greatest at the 2-position and charge densities greatest at the 3-position. Similar calculations were performed by the SCF MO method and duplicate results obtained for localization energies (Table V) and charge densities (Fig. 5).[3]

The localization energy difference among positions of 7,6-borazarothieno[3,2-*c*]pyridine, **97**, has been found to be the smallest of the borazaromatic compounds calculated (0.23 eV) and makes meaningful predictions difficult. Calculated charge densities for **97** are highest at the 3-position where nitration in sulfuric acid occurs. Neglecting other criteria, one would expect that from localization energies the 3-position of 7,6-borazarothieno[3,4-*c*]pyridine, **96**, and the 6-position of 3,2-borazaropyridine, **112**, would be most susceptible to electrophilic attack. However, there is good reason to suspect from experimental substitution patterns in acetic acid that the high bond orders at the carbon–nitrogen bond of the isoquinoline ring may have a strong influence on electrophilic substitution in these compounds (Fig. 6). While localization energies are good for predicting sites of electrophilic substitution for the simple borazaromatic compounds, this becomes one of several determining factors in the borazaroisoquinolines and pyridines and the charge densities and bond orders must be considered.

C. Aromaticity

Historically the term aromaticity has developed from various descriptive chemical criteria: the ease at undergoing electrophilic substitution rather than addition reactions; the enhanced stabilization of certain compounds resulting in resistance to oxidation and hydrolysis; and the property of abnormally high heats of formation and bonds intermediate between normal single and double bonds. However this descriptive type of definition gives only general comparative characteristics of compounds behaving like "aromatic" hydrocarbons in chemical and physical properties.

There has been recent attempts to enlarge the concept of aromaticity by giving a more explanatory definition based on internal criteria of the molecule rather than on comparative physical and chemical properties within a given class of compounds. An example of this type of definition is the ability of certain compounds to sustain an induced ring current.[112] However, since this type of definition is found to extend to some commonly conceived "nonaromatic" compounds, one is either forced to further modify it or to adopt another internal criterion such as magnetic properties. At the present moment no satisfactory explanatory definition of aromaticity is available, and thus it is best to limit a discussion of borazaromaticity to the more empirical and descriptive definitions, while still granting borazaromatic compounds the ability to sustain ring currents.

Borazaromatic compounds possess the unusual stability of the trigonal boron toward reagents with which they ordinarily undergo reactions and the ability to undergo electrophilic substitution reactions. These compounds have characteristic aromatic acid properties and produce uv, ir, and nmr spectra that are quite similar to those of common aromatic hydrocarbons. Thus borazaromaticity is a relative concept which is properly understood only when compared with normal aromatic compounds. Thus one can speak of borazarene rings being as "aromatic" as the benzene ring and discuss similarities and differences in charge densities, bond orders, heats of atomization, and such. There is a strong temptation to constitute aromaticity as an absolute concept and then measure existing compounds and compare them with purely hypothetical values.

There has been a great amount of confusion in the literature with regard to the measured differences such as "resonance energy values," either due to the arbitrary choice of the reference compound or the failure to allow for changes in bond properties with the state of hybridization of the atoms involved.[84] Our treatment of borazaromaticity is spared some of these pitfalls by the sheer lack of experimental values for heats of formation.

Even though one is unable to make precise quantitative comparisons between borazaromatic compounds and parent hydrocarbons, one can find several general qualitative trends which allows one to speak of degrees of aromaticity among boron compounds. From the many properties listed in previous sections the following can be said:

1. Cyclic boron-containing compounds with nitrogen in the conjugated ring are more aromatic than those heterocycles containing the more electronegative oxygen or sulfur, since these last two do not participate in the dipolar $-X^{+}{=}B^{-}$ structure as readily as the nitrogen atom.

2. The borazaromatic compound is more stable when the heteroatom is adjacent the boron, for example, the greater stability of the 2,1-borazarene system.

3. The aromaticity of a given system is reduced by replacing the imino hydrogen by electrophilic groups.

4. The aromaticity increases in compounds where boron is attached to a single heteroatom as opposed to where it is attached to two heteroatoms. 2,1-Borazaronaphthalene is more aromatic than are the borazaroquinazolines.

5. The aromaticity increases where boron is in a bridged position in a fused polycyclic system and thus attached to two carbons as opposed to where it is attached to a single carbon and a hydrogen; likewise B-aryl- and alkyl-substitutents tend to stabilize a given series. 10,9-Borazaronaphthalene appears to be one of the most aromatic compounds synthesized and shows very similar properties to those of naphthalene.

6. The more aromatic borazaromatic acids tend to act as protic acids while less aromatic ones and other boronic acids behave as Lewis acids.

VIII. Conclusions

The field of borazaromatic chemistry is only about 20 years old and has already opened many challenging problems for synthetic, analytic, and theoretical chemists. The unusual stabilities of borazaromatic compounds invite future investigations on a variety of reactions, rates of reactivity, and practical application of these compounds to the medical and drug fields. There has been almost no testing of these compounds for biochemical activity. The advent of new spectroscopic instrumentation and the development of more sophisticated molecular orbital methods offer new possibilities for theoretical investigations. These compounds have heteroatoms at key positions in aromatic systems which allow for a more thorough understanding of the parent compounds. Besides, the valuable tools of ^{14}N and ^{11}B nmr can be utilized in elucidation of borazaromaticity. These borazaromatic compounds which straddle the ever-fading boundaries separating organic and inorganic chemistry beckon investigators from every branch of chemistry.

At the present there are several obstacles to further progress in this field. The borazaromatic compounds require considerable time and effort in their preparation and the yields of many of the intermediates are disappointingly low. At present it is quite difficult to collect a sufficient quantity of the various compounds for extensive investigation. Likewise present methods for obtaining heats of formation are not adequate for use on the borazaromatic compounds. As these obstacles are overcome, one might anticipate more interesting results from the study of these compounds.

IX. Appendix

The nomenclature of borazaromatic compounds has been a point of controversy.[1,2,8,19] The authors in this field have generally followed the procedure of naming these compounds after the isoconjugate parent aromatic hydrocarbon in order to emphasize the aromaticity of these compounds and their similarity to the parent compound. The prefixes "bora" and "aza" indicate the replacement of a methine group by boron and nitrogen; the added prefix segment "aro" denotes the replacement of carbon atoms by isoelectronic ions rather than heteroatoms.[2]

Though the names listed in this review are given as secondary names in the subject indices of *Chemical Abstracts*, the principal names are derived from reference to the azaborine system: the commonly named 4,1-borazarobenzene is listed as 1,4-dihydro-1,4-azaborine; 4,3-borazaroisoquinoline as 1,2-dihydro-3,2,1-benzodiazaborine; 6,5-borazarochrysene as 5,6-dihydro-benz[*c*]naphth[2,1-*e*][1,2]azaborine. The nomenclature based on azaborine makes the visualizing of the borazaromatic compounds very difficult and so it has seemed more feasible to continue the practice of naming these compounds in the manner used by the principal investigators in the field.

X. Addendum

Some very recent work in the field of borazaromatic chemistry includes the following:

A series of substituted arylboratheophyllines, **143**, containing the mentioned diazaborolidine ring, **83**[113];

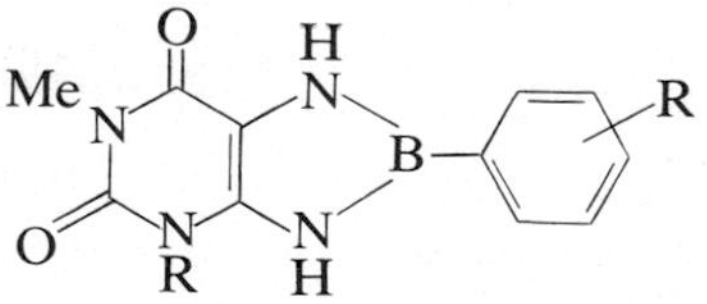

143: (R = H, *o*-, *m*-, or *p*-CH_3 or CH_3O, or *p*-Br)

Some substituted borazaroisoquinolines that have been proven active against Gram-negative bacteria, for example, *Escherichia coli* or *Proteus vulgaris*, and have diuretic and saluretic activities[114];

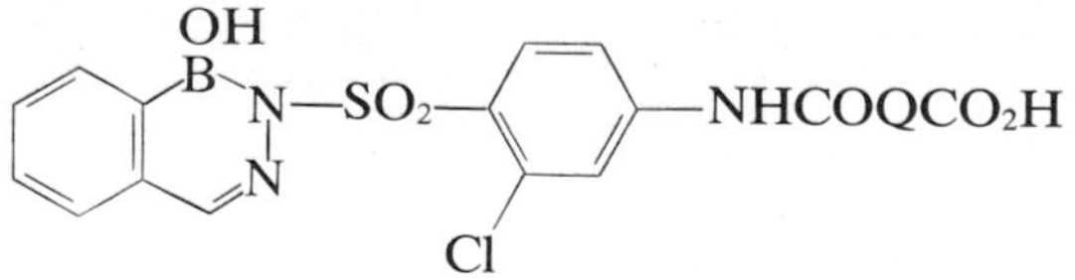

144: Q = $(CH_2)_2$, $(CH_2)_3$, *o*-phenylene, or γ-phthalimidopropylene

The borazarofuropyridine analogs of **94a–c** and **97a–c**.[115]

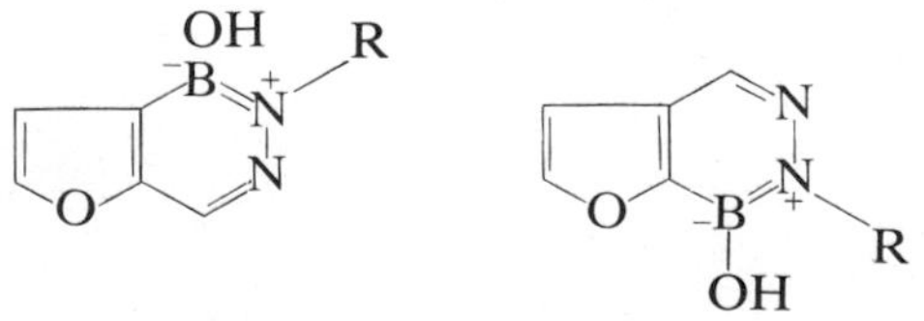

145: R = H, CH_3 or C_6H_5 **146:** R = H, CH_3, or C_6H_5

XI. Acknowledgment

I wish to thank Professor Michael J. S. Dewar for generously consenting to read this paper and for offering several very helpful suggestions.

XII. References

1. M. J. S. Dewar, *Prog. Boron Chem.*, **1,** 235 (1964).
2. M. J. S. Dewar, *Adv. Chem. Ser.*, **42,** 227 (1964).
3. M. J. S. Dewar, A. J. Fritsch, and M. J. Garcia, unpublished work, 1970.
4. M. J. S. Dewar and P. A. Marr, *J. Am. Chem. Soc.*, **84,** 3782 (1962).
5. D. G. White, *J. Amer. Chem. Soc.*, **85,** 3634 (1963).
6. G. C. Culling, M. J. S. Dewar, and P. A. Marr, *J. Am. Chem. Soc.*, **86,** 1125 (1964).
7. K. M. Davies, M. J. S. Dewar, and P. Rona, *J. Am. Chem. Soc.*, **89,** 6294 (1967).

7a. H. Wille and J. Goubeau, *Chem. Ber.*, **107,** 110 (1974).

8. M. J. S. Dewar and R. Dietz, *J. Chem. Soc.*, 2728 (1959).
9. M. J. S. Dewar, R. Dietz, V. P. Kubba, and A. R. Lepley, *J. Am. Chem. Soc.*, **83,** 1754 (1961).
10. M. J. S. Dewar and R. Dietz, *Tetrahedron*, **15,** 26 (1961).
11. H. R. Snyder, A. J. Reedy, and W. J. Lennarz, *J. Am. Chem. Soc.*, **80,** 835 (1958).
12. W. J. Lennarz and H. R. Snyder, *J. Am. Chem. Soc.*, **82,** 2172 (1960).
13. M. J. S. Dewar and R. Dietz, *J. Org. Chem.*, **26,** 3252 (1961).
14. M. J. S. Dewar, J. Hashmall, and V. P. Kubba, *J. Org. Chem.*, **29,** 1755 (1964).
15. M. J. S. Dewar, G. J. Gleicher, and B. P. Robinson, *J. Am. Chem. Soc.*, **86,** 5698 (1964).
16. M. J. S. Dewar and R. Jones, *J. Am. Chem. Soc.*, **90,** 2137 (1968).
17. M. J. S. Dewar and A. J. Fritsch, unpublished work, 1969.
18. F. A. Davis, M. J. S. Dewar, R. Jones, and S. D. Worley, *J. Am. Chem. Soc.*, **91,** 2094 (1969).
19. M. J. S. Dewar, V. P. Kubba, and R. Pettit, *J. Chem. Soc.*, 3073 (1958).
20. M. J. S. Dewar, R. B. K. Dewar, and Z. L. F. Gaibel, *Org. Syn.*, **46,** 65 (1966).
21. P. J. Grisdale and J. L. R. Williams, *J. Org. Chem.*, **34,** 1675 (1969).
22. M. J. S. Dewar and R. Dietz, *J. Chem. Soc.*, 1344 (1960).
23. M. J. S. Dewar and P. M. Maitlis, *Chem. Ind.*, 1626 (1960).
24. F. A. Davis and M. J. S. Dewar, *J. Am. Chem. Soc.*, **90,** 3511 (1968).
25. M. J. S. Dewar and V. P. Kubba, *Tetrahedron*, **7,** 213 (1959).
26. M. J. S. Dewar and R. H. Logan, *J. Am. Chem. Soc.*, **90,** 1924 (1968).
27. M. J. S. Dewar and V. P. Kubba, *J. Org. Chem.*, **25,** 1722 (1960).
28. M. J. S. Dewar and P. J. Grisdale, *J. Org. Chem.*, **28,** 1759 (1963).
29. M. J. S. Dewar and V. P. Kubba, *J. Am. Chem. Soc.*, **83,** 1757 (1961).
30. M. J. S. Dewar and P. M. Maitlis, *J. Am. Chem. Soc.*, **83,** 187 (1961).
31. M. J. S. Dewar and P. M. Maitlis, *Tetrahedron*, **15,** 35 (1961).
32. R. Huisgen, I. Ugi, I. Zeigler, and H. Huber, *Tetrahedron*, **15,** 44 (1961).
33. S. S. Chissick, M. J. S. Dewar, and P. M. Maitlis, *Tetrahedron Lett.*, **23,** 8 (1960).

34. M. J. S. Dewar, C. Kaneko, and M. K. Bhattacharjee, *J. Am. Chem. Soc.*, **84,** 4884 (1962).
35. M. J. S. Dewar and W. H. Poesche, *J. Org. Chem.*, **29,** 1757 (1964).
36. M. J. S. Dewar and W. H. Poesche, *J. Am. Chem. Soc.*, **85,** 2253 (1963).
37. M. J. S. Dewar and R. Jones, *Tetrahedron Lett.*, **22,** 2707 (1968).
38. J. Michl, private communication, 1970.
39. J. M. Davidson and C. M. French, *J. Chem. Soc.*, 191 (1960).
40. P. M. Maitlis, *J. Chem. Soc.*, 425 (1961).
41. M. J. S. Dewar and E. A. C. Lucken, *Chem. Soc. Spec. Publ.*, **12,** 223 (1958).
42. P. M. Maitlis, *Chem. Rev.*, **62,** 223 (1962).
43. R. L. Letsinger and S. B. Hamilton, *J. Org. Chem.*, **25,** 592 (1960).
44. J. Goubeau and A. Zappel, *Z. Anorg. Chem.*, **279,** 38 (1955).
45. D. Ulmschneider and J. Goubeau, *Chem. Ber.*, **90,** 2733 (1957).
46. M. J. S. Dewar, V. P. Kubba, and R. Pettit, *J. Chem. Soc.*, 3076 (1958).
47. M. F. Hawthorne, *J. Am. Chem. Soc.*, **81,** 5836 (1959).
48. M. F. Hawthorne, *ibid.*, **83,** 831 (1961).
49. E. Nyilas and A. H. Soloway, *ibid.*, **81,** 2681 (1959).
50. J. C. Catlin and H. R. Snyder, *J. Org. Chem.*, **34,** 1664 (1969).
51. L. F. Hohnstedt and A. M. Pellicciotto, U.S. Office of Naval Research, 2793(00), (1961).
52. C. A. Brown, U.S. Office of Naval Research, 1939(02), (1956).
53. H. Beyer, K. Niedenzu, and J. W. Dawson, *J. Org. Chem.*, **27,** 4701 (1962).
54. R. J. Brotherton and A. C. McCloskey, *Adv. Chem. Ser.*, **42,** 131 (1964).
55. K. Niedenzu and J. W. Dawson, "Boron-Nitrogen Chemistry," Academic Press, New York, 1965, p. 140.
56. J. E. Mulvaney, J. J. Bloomfield, and C. S. Marvel, *J. Polymer. Sci.*, **62,** 59 (1962).
57. H. Zimmer, E. R. Andrews, and A. D. Sill, *Arzneim.-Forsch.*, **17,** 607 (1967).
58. H. Noeth and G. Schmid, *Z. Anorg. Allgew. Chem.*, **345,** 69 (1966).
58a. W. L. Cook and K. Niedenzu, *Syn. Inorg. Metal-Org. Chem.*, **2,** 267 (1972)(for **84a, 84b, 84c**).
58b. J. S. Merriam and K. Niedenzu, *J. Organometal. Chem.*, **51,** C1–C2 (1973)(for **84d**).
58c. M. J. S. Dewar, R. Golden, and P. A. Spanninger, *J. Am. Chem. Soc.*, **93,** 3298 (1971).
58d. M. J. S. Dewar and P. A. Spanninger, *Tetrahedron*, **28,** 959 (1972).
59. M. J. S. Dewar and R. C. Dougherty, *J. Am. Chem. Soc.*, **84,** 2648 (1962).
60. M. J. S. Dewar and R. C. Dougherty, *J. Am. Chem. Soc.*, **86,** 433 (1964).
61. M. J. S. Dewar and J. L. von Rosenberg, Jr., *J. Am. Chem. Soc.*, **88,** 358 (1966).
62. J. Namtvedt and S. Gronowitz, *Acta Chem. Scand.*, **20,** 1448 (1966).
63. M. J. S. Dewar and P. M. Maitlis, *J. Chem. Soc.*, 2521 (1957).
64. S. Gronowitz and A. Bugge, *Acta Chem. Scand.*, **19,** 1271 (1965).
65. S. Gronowitz and J. Namtvedt, *Acta Chem. Scand.*, **21,** 2151 (1967).
65a. S. Gronowitz, T. Dahlgren, J. Namtvedt, C. Roos, G. Rosen, B. Sjoberg, and U. Forsgren, *Acta Pharm. Suecica*, **8,** 623 (1971).
66. R. Mayer, H. Kleinert, S. Richter, and K. Gewald, *Angew. Chem.*, **74,** 118 (1962).
67. J. Namtvedt, *Acta Chem. Scand.*, **22,** 1611 (1968).
68. M. J. S. Dewar and P. M. Maitlis, *J. Chem. Soc.*, 944 (1957).
69. S. Gronowitz and J. Namtvedt, *Tetrahedron Lett.*, **26,** 2967 (1966).
70. J. Namtvedt and S. Gronowitz, *Acta Chem. Scand.*, **22,** 1373 (1968).
70a. S. Gronowitz and A. Maltesson, *Acta Chem. Scand.*, **25,** 2435 (1971).
70b. S. Gronowitz and A. Maltesson, *Chem. Scr.*, **2,** 79 (1972).

70c. S. Gronowitz and C. Toresson, *ibid.*, 143.
71. S. S. Chissick, M. J. S. Dewar, and P. M. Maitlis, *J. Am. Chem. Soc.*, **81,** 6329 (1959).
72. S. S. Chissick, M. J. S. Dewar, and P. M. Maitlis, *J. Am. Chem. Soc.*, **83,** 2708 (1961).
73. H. L. Yale, F. H. Bergeim, F. A. Sowinski, J. Bernstein, and J. Fried, *J. Am. Chem. Soc.*, **84,** 688 (1962).
74. J. Fried, F. H. Bergeim, H. L. Yale, and J. Bernstein, U. S. Patent 3,293,252 (C1 260–268), December 20, 1966; *Chem. Abstr.*, **66,** 115783n (1967).
75. R. Koester and K. Iwasaki, *Adv. Chem. Ser.*, **42,** 148 (1964).
76. R. Koester, K. Iwasaki, S. Hattori, and Y. Morita, *Ann. Chem.*, **720,** 23 (1969).
76a. B. Frange, *Bull. Soc. Chim. Fr.*, (7–8)(Pt. 1), 2165 (1973).
77. R. K. Bartlett, H. S. Turner, R. J. Warne, M. A. Young, and I. J. Lawrenson, *J. Chem. Soc.* (*A*), 479 (1966).
78. H. Noeth, P. Fritz, and W. Meister, *Angew. Chem.*, **73,** 762 (1961).
79. A. H. Soloway, *J. Am. Chem. Soc.*, **82,** 2442 (1960).
80. J. Michl, R. Jones, and J. Janata, unpublished material.
81. M. J. S. Dewar, *J. Chem. Soc.*, 2329 (1950).
82. M. J. S. Dewar and A. R. Lepley, *J. Am. Chem. Soc.*, **83,** 4560 (1961).
83. M. J. S. Dewar and H. Rogers, *J. Am. Chem. Soc.*, **84,** 395 (1962).
84. M. J. S. Dewar, "The Molecular Theory of Organic Chemistry," McGraw-Hill, New York, 1969, p. 383.
85. F. A. Davis, M. J. S. Dewar, and R. Jones, *J. Am. Chem. Soc.*, **90,** 706 (1968).
86. W. D. Phillips, H. C. Miller, and E. L. Muetterties, *J. Am. Chem. Soc.*, **81,** 4496 (1959).
87. M. J. S. Dewar and R. Jones, *J. Am. Chem. Soc.*, **89,** 2408 (1967).
88. R. J. Boyd, D. H. Lo, and M. A. Whitehead, *Chem. Phys. Lett.*, **2,** 227 (1968).
89. D. A. Brown and C. G. McCormack, *Theor. Chim. Acta*, **6,** 350 (1966).
90. O. Chalvet, R. Daudel, and J. J. Kaufman, *Adv. Chem. Ser.*, **42,** 251 (1964).
91. O. Chalvet, R. Daudel, and J. J. Kaufman, *J. Am. Chem. Soc.*, **87,** 399 (1965).
92. D. W. Davies, *Trans. Faraday Soc.*, **56,** 1713 (1960).
93. D. W. Davies, *Trans. Faraday Soc.*, **64,** 2881 (1968).
94. M. Giambiagi, M. S. de Giambiagi, and E. Silberman, *Theor. Chim. Acta*, **5,** 435 (1966).
95. K. Hensen and K. P. Messer, *Theor. Chim. Acta*, **9,** 17 (1967).
96. R. Hoffmann, *Adv. Chem. Ser.*, **42,** 78 (1964).
97. R. Hoffmann, *J. Chem. Phys.*, **40,** 2474 (1964).
98. P. M. Kuznesof and D. F. Shriver, *J. Am. Chem. Soc.*, **87,** 399 (1965).
99. P. M. Kuznesof and D. F. Shriver, *J. Am. Chem. Soc.*, **90,** 1683 (1968).
100. J. F. Labarre, M. Graffaul, J. P. Faucher, M. Pasdeloup, and J. P. Laurent, *Theor. Chim. Acta*, **11,** 423 (1968).
101. J. C. Patel and S. Basu, *Naturwissenschaften.*, **47,** 302 (1960).
102. P. G. Perkins and D. H. Wall, *J. Chem. Soc. A*, 235 (1966).
103. C. C. J. Roothaan and R. S. Mulliken, *J. Chem. Phys.*, **16,** 118 (1948).
103a. S. D. Peyeumhoff and R. J. Buenker, *Theor. Chim. Acta*, **19,** 1 (1970).
104. N. C. Baird and M. A. Whitehead, *Can. J. Chem.* **45,** 2059 (1967).
105. J. J. Kaufman and J. R. Hamann, *Adv. Chem. Ser.*, **42,** 273 (1964).
106. R. Carbo, M. S. de Giambiagi, and M. Giambiagi, *Theor. Chim. Acta*, **14,** 147 (1969).
107. M. J. S. Dewar and A. J. Harget, *Proc. Roy. Soc. A*, **315,** 457 (1970).
108. F. A. Davis and M. J. S. Dewar, *J. Org. Chem.*, **33,** 3324 (1968).
109. M. J. S. Dewar, R. Jones, and R. H. Logan, *J. Org. Chem.*, **33,** 1353 (1968).
110. M. J. S. Dewar and C. C. Thompson, Jr., *J. Am. Chem. Soc.*, **87,** 4414 (1965).

111. H. C. Longuet-Higgins and C. A. Coulson, *Trans. Faraday Soc.*, **43,** 87 (1947).
112. J. March, "Advanced Organic Chemistry," McGraw-Hill, New York, 1968, p. 38.
113. R. Caujolle and D. Quoc Quan, *CR Acad. Sci. Ser. C.*, **271,** 754 (1970).
114. *S. Herrling, H. Huemer, and H. Mueckter, German Patent* 1,812,103; *Chem. Abstr.*, **73,** 120744z (1970).
115. S. Gronowitz and U. Michael, *Ark. Kemi*, **32,** 283 (1970).

CHAPTER VIII

Syntheses and Properties of Cyanine and Related Dyes

DAVID M. STURMER

Research Laboratories, Eastman Kodak Company, Rochester, New York

Dyes are used in photography for a number of distinct purposes, the most important of which are (1) as spectral sensitizers (sensitizing dyes), (2) as color images in color photography, (3) in antihalation and filter layers and in color filters, and (4) as desensitizers. Of these, dyes of categories 1 and 4 are the subject of this chapter. The first sections provide general descriptions of the historical development and generic concepts applied to sensitizers; the later sections detail important synthetic reactions and physical/chemical properties.

I. Historical Perspective

Comprehensive reviews of sensitizing dye literature were prepared by Leslie G. S. Brooker,[1–9] Frances M. Hamer,[10] G. E. Ficken,[11] and A. I. Kiprianov.[12–19] Published proceedings of various symposia on dyes provide additional reviews of color/constitution, physical/chemical properties, and spectral sensitization (see listing in Table I). The encyclopedic review by Hamer is outstanding and remains the best source for the heterocyclic chemistry of cyanine and related dyes. Based on the changing emphasis in the various reviews, five major time periods characterize important eras of investigations of sensitizing dyes.

TABLE I. DYE SYMPOSIA

Symposium title, location	Date	Proceedings: editor(s) publisher, date	Synthesis, color/ const.[a]	Physical, chemical properties[a]	Spectral sensitiz.[a]
Vogel centennial Putney, Vermont	Aug. 1973	Not published	MO	X, A, *p*K, R	L, EL AD
Photographic Sensitivity Cambridge, England	Sept. 1972	R. J. Cox, Ed., Academic Press, NY 1973		L, A, R	IP
Dye Sensitization Bressanone, Italy	Aug. 1967	G. Semerano, U. Mazzucato, W. F. Berg, and H. Meier, Eds., Focal Press, NY 1970	MO	L, A, R	DES IP
5th International Farben-symposium, Basel, Switzerland	Sept. 1973	*Chimia*, **27,** 604–608, 640–662 (1973)	NC PC MO		IS A
4th International Farben-symposium, Lindau	May 1970	Announced *Chimia*, **24,** 42 (1970)	Metal-complexed dyes		
3rd International Farben-symposium, Interlaken, Switzerland	May 1967	*Chimia Supplementum* (1968)	NC Syn. MB	Thermal cleavage of azacyanines	

TABLE I (*continued*)

Symposium title, location	Date	Proceedings: editor(s) publisher, date	Synthesis, color/ const.[a]	Physical, chemical properties[a]	Spectral sensitiz.[a]
2nd International Farben-symposium, Elmau, Switzerland, "Optische Anregung Organischer Systeme"	April 1964	M. Jung, Ed., Verlag Chem., Weinheim, 1966	MO, NC PC SS	L, A	
International Farbensymposium, Basel, Switzerland	June 1960	*Separatum aus, Chimia*, **15,** 1–227 (1961)	NC Syn., SS MB	L, MO	
Scientific Photography Liège, Belgium	1959	H. Sauvenier, Ed., Pergamon Press, NY 1962		R Aza	DES Aza
XIV International Congress Pure Applied Chemistry, Zurich Switzerland	July 1955	*Experientia Supplementum* II (1955)	Syn. SS, ST		
International Congress of Scientific and Applied Photography, London, England	Sept. 1953	R. S. Schultze, Ed., Royal Photogr. Soc., London, 1955	ST	Fading ($h\nu$)	AD DES

9th International Congress of Scientific and Applied Photography, Paris, France	July 1935	L. P. Clerc, Ed., Revue d'Optique Theor. Inst., Paris, 1936	Syn. Aza		DES Aza
8th International Congress of Scientific and Applied Photography, Dresden, Germany	1931	J. Eggert and A. V. Biehler, Eds., J. A. Barth, Leipzig, 1932		$AgNO_3$ Dyes	
7th International Congress of Scientific and Applied Photography, London, England	July 1928	W. Clark, T. Slater-Price, B. V. Storr, Eds., and W. Heffer, Cambridge, 1929	Aza		DES
"Photographic Science"		(all published by Focal Press, NY)			
Series: Paris, France;	1965	J. Pouradier, Ed., 1967			
Torino, Italy	1963	G. Semerano and U. Mazzucato, Eds., 1965			
Zurich, Switzerland	1961	W. F. Berg, Ed., 1963			

[a] Symbols used: A = aggregation, AD = dye adsorption, Aza = aza-derivatives of cyanines, DES = desensitization versus dye structure, EL = energy level determinations, IP = dye ionization potentials, IS = isomers, L = luminescence, MB = methylene bases; MO = molecular orbital calculations, NC = novel chromophores, PC = photochromics, pK = dye pKa values, R = (redox) electrochemical potential determinations, SS = solvent sensitivity, ST = steric effects, Syn. = reviews of synthetic methods and/or approaches to specific dye classes, X = x-ray crystal structure.

A. 1850–1910: Early Discoveries

Particularly knowledgeable accounts of these years were written by Brooker[1,4] and Hamer.[10] "The Dyes" were discovered in 1856 by the young chemist C. H. Greville Williams, while characterizing quinoline obtained by the distillation of alkaloids with caustic alkali.[20] After quaternization with alkyl iodides, these samples of quinoline gave reddish blue dyes by treatment with silver oxide. Although the dyes attracted little commercial attention for use as fabric dyes despite their superior color, Brooker[4] noted that "the dye from the amyl iodide salt was, however, manufactured in small quantities by the Paris firm of Menier who gave it the name of 'cyanine'," meaning blue.

Spectral sensitization was discovered in 1873 by H. W. Vogel,[21] during a study of the spectral sensitivity of silver halide dry collodion plates. Certain plates gave spectral sensitivity maxima (!) in the green. This spectral sensitivity was removed by washing the plate prior to exposure and regained by addition of other dyes, including Williams' original cyanine. Subsequent experiments were described by Vogel in a letter to *The Philadelphia Photographer* (written December 1, 1873), and this was republished recently[22] along with a review by W. West on "The First Hundred Years of Spectral Sensitization."[23]

An isocyanine condensation in the presence of formaldehyde, carried out by Homolka[24] around 1904, led to "Pinacyanol," the first sensitizer for red light. Furthermore, this then unknown structure was to be a key compound 12 years later, which unlocked a pattern of structural modifications for the next half century.

By 1910 spectral sensitivity was extended throughout the visible region with *isocyanine* dyes (from 2-methylquinoline salts), erythrosin, and hundreds of other dyes.

B. 1915–1925: The Necessity for Structural Definition

Most spectral sensitizers were made in Germany prior to 1915, but during World War I, W. J. Pope and W. H. Mills at Cambridge were asked to supply photographic sensitizers for Britain. The exact reagents for the most useful dyes were buried in German patents, and the Cambridge group, of necessity, prepared and carefully evaluated a great many dyes. At the same time, structural identities for the dyes were established and published after the war (Fig. 1).* Frances M. Hamer was one of Mills' students at Cambridge during this period (1917–1924), and

* In general, counterions for dyes are not specified in this chapter.

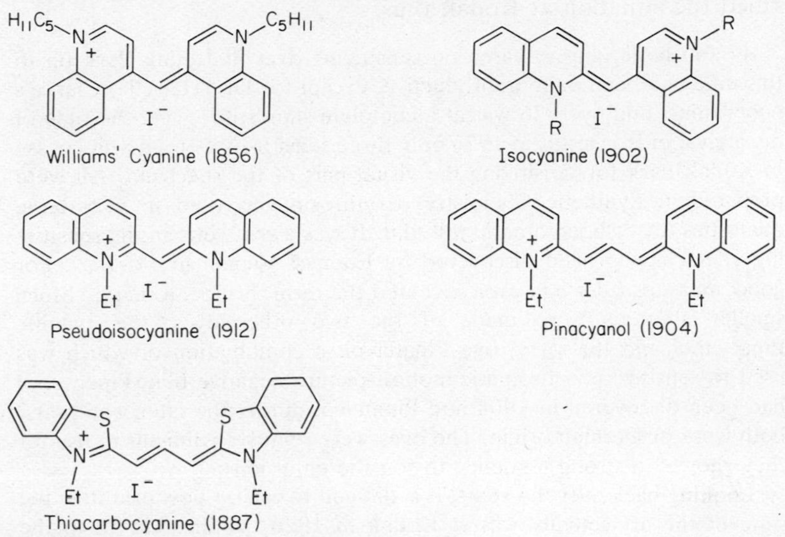

Williams' Cyanine (1856)

Isocyanine (1902)

Pseudoisocyanine (1912)

Pinacyanol (1904)

Thiacarbocyanine (1887)

Fig. 1. Structures for typical sensitizers as of 1920. Since the colors of pinacyanol and Williams' cyanine were nearly identical, both pinacyanol and the more obscure thiacarbocyanine were thought to be monomethine cyanines until 1920.

her account of this work is contained in the first chapter ("Mainly Introductory") of her review.[10]

The majority of structures were dyes with monomethine links between two heterocyclic nuclei, but the important discovery in this work was published by Mills and Hamer[25]: "Pinacyanol" was two quinoline rings linked not by one but by *three* methine carbons. The structure of the benzothiazole dye, thought since 1887 to be a dye with a monomethine link, was also found to have three methine carbon atoms.[26] Since the actual monomethine dyes related to these trimethine dyes absorbed at much shorter wavelengths,[25,26] one of the first important relations between dye color and constitution was established. The term *carbocyanine* was suggested for these trimethine dyes, the term *simple cyanine* being used to describe the monomethine dyes.

The publications of Pope and Mills' group[10] had little apparent effect until Dr. Hamer joined Ilford* in England around the beginning of 1925 and Dr. Leslie Brooker was employed by Kodak† in 1926: Brooker[27]

* Dr. Hamer later joined Kodak at Harrow, England (1930).

† Eastman Kodak Company, located at Rochester, N.Y. In 1926, film manufacture, preparation of the organic chemicals and sensitizers (Synthetic Chemistry Department), and the laboratory in which Brooker was employed (Organic Chemistry Research Laboratory) were located at Kodak Park, about 4 miles from Rochester city center.

described the situation at Kodak thus:

> By and large, any research on sensitizing dyes at Kodak Park up to this time had been quite unproductive, except for Dr. (Hans T.) Clarke's neocyanine, and by 1926 was at a complete standstill. . . . At the time of my arrival in Rochester in 1926 only three sensitizers were in regular use in Kodak Park for sensitizing the visual part of the spectrum. All were prepared at Synthetic Chemistry. Erythrosin was used in very large quantities for orthochromatic roll film. It was a real veteran, its sensitizing effect having been discovered by Eder of Vienna in 1884. . . . For good measure Eder had even invented the term 'orthochromatic.' Much smaller amounts were made of the two other dyes, the cyanine Pinacyanol and the styryl dye Pinaflavol, a combination of which was used to sensitize panchromatic motion-picture negative film. Pinacyanol had been discovered in 1904 and Pinaflavol during the later war years; both were of German origin. The dyes were relatively difficult to use for they showed a strong tendency to fog the emulsions.
>
> Looking back over the years, it is difficult to realize how primitive the state of the art actually was at Kodak in 1926. . . . In addition to the three dyes already mentioned, very small amounts of the infrared

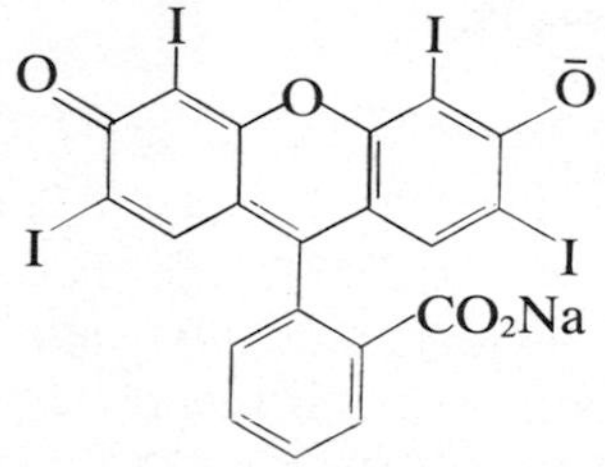

Erythrosin

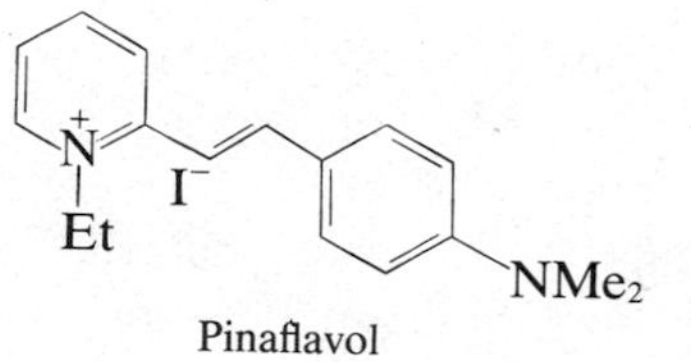

Pinaflavol

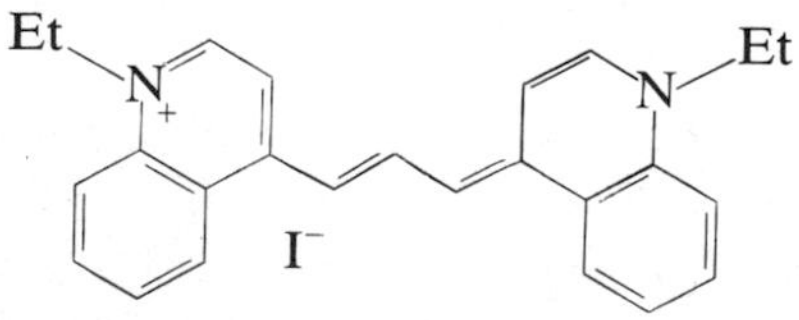

Kryptocyanine

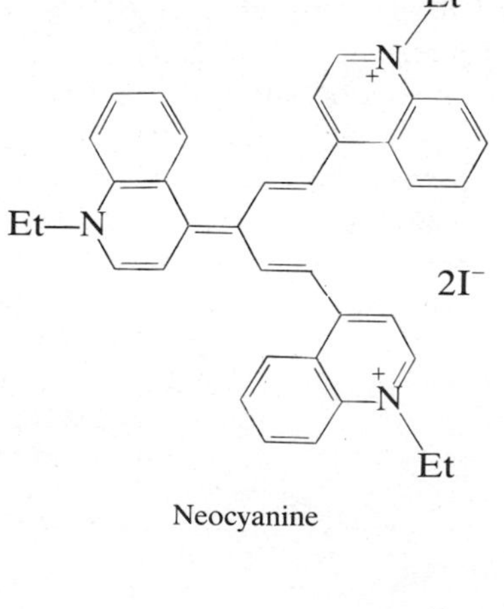

Neocyanine

sensitizers, kryptocyanine and neocyanine, were made for use in Wratten plates, and these five dyes constituted the total Kodak arsenal of sensitizers.

C. 1920–1962: The Development of Color and Constitution

Enormous synthetic efforts followed Pope and Mills' structural assignments.[10] These brought generalized synthetic procedures, new intermediates, a new class of dyes (merocyanines), and definitive theories on color-structure relations. Major synthetic advances were numerous: (1) W. Koenig (Dresden) used orthoesters and vinylogs of these compounds [$\mathrm{EtO{+}(CH{=}CH)_{n}{-}CH(OEt)_2}$, $n = 0,1$] to prepare symmetrical dyes.[28–30] Variations of these reagents led to powerful sensitizers[31] [from $\mathrm{RC(OEt)_3}$] and to extended-chain dyes with seven methine carbons.[32–35]

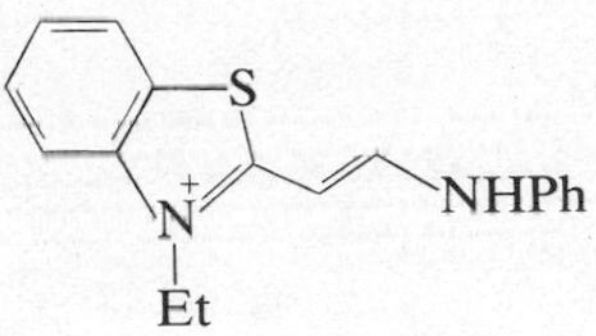

ICI intermediate

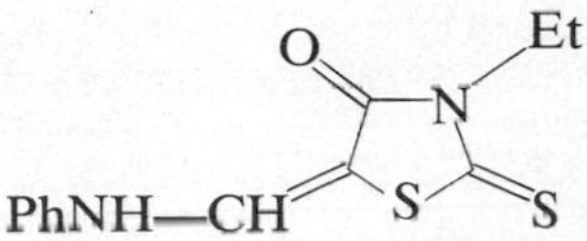

Dains intermediate

(2) Piggott and Rodd at ICI, Ltd., developed reagents for unsymmetrical dyes,[36,37] which are known as "ICI intermediates," and the analogous reagents from ketomethylene compounds were first made by Dains and co-workers.[38,39] (3) A large class of valuable sensitizers known as *merocyanines* was developed in the early 1930s. A comprehensive account of these dyes was subsequently published.[40,41] The dyes also served as intermediates for more complex dyes, usually incorporating two or more typical dye chromophores. (4) Extended-chain quaternary salts[42] provided important synthetic access to infrared sensitizers.

The color-structure relations for symmetrical dyes were essentially dictated by the early structural work on cyanines,[10] but unsymmetrical dyes were instrumental in understanding color and constitution. Much of the early experimental work on cyanine and merocyanine dyes was published by L. G. S. Brooker as a series of papers on "Color and Constitution,"[43–55] although independent work by Kiprianov used certain similar concepts.[56–59] Brooker's basic ideas, as contained in the 1942 edition of *The Theory of the Photographic Process*[1] and in the papers cited, emphasized the concept of continuous polarity among various dye systems. Contributing resonance structures were of different stabilities in highly unsymmetrical dyes. As one resonance structure became the most stable through predictable structural changes, the dyes absorbed at unexpectedly short wavelengths. Other factors, such as solvent, contributed to resonance structure stabilities, particularly for the merocyanines.

D. 1950–1966: Novel Chromophores[60]

Applications of the generalized syntheses of polymethine dyes now employed some rather novel reagents. In addition to the heterocycles in which the chromophoric nitrogens (typical of the original cyanines) are replaced by another heteroatom,[61–65] hydrocarbon analogs of cyanines (positive charge) and oxonols (negative charge) were prepared,[66–70] along with ferrocene-[71–73] and phosphorus-containing[74–81] dyes. Conjugated betaines[82,83] ("merocyanines") were prepared by oxidative syntheses, analogous to the many preparations of simple cyanines nearly a century earlier (1850–1920).

E. 1960–Present: Structure and Properties: A Closer Look

Quite detailed experimental and theoretical data for ground- and excited-state cyanine dyes have characterized many recent investigations, and progress has been made in synthesis as well. Ground-state dyes have been investigated by x-ray crystallography on dye single crystals and nuclear magnetic resonance for dyes in solution. Experimental probes for excited-state dyes have included luminescence at room temperature and low temperature as well as flash photolysis techniques for the detection of excited-state absorptions at room temperature. Molecular orbital calculations of varying sophistications have depicted both ground and excited states for molecular and aggregated dye molecules. These areas of investigation are reviewed in subsequent sections of this chapter.

II. Generic Concepts: Cyanine and Related Dyes

Most of the relationships between the color and structure of dye molecules were established before the widespread use of computers.[1,2,10,84,85] Brooker[86,†] and co-workers provided extensive documentation of the major trends with two series of publications: (1) *Color and*

[†] The late Dr. Leslie G. S. Brooker guided sensitizing dye research at the Research Laboratories of the Eastman Kodak Company (Rochester, NY) for over 40 years. "A Look at Leslie Brooker—His Life and Works" was presented at the Vogel Centennial (Putney, Vt., 1973) and published recently.[86]

Constitution[43–55] I-XII (1940–1951), XIII (1965); and (2) *Steric Hindrance*[6,8] (1947–1953). Kiprianov and co-workers published parallel work independently.[12–19] Resonance theories were developed to accommodate all the major trends in a highly successful manner. For an experienced dye chemist, these still represent a most useful approach to the design of dyes with a specified color, band shape, or solvent sensitivity. More recently, the cataloging of specific oxidation and reduction potentials (see Section III.D) has led to additional structure/properties relationships and the computer-assisted design of new molecules.

A. Chromophoric Systems

Sensitizing dyes are characterized by high extinction transitions in the visible or infrared regions of the spectrum. The primary types of chromophores for these molecules are the amidinium-ion system (**A**), the carboxyl-ion system (**B**), and the dipolar amidic system (**C**). For each system two extreme resonance structures are shown, where any of the formal charges are located at the ends of the chromophore. Intermediate resonance structures, with the charges nearer the center of the chromophore or with additional dipoles, were considered less important in the resonance picture of dyes. However, structural changes which favored intermediate forms had significant effects on the color of symmetrical dyes (**A**).[7,9,51] For the amidic dyes (**C**) structural features stabilizing

$$>\overset{+}{N}{=}CH\left(CH{=}CH\right)_n\ddot{N}< \quad\longleftrightarrow\quad >\ddot{N}{-}CH\left(CH{-}CH\right)_n\overset{+}{N}< \qquad \textbf{(A)}$$

$$O{=}\underset{|}{C}\left(CH{=}\underset{|}{C}\right)_n O^- \quad\longleftrightarrow\quad {}^-O{-}\underset{|}{C}\left(CH{-}\underset{|}{C}\right)_n O \qquad \textbf{(B)}$$

$$>\ddot{N}\left(CH{=}CH\right)_n\underset{|}{C}{=}O \quad\longleftrightarrow\quad >\overset{+}{N}\left(CH{-}CH\right)_n\underset{|}{C}{-}O^- \qquad \textbf{(C)}$$

both neutral and dipolar extreme resonance forms gave longer wavelength absorbing dyes.[52,53] The resonance concepts agree qualitatively with the results from quantum mechanical calculations (see Section V.A).

The important characteristics that influence the absorption wavelengths for these dyes are the length of the conjugated chain and the nature of the terminal group. Many of the early cyanine dyes comprised a chain with an odd number of methine carbon atoms (C–H) and two heterocyclic rings like quinoline or benzothiazole. Historically, the terms *simple cyanine, carbocyanine, dicarbocyanine,* and such, were used to designate both the specific dyes derived from quinoline and also generic dye structures **1** from other heterocycles with one, three, five, and so on,

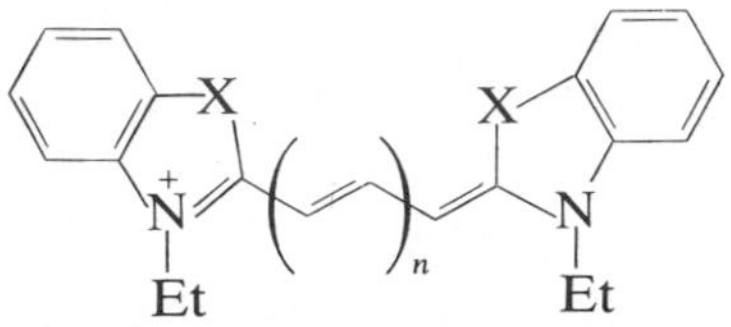

$n=0$, a simple cyanine
$n=1$, a carbocyanine
$n=2$, a dicarbocyanine
$n=3$, a tricarbocyanine

1

methine carbon atoms. For the dyes from quinoline, the ring position attached to the methine chain and the *N*-substituent are usually specified (e.g., 1,1′-diethyl-2,2′-cyanine is dye **1** with X = CH = CH and $n=0$).

Dyes derived from the primary chromophores (**B**) and (**C**) were subsequently designated *oxonols* and *merocyanines,* although the term *neutrocyanine* has also been used for (**C**). For certain cyanine dyes and for merocyanines "simple" designation refers to a dye with zero methines, that is, the shortest possible linkage that retains the chromophore between the terminal groups.

Dyes that differ only by the number of vinyl groups (CH = CH) in the conjugated chain are termed a *vinylogous series.* Absorption maxima for vinylogous series of dyes like **2** to **5** in Table II shift to longer wavelengths as the methine chain length increases. The shift approximates 100 nm per vinyl group in most symmetric chromophores like **2** (amidinium-ion system, **A** above) and **3** (carboxyl-ion system, **B** above), and these are termed *nonconverging* series. Less symmetric chromophores including those of the dipolar amidic systems (**4, 5, C** above), and the polyenes **6** show markedly reduced shifts as each vinyl unit is added (*converging* series). In the dyes the degree of asymmetry and the absorption shifts are both related to the structures of the heterocyclic terminal groups (see Section II.B).

A remarkably large group of dyes incorporate two or more of the chromophoric systems (**A**), (**B**), or (**C**). Typical combinations are shown in the structures **7** to **11** below. In all of these dyes the separate chromophores are conjugated through common rings. In the first two, the common ring serves as an electron acceptor for the merocyanine portion on the left side and an electron donor for the cyanine or merocyanine part on the right. In the second two the common rings serve as either electron acceptors (structure **9,** merocyanine) or as electron donors (structure **10,**

TABLE II. ABSORPTIONS OF VINYLOGOUS DYES[a]

Dye		λ_{max}, nm ($\epsilon \times 10^{-4}$, liters mole^{-1} cm^{-1})				Ref.
		$n = 0$	1	2	3	
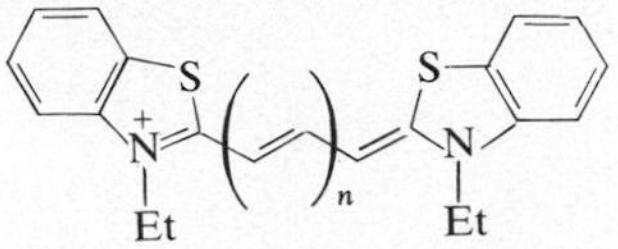 2		423 nm (8.5)	557 nm (15)	650 nm (23)	758 nm (25)	[b]
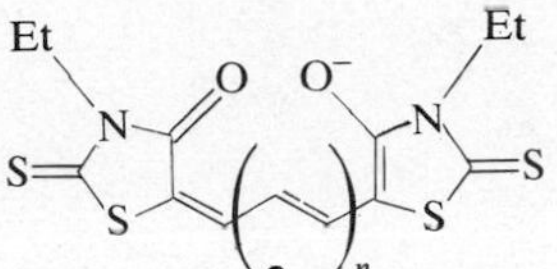 3		542 nm (8.2)	613 nm (11)	714 nm (14)	—	[c,d]
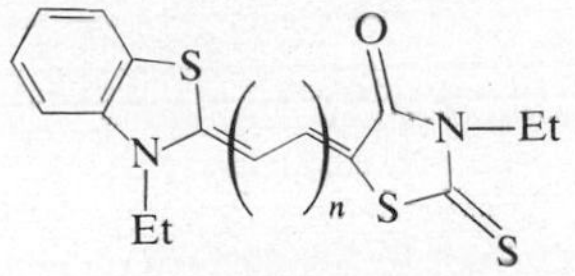 4	(pyridine)	432 nm (6.0)	528 nm (9.3)	605 nm (8.5)	635 nm (6.9)	[c]
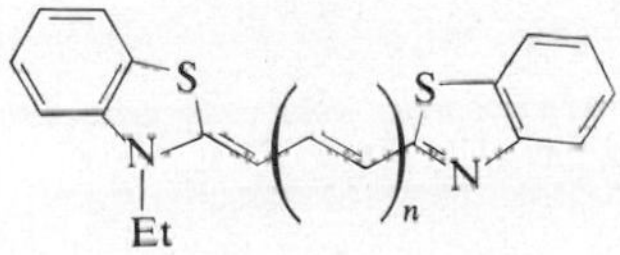 5		396 nm (5.9)	458 nm (5.7)	490 nm (6.4)	510 nm (—)	[b]
6: Ph(CH=CH)$_n$Ph		— —	328 nm (9)	357 nm (12)	383 nm (27)	[e]

[a] Spectra were obtained in methanol or ethanol for dye series **2, 3, 5,** pyridine for **4,** and ether–alcohol at −196°C for **6**;
[b] L. G. S. Brooker, R. H. Sprague, C. P. Smyth, and G. L. Lewis, *J. Am. Chem. Soc.*, **62,** 1116 (1940);
[c] L. G. S. Brooker, G. H. Keyes, R. H. Sprague, R. H. Van Dyke, E. Van Lare, G. Van Zandt, F. L. White, H. W. J. Cressman, and S. G. Dent, Jr., *J. Am. Chem. Soc.*, **73,** 5332 (1951);
[d] M. V. Deichmeister, I. I. Levkoev, and E. B. Lifshits, *Zh. Obshch. Khim.*, **23,** 1529 (1953);
[e] K. W. Hausser, R. Kuhn, and G. Seitz, *Z. Phys. Chem. B.*, **29,** 391 (1935).

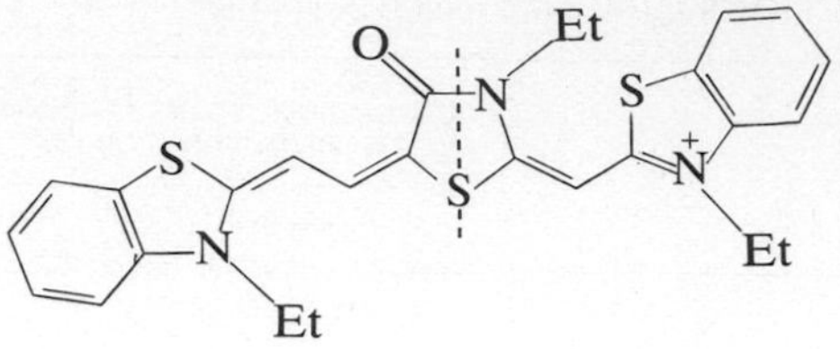

7 Merocyanine (**C**) Cyanine (**A**)

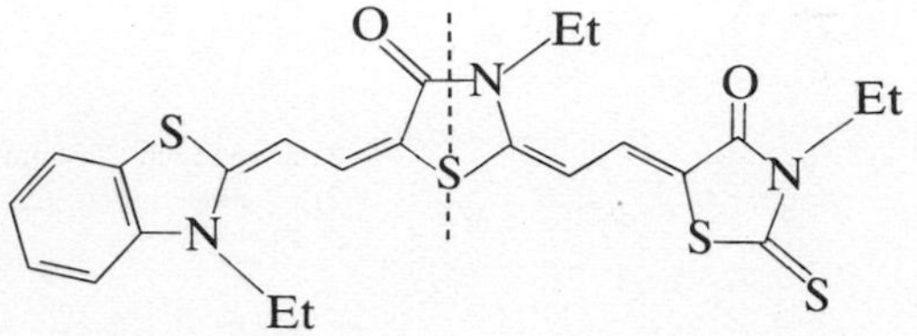

8 Merocyanine (**C**) Merocyanine (**C**)

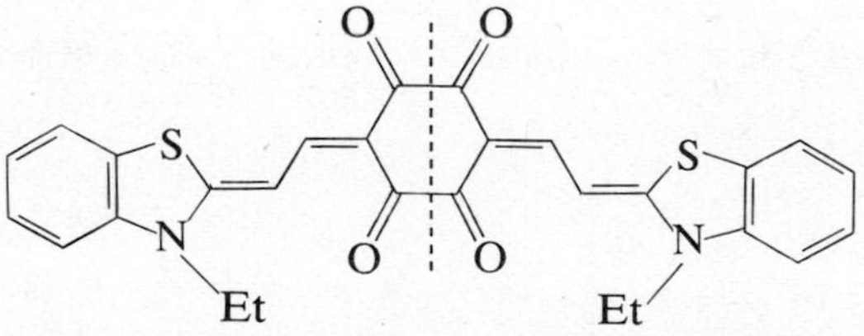

9 Merocyanine (**C**) Merocyanine (**C**)

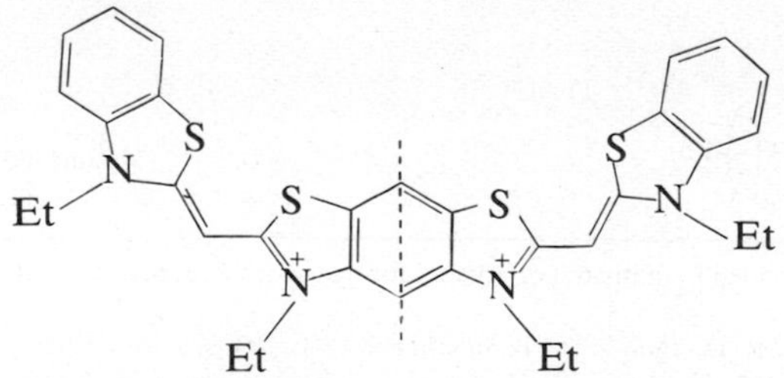

10 Cyanine (**A**) Cyanine (**A**)

cyanine). The last structure, **11,** incorporates an oxonol chromophore, which has common rings with two cyanine chromophores. Most dyes like these five are prepared using standard synthetic reactions[10] and multiple reactive sites on the common rings.

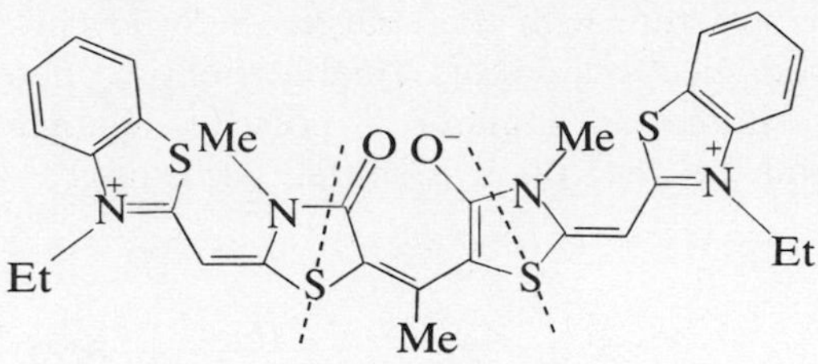

11 Cyanine (**A**) Oxonol (**B**) Cyanine (**A**)

B. Terminal Groups

Virtually any atom or group of atoms can function as terminal groups for dyes if the nitrogens and oxygens in the primary chromophores (**A**), (**B**), and (**C**) are replaced by electronically equivalent atoms. One example was given in formula **5** of Table II, where the $\rangle C{=}\ddot{N}-$ was substituted for the $\rangle C{=}O$ group in the dipolar amidic system (C). There are many other examples to be treated in detail later, but the fundamental concepts and perhaps the largest class of useful sensitizers are derived from dyes with heterocyclic terminal groups. The heterocycles are of two major types: (1) *basic* or electron donating and (2) *acidic* or electron accepting. These are treated separately below.

1. *Basic Heterocycles*

The quinoline dyes were studied extensively until 1920 when the work of Mills and Hamer focused attention on the superior benzothiazole dyes. Other heterocyclic thiazoles as well as the related oxazoles, pyrroles, and imidazoles were subsequently used for cyanines. In the late 1930s, it was clear to both Brooker and Kiprianov independently that certain unsymmetrical dyes absorbed at unexpectedly short wavelengths, whereas the absorption of others more closely approximated the mean wavelength for the related symmetrical dyes. These observations resulted in the concept of *deviation*, which related the absorption characteristics of unsymmetrical dyes to the electron-donating abilities ("basicities") of the various heterocycles. Brooker selected the *p*-dimethylaminophenyl group as a weakly basic terminal group and suggested that the importance of the two extreme amidinium resonance structures would depend on the "basicity" of the other terminal group. For the symmetrical dyes in the example below, the two structures are equivalent, but in the unsymmetrical "styryl" dye the structure with the charged heterocycle is favored, particularly for good electron donors (highly "basic" heterocycles). The resulting bond alternation induces a polyene character to the dye chromophore, and the absorption is shifted accordingly.

12

S N⁺ Et — NMe₂ ⟷ S N Et — N⁺Me₂

Unsymmetrical carbocyanine (a *styryl* dye)

13 $Me_2\overset{+}{N}$=⟨ ⟩=CH—⟨ ⟩—NMe_2 ⟷

Me_2N—⟨ ⟩—CH=⟨ ⟩=$\overset{+}{N}Me_2$

14

S N⁺ Et — S N Et ⟷

S N Et — S N⁺ Et

Related symmetrical dyes [Michler's hydrol blue (**13**) and Thiacarbocyanine (**14**)]

A quantitative expression of these observations was calculated by Eq. 1. The λ_{obs} is the observed absorption maximum for the unsymmetrical carbocyanine and λ_I is the arithmetic mean ("isoenergetic wavelength") for the absorption maxima of the related symmetrical dyes.[49,84] Many of the known deviations are listed in Table III, along with the structures of the heterocycles. The higher the deviation, the more readily the nucleus accommodates a positive charge.

$$\text{Deviation} = \Delta\lambda = \lambda_I - \lambda_{obs} \quad (1)$$

2. *Acidic Heterocycles*

A similar classification was made for the *acidic* electron-accepting terminal groups used in dipolar (merocyanine) chromophores.[52] The unsymmetrical dyes again incorporated the *p*-dimethylaminophenyl group, connected to the acidic group by one or three methine carbon atoms. For the unsymmetrical dye the nonpolar resonance form was expected to be the dominant one, except when highly electron accepting terminal groups were present. Nonpolar dyes exhibited the characteristics of polyenes, absorbing at much shorter wavelengths than expected from the arithmetic mean absorption for the two symmetrical dyes. Because of this, the least acidic (electron-accepting) groups showed the highest deviations. The least acidic heterocycles are characterized by rhodanines

15

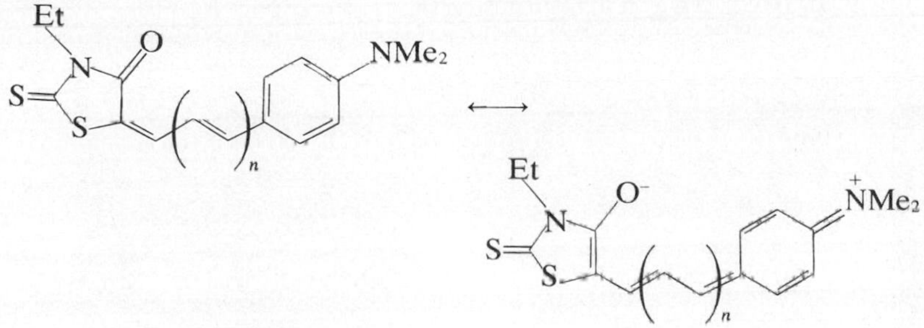

Unsymmetrical dye (a *benzylidene*, $n = 0$)

16

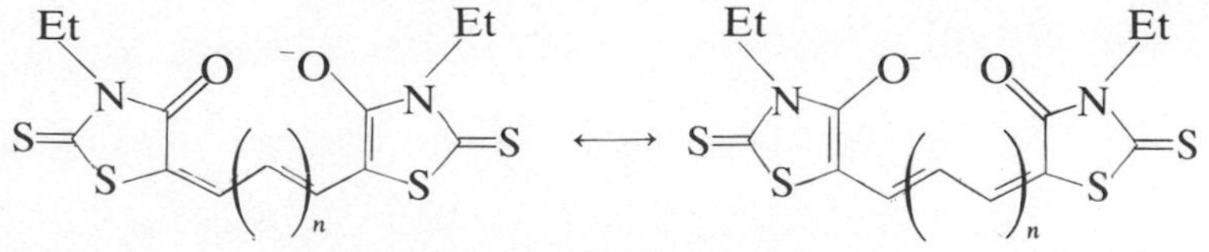

17

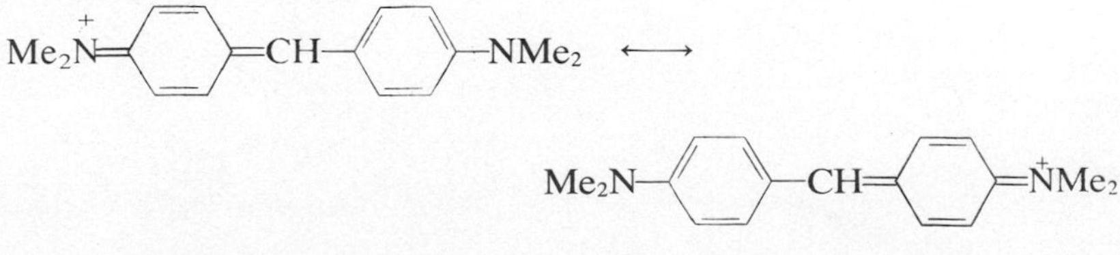

Related symmetrical dyes (an oxonol, $n = 0$, and Michler's hydrol blue)

TABLE III. Heterocycles and Deviations

A. Basic Heterocycles and B. Acidic Heterocycles

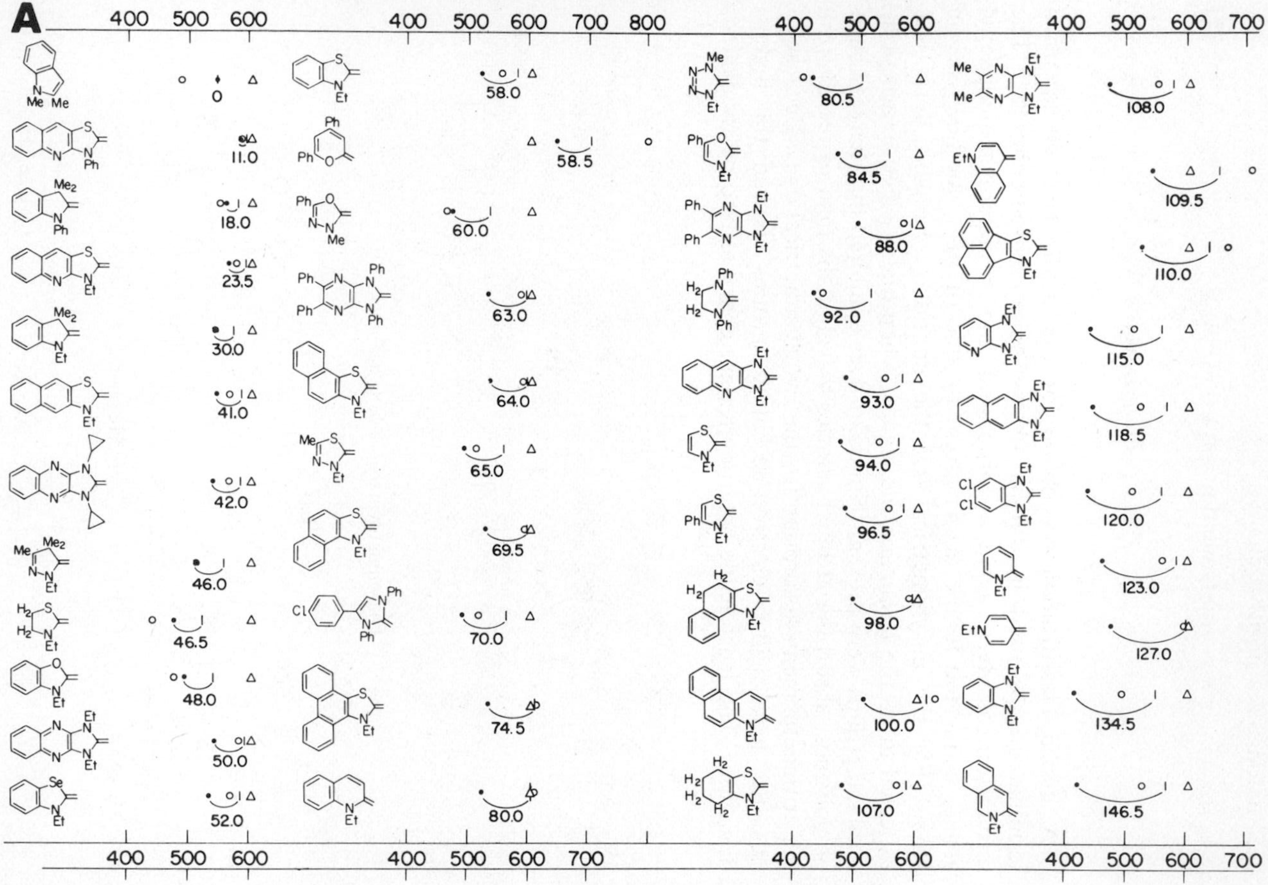

A

400 500 600	400 500 600 700 800	400 500 600	400 500 600 700
0	58.0	80.5	108.0
11.0	58.5	84.5	109.5
18.0	60.0	88.0	110.0
23.5	63.0	92.0	115.0
30.0	64.0	93.0	118.5
41.0	65.0	94.0	120.0
42.0	69.5	96.5	123.0
46.0	70.0	98.0	127.0
46.5	74.5	100.0	134.5
48.0	80.0	107.0	146.5
50.0			
52.0			

400 500 600 | 400 500 600 700 | 400 500 600 | 400 500 600 700

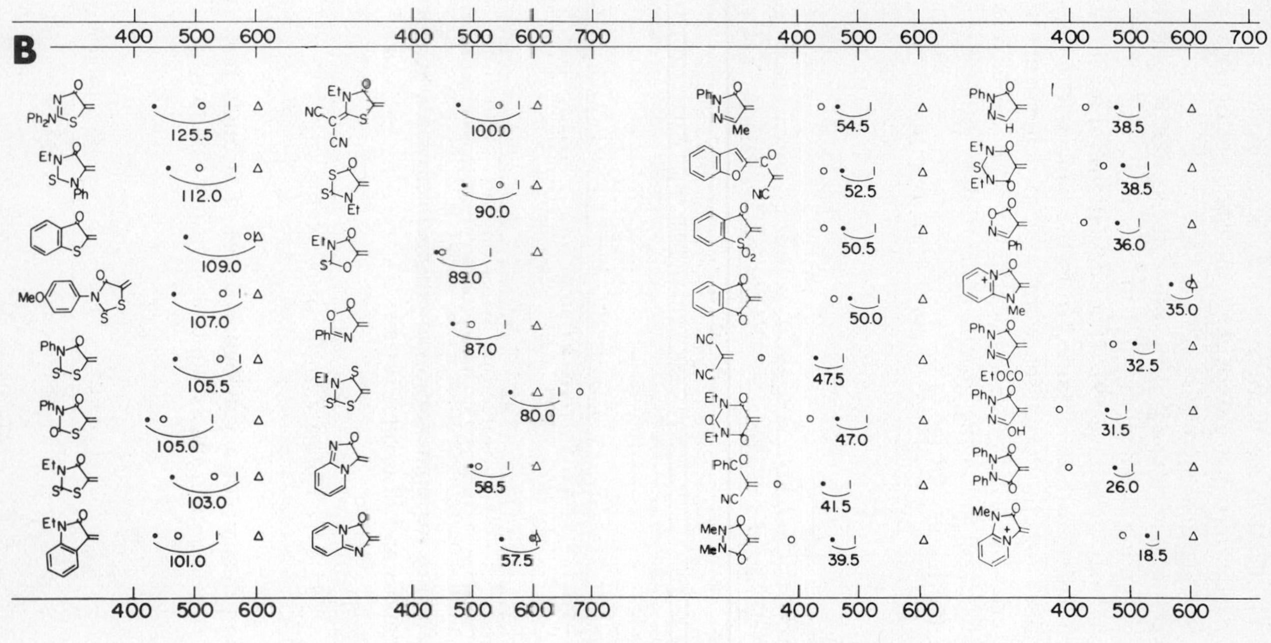

Symbols for absorption maxima: Δ = symmetrical *bis*(*p*-dimethylaminophenyl) cyanine ;
O = symmetrical carbocyanine (methine oxonol) from nucleus shown
● = (experimental) unsymmetrical styryl (benzylidene)
I = mean wavelength λ_I

and hydantoins (Table III); the more acidic are isoxazolones and pyrazolidinediones. Acidic terminal groups with intermediate deviations include indanediones and malononitrile.

Deviations were also estimated using other dyes, like the merocarbocyanines incorporating 3,3-dimethyl-3*H*-indoles (e.g., dye **18**) as well as the "cinnamylidenes" (e.g., dye **19**). These deviations for acidic groups were in a similar order to those obtained from the benzylidene dyes **15,** although the values for the deviations were solvent dependent. In the cinnamylidene series using dyes like **19,** deviations were less affected by the steric properties of the acidic groups compared to either the benzylidenes **15** or the indoles **18**.

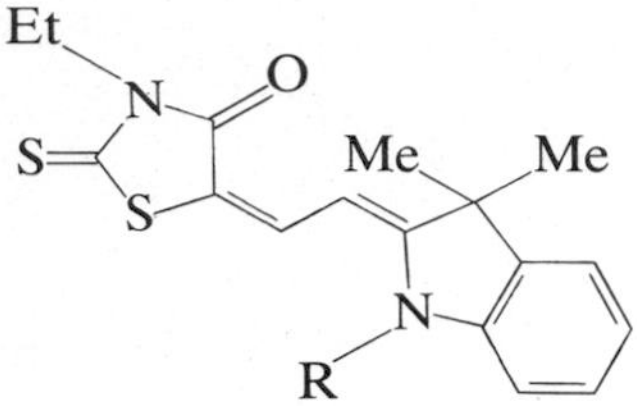

Et N O S S Me Me N R

18

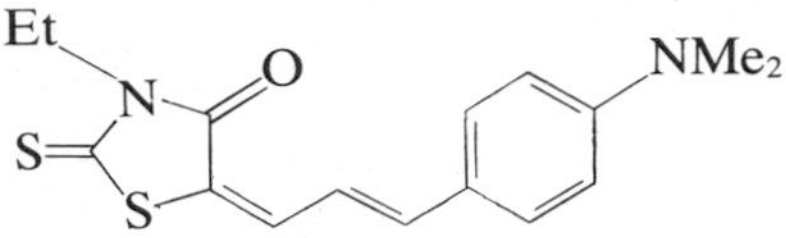

Et N O S S NMe_2

19

C. Solvent Effects

Solvent influences on absorption wavelength, extinctions, and bandshape are large for merocyanines and other dipolar dyes. Brooker summarized the major experimental observations in his later "Color and Constitution" papers and in certain reviews.

1. *Dye Polarity and Convergence*

The absorption maxima of a vinylogous series of strongly polar dyes (e.g., **20**) were convergent in highly aqueous pyridine solutions,

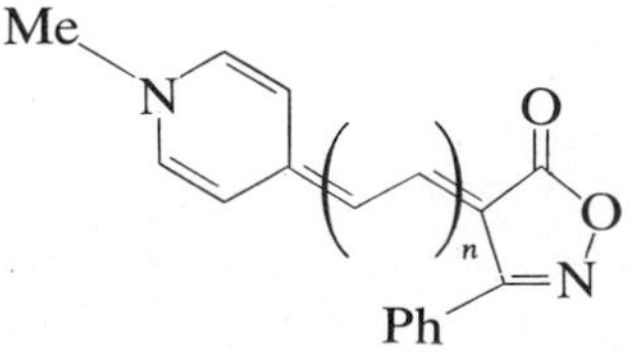

Me N O O N *n* Ph

20

	$n=0$	1	2	3	pyridine–water
λ_{max}	380	470	530	553	1:1
λ_{max}	400	515	610	710	1:0

but nonconvergent in pure pyridine.[52] A series of weakly polar dyes (e.g., **21**) showed the opposite behavior: nonconvergence in aqueous pyridine and convergence in anhydrous pyridine.[52] The nonconvergence typically shown by symmetrical dyes (e.g., cyanines and oxonols) has already been discussed (see Table II).

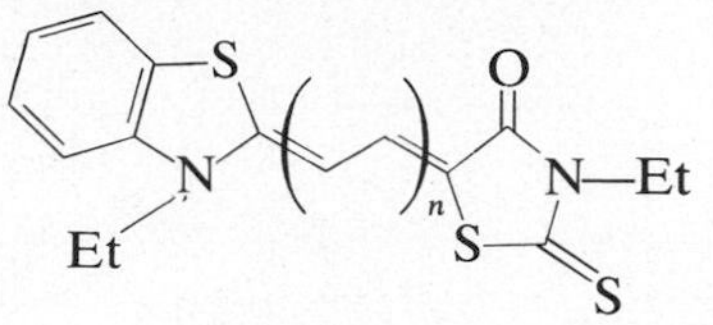

21

	$n = 0$	1	2	3	pyridine–water
λ_{max}	432	540	633	733	1:1
λ_{max}	432	528	605	635	1:0

2. *Dye Polarity and Extinction Coefficients*

Both wavelength and intensity are solvent dependent. For pyridine : water solutions, the solvent mixtures that gave the highest extinction coefficients for a dye were termed *isoenergetic points.*[53] Brooker

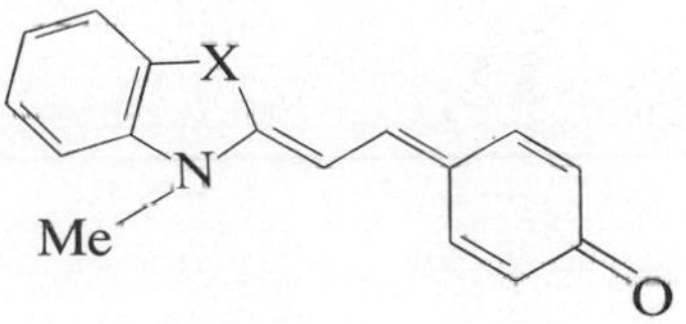

22

Isoenergetic points[53]

X = S, 92.5% pyridine–water (v/v)

X = CMe_2 50% pyridine–water (v/v)

assumed that the nonpolar and dipolar resonance structures in the amidic chromophore (**C**) had equal stabilities under these conditions. Lower extinctions in different pyridine–water mixtures were commonly accompanied by a shift of the maximum absorption to shorter wavelengths, particularly as the water content increased. For the two dyes listed (**22**; X = sulfur or CMe_2), the more basic nucleus (X = sulfur) requires more of the nonpolar solvent pyridine to reach the isoenergetic point.

3. *Dye Conformation and Solvent Polarity*[2]

The partly polar (allopolar) dyes like **23** exist in two distinct conformations (Fig. 2): one with complete charge separation (holopolar, **H**) and the other with at least one nonionized resonance form (meropolar, **M**). Major absorptions shown in Fig. 2 were assigned to these isomers, the holopolar form predominating in polar solvents. The totally planar structure **A** in

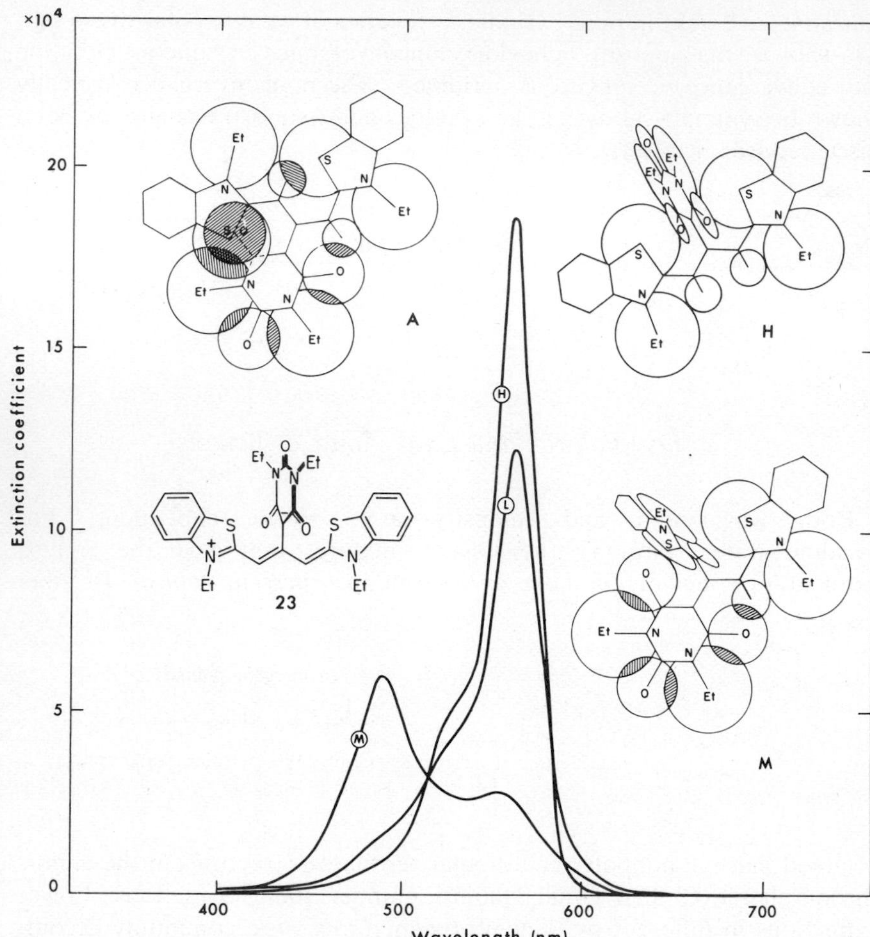

Fig. 2. Allopolar isomerism of Dye **23**. Projections **H** and **M** represent the holopolar and meropolar isomers, respectively. These isomers predominate in methanol and methylcyclohexane:lutidine (90:10) solutions, respectively, as indicated by the letters **H** and **M** on the absorption curves. The absorption curve in pure lutidine is marked L.

Fig. 2 is too crowded to exist, but Brooker postulated a propeller conformation, absorbing between the bands for the major isomers, in which all three ring systems were twisted.

D. Additional Structure-Property Relationships

The color-structure relations above are important in the design of sensitizers for specific wavelengths. In the normal visible spectrum (400–700 nm) compounds with a wide variety of structures will absorb at the

TABLE IV. SPECTRAL SENSITIZERS

Carbocyanine	Brooker deviation for heterocycle	Spectral sensitization
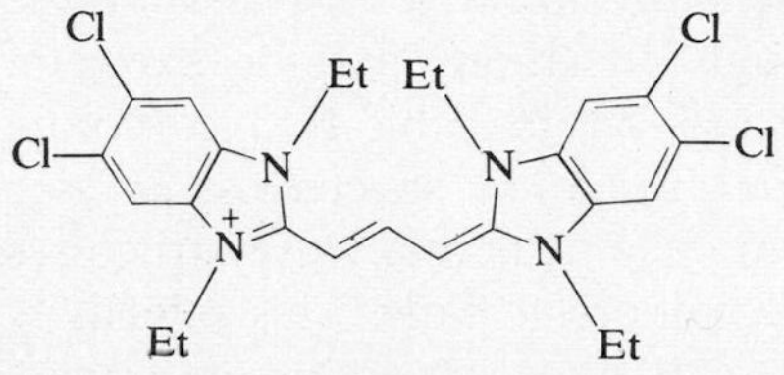 **24**	120	Good
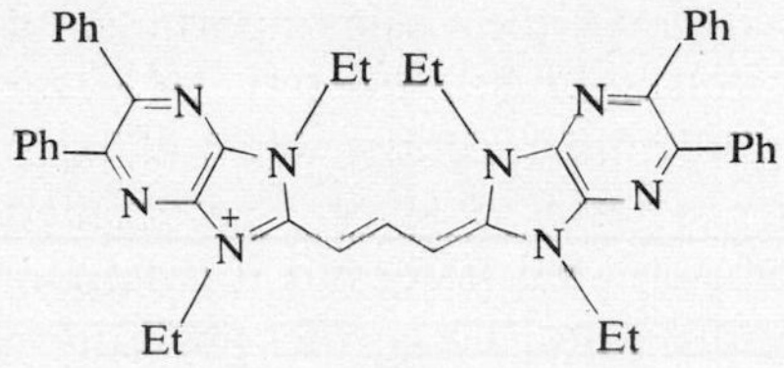 **25**	88	Moderate
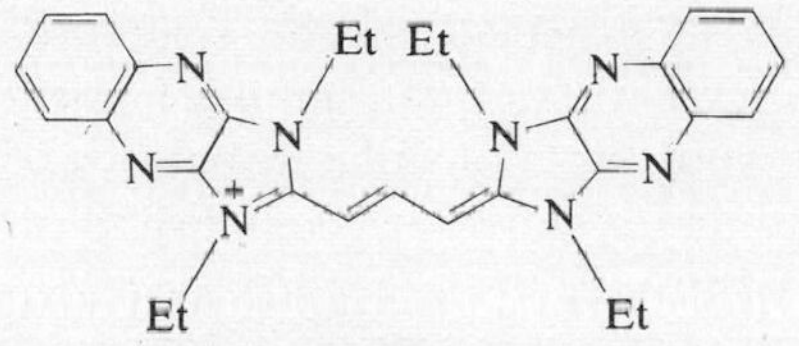 **26**	50	Very poor
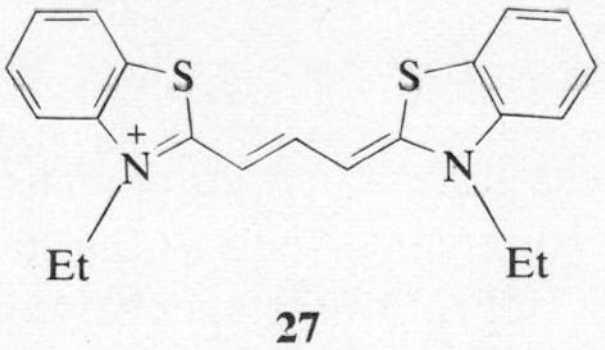 **27**	58	Moderate
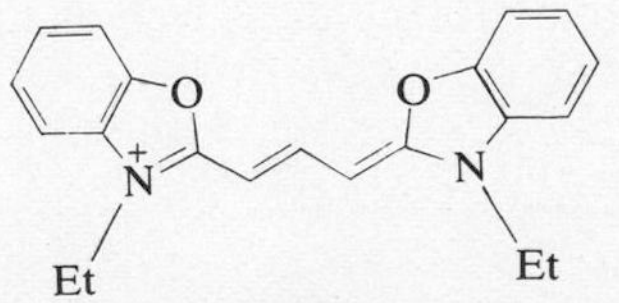 **28**	48	Good

wavelengths required for spectral sensitization. In the infrared prime sensitizers have come from the electronically symmetrical dyes that show no convergence in their absorptions as the length of the chromophoric chain is increased.

Brooker noted that within certain heterocyclic series increased deviation correlated with increased spectral sensitization (dyes **24–26** in Table IV), so that many heterocycles with high deviations provide excellent sensitizers. However, the last three dyes (**26–28**) in Table IV incorporate heterocycles with similar deviations, but range in spectral sensitizing power from very poor to good. Furthermore, of the many dye structures now available, only a few are used in commercial films. This selectivity suggests that properties other than color or Brooker deviation are important for optimum spectral sensitization.

In 1945 Sheppard and co-workers recognized the basic importance of the electrochemical properties of sensitizers (see Section III). However their significance was not fully appreciated until recently, when electrochemical potentials were available for large series of dyes. The detailed aspects of spectral sensitization will not be treated here, but it is important to note that effective sensitizers require both the correct absorption wavelengths and suitable electrochemical potentials (as well as certain desirable physical properties). The relations between structure and electrochemical potentials are an important new aspect of the structure-property study of sensitizing dyes. To the extent that quantum mechanical calculations can predict these potentials from the molecular structures of the dyes, the design of new sensitizers can be effectively accomplished on a computer, with subsequent syntheses of promising molecular systems.

III. Physical Characterization of Ground-State Dyes

A large number of cyanine and related dyes have been studied using x-ray crystallography and nuclear magnetic resonance as well as the more traditional electrochemical measurements, ir (vibrational) spectroscopy, protonation equilibria (pK_a), and dipole-moment determinations. Detailed information about the ground states of single dye molecules as well as the intermolecular arrangements has resulted.

A. X-Ray Crystallography

Crystal-structure analyses of cyanine and related dyes were reviewed by Smith.[87] Most typical sensitizers were nearly planar, with angles of less than 15° between planes defined by heterocyclic rings. These included: merocyanines from rhodanine,[88–92] thiacarbocyanines[93–96] (bromide,[93,94] *p*-toluenesulfonate,[94a] 7,7,8,8-tetracyanoquinodimethan complex,[95,96] or tetrachloroaurate[96a]), thiadicarbocyanine iodide,[97] thiacyanine **29** (bromide[98] or 7,7,8,8-tetracyanoquinodimethan complex[99]), phosphathiacyanine **30** perchlorate,[100] thiazolinocarbocyanine (iodide[101] or 7,7,8,8-tetracyanoquinodimethan complex[101a]), 5,5′,6,6′-tetrachlorobenzimidazolocarbocyanine **31** iodide,[102] the acetylenic dye **32** {5,6-dichloro-1,3-diethyl-2[(5,6-dichloro-1,3-diethyl-2-benzimidazolinylidene)-1-propynyl]benzimidazolium *p*-toluenesulfonate},[103] 5,5′-dichloro-3,3′,9-triethylthiacarbocyanine bromide,[104] 5,5′,7,7′-tetramethyl-3,3′,9-triethyl-8,9-*cis*-thiacarbocyanine perchlorate,[105] methylene blue,[106] 1,3-*bis*(dimethylamino)trimethinium perchlorate,[107,107a] 1,3-bis(pyrrolidino)trimethinium perchlorate,[107b] 1,5-bis(anilino)pentamethinium bromide,[108,108a] and 1,5-bis(dimethylamino)pentamethinium perchlorate[109] (chloride[109a]). The average bond lengths and bond angles for four of these dyes, **29–32,** and a nonplanar dye, **33,** are shown in Fig. 3. The bond lengths agreed with those observed for simpler molecules[103] and with theoretical bond orders.[102,110]

Certain other features of the structures should be noted. First, the short nonbonded S–S distances in the two simple cyanines from benzothiazole **29–30** suggested an attractive interaction. Second, considerable bond angle distortion occurred for the methine carbon atoms and the first ring carbon of the benzimidazole dyes **31–32**, contributing significantly to the relief of steric interactions between the heterocyclic ring systems. Third, distinct solvent of crystallization was present in most of the cationic dyes.

More highly twisted chromophores were observed by x-ray analyses of simple cyanines from quinoline[111–114] and imidazo[4,5-*b*]quinoxaline.[110] An unsymmetrical cyanine[111] from quinoline and benzothiazole with a methyl substituent on the methine carbon was twisted by about 60°. The symmetrical 1,1′-diethyl-4,4′-quinocyanine bromide,[112] 1,1′-diethyl-2,2′-quinocyanine (bromide,[112] chloride[113]) and 1,1′-diethylphospha-2,2′-quinocyanine perchlorate[114] were also twisted by 40 to 60° between the planes of the heterocyclic rings. A twist angle of 55° was reported for the imidazo[4,5-*b*]quinoxalinocyanine **33** iodide.[110] For most of these nonplanar dyes, steric interactions were also relieved by bond-angle distortion at the methine carbon. However nearly equivalent bond lengths in the two heterocyclic rings for each symmetrical cyanine suggested that the charge on the dye was delocalized over the chromophore in spite of the molecular distortions.[87]

X-Ray crystal analyses also provided intermolecular data. Because of photographic use of J-aggregated cyanine and carbocyanine dyes, the

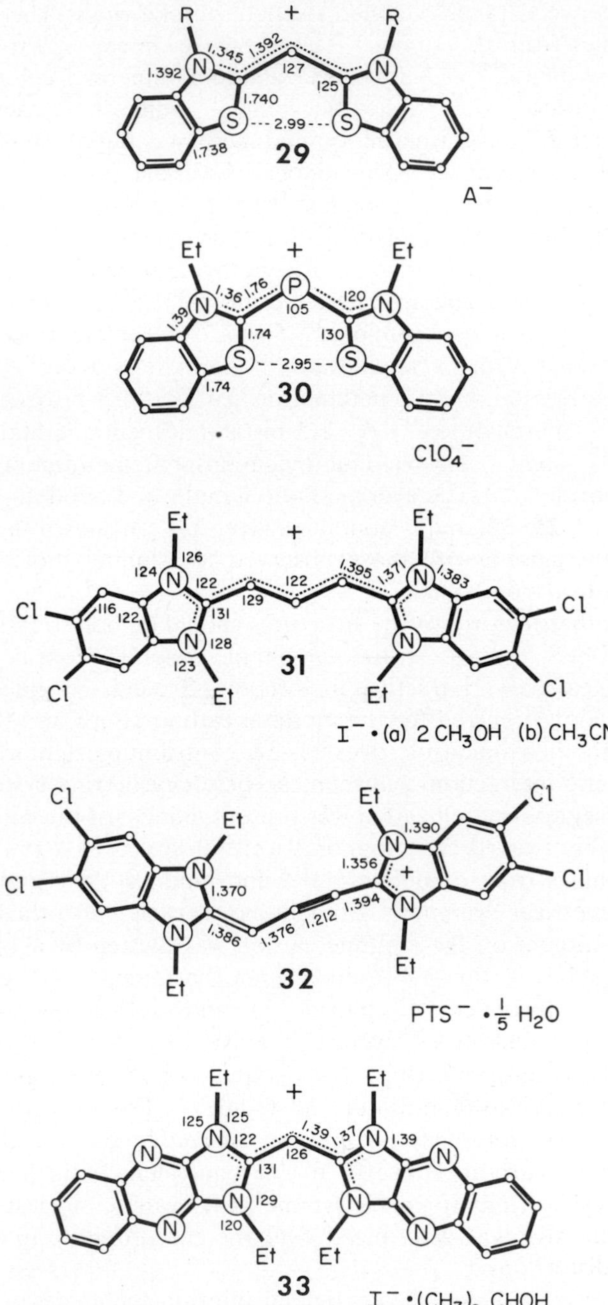

Fig. 3. x-Ray crystal structures[87] for planar simple cyanines **29–30**, a planar carbocyanine **31**, an acetylenic dye **32**, and a twisted simple cyanine **33**.

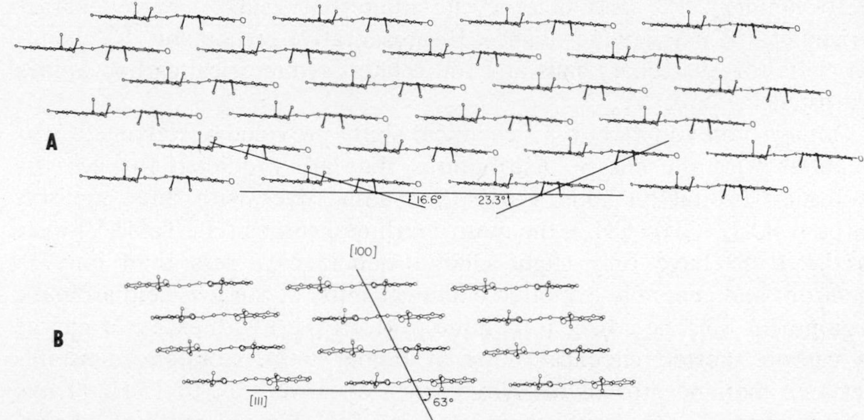

Fig. 4. Cation–cation arrangements[87] for the acetonitrile solvate of the tetrachlorobenzimidazolocarbocyanine **31** iodide (A) and the ethanol solvate of 3,3′-diethylthiacarbocyanine bromide (B).

cation–cation arrangements of most interest have been those for 1,1′-diethyl-2,2′-quinocyanine chloride,[113] 5,5′,6,6′-tetrachloro-1,1′,3,3′-tetraethylbenzimidazolocarbocyanine iodide,[102] and 5,5′-dichloro-3,3′,9-triethylthiacarbocyanine bromide.[104] The heterocyclic rings of these three dyes had plane-to-plane orientations, and interplanar distances of 3.3 to 3.6 Å. The lateral shift of adjacent dye chromophores in the benzimidazole dye was 12 to 13 Å (Fig. 4A). The atoms in the quinocyanine dye exhibited a more complex relationship, where the highly twisted chromophore introduced an inversion center relating atoms of adjacent chromophores. In contrast to this, 3,3′-diethylthiacarbocyanine[93,94] (Fig. 4B) with similar interplanar distances (~3.6 Å), exhibited a lateral shift of adjacent chromophores of only 4.05 Å. Smith[87] concluded that the intermolecular geometry for most planar dyes was a layered arrangement with a plane-to-plane stacking of the cations on edge and with a lateral displacement of adjacent chromophores that was highly influenced by substituents.

B. Nuclear Magnetic Resonance

Proton chemical shifts and coupling constants have been tabulated for many symmetrical cyanines and carbocyanines,[115–128] unsymmetrical carbocyanines,[127,128] merocyanines,[117,129–132] oxonols,[122] protonated dyes,[133,134] dye intermediates,[135–138] and europium complexes of merocyanines,[139] anils,[140] and cyanines.[141] Certain cyanines,[142–145] merocyanines,[142,143] and heterocyclic quaternary salts[146] were characterized by ^{13}C magnetic resonance. Representative proton and ^{13}C chemical shifts for quaternary salts and the related symmetrical carbocyanines are listed in Table V.

Both proton and carbon-13 chemical shifts provided approximate correlations with the charge distributions that are predicted for dyes by molecular orbital methods. For carbocyanine dyes with three methine carbons (CH_α–CH_β–CH_α), the meso-methine protons (H_β, Table V) were predicted to have only slight charge deficiencies (less than one π-electron) and generally exhibited chemical shifts in the low-field aromatic region near 8δ. The adjacent methine protons (H_α) with excess π-charge at carbon showed chemical shifts at higher field. Coupling constants between methine protons for symmetrical dyes were 12 to 15 Hz (Table V), consistent with trans conformations for the polymethine chains. However, merocyanines exhibited unequal coupling constants, and both coupling constants and chemical shifts changed as a function of solvent polarity.[117,130] The increased polyene character of merocyanines in CS_2

TABLE V. NUCLEAR MAGNETIC RESONANCE DATA FOR PROTONS[a] AND HETEROCYCLIC RING CARBONS[b] (AT C_2) FOR QUATERNARY SALTS AND SYMMETRICAL CARBOCYANINE DYES

Heterocycle (2-methyl-3-ethyl quaternary salt structure: X, CH_3, $\overset{+}{N}$, Et)	Brooker deviation (Table III)	Quaternary salts		Carbocyanine dyes			
		2-CH_3	$^{13}C_2$	$^{13}C_2$	H_α	H_β	$J_{\alpha\beta}$
X = CMe_2	30	—	190.7	173.34	—	—	—
X = O	48	3.12	166.7	161.19	6.13	8.30	13.5[c]
X = Se	52	3.14	187.7	168.68	6.69	7.14	12.5
X = S	58	3.22	176.3	163.98	6.64	7.79	13.0
X = CH = CH	80	3.10	160.3	151.77	—	—	—
X = N-Et-(5,6-dichloro-)	120	—	153.3	—	5.93	7.84	13.3[d]
X = N-Et-(5-cyano-)	—	3.06	153.9	—	6.04	8.13	13.3
X = N-Et	134	3.04	151.6	147.52	—	—	—

[a] Proton chemical shifts in δ (ppm) and coupling constants $J_{\alpha\beta}$ in hertz from E. Kleinpeter and co-workers, *J. Prakt. Chem.*, **315,** 587, 600 (1973).

[b] Carbon-13 chemical shifts adjusted to δ_{TMS} (ppm) from E. Kleinpeter and co-workers, *J. Prakt. Chem.*, **315,** 765 (1973), for quaternary salts and from P. M. Henrichs, *J. Chem. Soc. Perkin* **2,** 542 (1976) for carbocyanine dyes.

[c] The 5,5′-diphenyl analog.

[d] *N*-ethyl was replaced by *N*-butyl/*N*-octyl on each heterocycle.

(relative to D_2O) was reflected by larger coupling constant variation across the polymethine chain.[117] A general relation between bond length and coupling constant was also proposed.[117] Chemical shifts and coupling constants for the aromatic protons on the heterocyclic rings were also assigned for many dyes and quaternary salts.[115–117,126–128,131,137]

A number of other phenomena were analyzed via nuclear magnetic resonance. First, the sites of protonation for vinylogous cyanines were established as the first methine carbon atom (Eq. 2),[133] while a simple

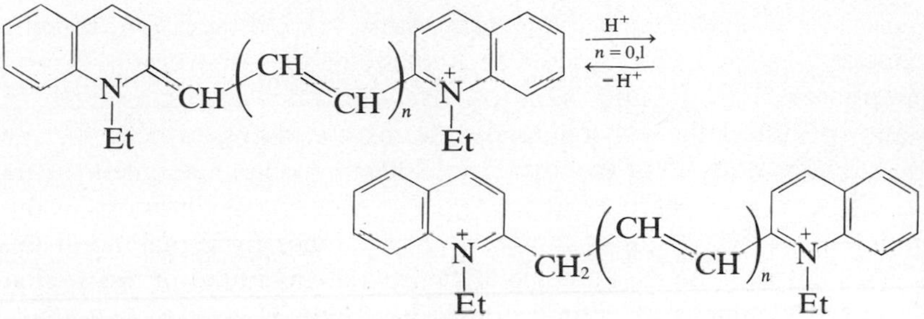

(2)

oxonol from rhodanine was suggested to protonate at a tertiary ring carbon (Eq. 3).[134] Second, N–Me groups for many symmetrical simple cyanines showed single absorptions, and symmetric conformations with the N–Me groups on the same side of the dye molecule (*syn*) were suggested.[125,126] Azacyanine analogs as well as pyridocyanines showed

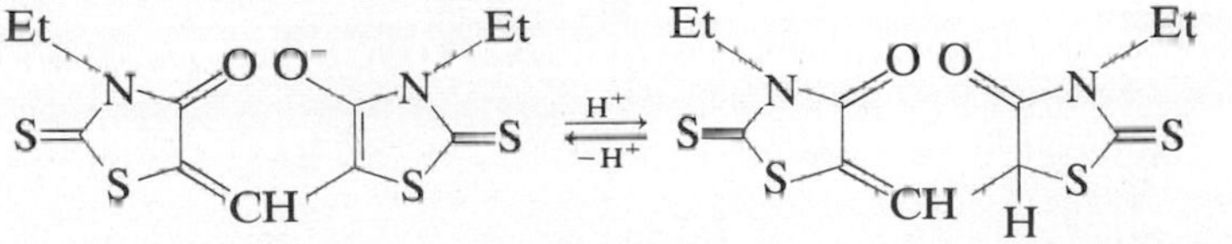

(3)

two N–Me absorptions (nonsymmetric conformations, *anti*)[125] but only one set of absorptions in the N–Et cyanines (symmetric conformations, *syn*).[126] Third, experimental activation energies for the rotation of simple amine groups R_2N were determined as a function of ring size (cyanines)[116,144], and solvent polarity (merocyanines).[129] Fourth, activation energies for proton-deuterium exchange and carbon-13 chemical shifts in methyl quaternary salts paralleled the calculated basicities of the heterocycles.[137,138,146] Fifth, europium shift reagents exhibited the expected binding to merocyanines[139] and anils.[140] In addition, these reagents showed distinct effects on the spectra of cationic species[141] like 1,1′-diethyl-2,2′-quinocyanine and the 1-ethylquinolinium ion, and the nature of the anion affected the magnitude of the induced shifts. Sixth, proton chemical shifts for dimerized 1,1′-diethyl-2,2′-quinocyanine were consistent with calculated chemical shifts if maximum overlap of the quinoline rings was assumed. A model for the dimer was proposed based on these results and a nonplanar geometry for each cyanine cation.[147]

C. Infrared Spectroscopy

Infrared vibrational spectra have been published for several cyanine dye series[148–154] and related quaternary salts.[153,155] In addition, specific frequencies have been, used to characterize "merocyaninelike" chromophores[148,156–159] and monitor dye-substrate interactions.[160–162] Only the spectra of the vinylogous dye series are discussed here.

An extensive study[148] of the 1400 to 1700 cm^{-1} region assigned strong vibrations between 1480 to 1590 cm^{-1} to C=N or C=C vibrations of the chromophores of vinylogous cyanines, merocyanines from malononitrile, and malononitrile oxonols. A single strong band was found in this region except for unsymmetrical carbocyanine dyes like **34** where two bands

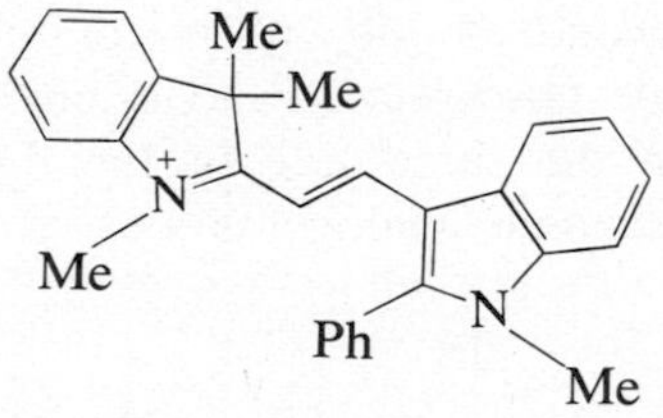

34

were noted. Other bands were assigned to aromatic C=C bonds (1590–1650 cm^{-1}) and to C–H vibrations (1400–1480 cm^{-1}). Other studies of this region were done for pyridocyanines.[154] The out-of-plane (γ) vibrations for aromatic hydrogens (700–900 cm^{-1}) for a number of quinoline dyes were also reported.[149]

Complete infrared vibrational spectra for vinylogous cyanine dyes from seven basic heterocycles were published by Leifer and co-workers.[150–152] Assignments of the bands were as follows: (1) resonant-conjugated unsaturated C=C (C=N) stretching modes in the chromophore (1580–1380 cm^{-1}), generally three bands except for the simple cyanines; (2) aromatic stretching (1590–1630 cm^{-1} and 1450–1490 cm^{-1}); (3) chromophoric CH out-of-plane bending (940–1030 cm^{-1}) except for simple cyanines; and (4) aromatic CH (730–790 cm^{-1}).

Polarized ir vibrations were determined for dyes in polymer films[163] and J-aggregates adsorbed to silver halide.[164]

D. Reduction and Oxidation

Early work by Sheppard and co-workers[165] related the redox properties of dyes to their photographic effects. This was followed by extensive investigations of electrochemical reduction and oxidation potentials as well as direct ionization-potential and electron-affinity determinations. All of these measurements essentially calibrated dyes as molecular probes for the study of solid substrates like silver halides as well as for studies on substituted carboxylate radicals from pulse radiolysis.[166,167]

Polarographic halfwave potentials for the reduction of dyes (E_{red}) were initially published for symmetrical cyanine dyes,[168–179] triphenylmethane dyes,[180–183] other cationic dyes[171–174,184] (e.g., phenosafranine and acridine orange), and protonated dyes.[185] Oxidation potentials (E_{ox}) were also published for symmetrical cyanines.[186–188] An extensive tabulation of reduction and oxidation potentials determined under constant experimental conditions was published by Large[189] (Table VI), and additional examples of substituent effects on reduction potentials were published.[190]

Although many correlations were established using electrochemical data, the review by Large suggested that completely reversible electrochemical potentials were rare. Some apparent deviations[189] were caused by (1) specific substituents like nitro, which did not follow an otherwise satisfactory relation between Hammett σ-constant and reduction potential for ring-substituted thiacarbocyanines; and (2) increased chain length for a vinylogous series of dyes. In the latter case, it was well known that the absorption maxima for dyes ($E_{h\nu}$ in electron volts) could be linearly related to the quantity [$\Delta E = (E_{ox} - E_{red})$ in volts], but the slope was significantly different from 1.0. Stability and absorption limits for sensitizers were suggested to be near 1300 nm based on this anomalous relation.[178,179] However, an inspection of the $E_{h\nu}$ and ΔE values in Table VI shows greater differences between $E_{h\nu}$ and [$E_{ox} - E_{red}$] as the chain length increases,[189] and for quinoline dyes this was ascribed to irreversible electrochemical behavior for the carbocyanine and dicarbocyanine during the oxidation process.[189] Thus available electrochemical data as measures of ground or excited states of dye molecules should be treated with some caution, and more reversible data would be desirable.

Other estimates of redox properties of dyes have resulted from ionization-potential and electron-affinity data. Indirect measures of these were calculated from oxidation potentials,[191,192] dye-quinone charge-transfer absorption wavelengths,[193] and reduction potentials.[171] More

TABLE VI. ELECTROCHEMICAL DATA ON HOMOLOGOUS SERIES OF CYANINE DYES[a,b]

Nuclei	n	E_{ox}(v)	E_{red}(v)	ΔE^c	λ_{max} (nm)	$E_{h\nu}$ (eV)
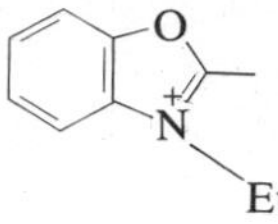 O, N+, Et	0	>1.0	−1.69	>2.69	372	3.33
	1	0.94	−1.26	2.20	485	2.56
	2	0.49	−1.16	1.65	580	2.14
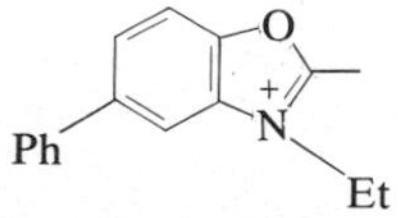 Ph, O, N+, Et	0	>1.0	>1.16	>2.2	387	—
	1	0.90	−1.10	2.00	496	2.50
	2	0.44	−1.06	1.50	593	2.09
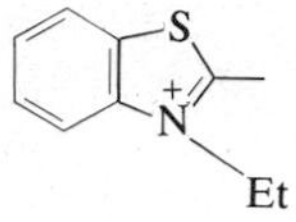 S, N+, Et	0	>1.2	−1.38	>2.38	422	2.94
	1	0.75	−1.00	1.75	557	2.23
	2	0.48	−0.83	1.31	650	1.91
	3	0.26	−0.72	0.98	762	1.63
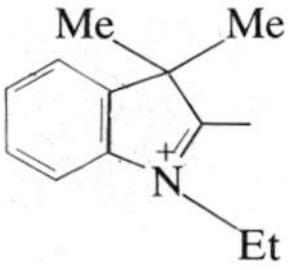 Me, Me, N+, Et	1	>1.0	−1.06	>2.06	545	2.27
	2	0.68	−0.87	1.55	636	1.95
	3	0.48	−0.66	1.14	745	1.66
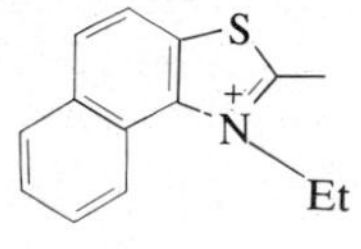 S, N+, Et	0	0.95	−1.60	2.55	456	2.72
	1	0.62	−1.12	1.74	597	2.08
	2	0.29	−0.99	1.28	691	1.79
	3	0.19	−0.77	0.96	800	1.55
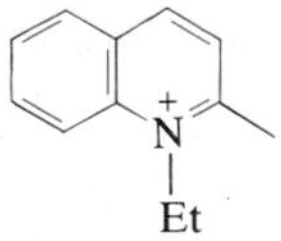 N+, Et	0	0.99	−1.03	2.02	522	2.34
	1	0.58	−1.10	1.68	605	2.05
	2	0.28	−0.98	1.26	708	1.75
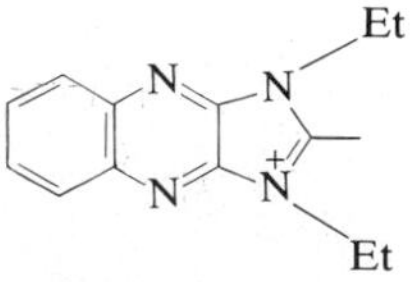 Et, N, N, N+, N, Et	0	>1.0	−0.90	>1.90	474	2.62
	1	1.00	−0.79	1.79	581	2.13
	2	0.65	−0.60	1.25	688	1.80
	3	0.41	−0.58	0.99	795	1.56
	1	1.00	−1.32	2.32	445	2.79

TABLE VI (*Continued*)

Nuclei	n	E_{ox}(v)	E_{red}(v)	ΔE^{c}	λ_{max} (nm)	$E_{h\nu}$ (eV)
Ph, N, Me	1	>1.0	−0.20	>1.20	587	2.11
S, +, N, Ph, Et	1	0.54	−1.35	1.89	559	2.22
Et, N, +, N, Et	0	1.00	−2.20	3.20	396	3.13
	1	0.37	−1.78	2.15	496	2.50
Cl, Cl, Et, N, +, N, Et	1	0.58	−1.50	2.08	514	2.41
	2	0.27	−1.28	1.55	610	2.06
Se, +, N, Et	0	>1.0	−1.58	>2.58	429	2.89
	1	0.62	−1.02	1.64	570	2.18
	2	0.46	−0.84	1.30	660	1.88
	3	0.30	−0.64	0.94	770	1.61
Et—N+	0	0.64	−1.14	1.78	593	2.09
	1	0.19	−0.99	1.18	704	1.76
Et—N+	1	0.13	−1.41	1.54	603	2.06

[a] R. F. Large, in "Photographic Sensitivity" (Cambridge), R. J. Cox, Ed., Academic Press, New York, 1973, p. 241. All data were obtained in methanol solutions. Cathodic data (E_{red}) were obtained at a dropping mercury electrode, and anodic data (E_{ox}) were obtained at a pyrolytic graphite electrode. The reference electrode was an aqueous saturated silver, silver chloride electrode. More negative E_{red} values indicate more difficult reduction of the ground-state dyes. More positive E_{ox} values indicate more difficult oxidation of the ground-state dyes. Each dye contains identical terminal nuclei, connected by one, three, five, or seven methine carbons ($n = 0$, 1, 2, 3, respectively).

[b] *N*-Ethyl derivatives of all dyes were not available and other derivatives have been substituted in some cases. The general conclusions from the data are not altered by such substitution of dyes.

[c] ΔE is the absolute difference between the oxidation and reduction potentials.

TABLE VII. IONIZATION POTENTIALS (IP) AND ELECTRON AFFINITIES (EA)

Dye structure	n	Threshold values			Shoulder	Dye Structure	Shoulder	Dye Complex[d]
		EA_{VAC}[a] (eV)	IP_{VAC}[b] (eV)	IP_{CdS}[b] (eV)	IP_{CdS}[c] (eV)		IP_{CdS}[c] (eV)	IP
A. Vinylogous dyes						B. Misc. dyes		
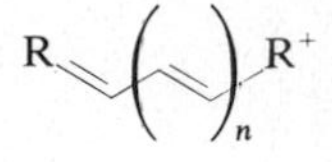	n							
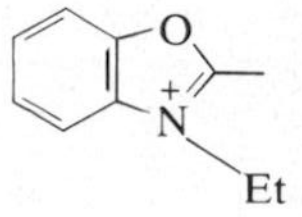	2	—	4.99	4.68	—	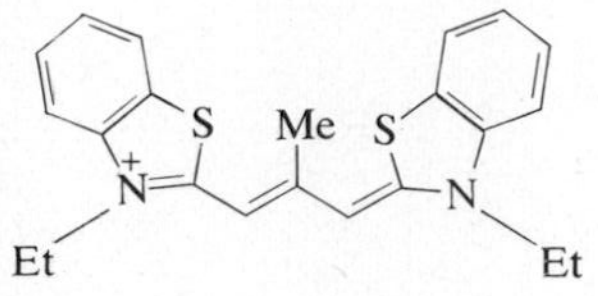	5.4	—
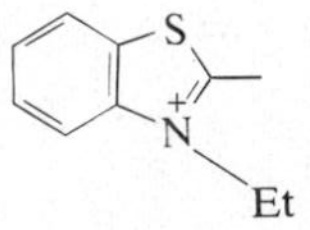	1	—	—	—	5.68			
	2	2.70	4.98	4.58	5.3	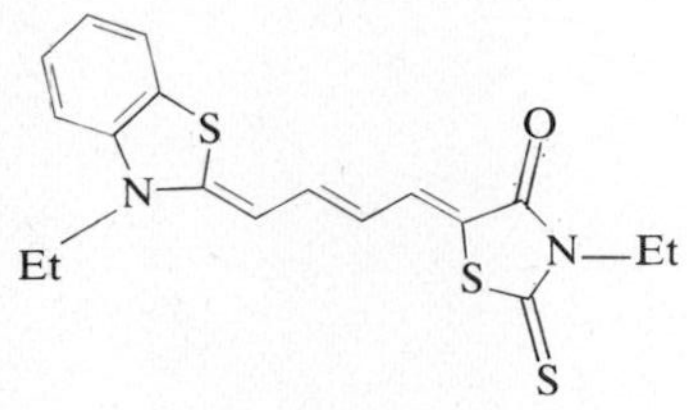	5.3	—
	3	2.73	4.84	4.50	5.05			

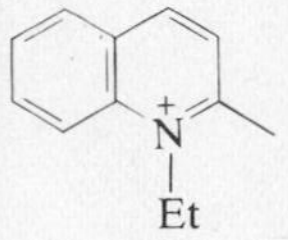	0	—	5.20	4.85	—
	1	2.74	4.96	4.63	5.35
	2	—	4.79	4.47	—
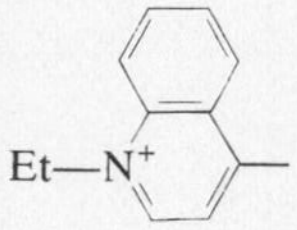	1	—	4.81	4.26	—

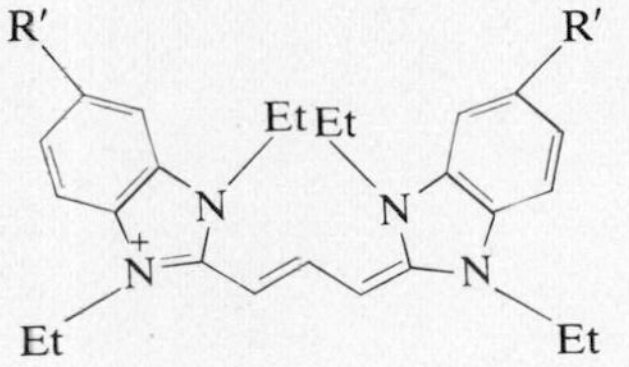

R′		
H	—	6.37
Cl	—	6.40
COMe	—	6.44
CN	—	6.46
SO_2Me	—	6.50
NO_2	—	6.50
SO_2CHF_2	—	6.52

[a] Electron-beam retardation, corrected to a vacuum (VAC) dielectric [J. W. Trusty and R. C. Nelson, *Photogr. Sci. Eng.*, **16**, 421 (1972)].
[b] External photoelectric effect for CdS substrate and corrected to a vacuum dielectric [R. G. Selsby and R. C. Nelson, *J. Mol. Spectrosc.*, **33,** 1 (1970)].
[c] External photoelectric effect [P. Yianoulis and R. C. Nelson, *Photogr. Sci. Eng.*, **18,** 94 (1974)].
[d] Absorption measurements on dye/quinone charge-transfer complexes [J. Nys and W. VandenHeuvel, *Photogr. Korresp.*, **102,** 37 (1966)].

TABLE VIII. DIPOLE MOMENTS (BENZENE, 25°C)

Dye structure		μ(D)	Ref.
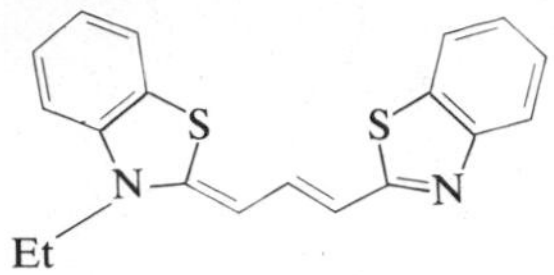		4.51	[a]
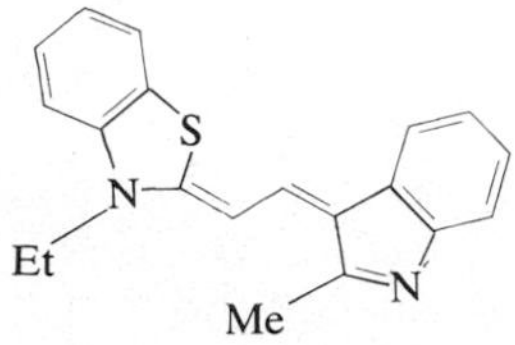		7.68	[a]
	$n=0$	2.37	[b]
	$n=1$	4.17	[b]
	$n=2$	5.32	[b]
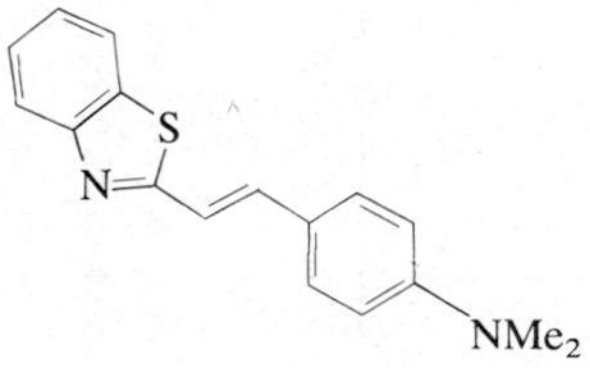		3.59	[c]
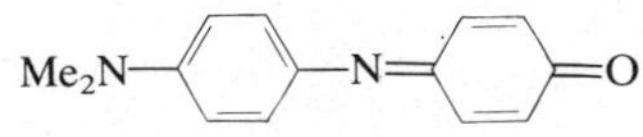		5.80	[d]
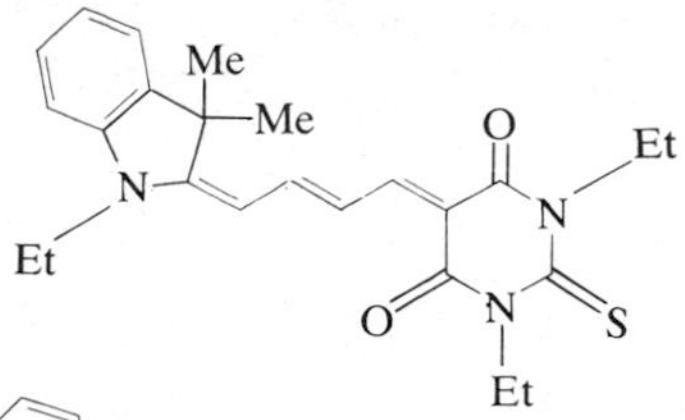		9.7	[e]
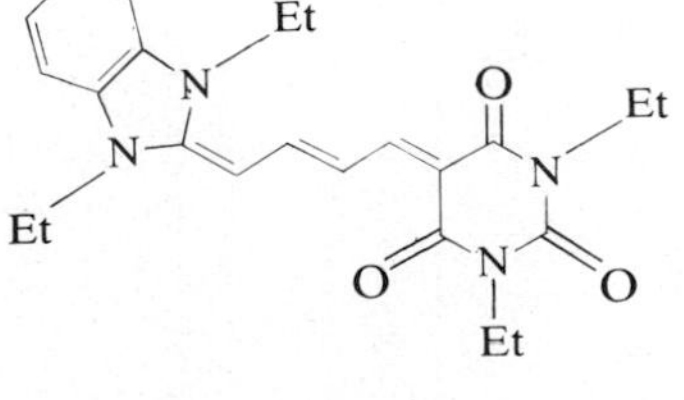		17.7 (dioxane)	[e]

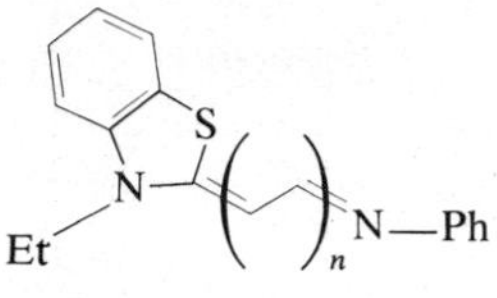

TABLE VIII (*Continued*)

Dye structure		μ(D)	Ref.
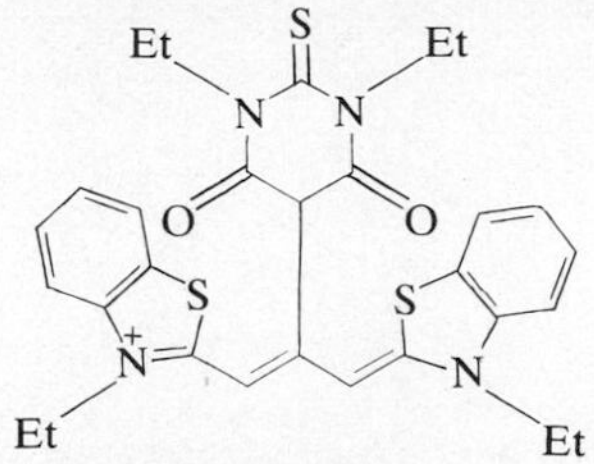 [Structure: merocyanine with N,N'-diethylthiobarbituric acid, two N-ethylbenzothiazole groups]		13.3 (dioxane)	f
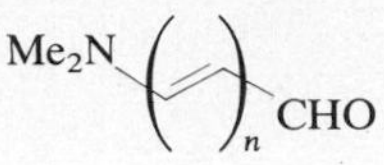 $Me_2N(CH=CH)_nCHO$	$n=0$	3.86	g
	$n=1$	6.24	g
	$n=2$	7.67	g
	$n=3$	8.24	g
	$n=4$	8.50	g
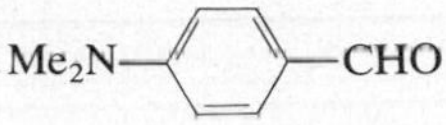 $Me_2N-C_6H_4-CHO$		5.60	g
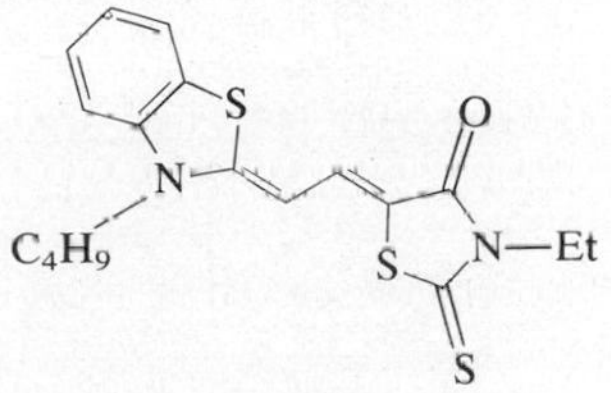 [Structure: N-C_4H_9 benzothiazole merocyanine with N-Et rhodanine]		7.68	h
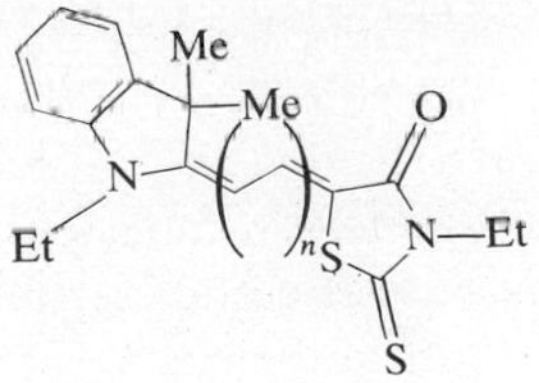 [Structure: N-Et 3,3-dimethylindoline merocyanine with N-Et rhodanine]	$n=1$	6.27	h
	$n=2$	6.91	e
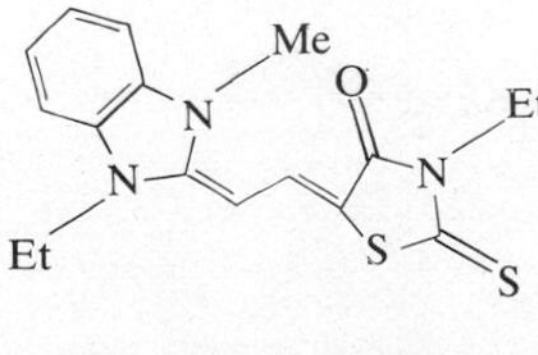 [Structure: 1-Me-3-Et benzimidazole merocyanine with N-Et rhodanine]		9.9	h
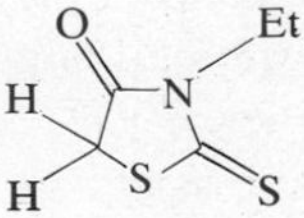 [Structure: N-ethylrhodanine]		1.75	h

TABLE VIII (*Continued*)

Dye structure	μ(D)	Ref.
1-ethyl-3,3-dimethyl-2-methyleneindoline (Me, Me, =CH$_2$, N–Et)	1.13	*h*
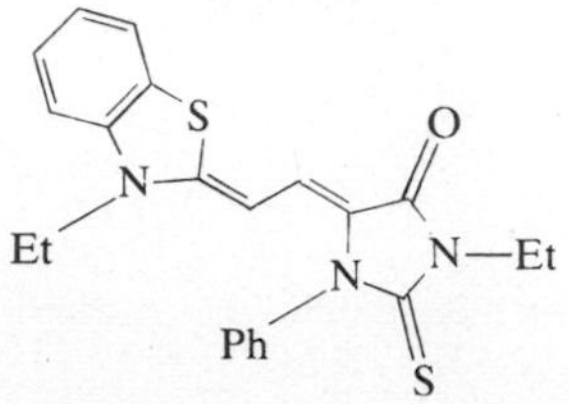 merocyanine: 3-ethylbenzothiazolinylidene-ethylidene linked to 1-phenyl-3-ethyl-2-thiohydantoin (S, N–Et, O, N–Et, N–Ph, =S)	7.35	*i*

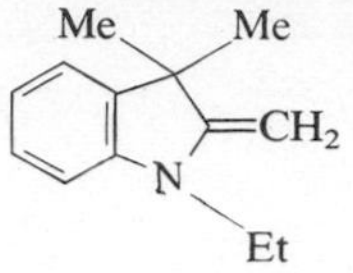

[a] L. G. S. Brooker, R. H. Sprague, C. P. Smyth, and G. L. Lewis, *J. Am. Chem. Soc.*, **62**, 1116 (1940).
[b] L. G. S. Brooker, F. L. White, G. H. Keyes, C. P. Smyth, and P. F. Oesper, *ibid.*, **63**, 3192 (1941).
[c] L. G. S. Brooker and R. H. Sprague, *ibid.*, **63**, 3203 (1941).
[d] L. G. S. Brooker and R. H. Sprague, *ibid.*, **63**, 3214 (1941).
[e] L. M. Kushner and C. P. Smyth, *ibid.*, **71**, 1401 (1949).
[f] A. J. Petro, C. P. Smyth, and L. G. S. Brooker, *ibid.*, **78**, 3040 (1956).
[g] M. Hely Hutchinson and L. E. Sutton, *J. Chem. Soc.*, 4382 (1958).
[h] E. A. Shott-L'vova, Ya. K. Syrkin, I. I. Levkoev, and Z. P. Sytnik, *Dokl. Akad. Nauk SSSR*, **116**, 804 (1957); *Chem. Abstr.*, **52**, 4361i (1958).
[i] E. A. Shott-L'vova, Ya. K. Syrkin, I. I. Levkoev, and M. V. Deichmeister, *Dokl. Akad. Nauk SSSR*, **145**, 1321 (1962); *Chem. Abstr.*, **58**, 8062c (1963).

direct determinations were published by Vilesov[194] and Nelson and co-workers.[195–206] The early work of Nelson provided the general shape of the photoelectric yield curve (from which threshold ionization potentials were taken),[196] the dependence of peak photoelectric current on dye concentration,[196] the influence of substrate (optical dielectric constant K)

$$IP_{sub} = IP_{vac} - \frac{3.6}{D} \cdot \frac{K-1}{K+1} \tag{4}$$

$$EA_{sub} = EA_{vac} + \frac{3.6}{D} \cdot \frac{K-1}{K+1} \tag{5}$$

on ionization energies (Eq. 4)[197] and the regeneration of oxidized dye.[198] More extensive dye series were subsequently examined, and the threshold ionization potentials[199,200] and electron affinities,[205,206] listed in Table VII, are corrected to a vacuum dielectric constant (K = 1 in Eqs. 4 and 5).

The influences of temperature and electric fields on the shape of the photoelectric yield curves were small.[201] Multiple dye states and imperfect crystal surfaces led to a spread in the ionization energies for dye molecules adsorbed to a real surface.[202–204] The ionization energies defined near the peak photocurrent[204] are also listed in Table VII. Ionization energy distributions, calculated using localized charges to approximate surface defects, showed good agreement with the experimental tail of the photoelectric yield curves.[204]

E. Dipole Moments

The dipole moments for many types of nonionic sensitizing dyes have been published. A list of representative examples is shown in Table VIII. These include the anhydronium dye bases studied early by Brooker, Smyth, and co-workers as well as data from later studies on holopolar dyes, heterocyclic merocyanines, and vinylogous dimethylformamides.

These measurements suggested significant dipolar resonance forms for the ground states of merocyanines and the electronically analogous anhydronium bases. As noted for the vinylogous formamides [$Me_2N(CH{=}CH)_nCHO$], the neutral resonance forms did not account for the measured dipole moment. The net charge transfer (between the donor and acceptor ends of the molecules) was about 0.49 electron ($n = 0$) and decreased asymptotically to 0.25 electron ($n = 4$). Most of the other dipole moments in Table VIII are also two- or threefold higher than the values expected for the compounds without dipolar resonance forms. Both higher basicity for the heterocyclic donor and increased acidity for the heterocyclic acceptor provided increased dipole moments.

F. Protonation of Cyanines

Protonation equilibria have been studied for a large number of cyanines[207–210] and carbocyanines,[190,210–213] as well as longer chain analogs,[210,214] vinylogous dye bases,[215] and diquinolyl methanes.[208,209] Analogous halogenation of dyes has been reviewed.[216] The reactivity of heterocyclic quaternary salts was related to rates of reaction with ethyl magnesium bromide,[217] pK_a values for the corresponding amines,[217] and nmr data (see Section III.B). Temperature jump studies on cyanines from quinoline provided rates of protonation as a function of structure, ionic strength, and dielectric constant.[218]

The most extensive literature on the relation between structure and protonation deals with simple cyanines (Eq. 6).[115,207–210,218]

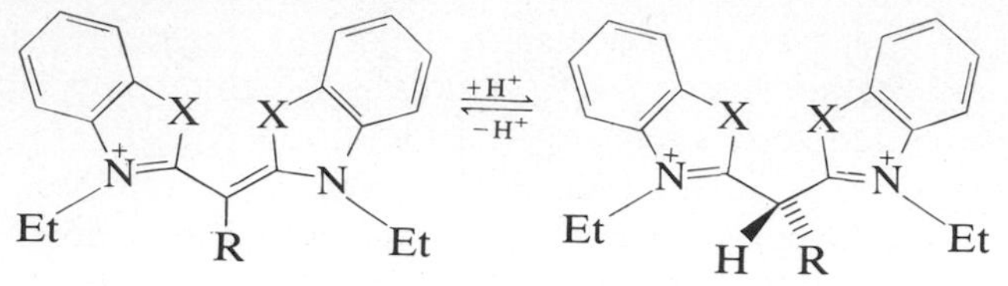

(6)

A relation between pK_a and the absorption energy of 17 simple cyanines[207] was reviewed recently,[210] and significant exceptions were noted in the cases of 3,3′,5,5′-tetramethyloxacyanine, 1,1′,3,3′-tetraethylbenzimidazolocyanines, and 2,2′-quinocyanines **37** with a rigidizing

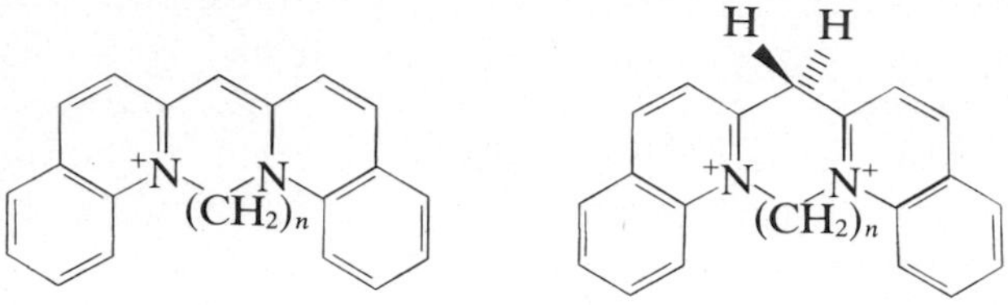

$(CH_2)_n$ group between the heterocyclic nitrogens. One of the factors influencing the observed pK_a values was the variable degree of steric strain among the dyes and their protonation products.[210] For the 2,2′-quinocyanines (Table IX), a 3,3′-dimethyl substitution on the quinoline rings increased steric hindrance in the protonated product more than in the dye, since the pK_a was decreased by 0.9 unit to 3.1 relative to 4.0 for the unsubstituted cyanine. Ethyl substitution on the central carbon raised the observed pK_a to 8.3. This suggested large steric interactions between the ethyl substituent and the quinoline rings in the dye form, which ideally should be planar. Protonation, which provides a more tetrahedral hybridization for the central carbon, decreased these steric interactions as evidenced by the high pK_a (8.3). Another factor influencing pK_a was suggested[210] to be the proximity of the positive charges in the protonated product, since a large decrease in pK_a was noted for the rigidized 1,1′-ethylene-2,2′-quinocyanine (Table IX, p$K_a = -1.0$) relative to the unrigidized 1,1′-diethyl-2,2′-quinocyanine (p$K = +4.0$).

Protonation equilibria for carbocyanines and longer chain dyes are less affected by steric interactions, and correlations between pK_a and other data are good. Brooker's deviation for 10 heterocycles was linearly related to the pK_a for the symmetrical carbocyanines (Fig. 5).[210] Increased chain length increases the pK_a.[214] Hammett substituent constants

TABLE IX. pK_a AND STERIC INTERACTIONS[a]

Dye **35**	X	R	Protonated dye **36**	pK_a
Least hindered[b]	—O	H	Least hindered[b]	−2.6
Moderately hindered	H, —H	H	Moderately hindered	+4.0
Moderately hindered	H, —Me	H	Strongly hindered	+3.1
Strongly hindered	H, —H	Et	Moderately hindered	+8.3
Strongly hindered[c]	N—Et	H	Strongly hindered[c]	+4.0
Dye **37**	$n=1$	—	Protonated dye **38**	−2.0
	$n=2$	—		−1.0

[a] Data and interpretations from A. H. Herz, *Photogr. Sci. Eng.*, **18**, 207 (1974) except for the pK_a of the benzoxazole dye [G. Scheibe, *Chimia*, **15**, 10 (1961)].
[b] 3,3′,5,5′-Tetramethyloxacyanine.
[c] 5,5′,6,6′-Tetrachloro dye.

are linearly related to changes in pK_a for substituted carbocyanines[190,211] as well as for 6,6′-substituted-2,2′-quinocyanines.[210] The pK_a values for substituted carbocyanines were also related by a family of lines to several quantities: polarographic reduction potentials (E_{red}), the quantity [$E_{red}+E_{max}$] (where E_{max} is the long-wavelength transition energy), ionization potentials, and electron affinities for vinylogous thiacarbocyanines.[190]

Cyanine aggregates are less basic than the corresponding monomers and require a more acidic pH for their protonation.[210] With basic benzimidazolocarbocyanines, aggregation to the J-state produced large effects on the protonation reaction. For example, a $2\times10^{-7}M$ solution of an *N,N*′-disulfobutyl-5,5′,6,6′-tetrachlorobenzimidazolocarbocyanine in its monomeric state (M) was half-protonated at pH 7.9 (pK_a 7.9) but the same dye in a $2\times10^{-5}M$ solution, where it existed solely as a J-aggregate (apparently a tetramer with $n=4$), was half-protonated at pH 5.8. Unless the concentration of the aggregate and its association number n are defined, it seems necessary and unambiguous to report basicity as the pH value for half-protonation of the dye. This value then becomes equivalent

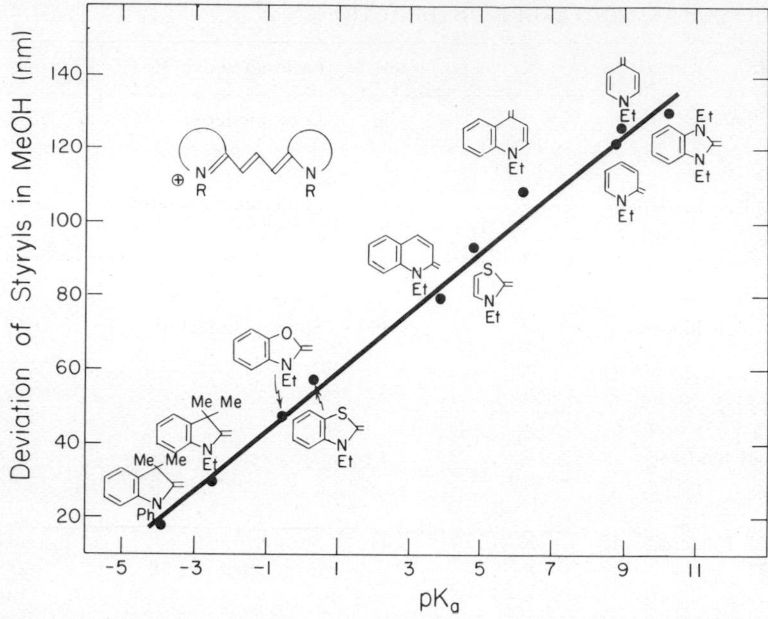

Fig. 5. Brooker deviation of styryl dyes (a spectral measure of basicity) correlated with pK_a values for symmetrical carbocyanine dyes (water, 25°C). The correlation coefficient was 0.9968. (Reproduced by permission from *Photogr. Sci. Eng.*, **18,** 207 (1974).)

to pK_a provided the reversible protonation reaction involves one H^+ per dye chromophore.

G. Base Adducts of Cyanine Dyes

Dye cations were shown to act as electron acceptors for a variety of Lewis bases, including oxygen, nitrogen, and carbanionic bases.[219] Certain dyes may be isolated as alkoxy adducts[220] whereas others such as **26** and **28** undergo irreversible degradation in base.

IV. Synthetic Approaches to Cyanine and Related Dyes

As noted in Sections I.C and D, quite general synthetic methods were developed after 1920 and extended to many new systems after 1950. Several reviews have covered the classes of sensitizers prepared by these methods.[7,8,40–42,76,221–228] Dye-forming reactions may be classed as oxidative or nonoxidative. The oxidative syntheses are primarily of historical interest, while nonoxidative syntheses are the most versatile and employ varied combinations of suitable nucleophilic and electrophilic reagents. The review by Hamer[10] lists the nearly infinite set of specific dye structures prepared before 1959, and Ficken's review[11] provides supplemental references to more recent compounds.

A. Oxidative Syntheses

Oxidative formation of cyanine dyes produced the first cyanine[20] and many of the other sensitizers[25,26] before 1920. Larivé and coworkers[135,136,229–234] defined the structures and reactivities of intermediates in these syntheses (typical reaction sequences are shown in Eq. 7). The quaternary salt **39** reacted with the methylene base **40** (see also Section IV.B.1) to provide the dihydroquinoline **41**. This dye precursor and other dihydroquinoline derivatives like **45** and **46** were characterized by nmr.[135,136] Oxidation of these initial products led to simple cyanines such as **42**, a common product of these reactions. A variation of this type of synthesis provided other simple cyanines[26,235,236] and pyridinium betaines.[82,83]

Carbocyanine dyes were also formed under certain conditions by oxidative reactions. In the presence of oxygen the methylene base **40** underwent oxidative cleavage to give formaldehyde, which reacted with two additional equivalents of methylene base to give the trimethylene compound **43**. Kryptocyanine (**44**) from oxidation of **43** was formed in 15 to 20% yield using oxygen-saturated ethanol, but the yield dropped to less than 2% in ethanol degassed with nitrogen. Since the original synthesis of a kryptocyanine chromophore resulted from a similar reaction of the *N*-isopropyl analogs of **39** and **40**, the influence of the *N*-alkyl substituents on the relative amounts of simple cyanine and krytocyanine was quite marked. A direct comparison of *N*-methyl and *N*-ethyl substituents gave ratios of simple cyanine/kryptocyanine yields (**42**/**44**) of 80/20 for *N*-methyl and 55/45 for *N*-ethyl.[229] Thiacarbocyanines (**1**; X = sulfur) and indocarbocyanines (**1**; $X = CMe_2$) were also prepared by this method.[237,238]

B. Nonoxidative Syntheses

Brooker[2] characterized dye syntheses as "the condensation type, two intermediates reacting under suitable conditions with elimination of some simple molecule." Most combinations of nucleophilic and electrophilic

Me

CH_2

H

Me

+

N

Me

N

Me

N

Me

N

Me

39

40

41

$-H^-$

O_2

Me

+

N

Me

N

Me

(7)

42

1-methyl-4-quinolone

+

CH_2O $\xrightarrow{+2\,\mathbf{40}}$ Me—N^+ —$CH_2CH_2CH_2$— ^+N—Me

43

[O]

Me

Me

N_+

N

44

Oxidative syntheses

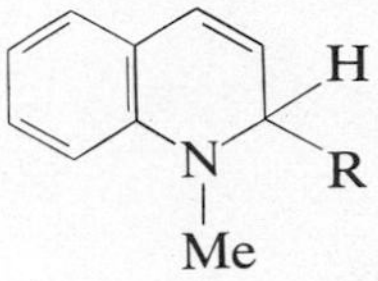

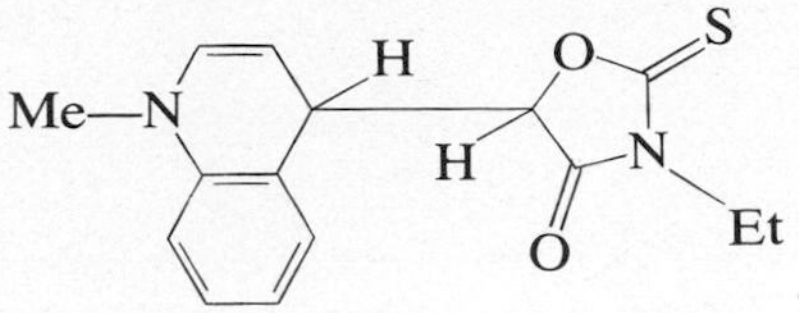

45 R = H, OH, OMe **46**

Dihydroquinolines

reagents are described in this way, including those necessary for the synthesis of complex dyes. In contrast to oxidative dye syntheses in which a methylene base and the related quaternary salt can combine to form a dye (Eq. 7), nonoxidative syntheses often combine reagents of quite different structure. For example, symmetrical carbocyanine (trimethine) dyes are typically prepared from two equivalents of an active methyl quaternary salt, which provides nucleophilic methylene bases (Section IV.B.1), and an orthoester as the electrophilic reagent. Most of the atoms in the resulting dye originate from the heterocyclic quaternary salt, and the orthoester provides only the central carbon of the trimethine chain.

Typical synthetic reactions are illustrated in Schemes 1 and 2. Many of the heterocycles and other structures used in these reagents are shown in Table III.A,B. Nucleophilic methylene bases derived from quaternary salts like 3-ethyl-2-methylbenzothiazolium (Scheme 1, **47**) reacted with a variety of electrophilic reagents to provide dyes in which the nucleophilic reagent **47** serves as the terminal heterocyclic group for the chromophore: (1) long-chain dyes **48** and **49**; (2) cyanines **50** and carbocyanines **51**; and (3) complex dyes **52** and **53**. The quaternary salts were also converted to generally useful electrophilic reagents like the ICI intermediate **54** shown in Scheme 2. Reaction with various nucleophilic reagents provided several types of dyes. Those with simple chromophores include: (1) hemicyanines (**55**), in which one of the terminal nitrogens is nonheterocyclic; (2) enamine tricarbocyanines (**56**) useful as laser dyes (see Section V.B); and (3) merocyanines (**57**). More complex polynuclear dyes from reagents with more than one reactive site include the trinuclear "B–A–B" dye **58**, containing basic-acidic-basic heterocycles, and the tetranuclear "B–A–A–B" dye **61**, containing basic-acidic-acidic-basic heterocycles. Structural variations of the reagents used in these reactions have been a primary source of progress in dye synthesis. Because of the many possible combinations of nucleophilic reagents with electrophilic reagents to form dye molecules, the following sections describe the structures and properties of these reagents separately: nucleophilic reagents in Section IV.B.1 and electrophilic reagents in Section IV.B.2. Some

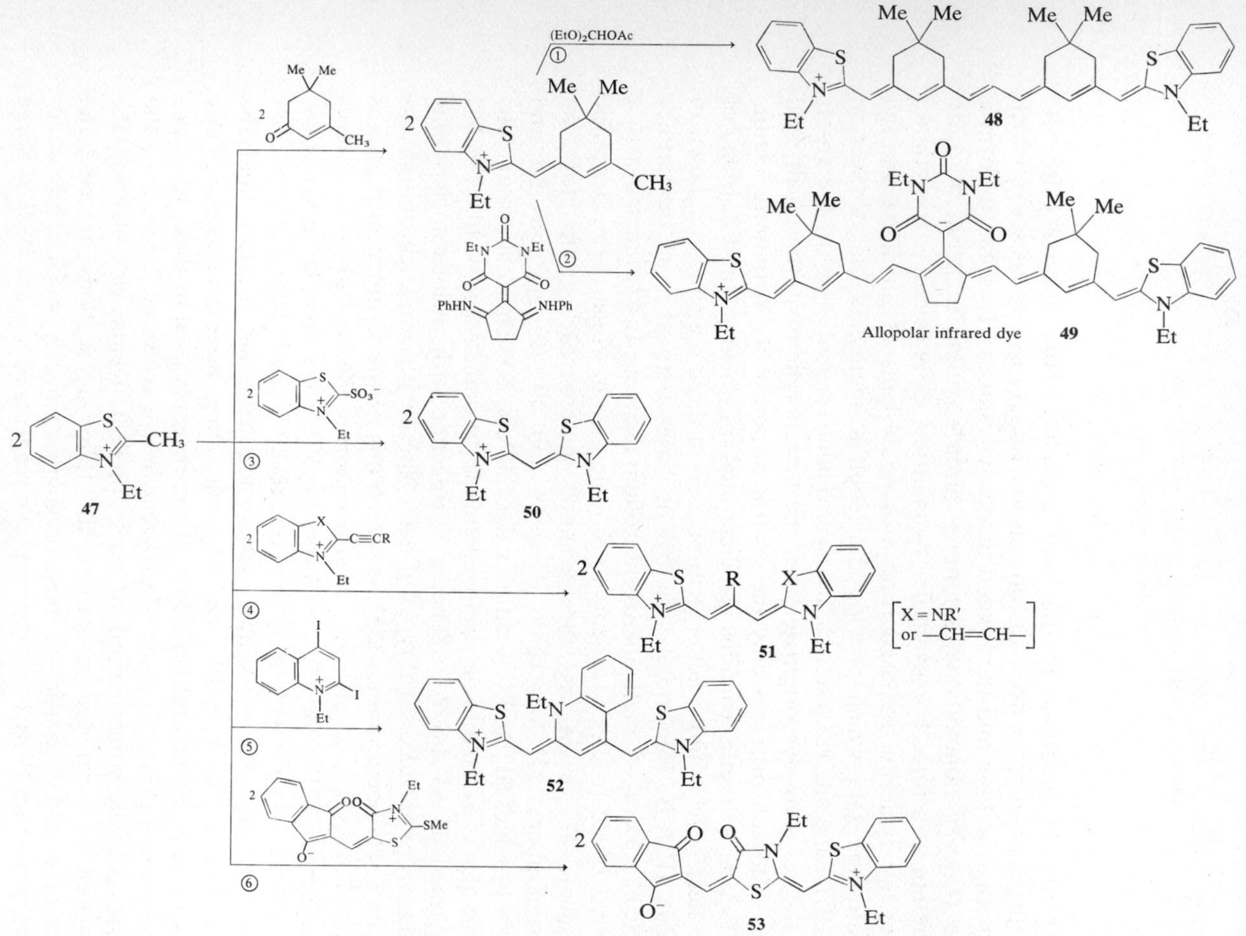

(EtO)$_2$CHOAc

Allopolar infrared dye **49**

47 **48** **50** **51** **52** **53**

[X = NR′ or —CH=CH—]

Scheme 1

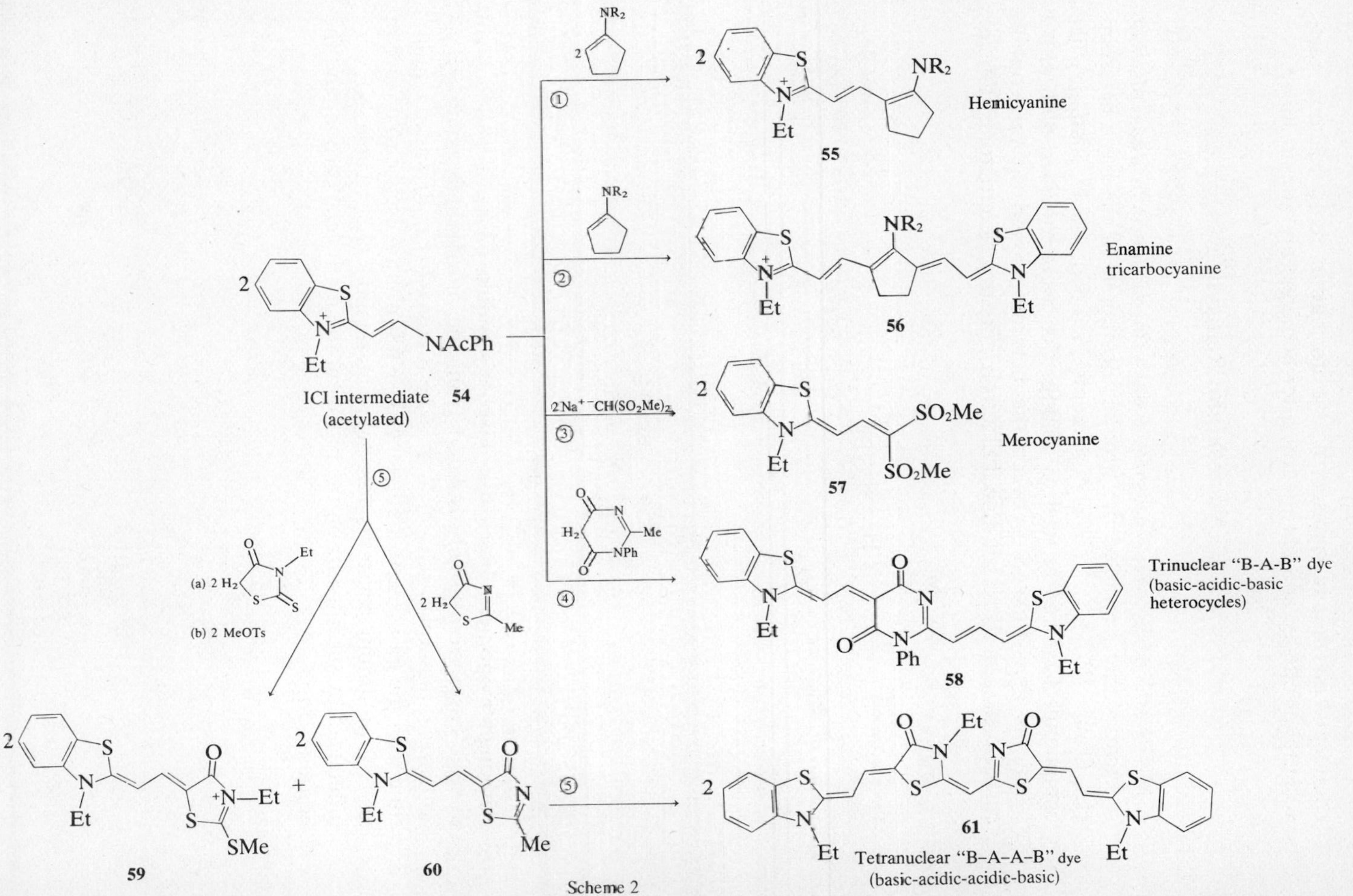

ICI intermediate (acetylated) **54**

① ② ③ $2 Na^+ {}^-CH(SO_2Me)_2$ ④ ⑤

Hemicyanine **55**

Enamine tricarbocyanine **56**

Merocyanine **57**

Trinuclear "B-A-B" dye (basic-acidic-basic heterocycles) **58**

(a) $2 H_2$ (b) 2 MeOTs

59 + **60**

Tetranuclear "B–A–A–B" dye (basic-acidic-acidic-basic) **61**

Scheme 2

comments on the mechanism of dye formation (e.g., reaction rates, substituent effects, and tetrahedral intermediates) are also included.

1. *Nucleophilic Reagents*

Methylene bases were recognized early as nucleophilic reagents with reactivities that paralleled the Brooker deviations of the basic heterocycles (Table III.A).[47,239,240] The reagents included simple enamines like compounds **62** to **64** as well as analogs that were modified with either additional vinyl linkages between the heteroatom and the nucleophilic carbon (**40**; Eq. 7) or additional electron-donating heteroatoms (**65**–**66**).

62 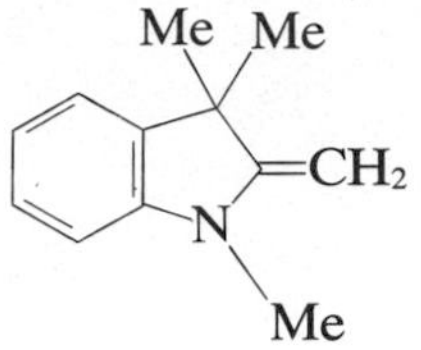 **63** 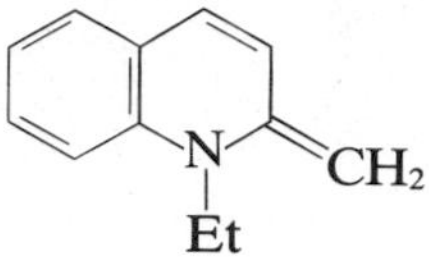 **64** 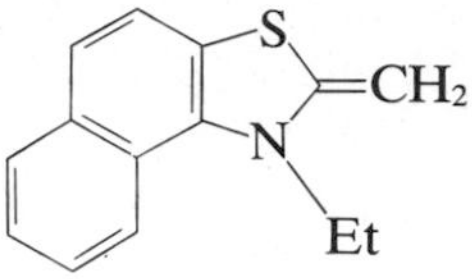 **65**

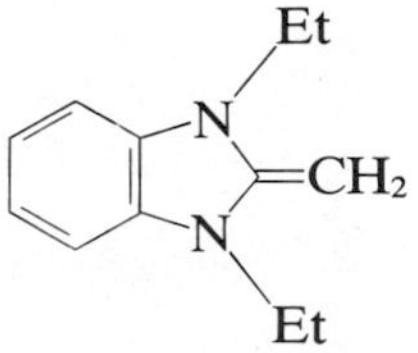

66

Although monomeric bases like Fischer's indole **63** were well known, dimeric bases were usually isolated from the more widely studied benzothiazole quaternary salts.[30,241–243] The dimer **67** (Eq. 8) was in equilibrium with the monomeric methylene base **68** in the presence of a trace of acid. Equilibrium constants for the dimerization (K_{eq}, Table X) were influenced by substituents on the nitrogen, the methylene carbon, and the benzene ring.

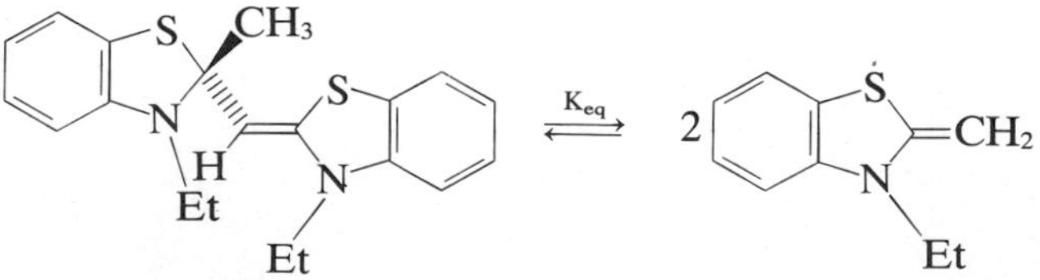

$$\mathbf{67} \underset{}{\overset{K_{eq}}{\rightleftharpoons}} 2\ \mathbf{68} \qquad (8)$$

67 **68**

TABLE X. EQUILIBRIUM CONSTANTS AND REACTION RATES FOR METHYLENE BASES FROM SUBSTITUTED BENZOTHIAZOLES[a]

R″ (positions 6, 5) — benzothiazole — S, N(R), =CHR′

69

Structure **69**	K_{eq} (mole/liter)[b] (5-Subst. only)	Rate of dye formation[c] (liters mole^{-1} sec^{-1}) 5 Subst.	6 Subst.
A. R = Et, R′ = H			
R″ = OMe	6.6×10^{-4}	0.422	0.735
H	6.9×10^{-4}	0.364	0.364
F	2.2×10^{-4}	0.152	0.262
Cl	2.0×10^{-4}	0.122	0.163
CF_3	0.8×10^{-4}	—	—
NO_2	—	0.031	0.012
B. R′ = R″ = H			
R = Me	0.70×10^{-4}	—	—
Et	6.9×10^{-4}	—	—
$CH_2C_6H_5$	4.6×10^{-4}	—	—

[a] J. R. Owen, *Tetrahedron Lett.*, 2709 (1969); *Eastman Org. Chem. Bull.*, (Eastman Kodak Co., Rochester, N.Y.), **43**(3), 3 (1971).
[b] The equilibrium was defined (Eq. 8) as $K_{eq} = [\text{monomer}]^2/[\text{dimer}]$.
[c] The rate constant for merocyanine formation (Eq. 9).

The utility of methylene bases as synthetic reagents[244–254] was greatly enhanced by Owen's preparation of monomeric methylene bases.[255,256] Under the appropriate conditions pure monomeric methylene bases from many heterocycles were prepared and isolated (**69–74**) or characterized by nmr spectroscopy (**64–66**). Merocarbocyanine dye formation from monomeric methylene bases and 1,3-diethyl-5-dimethylaminomethylene-2-thiobarbituric acid (Eq. 9) was first order in both reagents, indicating that the formation of the dye precursor **75** was the rate-determining step. The second-order rate constants for dye formation (Table X) were influenced more by substituents in the 6-position than the 5-position of the benzothiazole base **69**. Modified Hammett equations were used to analyze the substituent effects on both the equilibrium constants and reaction rate constants. These indicated that the transmission of inductive effects through the nitrogen was about twice that for the sulfur.[256] Related studies were carried out on the reaction rates of substituted quinoline quaternary salts with *p*-dimethylaminobenzaldehyde.[257–259]

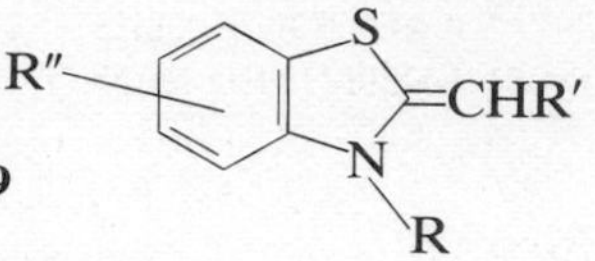

69

[No dimers, R′ = Me, Ph, CHO]

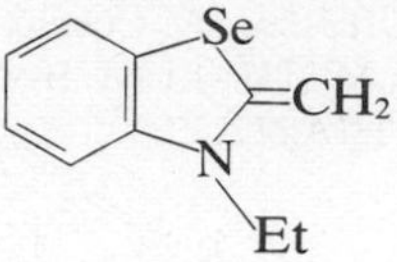

70

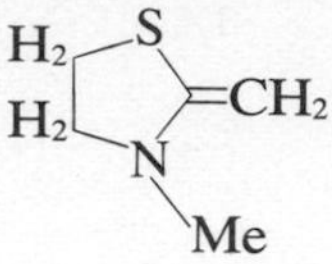

71

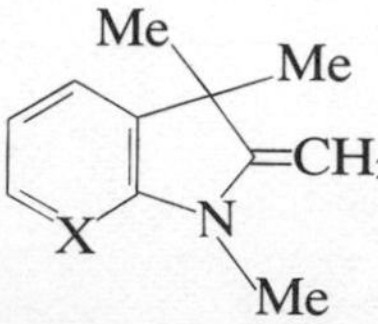

72 [X = CH, N]

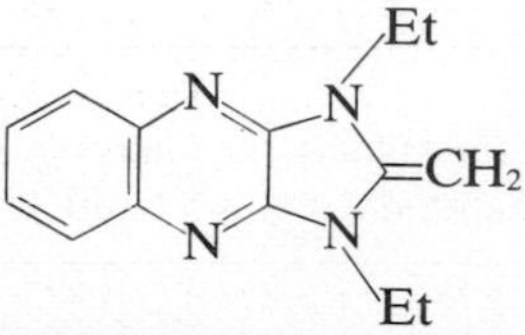

73

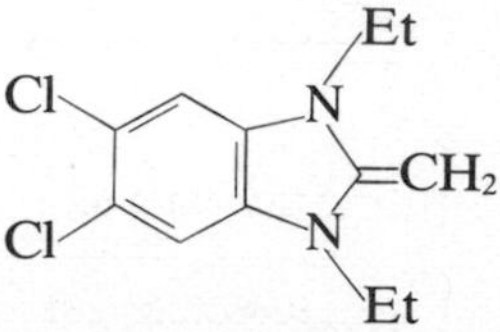

74

Isolated monomeric bases

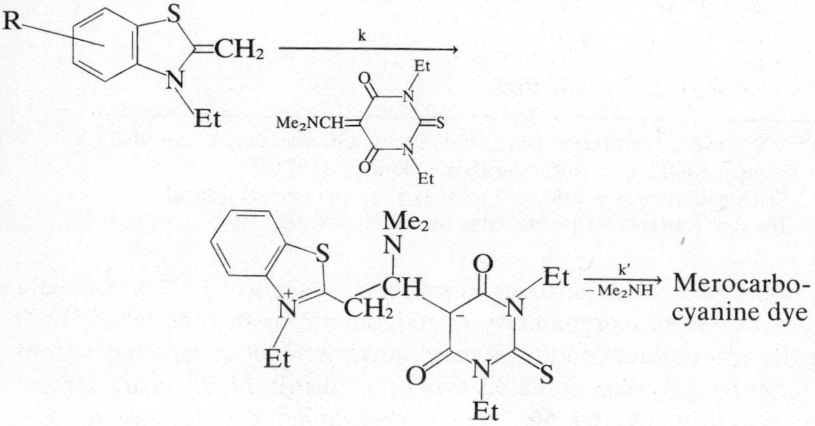

Merocarbocyanine dye (9)

75

Many basic heterocycles yield quaternary salts with one or two reactive methyl groups. Normal alterations[10,11,221,225] in the structure of monomethyl quaternary salts were illustrated in Table III.A. Heterocycles like imidazole, thiazole, pyridine, oxazole, and pyrrole were systematically modified by the use of additional fused rings. In the imidazole and thiazole series, for example, these fused rings included benzene, pyridine, pyrazine, isomeric naphthalenes, phenanthrene, quinoline, and quinoxaline. Most of these rings lowered the Brooker deviations from those observed from imidazole or thiazole, so that less basic conditions were necessary for synthetic reactions. All of these variations in structure lead to dyes with modifications outside the primary chromophore, and these are treated as substituent effects in Section VIII.A.

The wide range of Brooker deviations (Table III.A) suggested large differences among heterocycles for the stabilization of a positive charge, and both methylene-base reactivity and the pK_a of carbocyanines paralleled deviations. As a typical example of the sensitivity of Brooker deviations to small structural changes, consider the three isomeric naphthothiazoles (Table III.A) in which the orientation of the fused naphthalene ring is the major structural modification. The lowest deviation (least positive charge stabilization) was exhibited by the linear naphtho[2,3-*d*]thiazole where both the nitrogen and sulfur were bonded to β-positions of the fused naphthalene ring (β,β-naphthothiazole). The highest deviation (highest positive charge stabilization) was shown by the angular naphtho[1,2-*d*]thiazole (**65**) where the sulfur was bonded to the β-position of naphthalene (β-naphthothiazole). If the nitrogen in the thiazole ring was most important for the stabilization of the positive charge (as well as for the transmission of substituent effects in methylene bases), the β-naphthothiazole would be expected to have the highest deviation, since the positively charged nitrogen is directly bonded to the site which also dominates electrophilic substitution reactions in naphthalene (α-carbon).

More complex reagents than the quaternary salts in Table III.A included cyanine and merocyanine dyes with reactive methyl groups (**76**–**77**) and bisquaternary salts (**78**–**79**). The merocyanine **77** was the type of intermediate used in the synthesis of trinuclear and tetranuclear dyes

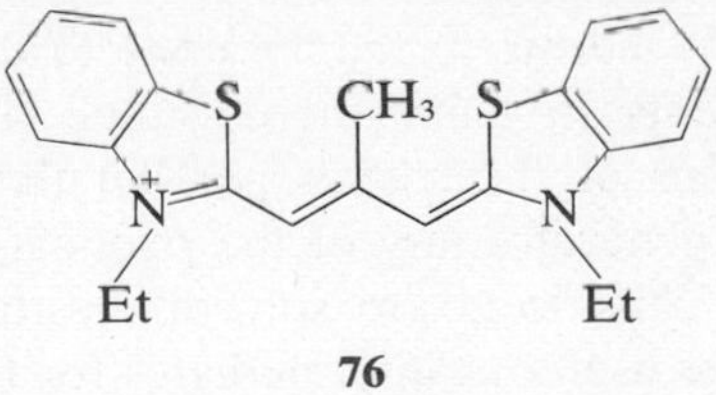

76 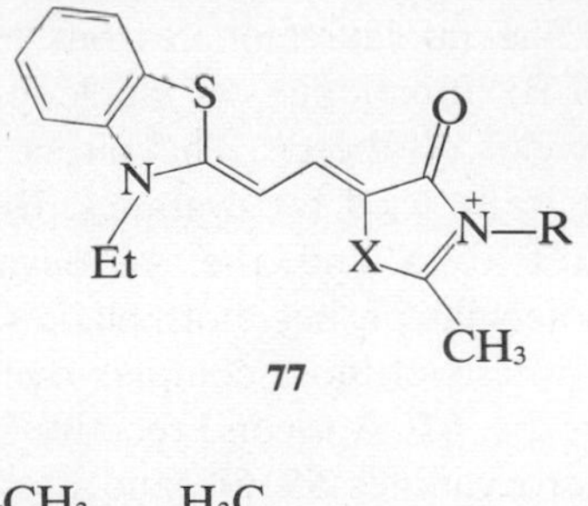 **77**

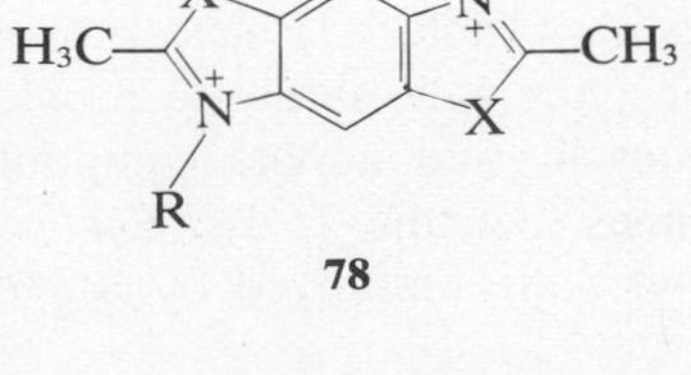

78 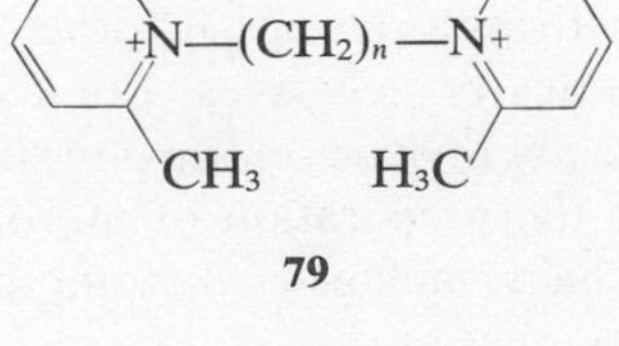 **79**

(Scheme 2; **58–61**). In Chapter XV of her review[10] Hamer listed examples of the many trinuclear and polynuclear dyes derived from such intermediates, and additional examples have been reviewed more recently.[11] The original syntheses of multichromophore dyes[260–263] from the bisquaternary salts **78–79** were extended recently with new examples of isomeric biscyanines[264–274] and bishemicyanines.[275–284]

Other nucleophilic reagents were also developed. Reactive quaternary salts with additional vinyl groups led to improved preparations of long-chain dyes.[42] Both tricarbocyanines and tetracarbocyanines were prepared from **80**. The related **81** was incorporated into pentacarbocyanines

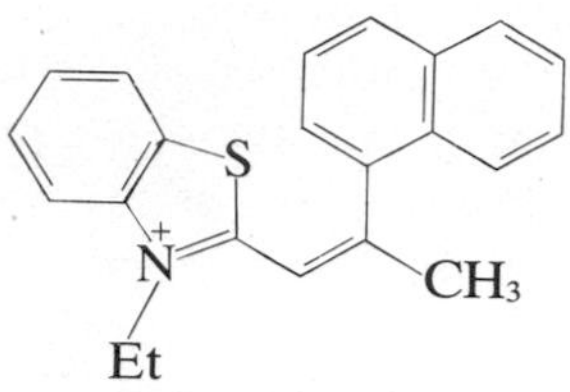

80

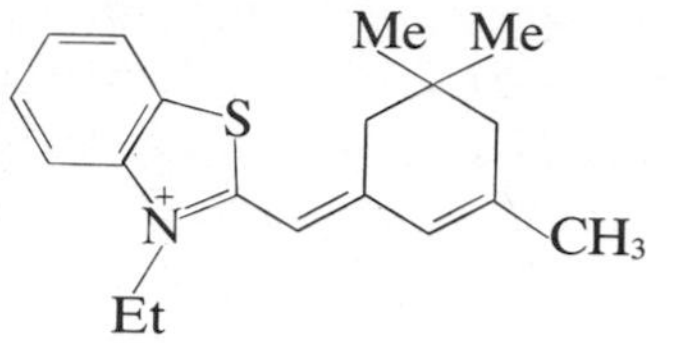

81

and other dyes (Scheme 1; **48–49**). Incorporation of the neopentylene group into the quaternary salt **81** was an important advance in the preparation of ir dyes because the rigidizing effect of this group led to relatively high extinction coefficients for the longest wavelength absorptions. Simple enamines with two reactive sites (**62**) led to chain-substituted ir dyes by alternative routes (Scheme 2; dye **56**). Bridged and heterocyclic analogs of these enamines are also known.[285–287]

The enolate anions from many active methylene compounds (Table III.B) functioned as nucleophilic reagents in early merocyanine and oxonol syntheses. The direct preparations of these dyes parallel those typically used for cyanines. Besides the general utility of the oxonols as filter dyes and the merocyanines as sensitizers or solvent-sensitive molecular probes, both classes of dyes are useful as intermediates for the synthesis of more complex dyes. A typical example is shown in Scheme 2 for dye **61**. A second reactive site was generated on the rhodanine ring of merocyanines **59–60**, and these reacted to produce the tetranuclear dye **61**.

Similar activating groups in other reagents (Table XI) have provided extensive series of new dyes. For example, the sulfone group provided more acidic methylene groups in heterocyclic and acyclic compounds (**82–86**) for the preparation of merocyanines (Scheme 2; dye **57**).[288–294] Extended chain analogs of ketomethylenes were employed by Brooker

TABLE XI. NUCLEOPHILIC REAGENTS FOR DYE SYNTHESIS: NUCLEOPHILES AND/OR PRECURSORS FOR NUCLEOPHILIC REAGENTS

A. Active methylene (methyl) compounds

O H₂ H₂ S O₂ — *a*; NR₂ S O₂ — *a*; O H₂ Me Me O — *b*; S OH H N Me — *c*; N⁺ CH₂ RC=O — *d*

O O H₂ H₂ O O — *e*; O H₂ H₂ O X O, X = NR, O — *f*; O H₂ O O O Me Me — *g*; O H₂ O N N R R′ — *h*

R O R N O N O CH₃ O R — *i*; Me Me O H₃C O Ph N — *j*; Me Me O Et H₃C N O N S Et — *j*; Et O H₂ N O H₂ N O Et — *k*

O ONa H₂ N N H O O — *l*

B. Novel nucleophiles and precursors for novel nucleophiles

+ H₂ — *m*; R CH₃ S S⁺ — *n*; + CH₃ — *o*

TABLE XI (*Continued*)

B. Novel nucleophiles and precursors for novel nucleophiles

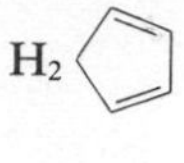

H_2 (cyclopentadiene) — *p*

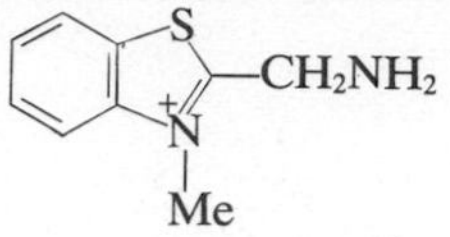

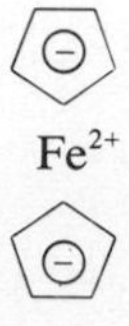

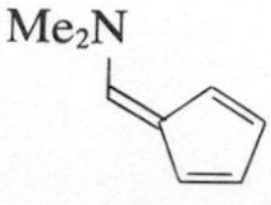

(3-methylbenzothiazolium-2-yl)$-CH_2NH_2$ (S, $\overset{+}{N}$, Me) — *q*

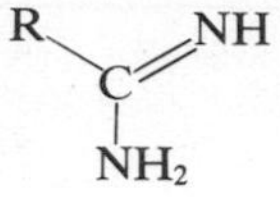

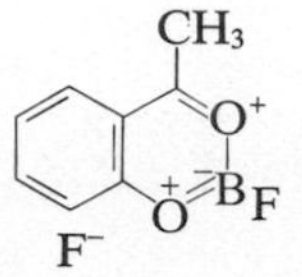

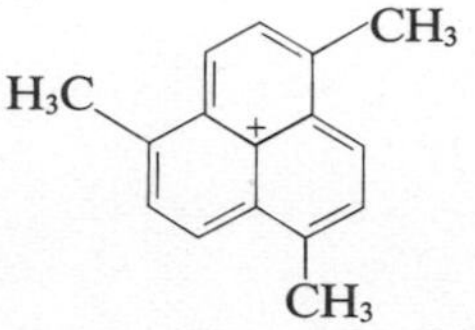

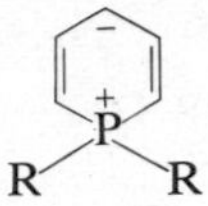

R–C(=NH)–NH_2 — *r*

Fe^{2+} (two cyclopentadienide ⊖ rings) — *s*

Me_2N–CH=(cyclopentadiene) — *t*

CH_3, O^+, $\overset{+}{O}$–$\overset{-}{B}$F, F^- — *u*

H_3C, CH_3, CH_3 (phenalenium +) — *v*

$R_3\overset{+}{P}$—$\overset{-}{C}HR$ — *w*

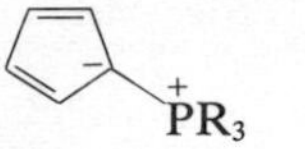

(cyclopentadienide −)–$\overset{+}{P}R_3$ — *x*

R–$\overset{+}{P}$–R (ring −) — *y*

$P(CH_2OH)_3$ — *y*

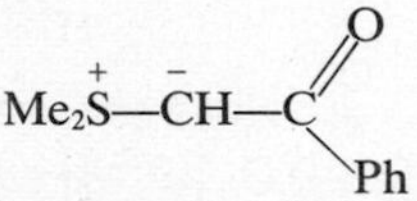

$Me_2\overset{+}{S}$—$\overset{-}{C}H$—C(=O)Ph — *z*

---

[a] A. C. Craig and L. G. S. Brooker, U.S. Patent 3,148,065 (1964); J. C. Martin, U.S. Patent 3,347,868 (1967).

[b] D. W. Heseltine and L. G. S. Brooker, U.S. Patent 3,140,182 (1964).

[c] E. B. Knott, *Chimia*, **15**, 106 (1961); *J. Chem. Soc.*, 4244 (1960).

[d] K. Dickoré and F. Kröhnke, *Chem. Ber.*, **93**, 1068 (1960); F. Kröhnke and H. G. Nordmann, *Justus Liebigs Ann. Chem.*, **730**, 158 (1969).

[e] D. W. Heseltine and L. G. S. Brooker, U.S. Patent 3,140,951 (1964).

[f] J. D. Mee, Research Disclosure 12529, volume 125 (1974).

[g] Belgian Patent 682,285 (1966).

[h] D. W. Heseltine and L. G. S. Brooker, U.S. Patents 2,927,026 (1960) and 3,213,089 (1965); G. de W. Anderson, British Patent 834,751 (1960).

[i] L. G. S. Brooker and F. G. Webster, U.S. Patent 2,965,486 (1960).

[j] L. G. S. Brooker et al., *J. Am. Chem. Soc.*, **87**, 2443 (1965).

[k] L. G. S. Brooker and F. L. White, U.S. Patent 2,955,939 (1960).

[l] P. W. Jenkins, French Patent 2,099,780 (1972).

[m] K. Hafner, *Angew. Chem.*, **70**, 413, 419 (1958); G. Bach, E. J. Poppe, and W. Treibs, *Naturwissenschaften*, **45**, 517 (1958); C. Jutz, *Angew. Chem.*, **71**, 380 (1959); E. C. Kirby and D. H. Reid, *J. Chem. Soc.*, 3579 (1961); *ibid.*, 163 (1961); A. Treibs, R. Zimmer-Galler, and Chr. Jutz, *Chem. Ber.*, **93**, 2542 (1960).

[n] E. Klingsberg, *Synthesis*, 29 (1972); *J. Heterocycl. Chem.*, **3**, 243 (1966).

and coworkers[55] (reagents *j*, Table XI) to prepare rigidized, solvent-sensitive dyes. Specific alterations in the structures of bisketomethylene compounds (reagents *e*, *f*, and *g*, Table XI)[295–297] led to multiple-chromophore dyes with variable intensities for secondary absorption maxima (see Section VII).

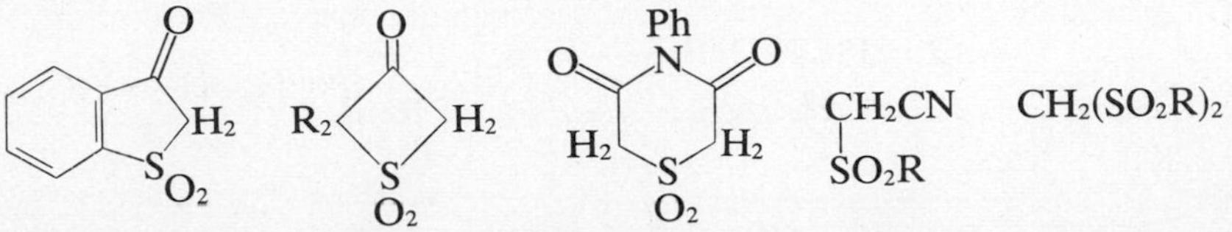

82 **83** **84** CH_2CN / SO_2R **85** $CH_2(SO_2R)_2$ **86**

Novel nucleophilic reagents from hydrocarbons, ferrocenes, boron heterocycles, and phosphorus are listed in Table XI.B. Extensive investigations[74–81] of phosphorus reagents (*w–y* Table XI) led to phosphocyanines (terminal R_3P group, dye **87**) and phosph*a*cyanines (chromophoric methine replaced by phosphorus). Simple phosphacyanines (**91**) were formed from **89** through a tetrahedral intermediate (**90**) (Eq. 10).[76] Arsenic and sulfur analogs of the phosphocyanines were also prepared.[298,299]

---

[o] K. Hafner, H. W. Riedel, and M. Danielisz, *Angew. Chem.*, **75**, 344 (1963); C. Jutz and F. Voithenleitner, *Chem. Ber.*, **97**, 29, 1590, 2050 (1964).

[p] E. D. Bergmann, *Chem. Rev.*, **68**, 41 (1968); C. H. Schmidt, *Angew. Chem. Int. Ed. Engl.*, **2**, 101 (1963); J. A. Berson, E. M. Evleth, Jr., and Z. Hamlet, *J. Am. Chem. Soc.*, **87**, 2887 (1965); Ch. Jutz and H. Amschler, *Angew. Chem.*, **73**, 806 (1961); for molecular orbital calculations on the dyes, see R. Zahradnik and J. Michl, *Collect. Czech. Chem. Commun.*, **30**, 1060 (1965); R. Kuhn and H. Fischer, *Angew. Chem. Int. Ed. Engl.*, **2**, 692 (1963); *Angew. Chem.*, **73**, 435 (1961).

[q] A. I. Kiprianov and T. M. Verbovs'kaya, *Zh. Obshch. Khim.*, **33**, 479 (1963).

[r] J. D. Kendall, British Patent 461,668 (1935).

[s] Ch. Jutz, *Tetrahedron Lett.*, No. 21, 1 (1959); A. Treibs and R. Zimmer-Galler, *Chem. Ber.*, **93**, 2539 (1960); P. M. Heertjes and L. T. Khoo, *Chim. Suppl.*, 157 (1968).

[t] K. Hafner, *Chimia*, **27**, 640 (1973).

[u] D. S. Daniel and D. W. Heseltine, U.S. Patent 3,567,439 (1971) and 3,745,160 (1973); H. Depoorter and J. R. Schellekens, German Patent 2,129,610 (1972).

[v] J. G. Dingwall, D. H. Reid, and K. Wade, *J. Chem. Soc. C*, 913 (1969).

[w] H. Depoorter, J. Nys, and A. van Dormael, *Bull. Soc. Chim. Belg.*, **73**, 921 (1964).

[x] A. van Dormael, *Chimia*, **15**, 67 (1961); H. Depoorter, J. Nys, and A. van Dormael, *Bull. Soc. Chim. Belg.*, **74**, 12 (1965).

[y] K. Dimroth, *Fortschr. Chem. Forsch.*, **38**, 5 (1973) and references therein.

[z] See references in J. Nys, *Chim. Suppl.*, 115 (1968).

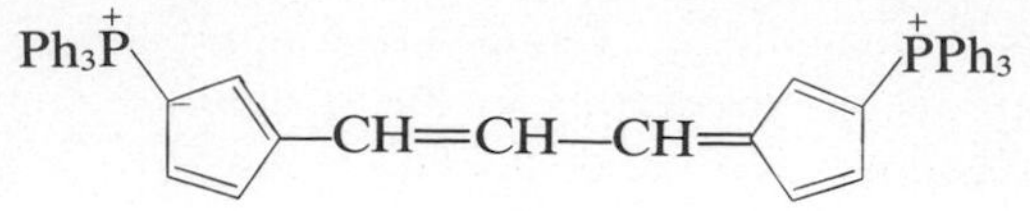

87

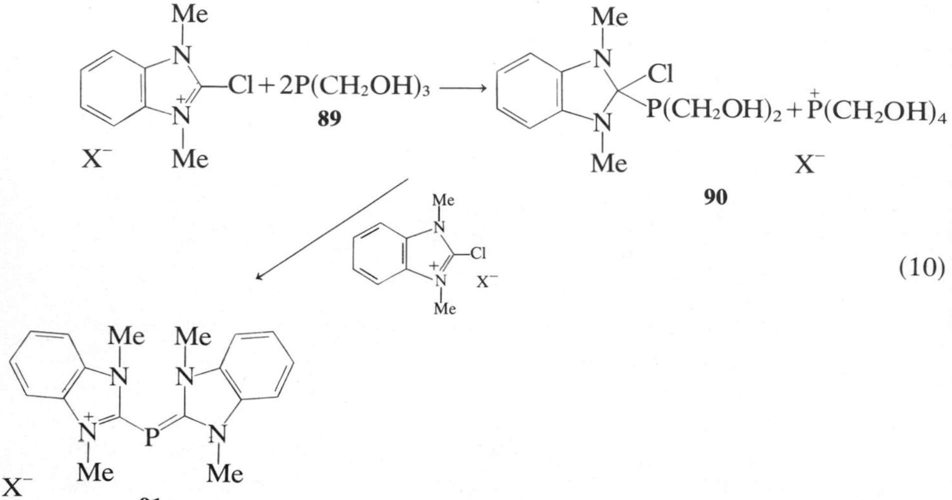

89 **90** (10)

91

2. *Electrophilic Reagents*

Three types of electrophilic reagents are useful for the synthesis of sensitizers: (1) heterocycles or aromatic rings with replaceable substituents, (2) functionally symmetric reagents like orthoesters, and (3) vinylogous amides and analogous reagents. The first type lead to short chromophoric linkages such as simple cyanines, whereas the latter two provide symmetrical and unsymmetrical dyes with longer chain lengths.

Aromatic substitution reactions occur readily on electrophilic reagents of the general structure **92**. Reactions of these with methylene bases and

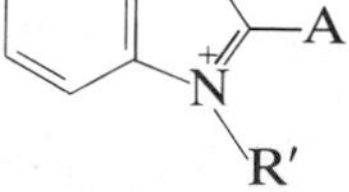

A = I, SR, SO_3^-, SO_2Me, OR, —CH=N—OR, CN

92

other nucleophiles lead to a variety of simple cyanines.[10,76] For leaving groups (A-groups in **92**) such as Cl^-, I^-, RS^-, SO_3^{2-}, and RO^-, the reaction pathway for the formation of simple cyanines may be an addition-elimination involving a tetrahedral intermediate such as **90**. In the synthesis of the simple phosphacyanine by aromatic substitution in **89**, the tetrahedral intermediate **90** was stable enough to be isolated. Related tetrahedral derivatives were noted as hydroxide adducts of quaternary salts,[243,254] intermediates in the oxidative syntheses of simple cyanines (Eq. 7), dimeric methylene bases (Eq. 8), and Lewis base adducts of cyanine dyes.[219–220] Similar displacements of aromatic substituents were important for the preparation of bridged dyes (Scheme 1; **52**)[300,301] and trinuclear dyes (Scheme 1; **53**).[302,303]

The oxime reagents (**92** with A = CH = NOR) were reviewed by Hamer as the nitrite method.[304] A variety of simple cyanines were made by this method, including the highly twisted cyanine **94** from imidazo[4,5-*b*]-quinoxaline and the unsymmetrical benzothiazole dye **95** (Eq. 11).[305] The

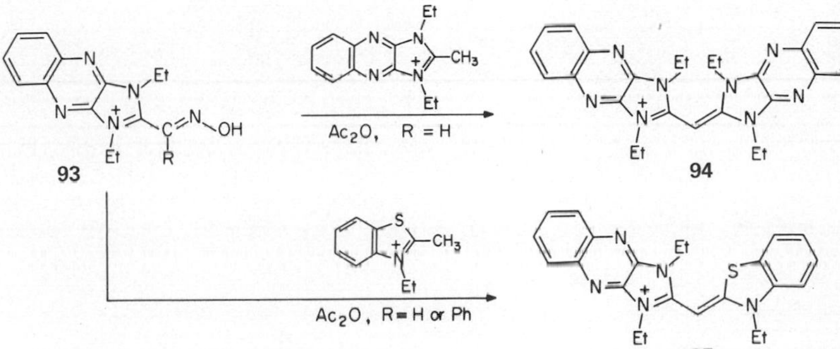

Ac_2O, R = H

Ac_2O, R = H or Ph

(11)

93 **94** **95**

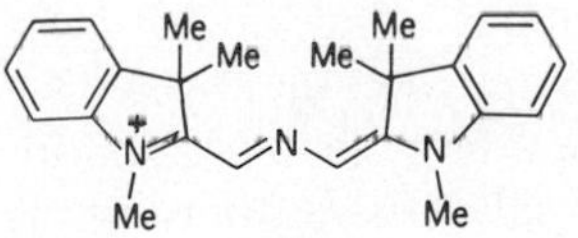

96 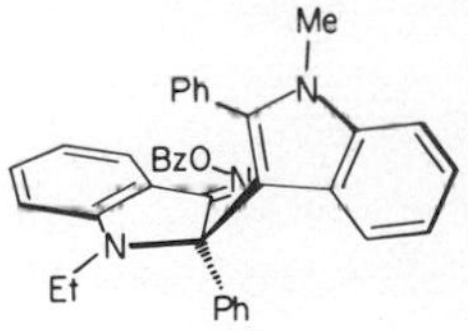 **97**

dye isolated from the first nitrite synthesis was originally suggested to be the β-azacarbocyanine **96**. However the correct structure for the product of the reaction was the related simple cyanine (**1**; $X = CMe_2$ with $n = 0$).[304] Several precursors to the simple cyanines have been suggested:[304] (1) cyano quaternary salts (**92**; A = CN), formed by dehydration of the oxime; (2) β-azacarbocyanines (e.g., **96**), formed by addition of a methylene base to the oxime; and (3) tetrahedral adducts of the oxime (e.g., **97**). The tetrahedral intermediate **97** was observed for the reaction between 1-methyl-2-phenylindole and 1-ethyl-2-phenyl-3-benzoyloxyiminoindolium chloride.[306]

Substitution on alkoxyphenalenium salts provided a variety of merocyanine, azo, and cyanine dyes[307,308] (Eq. 12). Both substitution for the ethoxy group and hydride loss occurred with an excess of the phenalenium salt to give dyes **99a–b**, and multiple substitution occurred with excess methylene base.[308] Hydrolysis of **99a** in hydrochloric acid

EtO OEt BF_4^- + S =CH_2 ⟶ N Et EtO R_1 R_2

98 (excess) **99**

(12)

(a) R_1 = CH= S N Et , R_2 = H

(b) R_1 = OEt, R_2 = CH= S N Et

replaced the ethoxy group by hydroxide, and subsequent proton loss gave the merocyanine dye containing the phenalenone ring. Reaction of **98** with 3-methyl-2-iminobenzothiazole (an aza-methylene base) gave only products derived from substitution for the ethoxy groups, but diisopropylethylamine reacted with **98** to place a substituent in position R_2 (**99**, R_1 = ethoxy and R_2 = 2-diisopropylaminovinyl).[308]

Carbocyanines and longer chain dyes have been prepared from many electrophilic reagents. The typical examples of these reagents, listed in Table XII, may be divided into three categories: (*a*) one-carbon reagents, which add a single carbon to the chromophore when reacted with two equivalents of nucleophilic reagents; (*b*) three- and five-carbon reagents, which usually add three and five carbons to a chromophore when reacted with two equivalents of nucleophilic reagents; and (C) unsymmetrical reagents, which form dyes by reaction with only one equivalent of a nucleophilic reagent. Some of the compounds in Table XII.A,B are structurally unsymmetrical but provided essentially symmetrical reactivity

TABLE XII. SYMMETRICAL AND UNSYMMETRICAL ELECTROPHILIC REAGENTS FOR DYE SYNTHESIS

A. One carbon reagents (reagents that provide one carbon to the methine chain of the resulting dye)

R—C(OR)$_3$ *a*

$RCO_2CH(OR)_2$ *b*

PhN=CH—NHPh *c*

S CH$_3$ N$^+$ R *d*

N N N *e*

$Me_2\overset{+}{N}$=CH Cl *f*

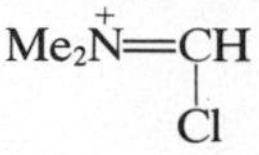

B. Three- and five-carbon reagents

HC≡C—CH(OEt)$_2$ *g*

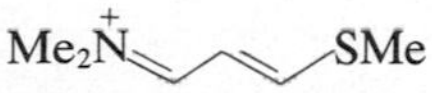

$Me_2\overset{+}{N}$ SMe *h*

N N NR$_2$ *i*

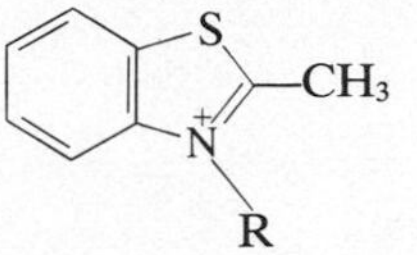

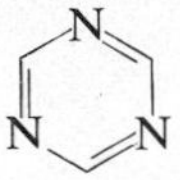

EtOCH=CH—CH(OEt)$_2$ *g*

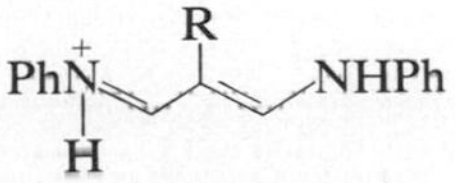

R PhN$^+$ H NHPh *j*

PhN$^+$ H ( )$_2$ NHPh *k*

X$^+$ X = NR O, S *l*

$Me_2\overset{+}{N}$ ( )$_n$ NMe$_2$ *m*

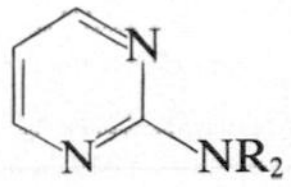

H CHO R CHO *n*

O O OR′ OR R = H, R′ = alkyl R = R′ = alkyl *o*

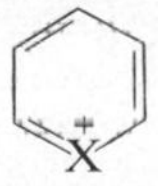

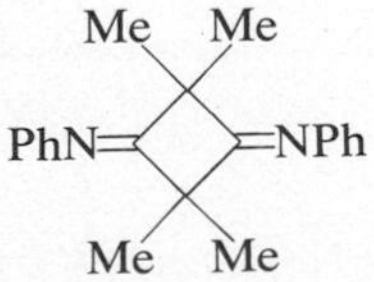

Me Me PhN= =NPh Me Me *p*

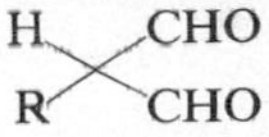

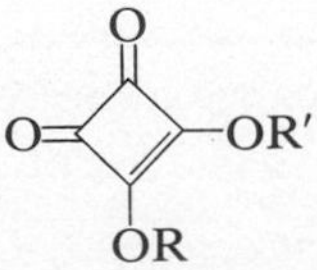

O ( )$_n$ CHO *q*

OH O OH O O *r*

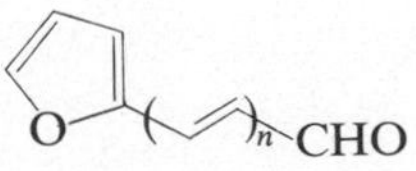

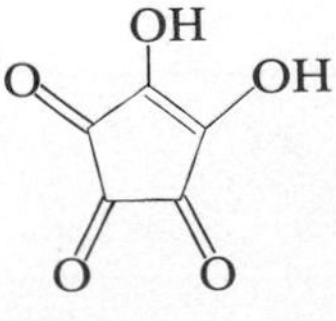

TABLE XII (*Continued*)

C. Unsymmetrical reagents

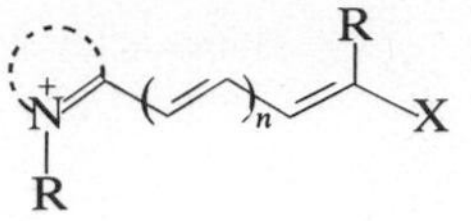

$X = OR, NHPh, SR, Cl, NR_2$

s

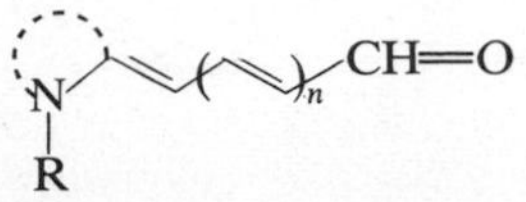

t

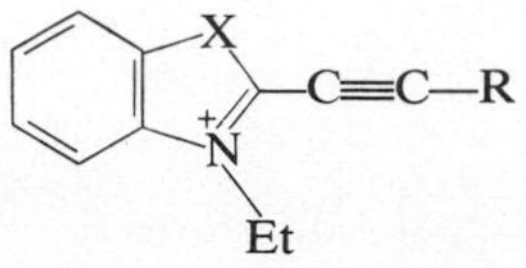

u

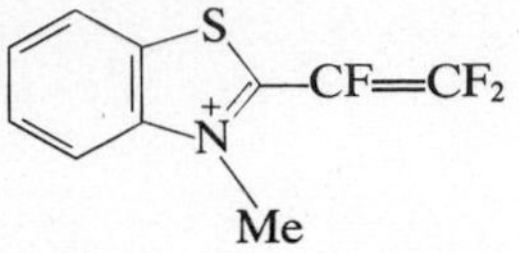

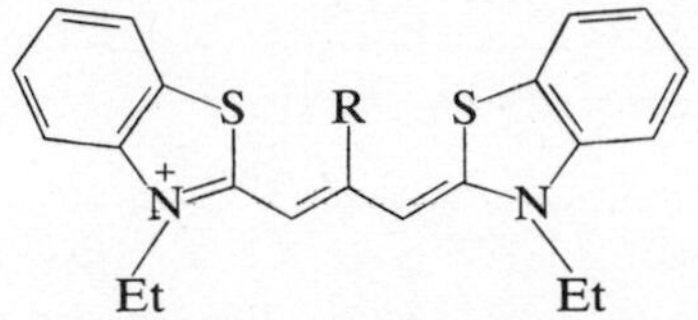

$R = Cl$ $R = SO_3^-$

v

w

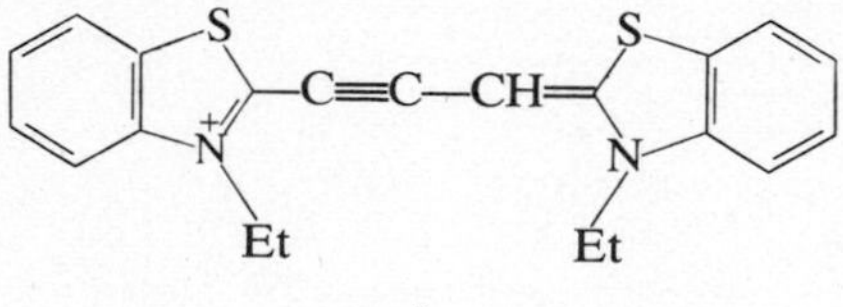

x

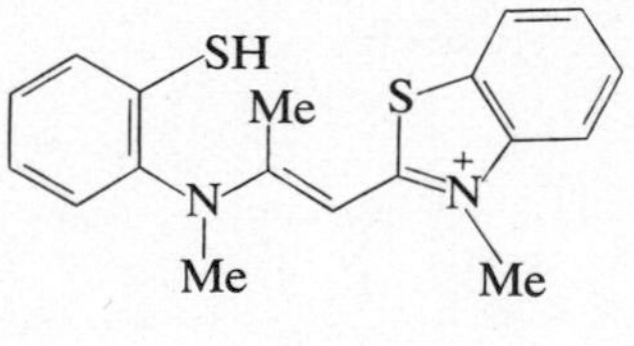

y

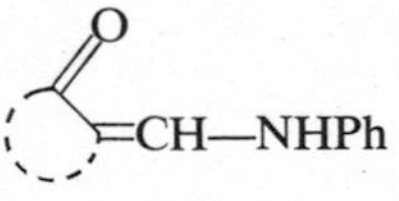

z

[a] L. G. S. Brooker and F. L. White, *J. Am. Chem. Soc.*, **57**, 2480 (1935) and references therein.

[b] S. G. Dent and L. G. S. Brooker, U.S. Patent 2,537,880 (1951).

[c] H. A. Piggott and E. H. Rodd, British Patent 344,409 (1929) and 354,898 (1930).

[d] L. G. S. Brooker and F. L. White, *J. Am. Chem. Soc.*, **57**, 547 (1935).

[e] A. Kreutzberger, *Arch. Pharm.*, **299**, 897, 984 (1966).

[f] CIBA, Ltd., British Patent 954,240 (1964).

[g] W. König, German Patent 410,487 (1922).

as electrophilic reagents for dye syntheses. These essentially symmetrical electrophilic reagents included: one-carbon analogs of orthoesters,[309–315] simple polymethines with displaceable terminal groups (vinylogs of orthoesters),[66–71,316–318] ring compounds like triazines and pyridiniums,[319–322] malonaldehydes,[323–327] malonaldehyde derivatives,[328–334] croconic acid,[335–337] and squaric acid (**100**) derivatives.[338–341] These squaric acid derivatives have provided a large series of novel dyes like **101** and **102** by electrophilic reaction at the 1,2- or 1,3-positions of the cyclobutane ring (Eq. 13). Dyes **101** and **102** are compared with a more typical bismerocarbocyanine **103** and other multichromophore dyes in Section VII.

Brooker[4] and Hamer[10] emphasized the importance of the orthoesters (reagents *a,b*, Table XII), their vinylogs (reagents *g*), and the dianilides

---

[h] B. Hirsch and E. Förster, *Chimia*, **20**, 126 (1966).
[i] R. G. Shepherd, *J. Chem. Soc.*, 4410 (1964).
[j] H. A. Piggott and E. H. Rodd, British Patent 355,693 (1930); S. Beattie, I. M. Heilbron, and F. Irving, *J. Chem. Soc.*, 260 (1932).
[k] N. I. Fisher and F. M. Hamer, *J. Chem. Soc.*, 189 (1933) and references therein.
[l] S. Creyf and L. Roosens, *Ind. Chim. Belge*, **23**, 853 (1958); *ibid.*, 967 (1958); J. Nys, *Chim. Suppl.*, 115 (1968).
[m] H. E. Nikolajewski, S. Daehne, B. Hirsch, and E.-A. Jauer, *Angew. Chem. Int. Ed. Engl.*, **5**, 1044 (1966).
[n] J. Clernik, *Collect. Czech. Chem. Commun.*, **37**, 2273 (1972); Ch. Reichardt and P. Miederer, *Justus Liebigs Ann. Chem.*, 750 (1973) and references therein.
[o] H. E. Sprenger and W. Ziegenbein, *Angew. Chem. Int. Ed. Engl.*, **7**, 530 (1968) and references therein.
[p] J. C. Martin, U.S. Patent 3,161,681 (1964).
[q] L. G. S. Brooker and G. H. Keyes, *J. Franklin Inst.*, **219**, 255 (1935).
[r] G. A. Rillaers and H. Depoorter, U.S. Patent 3,793,313 (1974); A. Treibs and L. Schulze, *Justus Liebigs Ann. Chem.*, 201 (1973).
[s] F. M. Hamer,[10] Chapters V, VII, VIII; (X=OR) H. Larivé, R. Dennilauler, and R. Baralle, *Chimia*, **19**, 238 (1965); ($n=0$, X=NHPh) ICI intermediate; (X=Cl) E. B. Rauch and J. A. Welsh, U.S. Patent 3,410,848 (1968).
[t] F. M. Hamer,[10] Chapters V, VII, VIII; J. D. Mee and D. W. Heseltine, French Patent 1,565,912 (1969).
[u] A. I. Kiprianov and G. G. Dyadyusha, *Zh. Obshch. Khim.*, **30**, 3654 (1960) and references therein; P. Collet and M. A. Compere, British Patents 1,093,938 and 1,093,940.
[v] M. M. Kul'chitskii, A. Ya. Il'chenko, and L. M. Yagupol'skii, *Zh. Org. Khim.*, **9**, 827 (1973) and references therein.
[w] G. L. Oliver, U.S. Patent 3,656,960 (1972); I. I. Levkoev, N. N. Sveshnikov, and N. I. Shirokova, *Dokl. Akad. Nauk. SSSR*, **153**, 350 (1963).
[x] J. D. Mee, *J. Am. Chem. Soc.*, **96**, 4712 (1974).
[y] J. Metzger, H. Larivé, R. J. Dennilauler, R. Baralle, and C. Gaurat, *Bull. Soc. Chim. Fr.*, 2868 (1964); L. Oliveros and H. Wahl, *ibid.*, 3204 (1971).
[z] F. B. Dains and E. W. Brown, *J. Am. Chem. Soc.*, **31**, 1148 (1909); F. B. Dains, R. Thompson, and W. F. Asendorf, *ibid.*, **44**, 2310 (1922).

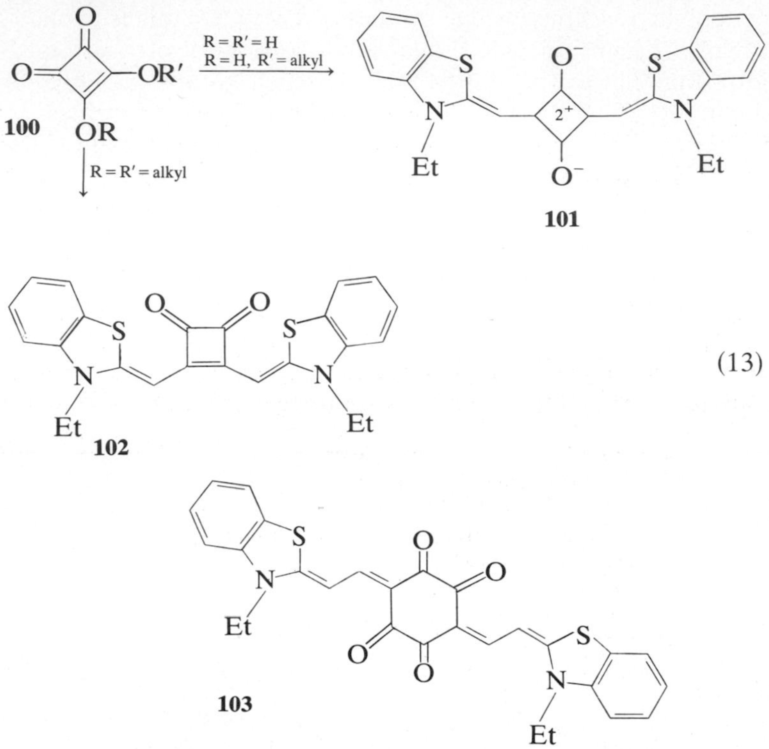

100 → 101 (R = R′ = H; R = H, R′ = alkyl); 100 → 102, 103 (R = R′ = alkyl) (13)

(reagents *j–l*) in the early development of general synthetic methods. The furfurals **104** (reagent *q*, $n=0-2$) reacted as cyclic analogs of orthoesters and in 1935[227] were shown to give tri-, tetra-, and pentacarbocyanines **106** via the bistetrahydroquinoline adduct **105** (Eq. 14).[227] Dyes of similar chain lengths **107** were also prepared from rigidized quaternary salts and standard electrophilic reagents (Scheme 1; dye **48**) and extended to allopolar heptacarbocyanines (Scheme 1; dye **49**). Absorption spectra for rigidized infrared dyes are characteristically broad (Fig. 6), due primarily to the presence of several isomeric forms. Pyridine tended to maximize the contribution of the long wavelength absorbing isomer.

Two of the most general unsymmetrical reagents are the diphenylformamidine derivatives of quaternary salts (ICI intermediates, reagent *s*, $n=0$ with X = NHPh, Table XII.C) and ketomethylene compounds (Dains intermediates, reagent *z*, Table XII.C) which were also described in a previous section. Analogous reagents are summarized in Table XII.C. The electrophilic reagents with displaceable groups attached to a vinylogous quaternary salt are represented by reagent *s* in Table XII, where the

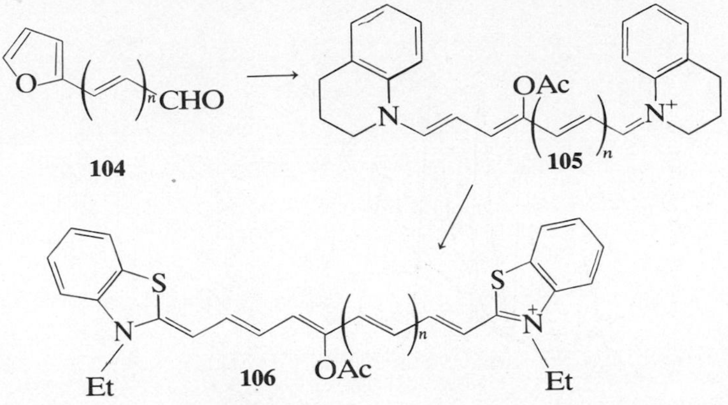

(14)

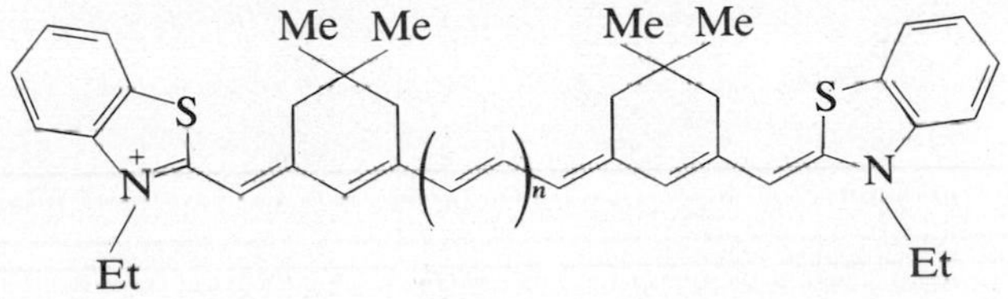

104

105

106

107

leaving group X is typically alkoxy, alkythio, chloro, anilino, and dialkylamino. Related compounds include reagent *v* with a fluoro leaving group and reagent *w* with the leaving group R as chloro or SO_3^{2-}. Uncharged analogs of these compounds are the formylmethylene compounds (reagent *t*, $n = 0$) and the Dains intermediate (reagent *z*).

The mechanism of dye formation was studied for merocyanines from analogs of the Dains intermediate (Eq. 9),[255,256] merocyanines from ICI intermediates (reagent *s* with X = NHPh or $N(COCH_3)Ph$),[138] and styryl dyes from an analog of the formylmethylene reagent *t* (*p*-dimethylaminobenzaldehyde).[257–259] Owen used a dimethylamino analog of the Dains intermediate in studies of the reactions of monomeric methylene bases.[255,256] The formation of merocyanine dye was first order in both the methylene base and the Dains analog (Eq. 9). The formation of a rhodanine merocyanine **109** from the benzimidazole ICI intermediate **108** (Eq. 15) gave zero-order rate constants.[138] The reaction in pyridine was over 40 times faster for the acetylated intermediate (**108**; R = $COCH_3$) than for the protonated compound **108**; R = H). Styryl dye formation from *p*-dimethylaminobenzaldehyde and quaternary salts was investigated extensively.[257–259] In aqueous alcoholic sodium hydroxide, condensation with *N*-methyl-4-picolinium iodide to give styryl dye **110**

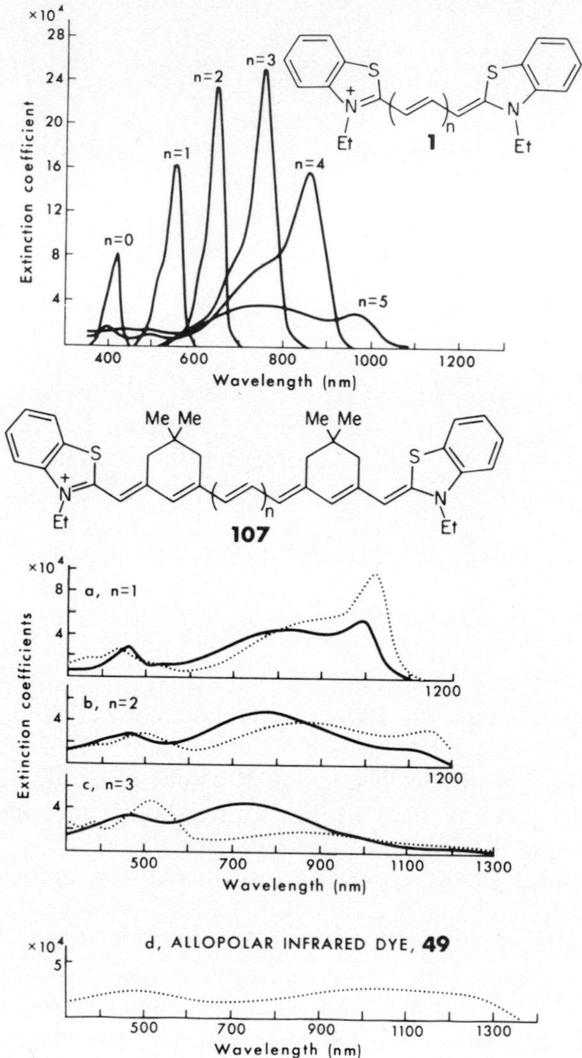

Fig. 6. Absorption spectra for vinylogous series of symmetrical cyanine dyes. For methanol solutions of thiacyanine dyes **1**, both λ_{max} and extinction coefficient increased as the methine chain was lengthened from $n = 0$ to $n = 3$, but the vibrational shoulder on the short-wavelength side was least prominent for $n = 3$. Longer chain thiacyanines ($n = 4$, 5) showed lower extinction coefficients for the longer wavelength band, and additional absorption bands in the 600 to 800 nm region. Rigidized long chain dyes (**107**; $n = 1$–3) and the allopolar infrared dye **49** (Scheme 1) showed increased absorption at long wavelength (curves **a**–**c** and **d**). Anhydrous pyridine (· · ·) maximized the long-wavelength peaks relative to the spectra in methanol (———). Brooker attributed this to a higher proportion of all-*trans* conformations for the dyes in pyridine.[3,4]

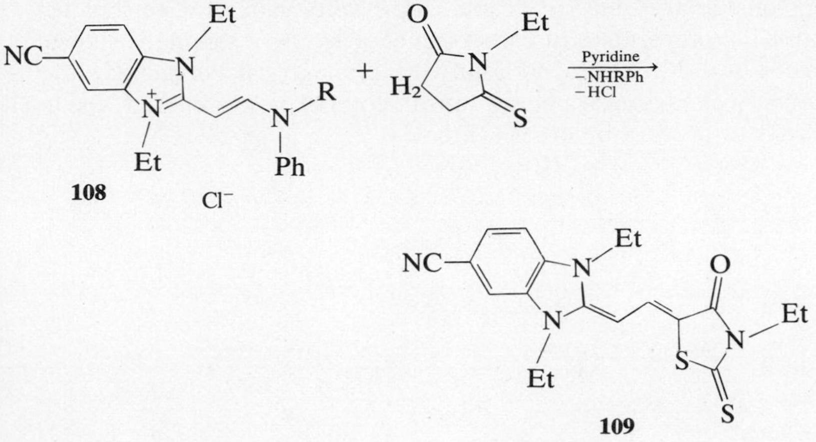

108 + → 109 (15)

Pyridine, −NHRPh, −HCl

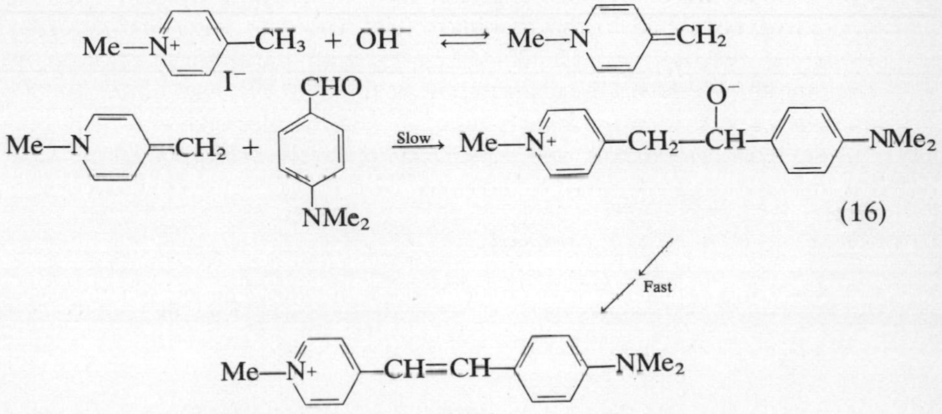

(16)

Slow, Fast

110

(Eq. 16) was first order in the benzaldehyde, quaternary salt, and base ($k = 1.7 \times 10^{-2}\ M^{-2}\ \text{sec}^{-1}$). The effective concentration of methylene base was low, but proton exchange was rapid for the quaternary salt. The kinetic data suggested that the rate-determining step was carbon–carbon bond formation between the aldehyde and methylene base rather than the subsequent dehydration.[259]

Acetylenic reagents for dye synthesis have been known for some time.[342–345] The 3,3-diethoxypropyne (reagent g, Table XII) is essentially a symmetrical electrophilic reagent, which was used to prepare a number of dicarbocyanine dyes. A vinylog of this reagent (HC≡C—CMe═CH—CHO) was used to prepare methyl-substituted tricarbocyanines.[345] The acetylenic quaternary salts *u* (Table XII) were general electrophilic reagents for the preparation of carbocyanine dyes. An example is shown in Scheme 1 for dye **51**. In addition, the symmetrical benzimidazolocarbocyanine with a central phenyl substituent **111** was prepared; these were difficult to prepare by other syntheses.

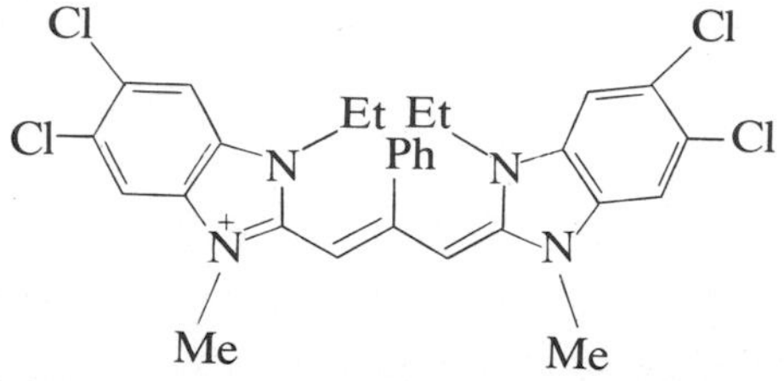

111

Acetylenic dyes, isolated recently from several synthetic approaches by Mee (Scheme 3), may be considered as electrophilic reagents for the synthesis of more complex dyes and also as an entirely new class of cationic dyes related to the cyanines.[346,347] The dyes were first detected from the reaction of **112** (X = CH=CH) with phosphoryl chloride in pyridine to give **113**. Direct formation of the dyes was also noted for the analogs of **112** incorporating imidazole heterocycles like 5,6-dichloro-1,3-diethylbenzimidazole and 1,3-diethylimidazo[4,5-*b*]quinoxaline. The

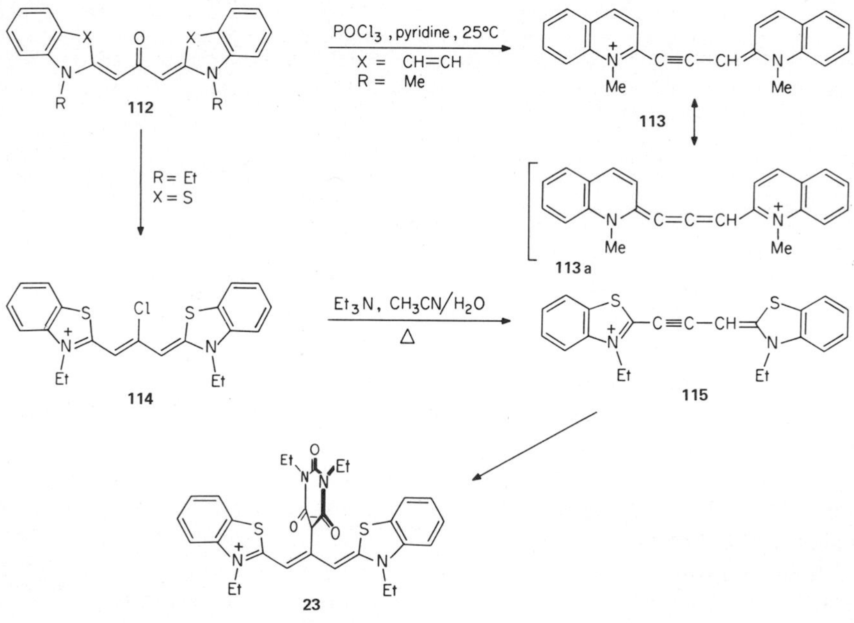

$POCl_3$, pyridine, 25°C; X = CH=CH; R = Me

R = Et; X = S

Et_3N, CH_3CN/H_2O; Δ

112 **113** **113a** **114** **115** **23**

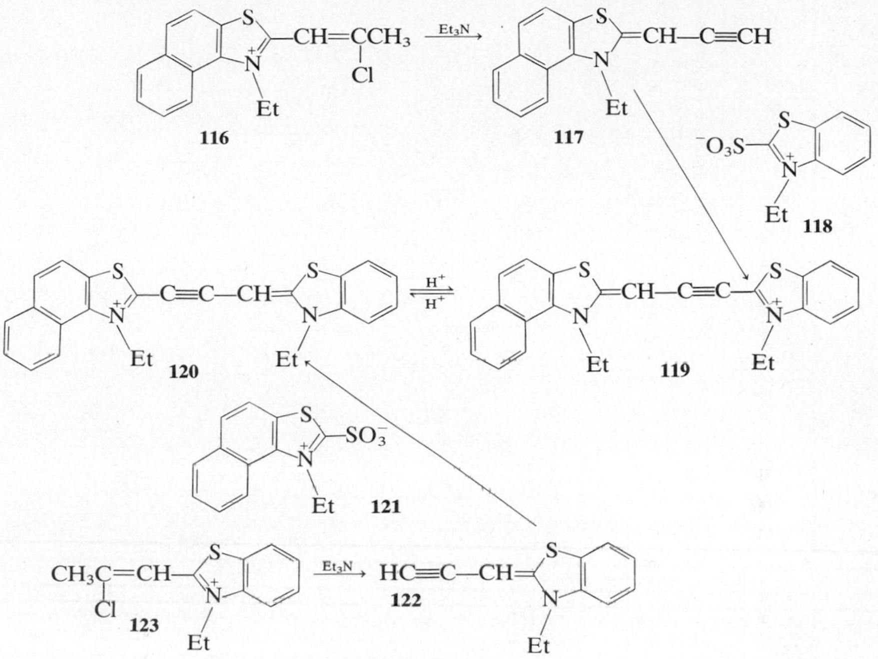

Scheme 3

benzothiazole analog (**112**; X = sulfur) gave the *meso*-chlorothiacarbocyanine **114** under the same conditions and required more forcing conditions (Scheme 3) to produce the acetylenic dye **115**. It was suggested that when the acetylenic dyes were formed directly, part of the driving force for their formation was derived from steric hindrance between the chlorine substituent and the heterocyclic rings. The acetylenic dyes were quite general electrophilic reagents, which reacted with the anions of either aryl thiols or ketomethylene compounds (e.g., 1,3-diethylbarbituric acid) to form chain-substituted dyes such as the allopolar dye **23**.

The acetylenic dyes are unsymmetrical by virtue of the triple bond in the chain, even though the heterocyclic terminal groups may be the same (dyes **113** and **115**). This asymmetry was reflected by the acetylenic dye absorptions, which were at shorter wavelengths than those of the corresponding (symmetrical) carbocyanines (Table XIII.A).[346–347a] Two possible resonance contributors for these dyes were the acetylenic form (structure **113**) and the cumulene form (**113a**). X-Ray crystallography on dye **32** (Fig. 3)[103] favored the acetylenic form as the major resonance contributor. The bond length between the "acetylenic" (sp-hybridized)

TABLE XIII. ACETYLENIC ANALOGS OF CYANINE DYES[a]

A. Acetylenic dyes containing identical heterocycles
$R_1^+—C\equiv C—CH=R_2$

B. Acetylenic dyes with two heterocycles
$R_1^+—C\equiv C—CH=R_2 \rightleftarrows R_1=CH—C\equiv C—R_2^+$
Tautomer **A** Tautomer **B**

R_1	R_2	$\lambda(\epsilon)$[a] (acetonitrile)	R_1	R_2	$\lambda(\epsilon)$[a] (acetonitrile)	Relative proportion of tautomers *A* and *B* at equilibrium(%)[c]
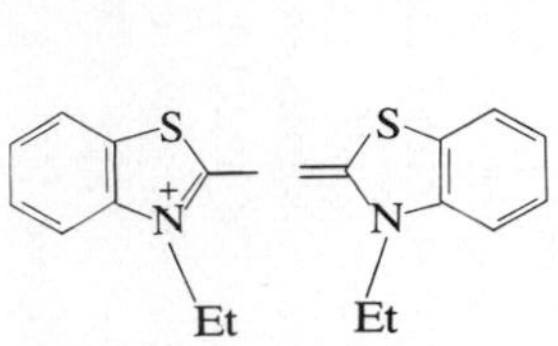		513 nm (9.9×10^4) [553 nm (16×10^4)][b]	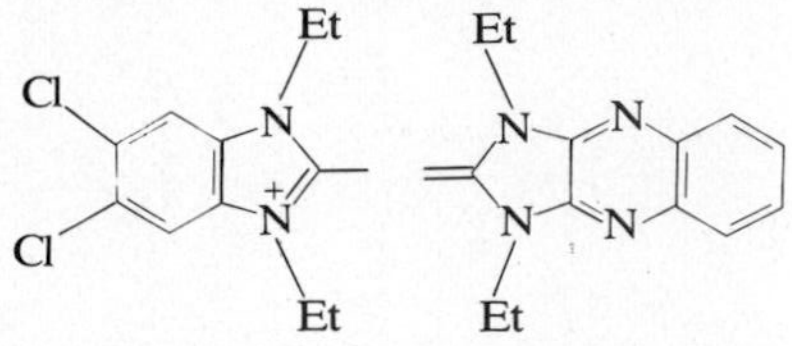		*A* 488 (13×10^4) *B* 532 ($>13\times10^4$)	>99 <1
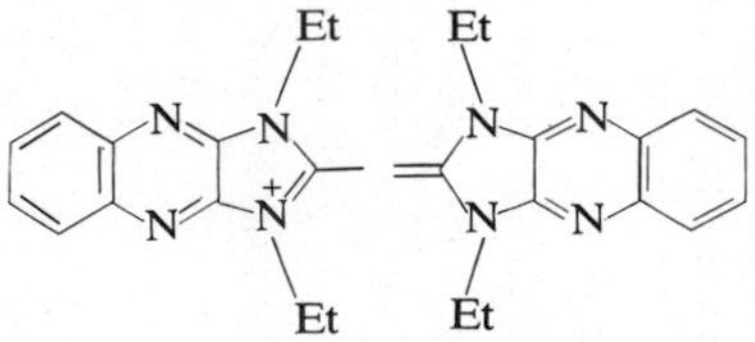		539 (21×10^4) [581 (32×10^4)][b]	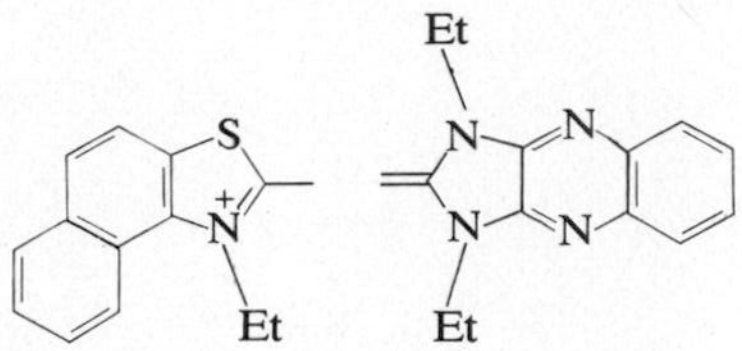		*A* 537 (14×10^4) *B* 556 (12×10^4)	75 25
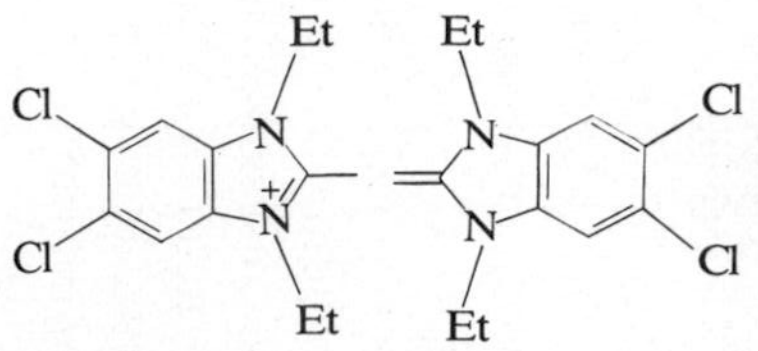		473 (14×10^4) [513 (21×10^4)][b]	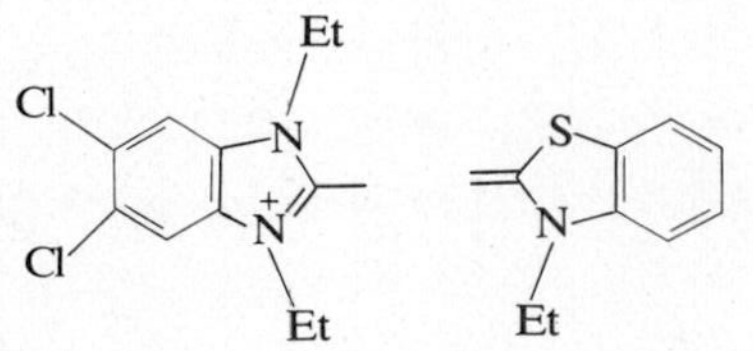		*A* 468 (9×10^4) *B* 510 (12×10^4)	>98 <2

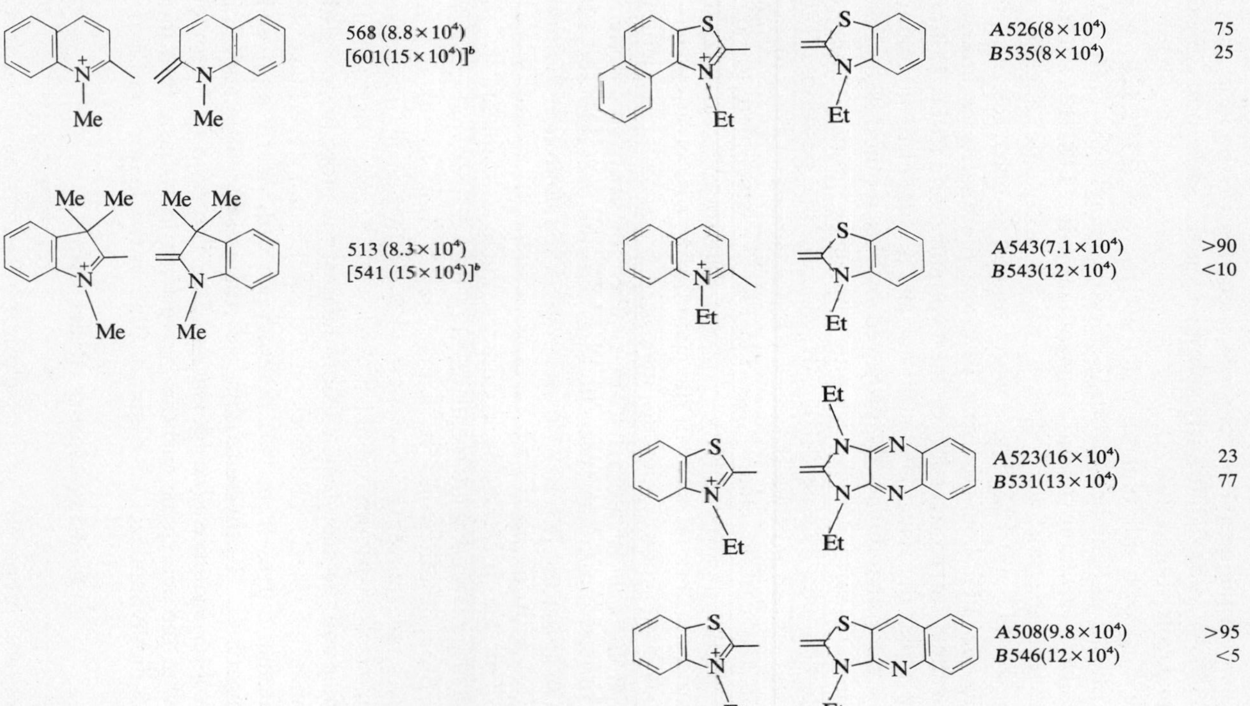

λ (ϵ)
568 (8.8 × 10^4) [601(15 × 10^4)][b]
513 (8.3 × 10^4) [541 (15 × 10^4)][b]

λ (ϵ)	Equilibrium
*A*526(8 × 10^4) *B*535(8 × 10^4)	75 25
*A*543(7.1 × 10^4) *B*543(12 × 10^4)	>90 <10
*A*523(16 × 10^4) *B*531(13 × 10^4)	23 77
*A*508(9.8 × 10^4) *B*546(12 × 10^4)	>95 <5

[a] J. D. Mee, *J. Am. Chem. Soc.*, **96,** 4712 (1972) and submitted for publication; absorption maxima (λ) in nm and extinction coefficients (ϵ) in liters mole^{-1} cm^{-1}.

[b] Absorption maxima (λ) and extinction coefficients (ϵ) for the related trimethine dyes.

[c] Equilibrium values ±5%.

carbons approached that for a normal acetylenic bond. In addition, the positive charge was localized to some degree on the benzimidazole heterocycle nearest the acetylenic carbons, since the CN bond lengths in that ring (1.356 Å) were significantly shorter than the same bonds in the other benzimidazole ring (1.370 Å).

Acetylenic dyes with different terminal heterocycles (**119**, **120**) were prepared as shown in Scheme 3 from 2-chloropropenyl salts (**116**, **123**) and 2-sulfobenzothiazolium (naphthothiazolium) salts (**118**, **121**). Intermediates in the reaction were shown to be acetylenic analogs of methylene bases, **117** and **122**. These were isolated and characterized for β-naphthothiazole **117**, benzothiazole **122**, and the analogous imidazo-[4,5-*b*]quinoxaline. Interconversion of the two tautomeric forms **119** and **120** was slow under the reaction conditions, and a number of tautomeric pairs of dyes were unambiguously prepared by this method. Equilibria between tautomeric forms were determined by acid-catalyzed equilibration of each tautomer (Table XIII). In general, the position of the equilibrium was influenced by a balance of three factors: (1) bond length alternation in the carbon chain (dye **32**, Fig. 3) which led to some localization of the positive charge in the heterocycle nearest the acetylenic carbons; (2) differences in the stabilization of a positive charge by the various basic heterocycles, as measured by Brooker deviations; and (3) minor steric interactions between either the acetylenic bond or the methine proton and the heterocyclic rings. For the dyes **119** and **120**, in which the two thiazole heterocycles were used, the dominant tautomer was **120** (75%) with the acetylenic carbons near the β-naphthothiazole heterocycle. For the preferred acetylenic resonance contributor, this tautomer **120** also favored some localization of the positive charge on the more basic β-naphthothiazole. A similar result was found for the tautomeric dyes incorporating 5,6-dichloro-1,3-diethylbenzimidazole and 1,3-diethylimidazo[4,5-*b*]quinoxaline heterocycles (Table XIII.B): the major tautomer had the acetylenic carbons nearest the more basic 5,6-dichlorobenzimidazole. Steric influences on the equilibrium were illustrated with the tautomeric dyes from two heterocycles with nearly equivalent Brooker deviations: 3-ethylbenzothiazole (deviation, **58**) and 1,3-diethylimidazo[4,5-*b*]quinoxaline (deviation, **50**). The ratio of tautomers was 77:23 with the acetylenic carbons nearest the least basic (but most hindered) 1,3-diethylimidazo[4,5-*b*]quinoxaline in the major tautomer.

C. Ring-Closure Reactions

Many dyes are prepared by ring-closure reactions at or near the dye-forming step of a synthetic sequence. The constitution of thiacyanine was originally established by the reaction of diethyl malonate and *o*-aminothiophenol.[26] Dithiolium, oxathiolium, and thiazolium quaternary salts and dyes were prepared in a similar manner,[348–353] including an optically active thiazoloquinocyanine (**124**; Eq. 17).[353] The vinylbenzothiazolium salt, reagent *y* in Table XII, was formally analogous to the ICI intermediates, but under certain reaction conditions the preferential electrophilic reaction was internal ring closure involving the *ortho*-thiol (or hydroxy) function. The compound formed by this ring closure was the methylene base dimer **67**, which provided nucleophilic reagents for dye syntheses.[247,248] Other ring closures in the final synthetic step include internal alkylation/acylation reactions[354–358] and bimolecular alkylations.[359–365]

$$\mathrm{Ph\overset{*}{C}HNH_2(Me)} \longrightarrow \mathrm{HNC(=S)Me\,(Ph\overset{*}{C}HMe)} \longrightarrow \text{thiazolium (2-Me, 4-Me, N-Ph}\overset{*}{C}\text{HMe)},\ \alpha_D = -68^\circ \longrightarrow \mathbf{124},\ \alpha_D = +1500^\circ \tag{17}$$

Four modifications of cyanine dyes were prepared through ring-closure reactions of a single intermediate **126** (Adduct A, Scheme 4).[354,355] Quaternary salts like **126** with reactive *N*-alkyl groups formed readily from methyl vinyl ketone and heterocyclic hydrobromide salts (**125**). This Michael addition reaction to the heterocyclic nitrogen occurred for benzothiazoles (Scheme 4) as well as for 2-benzyl-5-phenylbenzoxazole, 2-benzyl-1-methylimidazo[4,5-*b*]quinoxaline, and 2-benzyl-5,6-dichloro-1-ethylbenzimidazole. Cyclodehydration of these adducts by brief heating in dimethylacetamide provided dihydropyridinium salts (**127**) for all the heterocycles except the dichlorobenzimidazole, where reverse Michael addition was favored over cyclodehydration.[354] Dehydrogenation of the dihydropyridinium salts (Pd/C, dimethylacetamide) gave the related pyridinium salts (**128**). For the benzothiazolium salt **126** (R = Ph, R′ = methyl), cyclodehydration to **127** also occurred in dimethyl sulfoxide-d_6 after about 10 min (37°C) and warming to 65°C caused the subsequent dehydrogenation to the pyridinium salt **128**. Reaction of methyl vinyl

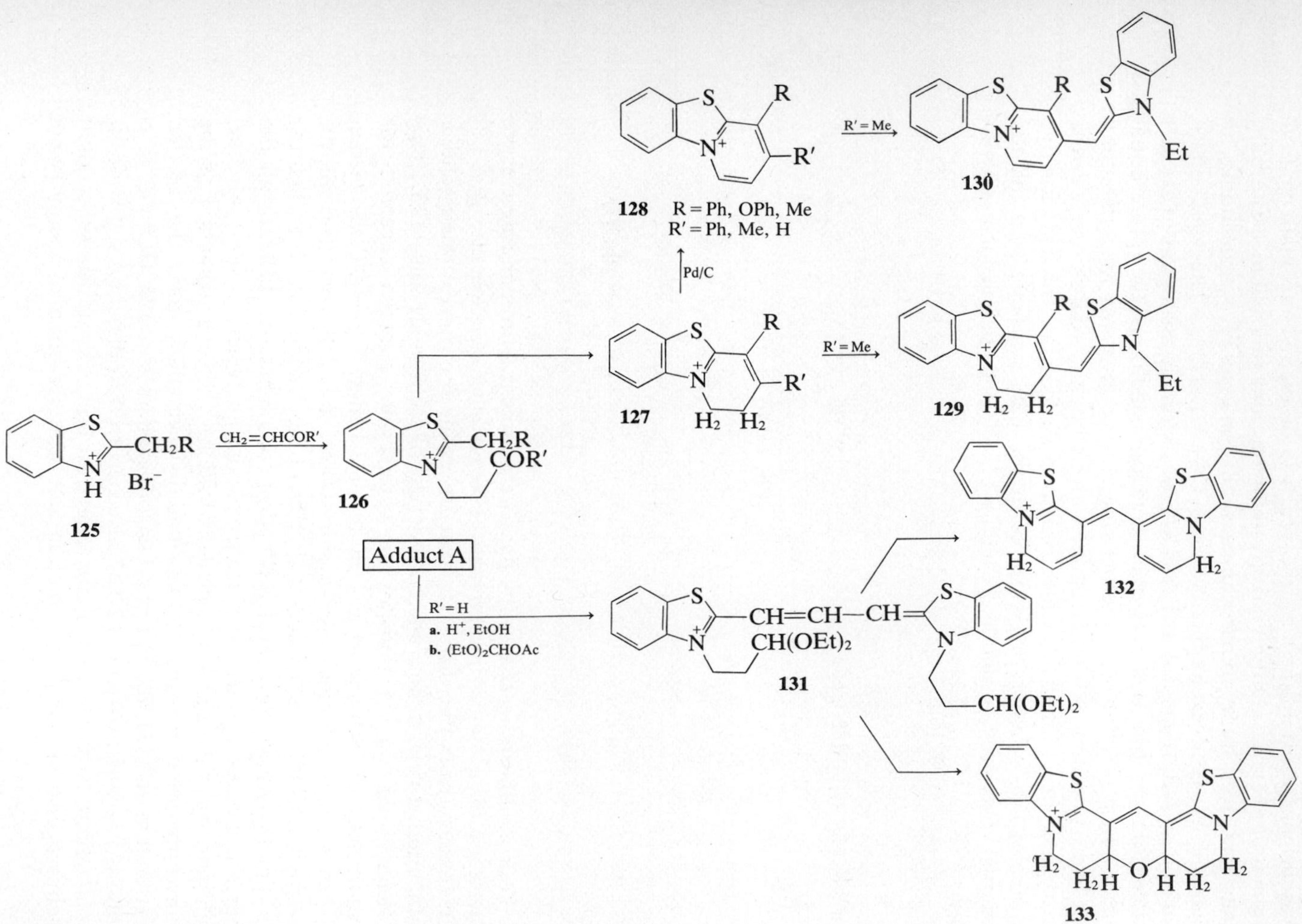

125

$CH_2=CHCOR'$

126

Adduct A

$R' = H$
a. H^+, EtOH
b. $(EtO)_2CHOAc$

127

Pd/C

128 $R = Ph, OPh, Me$
$R' = Ph, Me, H$

$R' = Me$

130

$R' = Me$

129

131

132

133

Scheme 4

ketone with hydrobromide salts of other heterocycles (2,3,3-trimethyl-3*H*-indole or 2-ethoxycarbonylmethyl-1-methylbenzimidazole) gave isomeric adducts corresponding to Michael addition to either the heterocyclic nitrogen or the activated 2-methylene (methyl) group. Cyclization and dehydrogenation led to the isomeric pyridinium salts **134** ($R' = CMe_2$ with $R = R'' = H$; $R' = NMe$ with $R,R'' = CO_2Et,H$) and **135** ($R' = CMe_2$ with $R = R'' = H$). The dihydropyridinium salt **127** gave novel rigidized dyes like **129**, and the related aromatic dyes (**130**) were formed from the pyridinium salts (**128**). Direct formation of dyes from the uncyclized adduct **126** led to the thiacarbocyanine **131**.[355] The reactive *N*-alkyl groups in this dye were ring closed to give the partially rigidized thiacarbocyanine **132** as well as the completely rigidized and highly fluorescent **133**.[355]

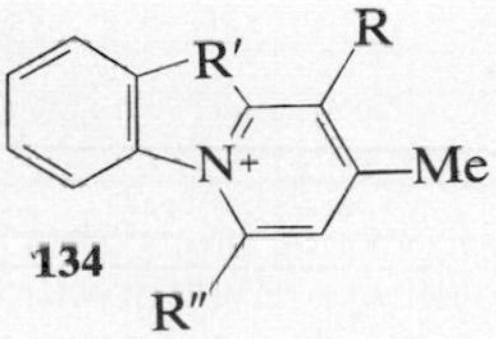

134 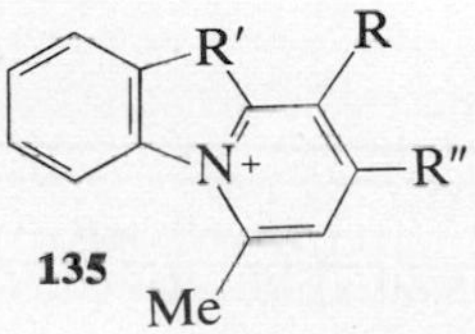 **135**

Alkylation positions for heterocycles with multiple reaction sites were reviewed by Duffin.[359] A number of pyrrolo-, oxazolo-, and thiazolopyridines were alkylated preferentially on the pyridine nitrogen. In contrast to this, isomeric thiazoloquinolines were originally presumed to be alkylated on the thiazole nitrogen.[362–364] Recent work[362] provided unambiguous syntheses of the two reactive quaternary salts from thiazolo[4,5-*b*]quinoline (**136**) with alkyl groups either on the quinoline nitrogen (for $R = Me$) or the thiazole nitrogen (for $R = SMe$). Comparisons of the resulting dyes with those produced from the directly alkylated intermediates suggested that the thiazolo[4,5-*b*]quinolines were also alkylated preferentially on the quinoline nitrogen. For other thiazoloquinoline isomers not all of the quaternary salts with alkyl groups on the quinoline nitrogen would have functioned as dye intermediates. The position of the *N*-methyl group in isomeric 1-methyl (or 3-methyl)imidazo[4,5-*b*]-pyridines influenced the course of quaternary salt formation by a subsequent alkylation.[365] The synthesis of rigidized thiacyanines **137** from bis-(benzothiazolyl)methanes involved double alkylation with 1,2-dibromoethane, and the related bis(3-ethylbenzothiazolium)butanes[301] with $MeCO_2CH(OEt)_2$ gave the rigidized carbocyanine **138**.

Shepherd reported that oxonols may be linked by a substituted alkyl chain to form dimeric dyes **139**.[360] Other dimeric dyes were also possible

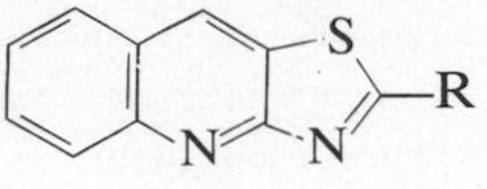

R = Me, SMe

136

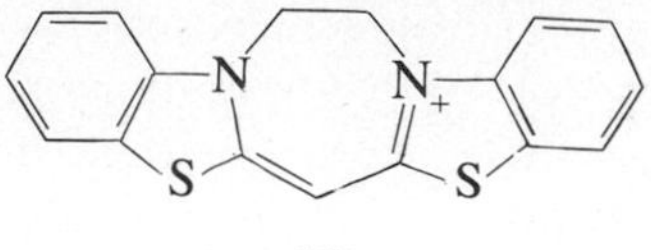

137

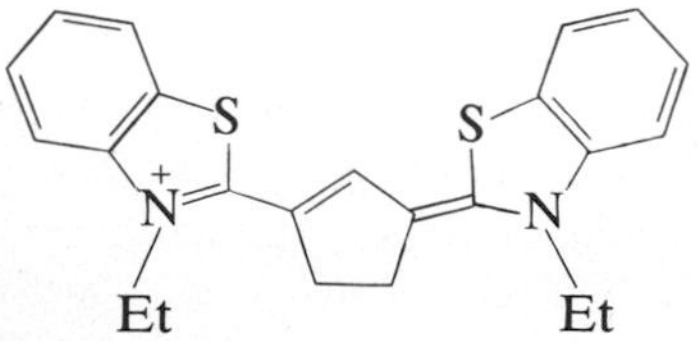

138

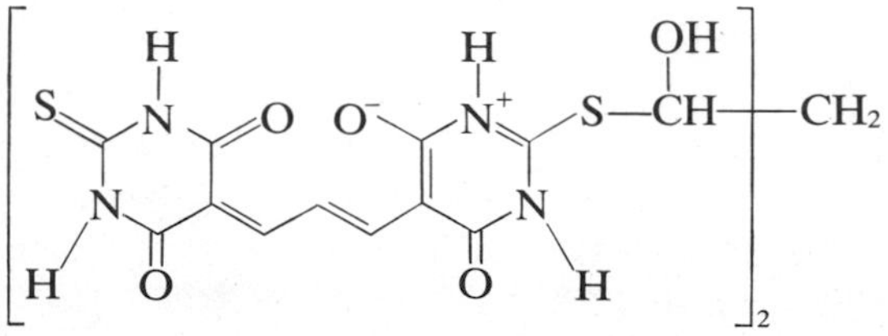

139

side products in the preparation of the rigidized cyanine **137**. Dimeric carbocyanine dyes of known structure were prepared directly using bis-quaternary salts **140**. Ring closure in the dye-forming step (Eq. 18) completed both carbocyanine chromophores to give dyes like **141** with several possible conformations (Fig. 7).

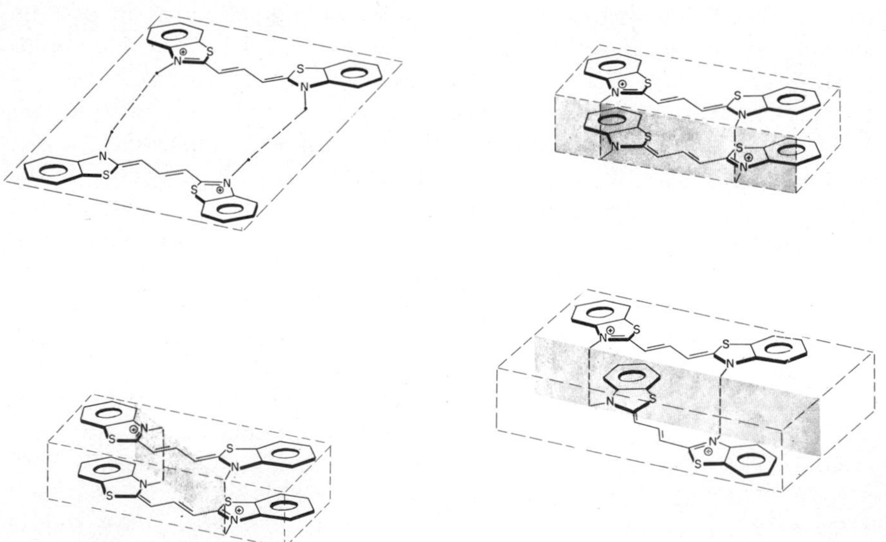

Fig. 7. Possible conformations of the macrocyclic *N,N'*-alkylene bisthiacarbocyanine **141**.

$$2\,[\mathbf{140}]\ (Br^-)_2(CH_2)_n \text{ bis(2-methylbenzothiazolium)} + 2\,CH(OEt)_2(OAc) \longrightarrow \mathbf{141}\ \text{(bis-carbocyanine bridged by } (CH_2)_n \text{ groups)} \qquad (18)$$

140 **141**

V. The Electronic Structure of Sensitizers

Large conjugated molecules like the cyanines and related dyes are described by general quantum theory as in Fig. 8. The filled orbitals of the ground states contain sigma (σ), pi (π), and lone-pair (n) electrons. Antibonding orbitals are typically pi (π^*) and sigma (σ^*). Although the relative energies of the orbitals will depend on the specific molecular system, orbitals of different types may have similar energies. For example, lone-pair orbitals and the least stable bonding π-orbitals are often of

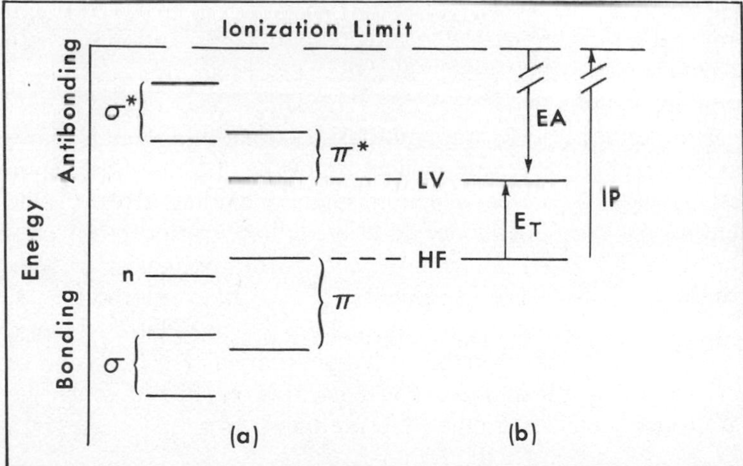

Fig. 8. Quantum mechanical orbital descriptions of cyanine and related dyes. (*a*) Bonding and antibonding orbitals for a typical dye molecule include: σ, π, n (bonding) and π^*, σ^* (antibonding). (*b*) Simplified energy diagram for the orbitals related to sensitizing activity: highest filled (HF) π-orbital and lowest vacant (LV) π-orbital. Combinations of these orbital energies with appropriate correction terms lead to transition energies E_T (singlet–singlet, singlet–triplet), electron affinities EA, and ionization potentials IP. Note that, in general, E_T is not equal to either $|IP-EA|$ or the analogous polarographic quantities $|E_{oxid}-E_{red}|$.

similar stability. The possible electronic transitions in the visible and infrared spectral regions include $\pi \rightarrow \pi^*$ and $n \rightarrow \pi^*$, but the latter are low extinction bands and can be obscured by the strong $\pi \rightarrow \pi^*$ transitions.

General reviews that treat dyes as well as less complex molecules have been published on absorption and electronic states,[366–368] classification of transition types and intensities,[369,370] molecular orbital descriptions of conjugated systems,[366–368,371–373] luminescence and intersystem crossing,[374,375] dye-laser activity,[376–378] and energy transfer.[379–381] Several dye properties can be related to spectral sensitization, including: (1) ionization potential, electron affinity, and related polarographic potentials; (2) absorption wavelength as related to dye structure and environment; (3) physical properties of excited states (relative stability, lifetime, decay pathways); and (4) chemical reactions of excited state dyes. Basic descriptions of absorption, structure, relative excited-state energies, and other photophysical properties are described in this section. Perturbations of absorption and excited-state properties, including induced color shifts, substituent effects, and photochemical reactions, are treated in later sections.

A. Absorption and Dye Structure: Quantitative Descriptions

Several theoretical approaches can predict dye absorption wavelengths. These include theories based on the essential dye chromophores (chromophore lengths, free electron), numerical descriptions of terminal groups (Brooker deviations), or combinations of atomic orbitals (molecular orbitals). Wavelength shifts are a function of the chromophore length (L) for most vinylogous series of compounds including dyes (Table II). Formulations for the dependence of λ on L have varied from the initial forms[382,383] ($\lambda \alpha L^a$, $a = 1$ for cyanines, $a = \frac{1}{2}$ for polyenes) to functions with continuously variable exponents.[384,385] Other methods assigned numerical values to specific terminal nuclei,[84,386] and Platt[84] systematized a large portion of Brooker's data for unsymmetrical dyes[43–55] and certain polyenes using Eqs. 19 and 20. The basicities (b_L, b_R) of more than 50 terminal groups were determined for methanol as a

$$E = [E_I^2 + (b_L - b_R)^2]^{1/2} \tag{19}$$

$$\frac{\epsilon_{\max}}{\epsilon_{I_{\max}}} = \frac{E_I}{E}\left[\frac{w_0}{w_0 + k|b_L - b_R|}\right] = \frac{E_I}{E}\left[\frac{w_0}{w_0 + k(E^2 - E_I^2)^{1/2}}\right] \tag{20}$$

solvent, relating the observed transition energy E to the "isoenergetic" energy E_I. The basicity difference ($b_L - b_R$) essentially corrected for the asymmetry of the chromophore in unsymmetrical dyes due to different (left and right) terminal groups. Since the chomophore asymmetry and thus the effective b_L, b_R values are a function of both structure and solvent, increased bandwidths and lowered ϵ_{max} values were found to be a function of the term $[w_0 + k|b_L - b_R|]$ in Eq. 20. This term, including the empirical constants w_0 and k, varies from about 1.0 kK for symmetrical cyanine dyes to about 5.0 kK for polyenes, and the striking variation of extinction coefficient (ϵ_{max}) with solvent composition for certain merocyanines was also reproduced. Cyanine dyes and polyenes were successfully described by modified versions of free-electron theory.[387–401] Variable shapes for the potential well of the free electron can reflect the single/double bond alternations in polyenes and unsymmetrical dyes[388–394] as well as differences among terminal groups.[397–400] Two- and three-dimensional potential wells were devised to treat polyenes, polyacetylenes, and cyanine dyes.[392–394]

Molecular orbital calculations applied to the cyanine and related dyes were generally those that give precise correlations with experimental data. Many investigations[171,402–410] using Hückel molecular orbital theory successfully correlated calculated molecular energy levels with appropriate physical data. The Hückel calculations proceeded from simple descriptions of atomic properties for aromatic carbons and heteroatoms (X): π-electronegativity (coloumb integral $\alpha_X = \alpha_{carbon} + h_X\beta_{C-C}$) and a bond strength between directly bonded atoms ($\beta_{C-X} = k_{C-X}\beta_{C-C}$) as well as some correction for σ-inductive effects of the heteroatoms. Necessary values for h_X and k_{C-X} were determined empirically. The resulting molecular orbitals related most directly to physical data were the highest filled orbital ϵ_{HF}, the lowest vacant orbital ϵ_{LV}, and the difference between these two $\epsilon_{HF} - \epsilon_{LV}$, defined as the calculated transition energy ϵ_T. All of these were defined in the arbitrary units β', but in a typical case (Fig. 9*a–c*) these were linearly related to the observed oxidation potential, reduction potentials, and transition energies, respectively.

The self-consistent-field-configuration-interaction (SCF CI) calculations like the Pariser–Parr–Pople method[410–427] have also provided good theoretical descriptions of transition energies (Fig. 9*d*), oscillator strengths, and other data. Additional SCF CI calculations were done for heterocyclic compounds related to the terminal groups in dyes.[428–432] Interactions between nonbonded atoms were utilized in the form of π-orbital overlap, electron repulsions, and σ-inductive effects,[412,416,417,427,429–431] so that differences among conformations of a dye molecule could be estimated.[424–426] All-valence-electron calculations defined the electronic ground states for model cyanine dyes[430,433–439] and molecular geometries for dye aggregates.[433–438]

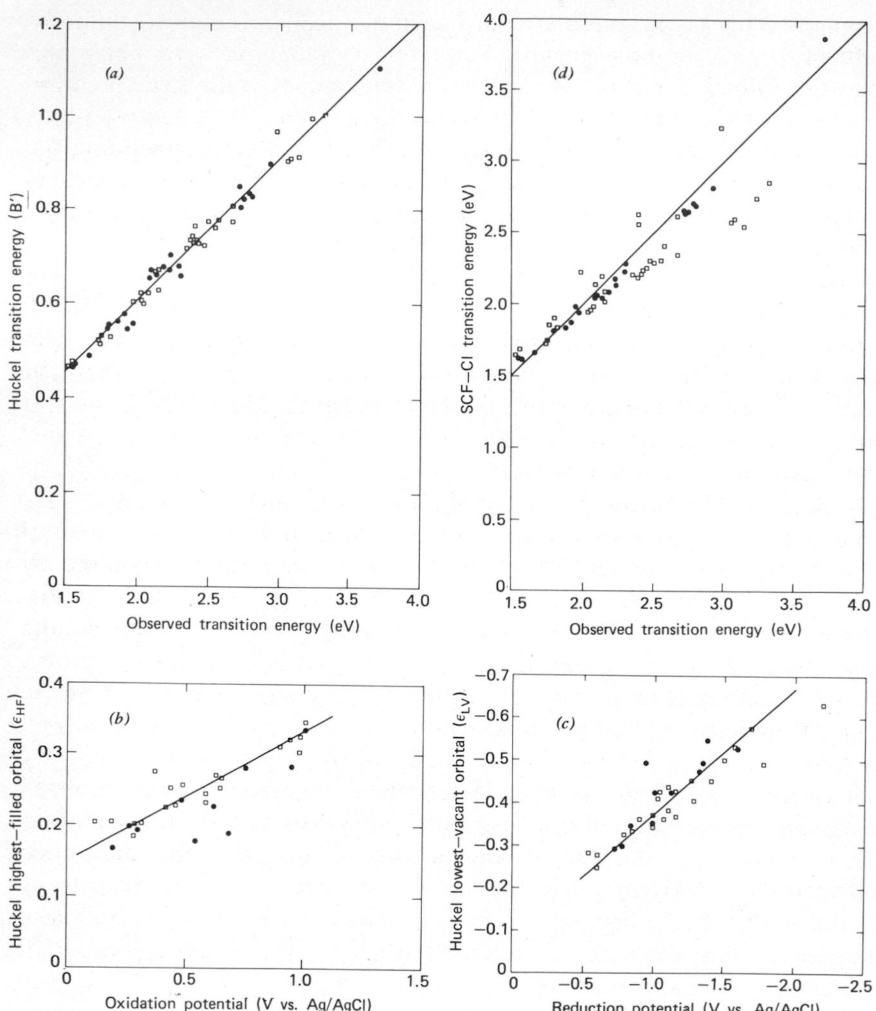

Fig. 9. Theoretical molecular orbital calculations on thiazole (●) and other dyes (□, oxazole, imidazole, quinoline, and selenazole) by the Hückel (HMO) and SCF CI (PPP) methods.[408,411] Calculated singlet–singlet transition energies from either method correlate with the observed absorption maxima for vinylogous cyanine dyes. Polarographic oxidation and reduction potentials from Table VI generally show linear relationships to the Hückel highest filled (ϵ_{HF}) and lowest vacant(ϵ_{LV}) orbitals, respectively.

Both the Hückel approach and the π-SCF CI methods required parameter calibration for precise applications to the molecular design of heterocyclic dyes.[411] The Hückel method was calibrated with transition energies for cyanine dyes,[408] where λ_{max} represented an accurate measure of the 0–0′ vibrational component of the electronic transition. The use of a vinylogous series of dyes from a single heterocycle (e.g., the thiacyanine series, **1**, X = sulfur with $n = 0$–5) led to poor calibrations, since any of a large, degenerate set of h_X, k_{C-X} parameters for sulfur provided equally precise linear correlations between calculated and observed transition energies. However variations in heterocyclic structure due primarily to use of additional fused benzene rings led to unique parameter sets that gave reasonably precise correlations not only with transition energies (Fig. 9*a*) but also with electrochemical potentials (Figs. 9*b*–*c*). The π-SCF CI methods were calibrated for uncharged heterocycles[412] as well as quaternary salts.[429–431] In these calculations as applied to dyes,[411,429–431] important variables included the molecular geometries, the extent of configuration interaction, and the size of the inductive effect for the heteroatoms.

In addition to the general color and constitution relations (previous section), certain other details of absorption spectra were investigated experimentally for model dyes: (1) transition polarizations,[440–453] (2) vibrational fine structure,[454–459] and (3) oscillator strengths.[457–463] The long-wavelength transitions were generally high extinction transitions involving the π-electrons and showed long-axis polarizations in cyanine dyes, as determined by dichroic absorption of polarized light (stretched polymer films with dye)[444–447,449] and by fluorescence polarization.[450–453] Shorter wavelength transitions were polarized both parallel and perpendicular to the long axis.[84,440–443]

Most cyanines showed prominent vibrational shoulders associated with the long wavelength electronic transition,[366–368,454] which included one or two vibrational quanta ($0 \rightarrow 1'$, $0 \rightarrow 2'$) relative to the absorption maximum ($0 \rightarrow 0'$). Intensity ratios for the $0 \rightarrow 0'/0 \rightarrow 1'$ bands increased as the chromophoric chain was lengthened (see thiacyanine series **1**, Fig. 6). Both Scheibe[455,456] and West[457] identified two distinct vibrational-frequency progressions in the low temperature absorption spectra of azacyanines[455,456] and several cyanine and carbocyanine dyes[457] (Fig. 10). The prominent progression of ~1370 cm^{-1} vibrational frequency in the excited state was suggested to correspond to the ~1500 cm^{-1} progression found in both the fluorescence emission spectra and the ir spectra of these dyes.[457] The second progression was ~550 cm^{-1} in the excited state and ~600 cm^{-1} in the ground state.[457] Bottger noted a strong Raman frequency at 600 cm^{-1} for a cyanine dye.[459] Less symmetric dyes (e.g., dye bases and merocyanines) exhibited long wavelength electronic absorptions where the $0 \rightarrow 1'$ vibrational component was the strongest (i.e., "λ_{max}"), and the $0 \rightarrow 0'$ and $0 \rightarrow 2'$ components were significantly weaker. This variation in the vibrational component producing the apparent

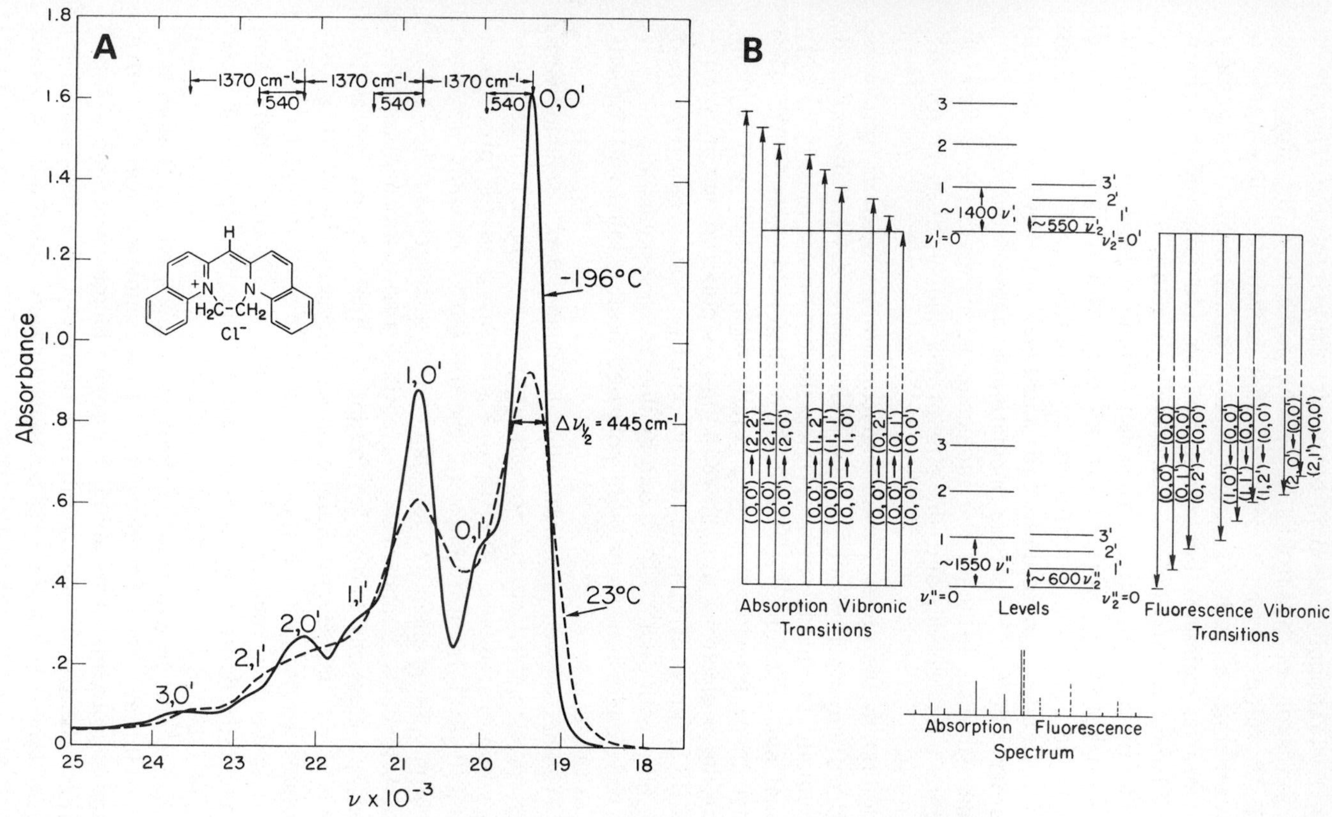

Fig. 10. Vibrational characteristics for a monomeric cyanine dye. (*a*) Absorption spectra in EPA at 23 and −196°C. The vibrational quantum numbers are denoted for the excited state only. (*b*) Vibrational spacings from the absorption spectra (~1400, ~550 cm^{-1}) are paralleled by slightly larger values from the fluorescence spectra (~1550, ~660 cm^{-1}.[457] (Reproduced by permission from *Photogr. Sci. Eng.*, **14,** 52 (1970).)

absorption maximum was analyzed by Scheibe and co-workers in terms of the variable shifts in the equilibrium distances for the ground and excited states.[455]

The oscillator strengths for typical cyanine and carbocyanine dyes increase for the longer chain dyes but are independent of temperature,[457] and Platt's successful analysis[84] of extinction-coefficient variations assumed a similar independence of merocyanine oscillator strengths in various solvents. Dye conformation, however, strongly affects absorption characteristics like bandwidth at low temperature,[457] extinction coefficient,[2,463] and oscillator strength.[460] Dipolar dyes with potentially attractive forces between the terminal chromophoric groups (merocyanines **4** and dye bases **5**) showed smaller increases in extinction coefficients than did symmetrical cyanines as the number of methine carbons increased. The vinylogous thiacyanine dyes showed maximum extinction coefficients for the thiatricarbocyanine (**2**, $n=3$ with seven methine carbons, Table II).[43] Related cyanines with longer methine chains showed broader absorption bands and lower extinctions (Fig. 6). For the polar merocyanines **20**, maximum extinctions were exhibited by the merodicarbocyanine (**20**; $n=2$ with four methine carbons) for a variety of pyridine : water mixtures.[52] Other series of merocyanines showed similar variations, but the extinction coefficients for the dye bases **5** (Table II) were not significantly increased by additional methine carbons.[43] Ground-state dipole moments for merocyanines also showed small increases for the longer chain members of vinylogous series (Table VIII). Both solvation and dye conformation were suggested to cause the lowered extinction coefficients of longer chain dyes (see Section VI.B.2).

B. Photophysical Behavior

Excited-state properties of the cyanine and related dyes have received much attention, particularly during the period 1960–present (see Section I.E). The partitioning of excited molecules (singlet versus triplet, bimolecular chemical reactions versus internal conversions) is important for sensitized processes as well as for dye-laser systems. Data on fluorescence, phosphorescence, transient absorption spectra, and dye lasers are discussed in this section.

The fluorescent properties of cyanine dyes have been known since the extensive study by Fisher and Hamer in 1936.[464] Eleven vinylogous series of symmetrical cyanine dyes were examined for fluorescence using both ultraviolet and visible exciting light. Except for the series containing quinoline, most of the dyes showed detectable fluorescence, which increased with an increase in chain length.[464,465] Fluorescence maxima for vinylogous cyanines, hemicyanines, and merocyanines were determined later.[466] Oxacarbocyanine iodides (**1**; X = oxygen, $n = 1,3$), 2,2′-quinocyanine iodides (**1**; X = CH=CH, $n = 0$–3), thiacarbocyanine iodides (**1**; X = sulfur, $n = 0$–3), and the hemicyanine iodides **142** ($n = 0$–3 for R = H

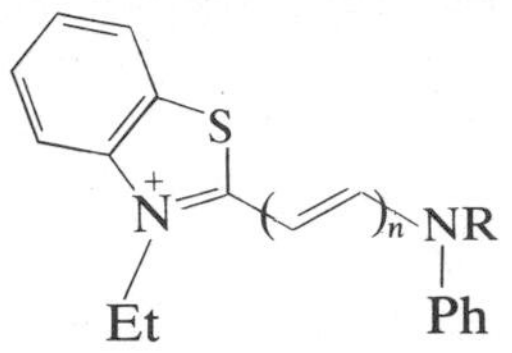

142

or methyl) gave detectable fluorescence in glycerol solution for all but the first member of each series. Most of the vinylogous dye series studied by Fisher and Hamer also included nonfluorescent simple cyanines.[464] Rhodanine merocyanines from 3-ethylbenzothiazole (**4**; $n = 0$–3) and the benzoxazole analogs were fluorescent for the dyes with $n > 0$, while the 4-quinoline/rhodanine dyes were fluorescent when $n = 0,1$.[466] The tendency for nonfluorescence by the first dye in each series was related to vibrational (twisting) motion induced by steric hindrance, as an extension of Lewis and Calvin's "Loose Bolt Theory" concerning the loss of fine structure in absorption spectra of substituted chromophores.[383,466,466a] Fluorescence maxima showed more constant Stokes shifts for the cyanine series than for the merocyanines. Both the absorption and fluorescence maxima for merocyanines shifted to longer wavelength as solvent dielectric constant increased, but the absorption was shifted the most. As the merocyanine chain length increased, the absorption wavelengths approached limiting values (convergent series like dye **21** in pyridine), but the fluorescence maxima formed a divergent series with the largest wavelength shifts occurring between merotricarbocyanines and dicarbocyanines.[466]

More recently investigated were general fluorescence behavior of other dyes[467–470] and dye crystals,[471,472] luminescence behavior of aggregates, monomers, and *cis-trans* isomers,[458,473–477] and the fluorescence lifetimes and radiationless transitions of the excited singlet states.[478–483] Specific details of excited-state reactions of dicarbocyanines have also been published.[484–493]

Most cyanine dyes exhibited small Stokes shifts for fluorescence maxima (Table XIV). Typical carbocyanines like oxacarbocyanine (O.C., **1**; X = oxygen with $n = 1$) and thiacarbocyanine (Th.C., **1**; X = sulfur with $n = 1$) showed 14 to 16 nm shifts in methanol with low quantum efficiencies for fluorescence ($\Phi_{Fl} < 0.05$). The dicarbocyanine analogs also showed small Stokes shifts but higher quantum yields ($\Phi_{Fl}=0.3$–0.5, Table XIV). Alkyl substitution on the center carbon of carbocyanines (*meso*-alkyl) lowered fluorescence quantum yields, but phenyl substitution, additional fused benzene rings, and rigidizing the methine chain raised Φ_{Fl}. This trend was most completely illustrated by a series of oxacarbocyanines in methanol (Table XIV):[483] oxacarbocyanine (O.C.) iodide or chloride ($\Phi_{Fl}=0.04$), 9-methyl-O.C. (0.005), 5,5′-diphenyl-O.C. (0.11), 9-ethyl-5,5′-diphenyl-O.C. (0.01), 5,6,5′,6′-dibenzo-O.C. **143** (0.26), rigidized 5,5′-diphenyl-O.C. **144a** (0.83), and rigidized 9-ethyl-5,5′-diphenyl-O.C. **144b** (0.77). The decreased quantum yields for chain-substituted nonrigidized dyes were attributed to nonfluorescent *cis* forms of the dye and enhanced vibrational deactivation of the normal dye conformation with a *trans* methine chain. In addition, higher Φ_{Fl} values for oxacarbocyanines

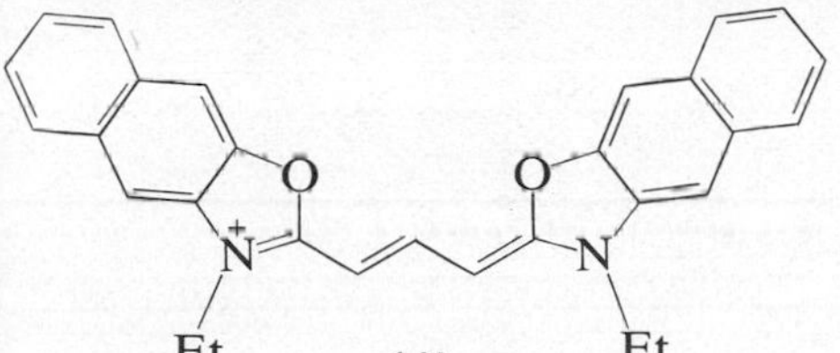

143

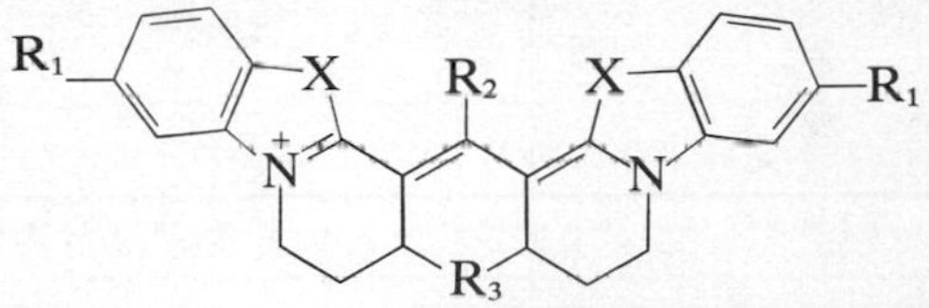

144a: X = O, R_1 = Ph, R_2 = H, R_3 =

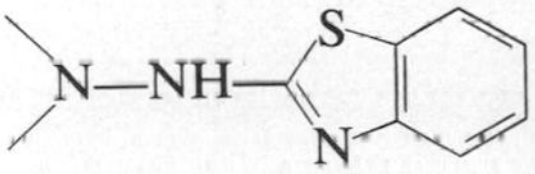

144b: X = O, R_1 = Ph, R_2 = Et, R_3 =

144c: X = S, R_1 = R_2 = H, R_3 =

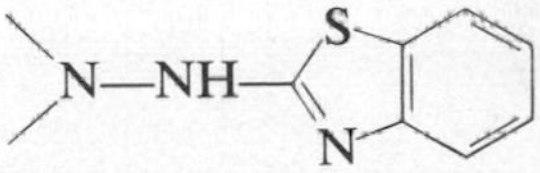

than for thiacarbocyanines, even for the rigidized dyes (**144a**, **144b** versus **144c**), paralleled an earlier ranking of general fluorescence intensities for carbocyanines: oxacarbocyanine > thiacarbocyanine > benzimidazolocarbocyanine > selenacarbocyanine.[468]

Fluorescence lifetimes[477,483] were generally less than 1 ns in solvents like methanol, chloroform, and dimethylformamide. Enhanced emission intensity and fluorescence lifetimes (Table XIV) were noted in viscous media (glycerol,[477] gelatin,[483] polymethylmethacrylate[483]). The fluroescence lifetime and quantum yield for oxacarbocyanine in a gelatin film were 2.9 nsec and 0.75, respectively, compared to less than 0.35 nsec and 0.04 in methanol solution.[483] However a monolayer film of a dye with the same chromophore, 3,3′-dioctadecyloxacarbocyanine, showed an approximate fluorescence quantum yield of 0.04 to 0.09, slightly higher than the

TABLE XIV. LUMINESCENCE DATA AND EXCITED STATE ABSORPTIONS FOR CYANINE DYES

Room temperature fluorescence data								Data for long-lived transients							
Dye	Matrix	λ_{max} (nm)	λ_{Fl} (nm)	Φ_{Fl}	τ_{Fl} (ns)	Maximum Φ_{ST}	Ref.	Dye	Matrix	T(°C)	Transient absorp-tions (nm)	λ Phosph. (nm)	Decay rates (sec^{-1})	ΔE_{ST}	Ref.
Oxacarbo-	MeOH	482	498	0.04	<0.35	0.01	[a]	Th.C.	Gelatin	25°C	640	—	1.8×10^2	—	[d]
cyanine															
(O.C.)	DMF	—	—	—	0.58	—	[b]		EtOH	23	550	—	1.7×10^2	—	[e]
	Glycerol	—	—	—	2.2	—	[b]		EtOH	−196	—	728	—	0.53eV	[f]
	Gelatin	492	510	0.75	2.9	—	[a]		EPA	−196	—	735	—	0.47	[g]
5,5′-Di-	MeOH	494	511	0.11	0.45	0.01	[a]	9-Methyl-	Gelatin	25	640	—	1.8×10^2	—	[d]
phenyl-								Th.C.			690		1.3×10^2	—	[d]
O.C.									EPA	−196	—	675	1.2×10^1	0.40	[h]
9-Methyl-	MeOH	497	514	0.01	—	0.01	[a]	Thia-	EtOH	−196	—	563	—	0.73	[f]
5,5′-Di-								cyanine	Alcohols	−196	—	535	$\sim 2.5 \times 10^0$	0.61	[i]
phenyl-O.C.															
"Rigidized"	MeOH	507	520	0.83	3.0	0.03	[a]	2,2′-	EtOH	−196	—	639	—	0.43	[f]
5,5′-Di-	Gelatin	535	545	—	2.3	—	[a]	Quino-	Alcohols	−196	—	640	5×10^0	0.43	[i]
phenyl-O.C.								cyanine	Gelatin	25	620	—	0.7×10^2	—	[d]
"Rigidized"	MeOH	511	523	0.77	—	0.03	[a]	O.C.	EtOH	−196	—	620	—	0.55	[f]
9-ethyl-									EPA	−196	—	614	—	0.47	[g]
5,5′-di-															
phenyl-O.C.															

Thiacarbo-cyanine (Th.C.)	MeOH	557	571	0.02	<0.35	0.01	[a]
9-Methyl-Th.C.	MeOH	543	567	0.001	—	0.01	[a]
"Rigidized" Th.C.	MeOH	571	581	0.52	5.7	0.03	[a]
	Gelatin	585	595	—	3.3	—	[a]
Phenosafra-nine	MeOH	525	555	—	2.6	—	[a]
	Gelatin	540	560	—	3.1	—	[a]
Oxadicarbo-cyanine	EtOH	580	605	0.49	1.2	<0.01	[c]
Thiadi-carbo-cyanine	EtOH	650	675	0.35	1.2	<0.01	[c]

9-ethyl-O.C.	EtOH	−196	—	620	—	0.52	[f]
	EPA	−196	—	634	—	0.47	[g]
5,5′-Dichloro-O.C.	EPA	−196	—	621	—	0.47	[g]
5,5′-Di-iodo-O.C.	EPA	−196	—	618	—	0.45	[g]
Oxadi-carbo-cyanine	EtOH	25	620 photoisomer	—	7.7×10^2	—	[c]
	EtOH–naph-thalene	25	650 $T_1 \leftarrow T_0$	—	$\sim2\times10^2$	—	[c]
Thiadi-carbo-cyanine	EtOH	25	705 photoisomer	—	3.1×10^3	—	[c]
	EtOH–naph-thalene	25	700–800 $T_1 \leftarrow T_0$	—	8.9×10^2	—	[c]

[a] D. F. O'Brien, T. M. Kelly, and L. F. Costa, *Photogr. Sci. Eng.*, **18**, 76 (1974).
[b] J. Q. Umberger, *Photogr. Sci. Eng.*, **11**, 392 (1967).
[c] D. N. Dempster, T. Morrow, R. Rankin, and G. F. Thompson, *J. Chem. Soc., Faraday Trans.* 2, **68**, 1479 (1972).
[d] S. H. Ehrlich, *Photogr. Sci. Eng.*, **18**, 179 (1974).
[e] P. J. McCartin, *J. Chem. Phys.*, **42**, 2980 (1965).
[f] W. West in "Scientific Photography," H. Sauvenier, Ed., Pergamon Press, Oxford, 1962, p. 557.
[g] U. Mazzucato, G. Favarro, and M. P. Mazzucato, *Ric. Sci. Parte 2: Sez. A*, **4**, 501 (1964).
[h] W. Cooper and K. A. Rome, *J. Phys. Chem.*, **78**, 16 (1974).
[i] R. A. Berg and A. Ron, *J. Chem. Phys.*, **59**, 3289 (1973).

methanol value for O.C. but significantly lower than the value for a gelatin film.[483] Although the dye was physically rigidized in the monolayer, which should increase Φ_{Fl}, it was also aggregated, and aggregation reduces Φ_{Fl}. Thiacarbocyanine in poly(methyl methacrylate) showed a tenfold increase in Φ_{Fl} (≈ 0.5) over the value for the dye in an unpolymerized methyl methacrylate solution, and chemically rigidized analogs (**144a**, **144b**, **144c**) exhibited high fluorescence yields and lifetimes of 2 to 6 nsec both in methanol solution and in gelatin films.[483]

The early observations of Buettner on 2,2′-quinocyanines[480] have been supported by more recent work on other cyanine dyes: the main nonfluorescent process is the radiationless deactivation $S_1 \rightarrow S_0$. Maximum singlet-triplet interconversion in methanol for carbocyanines was about 3% (Table XIV, max $\Phi_{ST} \leq 0.03$), so that the sum $[\Phi_{Fl} + \Phi_{ST}]$ was less than 0.10. In contrast to this, hydrocarbons and other classes of dyes exhibited significant triplet yields ($\Phi_{ST} = 0.2$–0.9), and the sum $[\Phi_{Fl} + \Phi_{ST}]$ was close to one.[483] Fluorescence quenching for cyanines was observed in the presence of excess iodide ion[477] and iodomethane,[483] but substitutents like meso-alkyl (in nonrigidized carbocyanines)[483] affected the fluorescence intensity more.

The fluorescence of oxadicarbocyanine was first described in 1935,[465] and since that time, this and other dicarbocyanines have been studied extensively[484–493] because of their importance to the dye-laser field. 3,3′-Diethyloxadicarbocyanine was suggested to convert to a long-lived photoisomer absorbing from 600 to 650 nm ($\lambda_{max} = 620$ nm, $\epsilon = 1.8 \times 10^5$ liters mole^{-1} cm^{-1}), and no absorption capable of being quenched by oxygen was found ($\Phi_{ST} \leq 0.01$).[484] A second dye (3,3′-diethylthiadicarbocyanine) behaved similarly. The characteristics reported for the first excited singlet states of these dyes (lifetimes $\cong 1$ nsec) do not suggest their known use for mode-locked lasers, but at the high photon densities in the laser cavities, the dyes could be completely converted to the photoisomers in less than 500 nsec. Thus the properties of the photoisomers (excited singlet-state lifetimes < 300 psec) were suggested to be critical for the mode-locked pulses that were observed experimentally.[484]

Transient photostates with lifetimes greater than fluorescence lifetimes were observed under a variety of conditions (Table XIV). Phosphorescence maxima at low temperature (−196°C) were compared with absorption maxima to estimate the energy difference between the first excited-singlet state (S_1) and the lowest triplet state (T_0).[375,468,494–502] Carbocyanine dyes generally showed 0.45 to 0.55 eV ($S_1 - T_0$).[375,468] For thia- and selenacarbocyanines this energy difference was decreased by *meso*-alkyl substitution. Simple cyanines from quinoline showed lower energy differences (0.43 eV), and thiacyanine was higher (0.73 eV).[375]

Singlet-triplet splittings for commonly used sensitizers for photooxygenation reactions were defined by the difference between the energies for the onset of singlet–singlet absorption and phosphorescence emission (xanthenes, 0.33 eV; isoalloxazines, 0.32 eV; and acridines, 0.31 eV).[494] For the xanthene and acridine dyes, where the absorption and phosphorescence curve shapes approached those for cyanines, energy differences between the maxima were approximately 0.45 to 0.55 eV.

Phosphorescence lifetimes at 77°K have been reported for simple cyanines,[475,501] a simple merocyanine,[500] and carbocyanines.[502] Lifetimes for the 1,1′-diethyl-2,2′-quinocyanine (0.2–0.3 sec)[475,501] were shorter than for the rigidized 1,1′-ethylene-2,2′-quinocyanine (0.59 sec).[475] Unusually short lifetimes were noted for 1,1′-methylene-2,2′-quinocyanine (0.06–0.10 sec).[475] Concentration-dependent lifetimes for thiacyanine varied from 0.35 sec at 10^{-6} *M* to a limiting 0.55 sec at 10^{-4} *M*.[501] The simple merocyanine 3-ethyl-5-(1,3-dithiolanylidene)rhodanine (**145**) gave lifetimes of 0.043 sec in 2-methyltetrahydrofuran and 0.032 sec in 2-methyltetrahydrofuran : ethyl iodide.[500] Single-crystal studies on this

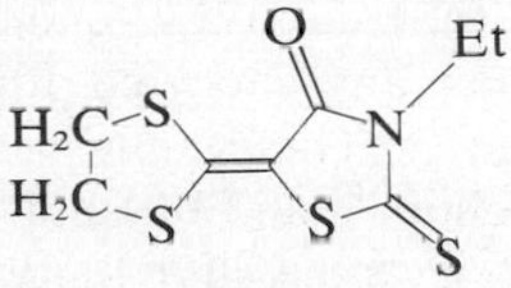

145

merocyanine at 4.2°K characterized the $S_0 \rightarrow T_0$ absorption and the $T_0 \rightarrow S_0$ emission.[500] The phosphorescence lifetime for 3,3′-diethyl-9-methylthiacarbocyanine bromide was 0.081 sec (77°K, ethyl ether–isopentane–ethyl alcohol) but with added ethyl iodide it decreased to 0.003 sec.[502]

Electron-spin resonance signals also resulted from light absorption at low temperatures.[503–506] In some cases delayed fluorescence[507] and excimer emission[473] were noted but the delayed fluorescence was suggested to result from photolytic degradation products.[474] Aggregation also increased phosphorescence from dyes.[508]

Flash photolysis of alcoholic dye solutions near room temperature produced transient photostates assigned to photoisomers[484,485,509–511a] (see next section), which revert to more stable forms via first-order processes [$E_{activation} = 12$ to 15 kcal/mole]. Triplet–triplet absorption spectra of organic molecules were recently reviewed, and reliable data through September 1971 were tabulated.[370] Although a number of organic dyes had been examined, only two series of cyanines were noted. In more recent studies including those noted in Table XIV as examples of long-lived dye transients, flash photolyses to produce both triplet and other excited states of dyes at room temperature were accomplished by two general methods: (1) the use of dyes in rigid films (polyvinyl alcohol,[480,512–514] gelatin,[512,515,516] or dye monolayers[496,517]) and (2) energy-transfer excitation from the triplet states of hydrocarbons like anthracene.[484,506,518,519]

Flash photolyses of simple cyanines have been reported by a number of investigators. The most widely studied was 1,1′-diethyl-2,2′-quinocyanine (**1**; X=CH=CH with $n = 0$, λ_{max} 523 nm). In poly(vinyl alcohol) or gelatin films at room temperature, this dye showed broad triplet-triplet absorption with a maximum near 640 nm.[480,513,515] Buettner compared this cyanine with its 6-bromo- and 6-iodo analogs.[480] All the cyanines showed triplet–triplet absorption maxima near 630 to 660 nm with high extinction coefficients ($3–4 \times 10^4$ liters mole^{-1} cm^{-1}). 2,2′-Quinocyanine and 6-bromo-2,2′-quinocyanine showed low quantum yields of triplet formation (0.02, 0.08), but 6-iodo-2,2′-quinocyanine had a significant triplet yield (0.26). Decay rates for the triplet absorptions showed both short-lived and longer lived components.[480,513,515] For the first-order long-lived component, the decay rate for 6-iodo-2,2′-quinocyanine ($\sim 4 \times 10^2$ sec^{-1}) was significantly higher than any reported for the unsubstituted 2,2′-quinocyanine ($4–8 \times 10^1$ sec^{-1}). Thiacyanine gave triplet-triplet absorption in both poly(vinyl) alcohol)[513] and butanol.[514]

Flash photolyses of carbocyanine and longer chain dyes, leading to facile isomer formation in nonviscous solutions,[484,485,509,510,514] gave transient absorptions extending well into the infrared for dyed poly(vinyl alcohol)[513] and gelatin films.[515,516] Excited-state lifetimes were monitored by both the regeneration rates of the bleached ground state and the relaxation rates for the induced absorptions. For typical carbocyanines in gelatin films, these rates gave similar lifetimes, although chain substitution (9-methyl) resulted in two sets of lifetimes: thiacarbocyanine, 5.2 msec (576 nm), 5.5 msec (640 nm), and 9-methylthiacarbocyanine, 5.9 msec (582 nm), 5.6 msec (640 nm), and 11.6 msec (530 nm), 8.0 msec (690 nm).[515] The examination of a series of dyes in gelatin films showed that the amount of bleached ground state depended more on reduction potential than structural features. Since the largest amount of bleaching occurred for the dyes with the most negative reduction potential, both dye radicals and triplets may contribute to observed transient absorption spectra.[515] An extensive study of methylene blue distinguished among monomer triplets, dimer triplets, and chemical intermediates as long-lived transients produced by flash or laser photolyses.[495]

The organic dye laser field has grown enormously in recent years, particularly with respect to continuously tunable lasers that operate at wavelengths from 340 to 1150 nm. Early emphasis on the dicarbocyanines and other dyes from quinoline[490–493,520–522] was augmented to include di-, tri-, and tetracarbocyanines from benzoxazole, benzothiazole, and related heterocycles.[523–539]

Snavely,[376,540] Strome,[541] and others[377,378,542–548] have reviewed the general subject of organic dye lasers. Their extensive work showed that cyanine, oxonol, and merocyanine chromophores were all suitable for dye laser applications. Specific reviews of the structures and efficiencies of polymethine laser dyes have been published recently.[376–378,542–548] In the best dyes the chromophore was not highly branched, and singlet-triplet intersystem crossing was negligible. The efficient laser dyes are not bleached irreversibly during excitation,[484,549,550] and the quenching of triplet states by oxygen and other materials (e.g., cycloheptatriene[551]) increased laser output significantly. The problem of triplet state interference in solid-state dye lasers was studied recently.[552]

The fluorescence output of these dyes can be tuned to specific wavelengths by several methods.[540] Broad, untuned fluorescence has bandwidths of about 4 nm, but the tuning element narrows the line in such a way that nearly all the energy contained in the broad line is condensed into the narrow line of about 0.7-nm bandwidth.[540,541]

VI. Induced Color Shifts in Cyanine and Merocyanine Dyes

Many classes of dyes exhibit reversible color shifts as a function of environmental changes. Photoinduced color changes at low temperature were studied for many years and recently, such changes have been more extensively studied at room temperature (see Section V.B). Color shifts were also induced by temperature, solvent, or ionic-strength changes. In general, the color shifts that have historically received the most emphasis may be divided into two categories: (*a*) color shifts and isomerism—usually involving two or more distinct species, each with an identifiable absorption wavelength (*syn*/*anti* isomerism, *cis*/*trans* isomerism, allopolar isomerism, merocyanine/spiropyran isomerism) and (*b*) color shifts and continuous environmental changes—usually a single chromophore showing a progressive shift in wavelength or extinction coefficient as a function of substituents (steric crowding of the chromophore environment), a change in solvation (solvent sensitivity), or the influence of neighboring chromophores (J- and H-aggregation).

A. Color Shifts and Isomerism

Typical isomeric forms of simple cyanine and carbocyanine dyes are shown in Eq. 21. The *syn/anti* terms refer to the relative orientation of the heterocyclic groups and *cis/trans* terms describe the methine chain. New interactions are labeled as *r* (1,4) and R (1,5), and the R-interaction depends on the nature of the heteroatom. Simple cyanines (**146** with $n=0$), investigated by spectral methods[440–442,460] and with nuclear magnetic resonance,[125,126] existed in the *syn* conformation (Eq. 21) for most dyes where X = CH except for the pyridocyanines. Simple aza-cyanines (X = –N=)[125,126] and certain hindered quinocyanines[460] existed

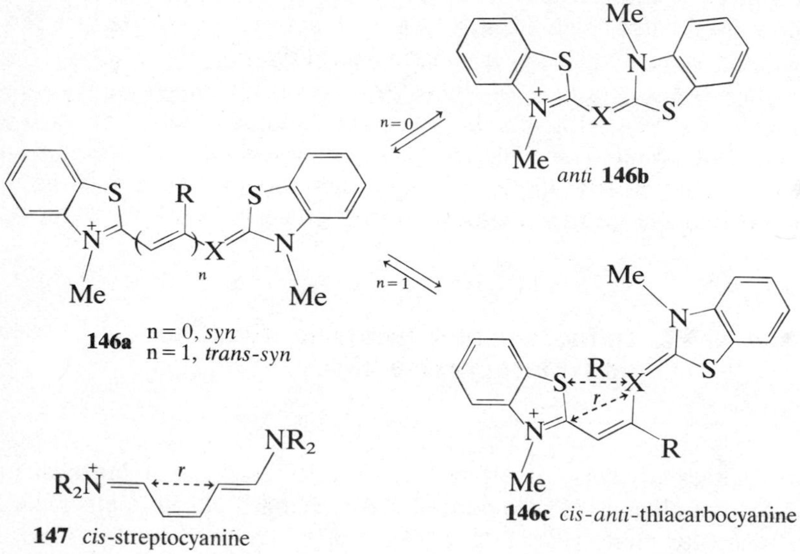

(21)

anti **146b**

146a n = 0, *syn*; n = 1, *trans-syn*

146c *cis-anti*-thiacarbocyanine

147 *cis*-streptocyanine

as the *anti*-conformers **146b**. Carbocyanines, on the other hand, existed primarily as the *trans-syn* isomers ($n=1$, R = H, Eq. 21) or mixtures of isomers (R = alkyl).[126a,554] Less stable dyes produced by photoisomerization were assigned *cis-anti* conformations (**146c**; Eq. 21).

Distinctly different wavelength shifts occurred in the photoisomerization of various dyes. West and others[473,509–511,553,554] studied thiacarbocyanines and related dyes where the unstable isomer (*cis-anti*) absorbed at shorter wavelength than the *trans-syn*. Scheibe and others[460,510,555–559] noted metastable (*cis*) isomers from the nonheterocyclic streptocyanines **147**, which absorbed at longer wavelength than the trans. Oscillator strengths for the metastable isomers were lower.

The different wavelength shifts for thiacarbocyanines and the streptocyanines may be related to two nonbonded interactions for the *cis* isomers (Eq. 21) that do not occur in the *trans* form: (1) the 1,4-interaction labeled *r* for both the streptocyanine and thiacarbocyanine and (2) the 1,5-interaction labeled R involving the sulfur atom in the thiacarbocyanine. Isomerism in longer chain dyes like rigidized thiapentacarbocyanines was used to explain distinct shifts between absorption maxima in methanol and benzene.[560]

The thermal *cis-trans* isomerizations were first order for the thiacarbocyanines and other dyes. Activation energies were 8 to 15 kcal/mole.[502,509,510,561] More complex kinetic behavior was found for 1,5-*N,N'*-dipyrrolidinylpentamethine perchlorate, and the "photoisomerization" was attributed to dimer formation.[562]

Separation or isolation of distinct isomers was noted in three cases. Chromatography of thiatricarbocyanine on $CaCO_3$ gave three zones, each of which gave three zones when rechromatographed.[563,564] An unsymmetrical carbocyanine from 3H-indole and benzimidazole gave four red color zones by chromatography on type II alumina, but each zone had an identical solution absorption spectrum after elution.[565] The photoisomerization of **148a** to **148b** (Eq. 22) coupled with a slow thermal return reaction and gave $H_\alpha - H_\beta$ coupling constants that were consistent with *trans* and *cis* isomers.[566] Irradiation of the *trans* isomer **148a** (λ_{max},

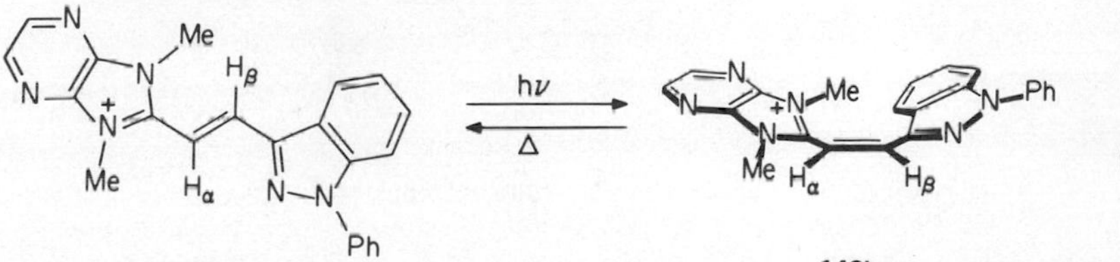

(22)

148a 148b

405 nm; $J_{\alpha\beta} = 17$ Hz) at 436 nm gave the *cis* isomer **148b** (λ_{max}, 350 nm; $J_{\alpha\beta} = 12$ Hz). The large absorption difference was ascribed to the nonplanarity of the *cis* isomer.[566] In addition, two crystalline forms were isolated, one of which was colorless.[566] Two forms (**149a**, **149b**) of a merocarbocyanine were also separated.[567] One isomer exhibited a metastable phase, which separated from supersaturated solution as helical ribbons up to 5 mm long with eight turns per crystal. The isomerism was attributed by nmr to *syn/anti* rotational forms involving the terminal thiohydantoin group, both isomers retaining a trans methine chain ($J_{\alpha\beta} = 14$–15 Hz). Absorption spectra in pyridine showed distinct peak absorptions for **149a** and **149b**, which were consistent with the

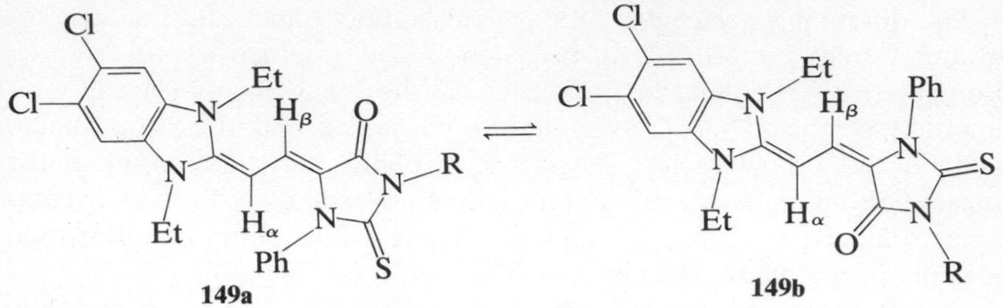

149a

149b

wavelengths calculated by molecular orbital theory (SCF CI[411]): **149a**, 513 nm (obs), 513 nm (calc) and **149b**, 527 nm (obs), 529 nm (calc). Extinction coefficients for the two isomers in pyridine were nearly identical (14×10^4 liters $mole^{-1}$ cm^{-1}). The thermodynamic equilibrium in chloroform was a 35 : 65 ratio for **149a** : **149b**, but photoequilibration in pyridine (fluorescent lights) produced more of the **149a** isomer.

A potentially large class of dyes that exhibit absorption bands for distinct isomers are the cyanine dyes with bulky groups R on the chromophoric chain (Eqs. 23–24).[2,7,8,42,568–574] For the dyes with an overall neutral charge, this was designated allopolar isomerism (see Fig. 2). In the holopolar (completely charged) or cyanine form **150a**, the group R is twisted and its π-electrons overlap very little with the chain.

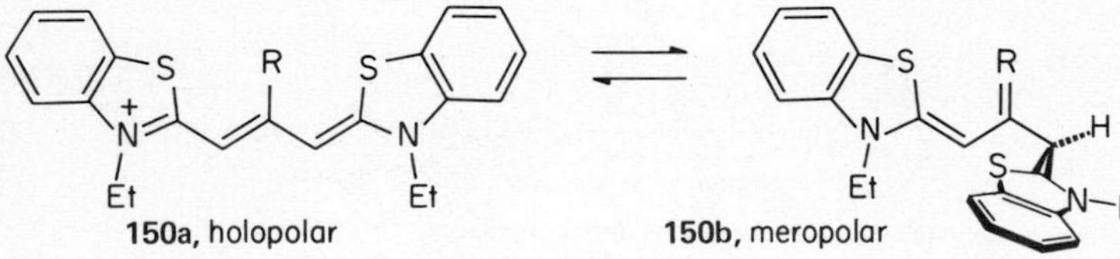

150a, holopolar **150b**, meropolar (23)

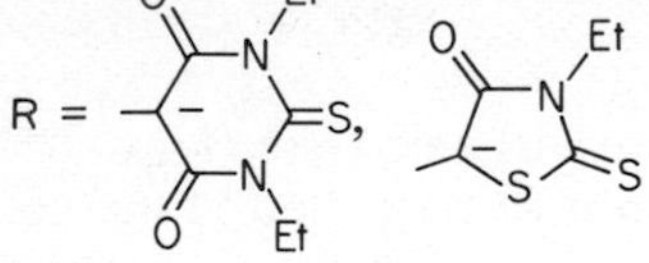

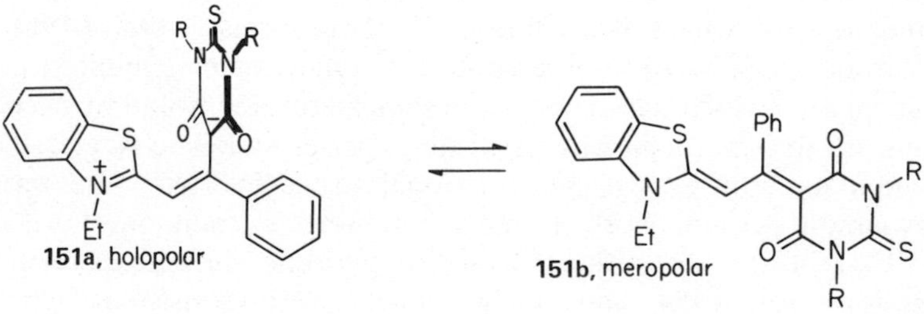

151a, holopolar **151b**, meropolar (24)

In the meropolar form **150b** one of the basic nuclei is twisted out of the plane. The other is conjugated with the group R so that the charge separation is decreased. The proportion of these two isomers changes as a function of solvent composition (Fig. 2). More detailed accounts of this isomerism were presented in early reviews of these dyes.[2,7,8,42] However the general phenomenon was probably first discovered because of the unusual solvent sensitivity of chain phenyl merocarbocyanines like **151a–b** (Eq. 24). Steric hindrance between the chain phenyl group and the 6-membered acidic nucleus (thiobarbituric acid) led to the nonplanar charge-separated form **151a** in which the chromophore is a simple vinyl quaternary salt. Other cases of isomerism were noted for uncharged substituents like −SMe in place of R in **150a-b**.[560]

Spiropyrans: **152** (25)

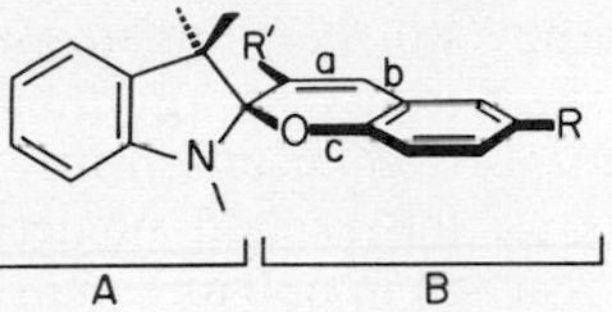

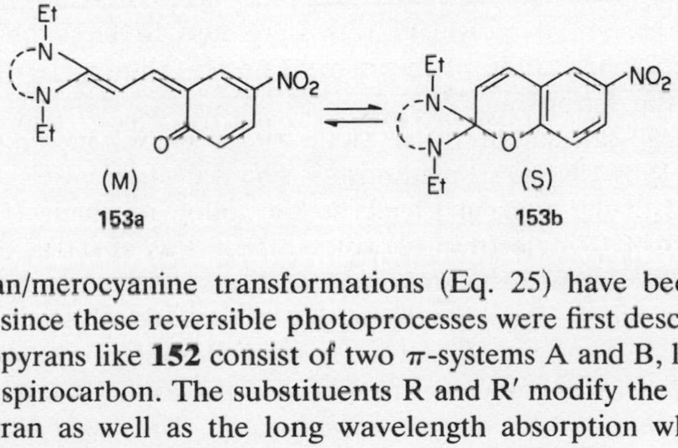

(M) **153a** ⇌ (S) **153b**

Spiropyran/merocyanine transformations (Eq. 25) have been studied extensively since these reversible photoprocesses were first described[575] in 1926. Spiropyrans like **152** consist of two π-systems A and B, linked by a tetrahedral spirocarbon. The substituents R and R′ modify the stability of the spiropyran as well as the long wavelength absorption which is associated with the π-system marked B. The equilibrium between the merocyanine and spiropyran forms is also influenced by the electronic character of the heterocyclic ring marked A. Many reviews of photoisomerisms, thermal processes, and equilibria are available,[576–580] and other types of photochromic processes have also been reviewed recently.[577]

The isomerization process[581–590] (Eq. 25) involves a colorless spiropyran form **153b** and several distinct colored species which were attributed to isomeric merocyanine forms (**153a**). The relative stabilities and interconversion rates for spiropyran and merocyanine forms depend on the substituents on the aromatic ring and vinyl carbons (see groups R, R′, in

152) as well as the nature of the heterocyclic nucleus. Many of the compounds that were studied extensively incorporated the weakly basic 1,3,3-trimethyl-3*H*-indole. For these spiropyrans electron-withdrawing substituents on the aromatic ring (e.g., $R = NO_2$ in **152**) provide increased stability for the merocyanine forms, and substituents on other aromatic carbons have additive effects. Alternatively the merocyanine forms are thermodynamically favored when highly basic heterocycles like benzimidazole are present. For example, a 1,3-diethylbenzimidazolospiropyran exists predominantly in the merocyanine form while incorporation of a sterically similar but less basic heterocycle in a 1,3-diethylimidazo[4,5-*b*]quinoxalinospiropyran favors the closed spiropyran form.

The photochemical coloration (spiropyran → merocyanine) and bleaching (merocyanine → spiropyran) reactions in solution were postulated to involve triplet states, since triplet energy donors sensitized both reactions.[591,592] Recent evidence for an unsensitized coloration reaction established an active role for the excited singlet state of a spiropyran as well.[593] The coloration reaction in polymer matrices was shown to involve an excited triplet state that was not the phosphorescent state.[594] The bleaching reaction in polymers was suggested to be a photochemical *trans* → *cis* isomerization of the merocyanine, followed by thermal ring closure to the spiropyran.[595]

Theoretical calculations were done on a model indolinospiropyran (**152**; R = H) by CNDO CI methods.[596] The transition $S_1 \leftarrow S_0$ (associated with part B of the molecule) leads to the following changes: (1) oxygen atom, loss of 0.073 electron; (2) spirocarbon, loss of 0.017 electron; (3) π-bond orders for bonds *b* and *c* (see **152**) increase in the excited singlet state (S_1) while *a* decreases to nearly two-thirds of its S_0 value; (4) π-bond orders for bonds *b* and *a* are equivalent in the S_1 state; and (5) the bond-order/charge-density pattern for the excited singlet state was similar to that for the lowest triplet state. The calculations support heterolytic cleavage of the spirocarbon–oxygen bond and emphasize large changes in π-bond orders for bonds *a* and *b* in both the excited-singlet and triplet states, resulting from simple $\pi \rightarrow \pi^*$ excitation of the aromatic part[597,598] of the spiropyran.

B. Color Shifts and Environment

Continuous changes of the environment for a dye chromophore generally produce gradual changes in absorption maxima, extinction coefficients, and other dye properties. The major factors responsible for these shifts include steric crowding, solvent sensitivity, and aggregation. Historically steric crowding was emphasized first. The understanding of solvent sensitivity and aggregation was advanced in several stages of synthetic and theoretical refinements, including recent quantum mechanical calculations. Before discussing these subjects, some less popular areas of investigation should be noted. Increased pressure shifted absorption maxima to the red for dyes dissolved in a cellulose acetate film and increased Davydoff splitting for dye crystals dispersed in NaCl.[599] Certain dye–dye interactions may optimize excitonic superconductivity,[600] and the detailing of photoconductivity and trap depths in dye crystals is well known.[601–603] Crystalline cationic dyes with complex anions exhibited some electrical conductivity.[604–607] Electric field effects on absorption spectra (electrochromism) received both theoretical and experimental attention.[608–615]

1. *Steric Crowding*

Brooker and co-workers reviewed[6,8] the early work on steric effects in substituted anilines and pyrrole dyes, but the prime documentation of steric influences was based on a large series of new dyes from quinolines.[6] For the symmetrical simple cyanines 3,3′-dimethyl substitution produced large steric interactions in the planar conformation (Fig. 11, dye A). The relief of these interactions by twisting about the central methine carbon lowered the extinction coefficients and shifted the maximum absorption to longer wavelength. Substitution by larger groups (e.g., N–Et for N–Me) shifted the absorption to even longer wavelength. For a variety of substituent patterns a linear relation between increased absorption wavelength and lowered extinction was shown. The steric crowding in these quinoline dyes is not totally relieved by increasing the chain length. As shown in the quinocarbocyanine B (Fig. 11), "residual" crowding will occur for R_1 substituents larger than hydrogen owing to the proximity of R_1 and the chain hydrogen atom (H_β). Significant effects on absorption wavelengths occurred for a variety of R_1 substituents. "Buttressing" effects were also noted in the quinocarbocyanine B. If $R_2 = \mathrm{Me}$ or Cl, the increased steric interaction between the R-groups on nitrogen and the chain hydrogens (H_α) lowered the extinction, shifted the absorption to longer wavelengths, and decreased the solution aggregation tendency of the carbocyanine dyes. Highly unsymmetrical styryl dyes containing quinolines generally showed decreased extinction coefficients as steric hindrance increased, but the absorption maxima were shifted to shorter wavelengths.

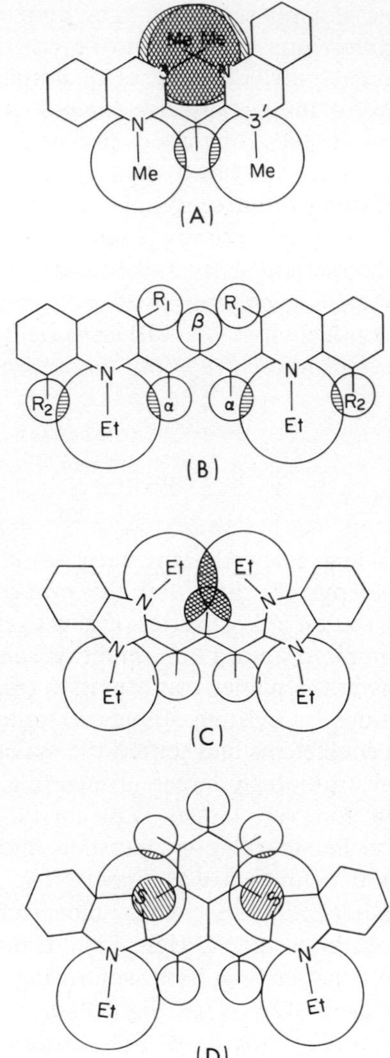

Fig. 11. Dye structures used to document steric influences on absorption maxima, extinction coefficients, and band shapes. The quinoline-related dyes like (A) and (B) were historically important whereas the benzimidazole simple cyanines and carbocyanines (C) provided less equivocal examples. The meso-phenylthiacarbocyanine (D), which shows an extremely unfavorable all-planar conformation, exhibited a narrow absorption band, suggesting that the phenyl group was rigidly oriented between the two sulfur atoms and at 90° to the thiacarbocyanine chromophore.

The quinoline dyes are perhaps not ideal substrates for the study of steric effects, since *syn/anti* isomerism of the quinoline rings in both the cyanines and the carbocyanines was suggested to be a function of the substituent pattern.[8] In addition, synthetic rigidizing of the simple 2,2-quinocyanine changes the placement of the quinoline rings, and although the rigidized dye may be more nearly planar, its extinction coefficient and oscillator strength were lower than for the nonrigidized analog.[475] Benzimidazole dyes, however, are symmetric with respect to the position of attachment to the chromophoric chain. In the symmetrical simple cyanine, twisting lowered the observed extinction coefficient and shifted the absorption maximum to longer wavelength. Some evidence for "buttressing" effects was found in the symmetrical benzimidazolocarbocyanines. As shown in dye C (Fig. 11), one ethyl group from each imidazole ring interacts moderately with the meso-hydrogen atom on the chain. Two or three chloro substituents on each benzimidazole ring provided high extinction carbocyanine dyes with relatively constant bandwidths. However, a fourth chloro substituent lowered the extinction coefficient and increased the bandwidth. This effect is attributable to a reinforcement (buttressing) of the meso-hydrogen-*N*-ethyl interaction by the fourth chlorine substituent.

The meso-phenyl thiacarbocyanines (Fig. 11, dye D) and related dyes exhibit large steric interactions in a totally planar conformation. Since these dyes show experimental absorption curves with exceptionally small bandwidths, the meso-phenyl was presumed to be perpendicular to the plane of the dye chromophore.

2. *Solvent Sensitivity*

The influence of solvents on absorption spectra for dyes is described in numerous publications since 1945. The classes of dyes used as experimental probes are, however, rather few, and certain reviews[12,55,616–619] treat these in some detail: (1) typical cyanine dyes,[620–623] (2) merocyanines,[52–55,624–633] (3) betaines,[82,83,634–637] (4) dye bases,[43,455] and (5) hydroxystyryls and related dyes.[46,53,616,638–642] Typical structures are shown in Fig. 12. Theoretical descriptions have been based on the McConnell relation (see previous section), the McRae equation or similar treatments,[643–647] variations in the vibrational band structure for an electronic transition,[455] and specific molecular orbital treatments.[132,427,627] As noted previously, ground-state dipoles for merocyanine dyes are high, and chemical shifts for the methine protons are solvent sensitive. Beyond this, the specific effects of solvents were considered either as the summation of several

Blue-shifting(χ_B) and red-shifting(χ_R) merocyanines Solvent Shifts

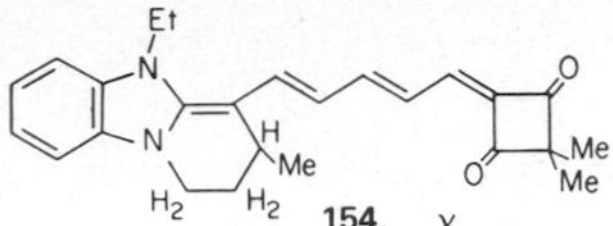

154, χ_B

	a	b	c	d	e
χ_B	-	53.7	50.1	43.2	63.0
λ_{max}	-	531	569	659	453

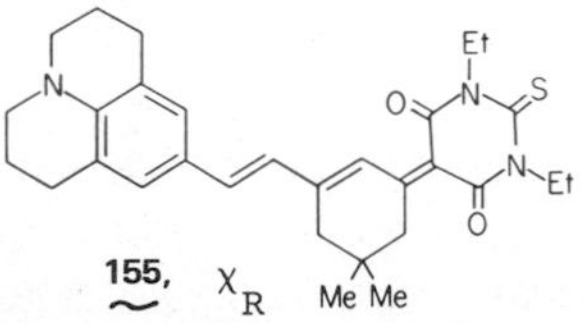

155, χ_R

	a	b	c	d	e
λ_{max}	570	623	623	637	662
χ_R	50.0	45.7	45.7	44.7	43.1

Three highly polar merocyanines[53,132,427,624,625,633,641]

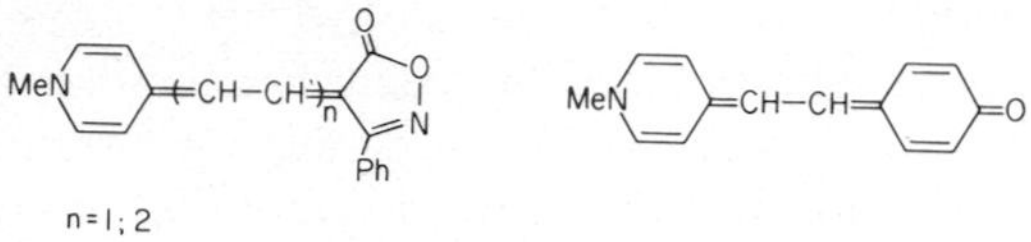

n = 1; 2

Other solvent-sensitive chromophores

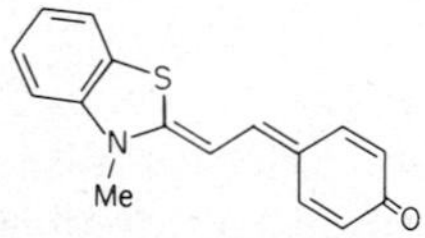

A "cusping" merocyanine[53,619,625] 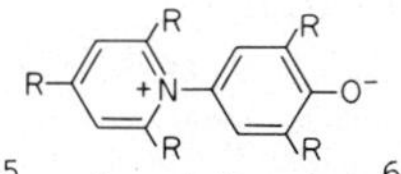 phenol betaines[617,636-637] 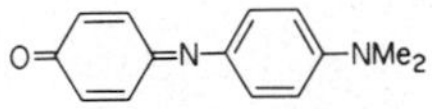 phenol blue[46,631,632]

Fig. 12. Typical solvent-sensitive dyes. Brooker and coworkers designed the highly sensitive blue-shifting dye labeled χ_B and red-shifting dye (χ_R) following extensive studies of highly polar merocyanines and other solvent-sensitive dyes. Note that as general solvent polarity increases in the series *a–e* (cyclohexane, acetonitrile, acetone, 2,6-lutidine, methanol), the χ_B and χ_R values are not parallel. This suggested that different features of solvent structure (refractive index, dielectric constant, hydrogen bonding) were responsible for the observed wavelength changes in the two dyes.

individual terms[643] or as the resultant of a single, empirically derived parameter.[55] A small red shift was generally observed for cyanine dyes in solvents with increasing refractive indices.[620–623] This shift was linearly related to dye-solvent dispersion forces, expressed as a simple function of the refractive index (n)[620]: $(n^2-1)/(2n^2+1)$. For merocyanines this general red shift was also expected, but it was usually superimposed on other, more powerful, effects involving dye-solvent permanent dipoles and hydrogen bonding.[132,629,631]

Brooker and co-workers constructed merocyanines with widely different terminal groups, so that the intrinsic polarity of the dyes ranged from

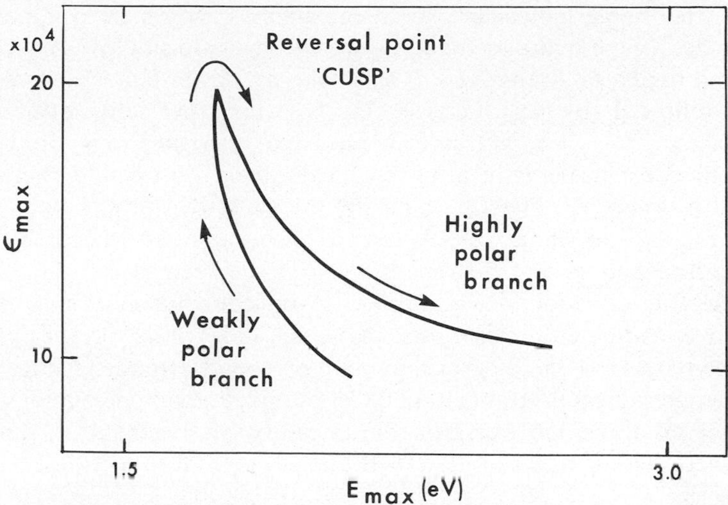

Fig. 13. "Cusping" merocyanines. Increased solvent polarity shifted the maximum absorption energy (E_{max}) and the extinction coefficients (ϵ) in the direction of the arrows.

weakly polar to highly polar.[52–55] Of the molecules that were extensively studied, a few serve to illustrate the general pattern of solvent effects on the absorption maxima and peak intensities (Fig. 13): (1) weakly polar dyes were red shifted (longer wavelength) and showed increased extinction coefficients (ϵ_{max}) as solvent polarity increased; (2) highly polar dyes were blue shifted and showed decreased ϵ_{max} values for increasingly polar solvents; and (3) moderately polar dyes exhibited a "cusp" in the plot of ϵ_{max} versus λ_{max}, where the effects of solvent polarity increases were reversed from red shifting to blue shifting.

The most successful correlations of absorption data and solvent properties were empirical. These used transition energies (so called E_T values) for model merocyanines[55] or phenol-betaines[617] to reflect gross solvent properties. Brooker suggested two dyes (Fig. 12) to characterize the properties of pure solvents and solvent mixtures: dye **154** (χ_B) is a highly polar, blue-shifting dye and dye **155** (χ_R) is a weakly polar, red-shifting dye. The absorption maxima (E_T in kcalories/mole), tabulated[55] as χ_B and χ_R, respectively, did not rank all solvents in the same order. However, the χ_B values correlated linearly with other spectral measures of solvent polarity while the χ_R values were related to certain reaction-rate data.

The striking variation in ϵ_{max} noted by Brooker[53] was not reflected in the observed oscillator strengths for three typical merocyanines.[625] This supported Platt's description (Eq. 20) of bandwidths and band intensities.

On the other hand, molecular orbital calculations for a merocyanine with variable degrees of hydrogen bonding showed a significant lowering of the calculated oscillator strength (f_{calc}) as strong hydrogen bonds were established, followed by an increase in f_{calc} for the fully protonated merocyanine.[132,427] For chloroform–methanol mixtures this prediction was verified experimentally using ϵ_{max} values (no observable change in bandwidth occurred). The ϵ_{max} data for the same dye in pyridine–water mixtures[53] showed similar trends except in pure water where the predicted increase in oscillator strength was not reflected in ϵ_{max}.

The McRae equation[643] was applied to typical solvent-sensitive dyes, primarily to estimate the absorption shifts due to intrinsic solvent polarity as defined by E_I (Eq. 26). For phenol blue and Brooker's suggested merocyanine indicator, **154** (χ_R), Eq. 26 provided good correlations with the observed transition energies for a variety of solvents.[631] Many E_I values were within 0.5 kcal/mole of the experimental transitions.

$$E_I = \left[\frac{n^2-1}{2n^2+1}\right]A + \left[\frac{D-1}{D+2} - \frac{n^2-1}{n^2+2}\right]B + C \tag{26}$$

where n = solvent refractive index
D = solvent dielectric constant
A, B, C = adjustable.

$$E_T = \text{observed transition} = E_I + \Delta E_H \tag{27}$$

The constants A, B, and C were approximately −33 kcal/mole, −4.4 kcal/mole, and 57.6 kcal/mole for phenol blue and −76.4 kcal/mole, −9.1 kcal/mole, and 65.6 kcal/mole for Brooker's indicator dye.[631] Significant deviations were noted for strongly hydrogen bonding media like *p*-cresol or mixtures of *p*-cresol with ōther solvents. The incorporation of a hydrogen-bonding correction term ΔE_H as in Eq. 27 has been suggested, but since substitutional patterns in phenol blue alter the sensitivity of the dye to hydrogen-bonding solvents,[631] a correction term characteristic of the solvent itself is difficult to evaluate.

3. *Aggregation*: *General Characteristics*

Interactions among dye molecules produce large spectral shifts and distinct changes in band shape. In aqueous solution at room temperature, these changes are most evident as the concentration of the dye increases or as dye is concentrated by adsorption to crystal surfaces or oppositely charged polymer sites. Relative to the absorptions for monomeric dye molecules (M-band, Fig. 14), progressive shifts of absorption maxima to

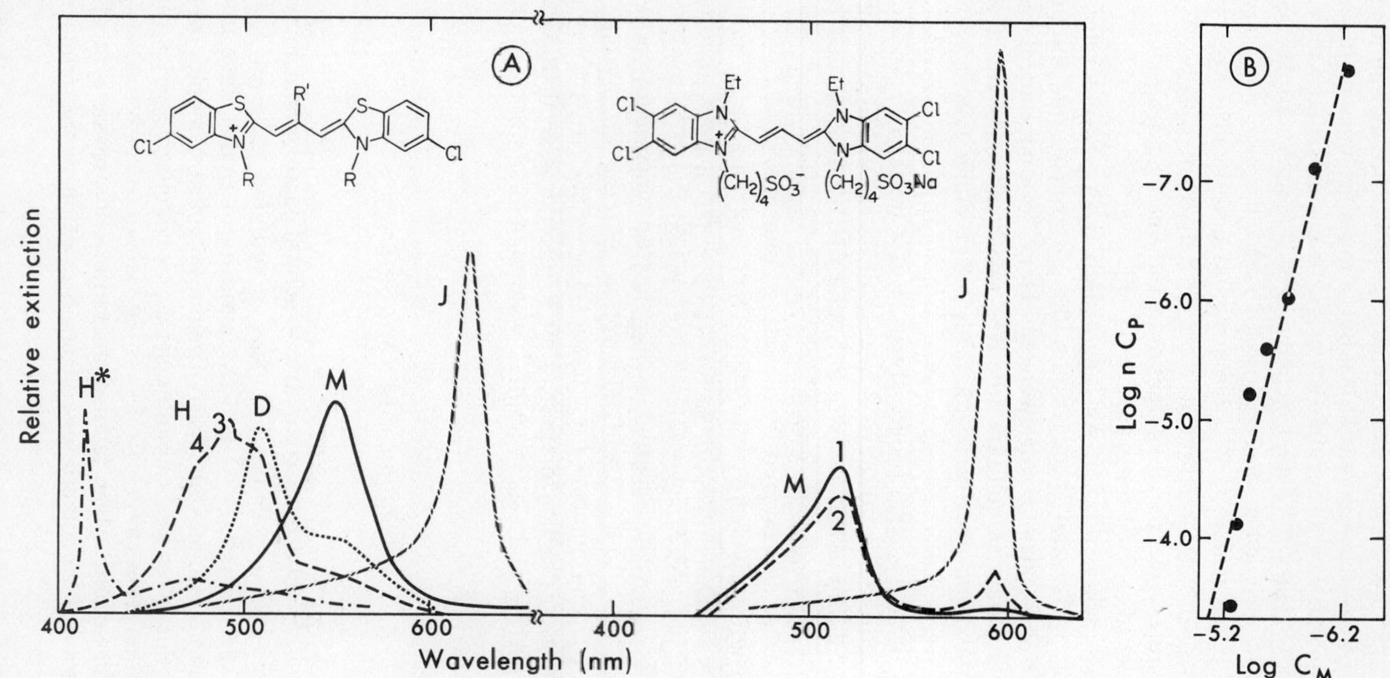

Fig. 14. Solution absorption characteristics of H- and J-aggregates.[660] (*a*) Thiacarbocyanines (**156** with X = chloro) and a benzimidazolocarbocyanine [**158** with R = $(CH_2)_4SO_3^-$]. The substituents R,R′ were (1) $C_4H_8SO_3^-$, Et for curves D, M, J; (2) $C_2H_4CO_2^-$, Me for curve H-3, 4; and (3) bridging alkyl groups in the dimeric thiacarbocyanine **141** for the unusually sharp H-aggregate absorption H*. (*b*) The aggregation number *n* was 4.0 for the benzimidazolocarbocyanine dye, where C_M and C_P were concentrations of monomeric and aggregated dye, respectively.

shorter wavelength are designated as H-bands (hypsochromic) although the term *metachromic* has been used also. Shifts to longer wavelength are less progressive. The new absorption is designated as a J-band and was first described by Jelley[648,649] and Scheibe.[650–652] Reviews of H- and J-aggregation are included in descriptions of metachromasy,[653–656] general aggregate phenomenon,[657—659] dye structure and detailed aggregation tendencies,[8,660–668] the relation of x-ray crystal structure to possible aggregate structures (see previous section), and the characteristics of aggregates absorbed to solid substrates.[664-677]

The degree of self-association for aggregated dyes was determined spectroscopically for several H-aggregates (dimers, trimers, and tetramers) and for J-aggregates as reviewed by Herz.[660] Under ideal circumstances aggregation in solution involves two-component equilibria (Fig. 14A): monomeric dye (curve **M**) with either dimer (curve **D**) or J-aggregate (curve **J**). Larger H-aggregates, formed at increasing concentration, shifted the wavelength progressively shorter (peaks H-3,4). Most H-bands were considered to have larger half band width than the monomer until recently, when an extremely sharp H-band designated H* was noted for two thiacarbocyanines.[457,660] On the other hand, most J-bands, which were also favored by increased concentrations of certain dyes, exhibited unusually sharp absorptions. Herz estimated a minimum of four monomer units per absorbing unit for the J-aggregate[660] although physically larger J- and H-aggregates were demonstrated to exist by solution centrifugation of J-aggregates,[660] electron microscopy of H-aggregate threads,[661] and J-band absorption spectra in crystalline dyes.[678–679] Association numbers for aggregates were also determined by fluorescence decay, extraction or partitioning procedures, evaluation of colligative properties, and polarographic evaluation of diffusion coefficients.[660]

4. *Aggregation and Dye Structure*

Aggregation is favored by certain dye structural features and matrix (solvent) compositions. Aggregation was noted for relatively few merocyanines and oxonols,[680–682] and the majority of studies have used symmetrical simple cyanines and carbocyanines. typical examples were from the benzothiazole series, quinoline series, benzimidazole series, and benzoxazole series (**156–159**).

The structural features which favor aggregation also provide a regular conformation for the monomeric dye. The observation of this tendency

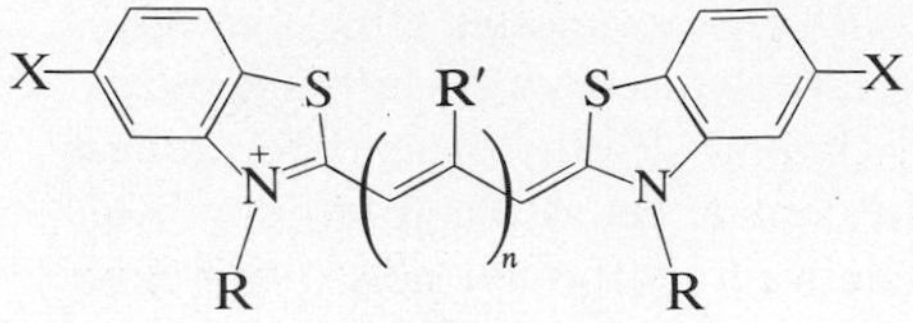

156 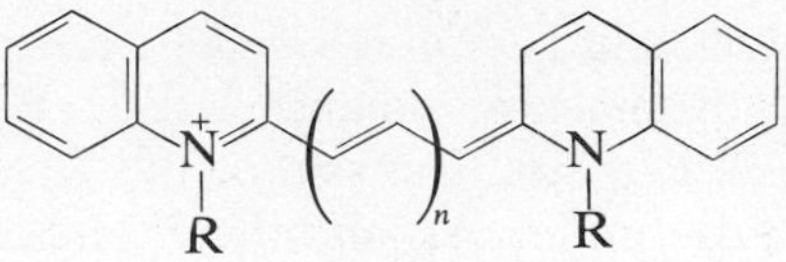 **157**

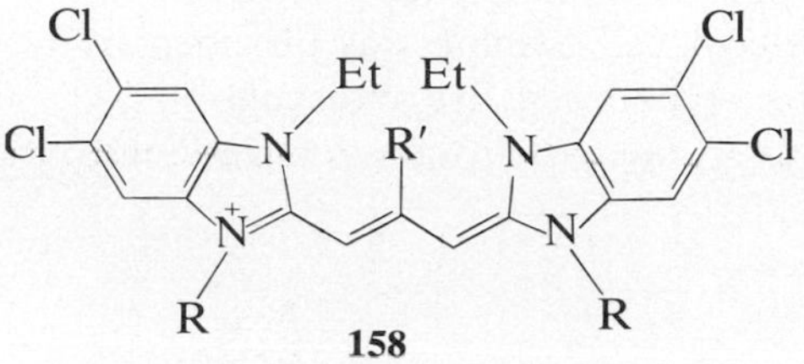

158 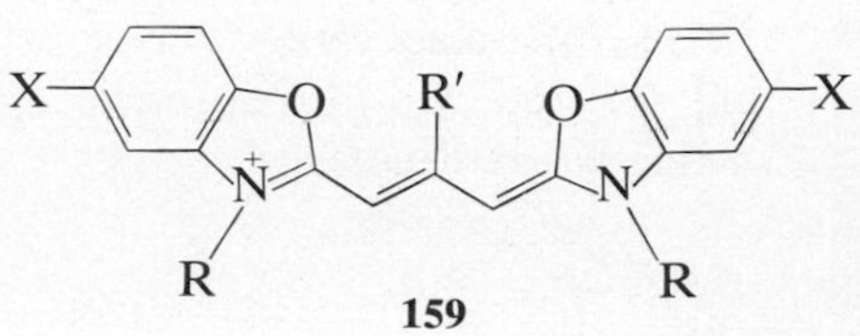 **159**

led Brooker to suggest that "compact" dyes aggregate more readily than "loose" dyes and "crowded" dyes.[8] The first two categories referred to planar dyes and the third to nonplanar dyes. Carbocyanine and dicarbocyanine dyes in the benzimidazole series aggregated readily in solution, but meso substitution on the chain (R = alkyl or phenyl) or a decrease in chain length to the simple cyanine led to "crowded" dyes and no observable aggregation.[660] However meso-chain substitution in the benzothiazolocarbocyanines markedly increases the aggregation tendency ("compact" dyes) even for large substituents such as ethyl, phenyl, or indenyl.[8] The meso placement of these chain substituents is critical since their most probable conformation is perpendicular to the plane of the chromophore. Careful study of the crystal structure in Fig. 4 will show that out-of-plane meso-ethyl groups might project between molecules of the adjacent layer and (along with other forces) influence the intermolecular angles (α, Fig. 4). Certain other out-of-plane substituents may be accommodated easily if they do not twist the dye molecule itself. The meso-alkyl, phenyl, or indenyl groups in thiacarbocyanines, for example, could fit between two other dye molecules of the adjacent layer and function as out-of-plane "posts" which fix the angle α.

Other substituent patterns influenced the type of aggregate formed or its wavelength. Two benzothiazolocarbocyanine dyes (in Fig. 14 with $R = (CH_2)_2CO_2^-$) exhibited H-bands (R′ = methyl) or J-bands (R′ = ethyl) as the first-formed species on silver halide surfaces, but the H-aggregate for the methyl dye was slowly converted to the thermodynamically more stable J-aggregate.[676] Substituent changes on nitrogen in 2,2′-quinocyanine dyes also favor different aggregates: H-aggregation for N–Me and

J-aggregation for N–Et. Ficken[665] recently reviewed the effects of substituents in N–Me and N–Et dyes for a large variety of X-substituents in 9-ethylthiacarbocyanines (**156**, R′ = Et) and 9-ethyloxacarbocyanines (**159**, R′ = Et). The size of the X substituent affected the total length of the dye molecules, and the differences between monomer absorption and J-aggregate wavelengths on silver halides were a function of these lengths. Similar results were found in several studies using monolayer systems of the Langmuir–Blodgett type.[683–688] The spacing required for J-aggregate absorption of 1,1′-dioctadecyl-2,2′-cyanine was obtained by the use of octadecane as an inert spacer between dye molecules[683–686] and was depicted in theoretical models as increasing the effective length of the dye molecules.

5. *Aggregate Properties*

Dye aggregates were shown to be dichroic by (1) the absorption of plane-polarized light by aggregates oriented on silver halide crystals[689–690] or as thin dye crystals[678,691] and (2) circular dichroism in the presence of optically active substrates.[692–698] This known interaction of dyes and macromolecular substrates[699–711] was extended with studies of dye-binding sites by either the reaction rates between bound dye molecules and hydrated electrons[712,713] or synthetic linking of dyes to porphyrin.[714]

The dependence of aggregation on temperature, solvent composition, and other factors was most thoroughly documented for small H-aggregates (dimers, trimers) of cyanines[656,660,715–717] and dimers of related dyes.[718–722] These aggregates for vinylogous thiacyanines, 9-ethylthiacarbocyanine, and *N*-substituted thiacarbocyanines exhibited the following general characteristics. First, dimers were readily formed in solution as the concentrations of the dyes increased. For the vinylogous thiacyanines, the free energies of dimerization ($\Delta G°$) were -5 to -7 kcal/mole, and the absorption-energy change (dimer-monomer) went from 1179 cm^{-1} (simple cyanine) to 2176 cm^{-1} (tricarbocyanine) (see Table XV). The ΔS and ΔH values for 9-ethyl-3,3′-dimethylthiacarbocyanine bromide dimerization were -1.9 eu and -5.8 kcal/mole, respectively, while trimerization showed -10.6 eu and -14.8 kcal/mole for the same quantities.[656] Dimer and trimer formation did not involve the anions, since the activities of free anions like bromide were not a function of the extent of the dye aggregation,[656] and internal anions like sulfoalkyl or carboxyalkyl did not affect the free energies of thiacarbocyanine dimerization (Table XV). Second, important features of the solvent

TABLE XV. MONOMER-DIMER EQUILIBRIA FOR AQUEOUS DYE SOLUTIONS

156

X	R	R′	E(cm^{-1}) [dimer-monomer]	$\Delta G°$ (kcal/mole)	Remarks	Ref.
H	Me	Et	1330	−5.2		[a]
				(−11.7 for trimer at 20°C)		[a]
Cl	Et	Et	1425	−6.7		[b]
Cl	$(CH_2)_2CO_2^-$	Et	1410	−6.4	Added KCl gave J-aggregate	[b]
Cl	$(CH_2)_2CO_2^-$	Me	1940	−6.6	Added KCl gave higher H-aggregate	[b]
H	Et	H	1525	−6.0		[c]
Simple thiacyanine			1179	−5.3		[c]
Thiadicarbocyanine			1839	−6.5		[c]
Thiatricarbocyanine			2176	−7.3		[c]

[a] J. F. Padday, *J. Phys. Chem.*, **72,** 1259 (1968);
[b] A. H. Herz, *Photogr. Sci. Eng.*, **18,** 323 (1974);
[c] W. West and S. Pearce, *J. Phys. Chem.*, **69,** 1894 (1965).

character were related qualitatively to the dielectric constant.[660] Polyhydroxylic solvents like water or glycols favored aggregation, and simple organic solvents required low temperature for aggregation to be observed.[660] The sharp H* aggregate was noted at low temperature in *n*-butanol.[457] The extent of aggregation in methanol–water mixtures was decreased at higher temperatures or with decreased water content. Increased ionic strength increased the degree of aggregation and changed the types of aggregates.[660] For the addition of KCl to solutions of dimerized 5,5′-dichloro-3,3′-dicarboxyethyl-9-alkylthiacarbocyanines, the 9-methyl dye showed increased H-aggregation including a small H* peak, and the 9-ethyl dye showed J-aggregate formation. The deaggregating influences of surfactants, methanol, acetone, phenol, and sodium 4-hydroxy-6-methyl-1,3,3*a*,7-tetraazaindene were reviewed recently.[660]

6. *Aggregate Luminescence*

The luminescence of dye aggregates paralleled distinctive features of the absorption spectra of J- and H-aggregates. J-Aggregates exhibited highly allowed transitions between the ground state and the first excited singlet state. Fluorescence from this state was generally strong. On the other hand, H-dimers and higher H-aggregates exhibited weak transitions between the ground state and first excited singlet state, and the highly allowed transitions of H-aggregates (Fig. 14) involved higher excited-singlet states. Excitation into these latter absorption bands produced a weak fluorescence and a *relatively* efficient phosphorescence. These observations suggested that the excited H-aggregates were more rapidly deactivated by internal conversion and some intersystem crossing than monomers or J-aggregates.

Resonance fluorescence was noted for J-aggregates studied in solution at low temperature and in monolayer films.[684,687,723–726] Weak emission at longer wavelengths was also detected for J-aggregates in solution.[723] J-aggregated 1,1′-diethyl-2,2′-cyanine bromide[725] showed two absorption and resonance fluorescence bands (J_1J_2) at 77°K, separated by about 151 cm^{-1} and with half-bandwidths of about 30 cm^{-1}. Although dyes in monolayers generally form H-aggregates, the 5,5′,6,6′-tetrachloro-1,1′-diethyl-3,3′-dioctadecylbenzimidazolocarbocyanine (**158**; R = octadecyl) formed a J-aggregate in a monolayer film, as evidenced by resonance fluorescence.[687] This aggregate absorbed at 595 nm regardless of substrate (including evaporated silver bromide). The tetraethyl analog (**158**; R = ethyl) also showed a 593-nm J-aggregate in aqueous solution, but both dyes showed a distinctly different J-aggregate absorption at 575 nm on silver halide microcrystals. This shift in absorption was suggested to result from epitaxial ordering of the dye by silver halide, leading to a different intermolecular arrangement than for the monolayer or solution J-aggregate.[687]

Luminescence from H-aggregates was observed for dimers of cyanines,[440] meso-substituted carbocyanines,[473] rigidized dyes,[457,458] and other dyes.[727–730] Cooper and co-workers[458] have shown that 3,8,3′-10-diethylenethiacarbocyanine bromide (monomer, 575 nm) gave several fluorescence peaks 490 to 600 nm for excitation in the H-bands (460 nm, 480 nm). The fluorescence intensity decreased markedly for excitation in the sharp H*-band (410 nm). In the presence of ethyl iodide, delayed luminescence for this same dye was observed at 645 nm. The fluorescence and phosphorescence action spectra exhibited similar peaks near 475 and 460 nm. The H-aggregated dimers of pyridocyanines[440] and meso-substituted carbocyanines[473] were studied at low temperature in typical alcoholic solvent mixtures. Meso-substituted oxa-, thia-, and selenacarbocyanines showed structured fluorescence from the *cis* and *trans* monomeric dyes, strong emission from the dimers, and eximer fluorescence from the dimers.[473] Wavelengths for monomer and dimer absorptions as well as dimer fluorescence for *cis* and *trans* isomers of each dye were assigned. The eximer stabilization energy was 2600 to 3100 cm^{-1}. Both phosphorescence ($\tau = 0.040$ sec) and delayed eximer fluorescence ($\tau = 0.081$ sec) were observed for 3,3′-diethyl-9-methyloxacarbocyanine, but the phosphorescence was primarily due to excitation of monomeric dye and not dimers.

7. *Aggregate Models*

Theoretical descriptions of aggregates utilized geometrical arrangements for dye molecules in close-packed monolayers,[683–688] x-ray crystal structures,[87] and oriented silver halide crystals.[674–677] Two basically different orientations are (1) Scheibe ladder or staircase arrangements[652] and variations (Fig. 15*a–c*, *e*) including the brickstone-work arrangement (Fig. 15*a*)[683–686] suggested recently by Kuhn and co-workers, and (2) the herringbone arrangement (Fig. 15*d*) suggested by Bird and co-workers.[674–677] The herringbone arrangement was suggested to be responsible for multibanded absorption spectra that could lead to sensitization of silver halides over a broad area of the visible spectrum.[677]

Models based primarily on analogs of the Scheibe ladder or staircase arrangements, where the long axes of neighboring dye molecules were colinear, were reviewed by Smith with respect to x-ray crystal data.[87] Close-packed monolayers of 1,1′-dioctadecyl-2,2′-quinocyanine showed a

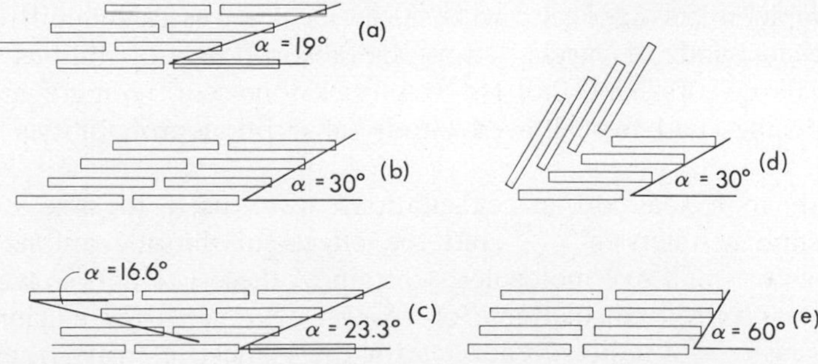

Fig. 15. Possible-structures for J-aggregates [(*a*) brickstone work,[684] (*b*) 30°-spaced aggregate,[674–677] (*c*) X-ray crystal structure model,[87] (*d*) herringbone model[674–677]] and H-aggregate structure (*e*).

J-aggregate at 552 nm, whereas the 6,6′-dimethyl analog did not.[685] This suggested that end-to-end contact of the dye molecules was important for this particular aggregate. A second J-aggregate at 578 nm similar to the one observed for dyed silver halides was ascribed to the brickstone-work arrangement (Fig. 15*a*). Theoretical calculations based on these geometries were computed with the extended-dipole model of Kuhn and coworkers.[683,731–733] Calculations using this theoretical approach indicated that the brickstone-work model adequately described J-aggregates of many dyes[733] and the card stack (Fig. 15*e*) was suitable for dimers, trimers, and such. Previous to this, the exciton (point dipole) approximation was used to describe dye aggregates in a qualitative manner,[440,508,734–739] but intermolecular terms at short distances (4–5 Å) were too large for quantitative predictions.[733] On the basis of spectral data it was suggested that the benzimidazolocarbocyanine **158** existed in a geometrical arrangement as in Fig. 15*c* for a J-aggregated dye monolayer (**158**; R = octadecyl) and a single crystal (**158**; R = ethyl).[687]

Two other theoretical models described absorption spectra of aggregated cyanine dyes. First, the transition density model[740–743] separated the shift from monomer to aggregates into two parts: (1) a monomer shift due to the "solvent" effect (dipole and Van der Waals forces) of the aggregate and (2) specific dipole splitting of excited state energy levels for identical monomer molecules in an aggregate. The results of the latter calculations, reported relative to an unspecified reference energy, showed that the shift was a continuous function of the angle α (see Fig. 15) and the size of the aggregate.[740–743] Maximum shifts to shorter wavelengths were at 90° (α) and to longer wavelengths at 18 to 25° (α), depending on the dye structure. Calculated absorption probabilities (for the N possible transitions of an aggregate with N molecules) indicated that more than one allowed absorption band was possible for a single aggregate geometry. For example, a 60° aggregate with 10 molecules was calculated to have five allowed bands of varying strengths (absorption probabilities 0.911, 0.068, 0.016, 0.004, and 0.001).[740] A brickstone work J-aggregate with four molecules had two allowed bands (absorption probabilities 0.022, 0.978).[741]

Second, molecular orbital calculations were used to assess stable ground-state geometries[433,436] and the effects of fluoride anions[438] for aggregates of small dye molecules. Certain of these geometries were the basis of π-electron calculations for several aggregates for a trimethine streptocyanine.[434] Predicted energies for the highest intensity transitions were consistent with the wavelengths found experimentally for various H-aggregates. Several geometrical arrangements and aggregate sizes gave similar results. The number of weakly allowed transitions (with energies near the normal H- and M-bands) was highly dependent on the specific geometry, and in some cases three to four bands were predicted for a dimeric species. J-Aggregate absorptions were predicted for a brickstone-work arrangement with four molecules. For three, five, or six molecules in a similar arrangement, two to three strong bands were predicted with up to two at longer wavelength than the monomer. Experimental spectroscopic studies of crystalline 1,1′-diethyl-2,2′-quinocyanine iodide (or chloride) showed parallel results.[691] Multiple absorption bands were observed for reflectance spectra of thick crystals, transmittance spectra of thin crystals, transmission spectra of aggregated-dye solution, and absorption spectra of dyed single grains of silver bromide. The relative orientations and positions of the bands suggested that typical J- and H- aggregate absorptions were actually crystal transitions and that the aggregated dye solutions contained microcrystals with structure similar to that of the bulk crystal.[691]

VII. Multiple Chromophore Dyes

Polynuclear dyes have been important sensitizers since the early use of neocyanine as an infrared sensitizer. The extensive structural variation in these dyes results primarily from the availability of two reactive groups in the central nuclei. In the more typical structures as reviewed by Hamer,[10] each central nucleus often functions as both an acidic and a basic terminal group for the chromophores.

Trinuclear dyes are represented by two types of structures: those with distinctly different chromophores such as dyes **7** and **8** and those with nearly identical chromophores such as **9** and **10**. Absorption wavelengths for several trinuclear dyes from benzothiazole are listed in Table XVI.

Brooker noted early that no formal resonance structures could be written between the chromophores of many of these dyes. Nevertheless distinct differences exist between the trinuclear dyes and the normal (dinuclear) analogs. Both Brooker and Kiprianov and their co-workers provided examples where the angle between identical chromophores was influential in determining the interaction. Examples of bismerocyanines, biscyanines, and bishemicyanines (Table XVI) showed that a decreased angle between the two chromophores increased the intensity of the short-wavelength band (λ_2) relative to the long-wavelength band (λ_1).

TABLE XVI. ABSORPTION CHARACTERISTICS OF TRINUCLEAR DYES

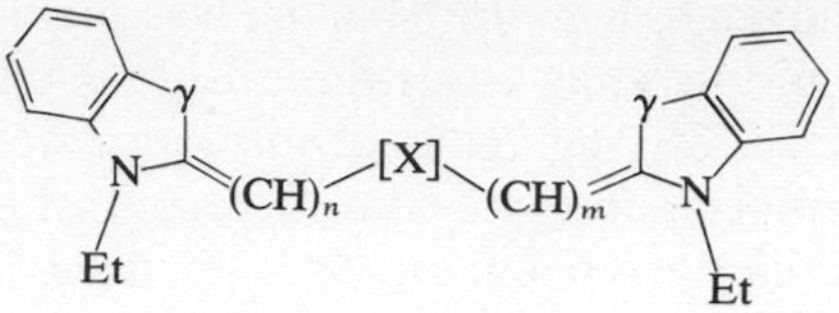

[X]	n	m	γ	λ_1 (nm)	λ_2 (nm)	Optical density ratio $[\lambda_1/\lambda_2]$	Ref.
A. Essentially linear dyes							
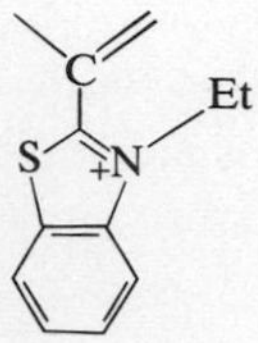	2	2	S	602	(340)[a]	(10)	[b]
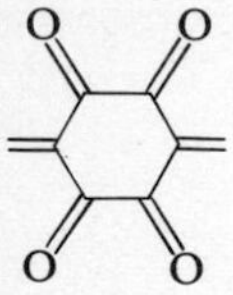	2	2	S	588	(350)[a]	(9.8)	[c]
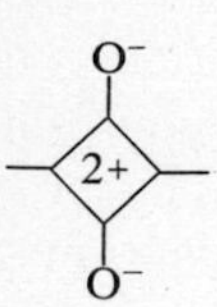	1	1	S	648	—	>10.0	[d]
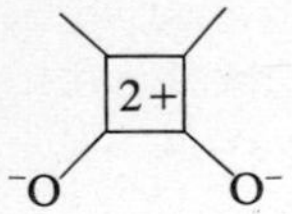	1	1	S	520	(338)[a]	(9.8)	[d]
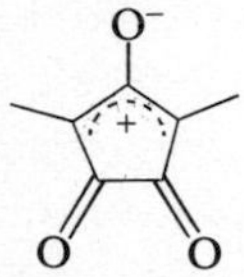	1	1	S	755	—	—	[e]
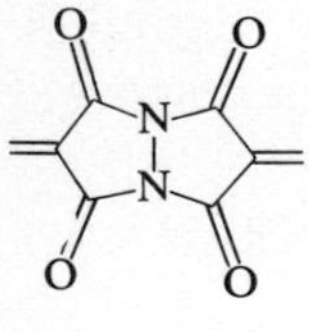	2	2	S	512	—	>10.0	[f]

TABLE XVI (*Continued*)

[X]	n	m	γ	λ_1 (nm)	λ_2 (nm)	Optical density ratio $[\lambda_1/\lambda_2]$	Ref.
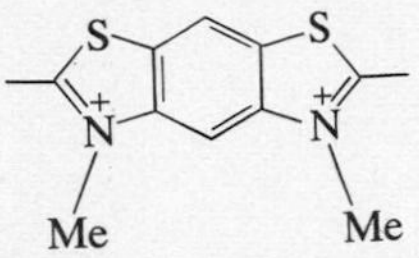 (benzobisthiazolium structure, N-Me, N-Me)	3	3	CMe_2	631	510	9.1	g
B. Nonlinear dyes with large angles between chromophores							
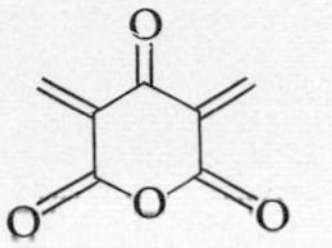 (pyran-trione structure)	2	2	S	548	432	2.6	h
	4	4	S	690	507	3.3	h
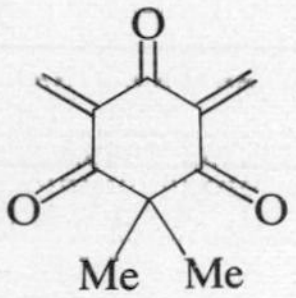 (cyclohexanetrione structure, Me Me)	2	2	S	570	440	2.5	h
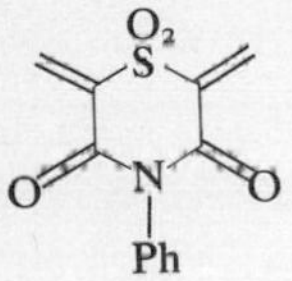 (O_2S, N-Ph structure)	2	2	S	495	430	2.9	h
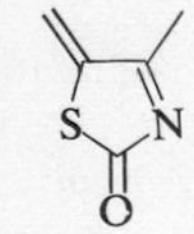 (thiazolone structure, S, N, O)	2	3	S	715	494	3.1	i
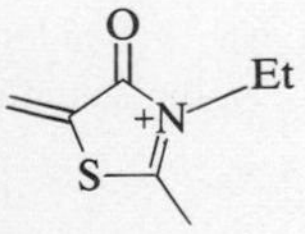 (thiazolium structure, O, +N-Et, S)	2	3	S	685	419 (broad)	1.3	j
	2	1	S	594	375 (broad)	3.1	j
	4	1	S	685	—	—	i
	6	1	S	822	—	—	j
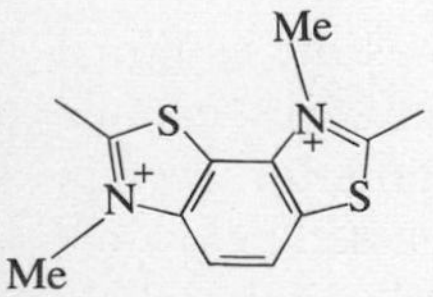 (benzobisthiazolium structure, N-Me, N-Me)	3	3	CMe_2	594	512	3.6	g

TABLE XVI (*Continued*)

[x]	n	m	γ	λ_1 (nm)	λ_2 (nm)	Optical density ratio [λ_1/λ_2]	Ref.
C. Nonlinear dyes with small angles between chromophores							
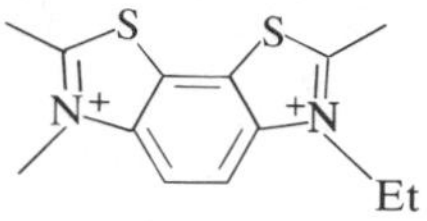 [structure: benzobisthiazolium, N-Et]	1	1	S	474	408	1.0	^{k}
	3	3	S	632	514	1.2	^{k}
	5	5	S	751	594	1.3	^{k}
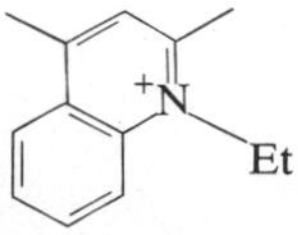 [structure: quinolinium, N-Et]	1	1	S	545	452	0.80	^{l}
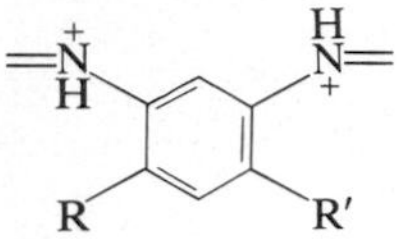 [structure: phenylenediamine, R, R′]	4 (R = R′ = H)	4	CMe_2	551	473	0.99	^{m}
	4 (R = H, R′ = Cl)	4	CMe_2	546	474	0.67	^{m}
	4 (R = R′ = Cl)	4	CMe_2	530	472	0.46	^{m}

[a] Secondary maxima at wavelengths larger than 400 nm were weak [$\lambda_1/\lambda_2 \gg 10$] and not well defined in these cases.

[b] Hamer,[10] p. 631.

[c] D. W. Heseltine and L. G. S. Brooker, U.S. Patent 3,140,951 (1964).

[d] H. E. Sprenger and W. Ziegenbein, *Angew. Chem. Int. Ed. Engl.*, **6**, 553 (1967).

[e] G. A. Rillaers and H. Depoorter, U.S. Patent 3,793,313 (1974).

[f] P. W. Jenkins, French Patent 2,099,780 (1972).

[g] A. I. Kiprianov and co-workers, *Zh. Org. Khim.*, **3,** 2036, 2041 (1967); spectra in nitromethane. The 3*H*-indole terminal nuclei have N-Me groups rather than the N-Et groups shown in the general structure.

[h] J. D. Mee, Research Disclosure 12529, Vol. 125 (1974).

[i] E. B. Knott, *J. Chem. Soc.*, 4244 (1960); spectra in pyridine/water (1/1).

[j] Hamer,[10] p. 661. Data for dyes with (n, m) values of (4,1) and (6,1) were for N-Ph rather than N-Et on the central rhodanine ring.

[k] A. I. Kiprianov, S. G. Fridman, G. G. Dyadyusha, and L. I. Kotova, *Zh. Org. Khim.*, **11,** 891 (1975).

[l] Hamer,[10] p. 624.

[m] A. I. Kiprianov and T. K. Nikolaenko, *Zh. Org. Khim.*, **8**, 1692 (1972). The 3*H*-indole terminal nuclei have N-Me groups rather than the N-Et groups shown in the general structure.

VIII. Substituent Effects

Starting with the methine carbons and terminal groups as the essential cyanine chromophore, the color and redox properties of dyes can be modified significantly by additional groups or substitution for certain atoms. The generic patterns for substituent effects are discussed below for three categories: ring substituents (added groups, fused rings); chain substituents (pendant groups, aza replacements, fused rings); and nitrogen substituents.

A. Ring Substituents

Substituents for the carbocyanine chromophore **160** have been studied extensively. Up to 40 groups in the 5,5′- and 6,6′-positions were tabulated by Ficken[11] for the benzothiazole and benzimidazole dyes. For the thiacarbocyanines, most substituents provided absorption maxima at

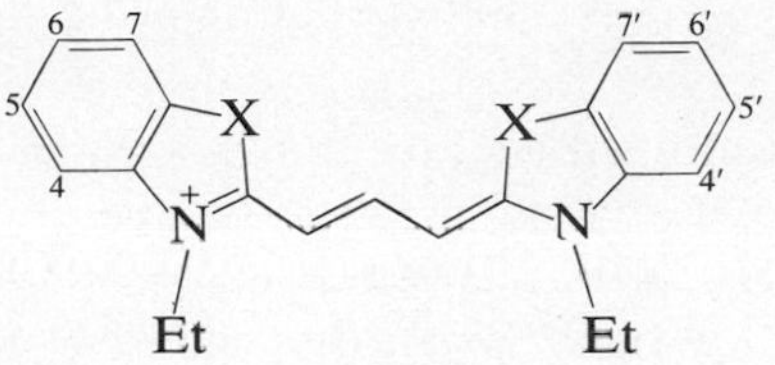

160

longer wavelengths. Halogens, alkoxy, and alkyl groups provided similar shifts in either the 5- or 6-positions. Electron-withdrawing groups and conjugated substituents, for example

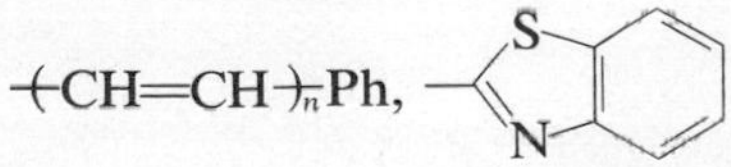

$-(CH{=}CH)_n Ph$,

showed larger effects in the 6,6′-positions than in the 5,5′-positions. Certain unsaturated substituents provided polymerizable dyes.[744,745] Direct ring sulfonation of carbocyanine and related dyes was successful.[746,747] Multiple substituents were introduced into thiacarbocyanines[748–756] and thiazolocarbocyanines.[757–759] Extensive work on multiple ring substituents in benzimidazolocarbocyanines[760–771] has provided good sensitizers containing many groups that significantly decreased the sensitizing power of thiacarbocyanines. This may be related to the large difference in reduction potentials for the unsubstituted dyes, which for the benzimidazolocarbocyanine is more negative by 0.7 V than for the thiacarbocyanine (see Table VI).

Hammett $\sigma\rho$ relations for many of these ring substituents correlate with data other than absorption maxima. Both pK_a[190,210–213] and polarographic reduction potential[190] were correlated with σ values for several dye series (X = N–R, O, S, and CMe_2). The R_f values for paper chromatography of thiacarbocyanines were a function of the Hammett σ

constant of the 5 (or 6) substituent, and 5-substituted dyes were the most strongly adsorbed.[772]

Additional benzene or heterocyclic rings were fused to the essential chromophores of thiazolocarbocyanine, imidazolocarbocyanine, and other dyes. The many variations in heterocyclic structure obtained in this way have been reviewed by Hamer[10] and Ficken.[11] Work on additional fused rings incorporating thiadiazoles,[773–775] thiophene (and benzo analogs),[776–782] naphthoquinones,[783–785] and aza heterocycles (quinoline, quinoxaline, pyrazine)[786–792] has continued. These new rings are similar to substituents in the sense that, although a similar color change was observed for many carbocyanines, the resulting redox properties (and, therefore, sensitizing power) depended strongly on the chromophore. As typical examples, consider the structures in Fig. 16, where the lowest vacant (ϵ_{lv}) and highest filled (ϵ_{hf}) orbitals for symmetrical carbocyanines are shown. The two patterns of substitution, linearly fused benzene rings and CH– replaced by N–, are additive, and the easily reduced dyes (desensitizers) are at the right. Experimentally, the thiazolocarbocyanines (shown in Fig. 16, X = sulfur) become desensitizers when a quinoline ring is fused to each thiazole ring (thiazolo[4,5-*b*]quinoline), but the imidazole analog from imidazo[4,5-*b*]quinoline provided sensitizing dyes. A second nitrogen in the center ring of the heterocycles is necessary in the imidazole dye series before the dyes become desensitizers (imidazo[4,5-*b*]-quinoxalines). This example and experience with substituent effects in

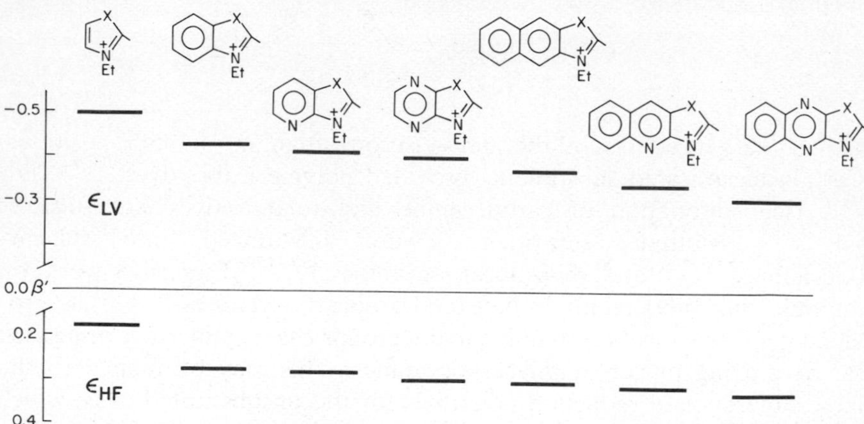

Fig. 16. Orbital shifts for benzo and aza derivatives of symmetrical trimethine dyes (X = sulfur). Replacement of aromatic C–H atoms by aza nitrogen augments the orbital shifts induced by the additional fused benzene rings. The arbitrary energy parameter β' was usually evaluated as the slope of linear relationships between these molecular orbitals and observed data (see Fig. 9). The least stable orbitals exhibit the most negative (least positive) β' values with respect to the arbitrary reference energy 0.0 β'. (Reproduced by permission from *Photogr. Sci. Eng.*, **18,** 56 (1974).)

benzimidazoles suggest that the synthetic design of effective sensitizers is perhaps limited most by the nature of the chromophore to be modified and not by inherent "desensitizing" properties of substituents or certain fused rings.

B. Chain Substituents

Dye modifications via chain substituents are historically most important (Hamer[10] devotes Chapter VI to chain-substituted carbocyanines). Effects on dye planarity and aggregation were noted in a previous section. The present section emphasizes the effects of chain substituents on color and redox properties. The two molecular orbitals that relate to these properties, ϵ_{lv} and ϵ_{hf}, differ in their sensitivity to substituents. These sensitivities, as a function of positions in various methine chains, are shown in Fig. 17. Electron-attracting substituents include groups attached to a methine carbon ($-CF_3$, $-CN$, $-NO_2$, $-CHO$, etc.), aza nitrogens that replace a methine carbon, and nonconjugated (twisted) atoms with higher electronegativity than carbon (hindered amines). In general, these substituents will stabilize both orbitals, but the net electronic effect on color and redox properties will depend on which orbital is most sensitive to substituents. Added to this type of electronic change are the effects of steric interactions and bond length changes, which vary with individual substituents.

Azacyanines are significantly lighter in color (shorter wavelength) than their methine analogs,[793–795] and related simple cyanines with phosphorus and arsenic atoms in the chain are also ligher.[298] Polyaza carbocyanines,[794–800] merocyanines,[798,799] and dicarbocyanines[801–804] with various patterns of aza substitution are known, but Kendall[805–808] first recognized the general importance of the position of the chain nitrogen in aza carbocyanines. Relative to the trimethine dyes, the α-aza dyes absorbed at shorter wavelength. Similar effects were found for carbocyanines with α-cyano, α-formyl, and α-nitro groups[809–811] since all of these stabilize primarily the highest filled orbital (ϵ_{hf} in Fig. 17). (Groups like α-alkyl, α-alkoxy, α-alkylthio, and α-fluoro generally provide longer wavelength absorption.[809,810,812–817]) On the other hand, β-aza[805–808] and β-cyano (nitro, C_3F_7, etc.)[818,819] carbocyanines absorbed at longer wavelengths (stabilized ϵ_{lv}, Fig. 17). The effects of meso substituents were also studied for di- and tricarbocyanines. First, for dicarbocyanines, central aza atoms,[807] nitro groups,[820] and substituted phenylazo groups[820,821] shifted the maxima to shorter wavelengths (see ϵ_{hf}, Fig. 17). Hammett σ-constants for substituents in the phenylazo groups were related to absorption maxima.[820,821] Second, absorptions of

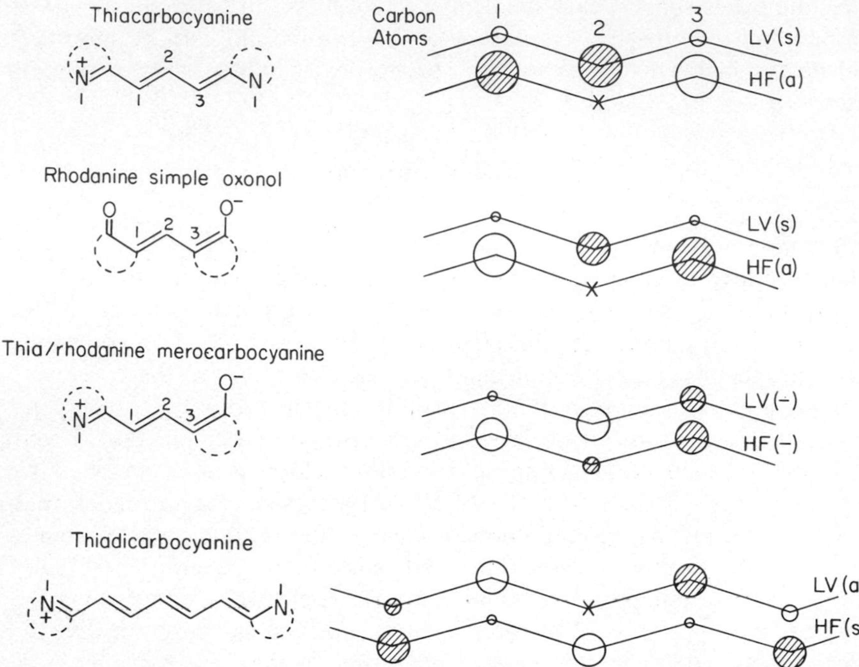

Fig. 17. Substituent inductive effects for thiacarbocyanine compared with a simple oxonol and merocarbocyanine of equivalent chromophore length and with thiadicarbocyanine. The molecular orbital coefficients for each carbon atom (C_{1j}, C_{2j}, . . .) are designated as either positive (open circles), negative (shaded circles), or zero (X), so that the orbital symmetry about carbon atom 2 is either symmetric (*s*) or antisymmetric (*a*). The size of the circles represents the values C_{ij}^2, and so on, for the lowest vacant (LV) and highest filled (HF) molecular orbitals. The value of C_{ij}^2 is a measure of the electron density at atom *i* in the molecular orbital *j* (HF or LV) in the diagram. A large circle suggests a high sensitivity of the orbital energy to substituent effects. If the sensitivities differ greatly for an atom in the HF and LV orbitals, significant color changes should result from substituents on that atom

rigidized tricarbocyanines with mesoamino substituents shifted to longer wavelength for less basic and crowded amines (e.g., Me_2N versus Ph_2N).[822–824]

Certain substituent patterns provide alternative routes for conjugation, like unsaturated bridges and heterocyclic rings incorporated into the methine chains. Many of these were reviewed by Ficken and Kendall,[807] Hamer (Chapters IX, XV),[10] and others.[825–831] Aromatic bridges[807] have included vinyl, allylidene, *o*-phenylene, and 1,8-naphthalene, but the effects on color depended on whether three or four methine carbons were bridged by the groups. Single heteroatom bridges utilized two or four chain carbons, effectively incorporating furan,[825,828,829] thiophene,[826,828] and similar heterocycles[807,827] in the chain. Dye **161** was prepared by nucleophilic additions to a substituted phenalene and is essentially a thiadicarbocyanine with three of the five methine carbons linked by the 1,8-positions of a naphthalene ring.[830,831] Isomeric dyes **162** and **163** were also isolated from similar nucleophilic additions. The large wavelength shifts for dyes **161** and **162** are related to sensitive chromophoric atoms in the ϵ_{lv} orbitals (Fig. 17): (1) dye **161**, a thiadicarbocyanine, atoms 2 and 4; and (2) dye **162**, a thiatricarbocyanine, atoms 2 and 6 but not 5. The tetracarbocyanine **163** has two positions where the aromatic substituent affects the ϵ_{lv} orbital (atoms 2,8) and one where it affects the ϵ_{hf} orbital (atom 5). The substituent R is attached to an atom where the ϵ_{lv} is sensitive, and the change from methyl to the more electron donating ethoxy shifts the dye to shorter wavelength.

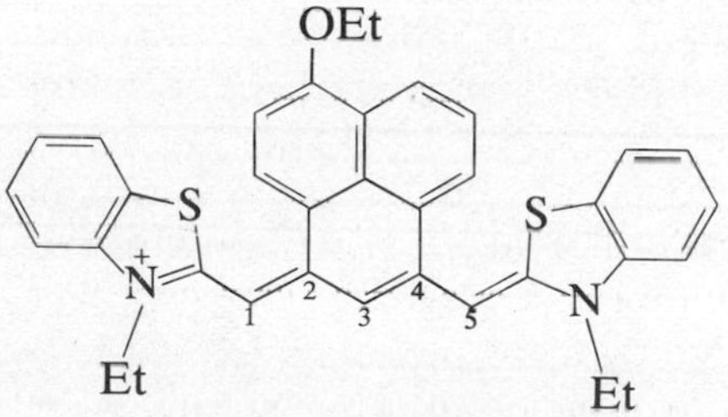

λ_{max} 757, 573 nm
(thiadicarbo, 650 nm)

161

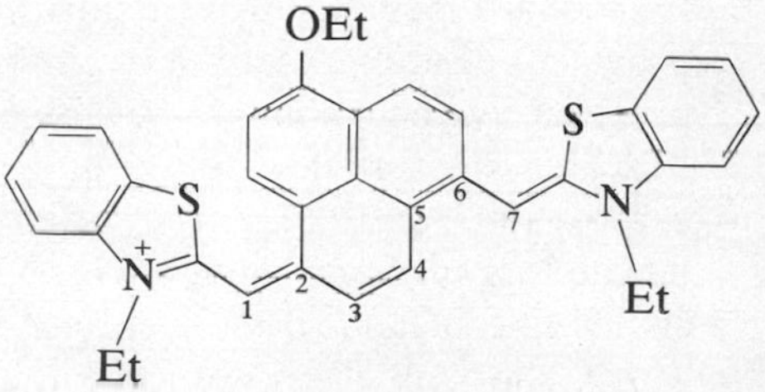

λ_{max} 845, 572 nm
(thiatricarbo, 750 nm)

162

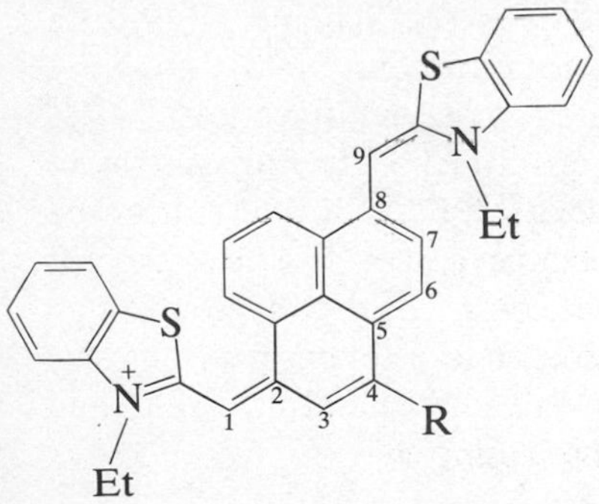

R = Me λ_{max} 905, 510 nm
R = OEt λ_{max} 840 nm
(thiatetracarbo, 850 nm)

163

C. Nitrogen Substituents

Several classes of nitrogen substituents have been incorporated into typical dye structures. The most useful class is substituted alkyl, as discussed below. Other classes should be noted, however. *N*-Aryl groups in benzimidazole,[832–834] thiazole,[814,835,836] and quinoline-related heterocycles[837–844] have been investigated extensively. Substituents on the aryl rings of the latter were related to dye color by Hammett σ-constants and to fluorescence intensity. Heterocyclic substituents on nitrogen included saturated and unsaturated groups.[845–854] Complexes of dye bases with metal ions[208,638,855] (iridium,[855] platinum,[855] zinc,[208] copper,[638] and nickel[638]) served as both colorimetric reagents[638] and potential superconducting systems.[855]

Nitrogen substituents related to alkyl groups are of primary importance for sensitizing dyes. Aside from the simple alkyl groups, there are five functional types of nitrogen substituents. First, the acidic *N*-substituents provided desirable solubility and adsorption characteristics for many practical cyanine and merocyanine sensitizers.[226,856–864] The patents in this area were reviewed recently by Poppe,[226] and the groups include carboxyalkyl, sulfoalkyl, thiosulfatoalkyl, sulfatoalkyl, and phosphoalkyl. Related groups are sulfoallyl and sulfobenzyl. Second, long *N*-alkyl chains like octadecyl gave surface-active dyes for monolayer studies.[865] Third, short alkyl groups containing electron-withdrawing groups were studied.[866–875] These included polyfluoroalkyls,[866,870,871] cyanoethyl,[867] $-CH_2CH_2CONR_2$,[868] polycyanoallyl (or vinyl),[869] $-(CH_2)_6OCOCMe{=}CH_2$ (polymerizable merocyanines),[869a] bromoalkyl,[872,873] thiosulfatoalkyl,[872–874] and acetals of alkylformyl groups.[875] The latter[875] are reactive and useful for the synthesis of rigidized dyes (see Scheme 4). Fourth, heteroatom substituents directly bonded to nitrogen ($N{-}O^-$, $N{-}NR_2$, N–OR) provided photochemically reactive dyes (see next section).[876–879] The subject of reactive fiber dyes (via groups attached to nitrogen) is to be reviewed and will not be covered here.[880] Fifth, cyclic saturated amines as the terminal groups for hemicyanines[44] and streptocyanines[881,882] significantly influenced the color of the dyes through variations in the "basicity" of the amino nitrogen as a function of ring size and structure. In addition, optically active thiazole quaternary salts and dyes were prepared from an optically active amine.[353]

IX. Photoreactions of Dyes

Photooxidation and photoreduction of dyes in solution have been studied extensively for the xanthene, thiazine, and acridine dye classes.

The pioneering work of Oster and co-workers[883–889] and some more recent results have been reviewed by Meier[890] and others.[891–893] Photoreactions of cyanines and related dyes have been studied less extensively, perhaps because of the low singlet-triplet intersystem crossing yields (Table XIV). Low quantum yield photofading reactions of cyanine dyes were observed under both oxidative conditions[894–897] (oxygen, reducible metal ions) and reductive conditions[898–900] (ascorbic acid, gelatin). Gerischer and co-workers have demonstrated photoconductively that cyanine dyes interact with oxidizing and reducing agents in solution.[901–903]

Both direct and sensitized photooxidations of cyanine dyes were investigated more recently.[904] The naphthothiacyanine **164** was photodegraded more readily as the H-aggregate, and triplet quenchers like 3,3,4,4-tetramethyl-1,2-diazetine dioxide give nearly diffusion-controlled quenching of the photooxidation. Since the quantum yields for direct photooxidation

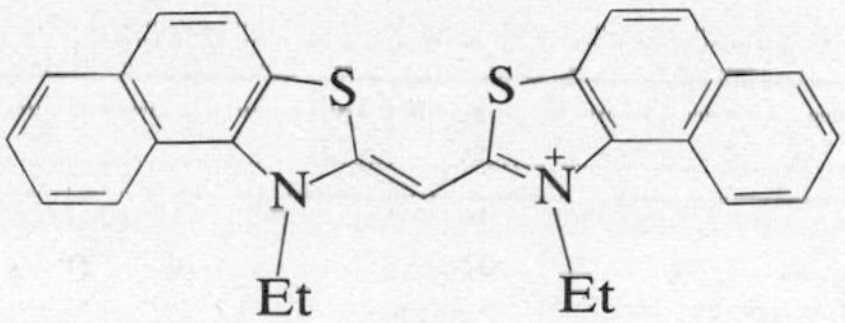

Photoxidation
Monomer, $\Phi \leq 10^{-6}$
H-aggregate, $\Phi \approx 10^{-2}$

164

were relatively low, sensitized photolyses using methylene blue (or 2-acetonaphthone) were used to determine the relative reactivity among a number of cyanine dyes toward singlet oxygen. Rate constants (k_{pr}) for dye bleaching, listed in Table XVII, are generally parallel to electrochemical oxidation potentials (i.e., thiatricarbocyanine reacted more readily than thiacarbocyanine), but other structural features like mesoalkyl enhanced the photooxidation rate about five times more than predicted from the electrochemical potentials.

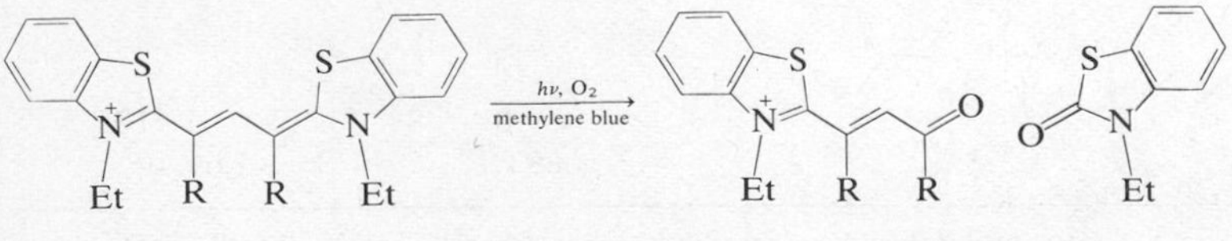

$\xrightarrow[\text{methylene blue}]{h\nu,\ O_2}$ (28)

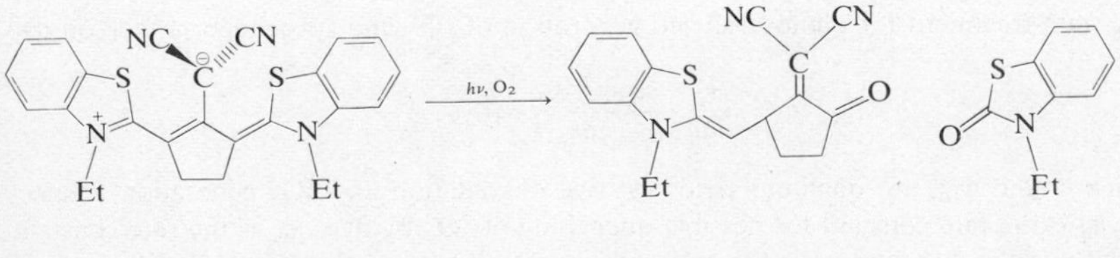

$\xrightarrow{h\nu,\ O_2}$ (29)

TABLE XVII. SENSITIZED PHOTOOXIDATIONS[a]

Dye structure	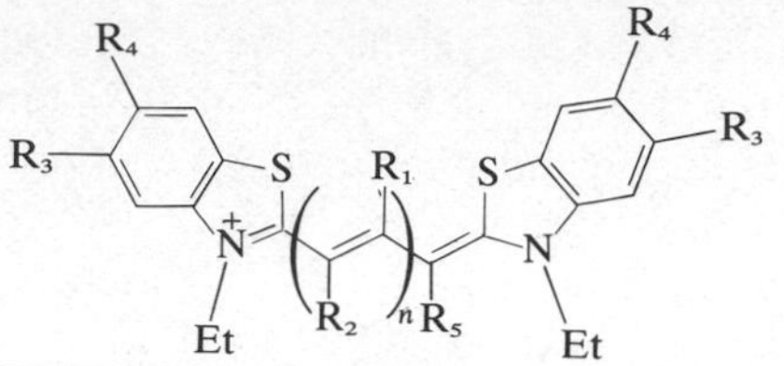	k_{pr} ($1\ m^{-1}\ sec^{-1}$)
A. Vinylogous thiacyanines (all R = hydrogen)	$n = 1$	0.35×10^7
	$n = 2$	1.3×10^7
	$n = 3$	5.1×10^7
B. Substituted thiacarbocyanines ($n = 1$) All R = hydrogen except:	$R_4 = OMe$	1.1×10^7
	$R_3 = OMe$	0.67×10^7
	$R_3 = Cl$	0.13×10^7
All R = hydrogen except R_1 = ethyl and	—	1.6×10^7
	$R_4 = OMe$	4.9×10^7
	$R_3 = OMe$	3.1×10^7
	$R_3 = Cl$	0.71×10^7
	$R_3 = CN$	0.33×10^7
All R = hydrogen except:	$R_5 = R_2 = Me$	28.8×10^7
	$(R_2, R_5) = —CH_2CH_2—$	11.8×10^7
	$(R_2, R_1) = —CH_2CH_2CH_2—$	14.8×10^7
C. Other dyes		
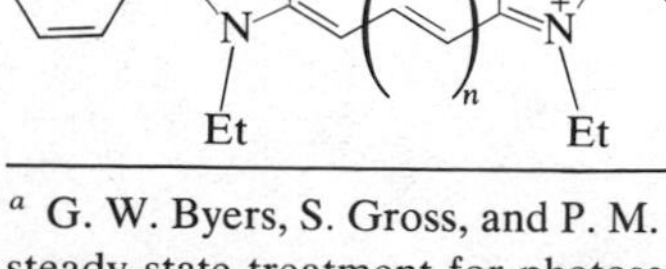 (quinoline cyanine, N-Et)	$n = 0$	$\approx 0.04 \times 10^7$
	$n = 1$	3.4×10^7
Kryptocyanine		10×10^7
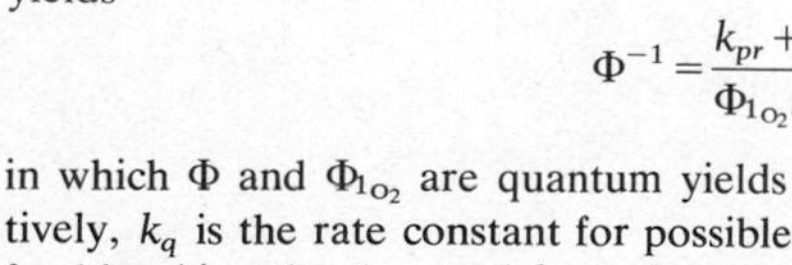 (naphthothiazole cyanine, N-Et)	$n = 0$	$\approx 0.04 \times 10^7$
	$n = 1$	1.6×10^7

[a] G. W. Byers, S. Gross, and P. M. Henrichs, *Photochem. Photobiol.*, **23**, 37 (1976). The usual steady-state treatment for photosensitized generation of $^1O_2$ and subsequent attack on dye yields

$$\Phi^{-1} = \frac{k_{pr} + k_q}{\Phi_{^1O_2} k_{pr}} + \frac{k_d}{\Phi_{^1O_2} k_{pr}} [\text{dye}]^{-1}$$

in which Φ and $\Phi_{^1O_2}$ are quantum yields for dye degradation and $^1O_2$ generation, respectively, k_q is the rate constant for possible quenching of $^1O_2$ by dye, k_{pr} is the rate constant for bleaching the dye, and k_d is the rate constant for $^1O_2$ natural decay ($1.9 \times 10^5\ sec^{-1}$ in methanol).

The products from these reactions included those shown in Eqs. 28–29, which were consistent with 1,2-addition of singlet oxygen followed by cleavage of the C–C bond between the first methine carbon and the heterocyclic ring.[567,904] These photooxidations (and photoreductions) are quite general, bimolecular reactions, which depend primarily on sufficient lifetimes for the reactive species to interact. In contrast to this, the two following photodegradations are examples of unimolecular reactions that depend on specific functional groups in the dyes.

Many nitro compounds of the structure **165** exhibited photofading reactions in solution and crystalline phases.[905–908] Substituted stilbenes ($R_1 = p\text{-}Me_2NC_6H_4$ and $R_2 = H$) as well as heterocyclic dyes ($R_1R_2C =$ a basic heterocycle, Table III.A) were reactive if the nitro group was ortho to the polymethine chain (Eq. 30). Photofragmentation caused by oxygen transfer from the nitro group gave products **167** and **168**. Photoisomerization led to structures like **166** for the heterocyclic dyes. Tautomers of these were postulated in 1955 by Splitter and Calvin[905] as unstable precursors to the final products (isatogens) from the stilbenes ($R_2 = H$). The relative importance of the photofragmentation and photoisomerization pathways was influenced by the nature of the solvent[908] as well as the crystalline phase.

$R_1R_2C{=}CH{-}(CH{=}CH)_n{-}C_6H_3(NO_2){-}R$ **165** $\xrightarrow{h\nu}$ **166** + $R_1R_2C{=}O$ **167** + **168** (30)

166 $\xrightarrow[n=0]{R_2 = H}$ isatogens

Photoreactive substituents on quaternary nitrogen atoms were characterized by Mee[876,877] and Mitchell.[878] Both N–OR and N–NR_2 for the N–X part of structure **169** (Eq. 31) provided photobleachable and thermally bleachable dyes. The photolytic reaction for the *N*-methoxy dyes led primarily to dye base **170**, chain-substituted analogs of **170**, and other heterocyclic products. Cleavage of the N–O bond in 1-methoxy-phenanthridinium perchlorate **171** (Eq. 32) occurred with a quantum yield of 0.25[910] and led to (1) the $CH_2(OMe)_2$ and CH_3OH products and (2) the isomeric methoxybenzonitriles.[909] The ratio of *ortho* : *meta* : *para* methoxybenzonitriles **173** (15 : 1.0 : 4) suggested homolytic cleavage of the N–O bond in **171** and a methoxy radical intermediate.

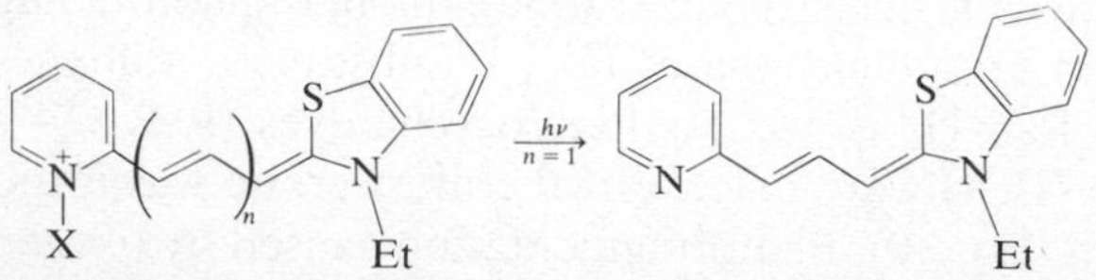

169 $\xrightarrow[n=1]{h\nu}$ **170** + [CH_2O for X = OMe] + other products (31)

171 $\xrightarrow[PhCN]{h\nu, CH_3CN}$ **172** + CH_3OH (25%) + $CH_2(OMe)_2$ (15%) + **173** (7.6%) (32)

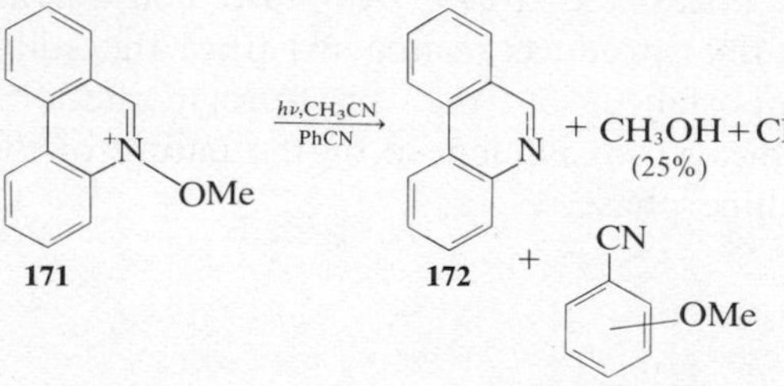

X. Acknowledgment

The author is indebted to D. W. Heseltine, D. L. Smith, J. D. Mee, P. W. Jenkins, G. W. Byers, P. M. Henrichs, D. F. O'Brien, J. E. Starr, D. D. Chapman, F. G. Webster, J. R. Nonnemacher, and A. Kocher of the Eastman Kodak Research Laboratories for much consultation and advice. Special thanks are due D. W. Heseltine, who, as current Head of the Sensitizing Dye Synthesis Laboratory, a long-time associate of L. G. S. Brooker, and a valuable consultant to many, has been responsible for much of the real progress in the cyanine dye field over the last decade.

XI. References

1. L. G. S. Brooker, "The Theory of the Photographic Process," C. E. K. Mees, Ed., Macmillan, New York, 1942, p. 987.
2. L. G. S. Brooker, "The Theory of the Photographic Process, " C. E. K. Mees and T. H. James, Eds., Macmillan, New York, 1966, p. 198.

3. L. G. S. Brooker, "Recent Progress in the Chemistry of Natural and Synthetic Coloring Matters," T. S. Gore, B. S. Joshi, S. V. Synthankar, and B. D. Tilak, Eds., Academic Press, New York, 1962, p. 573.
4. L. G. S. Brooker and P. W. Vittum, *J. Photogr. Sci.*, **5,** 71 (1957).
5. L. G. S. Brooker, *Rev. Mod. Phys.*, **14,** 275 (1942).
6. L. G. S. Brooker, F. L. White, R. H. Sprague, S. G. Dent, Jr., and G. Van Zandt, *Chem. Rev.*, **41,** 325 (1947).
7. L. G. S. Brooker and E. J. Van Lare, "Encyclopedia of Chemical Technology," Vol. 5, Wiley, New York, 1964, p. 763.
8. L. G. S. Brooker, F. L. White, D. W. Heseltine, G. H. Keyes, S. G. Dent, Jr., and E. J. Van Lare, *J. Photogr. Sci.*, **1,** 173 (1953).
9. L. G. S. Brooker and W. T. Simpson, *Ann. Rev. Phys. Chem.*, **2,** 121 (1951).
10. F. M. Hamer, "The Chemistry of Heterocyclic Compounds," Vol. 18, "The Cyanine Dyes and Related Compounds," A. Weissberger, Ed., Interscience, New York, 1964.
11. G. E. Ficken, "The Chemistry of Synthetic Dyes," Vol. 4, K. Venkataraman, Ed., Academic Press, New York, 1971, p. 211.
12. A. I. Kiprianov, *Usp. Khim.*, **29,** 1336 (1960).
13. A. I. Kiprianov, *Usp. Khim.*, **40,** 1283 (1971).
14. A. I. Kiprianov and L. M. Yagupol'skii, *Zh. Obshch. Khim.*, **20,** 2111 (1950).
15. A. I. Kiprianov and I. K. Ushenko, *Zh. Obshch. Khim.*, **20,** 134, 514 (1950).
16. A. I. Kiprianov and I. K. Ushenko, *Zh. Obshch. Khim.*, **15,** 207 (1945).
17. A. I. Kiprianov, I. K. Ushenko, and E. D. Sych, *Zh. Obshch. Khim.*, **15,** 200 (1945).
18. A. I. Kiprianov and G. T. Pilyugin, *Byull. Vsesoyuz. Khim. Obshch. Im. D. I. Mendeleev.*, 1939, see *Khim. Ref. Zh.*, **3,** 60 (1939); *Chem. Abstr.*, **34,** 4663 (1940).
19. A. I. Kiprianov, G. G. Dyadyusha, and F. A. Mikhailenko, *Usp. Khim.*, **35,** 823 (1966).
20. C. H. G. Williams, *Trans. Roy. Soc. Edinburgh*, **21,** 377 (1856).
21. H. W. Vogel, *Ber.*, **6,** 1302 (1873).
22. H. W. Vogel, *Phil. Photogr.*, **11,** 25 (1874); republished in *Photogr. Sci. Eng.*, **18,** 33 (1974).
23. W. West, *Photogr. Sci. Eng.*, **18,** 35 (1974).
24. B. Homolka, German Patent 172,118 (1905).
25. W. H. Mills and F. M. Hamer, *J. Chem. Soc.*, **117,** 1550 (1920).
26. W. H. Mills, *J. Chem. Soc.*, **121,** 455 (1922).
27. L. G. S. Brooker (Eastman Kodak Co., Rochester, N. Y.), retired, now deceased, private communication.
28. W. Koenig, *Ber.*, **55,** 3293 (1922).
29. W. Koenig, *Ber.*, **57,** 685 (1924).
30. W. Koenig and W. Meier, *J. Prakt. Chem.*, **109,** 324 (1925).
31. F. M. Hamer, *J. Chem. Soc.*, 3160 (1928).
32. O. Wahl, German Patent 499,967 (1928).
33. H. A. Piggott and E. H. Rodd, British Patent 355,693 (1930).
34. F. M. Hamer, British Patent 354,826 (1930).
35. T. Zincke, *Justus Liebigs Ann. Chem.*, **330,** 361 (1904).
36. H. A. Piggott and E. H. Rodd, British Patent 344,409 (1929).
37. H. A. Piggott and E. H. Rodd, British Patent 354,898 (1930).
38. F. B. Dains and E. W. Brown, *J. Am. Chem. Soc.*, **31,** 1148 (1909).
39. F. B. Dains, R. Thompson, and W. F. Asendorf, *J. Am. Chem. Soc.*, **44,** 2310 (1922).
40. L. G. S. Brooker, G. H. Keyes, R. H. Sprague, R. H. Van Dyke, E. Van Lare, G. Van Zandt, and F. L. White, *J. Am. Chem. Soc.*, **73,** 5326 (1951).

41. L. G. S. Brooker, G. H. Keyes, R. H. Sprague, R. H. Van Dyke, E. Van Lare, G. Van Zandt, F. L. White, H. W. J. Cressman, and S. G. Dent, Jr., *J. Am. Chem. Soc.*, **73,** 5332 (19[illegible]).
42. L. G. S. Brooker, *Exper. Suppl.*, **2,** 229 (1955).
43. L. G. S. Brooker, R. H. Sprague, C. P. Smyth, and G. L. Lewis, *J. Am. Chem. Soc.*, **62,** 1116 (1940).
44. L. G. S. Brooker, F. L. White, G. H. Keyes, C. P. Smyth, and P. F. Oesper, *J. Am. Chem. Soc.*, **63,** 3192 (1941).
45. L. G. S. Brooker and R. H. Sprague, *J. Am. Chem. Soc.*, **63,** 3203 (1941).
46. L. G. S. Brooker and R. H. Sprague, *J. Am. Chem. Soc.*, **63,** 3214 (1941).
47. L. G. S. Brooker, G. H. Keyes, and W. W. Williams, *J. Am. Chem. Soc.*, **64,** 199 (1942).
48. L. G. S. Brooker and R. H. Sprague, *J. Am. Chem. Soc.*, **67,** 1869 (1945).
49. L. G. S. Brooker, A. L. Sklar, H. W. J. Cressman, G. H. Keyes, L. A. Smith, R. H. Sprague, E. Van Lare, G. Van Zandt, F. L. White, and W. W. Williams, *J. Am. Chem. Soc.*, **67,** 1875 (1945).
50. L. G. S. Brooker, R. H. Sprague, and H. W. J. Cressman, *J. Am. Chem. Soc.*, **67,** 1889 (1945).
51. L. G. S. Brooker, F. L. White, and R. H. Sprague, *J. Am. Chem. Soc.*, **73,** 1087 (1951).
52. L. G. S. Brooker, G. H. Keyes, R. H. Sprague, R. H. Van Dyke, E. Van Lare, G. Van Zandt, F. L. White, H. W. J. Cressman, and S. G. Dent, Jr., *J. Am. Chem. Soc.*, **73,** 5332 (1951).
53. L. G. S. Brooker, G. H. Keyes, and D. W. Heseltine, *J. Am. Chem. Soc.*, **73,** 5350 (1951).
54. L. G. S. Brooker and G. H. Keyes, *J. Am. Chem. Soc.*, **73,** 5356 (1951).
55. L. G. S. Brooker, A. C. Craig, D. W. Heseltine, P. W. Jenkins, and L. L. Lincoln, *J. Am. Chem. Soc.*, **87,** 2443 (1965).
56. A. I. Kiprianov and I. K. Ushenko, *Zh. Obshch. Khim.*, **15,** 684 (1945).
57. A. I. Kiprianov and V. E. Petrun'kin, *Zh. Obshch. Khim.*, **10,** 600 (1940).
58. A. I. Kiprianov and V. E. Petrun'kin, *Zh. Obshch. Khim.*, **10,** 613 (1940).
59. A. I. Kiprianov and E. S. Timoshenko, *Zh. Obshch. Khim.*, **17,** 1468 (1947).
60. See reviews in *Opt. Anregung Org. Syst.* (*Int. Farbensymp. Elmau, 1964*), W. Jung, Ed., Verlag Chemie, Weinheim, 1966.
61. R. Gompper and E. Kutter, *Angew. Chem. Int. Ed. Engl.*, **2,** 687 (1963).
62. R. Wizinger-Aust, *Q. Rep. Sulfur Chem.*, **5,** 191 (1970).
63. R. Wizinger and D. Duerr, *Helv. Chim. Acta*, **46,** 2167 (1963).
64. U. Schmidt, R. Scheuring, and A. Lüttringhaus, *Justus Liebigs Ann. Chem.*, **630,** 116 (1960).
65. H. Prinzbach and E. Futterer, "Advances in Heterocyclic Chemistry," Vol. 7, A. R. Katrizky and A. J. Boulton, Eds., Academic Press, New York, 1966, p. 39.
66. C. Jutz and H. Amschler, *Angew. Chem.*, **73,** 806 (1961).
67. R. Kuhn and H. Fischer, *Angew. Chem.*, **73,** 435 (1961).
68. R. Kuhn, *Angew. Chem.*, **73,** 658 (1961).
69. R. Kuhn, *Angew. Chem.*, **74,** 721 (1962).
70. K. Hafner, H. W. Riedel, and M. Danielisz, *Angew. Chem. Int. Ed. Engl.*, **2,** 215 (1963).
71. I. Agranat and E. Aharan-Shalom, *J. Am. Chem. Soc.*, **97,** 3829 (1975).
72. F. D. Popp and E. B. Moynahan, "Advances in Heterocyclic Chemistry," Vol. 13, A. R. Katrizky and A. J. Boulton, Eds., Academic Press, New York, 1971, p. 1.

73. A. Treibs and R. Zimmer-Galler, *Chem. Ber.*, **93,** 2539 (1960).
74. K. Dimroth and P. Hoffmann, *Angew. Chem.*, **76,** 433 (1964); *Chem. Ber.*, **99,** 1325 (1966).
75. K. Dimroth, N. Grief, and A. Klapproth, *Justus Liebigs Ann. Chem.*, 373 (1975).
76. K. Dimroth, *Fortschr. Chem. Forsch.*, **38,** 5 (1973).
77. A. Van Dormael, *Chimia*, **15,** 67 (1961).
78. H. Depoorter, J. Nys, and A. Van Dormael, *Tetrahedron Lett.*, 199 (1961).
79. H. Depoorter, J. Nys, and A. Van Dormael, *Bull. Soc. Chim. Belg.*, **73,** 921 (1964).
80. H. Depoorter, J. Nys, and A. Van Dormael, *Bull. Soc. Chim. Belg.*, **74,** 12 (1965).
81. J. A. Ford, Jr., *Tetrahedron Lett.*, 815 (1968).
82. F. Kroehnke, K. Ellegast, and E. Bertram, *Justus Liebigs Ann. Chem.*, **600,** 176 (1956).
83. J. A. Berson, E. M. Evleth, Jr., and Z. Hamlet, *J. Am. Chem. Soc.*, **87,** 2887 (1965).
84. J. R. Platt, *J. Chem. Phys.*, **25,** 80 (1956).
85. W. T. Simpson, *J. Am. Chem. Soc.*, **73,** 5359 (1951).
86. J. Spence, "A Look At Leslie Brooker—His Life and Works" presented at the Vogel Centennial, Putney, Vt., 1973, published in *Photogr. Sci. Eng.*, **18,** 610 (1974).
87. D. L. Smith, *Photogr. Sci. Eng.*, **18,** 309 (1974).
88. G. Germain, C. Paternotte, P. Piret, and M. Van Meerssche, *J. Chim. Phys. Phys.-Chim. Biol.*, **61,** 1059 (1964).
89. G. Germain, P. Piret, M. Van Meerssche, and J. De Kerf, *Acta Crystallogr.*, **15,** 373 (1962).
90. G. Germain, P. Piret, M. Van Meerssche, and J. De Kerf, *Bull. Soc. Chim. Fr.*, 1407 (1961).
91. M. B. D'Enghien-Peteau, J. Meunier-Piret, and M. Van Meerssche, *J. Chim. Phys. Phys.-Chim. Biol.*, **65,** 1221 (1968).
92. H. R. Luss and D. L. Smith, *Acta Crystallogr.*, *Sect. B*, **29,** 998 (1973).
93. P. J. Wheatley, *J. Chem. Soc.*, 3245 (1959).
94. P. J. Wheatley, *J. Chem. Soc.*, 4096 (1959).
94a. B. Zeimer and S. Kulpe, *J. Prakt. Chem.*, **317,** 199 (1975).
95. V. F. Kaminskii, R. P. Shibaeva, and L. O. Atovmyan, *Zh. Strukt. Khim.*, **14,** 700, 1082 (1973).
96. R. P. Shibaeva, V. F. Kaminskii and L. O. Atovmyan, *Zh. Strukt. Khim.*, **15,** 720 (1974).
96a. W. Falkenberg, E. Hoehne, J. Siele, and R. Wagener, *Krist. Tech.*, **8,** 227 (1973).
97. J. E. Weidenborner, G. Castro, and N. R. Stemple, American Crystallographic Association Meeting, University of Calif., Berkeley, March, 1974, paper C–7.
98. K. Nakatsu, H. Yoshioka, and T. Aoki, *Chem. Lett.*, 339 (1972).
99. R. P. Shibaeva, L. O. Atovmyan, V. I. Ponomarjev, O. S. Philipenko, and L. P. Rozenberg, *Tetrahedron Lett.*, 185 (1973); *Krystallogr.*, **19,** 95 (1974).
100. R. Allmann, *Chem. Ber.*, **99,** 1332 (1966).
101. T. E. Borowiak, N. G. Bokii, and Yu T. Struchkov, *Zh. Strukt. Khim.*, **13,** 480 (1972).
101a. V. F. Kaminsky, R. P. Shibaeva, and L. O. Atovmyan, *Zh. Struct. Khim.*, **15,** 509 (1974).
102. D. L. Smith and H. R. Luss, *Acta Crystallogr.*, *Sect. B*, **28,** 2793 (1972).
103. D. L. Smith and H. R. Luss, *Acta Crystallogr.*, *Sect. B*, **31,** 402 (1975).
104. J. Potenza and D. Mastropaolo, *Acta Crystallogr.*, *Sect. B*, **30,** 2353 (1974).
105. H. Stoekli-Evans, *Helv. Chim. Acta*, **57,** 1 (1974).
106. H. E. Marr, III, J. M. Stewart, and M. F. Chiu, *Acta Crystallogr.*, *Sect. B*, **29,** 847 (1973).

107. B. W. Matthews, R. E. Stenkamp, and P. M. Colman, *Acta Crystallogr., Sect. B*, **29,** 449 (1973).

107a. K. Sieber, L. Kutschabsky and S. Kulpe, *Krist. Tech.*, **9,** 1101 (1974).

107b. A. Zedler and S. Kulpe, *J. Prakt. Chem.*, **317,** 38 (1975).

108. F. Brandl, H. J. Springer, P. Narrayanan, J. Preuss, W. Hoppe, and G. Scheibe, *J. Mol. Cryst. Struct.*, **4,** 391 (1974).

108a. J. Kroon and H. Krabbendam, *Acta Crystallogr., Sect. B*, **30,** 1463 (1974).

109. K. Sieber, L. Kutschabsky, and S. Kulpe, *Krist. Tech.*, **9,** 1111 (1974).

109a. B. Zeimer and S. Kulpe, *J. Prakt. Chem.*, **317,** 185 (1975).

110. D. L. Smith and E. K. Barrett, *Acta Crystallogr., Sect. B*, **27,** 969 (1971).

111. J. Effinger, G. Germain, J. Meunier, J. Vanderauwera, and M. Van Meerssche, *Bull. Soc. Chim. Belg.*, **69,** 387 (1960).

112. H. Yoshioka and K. Nakatsu, *Chem. Phys. Lett.*, **11,** 255 (1971).

113. B. Dammeier and W. Hoppe, *Acta Crystallogr., Sect. B*, **27,** 2364 (1971).

114. I. Kawada and R. Allmann, *Angew. Chem. Int. Ed. Engl.*, **7,** 69 (1968).

115. G. Scheibe, W. Seiffert, H. Wengenmayr, and C. Jutz, *Ber. Bunsenges. Phys. Chem.*, **67,** 560 (1963).

116. G. Scheibe, C. Jutz, W. Seiffert, and D. Grosse, *Angew. Chem.*, **76,** 270 (1964).

117. G. Scheibe, W. Seiffert, G. Hohlneicher, C. Jutz, and H. J. Springer, *Tetrahedron Lett.*, 5053 (1966).

118. S. Daehne and J. Ranft, *Z. Phys. Chem.* (*Leipzig*), **224,** 65 (1963).

119. S. Daehne and J. Ranft, *Angew. Chem.*, **75,** 1175 (1963).

120. S. Daehne and J. Ranft, *Z. Phys. Chem.* (*Leipzig*), **232,** 259 (1966).

121. R. Radeglia, S. Daehne, and H. Hartmann, *J. Prakt. Chem.*, **312,** 1081 (1970).

122. R. Radeglia, E. Gey, K. D. Nolte, and S. Daehne, *J. Prakt. Chem.*, **312,** 877 (1970).

123. R. Radeglia, H. Gey, H. Henning, W. Kuehnel, D. Labes, and H. G. Sieg, *Z. Chem.*, **13,** 182 (1973).

124. D. Lloyd, R. K. Mackie, H. McNab, and D. R. Marshall, *J. Chem. Soc., Perkin Trans. 2*, 1729 (1973).

125. H. J. Friedrich, *Angew. Chem.*, **75,** 298 (1963).

126. I. H. Leubner, *Org. Magn. Reson.*, **6,** 253 (1974).

126a. P. M. Henrichs and S. Gross, *J. Am. Chem. Soc.*, in press.

127. E. Kleinpeter, R. Borsdorf, G. Bach, and J. Von Grossmann, *J. Prakt. Chem.*, **315,** 587 (1973).

128. E. Kleinpeter, R. Borsdorf, and J. Von Grossmann, *J. Signalaufzeichnungsmaterialien*, **1,** 293 (1973).

129. R. Radeglia, *Z. Phys. Chem.* (*Leipzig*), **235,** 335 (1967).

129a. M. Waehnert and S. Daehne, *J. Prakt. Chem.*, **318,** 321 (1976).

130. R. Radeglia and S. Daehne, *J. Mol. Struct.*, **5,** 399 (1970).

131. E. Kleinpeter, R. Borsdorf, G. Bach, and J. Von Grossmann, *Z. Chem.*, **14,** 194 (1974).

132. H. G. Benson and J. N. Murrell, *J. Chem. Soc., Faraday Trans. 2*, 137 (1972).

133. L. H. Feldman, A. H. Herz, and T. H. Regan, *J. Phys. Chem.*, **72,** 2008 (1968).

134. E. A. Morris and E. F. Mooney, *Ind. Chim. Belge*, **32,** 86 (1967).

135. J. Metzger, H. Larivé, R. Dennilauler, R. Baralle, and C. Gaurat, *Ind. Chim. Belge*, **32,** 96 (1967).

136. J. Metzger, H. Larivé, R. Dennilauler, and E. J. Vincent, *Bull. Soc. Chim. Fr.*, 46 (1967).

137. E. Kleinpeter, R. Borsdorf, and F. Dietz, *J. Prakt. Chem.*, **315,** 600 (1973).

138. E. Kleinpeter, R. Borsdorf, G. Bach, and J. Von Grossmann, *J. Prakt. Chem.*, **316,** 761 (1974).
139. R. Radeglia and A. Weber, *J. Prakt. Chem.*, **314,** 884 (1972).
140. E. Kleinpeter and R. Borsdorf, *Z. Chem.*, **13,** 183 (1973).
141. R. E. Graves and P. I. Rose, *J. Chem. Soc., Chem. Commun.*, 630 (1973).
142. R. Radeglia, G. Engelhardt, E. Lippmaa, T. Pehk, K. D. Nolte, and S. Daehne, *Org. Magn. Reson.*, **4,** 571 (1972).
143. R. Radeglia, E. Gey, T. Steiger, S. Kulpe, R. Lueck, M. Ruthenberg, M. Stierl, and S. Daehne, *J. Prakt. Chem.*, **316,** 766 (1974).
144. R. Radeglia, *J. Prakt. Chem.*, **316,** 344 (1974).
144a. W. Grahn and C. Reichardt, *Tetrahedron*, **32,** 125, 1931 (1976).
145. P. M. Henrichs, *J. Chem. Soc. Perkin 2*, 542 (1976).
146. E. Kleinpeter and R. Borsdorf, *J. Prakt. Chem.*, **315,** 765 (1973).
147. R. E. Graves and P. I. Rose, *J. Phys. Chem.*, **79,** 746 (1975).
148. M. Pestemer, *Chimia*, **15,** 31 (1961).
149. H. J. Friedrich, *Z. Naturforsch., Teil B*, **18,** 635 (1963).
150. A. Leifer, D. Bonis, M. Boedner, P. Dougherty, M. Koral, and J. E. LuValle, *Appl. Spectrosc.*, **20,** 289 (1966).
151. A. Leifer, M. Boedner, P. Dougherty, A. Fusco, M. Koral, and J. E. LuValle, *Appl. Spectrosc.*, **20,** 150 (1966).
152. A. Leifer, D. Bonis, M. Boedner, P. Dougherty, A. J. Fusco, M. Koral, and J. E. LuValle, *Appl. Spectrosc.*, **21,** 71 (1967).
153. F. Zuccarello, A. Foffani, R. Cataliotti, and A. Poletti, *Ric. Sci.*, **37,** 1106 (1967).
154. R. Cataliotti, A. Poletti, F. Zuccarello, and A. Foffani, *Ric. Sci.*, **37,** 1115 (1967).
155. A. Leifer, D. Bonis, M. Collins, P. Dougherty, A. Fusco, M. Koral, and J. E. LuValle, *Spectrochim. Acta*, **20,** 909 (1964).
156. J. Nys, *Ind. Chim. Belge* (*Suppl., 1959*), **2,** 505 (1959).
157. S. Daehne, J. Ranft, and H. Paul, *Tetrahedron Lett.*, 3355 (1964).
158. Y. S. Moshkovskii, N. S. Spasokukotskii, M. V. Deichmeister, and L. D. Zhilina, *Zh. Obshch. Khim.*, **35,** 528 (1965).
159. M. Deichmeister, A. Z. Pinkhasova, and N. S. Spasokukotskii, *Zh. Prikl. Spektrosk.*, **14,** 932 (1971).
160. H. Tsukahara, *Nippon Shashin Gakkaishi*, **30,** 215 (1967); *Chem. Abstr.*, **69,** 72813q (1968).
161. F. Zuccarello, A. Foffani, and S. Fasone, *Ric. Sci.*, **39,** 632 (1969); *Chem. Abstr.*, **73,** 19943p (1970).
162. S. Boyer, B. Malingrey, and M. C. Preteseille, *C. R. Acad. Sci., Ser. C*, **268,** 1629 (1969).
163. H. Jakobi, A. Novak, and H. Kuhn, *Z. Elektrochem.*, **66,** 863 (1962).
164. W. E. Gray, W. R. Brewer, and G. R. Bird, *Photogr. Sci. Eng.*, **14,** 316 (1970).
165. S. E. Sheppard, R. H. Lambert, and R. D. Walker, *J. Phys. Chem.*, **50,** 210 (1946).
166. P. S. Rao and E. Hayon, *J. Am. Chem. Soc.*, **96,** 1287 (1974).
167. P. S. Rao and E. Hayon, *J. Am. Chem. Soc.*, **96,** 1295 (1974).
168. A. Stanienda, *Z. Phys. Chem.* (*Frankfurt am Main*), **32,** 238 (1962).
169. M. Tamura and H. Hada, "Scientific Photography" (*Proc. Int. Coll. Liège, 1959*), H. Sauvenier, Ed., Pergamon Press, London, 1962, p. 579.
170. H. Hada and M. Tamura, *Bull. Soc. Sci. Photogr. Japan*, **7,** 1 (1957).
171. T. Tani and S. Kikuchi, *Photogr. Sci. Eng.*, **11,** 129 (1967).
172. T. Tani, K. Honda, and S. Kikuchi, *J. Electrochem. Soc. Japan*, **37,** 17 (1969).
173. T. Tani, S. Kikuchi, and K. Honda, *Photogr. Sci. Eng.*, **12,** 80 (1968).

174. T. Tani, *J. Electrochem. Soc.*, **120,** 254 (1973).
175. Y. Hishiki and T. Ueda, *Nippon Shashin Gakkaishi*, **36,** 20 (1973); *Chem. Abstr.*, **79,** 99164x (1973).
176. O. Guertler, B. Konieczny, J. Von Grossmann, and G. Bach, *J. Prakt. Chem.*, **315,** 323 (1973).
177. O. Guertler, D. Fuertig, and J. Von Grossmann, *J. Signalaufzeichnungsmaterialien*, **2,** 33 (1974).
178. S. Daehne and O. Guertler, *J. Prakt. Chem.*, **315,** 786 (1973).
179. O. Guertler and S. Daehne, *Z. Phys. Chem. (Leipzig)*, **255,** 501 (1974).
180. S. Umano, *Nippon Kagaku Zasshi*, **77,** 796 (1956); *Chem. Abstr.*, **51,** 7916e (1957).
181. I. Nemcova and I. Nemec, *Chem. Zvesti*, **26,** 115 (1972); *Chem. Abstr.*, **77,** 69348b (1972).
182. W. Kemula and A. Axt, *Rocz. Chem.*, **43,** 199 (1969); *Chem. Abstr.*, **71,** 4495d (1969).
183. W. Kemula and A. Axt-Zak, *Rocz. Chem.*, **43,** 403 (1969); *Chem. Abstr.*, **71,** 66664s (1969).
184. S. Huenig, D. Scheutzow, H. Schlaf, and H. Puetter, *Justus Liebigs Ann. Chem.*, 1436 (1974).
185. J. M. Simson (Eastman Kodak Co., Rochester, N.Y.) and P. Zuman (Clarkson College, Pottsdam, N.Y.), private communication.
186. F. Doerr and G. Hohlneicher, *Tetrahedron Lett.*, 1297 (1964).
187. A. Stanienda, *Naturwissenschaften*, **47,** 353 (1960).
188. A. Stanienda, *Z. Wiss. Photogr., Photophys. Photochem.*, **59,** 76 (1966).
189. R. F. Large, "Photographic Sensitivity" (Cambridge, 1972), R. J. Cox, Ed., Academic Press, New York, 1973, p. 241.
190. P. Beretta and A. Jaboli, *Photogr. Sci. Eng.*, **18,** 197 (1974).
191. A. Stanienda, *Z. Naturforsch., Teil B*, **23,** 1285 (1968).
192. L. Gouveneur, G. Leroy, and I. Zador, *Electrochim. Acta*, **19,** 215 (1974).
193. J. M. Nys and W. Van den Heuvel, *Photogr. Korresp.*, **102,** 37 (1966).
194. F. I. Vilesov, *Dokl. Akad. Nauk SSSR*, **132,** 632 (1960).
195. R. C. Nelson, *J. Phys. Chem.*, **71,** 2517 (1967).
196. R. C. Nelson, *J. Opt. Soc. Am.*, **55,** 897 (1965).
197. R. C. Nelson, *J. Mol. Spectrosc.*, **23,** 213 (1967).
198. N. L. Petruzzella, R. G. Selsby, and R. C. Nelson, *J. Opt. Soc. Am.*, **59,** 112 (1969).
199. R. G. Selsby and R. C. Nelson, *J. Mol. Spectrosc.*, **33,** 1 (1970).
200. R. C. Nelson and R. G. Selsby, *Photogr. Sci. Eng.*, **14,** 342 (1970).
201. S. S. Choi and R. C. Nelson, *Photogr. Sci. Eng.*, **16,** 341 (1972).
202. R. C. Nelson, "Photographic Sensitivity" (Cambridge, 1972), R. J. Cox, Ed., Academic Press, New York, 1973, p. 177.
203. R. C. Nelson, *J. Photogr. Sci.*, **20,** 211 (1972).
204. P. Yianoulis and R. C. Nelson, *Photogr. Sci. Eng.*, **18,** 94 (1974).
205. J. W. Trusty and R. C. Nelson, *Photogr. Sci. Eng.*, **16,** 421 (1972).
206. J. W. Trusty, J. M. Ferrier, G. B. Freeman, and R. C. Nelson, *Photogr. Sci. Eng.*, **17,** 510 (1973).
207. G. Scheibe, *Chimia*, **15,** 10 (1961).
208. E. Daltrozzo, G. Hohlneicher, and G. Scheibe, *Ber. Bunsenges. Phys. Chem.*, **69,** 190 (1965).
209. G. Schiebe and E. Daltrozzo, "Advances in Heterocyclic Chemistry," Vol. 7, A. R. Katrizky and A. J. Boulton, Eds., Academic Press, New York, 1966, p. 153.
210. A. H. Herz, *Photogr. Sci. Eng.*, **18,** 207 (1974).

211. E. B. Lifshits, N. S. Spasokukotskii, L. M. Yagupol'skii, E. S. Kozlova, D. Y. Naroditskaya, and I. I. Levkoev, *Zh. Obshch. Khim.*, **38,** 2025 (1968).
212. B. S. Portnaya, I. I. Levkoev, and N. S. Spasokukotskii, *Dokl. Akad. Nauk SSSR*, **75,** 231 (1950).
213. E. B. Lifshits, I. I. Levkoev, L. M. Yagupol'skii, and N. S. Barvyn, *Zh. Nauchn. Prikl. Fotogr. Kinematogr.*, **11,** 175 (1966).
214. U. Mazzucato and P. Bassignana, *Ric. Sci. Parte 2: Sez. A*, **7,** 215 (1964); *Chem. Abstr.*, **63,** 4143d (1965).
215. A. I. Kiprianov and Y. L. Slominskii, *Zh. Org. Khim.*, **1,** 1321 (1965).
216. H. C. A. Van Beek, Ch. F. Hendriks, G. J. Van der Net, and L. Schaper, *Rec. Trav. Chim.*, **94,** 31 (1975).
217. G. Di Modica and E. Barni, *Gazz. Chim. Ital.*, **91,** 187 (1961).
218. P. J. Dynes, G. S. Chapman, E. Kebede, and F. W. Schneider, *J. Am. Chem. Soc.*, **94,** 6356 (1972).
219. G. Buttgereit and G. Scheibe, *Ber. Bunsenges. Phys. Chem.*, **69,** 301 (1965).
220. F. G. Webster, U. S. Patent 3,545,976 (1970).
221. E. J. Van Lare, *Kirk-Othmer Encycl. Chem. Technol.*, 2nd ed. (England), **6,** 605 (1965).
222. A. Van Dormael, *Chim. Ind.* (*Paris*), **90,** 619 (1963).
223. G. Bach, *Wiss. Z. Tech. Univ. Dresden*, **20,** 725 (1971).
224. J. Nys, *Chimia Suppl.*, 115 (1968).
225. H. Zwicky, *Chimia*, **19,** 416 (1965).
226. E. J. Poppe, *Z. Wiss. Photogr. Photophys. Photochem.*, **63,** 149 (1969).
227. L. G. S. Brooker and G. H. Keyes, *J. Franklin Inst.*, **219,** 255 (1935).
228. H. Nikolajewski, S. Daehne, B. Hirsch, and E. A. Janer, *Chimia*, **20,** 176 (1966).
229. J. Metzger, H. Larivé, R. Dennilauler, R. Baralle, and C. Gaurat, *Bull. Soc. Chim. Fr.*, 3156 (1969).
230. J. Metzger, H. Larivé, R. Dennilauler, R. Baralle, and C. Gaurat, *Bull. Soc. Chim. Fr.*, 1275 (1969).
231. J. Metzger, H. Larivé, R. Dennilauler, R. Baralle, and C. Gaurat, *Bull. Soc. Chim. Fr.*, 1266 (1969).
232. J. Metzger, H. Larivé, R. Dennilauler, R. Baralle, and C. Gaurat, *Bull. Soc. Chim. Fr.*, 57 (1967).
233. J. Metzger, H. Larivé, R. Dennilauler, R. Baralle, and C. Gaurat, *Bull. Soc. Chim. Fr.*, 40 (1967).
234. J. Metzger, H. Larivé, R. Dennilauler, R. Baralle, and C. Gaurat, *Bull. Soc. Chim. Fr.*, 30 (1967).
235. S. Petersen, C. Hackmann, and L. Juehling, Belgian Patent 650,455 (1964).
236. B. D. Tilak, *Ind. Chim. Belge*, **32,** 50 (1967).
237. A. I. Kiprianov and I. K. Ushenko, *Zh. Obshch. Khim.*, **17,** 1538 (1947).
238. H. Siegrist, *Chimia*, **20,** 295 (1966).
239. W. Mills and R. Raper, *J. Chem. Soc.*, 2460 (1925).
240. L. G. S. Brooker, *J. Am. Chem. Soc.*, **75,** 4335 (1953).
241. O. Mumm, H. Henz, and J. Diedrichsen, *Ber.*, **72,** 2107 (1939).
242. A. I. Kiprianov and F. S. Babichev, *Zh. Obshch. Khim.*, **20,** 145 (1950).
243. H. Larivé and R. Dennilauler, *Chimia*, **15,** 115 (1961).
244. R. Dennilauler and H. Larivé, French Patent 1,492,297 (1967).
245. J. R. Owen, German Patent 2,002,413 (1970).
246. J. R. Owen, German Patent 2,002,412 (1970).
247. H. Larivé and R. Baralle, French Patent 2,174,417 (1973).

248. H. Larivé and R. Baralle, French Patent 2,174,418 (1973).
249. H. R. Mueller and M. Seefelder, *Justus Liebigs Ann. Chem.*, **728,** 88 (1969).
250. H. Baumann, G. Hansen, H. R. Mueller, and M. Seefelder, *Justus Liebigs Ann. Chem.*, **717,** 124 (1968).
251. J. Nys and J. Libeer, *Sci. Inds. Photogr.*, **28,** 433 (1957).
252. A. Fabrycy and A. Kosmider, *Rocz. Chem.*, **48,** 603, 1069 (1974).
253. L. Oliveros and H. Wahl, *Bull. Soc. Chim. Fr.*, 3204 (1971).
254. J. Bourson, *Bull. Soc. Chim. Fr.*, 152, 3541 (1971).
255. J. R. Owen, *Tetrahedron Lett.*, 2709 (1969).
256. J. R. Owen, *Eastman Org. Chem. Bull.* (Eastman Kodak Co., Rochester, N.Y., **43**(3), 3 (1971).
257. I. N. Chernyuk, A. M. Ivanov, and V. E. Pridan, *Zh. Obshch. Khim.*, **44,** 1094 (1974).
258. P. Mastagli, H. Larivé, and P. Etevenon, *C.R. Acad. Sci.*, **252,** 3782 (1961).
259. D. N. Kramer, L. P. Bisauta, R. Bato, and B. L. Murr, Jr., *J. Org. Chem.*, **39,** 3132 (1974).
259a. T. J. Novak, D. N. Kramer, H. Klapper, L. W. Daasch, and B. L. Murr, Jr., *J. Org. Chem.*, **41,** 870 (1976).
260. S. R. H. Edge, *J. Chem. Soc.*, **123,** 2330 (1923).
261. C. D. Wilson, U.S. Patent 2,465,412 (1949).
262. C. D. Wilson, U.S. Patent 2,425,774 (1947).
263. F. L. White and G. H. Keyes, U.S. Patent 2,166,736 (1939).
264. A. I. Kiprianov and I. L. Mushkalo, *Zh. Org. Khim.*, **1,** 744 (1965).
265. A. I. Kiprianov, T. M. Verbovskaya, and I. L. Mushkalo, *Zh. Org. Khim.*, **3,** 2036 (1967).
266. A. I. Kiprianov, I. L. Mushkalo, and F. A. Mikhailenko, *Zh. Org. Khim.*, **3,** 2041 (1967).
267. S. G. Fridman and A. I. Kiprianov, *Zh. Org. Khim.*, **8,** 1289 (1972).
268. A. I. Kiprianov, S. G. Fridman, and F. A. Mikhailenko, *Zh. Org. Khim.*, **9,** 1253 (1973).
269. P. Meheux, P. Lochon, and J. Neel, *C.R. Acad. Sci., Ser. C*, **275,** 749 (1972).
270. A. M. Osman and S. A. Mohamed, *U.A.R.J. Chem.*, **14,** 475 (1971); *Chem. Abstr.*, **80,** 16426f (1974).
271. M. Y. Kornilov and E. M. Ruban, *Zh. Org. Khim.*, **9,** 2188 (1973).
272. A. I. Kiprianov, S. G. Fridman, G. G. Dyadyusha, and L. I. Kotova, *Zh. Org. Khim.*, **11,** 891 (1975) and references therein.
273. I. L. Mushkalo, E. D. Sych, and O. V. Moreiko, *Ukr. Khim. Zh.* (*Russ. Ed.*), **39,** 913 (1973); *Chem. Abstr.*, **80,** 16428h (1974).
274. F. A. Mikhailenko, A. N. Borguslavskaya, and V. V. Geiko, *Ukr. Khim. Zh.* (*Russ. Ed.*), **40,** 977 (1974); *Chem. Abstr.*, **82,** 45006r (1975).
275. A. I. Kiprianov and G. A. Lezenko, *Zh. Org. Khim.*, **8,** 1700 (1972).
276. A. I. Kiprianov and G. A. Lezenko, *Zh. Org. Khim.*, **8,** 1712 (1972).
277. A. I. Kiprianov and G. A. Lezenko, *Zh. Org. Khim.*, **9,** 2587 (1973).
278. A. I. Kiprianov and V. Y. Buryak, *Zh. Org. Khim.*, **8,** 1707 (1972).
279. A. I. Kiprianov and V. Y. Buryak, *Zh. Org. Khim.*, **9,** 1257 (1973).
280. A. I. Kiprianov and T. K. Nikolaenko, *Zh. Org. Khim.*, **8,** 1692 (1972).
281. A. I. Kiprianov and T. K. Nikolaenko, *Ukr. Khim. Zh.* (*Russ. Ed.*), **40,** 249 (1974); *Chem. Abstr.*, **81,** 93012r (1974).
282. A. I. Kiprianov, V. Y. Buryak, and P. V. Tarasenko, *Ukr. Khim. Zh.* (*Russ. Ed.*), **39,** 793 (1973); *Chem. Abstr.*, **80,** 49228w (1974).

283. H. Junek and W. Remp, *Monatsh. Chem.*, **104,** 433 (1973).
284. F. A. Mikhailenko and L. I. Schevchuk, *Ukr. Khim. Zh.* (*Russ. Ed.*), **40,** 885 (1974); *Chem. Abstr.*, **82,** 45002m (1975).
285. A. Fumia and L. G. S. Brooker, U.S. Patent 3,786,046 (1974).
286. K. Hafner, *Chimia,* **27,** 640 (1973).
287. A. G. Cook, "Enamines: Synthesis, Structure, and Reactivity," M. Dekker, New York, 1969.
288. L. G. S. Brooker and D. W. Heseltine, U.S. Patent 2,748,114 (1956).
289. A. C. Craig and L. G. S. Brooker, U.S. Patent 3,148,065 (1964).
290. J. C. Martin, U.S. Patent 3,347,868 (1967).
291. G. Pagani, *J. Chem. Soc., Perkin Trans. 2,* 1184 (1973).
292. G. Oliver, French Patent 2,022,536 (1970).
293. G. Oliver, U.S. Patent 3,723,154 (1973).
294. G. Oliver, British Patent 1,128,113 (1968).
295. P. W. Jenkins, French Patent 2,099,780 (1972).
296. D. W. Heseltine and L. G. S. Brooker, U.S. Patent 3,140,951 (1964).
297. J. D. Mee, *Res. Disclosures,* **125,** 32 (1974).
298. G. Maerkl and F. Lieb, *Tetrahedron Lett.*, 3489 (1967).
299. H. Depoorter, M. J. Libeer, and G. Van Mierlo, *Bull. Soc. Chim. Belg.*, **77,** 521 (1968).
300. L. G. S. Brooker and L. A. Smith, *J. Am. Chem. Soc.*, **59,** 67 (1937).
301. G. E. Ficken and J. D. Kendall, *Chimia,* **15,** 110 (1961).
302. E. B. Knott, *J. Chem. Soc., Suppl.*, 6204 (1964).
303. F. M. Hamer, ref. 10, Chapter XV.
304. F. M. Hamer, ref. 10, Chapter II.
305. W. S. Gaugh, D. M. Sturmer, and J. P. Freeman (Eastman Kodak Co., Rochester, N.Y.), private communication.
306. E. Foresti Serantoni, R. Mongiorgi, M. Colonna, L. Greci, L. Marchetti, and L. Andreetti, *Gazz. Chim. Ital.*, **104,** 1217 (1974).
307. S. Huenig and E. Woelff, *Chimia Suppl.*, 33 (1968).
308. J. K. Elwood, *J. Org. Chem.*, **38,** 2425 (1973).
309. H. W. Post, "Chemistry of Aliphatic Orthoesters" (A.C.S. Monograph Series No. 92), Reinhold, New York, 1943.
310. V. V. Mezheritskii, E. P. Olekhnovich, and G. N. Dorofeenko, *Usp. Khim.*, **42,** 896 (1973).
311. L. G. S. Brooker and F. L. White, *J. Am. Chem. Soc.*, **57,** 2480 (1935).
312. L. G. S. Brooker and F. L. White, *J. Am. Chem. Soc.*, **57,** 547 (1935).
313. Ciba-Ltd., British Patent 954,240 (1964).
314. G. Bach and W. Felgner, East German Patent 87,109 (1972); *Chem. Abstr.*, **78,** 73668u (1973).
315. J. Liebscher and H. Hartmann, *Z. Chem.*, **14,** 358 (1974).
316. H. Nikolajewski and S. Daehne, *Angew. Chem. Int. Ed. Engl.*, **5,** 1044 (1966).
317. M. Przbylska and J. Swiderski, *Rocz. Chem.*, **45,** 781 (1971); *Chem. Abstr.*, **75,** 152956c (1971).
318. M. Dabrowska, J. Oszczapowciz, and J. Swiderski, *Rocz. Chem.*, **36,** 475 (1962); *Chem. Abstr.*, **58,** 8062e (1963).
319. S. Creyf and L. Roosens, *Ind. Chim. Belge,* **23,** 853, 967 (1958).
320. A. Kreutzberger, *Arch. Pharm.* (*Weinheim, Ger.*), **299,** 984 (1966).
321. A. Kreutzberger, *Arch. Pharm.* (*Weinheim, Ger.*), **299,** 897 (1966).
322. G. E. Ficken and J. D. Kendall, *J. Chem. Soc.*, 3988 (1959).

323. C. Reichardt and K. Schagerer, *Angew. Chem.*, **85,** 346 (1973).
324. C. Reichardt and P. Miederer, *Justus Liebigs Ann. Chem.*, 750 (1973).
325. E. Shchelkina and A. V. Kazymov, *Zh. Org. Khim.*, **8,** 635 (1972).
326. C. Reichardt and M. Mormann, *Chem. Ber.*, **105,** 1815 (1972).
327. A. F. Vompe, N. V. Monich, and N. S. Spasokukotskii, *Zh. Org. Khim.*, **10,** 1980 (1974).
328. J. R. Owen, U.S. Patent Office Defensive Public., 890,002 (1971).
329. J. Ciernik, *Collect. Czech. Chem. Commun.*, **37,** 2273 (1972).
330. J. Ciernik and V. Hruskova, *Collect. Czech. Chem. Commun.*, **37,** 2771 (1972).
331. J. Ciernik and V. Hruskova, *Collect. Czech. Chem. Commun.*, **37,** 3800 (1972).
332. J. Ciernik and V. Vystavel, Czech. Patent 149,771–2 (1973); *Chem. Abstr.*, **80,** 28493p–94q (1974).
333. J. Ciernik and V. Hruskova, Czech. Patents 150,788–90 and 150,793–96 (1973); *Chem. Abstr.*, **81,** 65256g–60b, 93098y, 137609x (1974).
334. H. Bredereck, G. Simchen, and P. Speh, *Justus Liebigs Ann. Chem.*, **737,** 46 (1970).
335. G. A. Rillaers and H. Depoorter, German Patent 1,930,224 (1970).
336. A. Treibs and L. Shulze, *Justus Liebigs Ann. Chem.*, 201 (1973).
337. G. A. Rillaers and H. Depoorter, U.S. Patent 3,615,417 (1971).
338. H. E. Sprenger and W. Ziegenbein, *Angew. Chem. Int. Ed. Engl.*, **7,** 530 (1968).
339. G. Maahs and P. Hegenberg, *Angew. Chem. Int. Ed. Engl.*, **5,** 888 (1966).
340. H. Kampfer, Belgian Patent 709,463 (1968).
341. A. Treibs and K. Jacob, *Justus Liebigs Ann. Chem.*, **699,** 153 (1966).
342. A. I. Kiprianov and G. G. Dyadyusha, *Zh. Obshch. Khim.*, **30,** 3647 (1960).
343. A. I. Kiprianov and G. G. Dyadyusha, *Zh. Obshch. Khim.*, **30,** 3654 (1960).
344. P. D. Collet and M. A. Compere, French Patent 1,386,399 (1965).
345. E. R. H. Jones and K. J. Reed, British Patent 616,223 (1946).
346. J. D. Mee, *J. Am. Chem. Soc.*, **96,** 4712 (1974).
347. J. D. Mee (Eastman Kodak Co., Rochester, N.Y.), submitted for publication.
347a. J. Bourson, *Bull. Soc. Chim. Fr.*, 644 (1975).
348. E. Klingsberg, U.S. Patent 3,155,682 (1964).
349. H. Yamaguchi, *Chem. Pharm. Bull.* (*Tokyo*), **16,** 1451 (1968); *Chem. Abstr.*, **70,** 38878a (1969).
350. J. Liebscher and H. Hartmann, *Z. Chem.*, **14,** 189 (1974).
351. H. Hartmann, H. Schaefer, and K. Gewald, *J. Prakt. Chem.*, **315,** 497 (1973).
352. H. Hartmann and F. Mohn, *J. Prakt. Chem.*, **314,** 419 (1972).
353. J. Goetze, *Ber. Bunsenges. Phys. Chem.*, **71,** 2289 (1938).
354. D. D. Chapman, J. K. Elwood, D. W. Heseltine, H. M. Hess, and D. W. Kurtz, *J. Chem. Soc., Chem. Commun.*, 648 (1974).
355. L. L. Lincoln (Eastman Kodak Co., Rochester, N.Y.), private communication.
356. R. A. Jeffreys, *J. Chem. Soc.*, 5824 (1963).
357. H. Harnisch, *Justus Liebigs Ann. Chem.*, **751,** 155 (1971).
358. D. Schelz and M. Priester, *Helv. Chim. Acta*, **58,** 2529, 2536 (1975); **59,** 688, 692 (1976).
359. G. F. Duffin, "Advances in Heterocyclic Chemistry," Vol. 3, A. R. Katrizky, Ed., Academic Press, New York, 1964, p. 1.
360. R. G. Shepherd, *J. Chem. Soc.*, 4410 (1964).
361. R. E. Lyle and J. L. La Mattina, *J. Org. Chem.*, **40,** 438 (1975).
362. G. Rillaers and H. Depoorter, *Ind. Chim. Belge*, **32,** 61 (1967).
363. G. Di Modica and E. Barni, *Ann. Chim.* (*Rome*), **54,** 530 (1964); *Chem. Abstr.*, **61,** 16195g (1964).

364. G. Di Modica and E. Barni, *Gazz. Chim. Ital.*, **93,** 679 (1963).
365. T. N. Pliev, R. M. Bystrova, and Y. M. Yutilov, *Khim. Geterotsikl. Soedin.*, 1686 (1973).
366. S. F. Mason, "The Chemistry of Synthetic Dyes," Vol. 3, K. Venkataraman, Ed., Academic Press, New York, 1970, p. 169.
367. "The Chemistry of Synthetic Dyes, Vol. 1, K. Venkataraman, Ed., Academic Press, New York, 1952, p. 323.
368. S. F. Mason, *J. Soc. Dyers Colour.*, **84,** 604 (1968).
369. J. R. Platt, *J. Opt. Soc. Am.*, **43,** 252 (1953).
370. H. Labhart and W. Heinzelmann, *Org. Mol. Photophys.*, Vol. 1, J. B. Birks, Ed., Wiley, London, 1973, p. 297.
371. H. Suzuki, "Electronic Absorption Spectra and Geometry of Organic Molecules," Academic Press, New York, 1967, p. 364.
372. L. Salem, "Molecular Orbital Theory of Conjugated Systems," W. A. Benjamin, New York, 1966, pp. 367, 380.
373. R. Zahradnik and P. Carsky, "Progress in Physical Organic Chemistry," Vol. 10, A. Streitwieser and R. W. Taft, Eds., Interscience, New York, 1973, p. 327.
374. C. A. Parker, "Photoluminescence in Solutions," Elsevier, New York, 1968.
375. W. West, "Scientific Photography" (*Proc. Int. Coll. Liege, 1959*), H. Sauvenier, Ed., Pergamon Press, London, 1962, p. 557.
376. B. B. Snavely, *Org. Mol. Photophys.*, Vol. 1, J. B. Birks, Ed., Wiley, London, 1973, p. 239.
377. F. P. Schaefer, "Dye Lasers" (Topics in Applied Physics, Vol. 1), Springer-Verlag, New York, 1973; *Top. Curr. Chem.*, **61,** 1 (1976).
378. D. Leupold, R. Koenig, and S. Daehne, *Z. Chem.*, **10,** 409 (1970).
379. H. Kuhn and D. Moebius, *Angew. Chem. Int. Ed. Engl.*, **10,** 620 (1971).
380. K. H. Drexhage, M. M. Zwick, and H. Kuhn, *Ber. Bunsenges. Phys. Chem.*, **67,** 62 (1963).
381. P. Barth, K. H. Beck, K. H. Drexhage, H. Kuhn, D. Moebius, D. Molzahn, K. Roellig, F. P. Schaefer, W. Sperling, and M. M. Zwick, *Opt. Anregung Org. Syst.* (*Int. Farbensymp. Elmau*, 1964), W. Jung, Ed., Verlag Chemie, Weinheim, 1966, p. 639.
382. K. W. Hausser, R. Kuhn, and G. Seitz, *Z. Phys. Chem.* (*Leipzig*) *B*, **29,** 391 (1935).
383. G. N. Lewis and M. Calvin, *Chem. Rev.*, **25,** 273 (1939).
384. S. Daehne, *Z. Chem.*, **5,** 441 (1965).
385. S. Daehne and R. Radeglia, *Tetrahedron*, **27,** 3673 (1971).
386. A. Gasco, E. Barni, and G. Di Modica, *Tetrahedron Lett.*, 5131 (1968).
387. N. Mishra, L. N. Patnaik, and M. K. Rout, *J. Indian Chem. Soc., Sect. A*, **14,** 56 (1976).
388. H. Kuhn, *J. Chem. Phys.*, **17,** 1198 (1949).
389. H. Kuhn, *Helv. Chim. Acta*, **32,** 2247 (1949).
390. H. Kuhn, *Chimia*, **9,** 237 (1955).
391. H. Kuhn and W. Huber, *Helv. Chim. Acta*, **42,** 363 (1959).
392. H. Kuhn, W. Huber, G. Handschig, H. Martin, F. Schaefer, and F. Baer, *J. Chem. Phys.*, **32,** 467 (1960).
393. F. Baer, W. Huber, G. Handschig, H. Martin, and H. Kuhn, *J. Chem. Phys.*, **32,** 470 (1960).
394. H. D. Foersterling, W. Huber, H. Kuhn, H. H. Martin, A. Schweig, F. F. Seelig, and W. Stratmann, *Opt. Anregung Org. Syst.* (*Int. Farbensymp. Elmau, 1964*), W. Jung, Ed., Verlag Chemie, Weinheim, 1966, p. 55.
395. G. Araki and S. Fujinaga, *Progr. Theor. Phys.*, **15,** 307 (1956).

396. Y. Mizuno and T. Izuyama, *Progr. Theor. Phys.*, **21,** 593 (1959).
397. S. Olszewski, *J. Chem. Phys.*, **26,** 1205 (1957).
398. S. Olszewski, *Acta Phys. Pol.*, **16,** 211 (1957).
399. S. Olszewski, *Acta Phys. Pol.*, **18,** 107 (1959).
400. S. Olszewski, *Acta Phys. Pol.*, **18,** 121 (1959).
401. R. A. Berg, G. Moyano, and R. A. Pierce, *J. Phys. Chem.*, **71,** 3352 (1967).
402. D. Leupold, *Z. Phys. Chem.* (*Leipzig*), **223,** 405 (1963).
403. D. Leupold, and S. Daehne, *Theor. Chim. Acta*, **3,** 1 (1965).
404. D. Leupold, *Theor. Chim. Acta*, **9,** 336 (1968).
405. T. Tani, *Photogr. Sci. Eng.*, **16,** 258 (1972).
406. T. Tani, *J. Photogr. Sci.*, **19,** 161 (1971).
406a. J. Fabian, H. Hartmann, and N. Tyutyulkov, *J. Signalaufzeichnungsmaterialien*, **4,** 101 (1976).
407. T. Tani, *Bull. Soc. Photogr. Sci. Technol. Japan*, **21,** 15 (1971).
408. D. M. Sturmer and W. S. Gaugh, *Photogr. Sci. Eng.*, **17,** 146 (1973).
409. J. K. Elwood, *J. Org. Chem.*, **38,** 2430 (1973).
410. W. F. Smith, *Tetrahedron*, **20,** 671 (1964).
411. D. M. Sturmer, unpublished calculations using the methods of References 408 or 412.
412. H. A. Hammond, *Theor. Chim. Acta*, **18,** 239 (1970).
413. G. Leroy and J. Nys, *Bull. Soc. Chim. Belg.*, **73,** 673 (1964).
414. J. C. Rayez and O. Chalvet, *J. Chim. Phys. Phys.-Chim. Biol.*, **69,** 1537 (1972).
415. F. Doerr, "Dye Sensitization" (Bressanone, 1967), G. Semerano, U. Mazzucato, W. F. Berg, and H. Meier, Eds., Focal Press, New York, 1970, p. 265.
416. R. Zahradnik, "Advances in Heterocyclic Chemistry," Vol. 5, A. R. Katrizky and A. J. Boulton, Eds., Academic Press, New York, 1965, p. 1.
417. C. Parkanyi, "Mechanisms of Reactions of Sulfur Compounds," Vol. 4, Intra-Science Research Foundation, Santa Monica, Calif, 1970, p. 69.
418. F. Dietz and C. Glier, *J. Signalaufzeichnungsmaterialen*, **2,** 123 (1974).
419. M. Klessinger, *Theor. Chim. Acta*, **5,** 252 (1966).
420. M. Klessinger, *Tetrahedron*, **22,** 3355 (1966).
421. S. Daehne, S. Kulpe, K. D. Nolte, and R. Radeglia, *Photogr. Sci. Eng.*, **18,** 410 (1974).
422. H. J. Hofmann and M. Scholz, *J. Prakt. Chem.*, **313,** 349 (1971).
423. H. J. Hofmann, M. Scholz, and C. Weiss, *J. Prakt. Chem.*, **313,** 339 (1971).
424. J. Fabian and H. Hartmann, *Z. Chem.*, **12,** 349 (1972).
425. J. Fabian and H. Hartmann, *Tetrahedron*, **29,** 2597 (1973).
426. J. Fabian, H. Hartmann, and K. Fabian, *Tetrahedron*, **29,** 2609 (1973).
427. H. G. Benson and J. N. Murrell, *J. Chem. Soc., Faraday Trans.* **2,** 129 (1972).
428. G. Bethier and G. Del Re, *J. Chem. Soc.*, 3109 (1965).
429. Y. Ferre, E. J. Vincent, H. Larivé, and J. Metzger, *Bull. Soc. Chim. Fr.*, 3862 (1972).
430. Y. Ferre, E. J. Vincent, H. Larivé, and J. Metzger, *J. Chim. Phys. Phys.-Chim. Biol.*, **71,** 329 (1974).
431. Y. Ferre, H. Larivé, and E. J. Vincent, *Photogr. Sci. Eng.*, **18,** 457 (1974).
432. Y. Maroni-Barnaud, H. Wahl, and P. Maroni, *C. R. Acad. Sci.*, **251,** 1787 (1960).
433. F. Dietz and H. J. Koehler, *J. Prakt. Chem.*, **313,** 1101 (1971).
434. F. Dietz, *Tetrahedron*, **28,** 1403 (1972).
435. C. Honda and H. Hada, *Photogr. Sci. Eng.*, **20,** 15 (1976).
436. F. Dietz and C. Glier, *Z. Chem.*, **12,** 229 (1972).
437. F. Dietz and K. J. Paessler, *J. Signalaufzeichnungsmaterialien*, **1,** 57 (1973).
438. F. Dietz and C. Glier, *J. Signalaufzeichnungsmaterialien*, **1,** 221 (1973).

439. H. Weinstein, B. Apfelderfer, and R. A. Berg, *Photochem. Photobiol.*, **18,** 175 (1973).
440. G. S. Levinson, W. T. Simpson, and W. Curtis, *J. Am. Chem. Soc.*, **79,** 4314 (1957).
441. G. G. Dyadyusha, T. M. Verbovskaya, and A. I. Kiprianov, *Ukr. Khim. Zh.*, **32,** 357 (1966).
442. I. Leubner, J. Dehler, and G. Scheibe, *Ber. Bunsenges. Phys. Chem.*, **72,** 1133 (1968).
443. G. Scheibe, J. Kern, and F. Doerr, *Z. Elektrochem.*, **63,** 117 (1959).
444. R. Eckert and H. Kuhn, *Z. Elektrochem.*, **64,** 356 (1960).
445. H. Jakobi and H. Kuhn, *Z. Elektrochem.*, **66,** 46 (1962).
446. K. S. Lyalikov and I. V. Semenchenko, *Zh. Nauchn. Prikl. Fotogr. Kinematogr.*, **5,** 161 (1960).
447. Y. Tanizaki, *Bull. Chem. Soc. Japan*, **33,** 979 (1960).
448. F. Doerr, J. Kern, J. Thies, and V. Zanker, Z. *Naturforsch., Teil A*, **17,** 93 (1962).
449. Y. Tanizaki, *Int. Symp. Mol. Struct. Spectrosc.* (*Tokyo*), 1962, p. B222; *Chem. Abstr.*, **61,** 5082c (1964).
450. F. Doerr and M. Held, *Angew. Chem.*, **72,** 287 (1960).
451. F. Doerr, *Angew. Chem. Int. Ed. Engl.*, **5,** 478 (1966).
452. J. J. Dekkers, G. P. Hoornweg, C. MacLean, and N. H. Velthorst, *Chem. Phys. Lett.*, **19,** 517 (1973).
453. P. P. Feofilov, "The Physical Basis for Polarized Emission," Consultants Bureau, New York, 1961.
454. J. Pouradier, *J. Chim. Phys. Phys.-Chim. Biol.*, **61,** 1107 (1964).
455. G. Scheibe, E. Daltrozzo, O. Woerz, and J. Heiss, *Z. Phys. Chem., N. F.*, **64,** 97 (1969).
456. G. Scheibe, *Opt. Anregung Org. Syst.* (*Int. Farbensymp. Elmau*, 1964), U. Jung, Ed., Verlag Chemie, Weinheim, 1966, p. 109.
457. W. West, S. P. Lovell, and W. Cooper, *Photogr. Sci. Eng.*, **14,** 52 (1970).
458. W. Cooper, S. P. Lovell, and W. West, *Photogr. Sci. Eng.*, **14,** 184 (1970).
459. G. L. Bottger, unpublished observations quoted in Ref. 457.
460. G. Scheibe, H. J. Friedrich, and G. Hohlneicher, *Angew. Chem.*, **73,** 383 (1961).
461. J. Rohleder and A. Olszowski, *Rocz. Chem.*, **34,** 1033 (1960).
462. J. N. Murrell, *J. Chem. Soc.*, 3779 (1956).
463. M. J. S. Dewar, "Steric Effects in Organic Systems" (*Proc. Symp. Hull, 1958*), G. W. Gray, Ed., Butterworths, London, 1958, p. 46.
464. N. I. Fisher and F. M. Hamer, *Proc. R. Soc. London, Ser. A*, **154,** 703 (1936).
465. W. Konig, *Z. Wiss. Photogr. Photophys. Photochem.*, **34,** 15 (1935).
466. L. J. E. Hofer, R. J. Grabenstetter, and E. O. Wiig, *J. Am. Chem. Soc.*, **72,** 203 (1950).
466a. D. M. Sturmer and W. S. Gaugh, *Photogr. Sci. Eng.*, **19,** 273 (1975).
467. G. E. Ficken and J. D. Kendall, *J. Chem. Soc.*, 3202 (1959).
468. U. Mazzucato, G. Favaro, and M. P. Mazzucato, *Ric. Sci. Parte 2: Sez. A*, **4,** 501 (1964).
469. N. J. L. Roth and A. C. Craig, *J. Phys. Chem.*, **78,** 1154 (1974).
470. R. Koenig, D. Leupold, B. Voigt, and S. Daehne, *J. Lumin.*, **9,** 113 (1974).
471. E. I. P. Walker, A. P. Marchetti, and C. D. Salzberg, *J. Chem. Phys.*, **63,** 2090 (1975).
472. I. A. Akimov, *Opt. Spektrosk.*, **32,** 317 (1972).
473. W. Cooper and N. B. Liebert, *Photogr. Sci. Eng.*, **16,** 25 (1972).
474. A. A. Muenter and W. Cooper, *Chem. Phys. Lett.*, **22,** 212 (1973).
475. W. Cooper, *Photogr. Sci. Eng.*, **17,** 3 (1973).
476. J. T. Knudtson and E. M. Eyring, *J. Phys. Chem.*, **78,** 2355 (1974).

477. J. Q. Umberger, *Photogr. Sci. Eng.*, **11,** 392 (1967).
478. H. R. Stadelmann, *J. Lumin.*, **3,** 143 (1970).
479. H. Rammensee and V. Zanker, *Z. Angew. Phys.*, **12,** 237 (1960).
480. A. V. Buettner, *J. Chem. Phys.*, **46,** 1398 (1967).
481. O. Inaker and H. Kuhn, *Chem. Phys. Lett.*, **28,** 15 (1974).
482. Y. I. Lifanov, V. A. Kuz'min, A. K. Chibisov, I. I. Levkoev, and A. V. Karyakin, *Zh. Prikl. Spektrosk.*, **20,** 221 (1974).
483. D. F. O'Brien, T. M. Kelly, and L. F. Costa, *Photogr. Sci. Eng.*, **18,** 76 (1974). The instrumentation was described by T. M. Kelly and D. F. O'Brien, *Photogr. Sci. Eng.*, **18,** 68 (1974).
484. D. N. Dempster, T. Morrow, R. Rankin, and G. F. Thompson, *J. Chem. Soc., Faraday Trans. 2*, **68,** 1479 (1972).
485. E. G. Arthurs, D. J. Bradley, and A. G. Roddie, *Chem. Phys. Lett.*, **22,** 230 (1973).
486. M. M. Malley and P. M. Rentzepis, *Chem. Phys. Lett.*, **3,** 534 (1969).
487. J. C. Mialocq, A. W. Boyd, J. Jaraudias, and J. Sutton, *Chem. Phys. Lett.*, **37,** 236 (1976).
488. K. R. Naqvi, D. K. Sharma, and G. J. Hoytink, *Chem. Phys. Lett.*, **22,** 5, 222 (1973).
489. K. R. Naqvi, D. K. Sharma, and G. J. Hoytink, *Chem. Phys. Lett.*, **22,** 226 (1973).
490. G. Mourou, B. Drouin, M. Bergeron, and M. M. Denariez-Roberge, *IEEE J. Quantum Electron.*, **9,** 745 (1973).
491. G. Mourou and M. M. Denariez-Roberge, *IEEE J. Quantum Electron.*, **9,** 787 (1973).
492. A. M. Bonch-Bruevich, T. Razumova, and G. M. Rubanova, *Opt. Spektrosk.*, **34,** 305 (1973).
493. D. W. Vahey and A. Yariv, *Phys. Rev. A*, **10,** 1578 (1974).
494. R. W. Chambers and D. R. Kearns, *Photochem. Photobiol.*, **10,** 215 (1969).
495. R. Nilsson, P. B. Merkel, and D. R. Kearns, *Photochem. Photobiol.*, **16,** 109 (1972).
496. O. Inacker and H. Kuhn, *Chem. Phys. Lett.*, **27,** 471 (1974).
497. L. G. Gross and N. N. Gabitova, *Zh. Nauchn. Prikl. Fotogr. Kinematogr.*, **14,** 385 (1969).
498. V. I. Permogorov, L. A. Serdyukova, and M. D. Frank-Kamenetskii, *Opt. Spektrosk.*, **22,** 979 (1967).
499. U. Mazzucato, G. Favaro, and F. Masetti, *Ric. Sci. Parte 2: Sez. A*, **8,** 1435 (1965).
500. A. P. Marchetti, J. F. Landry, and D. S. Tinti, *J. Chem. Phys.*, **61,** 1086 (1974).
501. R. A. Berg and A. Ron, *J. Chem. Phys.*, **59,** 3289 (1973).
502. W. Cooper and K. A. Rome, *J. Phys. Chem.*, **78,** 16 (1974).
503. W. C. Needler, R. L. Griffith, and W. West, *Nature*, **191,** 902 (1961).
504. H. Schmidt and R. Zellhofer, *Z. Phys. Chem.* (*Frankfurt am Main*), **91,** 204 (1974).
505. R. A. Pierce and R. A. Berg, *J. Chem. Phys.*, **56,** 5087 (1972).
506. Y. I. Lifanov and A. K. Chibisov, *Khim. Vys. Energ.*, **8,** 418 (1974).
507. J. Kern, F. Doerr, and G. Scheibe, *Z. Elektrochem.*, **66,** 462 (1962).
508. E. G. McRae and M. Kasha, *J. Chem. Phys.*, **28,** 721 (1958).
509. P. J. McCartin, *J. Chem. Phys.*, **48,** 2980 (1965).
510. F. Doerr, J. Kotschy, and H. Kausen, *Ber. Bunsenges. Phys. Chem.*, **69,** 11 (1965).
511. Y. I. Lifanov, V. A. Kuz'min, A. V. Karyakin, A. K. Chibisov, and I. I. Levkoev, *Izv. Akad. Nauk SSSR, Ser. Khim.*, 787 (1973)
511a. A. M. Vinogradov, V. A. Kuz'min, M A. Al'perovich, I. I. Levkoev, and I. I. Zil'berman, *Dokl. Akad. Nauk SSSR*, **227,** 642 (1976).
512. A. V. Buettner, *J. Phys. Chem.*, **68,** 3253 (1964).
513. R. A. Pierce and R. A. Berg, *J. Chem. Phys.*, **51,** 1267 (1969).

514. Y. I. Lifanov, V. A. Kuz'min, A. K. Chibisov, and I. I. Levkoev, *Khim. Vys. Energ.*, **7,** 321 (1973).
515. S. H. Ehrlich, *Photogr. Sci. Eng.*, **18,** 179 (1974); **20,** 5 (1976).
516. C. F. V. Jessup and F. W. Willets, *J. Photogr. Sci.*, **23,** 203, 206 (1975).
517. K. H. Drexhage and M. Fleck, *Ber. Bunsenges. Phys. Chem.*, **72,** 330 (1968).
518. A. K. Chibisov, V. A. Kuz'min, G. P. Roitman, I. I. Levkoev, and A. V. Karyakin, *Izv. Akad. Nauk SSSR, Ser. Fiz.*, **34,** 1288 (1970).
519. A. K. Chibisov, V. A. Kuz'min, and Y. I. Lifanov, *Opt. Spektrosk.*, **36,** 919 (1974).
520. M. L. Spaeth and D. P. Bortfeld, *Appl. Phys. Lett.*, **9,** 179 (1966).
521. M. I. Dzyubenko, I. G. Naumenko, I. N. Chernyuk, and G. T. Pilyugin, *Ukr. Fiz. Zh.* (*Russ. Ed.*), **14,** 735 (1969).
522. N. S. Kozlov, O. D. Zhikareva, A. N. Rubinov, V. A. Mostovnikov, and S. A. Batishchev, *Zh. Obshch. Khim.*, **46,** 395 (1976).
523. O. L. Lebedev, Y. M. Gryaznov, A. A. Chastov, and A. V. Kazymov, *Zh. Prikl. Spektrosk.*, **6,** 261 (1967).
524. P. P. Sorokin, W. H. Culver, E. C. Hammond, and J. R. Lankard, *IBM J. Res. Dev.*, **10,** 401 (1966).
525. F. P. Schaefer, W. Schmidt, and K. Marth, *Phys. Lett. A*, **24,** 280 (1967).
526. P. P. Sorokin, J. R. Lankard, V. L. Morruzzi, and E. C. Hammond, *J. Chem. Phys.*, **48,** 4726 (1968).
527. Y. Tamoto, M. Ohtsuka, and K. Egami, *Bull. Soc. Photogr. Sci. Technol. Japan*, 19 (1974).
528. Y. Miyazoe and M. Maeda, *Appl. Phys. Lett.*, **12,** 206 (1968).
529. L. D. Derkacheva and A. I. Krymova, *Dokl. Akad. Nauk SSSR*, **178,** 581 (1968).
530. G. Yamaguchi, S. Murakawa, and C. Yamanaka, *Appl. Phys. Lett.*, **13,** 134 (1968).
531. J. I. Steinfeld, *Int. J. Quantum Chem., Symp.*, 325 (1969).
532. J. P. Hermann, *Opt. Commun.*, **12,** 102 (1973).
533. D. J. Bradley, M. H. R. Hutchinson, and H. Koester, *Proc. R. Soc. London, Ser. A*, **329,** 105 (1972).
534. M. Maeda and Y. Miyazoe, *Japan. J. Appl. Phys.*, **11,** 692 (1972).
535. A. Hirth, K. Vollrath, J. Faure, and D. Lougnot, *Opt. Commun.*, **7,** 339 (1973).
536. V. A. Mostovnikov, A. N. Rubinov, M. A. Al'perovich, V. I. Avdeeva, I. I. Levkoev, and M. M. Loiko, *Zh. Prikl. Spektrosk.*, **20,** 42 (1974).
537. L. D. Derkacheva, A. I. Krymova, V. I. Malyshev, and A. S. Markin, *Zh. Eksp. Teor. Fiz., Pis'ma Red.*, **7,** 468 (1968).
538. B. H. Soffer and R. H. Hoskins, U.S. Patent 3,470,492 (1969).
539. J. L. R. Williams and G. A. Reynolds, *J. Appl. Phys.*, **39,** 5327 (1968).
540. B. B. Snavely, *Electro-Opt. Syst. Des.*, **5,** 30 (1973).
541. F. C. Strome, *Eastman Org. Chem. Bull.* (Eastman Kodak Co., Rochester, N.Y.), **46**(2), 1 (1974).
542. J. P. Webb, F. G. Webster, and B. E. Plourde, *Eastman Org. Chem. Bull.* (Eastman Kodak Co., Rochester, N.Y.), **46**(3), 1 (1974).
543. F. G. Webster, U.S. Patent 3,831,105 (1974).
544. F. P. Schaefer, *Angew. Chem. Int. Ed. Engl.*, **9,** 9 (1970).
545. A. Dienes, C. V. Shank, and A. M. Trozzolo, *Creat. Detect. Excited State*, **2,** 149 (1974).
546. A. Dienes, C. V. Shank, and R. L. Kohn, *IEEE J. Quantum Electron.*, **9,** 833 (1973).
547. M. Nakashima, R. C. Clapp, and J. A. Sousa, *Nature* (*London*), *Phys. Sci.*, **245,** 124 (1973).

548. K. H. Drexhage, G. R. Erikson, G. H. Hawks, and G. A. Reynolds, *Opt. Commun.*, **15,** 399 (1975).
549. D. Beer and J. Weber, *Opt. Commun.*, **5,** 307 (1972).
550. R. C. Pastor, B. H. Soffer, and H. Kimura, *J. Appl. Phys.*, **43,** 3530 (1972).
551. C. Brecher, R. Pappalardo, and H. Samelson, U.S. Patent 3,677,959 (1972).
552. D. L. Hecht, *Diss. Abstr. B*, **32,** 5791 (1972).
553. W. West, S. Pearce, and F. Grum, *J. Phys. Chem.*, **71,** 1316 (1967).
554. D. L. Ross and J. Blanc, "Techniques of Chemistry," A. Weissberger, Ed., Vol. 3; "Photochromism," G. H. Brown, Ed., Interscience, New York, 1971, p. 471.
555. G. Scheibe, J. Heiss, and K. Feldmann, *Angew. Chem.*, **77,** 545 (1965).
556. J. Heiss and K. Feldmann, *Angew. Chem.*, **77,** 546 (1965).
557. G. Scheibe, J. Heiss, and K. Feldmann, *Ber. Bunsenges. Phys. Chem.*, **70,** 52 (1966).
558. H. Kuhn, *Angew. Chem.*, **74,** 74 (1962).
559. H. Nikolajewski, S. Daehne, D. Leupold, and B. Hirsch, *Tetrahedron*, **24,** 6685 (1968).
560. L. G. S. Brooker, D. W. Heseltine, and L. L. Lincoln, *Chimia*, **20,** 327 (1966).
561. E. Fischer and Y. Frei, *J. Chem. Phys.*, **27,** 808 (1957).
562. D. Leupold, H. Kobischke, and U. Geske, *Tetrahedron Lett.*, 3287 (1967).
563. L. Zechmeister and J. H. Pinckard, *Experientia*, **9,** 16 (1953).
564. R. A. Jeffreys, *Wissenschaftliche Photographie* (*Int. Conf. Koln, 1956*), O. Helwich, Ed., Verlag Chemie, Darmstadt, 1958, p. 123.
565. W. Seiffert, unpubl. results, quoted Ref. 460.
566. W. S. Gaugh and F. G. Webster (Eastman Kodak Co., Rochester, N.Y.), private communication.
567. F. J. Sauter (Eastman Kodak Co., Rochester, N.Y.), private communication.
568. A. Van Dormael and J. Nys, *Ind. Chim. Belge, Suppl.*, **2,** 512 (1959).
569. P. W. Jenkins and L. G. S. Brooker, U.S. Patent 3,681,081 (1972).
570. L. G. S. Brooker and F. G. Webster, British Patent 988,627 (1965).
571. P. W. Jenkins and L. G. S. Brooker, British Patent 1,075,391 (1967).
572. D. W. Heseltine, French Patent 1,500,025 (1967).
573. J. V. Holtzclaw and L. G. S. Brooker, U. S. Patent 3,440,053 (1969).
574. P. W. Jenkins and L. G. S. Brooker, British Patent 1,192,334 (1970).
575. A. Lowenbein and W. Katz, *Ber. Dtsch. Chem. Ges.*, 1377 (1926).
576. R. C. Bertelson, "Techniques of Chemistry," A. Weissberger, Ed., Vol. 3; "Photochromism," G. H. Brown, Ed., Interscience, New York, 1971, p. 45.
577. G. Eigemann, "Techniques of Chemistry," A. Weissberger, Ed., Vol. 3; "Photochromism," G. H. Brown, Ed., Interscience, New York, 1971, p. 433.
578. R. Dessauer and J. P. Paris, *Adv. Photochem.*, **1,** 275 (1963).
579. R. Exelby and R. Grinter, *Chem. Rev.*, **65,** 247 (1965).
580. J. H. Day, *Chem. Rev.*, **63,** 65 (1963).
581. R. Heiligman-Rim, Y. Hirshberg, and E. Fischer, *J. Phys. Chem.*, **66,** 2465 (1962).
582. T. Bercovici, R. Heiligman-Rim, and E. Fischer, *Mol. Photochem.*, **1,** 23 (1969).
583. R. Guglielmetti, *J. Photogr. Sci.*, **22,** 77 (1974).
584. J. R. Haase (Eastman Kodak Co., Rochester, N.Y.), private communication.
585. F. Mentienne, A. Samat, R. Guglielmetti, F. Garnier, J. E. Dubois, and J. Metzger, *J. Chim. Phys. Phys.-Chim. Biol.*, **70,** 544 (1973).
586. P. Appriou and G. Guglielmetti, *Bull. Soc. Chim. Fr.*, 510 (1974).
587. H. Pommier and J. Metzger, French Patent 2,181,208 (1974).
588. M. Kryszewski and B. Nadolski, *Advan. Mol. Relaxation Processes*, **5,** 115 (1973).
589. M. A. Gal'bershtam and N. P. Samoilova, *Khim. Geterotsikl. Soedin.*, 1209 (1973).
590. E. Klemm, D. Klemm, J. Reichardt, and H. Hoerhold, *Z. Chem.*, **13,** 375 (1973).

591. T. Bercovici and E. Fischer, *J. Am. Chem. Soc.*, **86,** 5687 (1964).
592. G. I. Lashkov, V. L. Ermolaev, and A. V. Shablya, *Opt. Spektrosk.*, **21,** 305 (1965).
593. H. Bach and J. G. Calvert, *J. Am. Chem. Soc.*, **92,** 2608 (1970).
594. D. A. Reeves and F. Wilkinson, *J. Chem. Soc., Faraday Trans. 2*, 1381 (1973).
595. M. R. V. Sahyun, *Photogr. Sci. Eng.*, **16,** 63 (1972).
596. B. Tinland, R. Guglielmetti, and O. Chalvet, *Tetrahedron*, **29,** 665 (1973).
597. N. W. Tyler and R. S. Becker, *J. Am. Chem. Soc.*, **92,** 1289 (1970).
598. N. W. Tyler and R. S. Becker, *J. Am. Chem. Soc.*, **92,** 1295 (1970).
599. G. A. Samara, B. M. Riggleman, and H. G. Drickamer, *J. Chem. Phys.*, **37,** 1482 (1962).
600. W. A. Little and H. Gutfreund, *Phys. Rev. B*, **4,** 817 (1971).
601. R. C. Nelson, *J. Chem. Phys.*, **47,** 4451 (1967).
602. H. Nakatsui and Y. Hishiki, *Photogr. Sci. Eng.*, **15,** 394 (1971).
603. L. D. Rozenshtein, V. V. Sinitskii, and Y. A. Vidadi, *Fiz. Tekh. Poluprovodn.*, **3,** 118 (1969); *Chem. Abstr.*, **71,** 4494c (1969).
604. B. H. Klanderman and D. C. Hoesterey, *J. Chem. Phys.*, **51,** 377 (1969).
605. V. A. Benderskii, L. A. Blyumenfel'd, P. A. Stunzhas, and E. A. Sokolov, *Nature*, **220,** 365 (1968).
606. M. Heider and J. Neel, *J. Chim. Phys. Phys.-Chim. Biol.*, **70,** 547 (1973).
607. M. Heider and J. Neel, *J. Chim. Phys. Phys.-Chim. Biol.*, **70,** 553 (1973).
608. J. R. Platt, *J. Chem. Phys.*, **34,** 862 (1961).
609. H. Labhart, *Helv. Chim. Acta*, **44,** 447 (1961).
610. H. Labhart, *Helv. Chim. Acta*, **44,** 457 (1961).
611. H. Labhart, "Advances in Chemical Physics," Vol. 13, I. Prigogine, Ed., Interscience, New York, 1967, p. 179.
612. V. A. Krongauz and A. A. Parshutkin, *Photochem. Photobiol.*, **15,** 503 (1972).
613. M. Kryszewski, *Acta Phys. Pol.*, **34,** 695 (1968).
614. H. Buecher, J. Wiegand, B. B. Snavely, K. H. Beck, and H. Kuhn, *Chem. Phys. Lett.*, **3,** 508 (1969).
615. L. M. Blinov, N. A. Kirichenko, B. P. Bespalov, and V. G. Rumyantsev, *Zh. Strukt. Khim.*, **14,** 662 (1973).
616. C. Reichart and K. Dimroth, *Fortschr. Chem. Forsch.*, **11,** 1 (1968).
617. C. Reichart, *Angew. Chem. Int. Ed. Engl.*, **4,** 29 (1965).
618. K. Dimroth, *Chimia*, **15,** 80 (1961).
619. L. G. S. Brooker, *Chimia*, **15,** 87 (1961).
620. W. West and A. L. Geddes, *J. Phys. Chem.*, **68,** 837 (1964).
621. E. B. Lifshits, *Dokl. Akad. Nauk SSSR*, **179,** 596 (1968).
622. S. E. Sheppard, P. T. Newsome, and H. R. Brigham, *J. Am. Chem. Soc.*, **64,** 2923 (1942).
623. S. E. Sheppard, *Rev. Mod. Phys.*, **14,** 303 (1942).
624. N. Bayliss and E. G. McRae, *J. Am. Chem. Soc.*, **74,** 5803 (1952).
625. E. G. McRae, *Spectrochim. Acta*, **12,** 192 (1958).
626. S. Daehne, H. J. Rauh, R. Schnabel, and G. Geiseler, *Z. Chem.*, **13,** 70 (1973).
627. S. Daehne, F. Schob, and K. D. Nolte, *Z. Chem.*, **13,** 471 (1973).
628. N. Spasokukotski, Y. S. Moshkovskii, M. V. Deichmeister, and L. D. Zhilina, *Zh. Obshch. Khim.*, **34,** 3259 (1964).
629. R. W. Taft and M. J. Kamlet, *J. Am. Chem. Soc.*, **98,** 2886 (1976) and references therein.
630. S. Huenig, G. Bernhard, W. Liptay, and W. Brenninger, *Justus Liebigs Ann. Chem.*, **690,** 9 (1965).
631. J. Figueras, *J. Am. Chem. Soc.*, **93,** 3255 (1971).

632. O. W. Kolling and J. L. Goodnight, *Anal. Chem.*, **45,** 160 (1973).
633. S. J. Davidson and W. P. Jencks, *J. Am. Chem. Soc.*, **91,** 225 (1969).
634. J. Nys, A. Van Dormael, and G. Verbeke, *Chimia*, **19,** 315 (1965).
635. K. Dimroth and C. Reichardt, *Justus Liebigs Ann. Chem.*, **727,** 93 (1969).
636. A. Schweig, *Z. Naturforsch., Teil A*, **22,** 724 (1967).
637. K. Dimroth, C. Reichardt, and A. Schweig, *Justus Liebigs Ann. Chem.*, **669,** 95 (1963).
638. J. W. Faller, A. Mueller, and J. P. Phillips, *J. Org. Chem.*, **29,** 3450 (1964).
639. H. W. Gibson and F. C. Bailey, *Tetrahedron*, **30,** 2043 (1974).
640. H. W. Gibson, *Can. J. Chem.*, **51,** 3065 (1973).
641. J. E. Kuder and D. Wychick, *Chem. Phys. Lett.*, **24,** 69 (1974).
642. A. Blazsek-Bodo, M. Vernes, and A. Szurkos, *Farmicia (Bucharest)*, **22,** 345 (1974); *Chem. Abstr.*, **81,** 171320f (1974).
643. E. G. McRae, *J. Phys. Chem.*, **61,** 562 (1957).
644. M. E. Baur and M. Nicol, *J. Chem. Phys.*, **44,** 3337 (1966).
645. W. Liptay, *Z. Naturforsch., Teil A*, **20,** 1441 (1965).
646. W. Liptay, *Opt. Anregung Org. Syst. (Int. Farbensymp. Elmau, 1964)*, W. Jung, Ed., Verlag Chemie, Weinheim, 1966, p. 263.
647. F. J. Kampas, *Chem. Phys. Lett.*, **26,** 334 (1974).
648. E. E. Jelley, *Nature*, **138,** 1009 (1936).
649. E. E. Jelley, *Nature*, **139,** 631 (1937).
650. G. Scheibe, *Angew. Chem.*, **49,** 563 (1936).
651. G. Scheibe, *Angew. Chem.*, **52,** 631 (1939).
652. G. Scheibe, A. Mareis, and R. Schiffmann, *Z. Phys. Chem. B*, **49,** 324 (1941).
653. J. F. Padday, *Chem. Biol. Intercell. Matrix, Advan. Study Inst.*, **2,** 1007 (1970).
654. J. Nys, "Dye Sensitization" (Bressanone, 1967), G. Semerano, U. Mazzucato, W. F. Berg, and H. Meier, Eds., Focal Press, New York, 1970, p. 26.
655. J. A. Bergeron and M. Singer, *J. Biophys. Biochem. Cytol.*, **4,** 433 (1958).
656. J. F. Padday, *J. Phys. Chem.*, **72,** 1259 (1968).
657. S. E. Sheppard and A. L. Geddes, *J. Am. Chem. Soc.*, **66,** 2003 (1944).
658. F. Dietz, *J. Signalaufzeichnungsmaterialien*, **1,** 157–180, 237–252, 381–382 (1973).
659. E. Coates, *J. Soc. Dyers Colour.*, **85,** 355 (1969).
660. A. H. Herz, *Photogr. Sci. Eng.*, **18,** 323 (1974).
661. E. S. Emerson, M. A. Conlin, A. E. Rosenoff, K. S. Norland, H. Rodriguez, D. Chin, and G. R. Bird, *J. Phys. Chem.*, **71,** 2397 (1967).
662. S. V. Natanson, N. S. Spasokukotskii, and E. S. Kozlova, *Dokl. Akad. Nauk SSSR*, **157,** 1455 (1964).
663. E. Daltrozzo, G. Scheibe, K. Gschwind, and F. Haimerl, *Photogr. Sci. Eng.*, **18,** 441 (1974).
664. B. Levy and N. Mattucci, *Photogr. Sci. Eng.*, **14,** 308 (1970).
665. G. E. Ficken, *J. Photogr. Sci.*, **21,** 11 (1973).
666. I. I. Levkoev, E. B. Lifschitz, S. V. Natanson, N. N. Sveschnikov, and Z. P. Sitnik, "Wissenschaftliche Photographie" (*Int. Conf. Köln, 1956*) O. Helwich, Ed., Verlag Chemie, Darmstadt, 1958, p. 109.
667. B. Grimm and S. Daehne, *J. Signalaufzeichnungsmaterialien*, **1,** 339 (1973).
668. B. Grimm, *Z. Chem.*, **14,** 277 (1974).
669. H. O. Dickinson, *J. Photogr. Sci.*, **2,** 50 (1954).
670. E. Clementi and M. Kasha, *J. Chem. Phys.*, **26,** 956 (1957).
671. A. H. Herz, R. P. Danner, and G. A. Janusonis, "Adsorption from Aqueous Solution" (*Advan. Chem. Ser. No.* 79), American Chemical Society, Washington, D.C., 1967, p. 173.

672. H. Philipaerts, W. Vanassche, F. H. Claes, and H. Borginon, *J. Photogr. Sci.*, **20,** 215 (1972).
673. J. M. Simson, *Photogr. Sci. Eng.*, **18,** 302 (1974).
674. G. R. Bird, B. Zuckerman, and A. E. Ames, *Photochem. Photobiol.*, **8,** 393 (1968).
675. V. K. Walworth, A. E. Rosenoff, and G. R. Bird, *Photogr. Sci. Eng.*, **14,** 321 (1970).
676. A. E. Rosenoff, V. K. Walworth, and G. R. Bird, *Photogr. Sci. Eng.*, **14,** 328 (1970).
677. C. Reich, *Photogr. Sci. Eng.*, **18,** 335 (1974).
678. H. Hada, M. Matsusaki, and M. Tamura, *Nippon Kagaku Zasshi,* **90,** 115 (1969); *Chem. Abstr..* **70,** 72505m (1969).
679. K. Yamaoka, Y. Matsuoka, and M. Miura, *J. Phys. Chem.*, **78,** 1040 (1974).
680. W. West, B. H. Carroll, and D. Whitcomb, *J. Phys. Chem.*, **56,** 1054 (1952).
681. G. W. Anderson, "Scientific Photography" (*Proc. Int. Coll. Liège, 1959*), H. Sauvenier, Ed., Pergamon Press, London, 1962, p. 487.
682. E. B. Lifschitz, *Zh. Nauchn. Prikl. Fotogr. Kinematogr.* **6,** 64 (1961).
683. V. Czikkely, G. Dreizler, H. D. Foersterling, H. Kuhn, J. Sondermann, P. Tillmann, and J. Wiegand, *Z. Naturforsch., Teil A,* **24,** 1821 (1969).
684. H. Buecher and H. Kuhn, *Chem. Phys. Lett.*, **6,** 183 (1970).
685. R. Steiger, R. Kitzing, and P. Junod, *J. Photogr. Sci.*, **21,** 107 (1973).
686. H. Kuhn, D. Moebius, and H. Buecher, "Techniques of Chemistry," Vol. 1, A. Weissberger and B. W. Rossiter, Eds., Part 3B, Interscience, New York, 1972, p. 577.
687. D. F. O'Brien, *Photogr. Sci. Eng.*, **18,** 16 (1974).
688. P. Fromherz, *Z. Naturforsch., Teil C,* **28,** 144 (1973).
689. W. West and V. I. Saunders, "Wissenschaftliche Photographie" (*Int. Conf. Köln, 1956*), O. Helwich, Ed., Verlag Chemie, Darmstadt, 1958, p. 48.
689a. A. M. Yacynych, H. B. Mark, Jr., and C. H. Giles, *J. Phys. Chem.* **80,** 839 (1976).
690. W. E. Gray, W. R. Brewer, and G. R. Bird, *Photogr. Sci. Eng.*, **14,** 316 (1970).
691. A. P. Marchetti, C. D. Salzberg, and E. I. P. Walker, *Photogr. Sci. Eng.*, **20,** 107 (1976); *J. Chem. Phys.*, **64,** 4693 (1976).
692. L. Stryer and E. R. Blout, *J. Am. Chem. Soc.*, **83,** 1411 (1961).
693. D. F. Bradley, *Trans. N.Y. Acad. Sci.*, **24,** 64 (1961)
694. R. A. Berg and B. A. Haxby, *Molec. Cryst. Liq. Cryst.*, **12,** 93 (1970).
695. V. I. Permogorov, *Izv. Akad. Nauk SSSR, Ser. Fiz.*, **34,** 1335 (1970); *Chem. Abstr.*, **73,** 99988v (1970).
696. G. Scheibe, O. Woerz, F. Haimerl, W. Seiffert, and I. Winkler, *J. Chim. Phys. Phys.-Chim. Biol.*, **65,** 146 (1968).
697. G. Scheibe, F. Haimerl, and W. Hoppe, *Tetrahedron Lett.*, 3067 (1970).
697a. C. Honda and H. Hada, *Tetrahedron Lett.* 177 (1976).
698. J. K. Maurus and G. R. Bird, *J. Phys. Chem.*, **76,** 2982 (1972).
699. W. Appel and G. Scheibe, *Z. Naturforsch., Teil B,* **13,** 359 (1958).
700. O. Woerz and G. Scheibe, *Z. Naturforsch., Teil B,* **24,** 381 (1969).
701. P. J. Hillson and R. B. McKay, *Nature,* **210,** 296 (1966).
702. R. B. McKay and P. J. Hillson, *Trans. Faraday Soc.*, **63,** 777 (1967).
703. R. E. Kay, E. R. Walwick, and C. K. Gifford, *J. Phys. Chem.*, **68,** 1896, 1907 (1964).
704. R. C. Bean, W. C. Shepherd, R. E. Kay, and E. R. Walwick, *J. Phys. Chem.*, **69,** 4368 (1965).
705. R. D. Edstrom and M. M. Koenst, *Arch. Biochem. Biophys.*, **155,** 307 (1973).
706. V. I. Pemogorov, *Opt. Spektrosk.*, **34,** 298 (1973).
707. D. F. Bradley and M. K. Wolf, *Proc. Natl. Acad. Sci. U.S.A.*, **45,** 944 (1959).

708. M. Abe, K. Miyaki, D. Mizuno, N. Narita, T. Takeuchi, T. Ukita, and T. Yamamoto, *Japan. J. Med. Sci. Biol.*, **12,** 175 (1959); *Chem. Abstr.*, **54,** 13230a (1960).
709. H. Ogasawara, *Kanko Shikiso,* **73,** 12 (1968); *Chem. Abstr.*, **69,** 94636m (1968).
710. J. Janda, *Biochem. J.*, **124,** 73 (1971); *Chem. Abstr.*, **76,** 11712d (1972).
711. I. Tyuma, *Nara Igaku Zasshi,* **8,** 296 (1957); *Chem. Abstr.*, **52,** 9267i (1958).
712. J. S. Moore, G. O. Phillips, D. M. Power, and J. V. Davies, *J. Chem. Soc., A,* 1155 (1970).
713. E. A. Balazs, J. V. Davies, G. O. Phillips, and D. S. Scheulfe, *J. Chem. Soc., C,* 1420, 1424, 1429 (1968).
714. F. J. Kampas, *J. Polym. Sci., Part C,* 81 (1970).
715. W. West and S. Pearce, *J. Phys. Chem.*, **69,** 1894 (1965).
716. G. Hui-Bon-Hoa, J. Langlet, and C. Balny, *J. Chim. Phys. Phys.-Chim. Biol.*, **65,** 372 (1968).
717. A. V. Borin and I. A. Pobedenostseva, *Zh. Nauchn. Prikl. Fotogr. Kinematogr.*, **3,** 256 (1958).
718. K. K. Rohatgi and A. K. Mukhopadhyay, *Chem. Phys. Lett.*, **12,** 259 (1971).
719. B. H. Robinson, A. Loeffler, and G. Schwarz, *J. Chem. Soc., Faraday Trans. 1,* 56 (1973).
720. K. L. Arvan and N. E. Zaitseva, *Opt. Spektrosk.*, **10,** 272 (1961).
721. M. Hida, A. Yabe, H. Murayama, and M. Hayashi, *Bull. Chem. Soc. Japan,* **41,** 1776 (1968).
722. K. K. Rohatgi and G. S. Singhal, *J. Phys. Chem.*, **70,** 1695 (1966).
723. E. Daltrozzo, discussion section published for Vogel Centennial as *Photogr. Sci. Eng.*, **18,** 335 (1974).
724. S. S. Collier, *Photogr. Sci. Eng.*, **18,** 430 (1974).
725. W. Cooper, *Chem. Phys. Lett.*, **7,** 73 (1970).
726. W. Cooper and A. A. Muenter, *Photogr. Sci. Eng.*, **20,** 117 (1976).
727. T. Foerster and E. Koenig, *Z. Elektrochem.*, **61,** 344 (1957).
728. E. D. Owen and Q. Sultana, *J. Appl. Chem. Biotechnol.*, **22,** 1043 (1972); *Chem. Abstr.*, **78,** 31391t (1973).
729. G. L. Zarur, *Diss. Abstr.*, **34,** 1062 (1973).
730. R. W. Chambers, T. Kajiwara, and D. R. Kearns, *J. Phys. Chem.*, **78,** 380 (1974).
731. H. J. Nolte and V. Buss, *Chem. Phys. Lett.*, **19,** 395 (1973).
732. V. Czikkely, H. Foersterling, and H. Kuhn, *Chem. Phys. Lett,* **6,** 11 (1970).
733. V. Czikkely, H. Foersterling, and H. Kuhn, *Chem. Phys. Lett.*, **6,** 207 (1970).
734. E. G. McRae and M. Kasha, "Physical Processes in Radiation Biology," L. Augenstein, R. Mason, and B. Rosenberg, Eds., Academic Press, New York, 1964, p. 23.
735. M. Kasha, H. R. Rawls, and M. A. El-Bayoumi, *Pure Appl. Chem.*, **11,** 371 (1965).
736. R. E. Ballard and B. J. Gardner, *J. Chem. Soc., B,* 736 (1971).
737. H. DeVoe, *J. Chem. Phys.*, **37,** 1534 (1962).
738. H. DeVoe, *J. Chem. Phys.*, **41,** 393 (1964).
739. K. Rohatgi, *Physicochem. Aspects Interaction Dyes in Solution Fibre Syst.* (*Symp., 1969*), p. 13.
740. K. Norland, A. Ames, and T. Taylor, *Photogr. Sci. Eng.*, **14,** 295 (1970).
741. G. Bird and C. Reich, discussion section published for Vogel Centennial as *Photogr. Sci. Eng.*, **18,** 335 (1974).
742. C. Reich, W. D. Pandolfe, and G. R. Bird, *Photogr. Sci. Eng.*, **17,** 334 (1973).
743. W. D. Pandolfe and G. R. Bird, unpublished work cited in footnote on p. 337 of *Photogr. Sci. Eng.*, **18,** 335 (1974).
744. A. N. Chigir, M. I. Cherkashin, and A. A. Berlin, *Izv. Akad. Nauk SSSR, Ser. Khim.*, 1199 (1969); *Chem. Abstr.*, **71,** 71892u (1969).

745. E. I. du Pont de Nemours and Co., British Patent 903,268 (1962).
746. D. W. Heseltine, Belgian Patent 627,309 (1963).
747. D. W. Heseltine and L. L. Lincoln, Belgian Patent 669,003 (1965).
748. I. I. Levkoev, Z. P. Sytnik, S. V. Natanson, V. V. Durmashkina, T. V. Krasnova, and S. R. Shuser, *Zh. Obshch. Khim.*, **24,** 1999 (1954).
749. I. I. Levkoev, N. N. Sveshnikov, and S. A. Heifetz, *Zh. Obshch. Khim.*, **16,** 1489 (1946).
750. I. I. Levkoev, N. N. Sveshnikov, N. S. Barvyn, and T. V. Krasnova, *Zh. Obshch. Khim.*, **30,** 291 (1960).
751. L. M. Yagupol'skii and V. I. Troitskaya, *Zh. Obshch. Khim.*, **31,** 628 (1961).
752. A. I. Kiprianov and I. N. Zhmurova, *Zh. Obshch. Khim.*, **23,** 874 (1953).
753. A. I. Kiprianov and I. N. Zhmurova, *Zh. Obshch. Khim.*, **23,** 493 (1953).
754. A. I. Kiprianov and I. N. Zhmurova, *Usp. Khim.*, **22,** 1246 (1953).
755. V. M. Zubarovskii and G. P. Khodot, *Khim. Geterotsikl. Soedin.*, 1037 (1973).
756. I. A. Ol'shevskaya and V. Y. Pochinok, *Khim. Geterotsikl. Soedin.*, 640 (1974).
757. E. D. Sych and L. P. Umanskaya, *Zh. Obshch. Khim.*, **34,** 2068 (1964).
758. T. G. Gnevysheva and I. I. Levkoev, *Dokl. Akad. Nauk SSSR*, **146,** 1081 (1962).
759. H. Ohlschaeger, O. Riester, J. Goetze, and A. Dorlars, German Patent 1,934,891 (1971).
760. E. B. Lifshits, I. I. Levkoev, L. M. Yagupol'skii, and N. S. Barvyn, *Zh. Nauchn. Prikl. Fotogr. Kinematogr.*, **11,** 175 (1966).
761. V. I. Troitskaya, A. I. Burmakov, L. S. Kudryavtseva, and L. M. Yagupol'skii, *Ukr. Khim. Zh.*, **37,** 808 (1971); *Chem. Abstr.*, **76,** 47370e (1972).
762. J. Ciernik, *Chem. Listy*, **55,** 44 (1961).
763. A. V. Stetsenko and Y. I. Bogodist, *Ukr. Khim. Zh.*, **26,** 92 (1960); *Chem. Abstr.*, **54,** 15938b (1960).
764. V. I. Troitskaya and L. M. Yagupol'skii, *Khim. Geterotsikl. Soedin.*, 275 (1974).
765. E. J. Van Lare, U.S. Patent 2,739,149 (1956).
766. Gevaert Photo-Producten N.V., Belgian Patent 640,453 (1964).
767. Gevaert Photo-Producten N.V., Belgian Patent 648,991 (1964).
768. H. Depoorter and O. Riester, British Patent 1,161,797 (1969).
769. G. Ficken, D. J. Fry, and K. J. Bannert, British Patent 1,132,528 (1968).
770. D. J. Fry and G. E. Ficken, British Patent 1,132,529 (1968).
771. L. Yagupol'skii, V. I. Troitskaya, I. I. Levkoev, E. B. Lifshits, P. A. Yufa, and N. S. Barvyn, *Zh. Obshch. Khim.*, **37,** 191 (1967).
772. G. Di Modica and S. Tira, *Boll. Sci. Fac. Chim. Ind. Bologna.* **22,** 19 (1964); *Chem. Abstr.*, **61,** 10806d (1964).
773. E. Barni, *Tinctoria*, **69,** 309 (1972); *Chem. Abstr.*, **78,** 5375w (1973).
774. S. G. Fridman and L. I. Kotova, *Khim. Geterotsikl. Soedin.*, 497 (1967).
775. S. G. Fridman, *Zh. Obshch. Khim.*, **35,** 1364 (1965).
776. Z. Moskalenko, *Khim. Geterotsikl. Soedin.*, 1189 (1972).
777. P. I. Abramenko, *Khim. Geterotsikl. Soedin.*, 459 (1971).
778. P. I. Abramenko and T. K. Ponomareva, *Khim. Geterotsikl. Soedin.*, 1606 (1972).
779. P. I. Abramenko, V. G. Zhiryakov, and T. K. Ponomareva, *Khim. Geterotsikl. Soedin.*, 1603 (1975); 56 (1976) and references therein.
780. H. P. Kuehlthau, German Patent, 2,211,959 (1973).
781. E. D. Sych, V. N. Bubnovskaya, L. T. Gorb, and M. Y. Kornilov, *Khim. Geterotsikl. Soedin.*, 1254 (1973).
782. S. I. Shul'ga and V. A. Chuiguk, *Ukr. Khim. Zh.*, (*Russ. Ed.*), **39,** 1151 (1973); *Chem. Abstr.* **80,** 49219u (1974).

783. A. V. Stetsenko and M. M. Kul'chitskii, *Ukr. Khim. Zh.*, **35,** 288 (1969); *Chem. Abstr.*, **71,** 22892k (1969).
784. A. V. Stetsenko and V. A. Korinko, *Ukr. Khim. Zh.*, **30,** 944 (1964); *Chem. Abstr.*, **62,** 11942f (1965).
785. G. Di Modica and S. Pasquino, *Ann. Chim.* (*Rome*), **63,** 509 (1973); *Chem. Abstr.*, **82,** 45007s (1975).
786. W. S. Gaugh, D. W. Heseltine, D. M. Sturmer, and J. P. Freeman, U.S. Patent 3,936,308 (1976).
786a. T. D. Weaver, U.S. Patent 3,898,216 (1975).
787. Eastman Kodak Co., Belgian Patent 660,253 (1965).
788. L. G. S. Brooker and E. J. Van Lare, U.S. Patent 3,431,111 (1969).
789. V. A. Gladkaya and Y. S. Rozum, *Ukr. Khim. Zh.*, **33,** 1287 (1967); *Chem. Abstr.*, **69,** 3701q (1968).
790. Y. S. Rozum and V. A. Gladkaya, *Ukr. Khim. Zh.*, **32,** 1200 (1966); *Chem. Abstr.*, **66,** 66722f (1967).
791. M. Gandino and A. R. Katritzky, *Ann. Chim.* (*Rome*), **60,** 462 (1970); *Chem. Abstr.*, **73,** 110855x (1970).
792. J. W. Carpenter, J. D. Mee, and D. W. Heseltine, U.S. Patent 3,809,691 (1974).
793. F. M. Hamer, *J. Chem. Soc.*, **125,** 1348 (1924).
794. N. I. Fisher and F. M. Hamer, *J. Chem. Soc.*, 907 (1937).
795. H. Balli and F. Kersting, *Chimia*, **20,** 318 (1966).
796. A. I. Kiprianov and T. M. Verbovskaya, *Zh. Org. Khim.*, **2,** 1848 (1966).
797. H. Balli and R. Loew, *Helv. Chim. Acta*, **59,** 155 (1976).
797a. H. Balli and R. Maul, *Helv. Chim. Acta*, **59,** 148 (1976).
798. K. Parija, A. Nayak, and M. K. Rout, *J. Indian Chem. Soc.*, **47,** 1129 (1970).
799. S. Huenig and H. Herrmann, *Justus Liebigs Ann. Chem.*, **636,** 32 (1960).
800. H. Schelz and H. Balli, *Helv. Chim. Acta*, **53,** 1913 (1970).
801. S. Huenig, and K. H. Fritsch, *Justus Liebigs Ann. Chem.*, **609,** 172 (1957).
802. S. Huenig, H. Balli, and H. Quast, *Angew. Chem.*, **74,** 28 (1962).
803. S. Huenig and H. Quast, *Justus Liebigs Ann. Chem.*, **711,** 139 (1968).
804. S. Huenig and H. Quast, *Opt. Anregung Org. Syst.* (*Int. Farbensymp. Elmau, 1964*), W. Jung, Ed., Verlag Chemie, Weinheim, 1966, p. 184.
805. J. D. Kendall, British Patent 447,109 (1936).
806. J. D. Kendall, "Scientific and Applied Photography" (*Int. Congr. Paris, 1935*), Revue D'Optique, Paris, 1936, p. 227.
807. G. E. Ficken and J. D. Kendall, *Chimia*, **15,** 110 (1961).
808. M. Tamura and H. Hada, "Scientific Photography" (*Proc. Int. Coll. Liège, 1959*), H. Sauvenier, Ed., Pergamon Press, London, 1962, p. 572.
809. L. Yagupol skii, L. I. Trushanina, and A. Y. Il'chenko, *Zh. Org. Khim.*, **11,** 2168 (1975).
810. M. M. Kul'chitskii, Y. L. Yagupol'skii, and L. M. Yagupol'skii, *Ukr. Khim. Zh.* **42,** 204 (1976).
811. E. D. Sych and L. P. Umanskaya, *Ukr. Khim. Zh.*, **34,** 604 (1968); *Chem. Abstr.*, **70,** 20966t (1969).
812. F. S. Babichev and E. Shchetsinskaya, *Zh. Obshch. Khim.*, **34,** 2441 (1964).
813. L. L. Lincoln and D. W. Heseltine, U.S. Patent 3,679,427 (1972).
814. A. I. Kiprianov and V. P. Khilya, *Zh. Org. Khim.*, **2,** 1478 (1966).
815. V. P. Khilya and G. I. Galatsan, *Khim. Geterotsikl. Soedin.*, 1282 (1973).
816. V. P. Khilya and G. I. Galatsan, *Khim. Geterotsikl. Soedin.*, 1258 (1973).
817. A. I. Kiprianov, Z. M. Ivanova, and S. G. Fridman, *Ukr. Khim. Zh.*, **20,** 641 (1954); *Chem. Abstr.*, **49,** 14739a (1955).

818. A. Y. Il'chenko, L. I. Trushanina, and L. M. Yagupol'skii, *Zh. Org. Khim.*, **8,** 1279 (1972).
819. L. M. Yagupol'skii, L. I. Trushanina, and A. Y. Il'chenko, *Zh. Org. Khim.*, **11,** 2168 (1975) and references therein.
820. C. Reichardt and K. Halbritter, *Chem. Ber.*, **104,** 822 (1971) and references therein.
821. C. Reichardt, *Angew. Chem.*, **77,** 508 (1965).
822. A. Fumia and L. G. S. Brooker, German Patent 2,046,672 (1971).
823. A. S. Kheinman, S. V. Natanson, N. S. Spasokukotskii, A. F. Vompe, and R. V. Karaul'shchikova, *Zh. Nauchn. Prikl. Fotogr. Kinematogr.*, **13,** 44 (1968).
824. Z. P. Dokuchaeva and A. A. Sadykova, *Zh. Nauchn. Prikl. Fotogr. Kinematogr.*, **11,** 350 (1966).
825. A. I. Kiprianov and A. A. Shulezko, *Zh. Obshch. Khim.*, **34,** 3932 (1964).
826. A. A. Shulezhko, *Zh. Org. Khim.*, **4,** 2207 (1968).
827. A. A. Shulezhko, *Zh. Org. Khim.*, **6,** 2122 (1970).
828. A. A. Shulezhko, I. T. Rozhdestvenskaya, and A. I. Kiprianov, *Zh. Org. Khim.*, **6,** 2118 (1970).
829. F. A. Mikhailenko and L. I. Shevchuk, *Synthesis*, 621 (1973).
830. S. Huenig and E. Wolff, *Justus Liebigs Ann. Chem.*, **732,** 26 (1970).
831. J. K. Elwood, *J. Org. Chem.*, **38,** 2425 (1973).
832. O. Neunhoeffer and J. A. Keiler, *Z. Wiss. Photogr. Photophys. Photochem.*, **58,** 142 (1964).
833. O. Neunhoeffer and J. A. Keiler, *Z. Wiss. Photogr. Photophys. Photochem.*, **58,** 147 (1964).
834. T. Tani, *Bull. Soc. Photogr. Sci. Technol. Japan*, **21,** 15 (1971).
835. R. G. Dubenko, I. M. Bazavova, and P. S. Pel'kis, *Khim. Geterotsikl. Soedin.*, 603 (1971).
836. V. N. Bubnovskaya, V. S. Shvarts, and F. S. Babichev, *Ukr. Khim. Zh.*, **33,** 924 (1967); *Chem. Abstr.*, **69,** 11405m (1968).
837. S. V. Lepikhova and G. T. Pilyugin, *Zh. Obshch. Khim.*, **39,** 1829 (1969).
838. G. T. Pilyugin, S. V. Shinkorenko, and L. E. Zhivoglazova, *Khim. Geterotsikl. Soedin.*, 321 (1968).
839. E. P. Opanasenko, V. I. Skalkina, and G. I. Kozhushko, *Zh. Obshch. Khim.*, **40,** 2722 (1970).
840. G. T. Pilyugin and S. V. Shinkorenko, *Zh. Obshch. Khim.*, **33,** 3223 (1963).
841. G. T. Pilyugin and S. V. Shinkorenko, *Zh. Obshch. Khim.*, **28,** 1313 (1958).
842. B. M. Gutsulyak and P. D. Romanko, *Zh. Obshch. Khim.*, **43,** 878 (1973).
843. G. T. Pilyugin, Y. O. Gorichok, B. M. Gutsulyak, and S. I. Gorichok, *Khim. Geterotsikl. Soedin.*, 896 (1965).
844. G. T. Pilyugin, V. I. Skalkina, E. P. Opanasenko, P. V. Prisyazhnyuk, and G. I. Kozhushko, *Zh. Vses. Khim. Ova.*, **17,** 116 (1972); *Chem. Abstr.*, **77,** 7273q (1972).
845. H. G. Derbyshire, U.S. Patent 3,136,772 (1964).
846. M. Gadino, P. Merli, and O. Turilli, French Patent 1,491,399 (1967).
847. Minnesota Mining and Manufacturing Co., French Demande 2,128,737 (1972).
848. O. V. Moreiko and E. D. Sych, *Khim. Geterotsikl. Soedin.*, 749 (1973).
849. E. D. Sych and O. V. Moreiko, *Khim. Geterotsikl. Soedin.*, 933 (1972); 1186 (1973).
850. E. D. Sych, O. V. Moreiko, and A. Y. Ilchenko, *Khim. Geterotsikl. Soedin.*, 1609 (1975).
851. E. A. Gloag and R. A. Jeffreys, British Patent 1,270,948 (1972).
852. R. C. Taber and L. G. S. Brooker, British Patent 1,112,034 (1968).
853. B. D. Illingsworth and E. J. Van Lare, French Patent 1,522,355 (1968).

854. V. M. Zubarovskii, *Khim. Geterotsikl. Soedin.*, 1542 (1973).
855. T. Winkler and C. Mayer, *Helv. Chim. Acta,* **55,** 2351 (1972).
856. D. S. Brown, G. Ficken, and E. J. Treherne, British Patent 1,309,274 (1973).
857. F. G. Webster and L. G. S. Brooker, U.S. Patent 3,743,638 (1973).
858. F. G. Webster, French Patent, 1,597,509 (1970).
859. H. Larivé and R. Baralle, French Patent 2,182,329 (1974).
860. R. M. Baralle, French Patent 2,199,547 (1974).
861. K. Kueffner and H. Glockner, German Patent 2,236,077 (1974).
862. L. I. Kirenskaya, N. I. Smol'yaninova, and I. I. Levkoev, U.S.S.R. Patent 412,218 (1974); *Chem. Abstr.*, **81,** 79392h (1974).
863. L. I. Kirenskaya, N. I. Smol'yaninova, A. V. Kazymov, T. N. Skryabneva, and I. I. Levkoev, U.S.S.R. Patent 432,166 (1974); *Chem. Abstr.*, **82,** 59924a (1975).
864. L. G. S. Brooker, French Patent 1,470,163 (1967).
865. J. Sondermann, *Justus Liebigs Ann. Chem.*, **749,** 183 (1971).
866. V. I. Troitskaya, V. I. Rudyk, E. B. Lifshits, and L. M. Yagupol'skii, *Zh. Org. Khim.*, **9,** 1051 (1973).
867. W. S. Gaugh, U.S. Patent 3,687,675 (1972).
868. G. E. Ficken and E. J. Squire, British Patent 1,001,480 (1965).
869. R. Liepins and C. Walker, *Ind. Eng. Chem., Prod. Res. Develop.*, **10,** 401 (1971); *Chem. Abstr.*, **76,** 73698n (1972).
869a. R. H. Sprague, J. Keller, P. Rorke, J. Russell, and W. Cumming, "Abstracts for the 28th Annual Conference of the Society of Photographic Scientists and Engineers," Denver, Colorado, May, 1975, papers VI-2, VI-3.
870. M. I. Dronkina and L. M. Yagupol'skii, *Zh. Org. Khim.*, **9,** 2167 (1973).
871. V. I. Troitskaya, V. I. Rudyk, E. V. Konovalov, and L. M. Yagupol'skii, *Zh. Org. Khim.*, **10,** 1524 (1974).
872. M. Yamamoto, S. Nakamura, K. Yoshimura, M. Yuge, S. Morosawa, and A. Yokoo, *Bull. Chem. Soc. Japan*, **46,** 1509 (1973).
873. O. Riester, German Patent 1,916,845 (1970).
874. C. D. Weston, "The Chemistry of Synthetic Dyes," Vol. 7, K. Venkataraman, Ed., Academic Press, New York, 1974, p. 35.
875. L. L. Lincoln and D. W. Heseltine, Canadian Patent 938,152 (1973).
876. J. D. Mee, P. W. Jenkins, and D. W. Heseltine, German Patent 1,950,779 (1970).
877. J. D. Mee, P. W. Jenkins, and D. W. Heseltine, German Patent 1,950,785 (1970).
878. G. F. Mitchell, Canadian Patent 938,151 (1973).
879. H. Bauman and J. Dehnert, U.S. Patent 3,102,878 (1963).
880. See Chapters I-III in "The Chemistry of Synthetic Dyes," Vol. 6, K. Venkataraman, Ed., Academic Press, New York, 1972.
881. G. Scheibe, D. Grosse, and J. Heiss, *Angew. Chem.*, **76,** 187 (1964).
882. D. Leupold and S. Daehne, *Z. Phys. Chem.* (*Frankfurt am Main*) **48,** 24 (1966).
883. G. Oster, *Trans. Faraday Soc.*, **47,** 660 (1951).
884. G. K. Oster and G. Oster, *J. Am. Chem. Soc.*, **81,** 5543 (1959).
885. F. Millich and G. Oster, *J. Am. Chem. Soc.*, **81,** 1357 (1959).
886. G. Oster, J. S. Bellin, R. W. Kimball, and M. E. Schrader, *J. Am. Chem. Soc.*, **81,** 5095 (1959).
887. B. Broyde and G. Oster, *J. Am. Chem. Soc.*, **81,** 5099 (1959).
888. G. Oster and N. Wotherspoon, *J. Am. Chem. Soc.*, **79,** 4836 (1957).
889. G. K. Oster and G. Oster, U.S. Patent 3,097,097 (1963).
890. H. Meier, "The Chemistry of Synthetic Dyes," Vol. 4, K. Venkataraman, Ed., Academic Press, New York, 1971, p. 389.

891. J. Bourdon and B. Schnuriger, "Physics and Chemistry of Organic Solid State," Vol. 3, D. Fox, M. M. Labes, and A. Weissberger, Eds., Interscience, New York, 1967, p. 60.
892. G. S. Egerton and A. G. Morgan, *J. Soc. Dyers Colour.*, **87,** 223 (1971).
893. R. M. Hochstrasser and G. B. Porter, *Q. Rev., Chem. Soc.*, **14,** 146 (1960).
894. C. Winther, *Z. Wiss. Photogr. Photophys. Photochem.*, **11,** 92 (1911).
895. P. Lasareff, *Z. Physik. Chem.*, **78,** 558 (1912).
896. A. M. Timber and E. C. Lingafelter, *J. Am. Chem. Soc.*, **71,** 4155 (1949).
897. R. A. Jeffreys, *Ind. Chim. Belge, Suppl.*, **2,** 495 (1958).
898. J. Bourdon and M. Durante, *Bull. Soc. Chim. Belg.*, **71,** 907 (1962).
899. A. Krasnovskii and N. N. Drozdova, *Dokl. Akad. Nauk SSSR*, **145,** 129 (1962).
900. G. Gotoh, N. Kimura, and S. Hayashi, *Bull. Inst. Chem. Res., Kyoto Univ.*, **47,** 340 (1969); *Chem. Abstr.*, **72,** 126773j (1970).
901. H. Gerischer, *Photochem. Photobiol.*, **16,** 243 (1972).
902. H. Gerischer and H. Selze, *Electrochim. Acta*, **18,** 799 (1973).
903. H. Tributsch and H. Gerischer, *Ber. Bunsenges Phys. Chem.*, **73,** 251 (1969).
904. G. W. Byers, S. Gross, and P. M. Henrichs, *Photochem. Photobiol.*, **23,** 37 (1976).
905. J. S. Splitter and M. Calvin, *J. Org. Chem.*, **20,** 1086 (1955).
906. D. M. Sturmer, Belgian Patent 788,279 (1972).
907. J. A. Van Allan, S. Farid, G. A. Reynolds, and S. C. Chang, *J. Org. Chem.*, **38,** 2834 (1973).
908. S. M. Krueger (Eastman Kodak Co., Rochester, N.Y.), private communication.
909. J. D. Mee, D. W. Heseltine, and E. C. Taylor, *J. Am. Chem. Soc.*, **92,** 5814 (1970).
910. D. F. O'Brien and P. W. Jenkins (Eastman Kodak Co., Rochester, N.Y.), private communication.

Index